Advances in Intelligent Systems and Computing

Volume 1309

AF880423

The series "Advances in Intelligent Systems and Computing" contains publications on theory, applications, and design methods of Intelligent Systems and Intelligent Computing. Virtually all disciplines such as engineering, natural sciences, computer and information science, ICT, economics, business, e-commerce, environment, healthcare, life science are covered. The list of topics spans all the areas of modern intelligent systems and computing such as: computational intelligence, soft computing including neural networks, fuzzy systems, evolutionary computing and the fusion of these paradigms, social intelligence, ambient intelligence, computational neuroscience, artificial life, virtual worlds and society, cognitive science and systems, Perception and Vision, DNA and immune based systems, self-organizing and adaptive systems, e-Learning and teaching, human-centered and human-centric computing, recommender systems, intelligent control, robotics and mechatronics including human-machine teaming, knowledge-based paradigms, learning paradigms, machine ethics, intelligent data analysis, knowledge management, intelligent agents, intelligent decision making and support, intelligent network security, trust management, interactive entertainment, Web intelligence and multimedia.

The publications within "Advances in Intelligent Systems and Computing" are primarily proceedings of important conferences, symposia and congresses. They cover significant recent developments in the field, both of a foundational and applicable character. An important characteristic feature of the series is the short publication time and world-wide distribution. This permits a rapid and broad dissemination of research results.

Indexed by SCOPUS, DBLP, EI Compendex, INSPEC, WTI Frankfurt eG, zbMATH, Japanese Science and Technology Agency (JST), SCImago.

All books published in the series are submitted for consideration in Web of Science.

More information about this series at http://www.springer.com/series/11156

M. Shamim Kaiser · Anirban Bandyopadhyay ·
Mufti Mahmud · Kanad Ray
Editors

Proceedings of International Conference on Trends in Computational and Cognitive Engineering

Proceedings of TCCE 2020

Springer

Editors
M. Shamim Kaiser
Jahangirnagar University
Dhaka, Bangladesh

Mufti Mahmud
Nottingham Trent University
Nottingham, UK

Anirban Bandyopadhyay
National Institute for Materials Science
Tsukuba, Japan

Kanad Ray
Amity University
Jaipur, Rajasthan, India

ISSN 2194-5357 ISSN 2194-5365 (electronic)
Advances in Intelligent Systems and Computing
ISBN 978-981-33-4672-7 ISBN 978-981-33-4673-4 (eBook)
https://doi.org/10.1007/978-981-33-4673-4

This Springer imprint is published by the registered company Springer Nature Singapore Pte Ltd.
The registered company address is: 152 Beach Road, #21-01/04 Gateway East, Singapore 189721, Singapore

Organization

Chief Patron

Farzana Islam, Jahangirnagar University, Bangladesh

Conference Co-chairs

Dr. Chi-Sang Poon, Massachusetts Institute of Technology, USA
Dr. Anirban Bandyopadhyay, National Institute for Materials Science, Japan
Dr. Kanad Ray, Amity University, Jaipur

Steering Committee

Robert Bai, University of Cambridge, UK
Anirban Bandyopadhyay, National Institute for Materials Science, Japan
Kanad Ray, Amity University, Jaipur, India
J. E. Lugo, University of Montreal, Canada
Subrata Ghosh, CSIR NIST, Jorhat, India
Chi-Sang Poon, MIT, USA
Jocelyn Faubert, University of Montreal, Canada
Mufti Mahmud, Nottingham Trent University, UK

Advisory Committee

Alamgir Hossain, Teesside University, UK
Yoshinori Kuno, SU, Japan
Yoshinori Kobayashi, SU, Japan

Azizur Rahman, City University London, UK
Satya Prashad Mazumder, BUET, Bangladesh
Md. Hanif Ali, Jahangirnagar University, Bangladesh
M. Atiq R. Ahad, OU, Japan
Md. Abdur Razzaque, University of Dhaka, Bangladesh
M. Sohel Rahman, BUET, Bangladesh
Md. Saidur Rahman, BUET, Bangladesh
Joarder Kamruzzaman, Federation University Australia
Subrata Kumar Aditya, University of Dhaka, Bangladesh
Md. Saiful Islam, BUET, Bangladesh
Md. Roshidul Hasan, BSMRAU, Bangladesh
Md. Obaidur Rahman, DUET, Bangladesh
Kazi M. Ahmed, UAP, Bangladesh
Sarwar Morshed, AUST, Bangladesh
Md. Nazrul Islam, MIST, Bangladesh
Md. Abu Taher, BUP, Bangladesh

Organizing Committee

Mesbahuddin Sarker (Chair), Jahangirnagar University, Bangladesh
Md. Abu Yousuf (Co-chair), Jahangirnagar University, Bangladesh
Shamim Al Mamun (Secretary), Jahangirnagar University, Bangladesh
Fazlul Karim Patwary, Jahangirnagar University, Bangladesh
Risala T. Khan, Jahangirnagar University, Bangladesh
K. M. Akkas Ali, Jahangirnagar University, Bangladesh
Jesmin Akhter, Jahangirnagar University, Bangladesh
Fahima Tabassum, Jahangirnagar University, Bangladesh
Md. Wahiduzzaman, Jahangirnagar University, Bangladesh
Mohammad Shahidul Islam, Jahangirnagar University, Bangladesh
Nusrat Zerin, Jahangirnagar University, Bangladesh
Manan Binte Taj Noor, Jahangirnagar University, Bangladesh
Rashed Mazumder, Jahangirnagar University, Bangladesh
Sazzadur Rahman, Jahangirnagar University, Bangladesh
Zamshed Iqbal Chowdhury, Jahangirnagar University, Bangladesh

Technical Program Committee Chairs

Anirban Bandyopadhyay (Chair), National Institute for Materials Science, Japan
M. Shahadat Hossain (Co-chair), University of Chittagong, Bangladesh
Nilanjan Dey (Co-chair), JIS University, Kolkata, India

A. A. Maman (Co-chair), Jahangirnagar University, Bangladesh
M. Shamim Kaiser (Secretary), Jahangirnagar University, Bangladesh

Technical Program Committee

Muhammad Arifur Rahman, Jahangirnagar University, Bangladesh
Sajjad Waheed, MBSTU, Bangladesh
Md. Zahidur Rahman, GUB, Bangladesh
Muhammad Golam Kibria, ULAB, Bangladesh
Md. Majharul Haque, Deputy Director, Bangladesh Bank
Samsul Arefin, CUET, Bangladesh
Md. Obaidur Rahman, DUET, Bangladesh
Mustafa Habib Chowdhury, IUB, Bangladesh
Marzia Hoque-Tania, Oxford University, UK
Antesar Shabut, CSE, Leeds Trinity University, UK
Md. Khalilur Rhaman, BRAC University, Bangladesh
Md. Hanif Seddiqui, University of Chittagong, Bangladesh
M. M. A. Hashem, KUET, Bangladesh
Tomonori Hashiyama, The University of Electro-Communications, Japan
Wladyslaw Homenda, Warsaw University of Technology, Poland
M. Moshiul Hoque, CUET, Bangladesh
A. B. M. Aowlad Hossain, KUET, Bangladesh
Sheikh Md. Rabiul Islam, KUET, Bangladesh
Manohar Das, Oakland University, USA
Kaushik Deb, CUET, Bangladesh
Carl James Debono, University of Malta, Malta
M. Ali Akber Dewan, Athabasca University, Canada
Belayat Hossain, Loughborough University, UK
Khoo Bee Ee, Universiti Sains Malaysia, Malaysia
Ashik Eftakhar, Nikon Corporation, Japan
Md. Tajuddin Sikder, Jahangirnagar University, Bangladesh
Mrs. Shayla Islam, UCSI, Malaysia
Antony Lam, Mercari Inc., Japan
Ryote Suzuki, Saitama University, Japan
Hishato Fukuda, Saitama University, Japan
Md. Golam Rashed, Rajshahi University, Bangladesh
Md. Sheikh Sadi, KUET, Bangladesh
Tushar Kanti Shaha, JKKNIU, Bangladesh
M. Shazzad Hosain, NSU, Bangladesh
M. Mostafizur Rahman, AIUB, Bangladesh
Tabin Hassan, AIUB, Bangladesh
Aye Su Phyo, Computer University Kalay, Myanmar
Md. Shahedur Rahman, Jahangirnagar University

Lu Cao, Saitama University, Japan
Nihad Adnan, Jahangirnagar University
Mohammad Firoz Ahmed, Jahangirnagar University
A. S. M. Sanwar Hosen, JNU, South Korea
Mahabub Hossain, ECE, HSTU, Bangladesh
Md. Sarwar Ali, Rajshahi University, Bangladesh
Risala T. Khan, Jahangirnagar University, Bangladesh
Mohammad Shahidul Islam, Jahangirnagar University, Bangladesh
Manan Binte Taj Noor, Jahangirnagar University, Bangladesh
Md. Abu Yousuf, Jahangirnagar University, Bangladesh
Md. Sazzadur Rahman, Jahangirnagar University, Bangladesh
Rashed Mazumder, Jahangirnagar University, Bangladesh
Md. Abu Layek, Jagannath University, Bangladesh
Saiful Azad, Universiti Malaysia Pahang, Malaysia
Mostofa Kamal Nasir, MBSTU, Bangladesh
Mufti Mahmud, NTU, UK
A. K. M. Mahbubur Rahman, IUB, Bangladesh
Al Mamun, Jahangirnagar University, Bangladesh
Al-Zadid Sultan Bin Habib, KUET, Bangladesh
Anup Majumder, Jahangirnagar University, Bangladesh
Atik Mahabub, Concordia University, Canada
Bikash Kumar Paul, MBSTU, Bangladesh
Md. Obaidur Rahman, DUET, Bangladesh
Nazrul Islam, MIST, Bangladesh
Ezharul Islam, Jahangirnagar University, Bangladesh
Farah Deeba, DUET, Bangladesh
Md. Manowarul Islam, Jagannath University, Bangladesh
Md. Waliur Rahman Miah, DUET, Bangladesh
Rubaiyat Yasmin, Rajshahi University, Bangladesh
Sarwar Ali, Rajshahi University, Bangladesh
Rabiul Islam, Kulliyyah of ICT, Malaysia
Dejan C. Gope, Jahangirnagar University, Bangladesh
Sk. Md. Masudul Ahsan, KUET, Bangladesh
Mohammad Shahriar Rahman, ULAB, Bangladesh
Golam Dastoger Bashar, Boise State University, USA
Md. Hossam-E-Haider, MIST, Bangladesh
H. Liu Wayne, State University, USA
Imtiaz Mahmud, Kyungpook National University, Korea
Kawsar Ahmed, MBSTU, Bangladesh
Kazi Abu Taher, BUP, Bangladesh
Linta Islam, Jagannath University, Bangladesh
Md. Musfique Anwar, Jahangirnagar University, Bangladesh
Md. Sanaul Haque, University of Oulu, Finland
Md. Ahsan Habib, MBSTU, Bangladesh
Md. Habibur Rahman, MBSTU, Bangladesh

M. A. F. M. Rashidul Hasan, Rajshahi University, Bangladesh
Md. Badrul Alam Miah, UPM, Malaysia
Mohammad Ashraful Islam, MBSTU, Bangladesh
Mokammel Haque, CUET, Bangladesh
Muhammad Ahmed, ANU, Australia
Nazia Hameed, University of Nottingham, UK
Partha Chakraborty, CoU., Bangladesh
Kandrapa Kumar Sarma, Gauhati University, India
Vaskar Deka, Gauhati University, India
K. M. Azharul Islam, KUET, Bangladesh
Tushar Sarkar, RUET, Bangladesh
Surapong Uttama, Mae Fah Luang University, Thailand
Sharafat Hossain, KUET, Bangladesh
Shaikh Akib Shahriyar, KUET, Bangladesh
A. S. M. Sanwar Hosen, Jeonbuk National University, Korea

Preface

TCCE 2020, the 2nd International Conference on Trends in Computational and Cognitive Engineering, took place on December 17 and 18, 2020. The conference is hosted by the Institute of Information Technology of Jahangirnagar University, Savar, Dhaka-1342, Bangladesh, and is one of the annual events of the International Invincible Rhythm Institute (IIoIR), Shimla, India. This series's first event was held at the Central University of Haryana, Mahendergarh, India, during November 28–30, 2019. TCCE focuses on experimental, theoretical, and application of computational and cognitive engineering. Computational and cognitive engineering consists of computer and mathematical methods commonly used in all fields of science, engineering, technology, and industry and analyzes diseases and behavioral disorders. The conference aims to provide a platform for international relationships among researchers, academia, business professional working in related field. This book encapsulates the peer-reviewed research papers presented at the meeting.

The conference on TCCE 2020 attracted 146 full papers from 11 countries in five tracks. These tracks include—artificial intelligence and soft computing; cognitive science and computational biology; IoT and data analytics; network and security; and computer vision. The submitted papers underwent a single-blind review process, soliciting expert opinion from at least two experts: at least two independent reviewers, the track co-chair, and the respective track chair. After receiving review reports from the reviewers and the track chairs, the technical program committee has selected 58 high-quality full papers from ten countries which were accepted for presentation at the conference. Consequently, this volume of the TCCE 2020 conference proceedings contains those 58 full papers presented on December 17–18, 2020. Due to the COVID-19 pandemic, the organizing committee decided to host the event virtually. However, the research community reacted amazingly in this challenging time.

The book series will be insightful and fascinating for those interested in learning about computational intelligence and cognitive engineering that explores the dynamics of exponentially increasing knowledge in core and related fields. We are thankful to the authors who have made a significant contribution to the conference and have developed relevant research and literature in computer and cognitive engineering.

We would like to express our gratitude to the organizing committee and the technical committee members for their unconditional support, particularly the chair, the co-chair, and the reviewers. TCCE 2020 could not have taken place without the tremendous work of the team and the gracious assistance. We would like to thank Dr. Md. Sazzad Hossain, Hon'ble member of UGC, and University Grants Commission for their support. Special thanks to Chief Patron Prof. Dr. Farzana Islam, Honorable Vice-chancellor, for her guidance and visionary support throughout the whole journey of TCCE 2020. We are grateful to Mr. Aninda Bose, Ms. Deenamaria Bonaparte, Mr. Parimelazhagan Thirumani, and other members of Springer Nature for their continuous support in coordinating this volume publication. Our special thanks to Editor(s)/Guest Editor(s) of the Studies in Rhythm Engineering, Springer Nature; Entropy, MDPI; Big Data Analytics, BioMed Central and International Journal of Ambient Computing and Intelligence, IGI Global for considering the extended version of the selected papers. We would also like to thank Dr. Nilanjan Dey of JIS University, Kolkata, India, and Mr. Mahfujur Rahman of DIU for continuous support. Last but not least, we thank all of our contributors and volunteers for their support during this challenging time to make TCCE 2020 a success.

Dhaka, Bangladesh — M. Shamim Kaiser
Nottingham, UK — Mufti Mahmud
Jaipur, India — Kanad Ray
Tsukuba, Japan — Anirban Bandyopadhyay
December 2018

Contents

About the Editors

Dr. M. Shamim Kaiser is currently working as Professor at the Institute of Information Technology of Jahangirnagar University, Savar, Dhaka-1342, Bangladesh. He received his Bachelor's and Master's degrees in Applied Physics Electronics and Communication Engineering from the University of Dhaka, Bangladesh, in 2002 and 2004, respectively, and the Ph.D. degree in Telecommunication Engineering from the Asian Institute of Technology, Thailand, in 2010. His current research interests include data analytics, machine learning, wireless network & signal processing, cognitive radio network, big data and cyber security, and renewable energy. He has authored more than 100 papers in different peer-reviewed journals and conferences. He is Associate Editor of the IEEE Access Journal, and Guest Editor of Brain Informatics Journal, and Cognitive Computation Journal. Dr. Kaiser is a life member of Bangladesh Electronic Society; Bangladesh Physical Society. He is also a senior member of IEEE, USA, and IEICE, Japan, and an active volunteer of the IEEE Bangladesh Section. He is Founding Chapter Chair of the IEEE Bangladesh Section Computer Society Chapter. Dr. Kaiser organized various international conferences such as ICEEICT 2015–2018, IEEE HTC 2017, IEEE ICREST 2018, and BI2020.

Anirban Bandyopadhyay is Senior Scientist in the National Institute for Materials Science (NIMS), Tsukuba, Japan., and completed his Ph.D. from Indian Association for the Cultivation of Science (IACS), Kolkata 2005, December, on supramolecular electronics. During 2005–2007: he has been ICYS Research Fellow NIMS, Japan, and from 2007 to now, he is Permanent Scientist in NIMS, Japan. He has 10 patents on building artificial organic brain, big data, molecular bot, cancer Alzheimer drug, fourth circuit element, etc. During 2013–2014, he has been Visiting Scientist in MIT, USA, on biorhythms. He is World Technology Network (WTN) Fellow, (2009–continued); he has been awarded Hitachi Science and Technology Award 2010, Inamori Foundation Award 2011–2012, Kurata Foundation Award, and SSI Gold medal (2017)., He is Inamori Foundation Fellow (2011–), and Sewa Society International SSS Fellow (2012–), Japan.

Dr. Mufti Mahmud is Senior Lecturer of Computing at the Nottingham Trent University, UK. He received Ph.D. degree in Information Engineering from the University of Padova – Italy, in 2011. A recipient of the Marie-Curie postdoctoral fellowship, he served at various positions in the industry and academia in India, Bangladesh, Italy, Belgium, and the UK since 2003. An expert in computational intelligence, data analysis, and big data technologies, Dr. Mahmud has published over 80 peer-reviewed articles and papers in leading journals and conferences. Dr. Mahmud serves as Associate Editor to the Cognitive Computation, IEEE Access, Big Data Analytics, and Brain Informatics journals. Dr. Mahmud is a senior member of IEEE and ACM, a professional member of the British Computer Society, and Fellow of the higher education academy – UK. During the year 2020–2021, he is serving as Vice Chair of the Intelligent System Application Technical Committee of IEEE CIS, a member of the IEEE CIS Task Force on Intelligence Systems for Health and the IEEE R8 Humanitarian Activities Subcommittee, and Project Liaison Officer of the IEEE UK and Ireland SIGHT committee. Dr. Mahmud is also serving as Local Organizing Chair of IEEE-WCCI2020; General Chair of BI2020 and BI2021; and Program Chair of IEEE-CICARE2020.

Kanad Ray senior member, IEEE, received the M.Sc. degree in physics from Calcutta University and the Ph.D. degree in physics from Jadavpur University, West Bengal, India. He has been Professor of Physics and Electronics and Communication, and is presently working as Head of the Department of Physics, Amity School of Applied Sciences, Amity University Rajasthan (AUR), Jaipur, India. His current research areas of interest include cognition, communication, electromagnetic field theory, antenna and wave propagation, microwave, computational biology, and applied physics. He has been serving as Editor for various Springer book series. He was Associate Editor of the Journal of Integrative Neuroscience (The Netherlands: IOS Press). He has been Visiting Professor to UTM & UTeM, Malaysia, and Visiting Scientist to NIMS, Japan. He has established MOU with UTeM Malaysia, NIMS Japan, and the University of Montreal, Canada. He has visited several countries such as Netherlands, Turkey, China, Czechoslovakia, Russia, Portugal, Finland, Belgium, South Africa, Japan, Singapore, Thailand, and Malaysia for various academic missions. He has organized various conferences such as SoCPROS, SoCTA, ICOEVCI, and TCCE as General Chair and a steering committee member.

Artificial Intelligence and Soft Computing

Bangla Real-Word Error Detection and Correction Using Bidirectional LSTM and Bigram Hybrid Model

Mir Noshin Jahan, Anik Sarker, Shubra Tanchangya, and Mohammad Abu Yousuf

Abstract Real-word error detection and correction in Bangla sentence is now more relevant topic of enormous interest nowadays. The complex character structure and grammatical rules in Bangla yield difficulty in processing the language. This paper proposes a hybrid method for real-word error detection and correction in Bangla which involves a combination of two different approaches like N-gram language model such as bigram and bidirectional long short-term memory (LSTM), a special type of recurrent neural network (RNN). We have initially collected Bangla dataset from different sources which is further reprocessed as the available datasets have valid output but does not provide input set with real-word error. The proposed model generates bigram sequences from the data corpus after accomplishing tokenization. Afterward, the sequences are fed into bidirectional LSTM model to predict the accurate word in order to replace inconsistent word. An advantage of this proposed procedure is that bidirectional LSTM remembers forward and backward relationship among words which creates better understanding of context for the network. Moreover, the system works on word length matching so that the output word length does not vary from the original one. The system shows significantly promising performance conveying 82.86% accuracy.

Keywords NLP · RNN · Bidirectional LSTM · Bigram · Real-word error

M. Noshin Jahan (✉) · A. Sarker · S. Tanchangya · M. Abu Yousuf
Institute of Information Technology, Jahangirnagar University, Savar, Dhaka 1342, Bangladesh
e-mail: noshinmir003@gmail.com

A. Sarker
e-mail: aniks2645@gmail.com

S. Tanchangya
e-mail: tanchangyashuv01@gmail.com

M. Abu Yousuf
e-mail: yousuf@juniv.edu

M. S. Kaiser et al. (eds.), *Proceedings of International Conference on Trends in Computational and Cognitive Engineering*, Advances in Intelligent Systems and Computing 1309, https://doi.org/10.1007/978-981-33-4673-4_1

1 Introduction

As our textual communication is increasing day by day with the upgrading of technology, a huge number of people share their opinion across Internet in Bangla. Real-word error detection and correction in Bangla sentence is an important and complicated task in natural language processing (NLP) that helps to predict the appropriate word to complete a sentence in a meaningful way. In textual documents, there are two types of errors such as real-word error and non-word error while typing. Non-word error occurs when a word makes no sense. Unlike non-word error, real-word error occurs in a sentence when the word is grammatically correct but is not compatible in obedience to the sentence. Bangla language has complex orthographic rules and many critical grammatical provision that are troublesome to maintain. While detecting spelling error in Bangla is cumbersome, checking real-word error in a Bangla sentence brings forward exceeding difficulty.

While processing data in software or chatting or typing in our daily working life, there may occur some real-word error where the word is correct on dictionary but has changed the original significance of the sentence which should not be ignored. An example can help us to comprehend the error such as "বাতাসে গাছের পাখা নড়ে উঠলো". Here, though the word "পাখা" is a correct word, still it is inappropriate for the sentence as it is not meaningful. "পাখা" can be replaced by "পাতা" which is appropriate and makes a sense. This matter enforces us to establish a hybrid model to analyze real-word error problem in Bangla sentence using N-gram language model and bidirectional long short-term memory.

Despite being a relevant topic of enormous interest nowadays, there is no satisfactory analysis on Bangla language in detecting and correcting real-word error in sentences. In this paper, a method is proposed to address the real-word error detection and correction in Bangla sentences which provides maximum matching with the original word. There is lack of publicly available standard datasets in Bangla language while there are lots of them in other languages. Researcher has to work on small dataset and most of them are not appropriate for use in future. Though large amount of research work has been accomplished in detecting and correcting error in English as well as other languages, very few research works have been found that deals with detecting and correcting real-word error in Bangla. Existing systems have drawbacks like low accuracy rate, lack of appropriate Bangla dataset, difficulty in detecting errors of an entire sentence and many others. The proposed technique aids in replacing inconsistent word in a Bangla sentence with a word having higher probability and maximum matching with the original word.

The rest of the paper is organized as follows. Section 2 covers overview of the existing works. Section 3 represents methodology of proposed procedure. Section 4 highlights used tools and technologies and Sect. 5 talks about experiment result. Lastly, conclusion is drawn in Sect. 6.

2 Related Work

Throughout all these years, many precious research works relevant to this research have been conducted in English as well as other languages. But yet a little progress is made in Bangla language. Hasan et al. proposed a model for detection of semantic error in Bangla sentences [1]. Their methodology is for simple sentences in the form of subject, object and verb based on subject-verb and object-verb relation.

Kundu et al. utilized a natural language generation model for correcting Bangla grammatical error that does not solve the problem of real-word error [2]. Yuan and Briscoe used neural machine translation system for grammatical error correcting for English sentences [3]. They applied RNN search model that contains bidirectional RNN and an encoder to capture information. Samanta and Chaudhuri proposed a method that tries to detect error noting bigram and trigram constituted by outright left and right neighbor of candidate word and generates suggestion according to the ranks of elements of confusion set [4]. Although their model showed good performance for only moderate test sets, this can be considered less accurate system because of small database.

Sharma and Gupta introduced a system where trigram and Bayesian methods are applied to resolve real-word errors [5]. Unlike other methods that use two or three features, their proposed system uses all the features of the sentences. Haque et al. designed a stochastic language models where they used unigram, bigram, trigram, back-off and deleted interpolation for single word prediction [6].

Rana et al. demonstrated a methodology using bigram, trigram and Markov assumption in order to find out and fix homophone error in real-word error in Bangla language [7]. By using bigram and trigram combination and extracting context features, they detected the error and produced decision list against the candidate word based on probability calculations. Despite the frequent occurrences of bigram, trigram is given first priority in their model as it extracts more features about the context.

An N-gram and semantic gram approach-based system was introduced by Wiegand and Patel to predict non-syntactic word for augmentative and alternative communication [8]. They evaluated and showed the execution of four algorithms. Jain and Jain suggested a model based on n-gram and dictionary lookup method for Hindi non-word spelling errors [9]. They showed that available methods are not suitable for applying in Hindi language which is similar to proposed case.

Mridha et al. focused on detecting unintentionally missed word while typing a sentence using bigram and provided suggestion using trigram [10]. Assuming bigram's foremost and endmost word to be trigram's foremost and endmost word consecutively, a suggestion list is generated for the missing word. Their method fails in some cases where even though a word is missed yet the sentence appears as correct. Stehouwer and van Zaanen concentrated on the problem of confused words where set of words are similar and often used incorrectly in context [11]. They used a generic classifier based on n-gram language model to predict correct word in context but its accuracy is expected to be less than any specific approaches.

Rakib et al. proposed a model that applies gated recurrent unit (GRU)-based RNN approach on a dataset where they used unigram, bigram, trigram, 4, 5-g datasets [12]. Their model can suggest not only the next most likely word but also suggest complete sentences simultaneously from given word sequence in Bangla providing 78.15% accuracy. Barmana and Boruah designed a system that predicts next Assamese word using long short-term memory (LSTM) with 88.20% accuracy [13]. Islam et al. used LSTM model for generating Bangla sentence where they made their model noise free by removing punctuation marks, additional spaces and new line [14].

Bangla text generation by utilizing bidirectional RNN on N-gram dataset was proposed by Abujar et al. which cannot detect error [15]. They focused on constructing a bidirectional RNN for preparing their model. They worked with a fixed length content and could not create arbitrary length content. Ghosh et al. analyzed and showed comparison between different techniques of Bangla handwritten character recognition to highlight limitation and future scopes of existing methods [16]. Santos et al. talked about a model that identifies and fixes syntax errors in English language [17]. They applied n-gram models and LSTM for modeling source code in order to find syntax errors and to synthesize the fixes. Islam et al. designed a RNN model which provides solution for three types of sentence corrections such as autocompletion, wrong arrangement and missing word [18].

3 Methodology

3.1 Dataset Details

Collecting data with real-word error was extremely challenging since there were no data available similar to our requirement. Henceforth, we had to generate it from available data. We have initially collected two types of data from website [19–21]. One of them is Bangla dictionary words and another one is Bangla text dataset. The text dataset was processed in such a way that all Bangla punctuation marks were omitted from the dataset. Then each sentence was extracted as individual lines in a corpus. The problem was though there were valid output set, still no input set with real-word error. Consequently, we had to generate an input set where sentences had real-word error. For this purpose, we generated homophones of a word in the sentence and replaced the original word with its homophone. To generate homophones, we manipulated each word in sentence through replacement and produced all possible new words. Thus, real-word error inclusive sentences were produced. We have generated four training and testing corpora. The accuracy of the corpus majorly depends of diversity and size of words in it. The corpus details are explained in Table 1.

We worked with dictionary lookup technique to check validity of a word where previously stated dictionary data had been used. At last, both input and output corpus were tokenized. Tokenization is a process of mapping string with an integer value

Table 1 Corpus specification

Corpus	Words in corpus	Training set (%)	Testing set (%)
1	321,381	70	30
2	238,643	70	30
3	212,955	70	30
4	235,780	70	30

to normalize the processing of analyzing strings. These tokenized values are used to read lines sequence by sequence in corpus and for bigram generation.

3.2 Bigram Generation and Probability Estimation

N-gram language model is a probabilistic model for predicting the probability of next word within any sequence of words. Bigram is a type of n-gram where n is referred as two. Bigram only predicts the probability of one word from previous one word. Once the tokenization procedure is accomplished, the model generates bigram sequences from the corpus. After that, the probability of the bigram sequences is calculated using a probability equation formulated with the help of general chain rule and Markov assumption which is further maximized by maximum likelihood estimation.

Conditional probability is one of the most fundamental and most important concepts in probability theory. To find joint probability distribution of two random events, conditional probability formula states,

$$P(XY) = P(X) \times P(Y|X) \tag{1}$$

Here, X and Y are considered random events.

General chain rule is applied with conditional probability formula (1) when distribution of probability is calculated for more than two random events. For n random words $word_1$, $word_2$, $word_3 \ldots$, $word_n$, this rule generates the following equation,

$$P(word_1 word_2 \ldots word_n) = \prod_{i=1}^{n} P(word_i | word_1 \ldots word_{i-1}) \tag{2}$$

According to Markov assumption equation, conditional probabilities of Eq. 2 can be approximately written as,

$$P(word_n | word_1 \ldots word_{n-1}) \approx P(word_n | word_{n-1}) \tag{3}$$

As observed, N-gram model is most efficient for predicting the probability of next item within any sequence of words. For simplicity, the proposed system focuses on using a type of n-gram such as where n is referred as two where bigram only predicts the probability of one word from prior one word. Equation 4 shows the formulation of bigram based on general chain rule (2) and Markov assumption (3),

$$P(word_1 word_2 \ldots word_n) = \prod_{i=1}^{n} P(word_i|word_{i-1}) \tag{4}$$

Then in order to maximize the probability of observed data, the system involves maximum likelihood estimation stated formally as Eq. 5

$$P(word_i|word_{i-1}) = \frac{Count(word_{i-1} word_i)}{Count(word_{i-1})} \tag{5}$$

By assigning $n = 1, 2, 3$ in Eqs. 5, 6, 7 and 7 are obtained as below.

$$P(word_1) = P(word_1) \tag{6}$$

$$P(word_1 word_2) = P(word_1) \times P(word_2|word_1) \tag{7}$$

$$P(word_1 word_2 word_3) = P(word_1) \times P(word_2|word_1) \times P(word_3|word_1 word_2) \tag{8}$$

Finally, after analyzing the Eqs. (6), (7) and (8) a summarized calculation can be found in Eq. 9,

$$P(word_1 word_2 \ldots word_n) = P(word_1 word_2 \ldots word_{n-1}) \times P(word_n|word_{n-1}) \tag{9}$$

This resulting equation is used in the proposed method for the purpose of calculating the probability of current processing word in a sentence and generate a probability score matrix (P) which contains probabilities of the fed inputs. The model also works on the output word's length so that the length does not differ much from the original word's length.

3.3 Bidirectional LSTM Approach

Bidirectional long short-term memory (LSTM) networks are special kind of recurrent neural network that were designed to avoid long-term dependencies problem. Bidirectional LSTM allows the model to understand a text better by feeding it an input

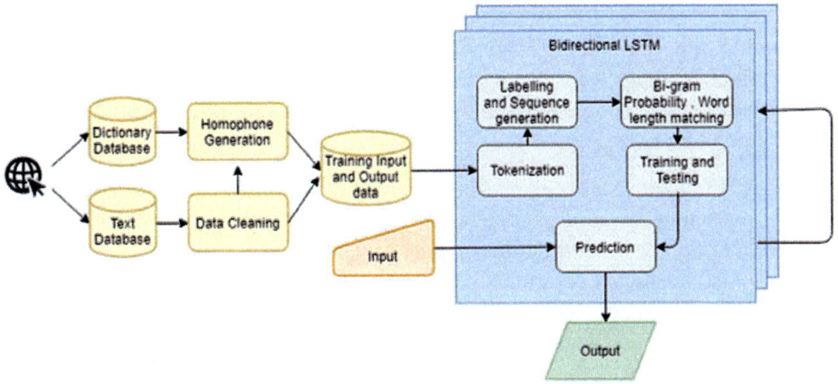

Fig. 1 System architecture for Bangla real-word error detection and correction

sequence into two directions, from beginning to the end and once again in reverse direction from end to beginning. There may arise some cases where we need more context to comprehend a text as the beginning of the sentence might have an impact on the last word of a paragraph. Like in this example, "অনিক মাঠে ফুটবল খেলছে" here, generating the third word might be disturbing as the word can be the both "ফুটবল" as well as "ক্রিকেট". Further proceeding into the context using LSTM can help to understand the dependencies of generated sequences and choose the correct word. In this research, we have used a sequential model of bidirectional LSTM with one bidirectional LSTM layer of 150 neurons and relu activation function. Following this layer, there is one dense layer consisting of neuron equal to the number of total words in training set and softmax activation function. We have trained our model with 25 epochs with a batch size of 64. Then the model was trained over four corpora each containing average 1 million words and prepared for testing. Traversing a whole sentence, the network that has been trained will predict the correction of current word in a sentence. It will automatically detect whether the word should be replaced or not, thus handling both error detection and correction of the word with most probable word with the help of previously fed training dataset. The workflow of the research work is demonstrated in Fig. 1.

4 Tools and Technology

The tools and techniques used for this research are-

(i) The data used for this work is collected from SUST CSE Developer Network Open Source Bengali Corpus [19], Bangla Article Dataset (BARD) [20] and Bangla Dictionary Word [21].
(ii) Google Drive to store the collected and processed data for the research.

(iii) Google Colab, an online platform which facilitates to combine executable code and to run code entirely in the cloud.
(iv) Python version 3.7 for coding.
(v) TensorFlow version 2 to build the neural network and train the model.
(vi) Python Libraries:

(a) Tokenizer: A Tokenizer allows to map each word to a single token which is generally represented as an integer value.
(b) Numpy: It is a multidimensional array and matrix structure where various mathematical operations like trigonometric, statistical, algebraic, etc. can be performed.
(c) Set: It is a built-in Python library which is used for finding a word or sequence from the corpus efficiently.

5 Experimental Result

In order to understand pertinence of proposed system, we have delved into some examples as shown in Table 2 where "গাছে", "তখন" and "মানা" are inappropriate words in conformity with the given sentences. The system checked each sentence by bigram probability calculation and minimum length difference to find out dependence of the sentences through this bidirectional LSTM network. Though numerous homophones words are available that match with those real-word errors , the system extracted most matching words "গেছে", "এখন" and "জানা" successively having minimum length difference with respect to those real-word error.

Our model showed promising accuracy level for training dataset. We tried to merge dataset in the corpus according to diverse context and situations. Our corpus consisted of text about education, arts, news, politics, economy, entertainment, sports, accident, etc. Table 3 shows details of result analysis of all the four corpora that have been used to train the hybrid LSTM model with a comparison to a bigram and RNN model and bigram and LSTM model with same dataset.

Figure 2a shows the accuracy of training corpus 1 (see Table 3) which is an average of 82.86%. The blue line depicts accuracy of training dataset. In Fig. 2b, the blue line depicts the loss of training corpus 1.

Table 2 Input/output assessment

Example	Input	Output
1	পুলিশ আসার আগেই ডাকাত পালিয়ে গাছে	পুলিশ আসার আগেই ডাকাত পালিয়ে গেছে
2	তুমি তখন কোথায় আছ	তুমি এখন কোথায় আছ
3	নিহত ব্যক্তির পরিচয় মানা যায়নি	নিহত ব্যক্তির পরিচয় জানা যায়নি

Table 3 Result analysis

Corpus	Hybrid model (%)	Bigram and RNN (%)	Bigram and LSTM (%)
1	81.96	59.97	60.96
2	80.98	56.51	57.93
3	81.22	57.19	58.40
4	82.86	57.81	59.16

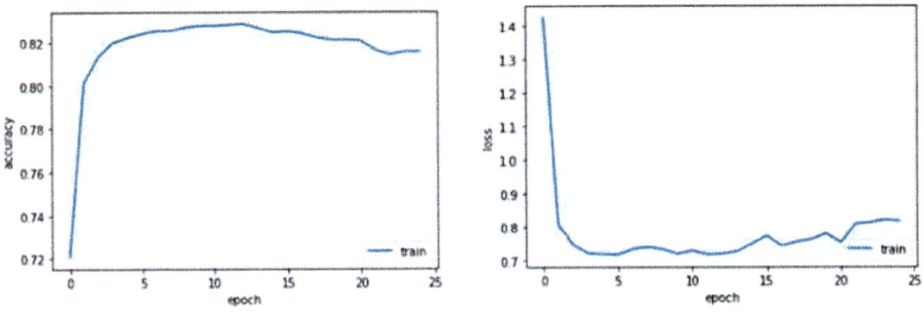

(a) Accuracy (Corpus 1) (b) Loss (Corpus 1)

Fig. 2 Performance evaluation

6 Conclusion

In this article, we have proposed a method for detection and correction of Bangla real-word error based on bigram and bidirectional LSTM. The dataset used here holds a major impact on training the model because of inadequacy of available standard resources. However, this proposed technique has achieved 82.86% accuracy in fixing the real-word error where the word is grammatically correct but has changed the original meaning of a sentence. Though the model was trained over four datasets each consisting of 1 million words beyond on average, we look forward to research more with variety of context oriented and further balanced data with an anticipation of achieving more fascinating performance.

References

1. Hasan, K.M.A, Hozaifa, M., Dutta., S.: Detection of semantic errors from simple Bangla sentences. In: 17th International Conference on Computer and Information Technology, Daffodil International University, Dhaka, Bangladesh (2014). https://doi.org/10.1016/0022-2836(81)90087-5
2. Kundu ,S. Chakraborti, S., Choudhury, K.: NLG approach for Bangla grammatical error correction. In: 9th International Conference on Natural Language Processing Macmillan Publishers, India (2011)

3. Yuan, Z., Briscoe, T.: Grammatical error correction using neural machine translation. In: Proceedings of the 2016 Conference of the North American Chapter of the Association for Computational Linguistics: Human Language Technologies (2016). https://doi.org/10.18653/v1/N16-1042
4. Samanta, P., Chaudhuri, B.B.: A simple real-word error detection and correction using local word bigram and trigram. In: Proceedings of the Twenty-Fifth Conference on Computational Linguistics and Speech Processing (2013)
5. Sharma, S., Gupta, S.: A correction model for real-word errors. In: 4th International Conference on Eco-friendly Computing and Communication Systems (2015). https://doi.org/10.1016/j.procs.2015.10.047
6. Haque, M.M., Habib, M.T., Rahman, M.M.: Automated word prediction in Bangla language using stochastic language models. In: International Journal in Foundations of Computer Science & Technology (2015). https://doi.org/10.5121/ijfcst.2015.5607
7. Rana, M.M., Khan, M.E.A., Sultan, M.T., Ahmed, M.M., Mridha, M.F., Hamid, M.A.: Detection and correction of real-word errors in Bangla language. In: International Conference on Bangla Speech and Language Processing(ICBSLP) (2018). https://doi.org/10.1109/ICBSLP.2018.8554502
8. Wiegand, K., Patel, R.: Non-syntactic word prediction for AAC. In: NAACL-HLT 2012 Workshop on Speech and Language Processing for Assistive Technologies (SLPAT). Montreal, Canada (2012)
9. Jain, A., Jain, M.: Detection and correction of non word spelling errors in Hindi language. In: International Conference on Data Mining and Intelligent Computing (ICDMIC) (2014). https://doi.org/10.1109/ICDMIC.2014.6954235
10. Mridha, M.F., Khan, M.E.A., Rana, M.M., Ahmed, M.M., Hamid, M.A., Sultan, M.T.: An approach for detection and correction of missing word in Bengali sentence. In: International Conference on Electrical, Computer and Communication Engineering (ECCE) (2019). https://doi.org/10.1109/ECACE.2019.8679416
11. Stehouwer, H., van Zaanen, M.: Language models for contextual error detection and correction. In: Proceedings of the EACL 2009 Workshop on Computational Linguistic Aspects of Grammatical Inference. Athens, Greece (2009)
12. Rakib, O.F., Akter, S., Khan, M.A., Das, A.K., Habibullah, K.M.: Bangla word prediction and sentence completion using GRU: an extended version of RNN on N-gram language model. In: International Conference on Sustainable Technologies for Industry 4.0 (STI). Dhaka, Bangladesh (2019). https://doi.org/10.1109/STI47673.2019.9068063
13. Barmana, P.P., Boruaha, A.: A RNN based approach for next word prediction in Assamese phonetic transcription. In: 8th International Conference on Advances in Computing and Communication (2018). https://doi.org/10.1016/j.procs.2018.10.359
14. Islam, M.S., Mousumi, S.S.S., Abujar, S., Hossain, S.A.: Sequence-to-sequence Bangla sentence generation with LSTM recurrent neural networks. In: International Conference on Pervasive Computing Advances and Applications—PerCAA (2019). https://doi.org/10.1016/j.procs.2019.05.026
15. Abujar, S., Masum, A.K.M., Mazharul Hoque Chowdhury, S.M., Hasan, M., Hossain, S.A.: Bengali text generation using bi-directional RNN. In: 10th International Conference on Computing, Communication and Networking Technologies (ICCCNT). At Kanpur, India (2019). https://doi.org/10.1109/ICCCNT45670.2019.8944784
16. Ghosh, T., Abedin, M.M., Chowdhury, S.M., Yousuf, M.A.: A comprehensive review on recognition techniques for Bangla handwritten characters. In: International Conference on Bangla Speech and Language Processing (ICBSLP) (2019). https://doi.org/10.1109/ICBSLP47725.2019.202051
17. Santos, E.A., Campbell, J.C., Patel, D., Hindle, A., Amaral, J.N.: Syntax and sensibility: using language models to detect and correct syntax errors. In: 25th IEEE International Conference on Software Analysis, Evolution and Reengineering, Campobasso, Italy (2018). https://doi.org/10.1109/SANER.2018.8330219

18. Islam, S., Farhana Sarkar, M., Hussain, T., Mehedi Hasan, M., Farid, D.M., Shatabda, S.: Bangla sentence correction using deep neural network based sequence to sequence learning. In: 21st International Conference of Computer and Information Technology (ICCIT) (2018). https://doi.org/10.1109/ICCITECHN.2018.8631974
19. SUST CSE Developer Network. https://scdnlab.com/corpus
20. tanvirfahim15/BARD-Bangla-Article-Classifier. https://github.com/tanvirfahim15/BARD-Bangla-Article-Classifier
21. MinhasKamal/BengaliDictionary. https://github.com/MinhasKamal/BengaliDictionary

Quantitative Analysis of Deep CNNs for Multilingual Handwritten Digit Recognition

Mohammad Reduanul Haque, Md. Gausul Azam, Sarwar Mahmud Milon, Md. Shaheen Hossain, Md. Al-Amin Molla, and Mohammad Shorif Uddin

Abstract Indian subcontinent is a birthplace of multilingual people, where documents such as job application form, passport, number plate identification, and so forth are composed of text contents written in different languages or scripts. These scripts consist of different Indic numerals in a single document page. Recently, deep convolutional neural networks (CNN) have achieved favorable result in computer vision problems, especially in recognizing handwritten digits but most of the works focuses on only one language, i.e., English or Hindi or Bangla, etc. However, developing a language-invariant method is very important as we live in a global village now. In this work, we have examined the performance of the ten state-of-the-art deep CNN methods for the recognition of handwritten digits using four most common languages in the Indian sub-continent that creates the foundation of a script invariant handwritten digit recognition system. Among the deep CNNs, Inception-v4 performs the best based on accuracy and computation time. Besides, it discusses the limitations of existing techniques and shows future research directions.

M. R. Haque (✉) · Md. G. Azam · S. M. Milon · Md. S. Hossain
Department of Computer Science and Engineering, Daffodil International University, Dhaka, Bangladesh
e-mail: reduan.cse@diu.edu.bd

Md. G. Azam
e-mail: mr.azam.cse@gmail.com

S. M. Milon
e-mail: sarwar15-1072@diu.edu.bd

Md. S. Hossain
e-mail: shaheen15-1096@diu.edu.bd

Md. A.-A. Molla
Department of General Education Development, Daffodil International University, Dhaka, Bangladesh
e-mail: alamin.ged@diu.edu.bd

M. S. Uddin
Department of Computer Science and Engineering, Jahangirnagar University, Dhaka, Bangladesh
e-mail: shorifuddin@gmail.com

M. S. Kaiser et al. (eds.), *Proceedings of International Conference on Trends in Computational and Cognitive Engineering*, Advances in Intelligent Systems and Computing 1309, https://doi.org/10.1007/978-981-33-4673-4_2

Keywords Digit recognition · Indic digits · Language-invariant system · Deep CNN

1 Introduction

Handwritten digit recognition is the process of recognizing handwritten digits without human interaction which is a classical problem in computer vision [1]. Although many researches have been done in this field, but till now the researches continue with the aim of improving the accuracy further due to similarities of shapes, writing styles, and the interconnections between the adjacent numerals. Moreover, very few work have been developed to recognize multilingual digit recognition systems. So, this work tries to find the gap and seeks a good comparative understanding of the available deep CNN techniques using four most used handwritten Indic digit datasets, i.e., Hindi, Oriya, Bangla, and English in Indian subcontinent.

Recently, deep learning models have been shown significant improvement that successfully to overcome these challenges. Also a lot of work has already done with English [2], Hindi [3], Bengali [4], Oriya [5] handwritten digit recognition.

Wakabayashi [5] worked for recognition of offline Oriya handwritten. They utilized curvature feature to recognize numerals as almost all the Oriya characters have curve line strokes. They tested 18,190 samples of Oriya handwritten digits and obtain 94.60% accuracy. But their system failed with some particular character. Chawpatnaik [6] improved the performance by applying iterative parallel thinning approach which is well suited for shaped complex character skeletonization, particularly Oriya digit. Some fundamental needs of thinning, like stroke connectivity preservation, spurious strokes removal have been in this method. But they considered only isolated character images. Tripathy [7] presented a new approach by which a Hopfield neural network design is used to solve the printed Oriya character recognition and achieved an accuracy of 96.7%. However, they failed to recognize same shape characters and some features classification. Pattayanak [8] aimed to develop a database which is consist of complete set of Oriya digit and by using SVM classifier and graphical user interface, they have been recognized, and they used DWT for features extraction from character image in scientific way.

Hindi is one of the most popular and usable languages in subcontinent. Many research works have been done by researchers. Vohra et al. [9] aimed to recognize multilingual handwritten characters with better accuracy, and for this purpose, they used support vector machine. They applied the method on English and Hindi language datasets, and their overall accuracy was 95.048%. In their experiment, very few unknown characters recognized as noise. Reddy [10] proposed a CNN with optimizer root mean square propagation and deep feed-forward neural network. The system took test and training dataset to measure accuracy, and it was 97.33%. They worked for some specific method and took user defined dataset, and their accuracy is not belong to other method. Several pooling technique is used here [11] to recognize any speech, and max pooling is used. The aim of the work is to develop efficient entity and recognition system from given text that used in natural language processing.

English is the most popular and usable language in the world, and many researchers work with MNIST dataset. Yadav [12] proposed a reconstruction that is accurate enough to serve as a replacement for the MNIST dataset, and they reconstruct the complete MNIST test set with 60,000 samples, and their result confirmed the trend observed. Baldominos [13] work reported good accuracy on the MNIST dataset for handwritten digit recognition but their work did not perform well when they use Arabic digit dataset. Alvear-Sandoval [14] applied stacked de-noising autoencoder for improving the performance of CNN classifier. Shamsuddin [15] got 99.4% accuracy using MNIST dataset but binary and normalization data dropped the accuracy.

Many research work were being conducted by the researchers on Bangla language which mother tongue of Bangladesh as well as seventh most widely spoken language. Alom [16] represents of many perplexing characters and excessive cursive in Bangla handwriting and achieve 98.78% recognition rate using the deep CNN method but CNN with dropout Gaussian filters and Gabor filters. Rabby [17] developed and implemented of a lightweight CNN model; it can be used on a low-processing device and get 99.74 and 98.93% validation accuracy but their proposed model confusion to overwritten. Sufian [18] proposed a task oriented model based on densely connected network along with CNN (BDNet) and got 99.78% accuracy but BDNet gives 47.62% error. Maity [19] focused specifically on offline handwritten character recognition and got handwritten text with 65% accuracy and alphabets with 99.5% accuracy. Pias [20] detecting the number plate of Bangladeshi vehicles and achieve 99.60% test accuracy but they also got 50.52% accuracy in their positive and negative image.

All the approaches hold a certain degree of truth, but factors, i.e., choosing of non-standardized databases, and dimensionality retention, etc., create confusion and debate over the years. For these reasons, those who are new in this field seeking a good comparative understanding of the available techniques surrounding each conclusion are in a dilemma. So we have experimented ten deep CNN architectures such as LeNet-5, VGG16, Inception-v1, ResNet50, Xception, Inception-v3, Inception-v4, Inception-ResNet, ResNeXt50, AlexNet. Moreover, this research work has following contributions:

i. Investigate the performance of the state-of-the-art deep CNN techniques to find the best one for the recognition of multilingual handwritten digits which utilizes MNIST, Hindi, Oriya, and Bangla whose are versatile and somehow challenging handwritten digit dataset for testing the deep CNN models.
ii. Quantitatively analyze and investigate the effects of deep CNNs in multilingual handwritten digit recognition.

The reminder of the paper is organized as follows: Sect. 2 consists of general architecture and recognition methods; Sect. 3 depicts experimental results and discussions, and conclusion and the future scope are drawn in Sect. 4.

2 Recognition Methods

Several methods exist to solve handwritten digit recognition tasks including SVM, KNN, CNN and deep CNNs. Among them deep Convolutional Neural Network has exhibited an outstanding performance on the area. This research focuses on the evaluation of the individual performance of ten widely used deep CNN models (i.e., LeNet-5, VGG16, Inception-v1, ResNet50, Xception, Inception-v3, Inception-v4, Inception-ResNet, ResNeXt50, AlexNet) and provides a quantitative evaluation among these models through extensive experimentation. A brief architectural overview of these four deep CNN models is summarized in Table 1.

2.1 *Convolutional Neural Network (CNN)*

CNN has become exoteric in the field of image processing and machine learning. It was first introduced by Fukushima, where images are feed into the network, process it and classify it in different categories.

The working procedure of convolutional neural networks consists of convolutional layer, pooling layer, and fully connected layer. A generic schematic flow diagram of a handwritten digit recognition algorithm using deep CNN is shown in Fig. 1.

2.2 *Deep CNN*

Deep CNN follows the basic architecture of convolutional neural networks with some extra working procedure and strategy. For that reason, we can some extra benefit

Table 1 Brief architecture of deep CNN models

Methods/parameters	Input size	Conv. layer	Filter size	Stride	Parameter (m)	Fc. layer
LeNet-5	32*32	8	5,5	1	23	1
AlexNet	256*256	8	11,11	4	61	2
VGG-16	224*224	19	3,3	2,2	138	3
Inception-v1	299*299	21	3,5	1,1	24	2
Inception-v3	299*299	21	3,5	1,1	24	2
ResNet-50	224*224*3	22	1,3,5	2	25	2
Xception	299*299	64	5,7,9	2	22	1
Inception-v4	299*299	21	3,5	1,1	42	2
Inception-ResNets	299*299	22	5,3	2	56	3
ResNext-50	224*224	22	1,3,5	2	23	3

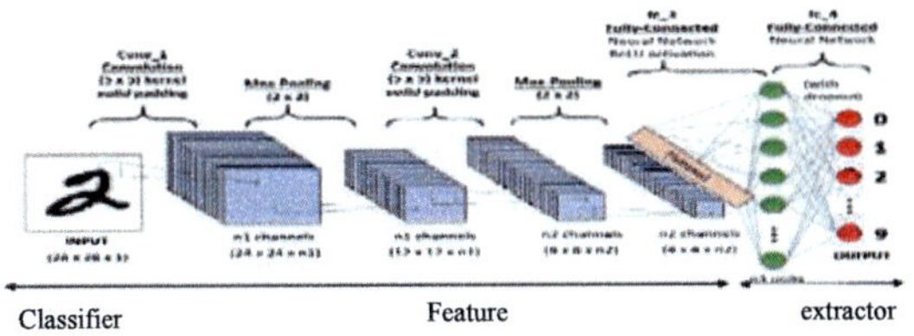

Fig. 1 Basic CNN architecture

from it. Basically, Deep CNN is an artificial function that emulates the human brain procedure and processing data creating patterns and gathering them to make some important decisions. We know, neural networks are a collection of neurons. Here, neurons are different kind of functions. These functions are taken input and give some efficient outputs.

A brief explanation of the investigated handwritten digit recognition deep CNN algorithms is given below.

LeNet-5

LeNet-5 [21] was proposed by Yann LeCun et al., where a convolutional neural network is used and trained by simple backpropagation algorithm. The method was successfully applied in handwritten zip code numbers provided by the US postal service.

AlexNet

AlexNet [22] was first used by Alex Krizhevsky. It has five convolutional layers with three fully connected layers. It used a large number of parameters (about 610,000) and used RGB image as an input.

VGG16

VGG16 ("Visual Geometry Group"- university of Oxford) was first proposed by K. Simonyan and A. Zisserman [23]. It has 16 layers with 13 convolutional and three fully connected layers. About 138 M parameters are used in this architecture. Most significant things are the accuracy of this architecture was 97.2% in ImageNet.

Inception-v1

Inception-v1 [24] achieved a milestone in CNN classifiers in compare to previous models that focus not only the performance and accuracy but also computational cost. The Inception-v1 network uses a lot of tricks to push performance, both in terms of speed and accuracy. The architecture has 22 (27, including the pooling layers) layers. It also known as GoogleNet and has nine such inception modules that were linearly fitted. Moreover, the method also reduces the total number of parameters.

Inception-v3

With 48 deep layers Inception-v3 [25] is the improvement version of Inception-v1. About 24 M parameters are used here. Instead of 5 $\times$ 5 convolution which was used in Inception-v1, here used 3 $\times$ 3 convolution.

ResNet50
With a convolution block and identity block ResNet50 [26] has been performed of five stages. Convolution block and identity block both are consist of three convolution layers. It has huge trainable parameters. About 23 million parameters are included in here.

Xception
Xception [27] is the conformation of inception. It has the same parameters like Inception-v1 (23 M). Xception maintains the 3 × 3 or 5 × 5 convolution.

Inception-v4
Inception-v4 [28] model is quite different from Inception-v4. It has 43 M parameters. Main difference from Inception-v4 is stem.

Inception-ResNet
Inception-ResNet [29] known as known as "Tiny Inception-ResNet-v2. Inception-ResNet as the same as Inception-v4. But it has 56 M parameters. This architecture was first used for the purpose of eliminate bonded labor to identify brick kilns.

ResNeXt-50
About 25 M parameters ResNeXt [30] is the special form of ResNet50. Instead of linear combination this architecture is used a hidden neural network which is being made this architecture which is more efficient.

3 Experimental Results and Discussion

3.1 Dataset Description

In this paper, MNIST, Devanagri, Oriya, and NumtaDB numerals datasets are used for the experiment. These are large, unbiased, unprocessed, and highly augmented datasets. All the datasets have been separated into training and testing sets with individual subsets.

MNIST
MNIST is a large database of handwritten digit dataset. Most of the time it is used for image processing. In machine learning field, the database is used for training and testing of in this field. It contains 60,001 image for training and 10,001 image for testing set. The MNIST database is combination of two subsets that combination create by MNIST's database.

NumtaDB
NumtaDB is a large, publicly available, and free usable datasets. It contains 85,000 image. It highly unprocessed and augment data because of its diverse shape. In

experiment, 85% data are used as training data, and 15% data are used for testing set.

Oriya
Oriya script recently developed from Kalinga script. It has 5970 sample image, and 17000 image are used for training set, and 3000 sample image are used for testing image.

Devnagari
Devnagari is contain different type of dataset including Hindi. It contains 46 classes of character that is huge, and each character contains 2000 example. The dataset is splitted into training (85%) and testing (15%) set. Here, most of the dataset are grayscale image that is used for benchmark classification algorithm for OCR system.

3.2 Discussion

Figures 2, 3, 4, and 5 represent some misclassified images that failed to recognize by all the above CNNs. Datasets statistics are shown in Table 2. Table 3 presents the recognition accuracy after applying different methods. In almost all the cases, Inception-v3 has shown better accuracy. Moreover, confusion matrix for Inception-v3 using NumtaDB dataset is described in Table 4.

Fig. 2 Sample image from different training sets from MNIST dataset

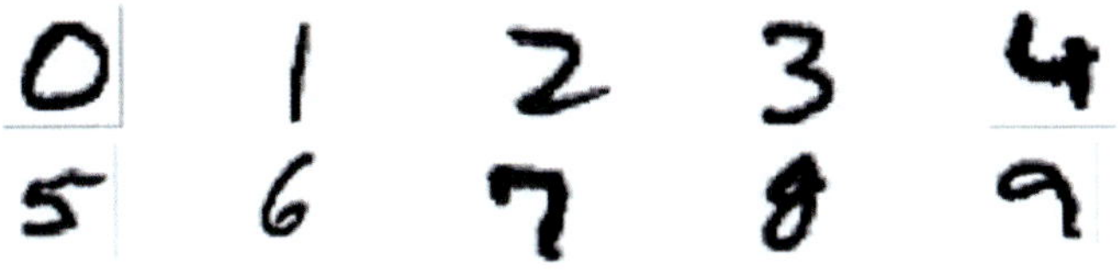

Fig. 3 Sample images from Hindi dataset

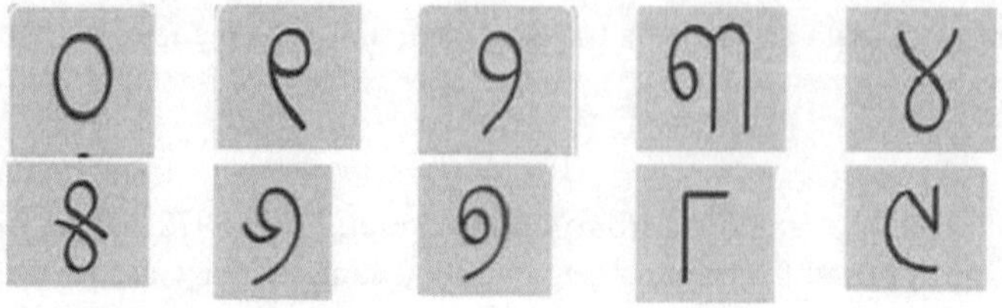

Fig. 4 Sample images from Oriya dataset

Fig. 5 Sample images from Bangla dataset

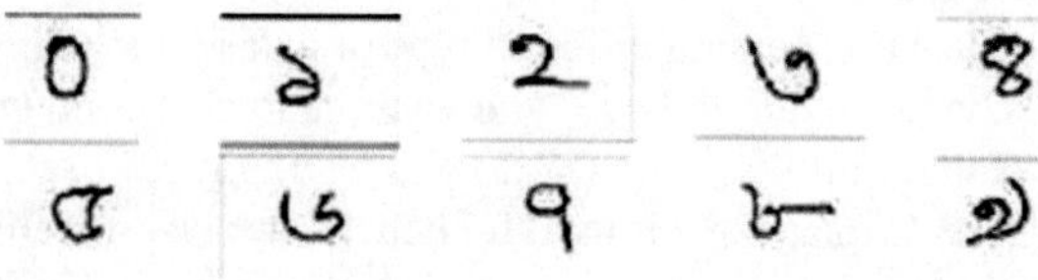

Table 2 Dataset statistics (for MNIST, Devanagri, Oriya, and NumtaDB)

Dataset	Number of training data	Number of testing data
MNIST	60,001	10,001
Devanagri	16,000	4000
Oriya	17,000	3000
NumtaDB	72,045	17,626

Table 3 Recognition accuracy under different deep CNNs

	Accuracy (in percentage)			
Method	MNIST	Devanagri	ORIYA	NumtaDB
LeNet-5	97	95.8	98	93
VGG16	92.1	94	98	99.80
Inception-v1	89.3	96	91.2	96
ResNet50	81.8	98	96	89
Xception	94.5	89.9	97	91.68
Inception V3	90.1	98	99	98
Inception-v4	97	98.3	92.1	78.38
Inception-ResNet	84.65	96	97.6	81
ResNeXt50	97	98.3	92.1	86.43
AlexNet	99.3	98	99	98.48

Table 4 Confusion matrix for Inception-v3 using NumtaDB dataset

	Predicted class										
Actual class		0	1	2	3	4	5	6	7	8	9
	0	**100**	0	0	0	0	0	0	0	0	0
	1	0	**98**	0	0	0	0	0	0	0	0
	2	0	0	**96**	0	0	0	0	1	0	0
	3	0	0	0	**100**	0	0	0	0	1	1
	4	0	0	0	0	**100**	0	0	0	0	0
	5	0	0	0	0	0	**100**	0	0	0	0
	6	0	0	0	0	0	0	**100**	0	0	2
	7	0	2	0	0	0	0	0	**95**	0	0
	8	0	0	4	0	0	0	0	0	**97**	3
	9	0	0	0	0	0	0	0	5	2	**94**
Accuracy = 98.00%											

Fig. 6 Some misclassified images from Bangla Digit dataset

Fig. 7 Some misclassified images from Hindi Digit dataset

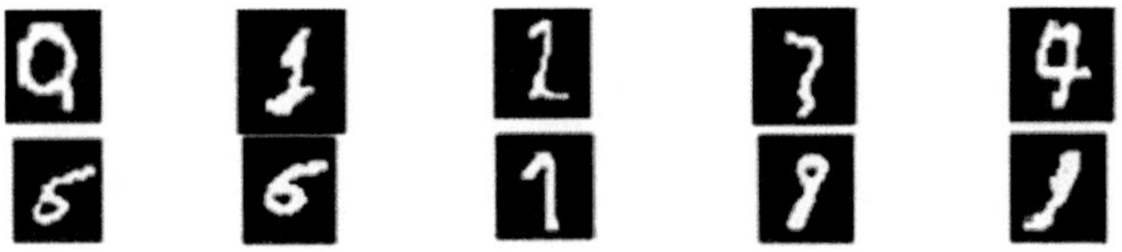

Fig. 8 Some misclassified image from MNIST dataset

Fig. 9 Some misclassified images from Oriya dataset

4 Conclusion

As a citizen of global village, we cannot avoid the importance of multilingual handwritten digit recognition. So, in this work, we have implemented ten deep CNN models as well as test their accuracy on multilingual digits datasets (i.e., MNIST, Bangla, Hindi, and Oriya and finally, examined which deep CNN performed well based on accuracy. This experimental result showed that Inception V4 performed better than other methods in all the datasets. Thus, this research can reveal a way for finding the method that is best performed in recognizing multilingual handwritten digit dataset. The future endeavor for this research is to add more languages so that a more robust generic script recognizer can be developed.

References

1. Pal, U., Chaudhuri, B.B.: Indian script character recognition: a survey. Pattern Recogn. **37**(9), 1887–1899 (2004)
2. Lopez, B., Nguyen, M.A., Walia, A.: Modified mnist (2019)
3. Chaudhary, M., Mirja, M.H., Mittal, N.: Hindi numeral recognition using neural network. Int. J. Sci. Eng. Res. **5**(6), 260–268 (2014)
4. Pal, U., Chaudhuri, B.B.: Automatic recognition of unconstrained off-line Bangla handwritten numerals. In: International Conference on Multimodal Interfaces. Springer, Berlin, Heidelberg (2000)
5. Pal, U., Wakabayashi, T., Kimura, F.: A system for off-line Oriya handwritten character recognition using curvature feature. In: 10th International Conference on Information Technology (ICIT), pp. 227–229 (2007)
6. Bag, S., Chawpatnaik, G.: A modified parallel thinning method for handwritten Oriya character images. In: Proceedings of the 4th International Conference on Frontiers in Intelligent Computing: Theory and Applications (FICTA) 2015. Springer, New Delhi (2016)
7. Jena, O.P., Pradhan, S.K., Biswal, P.K., Tripathy, A.R.: Odia characters and numerals recognition using hopfield neural network based on Zoning features. Int. J. Recent Technol. Eng. **8**(2), 4928–4937 (2019)
8. Pattanayak, Sudha, S., Pradhan, S.K., Mallik, R.C.: Printed Odia symbols for character recognition: a database study. In: Advanced Computing and Intelligent Engineering, pp. 297–397. Springer, Singapore (2020)
9. Vohra, U.S., Dwivedi, S.P., Mandoria, H.L.: Study and analysis of multilingual hand written characters recognition using SVM classifier. Oriental J. Comput. Sci. Technol. **9**(2), 109–114 (2016)
10. Reddy, R., Kumar, V., Babu, U.R.: Handwritten Hindi character recognition using deep learning techniques. Department of CSE, Acharya Nagarjuna University, Guntur, India. Int. J. Comput. Sci. Eng. (2019)
11. Fujisawa, H.: Forty years of research in character and document recognition—an industrial perspective. Pattern Recogn. **41**(8), 2435–2446 (2008)
12. Yadav, C., Bottou, L.: Cold case: the lost mnist digits. In: Advances in Neural Information Processing Systems (2019)
13. Baldominos, Alejandro, Saez, Y., and Pedro I.: A survey of handwritten character recognition with mnist and emnist. Applied Sciences 9.15, 3169, (2019).
14. .Alvear-Sandoval, R.F., Sancho-Gómez, J.L., Figueiras-Vidal, A.R.: On improving CNNs performance: the case of MNIST. Information Fusion. **52**, 106–109 (2019)

15. Shamsuddin, M., Razif, S., Abdul-Rahman, Mohamed, A.: Exploratory analysis of MNIST handwritten digit for machine learning modelling. In: International Conference on Soft Computing in Data Science. Springer, Singapore (2018)
16. Alom, M., Sidike, P., Taha, T., Asari, V.: Handwritten Bangla Digit Recognition Using Deep Learning. *ArXiv* abs/1705.02680 (2017)
17. Rabby, Azad, AKM.S.: et al.: Bangla handwritten digit recognition using convolutional neural network. In: Emerging Technologies in Data Mining and Information Security, pp. 111–112. Springer, Singapore (2019)
18. Sufian, A., et al.: Bdnet: Bengali handwritten numeral digit recognition based on densely connected convolutional neural networks. J. King Saud Univ.-Comput. Inf. Sci. (2020)
19. Maity, S., et al.: Handwritten Bengali character recognition using deep convolution neural network. In: International Conference on Machine Learning, Image Processing, Network Security and Data Sciences. Springer, Singapore (2020)
20. Pias, M., Mutasim, A.K., Amin, M.A.: Bangladeshi number plate detection: cascade learning versus deep learning. In: Proceedings of the 15th International Workshop on Content-Based Multimedia Indexing (2017)
21. Lecun, Y., Bottou, L., Bengio, Y., Haffner, P.: Gradient-based learning applied to document recognition. Proc. IEEE **86**(11), 2278–2324 (1998)
22. Kienzle, W., Chellapilla, K.: Personalized handwriting recognition via biased regularization. In: Proceedings of the 23rd International Conference on Machine Learning (2006)
23. Simonyan, K., Zisserman, A.: Very Deep Convolutional Networks for Large-Scale Image Recognition. arXiv preprint arXiv: 1409.1556 (2014)
24. Mohd, S.S., et al.: Offline Signature Verification using Deep Learning Convolutional Neural Network (CNN) Architectures GoogLeNet Inception-v1 and Inception-v3. Procedia Comput. Sci. **161**, 475–483 (2019)
25. Cheng, W., et al.: Pulmonary image classification based on inception-v3 transfer learning model. IEEE Access **7**, 146533–146541 (2019)
26. Wen, L., Xinyu, L., Liang, G.: A transfer convolutional neural network for fault diagnosis based on ResNet-50. Neural Comput. Appl. 1–14 (2019)
27. Chollet, F.: Xception: Deep learning with depthwise separable convolutions. In: Proceedings of the IEEE Conference on Computer Vision and Pattern Recognition (2017)
28. Sevilla, A., Glotin, H.: Audio Bird Classification with Inception-v4 extended with Time and Time-Frequency Attention Mechanisms. CLEF (Working Notes), (2017).
29. Chen, X., et al.: Visual crowd counting with improved inception-ResNet-A module. In: 2018 IEEE International Conference on Robotics and Biomimetics (ROBIO). IEEE (2018)
30. Qianqian, Z., Sen, L., Weiming, G.: Research on Vehicle Appearance Component Recognition Based on Mask R-CNN. In: Journal of Physics: Conference Series, vol. 1335, no. 1. IOP Publishing (2019)

Performance Analysis of Machine Learning Approaches in Software Complexity Prediction

Sayed Moshin Reza, Md. Mahfujur Rahman, Hasnat Parvez, Omar Badreddin, and Shamim Al Mamun

Abstract Software design is one of the core concepts in software engineering. This covers insights and intuitions of software evolution, reliability, and maintainability. Effective software design facilitates software reliability and better quality management during development which reduces software development cost. Therefore, it is required to detect and maintain these issues earlier. Class complexity is one of the ways of detecting software quality. The objective of this paper is to predict class complexity from source code metrics using machine learning (ML) approaches and compare the performance of the approaches. In order to do that, we collect ten popular and quality maintained open source repositories and extract 18 source code metrics that relate to complexity for class-level analysis. First, we apply statistical correlation to find out the source code metrics that impact most on class complexity. Second, we apply five alternative ML techniques to build complexity predictors and compare the performances. The results report that the following source code metrics: Depth inheritance tree (DIT), response for class (RFC), weighted method count (WMC), lines of code (LOC), and coupling between objects (CBO) have the most impact on class complexity. Also, we evaluate the performance of the techniques, and results show that random forest (RF) significantly improves accuracy without providing additional false negative or false positive that work as false alarms in complexity prediction.

S. Moshin Reza · O. Badreddin
University of Texas, Austin, TX, USA
e-mail: sreza3@miners.utep.edu

O. Badreddin
e-mail: obbadreddin@utep.edu

Md. Mahfujur Rahman (✉)
Daffodil International University, Dhaka, Bangladesh
e-mail: mrrajuiit@gmail.com

H. Parvez · S. Al Mamun
Jahangirnagar University,Dhaka, Bangladesh
e-mail: hasnatiit847@gmail.com

S. Al Mamun
e-mail: shamim@juniv.edu

M. S. Kaiser et al. (eds.), *Proceedings of International Conference on Trends in Computational and Cognitive Engineering*, Advances in Intelligent Systems and Computing 1309, https://doi.org/10.1007/978-981-33-4673-4_3

Keywords Software complexity · Software quality · Machine learning · Software design · Software reliability

1 Introduction

Software design is a process of creating software artifacts, primitive components, and constraints. Effective software design with object oriented structures facilitates better software quality, reusability, and maintainability [1]. One of the quality factors is complexity. This quality attribute is determined by many factors related to code structures, object-oriented properties, and source code metrics [2]. The less the complexity of a software, the less the cost of software development will be [3, 4]. This motivates us to research on software complexity prediction.

In software life cycle, the more the complexity is, maintenance becomes costly, unpredictable, human-intensive activity [2]. Moreover, high maintenance efforts often affect the software sustainability that many software systems become unsustainable over time [5, 6]. Therefore, software redesign becomes an essential step where complexity of the software needs to be reduced. Such action will enhance software maintainability and reduce the associated costs [7, 8]. Having set the importance of complexity detection for software redesign, we are motivated to predict class-level complexity from source code metrics.

Some studies introduced McCabe complexity, a widely accepted metrics developed by Thomas McCabe to show the level of software complexity [9]. Another approach on calculation of software complexity was based on counting number of operators and operands in software. But the calculation and counting process of total operators and operands are tedious [10].

In this paper, we use machine learning techniques to build complexity predictor. The reason behind using machine learning to get rid of manual process or code rules to detect class complexity. Also, successful research on detecting software defect, vulnerability using ML techniques motivate us [11, 12]. We use five ML classifiers, analyze the performance of the classifiers, and report the best technique in complexity prediction.

The rest of the paper is organized as follows. We present literature reviews in Sect. 2. We describe research methodology in Sect. 3. Results and evaluation are discussed in Sect. 4 and finally, we conclude the paper in Sect. 5.

2 Literature Review

Several research on code quality from source code metrics includes fault-prone modules detection [1], early detection of vulnerabilities [11], improvement of network software security [13, 14], software redesign [9], etc. All of these researches are targeted to reduce the maintenance effort and cost during the software development.

Chowdhury et al. investigate the efficacy of applying cohesion, complexity, and coupling metrics to automatically predict vulnerability and complexity entities [11]. This study used machine learning and statistical approaches to predict vulnerability that learn from the cohesion, complexity, and coupling metrics. The results indicate that structural information from the non-security realm such as cohesion, complexity, and coupling is useful in vulnerability prediction which minimize the maintenance effort.

Another study proposed by Briand et al. [15] analyzed correspondence between object-oriented metrics and fault proneness. This research results are created based upon few number of classes analysis. Gegick et al. [16] developed a heuristic model to predict vulnerable components and complexity. The model was successful on a large commercial telecommunications software and predicted vulnerable components with 8% false positive rate and 0% false negative rate.

In this research, we analyze source code metrics in relation to complexity. Also, we apply ML techniques to predict complexity from source code metrics.

3 Research Methodology

This research has two main goals. First, analyze source code metrics to what extent it is possible to predict complexity. Second, report the best ML approaches evaluating relative effectiveness in prediction of complexity from source code metrics. The details of our research questions, datasets, and machine learning approaches are discussed in the following subsections.

3.1 Research Questions

This research is focused on answering two primary research questions.

Research Question 1: How source code metrics are correlated with quality attribute: class complexity?

This question reveals the relationships between complexity and source code metrics, such as number of attributes, and lines of code. To answer this question, we apply statistical correlation on 18 source code metrics and complexity collected from ten different source code repositories to find out the relationship.

Research Question 2: How accurately can machine learning approaches predict class complexity from source code metrics?

This question is targeted to find out the accuracy of machine learning approaches in class-level complexity detection. We apply five machine learning techniques and evaluate the performance. This question reveals the best technique in detecting class complexity from source code metrics.

3.2 Proposed Research Framework

The proposed research is build upon three steps. First, extracting source code metrics and complexity from classes of large code bases. Second, prepare the dataset for complexity prediction by applying data cleaning process. Third, apply ML techniques and evaluate to find out the best one.

For the first step, we extract source code metrics and quality feature: complexity from a large number of classes. The details of dataset creation process are discussed in Sect. 3.3. In the second step, we apply data cleaning process to get better learned ML model. Uncleaned data fed into machine learning techniques may result to a bad model creation [17]. The details of the process are discussed in Sect. 3.4. For the final step, we select several ML techniques and train the dataset to detect highly complex classes. We also assess ML prediction effectiveness using performance metrics. The detailed picture of the study is shown in Fig. 1.

3.3 Dataset Collection

Dataset for complexity prediction needs diverse set of repositories. We search code-base repositories using ModelMine tool [18] with the following criteria; a repository with primary language Java, a minimum of 5000 commits, at least 100 active contributors, a minimum of 3000 stars and 500 forks. The selected repositories are shown in Table 1 with repository metadata information.

To validate the diversity of repositories, we consider high number of stars and forks as a proxy for popularity of repositories and high number of commits as a proxy of maintenance. Also, we consider repository size as follows: low (1–1000

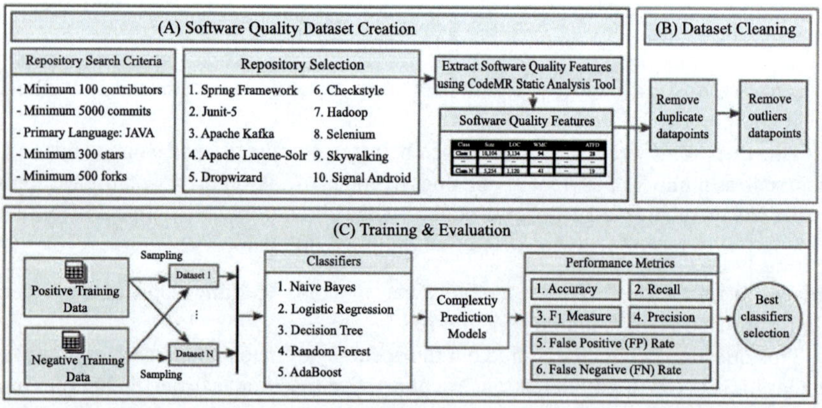

Fig. 1 Proposed methodology

Table 1 Selected repositories with metadata information

Serial	Repository name	Repository link	Commits	Contributors	Stars	Forks	Lines of code	Classes
1	Spring framework	https://github.com/spring-projects/spring-framework	21,154	491	38,200	25,800	232,447	5628
2	Junit5	https://github.com/junit-team/junit5/	6286	146	4000	899	16,856	659
3	Apache Kafka	https://github.com/apache/kafka	7787	691	16,300	8700	119,299	2463
4	Apache Lucene-Solr	https://github.com/apache/lucene-solr	33,899	194	3600	2500	602,185	8850
5	Dropwizard	https://github.com/dropwizard/dropwizard	5448	345	7700	3200	14,268	508
6	Checkstyle	https://github.com/checkstyle/checkstyle	9408	232	5400	7400	26,030	454
7	Hadoop	https://github.com/apache/hadoop	24,001	280	10,600	6600	695,992	10,496
8	Selenium	https://github.com/SeleniumHQ/selenium	25,354	518	18,100	5800	36,031	1175
9	Skywalking	https://github.com/apache/skywalking	5753	245	14,000	4100	61,588	2531
10	Signal-Android	https://github.com/signalapp/Signal-Android	5777	206	13,400	3400	116,268	2861

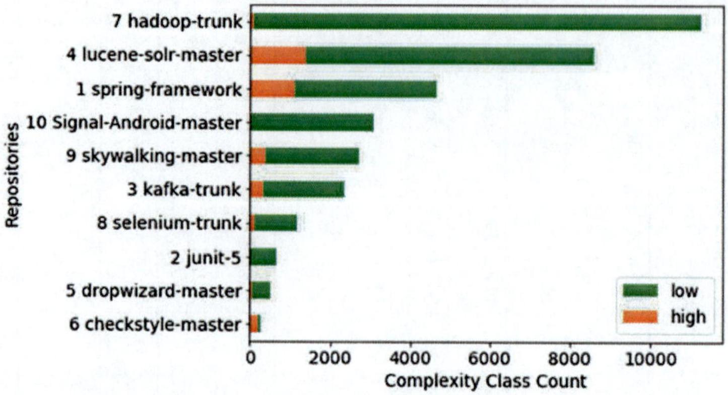

Fig. 2 Complexity distribution among repositories

classes), medium (1001–5000 classes), and high (more than 5000 classes) in size. This selection implies diversity in complexity of classes. Figure 2 shows number of complexity classes against each selected repository where three of them are selected from low, four of them are selected from medium, and rest of them are selected from high volume of category.

After extracting code repositories, we extract source code metrics for each class in the repository using CODEMR tool [19]. The tool provides 18 unique source code metrics for each class. The details of the source code metrics are described in Table 2. The target variable data is collected also for each class using same tool with different process. The data is then combined using the class file name for training and testing purpose.

3.4 *Dataset Cleaning and Analysis*

Data cleaning is critically important step for the complexity prediction. To get optimistic performance result of ML approaches, we clean the data in two stages. First, by identifying column variables that have single value or very few unique values. In this stage, we also remove the duplicate observations. In second stage, we apply box plot for each source code metrics and find the outliers. This technique helps to remove the bias datapoints from the dataset.

After cleaning the dataset, we have come up with much more differential and clear dataset for complexity prediction. Figure 3a visualizes the relationship between weighted method count, lines of code, and complexity. Figure 3b visualizes the relationship between response for class, method lines of code, and complexity.

Table 2 Source code metrics

No	Source code metric name	Description
1	Class lines of code (CLOC)	The number of all non-commented and nonempty lines of a class
2	Weighted method count (WMC)	The weighted sum of all class' methods
3	Depth of inheritance tree (DIT)	The location of a class in the inheritance tree
4	Number of children (NOC)	The number of associated sub-classes of a class
5	Coupling between object classes (CBO)	The number of classes that another class is coupled to
6	Response for a class (RFC)	The number of the methods that can be potentially invoked in response by an object of a class
7	Simple response for a class (SRFC)	The number of the methods that can be potentially invoked in response by an object of a particular class
8	Lack of cohesion of methods (LCOM)	Measure how methods of a class are related to each other
9	Lack of cohesion among methods (LCAM)	Measure cohesion based on parameter types of methods
10	Number of fields (NOF)	The number of fields (attributes) in a class
11	Number of methods (NOM)	The number of methods in a class
12	Number of static fields (NOSF)	The number of static fields in a class
13	Number of static methods (NOSM)	The number of static methods in a class
14	Specialization index (SI)	Measures the extent to which sub-classes override their ancestor's classes
15	Class-methods lines of code (CMLOC)	Total number of all nonempty, non-commented lines of methods inside a class
16	Number of overridden methods (NORM)	The number of methods that are inherit from a super-class and has return type as the method that it overrides
17	Lack of tight class cohesion (LTCC)	Measures cohesion between the public methods of a class and subtract from 1
18	Access to foreign data (ATFD)	The number of classes whose attributes are directly or indirectly reachable from the a class

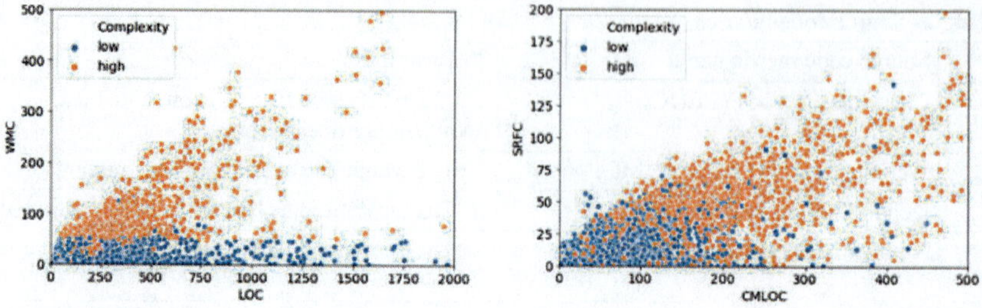

(a) WMC, LOC vs Complexity (b) SRFC, CMLOC vs Complexity

Fig. 3 Relationship of input variables with target variable

3.5 Machine Learning Classifiers and Evaluation Metrics

This subsection provides a brief overviews of five alternative machine learning classifiers used to build class complexity predictors. The machine learning classifiers are as follows: (1) Naive Bayes (NB), (2) Logistic Regression (LR), (3) Decision Tree (DT), (4) Random Forest (RF), and (5) AdaBoost (AB). These classifiers are well-known classifiers in building vulnerability predictors and used in several similar research [11, 20, 21]. The statistical performance of selected ML classifiers is calculated by performing ten-fold cross-validation technique. Cross-validation is a technique for assessing how accurately a predictive model will perform in practice after generating the model [22]. The objective of such operation is to reduce the variability of the results.

4 Result and Discussion

This section describes the results of correlation analysis, complexity prediction using ML models and compares the performance of ML classifiers.

4.1 Correlation Results

The results of Pearson correlation reveal the impact of source code metrics on quality attribute: complexity. Figure 4 visualizes the correlation between source code metrics and complexity. It is clear in the figure that not any single metric highly impact on complexity. This quality attribute is formed based on a combined behavior of source code metrics. Among the code metrics, DIT, SRFC, RFC, WMC, CMLOC, and CBO have moderately high impact on complexity. Generally, classes with higher number of WMC, LOC, or DIT associated with high number of defect in the software,

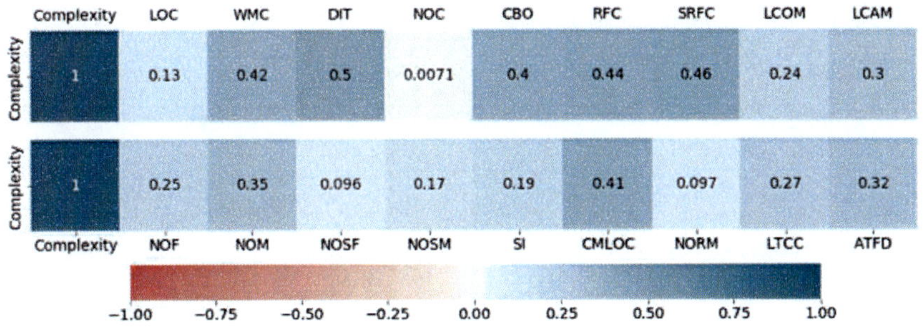

Fig. 4 Correlation among source code metrics and quality attribute

and it becomes hard to maintain over time [12]. This issue is also mentioned by Subramanyam et al. that DIT and CBO have influenced class complexity [12]. In another research, Chowdhury et al. experimentally showed that WMC, DIT, RFC, and CBO code-level metrics are strongly correlated to vulnerabilities which are directly generated from file complexity [23]. This answers research question 1.

4.2 *Performance Results*

In this subsection, we discuss the performance of ML complexity predictors. We use the following evaluation metrics: accuracy, precision, recall, F_1 score, FP rate, and FN rate to compare the performances. At first, we generate confusion matrices from the validation set. Table 3 visualizes the confusion matrices of the classifiers for predicting software complexity.

We evaluate the techniques using following metrics: accuracy, precision, recall, F_1 score, FP and FN rate, and the results are visualized in Table 4. Accuracy and precision are most used measurement in comparing the performance. Table 4 and Fig. 5 shows the accuracy and precision value of the selected classifiers. The result implies decision tree and random forest classifier have the highest accuracy and precision than other classifiers. We also observe random forest has highest recall and F_1 score.

Table 3 Confusion matrices of classifiers for predicting software complexity

Classifier names	Naive Bayes		Logistic regression		Decision tree		Random forest		Ada Boost	
	Predicted									
Actual	Low	High	Low	High	Low	High	Low	High	Low	High
Low	6416	475	6766	125	6832	59	6820	71	6813	78
High	330	434	173	591	223	541	58	706	82	682

Table 4 Prediction performance of machine learning models

Serial	Classifier name	Accuracy	Precision	Recall	F1 score	FP rate	FN rate
1	Naive Bayes	89	71	75	73	6.88	42.11
2	Logistic regression	96	91	86	88	1.44	26.08
3	Decision tree	98	95	96	96	0.90	7.53
4	Random forest	98	95	99	97	1.00	1.95
5	Ada Boost	97	94	93	94	1.13	12.27

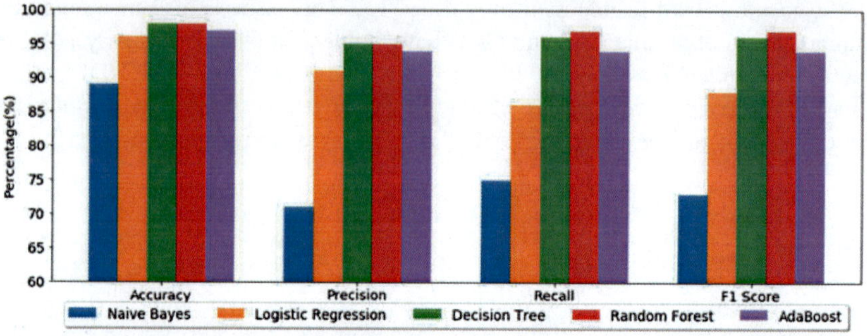

Fig. 5 Relative performance of ML classifiers

However, we evaluate the classifiers with another set of metrics: false positive rate and false negative rate. The higher the FN rate, the model generates more false alarms. This implies high complex classes are detected as low complex classes which are very risky. Figure 6 shows the relative performance of classifiers in terms of false positive rate and false negative rate. One may have to tolerate many false positives to ensure reduced number of complex classes left undetected. As such, if the target is to predict a larger percentage of high complexity class files, then Naive Bayes classifier can be evaluated favorably although in overall prediction, random forest and decision tree classifier'performance are better.

On the other hand, if the target is to predict a fewer percentage of high complex files as low to avoid risk, then obviously random forest might be the good choice as it has the lowest false negative rate. We focus more on false negative rate to reduce the risk of detecting high complex class as low. RF results indicate that it is much better model in prediction of complexity because of its bootstrapping random re-sample technique and working with significant elements. On the other hand, DT is working with all elements, and as a result, it creates more false alarms than RF. Therefore, random forest is the best complexity predictor among selected ML techniques.

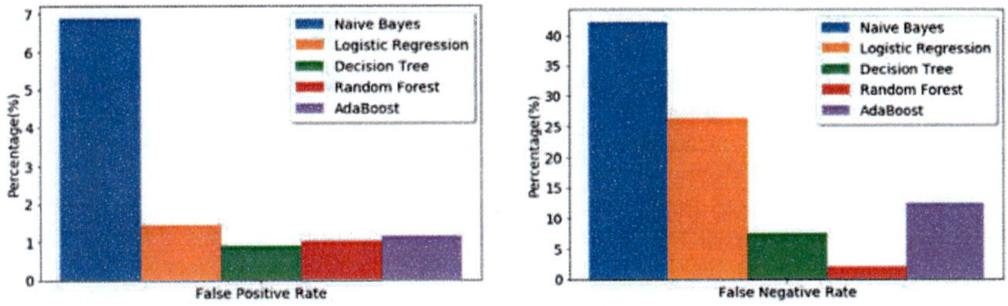

Fig. 6 Relative FP and FN rate of ML classifiers

5 Conclusion

In this study, we analyze the software source code metrics which are mostly impacted the class complexity. It is undoubtedly necessary to take proper action before classes become more complex. Otherwise, it will become more expensive to test and fix if large number of classes become highly complex. To reduce such risk and cost, it is necessary to build complexity predictor.

We start with extracting 38,778 classes of dataset with 18 source code metrics, we use five different machine learning approaches to train the dataset to classify high or low complex classes. In evaluation, we compare the performance of the approaches using the evaluation metrics. The result shows that RF classifier predicts high complexity classes with an accuracy of 98% and also having lowest FN rate of 1.95. Therefore, random forest is considered as best classifier to detect class complexity. In summary, we have made the following observation from our study. First, cross-validation implies low variance of performance metrics detecting software complexity. Second, FN rate needs to be reduced as much as possible to avoid the risk of detecting high complex class as low complex class.

Finally, the observations and results from this study can be useful in software quality research. Using ML automatic prediction on code quality will allow quality managers, practitioners to take preventive actions against bad quality, faults, and errors. Such proactive actions will allow software redesign and maintenance which ensure better software quality during the development.

References

1. Alakus, T.B., Das, R., Turkoglu, I.: An overview of quality metrics used in estimating software faults. In: 2019 International Artificial Intelligence and Data Processing Symposium (IDAP), pp. 1–6. IEEE (2019)
2. Ogheneovo, E.E., et al.: On the relationship between software complexity and maintenance costs. J. Comput. Commun. **2**(14), 1 (2014)

3. Yu, S., Zhou, S.: A survey on metric of software complexity. In: 2010 2nd IEEE International Conference on Information Management and Engineering, pp. 352–356. IEEE (2010)
4. Reza, S.M., Rahman, M.M., Parvez, M.H., Shamim Kaiser, M., Mamun, S.A.: Innovative approach in web application effort & cost estimation using functional measurement type. In: 2015 International Conference on Electrical Engineering and Information Communication Technology (ICEEICT), pp. 1–7. IEEE (2015)
5. Durdik, Z., Klatt, B., Koziolek, H., Krogmann, K., Stammel, J., Weiss, R.: Sustainability guidelines for long-living software systems. In: 2012 28th IEEE International Conference on Software Maintenance (ICSM), pp. 517–526. IEEE (2012)
6. Reza, S.M., Rahman, M.M., Mamun, S.A.: A new approach for road networks-a vehicle xml device collaboration with big data. In: 2014 International Conference on Electrical Engineering and Information & Communication Technology, pp. 1–5. IEEE (2014)
7. Bhattacharya, P., Iliofotou, M., Neamtiu, I., Faloutsos, M.: Graph-based analysis and prediction for software evolution. In: 2012 34th International Conference on Software Engineering (ICSE), pp. 419–429. IEEE (2012)
8. Paul, M.C., Sarkar, S., Rahman, M.M., Reza, S.M., Shamim Kaiser, M.: Low cost and portable patient monitoring system for e-health services in Bangladesh. In: 2016 International Conference on Computer Communication and Informatics (ICCCI), pp. 1–4. IEEE (2016)
9. Moreno-León, J., Robles, G., Román-González, M.: Comparing computational thinking development assessment scores with software complexity metrics. In: 2016 IEEE Global Engineering Education Conference (EDUCON), pp. 1040–1045. IEEE (2016)
10. Singh, G., Singh, Dilbag, Singh, V.: A study of software metrics. IJCEM Int. J. Comput. Eng. Manage. **11**, 22–27 (2011)
11. Chowdhury, I., Zulkernine, M.: Using complexity, coupling, and cohesion metrics as early indicators of vulnerabilities. J. Syst. Arch. **57**(3), 294–313 (2011)
12. Subramanyam, R., Krishnan, Mayuram S.: Empirical analysis of ck metrics for object-oriented design complexity: implications for software defects. IEEE Trans. Softw. Eng. **29**(4), 297–310 (2003)
13. Moshtari, S., Sami, A., Azimi, M.: Using complexity metrics to improve software security. Comput. Fraud Sec. **2013**(5), 8–17 (2013)
14. Rahman, S., Sharma, T., Reza, S.M., Rahman, M.M., Kaiser, M.S., et al.: Pso-nf based vertical handoff decision for ubiquitous heterogeneous wireless network (uhwn). In: 2016 International Workshop on Computational Intelligence (IWCI), pp. 153–158. IEEE (2016)
15. Briand, L.C., Wüst, J., Daly, J.W., Victor Porte, D.: Exploring the relationships between design measures and software quality in object-oriented systems. J. Syst. Softw. **51**(3), 245–273 (2000)
16. Gegick, M., Williams, L., Osborne, J., Vouk, M.: Prioritizing software security fortification throughcode-level metrics. In: Proceedings of the 4th ACM workshop on Quality of protection, QoP '08, pp. 31–38. Association for Computing Machinery, New York, NY, USA, Oct 2008
17. Munappy, A., Bosch, J., Olsson, H.H., Arpteg, A., Brinne, B.: Data management challenges for deep learning. In: 2019 45th Euromicro Conference on Software Engineering and Advanced Applications (SEAA), pp. 140–147. IEEE (2019)
18. Reza, S.M., Badreddin, O., Rahad, K.: Modelmine: a tool to facilitate mining models from open source repositories. In: 2020 ACM/IEEE 23rd International Conference on Model Driven Engineering Languages and Systems (MODELS). ACM (2020)
19. Shaheen, A., Qamar, U., Nazir, A., Bibi, R., Ansar, M., Zafar, I.: Oocqm: object oriented code quality meter. In: International Conference on Computational Science/Intelligence & Applied Informatics, pp. 149–163. Springer (2019)
20. Zhang, Y., Lo, D., Xia, X., Xu, B., Sun, J., Li, S.: Combining software metrics and text features for vulnerable file prediction. In: 2015 20th International Conference on Engineering of Complex Computer Systems (ICECCS), pp. 40–49. IEEE (2015)
21. Jimenez, M., Rwemalika, R., Papadakis, M., Sarro, F., Traon, Y.L., Harman, M.: The importance of accounting for real-world labelling when predicting software vulnerabilities. In: Proceedings of the 2019 27th ACM Joint Meeting on European Software Engineering Conference and Symposium on the Foundations of Software Engineering, pp. 695–705 (2019)

22. Yadav, S., Shukla, S.: Analysis of k-fold cross-validation over hold-out validation on colossal datasets for quality classification. In: 2016 IEEE 6th International Conference on Advanced Computing (IACC), pp. 78–83. IEEE (2016)
23. Chowdhury, I., Zulkernine, M.: Can complexity, coupling, and cohesion metrics be used as early indicators of vulnerabilities? In: Proceedings of the 2010 ACM Symposium on Applied Computing, pp. 1963–1969 (2010)

Bengali Abstractive News Summarization (BANS): A Neural Attention Approach

Prithwiraj Bhattacharjee, Avi Mallick, Md. Saiful Islam, and Marium-E-Jannat

Abstract Abstractive summarization is the process of generating novel sentences based on the information extracted from the original text document while retaining the context. Due to abstractive summarization's underlying complexities, most of the past research work has been done on the extractive summarization approach. Nevertheless, with the triumph of the sequence-to-sequence (seq2seq) model, abstractive summarization becomes more viable. Although a significant number of notable research has been done in the English language based on abstractive summarization, only a couple of works have been done on Bengali abstractive news summarization (BANS). In this article, we presented a seq2seq based Long Short-Term Memory (LSTM) network model with attention at encoder-decoder. Our proposed system deploys a local attention-based model that produces a long sequence of words with lucid and human-like generated sentences with noteworthy information of the original document. We also prepared a dataset of more than 19 k articles and corresponding human-written summaries collected from bangla.bdnews24.com (https://bangla.bdnews24.com/) which is till now the most extensive dataset for Bengali news document summarization and publicly published in Kaggle (https://www.kaggle.com/prithwirajsust/bengali-news-summarization-dataset) We evaluated our model qualitatively and quantitatively and compared it with other published results. It showed significant improvement in terms of human evaluation scores with state-of-the-art approaches for BANS.

P. Bhattacharjee (✉) · A. Mallick · Md. Saiful Islam · Marium-E-Jannat
Department of Computer Science and Engineering, Shahjalal University of Science and Technology, Sylhet 3114, Bangladesh
e-mail: prithwiraj12@student.sust.edu

A. Mallick
e-mail: avimallick999@gmail.com

Md. Saiful Islam
e-mail: saiful-cse@sust.edu

Marium-E-Jannat
e-mail: jannat-cse@sust.edu

M. S. Kaiser et al. (eds.), *Proceedings of International Conference on Trends in Computational and Cognitive Engineering*, Advances in Intelligent Systems and Computing 1309, https://doi.org/10.1007/978-981-33-4673-4_4

Keywords Attention · Abstractive summarization · BLEU · ROUGE · Dataset · seq2seq · LSTM · Encoder–decoder · Bengali

1 Introduction

Text or document summarization is the process of transforming a long document or documents into one or more short sentences which contain the key points and main contents. Automatic summarization became vital in our daily life in order to minimize the effort and time for finding the condensed and relevant delineation of an input document that captures the necessary information of that document. Despite different ways to write the summary of a document, the summarization can be categorized into two classes based on the content selection and organization: **Extractive** and **Abstractive** approach. Extractive Summarization basically finds out the most important sentences from the text using features and grouped to produce the summary. It is like highlighting a text through a highlighter. In contrast, abstractive summarization is a technique that generates new sentences instead of selecting the essential sentences of the original document that contain the most critical information. Like a human being, writing a summary from his thinking with a pen. Machine Learning-based summarizing tools are available nowadays. But the language-specific models are hard to find.

Although a notable number of works have been done on Bengali extractive summarization, only a few abstractive summarizations are available. The majority of the available works are based on the basic Machine Learning (ML) techniques and the dataset was too small. Due to the lack of standard datasets, no significant work is available on encoder-decoder based summarization systems. So, the most challenging part for BANS is to prepare a standard and clean dataset. To build a Bengali news summarization dataset, a crawler has been made to crawl data from online resources like a daily newspaper. We have collected more than 19 k data from bangla.bdnews24.com[1] online portal. The dataset represents the article and its corresponding summary. In this paper, a sequence to sequence LSTM encoder-decoder architecture with attention has been presented for Bengali abstractive news summarization. Figure 1 illustrates the proposed model. The source code and other details of the model already uploaded to Github.[2] Then the dataset of size 19096 has also been prepared which is till now the largest one and published it in Kaggle.[3] The word embedding layer has been used to represent the words in numbers and fed them into the encoder. Moreover, both the encoder and decoder parts are associated with some attention mechanisms. We got a notable improvement in terms of human assessment compared to other available Bengali abstractive summarization methods. We also evaluated ROUGE and BLEU scores. In short, our contribution to this work is threefold. They are:

[1]https://bangla.bdnews24.com/.

[2]https://github.com/Prithwiraj12/Bengali-Deep-News-Summarization.

[3]https://www.kaggle.com/prithwirajsust/bengali-news-summarization-dataset.

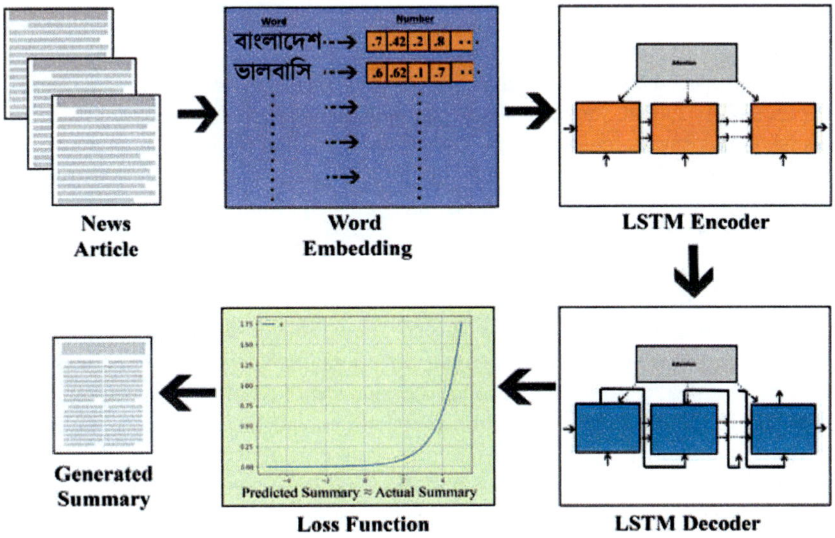

Fig. 1 Illustration of our neural attention model for abstractive summarization of Bengali news incorporates a set of LSTM encoder-decoder on top of a standard word embedding

- Preparation of till now the largest Bengali news summarization dataset of size 19,096 documents with its summary and published it in Kaggle[2].
- Presenting the encoder-decoder architecture with the attention mechanism for Bengali abstractive news summarization (BANS) in an efficient way.
- Evaluation of the model both qualitatively and quantitatively and the presented approach outperforms Bengali state-of-the-art approaches.

2 Related Work

There are different kinds of abstractive text summarization approaches that exist. We found that Yeasmin et al. [1] have described the different techniques regarding abstractive approaches. Then as we decided to focus on abstractive text summarization approaches on the Bengali language context, we covered Haque et al. [2] where 14 approaches of Bengali text summarization regarding both extractive and abstractive approaches are described. In 2004, Islam et al. [3] first introduced Bengali extractive summarization based on document indexing and keyword-based information retrieval. Then techniques of English extractive text summarization were applied for Bengali by Uddin et al. [4]. In 2010, Das et al. [5] used theme identification, page rank algorithms, etc. for extractive summarization. Sentence ranking and stemming process-based Bengali extractive summarization were first proposed by a researcher named Kamal Sarkar [6] and later in a better way by Efat et al. [7].

Haque et al. [8, 9] respectively proposed a key-phrase based extractive approach and a pronoun replacement based sentence ranking approach. In 2017, the heuristic approach proposed by Abujar et al. [10], K-means clustering method of Akther et al. [11] and LSA (Latent Semantic Analysis) method stated in Chowdhury et al. [12] became popular techniques for Bengali extractive summarization. The graph-based sentence scoring feature for Bengali summarization was first used by Ghosh et al. [13]. Moreover, Sarkar et al. [14] and Ullah et al. [15] proposed term frequency and cosine similarity based extractive approach respectively.

Recently, Munzir et al. [16] instigated a deep neural network-based Bengali extractive summarization. Again Abujar et al. [17] introduced Word2Vec based word embedding for Bengali text summarization. Then Talukder et al. [18] proposed an abstractive approach for Bengali where bi-directional RNNs with LSTM are used at the encoder and attention at the decoder. We also used LSTM-RNN based attention model like [18] but we applied attention to both the encoder and the decoder layer and did some comparative study with the corresponding result part and dataset part with the existing one. Another LSTM-RNN based text generation process is introduced by Abujar et al. [19] for Bengali abstractive text summarization.

We used the concept stated in Lopyrev et al. [20] for our system. The seq2seq model and the LSTM encoder-decoder architecture we used, was introduced by Sutskever et al. [21] and Bahdanau et al. [22] respectively. Again, the decoder and encoder part's attention technique is the concept stated in Luong et al. [23] and Rush et al. [24] respectively. Furthermore, the LSTM concept-based language parsing method has been adopted from Vinyals et al. [25].

3 Dataset

A standard dataset is a vital part of text summarization. We gathered a conceptual idea of preparing a standard dataset from Hermann et al. [26] and also observed some of the existing public English datasets like CNN-Daily Mail[4] dataset. We need a vast amount of data for training but no significant standard public dataset is available for Bengali summarization.

So, we collected news and its summary from the online news portal bangla. bdnews24.com (https://bangla.bdnews24.com/) as it had both the article and its summary. We made a crawler and crawled 19352 news articles and their summaries from different categories like sports, politics, economics, etc. Online news contains lots of garbage like advertisements, non-Bengali words, different websites' links, etc. So, we started preprocessing by making a data cleaning program that eliminates all kinds of garbage from the dataset. We uploaded data crawling, cleaning, and analysis source code[5] and their working details to Github and publicly published our dataset in Kaggle (https://www.kaggle.com/prithwirajsust/bengali-news-summarization-dataset).

[4]https://cs.nyu.edu/~kcho/DMQA/.

[5]https://github.com/Prithwiraj12/Data-Manipulation.

Table 1 Statistics of the dataset

Total no of articles	19, 096
Total no of summaries	19, 096
Maximum no of words in an article	76
Maximum no of words in a summary	12
Minimum no of words in an article	5
Minimum no of words in a summary	3

Table 2 Comparison of our standard dataset with BNLPC dataset

Source	Total articles	No of summary (per article)	Total summaries
BNLPC (http://www.bnlpc.org/research.php) dataset	200	3	600
Our dataset (https://www.kaggle.com/prithwirajsust/bengali-news-summarization-dataset)	19,096	1	19,096

A tabular representation of our processed data is shown in Table 1. The significance and comparison of our dataset with only publicly available Bangla Natural Language Processing Community (BNLPC[6]) summarization dataset has been shown in Table 2.

4 Model Architecture

By observing the significant performance of LSTM encoder-decoder with the attention mechanism described in Lopyrev et al. [20], we've used a similar neural attention model architecture. It has an LSTM Encoder part and an LSTM Decoder part. Both of the parts are associated with some attention mechanisms. Tensorflow's embedding layer embedding_attention_seq2seq has been used to represent the words in numbers to feed into encoders. After generating the decoder's output, a comparison between the actual and predicted summary has been done using the softmax loss function, and for minimizing the loss, the network started back-propagating. Lastly, a summary has been generated with minimal loss. The whole process works as a seq2seq approach and can be visualized by Fig. 1. Let's describe the major two components of our model.

[6]http://www.bnlpc.org/research.php.

Firstly, an input sequence is encoded to numbers via word embedding layer and fed into the LSTM encoder in reverse order. Sutskever et al. [21] proposed that because of calculating short term dependencies, the first few words of both the input sequence and output sequence must be closer to each other and it can be achieved by feeding input in reverse order and thus the result can be significant. That means Bengali sentence like "আজকের সংবাদ" is fed into each encoder cell reversely as individual word "সংবাদ" and "আজকের" respectively. Attention is also used to the encoder part as mentioned by Rush et al. [24].

Secondly, we used a greedy LSTM decoder which is different from a beam search decoder. Firstly, encoder output is fed into the first decoder cell. Then the output of the current decoder cell is fed into the next decoder cell along with the attention as well as the information from the previous decoder cell and continued the process till the last decoder cell. That means if the first generated word in the decoder cell is "সংবাদের" then this word will help to predict the next word suppose "সারাংশ" for the next decoder cell combining with attention and continued the process till the end. The decoder attention mechanism is implemented as stated in [21].

Before training, we made a vocabulary of the most frequent 40k words both from articles and summaries. The out of vocabulary words are denoted by _UNK token. _PAD token is used for padding the article and its summary to the bucket sizes. A bucket is nothing but an array where we define how many words an article and its summary can hold while training. We used five encoder-decoder LSTM models for training. Now, the trained model also padded the words of the given input sentences to the bucket sizes. So the model can well summarize the articles containing the number of words in all sentences equal to the largest bucket size and in our case it was (50, 20) for article and summary respectively.

5 Result and Discussion

We assessed our model based on two types of evaluation matrices for analyzing the result: They are **Quantitative Evaluation** and **Qualitative Evaluation**. Both of the evaluation methods are mandatory for checking how much the summary system is suitable for generating a summary. 70% of our data was used for training, 20% for validating, and 10% was used for testing. The system was trained three times with different parameter specifications. After the evaluation, we found that the system has the best output when the vocabulary size was set to 40 k, hidden unit to 512, learning rate to 0.5, and steps per checkpoint to 350. Table 3 shows some generated examples of our best model.

We showed two good quality as well as two poor quality predictions in Table 3 from our system. Here, the first two predictions are well summarised by our model and sometimes the new word has also been generated like "পুকুরে" in the second example. On the other hand, from the last two predictions on the Table 3 we found that repetition of words like "দগ্ধ" in the third example and "লাশ" in the fourth example

Table 3 Illustrates some predictions of our BANS system showing the input news article, actual summary and BANS predicted summary

New Article	Actual Summary	Predicted Summary
Bengali: বিএনপি জোটের হরতাল-অবরোধের মধ্যে চট্টগ্রাম নগরীর স্টিল মিল এলাকায় একটি বাসে আগুন দিয়েছে দুর্বৃত্তরা। **English:** The miscreants set fire to a bus in Steel Mill area of Chittagong city during the strike and blockade of the BNP alliance.	**Bengali:** চট্টগ্রামে বাসে আগুন। **English:** Fire on bus in Chittagong.	**Bengali:** চট্টগ্রামে বাসে বাসে আগুন **English:** Fire on the buses of Chittagong
Bengali: জামালপুরের মেলান্দহে পানিতে ডুবে দুই শিশুর মৃত্যু হয়েছে। **English:** Two children drowned at Melandho in Jamalpur.	**Bengali:** জামালপুরে পানিতে ডুবে দুই শিশুর মৃত্যু **English:** Two children drowned in Jamalpur	**Bengali:** পুকুরে পানিতে ডুবে শিশুর মৃত্যু **English:** The child drowned in the pond
Bengali: নরসিংদীর পলাশে অবরোধের পেট্রোলবোমায় দগ্ধ হওয়ার আট দিন পর হাসপাতালে চিকিৎসাধীন অবস্থায় মারা গেলেন ট্রাকচালক জাহিদ আহমেদ। **English:** Truck driver Zahid Ahmed died in a hospital after eight days being burnt by a petrol bomb in a blockade at Polash in Narsingdi.	**Bengali:** নরসিংদীতে বোমায় দগ্ধ ট্রাকচালকের মৃত্যু **English:** Truck driver killed in bomb blast in Narsingdi	**Bengali:** অবরোধের দগ্ধ দগ্ধ আরও মৃত্যু **English:** More death from burns burns in the siege
Bengali: ঠাকুরগাঁও সদরে নদীতে পড়ে নিখোঁজের একদিন এক যুবকের লাশ উদ্ধার করা হয়েছে। **English:** The body of a young man who went missing after falling into the river at Thakurgaon Sadar has been recovered.	**Bengali:** ঠাকুরগাঁওয়ে নিখোঁজ যুবকের লাশ উদ্ধার। **English:** The dead body of a missing young boy was recovered in Thakurgaon.	**Bengali:** কুষ্টিয়ায় নিখোঁজ যুবকের লাশ লাশ উদ্ধার **English:** The dead body dead body of a missing young boy was recovered in Kushtia

occurred twice. Further from the third example, we can see inaccurate reproduction of factual details. That means word "কুষ্টিয়া" has been produced by the model rather than predicting the word "ঠাকুরগাঁও" in the fourth example. Moreover, due to bucketing issues, some summaries are forcefully stopped before hitting the end token of the sentence which can be shown in third predictions on Table 3.

5.1 Quantitative Evaluation

Quantitative evaluation is a system-oriented evaluation. In this evaluation process, both the actual and predicted summaries are given as input to a program and the program generates a score comparing how much the predicted summary deviates from the actual summary. We found that Recall-Oriented Understudy for Gisting Evaluation (ROUGE) [27] and Bilingual Evaluation Understudy (BLEU) [28] are two standard quantitative evaluation matrices. As far as our knowledge, quantitative evaluation of the existing Bengali abstractive text summarization techniques [18, 19] is not mentioned or publicly available. So we could not compare our evaluation with them. But as per standard scoring mentioned in the papers [27, 28], our achieved

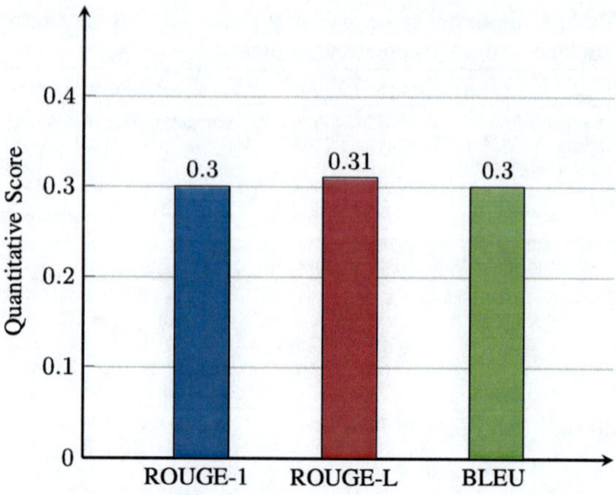

Fig. 2 Illustrates the quantitative analysis of our proposed model based on ROUGE-1, ROUGE-L and BLEU scores

score was also significant. There are different variants of ROUGE calculation exist. ROUGE-1, ROUGE-2, ROUGE-L, ROUGE-N, etc are some of them. Here, we computed the most adapted ROUGE-1, ROUGE-L, and measured the BLEU score as well. Firstly, We took 100 generated summaries and corresponding actual summaries and calculated the average BLEU score. Again for ROUGE calculation, we first calculated the Precision and Recall. Then using these two measurements calculated the average F1 score for that 100 examples. The bar diagram of Fig. 2 denotes ROUGE and BLEU scores of the best model.

5.2 *Qualitative Evaluation*

Qualitative evaluation is the user-oriented evaluation process. Here some users of different ages take part in rating the generated summary on a scale of 5 compared with the actual one. For the qualitative evaluation, we took some examples from our system and some from the existing one [18].

As far as our knowledge, qualitative evaluation of the existing method [18] is not publicly available. So for comparison, we also had to calculate the rating for [18]. We provided the examples of both the systems to the users via a google form[7] survey. A total of 20 users participated in a rating on a scale of 5. Among the users 45% were female and 55% were male. Moreover, all the users were from the educational background with an average age of 24. Again 45% were from linguistic faculty, 35% were from engineering faculty and 25% were from other faculties. We calculated the average rating regarding each of the models and found that our system outperforms the existing system based on human assessment. The qualitative rating of the systems is shown in Table 4.

[7] https://forms.gle/r9Mu5NEpVkMcSXbD9.

Table 4 Qualitative evaluation of existing system and the proposed system

System	Average rating (Out of 5)
Proposed system	2.80
Existing system [18]	2.75

6 Conclusion

To recapitulate, the development of the standard summarization dataset of 19,096 Bengali news has been one of our pioneering accomplishments, especially since it is the largest publicly published dataset in this field. Here a neural attention-based encoder-decoder model for abstractive summarization of Bengali news has been presented, which generates human-like sentences with core information of the original documents. Along with that, a large-scale experiment was conducted to investigate the effectiveness of the proposed BANS. From the qualitative evaluation, we have found that the proposed system generates more humanoid output than all other existing BANS. Indeed, the LSTM-based encoder-decoder has been exceptionally successful, nonetheless, the model's performance can deteriorate quickly for long input sequences. Repetition of summaries and inaccurate reproduction of factual details are two significant problems. To fix these issues, we plan to drive our efforts on modeling hierarchical encoder based on structural attention or pointer-generator architecture and developing methods for multi-document summarization.

Acknowledgements We would like to thank Shahjalal University of Science and Technology (SUST) research center and SUST NLP research group for their support.

References

1. Yeasmin, S., Tumpa, P.B., Nitu, A.M., Uddin, M.P., Ali, E., Afjal, M.I.: Study of abstractive text summarization techniques. Am. J. Eng. Res. **6**(8), 253–260 (2017)
2. Haque, M.M., Pervin, S., Hossain, A., Begum, Z.: Approaches and trends of automatic Bangla text summarization: challenges and opportunities. Int. J. Technol. Diffusion (IJTD) **11**(4), 1–17 (2020)
3. Islam, M.T., Al Masum, S.M.: Bhasa: a corpus-based information retrieval and summariser for Bengali text. In: Proceedings of the 7th International Conference on Computer and Information Technology (2004)
4. Uddin, M.N., Khan, S.A.: A study on text summarization techniques and implement few of them for Bangla language. In: 2007 10th International Conference on Computer and Information Technology, pp. 1–4. IEEE (2007)
5. Das, A., Bandyopadhyay, S.: Topic-based Bengali opinion summarization. In: Coling 2010: Posters, pp. 232–240 (2010)
6. Sarkar, K.: Bengali text summarization by sentence extraction. arXiv:1201.2240 (2012)

7. Efat, M.I.A., Ibrahim, M., Kayesh, H.: Automated Bangla text summarization by sentence scoring and ranking. In: 2013 International Conference on Informatics, Electronics and Vision (ICIEV), pp. 1–5. IEEE (2013)
8. Haque, M.M., Pervin, S., Begum, Z.: Enhancement of keyphrase-based approach of automatic Bangla text summarization. In: 2016 IEEE Region 10 Conference (TENCON), pp. 42–46. IEEE (2016)
9. Haque, M., Pervin, S., Begum, Z., et al.: An innovative approach of Bangla text summarization by introducing pronoun replacement and improved sentence ranking. J. Inf. Process. Syst. **13**(4) (2017)
10. Abujar, S., Hasan, M., Shahin, M., Hossain, S.A.: A heuristic approach of text summarization for Bengali documentation. In: 2017 8th International Conference on Computing, Communication and Networking Technologies (ICCCNT), pp. 1–8. IEEE (2017)
11. Akter, S., Asa, A.S., Uddin, M.P., Hossain, M.D., Roy, S.K., Afjal, M.I.: An extractive text summarization technique for Bengali document (s) using k-means clustering algorithm. In: 2017 IEEE International Conference on Imaging, Vision & Pattern Recognition (icIVPR), pp. 1–6. IEEE (2017)
12. Chowdhury, S.R., Sarkar, K., Dam, S.: An approach to generic Bengali text summarization using latent semantic analysis. In: 2017 International Conference on Information Technology (ICIT), pp. 11–16. IEEE (2017)
13. Ghosh, P.P., Shahariar, R., Khan, M.A.H.: A rule based extractive text summarization technique for Bangla news documents. Int. J. Modern Educ. Comput. Sci. **10**(12), 44 (2018)
14. Sarkar, A., Hossen, M.S.: Automatic Bangla text summarization using term frequency and semantic similarity approach. In: 2018 21st International Conference of Computer and Information Technology (ICCIT), pp. 1–6. IEEE (2018)
15. Ullah, S., Hossain, S., Hasan, K.A.: Opinion summarization of Bangla texts using cosine similarity based graph ranking and relevance based approach. In: 2019 International Conference on Bangla Speech and Language Processing (ICBSLP), pp. 1–6. IEEE (2019)
16. Al Munzir, A., Rahman, M.L., Abujar, S., Hossain, S.A., et al.: Text analysis for Bengali text summarization using deep learning. In: 2019 10th International Conference on Computing, Communication and Networking Technologies (ICCCNT), pp. 1–6. IEEE (2019)
17. Abujar, S., Masum, A.K.M., Mohibullah, M., Hossain, S.A., et al.: An approach for Bengali text summarization using word2vector. In: 2019 10th International Conference on Computing, Communication and Networking Technologies (ICCCNT), pp. 1–5. IEEE (2019)
18. Talukder, M.A.I., Abujar, S., Masum, A.K.M., Faisal, F., Hossain, S.A.: Bengali abstractive text summarization using sequence to sequence rnns. In: 2019 10th International Conference on Computing, Communication and Networking Technologies (ICCCNT), pp. 1–5. IEEE (2019)
19. Abujar, S., Masum, A.K.M., Islam, M.S., Faisal, F., Hossain, S.A.: A Bengali text generation approach in context of abstractive text summarization using rnn. In: Innovations in Computer Science and Engineering, pp. 509–518. Springer (2020)
20. Lopyrev, K.: Generating news headlines with recurrent neural networks. arXiv:1512.01712 (2015)
21. Sutskever, I., Vinyals, O., Le, Q.V.: Sequence to sequence learning with neural networks. In: Advances in Neural Information Processing Systems, pp. 3104–3112 (2014)
22. Bahdanau, D., Cho, K., Bengio, Y.: Neural machine translation by jointly learning to align and translate. arXiv:1409.0473 (2014)
23. Luong, M.T., Pham, H., Manning, C.D.: Effective approaches to attention-based neural machine translation. arXiv:1508.04025 (2015)
24. Rush, A.M., Chopra, S., Weston, J.: A neural attention model for abstractive sentence summarization. arXiv:1509.00685 (2015)
25. Vinyals, O., Kaiser, Ł., Koo, T., Petrov, S., Sutskever, I., Hinton, G.: Grammar as a foreign language. In: Advances in Neural Information Processing Systems, pp. 2773–2781 (2015)
26. Hermann, K.M., Kocisky, T., Grefenstette, E., Espeholt, L., Kay, W., Suleyman, M., Blunsom, P.: Teaching machines to read and comprehend. In: Advances in Neural Information Processing Systems, pp. 1693–1701 (2015)

27. Lin, C.Y.: Rouge: a package for automatic evaluation of summaries. In: Text Summarization Branches Out, pp. 74–81 (2004)
28. Pastra, K., Saggion, H.: Colouring summaries bleu. In: Proceedings of the EACL 2003 Workshop on Evaluation Initiatives in Natural Language Processing: Are Evaluation Methods, Metrics and Resources Reusable? pp. 35–42 (2003)

Application of Feature Engineering with Classification Techniques to Enhance Corporate Tax Default Detection Performance

Md. Shahriare Satu, Mohammad Zoynul Abedin, Shoma Khanom, Jamal Ouenniche, and M. Shamim Kaiser

Abstract The objective of this work is to propose a methodology that is helpful in analyzing tax data and predict significant features that cause tax defaulting. In this work, we gathered a Finnish tax default data of different firms and then split it according to primary and transformed feature sets. Different feature selection techniques were used to explore significant feature sets. After that, we applied various classification techniques into primary and transformed data sets and analyzed experimental outcomes. Besides, almost all classification techniques are represented the highest results for correlation-based feature selection subset evaluation, information gain feature selection and gain ratio attribute evaluation techniques. But, information gain feature selection is found as the most reliable feature selection method in this work. This analysis can be useful as a complementary tool to assess tax default factors in corporate sectors.

Keywords Tax default detection · Feature selection · Classification techniques

Md. S. Satu
Department of MIS, Noakhali Science and Technology University, Noakhali, Bangladesh

M. Zoynul Abedin (✉)
Department of Finance and Banking, Hajee Mohammad Danesh Science and Technology University, Dinajpur 5200, Bangladesh
e-mail: abedinmz@yahoo.com

S. Khanom
Department of Electronics and Communication Engineering, Institute of Science Trade and Technology, Dhaka, Bangladesh

J. Ouenniche
University of Edinburgh, Business School, 29 Buccleuch Place, Edinburgh EH8 9JS, UK

M. Shamim Kaiser
Institute of Information Technology, Jahangirnagar University, Dhaka, Bangladesh

M. S. Kaiser et al. (eds.), *Proceedings of International Conference on Trends in Computational and Cognitive Engineering*, Advances in Intelligent Systems and Computing 1309, https://doi.org/10.1007/978-981-33-4673-4_5

1 Introduction

Tax is a primary source of government earning that allocates funds to the various social plans, infrastructure and services. According to the World Bank studies, the average universal paid taxes are almost 40% where taxes of the default firms might not be recovered in the forthcoming years. But, only a small amount of works were happened in view of unpaid tax statistics. Now-a-days every country gives attention to obtain tax revenue because of planning budget [7, 10]. In Finland, 12% active firms were not paid any taxes at the end of 2015 [6]. However, tax audits were assessed the risk of different firms using the likelihood of tax default. Alongside, inability of paying tax may decline the economical state that causes raising of workless people, pretentious micro and macroeconomic consequences, communal displacement, monetary depressions and collapses. Recently, machine learning is an emergent field that resolves various forecasting issues in the finance and accounting sector. It produces laborious, potentially novel and comprehensive findings to make decision. Therefore, the current work is focused on machine learning based modeling that identify tax default at different firms.

There were happened a small amount of works in tax default. In the previous studies [6] they employed machine learning to analyze tax default conditions and focused only single feature selection and classification methods such as genetic algorithm (GA) and linear discriminant analysis (LDA) respectively. However, we extended this work with more machine learning methods and explored significant features about tax default. In the empirical point of view, machine learning was used massively to explore this issue that was not happened in previous studies. Again, feature transformation and selection methods were provided more appropriate tax default prediction. Moreover, non parametric statistical test is re-evaluated the findings about this work. In the managerial perspective, it is useful to reduce their work for the tax authorities and stakeholders. This automatic solution can forecast which organization will be failed to give taxes in the upcoming year. Tax related financial statements are automatically justified based on predicted results. So, this work enriches the corporate productivity and profitability about tax default.

The organization of this paper is given as follows: Sect. 2 represents the research methodology that contains dataset description, preprocessing, and classification process briefly. In Sect. 3, the experimental result about tax default is represented for both primary and transformed dataset and discussion section is given at Sect. 4. Then, we conclude this work by summarizing and indicating some future direction about how to analyze tax default data in Sect. 5.

2 Proposed Methodology

Most of the financial sector is concerned about their cost-benefit state with corresponding taxes for the government or any private organization. Almost all of the aspects in professional services are required to pay taxes. Hence, authorities should

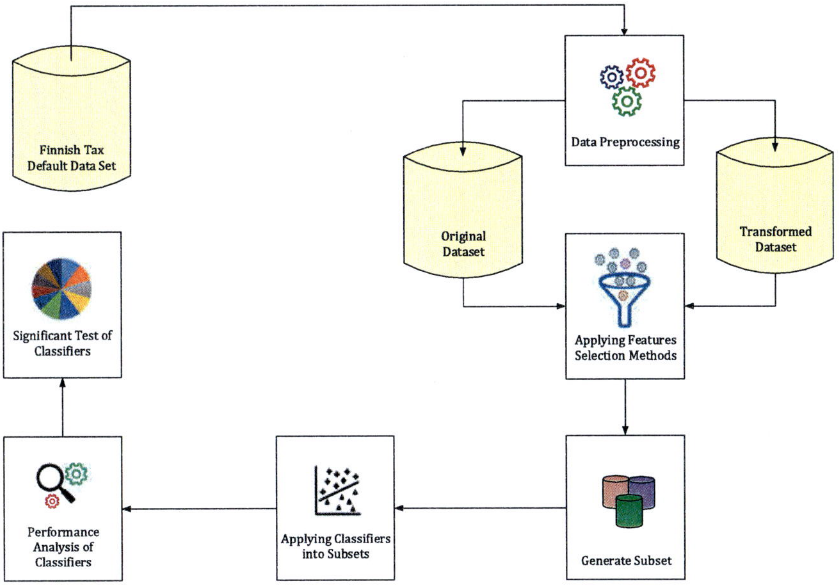

Fig. 1 Proposed methodology

be taken more steps so that no one can avoid it. But, it is too much tedious and costly to monitor tax default persons/firms by only increasing manpower. Therefore, an autonomous machine learning model is proposed that investigates tax default instances and explores related features using various classifiers. In the beginning, Finnish tax default dataset was split into primary and transformed dataset. Then, different feature selection and classification methods were implemented into these datasets to identify informative feature sets. Figure 1 shows the entire methodology of the tax default analysis.

2.1 Dataset Description and Transformation

The primary tax defaults was gathered from [6]. In the Finnish limited liability firms, corporate tax is provided 20% on their taxable income using financial reports and generated from annual statements. Various firms had failed to pay on employer contribution or value added taxes in 2014. In that year, 3036 firms were found whose have defaulted taxes and 1118 firms (Out of 3036 firms) had submitted financial statements of 2012 and 2013 respectively. Besides, 545 firms were discarded for the missing values. Furthermore, 161 firms were dropped because their sales or total assets were found below 10,000 €. However, 384 defaulting firms were considered where they were matched with non-defaulting firms. So, total 768 firms data were

Table 1 Feature description

S/N	Features	Period			
		t	t-1	$\pm\sqrt{t}$	$\pm\sqrt{t-1}$
1	Industry risk of payment defaults	×		×	
2	Industry risk of bankruptcy	×		×	
3	Sales/total assets	×	×	×	×
4	Total assets	×	×	×	×
5	Change in sales	×	×	×	×
6	Gross result/sales	×	×	×	×
7	Operating margin	×	×	×	×
8	Operating income	×	×	×	×
9	Quick ratio	×	×	×	×
10	Current ratio	×	×	×	×
11	Return on investment	×	×	×	×
12	Return on assets	×	×	×	×
13	Equity ratio	×	×	×	×
14	Net gearing	×	×	×	×
15	Debt/sales	×	×	×	×
16	Working capital /sales	×	×	×	×
17	Inventory turnover ratio	×	×	×	×
18	Collection period of trade receivables	×	×	×	×
19	Payment period of trade payable	×	×	×	×

In this table, the primary dataset represents 19 features for t = 2014 and 17 features for $t - 1$ = 2013. Transformed features are signed squared transformation of original features. × indicates the occurrence of individual features in a particular periods

combined for tax default and non-default data. This dataset was published at the official journal of Finland and the financial statement that is found at the Voitto + database (this dataset is downloaded from https://goo.gl/5-2cK41). A brief description of tax default dataset is shown in Table 1. In this dataset, 17 financial statement ratios, industrial payment default, and bankruptcy risks (total 36 variables) were used whereas financial statement ratios were not normally distributed in previous studies [4]. This might lessen the performance of tax default prediction model [6]. Hence, the signed square method was applied into the primary 36 variables and transformed it into another format. Now, the working steps of this work are described as follows.

2.2 Feature Selection

Different methods such as Correlation-based Feature Selection Subset Evaluation (CFSSE) [9], Information Gain Attribute Evaluation (IGAE) [8], Gain Ratio Attribute Evaluation (GRAE) [12], and Correlation Attribute Evaluation (CAE) [13] were used to explore several feature subsets from both primary and transformed dataset. However, a few number of tax defaulting works were happened where most of them were not explored significant features. In previous works, wrapper and embedded methods were used to identify features in the tax default dataset. So, filter approaches were emphasized in this work.

2.3 Classification

Classification is a supervised learning approach that assigns instances into a number of target categories or classes. Different methods such as Naïve Bayes (NB), Bayes Net (BN), Multi-Layer Perceptron (MLP), Sequential Minimal Optimization (SMO), Simple Logistic (SL), Instance-based k-nearest Neighbors (IBk), J48 Decision Tree, Random Forest (RF) and Reduces Error Pruning Tree (REPTree) were applied into all feature subsets of primary and transformed dataset respectively. These classifiers were widely used in many tax analysis [7, 14] and related financial applications [1, 2]. Each classifier was used tenfold cross validation in this work. Besides, these classification criteria differ from one classifier to another.

2.4 Evaluation Metrics

To analyze the performance of different classifiers, we used confusion matrix that contains true positive (TP), false positive (FP), true negative (TN) and false negative (FN) values. Several evaluation metrics were used to investigate the performance of different classifiers which are described briefly as follows:

- **Accuracy**: is measured by the percentage of corrected instances. It is a statistical measure that manipulates how well we identify correct samples.

$$\text{Accuracy} = \frac{\text{TP} + \text{TN}}{\text{TP} + \text{TN} + \text{FP} + \text{FN}} \tag{1}$$

- **F-Measure**: The harmonic mean of precision and recall is called F-Measure. In this case, precision indicates the percentage of relevant result and recall defines the percentage of correctly classified relevant result. So, F-Measure is manipulated as follows:

$$\text{Precision} = \frac{\text{TP}}{\text{TP} + \text{FP}} \quad (2)$$

$$\text{Recall} = \frac{\text{TP}}{\text{TP} + \text{FN}} \quad (3)$$

$$\text{F-Measure} = \frac{2 \times \text{Precision} \times \text{Recall}}{\text{Precision} + \text{Recall}} \quad (4)$$

- **AUROC**: Area under receiver operating characteristic (AUROC) demonstrates how well the probabilities of positive classes are isolated from negative classes. It can be calculated by following equation:

$$\text{AUROC} = \frac{\text{TP rate} + \text{TN rate}}{2} \quad (5)$$

3 Experimental Result

In this experiment, tax default dataset was explored and analyzed significant features for the failure to pay taxes. Furthermore, relevant classifiers were used to find out significant feature set that gives high performance to predict tax evasion. To convey this work, different data analytical tools such as Microsoft Excel [11], WEKA [5] and KEEL [3] were used. In the beginning, Finnish tax default dataset was split into primary and transformed dataset where each dataset contains 36 features respectively. Besides, feature engineering is reduced the computational cost for associated stakeholders (e.g. tax payers and authorities). Consequently, different feature selection methods were applied into primary and transformed dataset. In primary dataset, 12, 13, 15 and 21 features were generated using CFSSE, IGAE, GRAE, and CAE respectively. Later, 12, 13, 15 and 26 features were extracted from transformed dataset employing CFSSE, IGAE, GRAE, and CAE respectively. Subsequently, NB, BN, MLP, SMO, SL, IBk, J48, RF and REPTree were implemented into all generated datasets.

3.1 Experimental Analysis of Primary Tax Default Dataset

In Table 2, the performance of the classifiers are provided after implementing feature selection methods in the primary tax default dataset. When the performance of individual classifiers are observed, RF shows the highest 73.8% accuracy and F-measure (0.738). BN and J48 show the second highest accuracy 73.7% and F-measure (0.737) in this work. Nevertheless, BN shows AUROC (0.816) than other classifiers. Again, Fig. 2a show BN is the best classifier to analyze tax default data on average. How-

Table 2 Experimental result of primary and transformed tax default data

CT	FST	Fs	Acc (%)	F-M	AUROC	Fs	Acc (%)	F-M	AUROC
		Primary dataset				Transformed dataset			
NB	CFSSE	12	60.156	0.550	0.740	12	72.656	0.726	**0.790**
	IGAE	13	*65.234*	*0.635*	0.736	13	*72.787*	*0.727*	0.782
	GRAE	15	61.979	0.590	0.726	15	72.266	0.722	0.783
	CAE	21	63.802	0.610	*0.742*	26	71.094	0.707	0.782
	Baseline	36	56.771	0.523	0.686	36	71.484	0.711	0.782
BN	CFSSE	12	*73.698*	*0.737*	*0.816*	12	73.828	0.738	*0.817*
	IGAE	13	72.917	0.728	0.809	13	72.787	0.727	0.809
	GRAE	15	*73.698*	0.736	0.812	15	73.568	0.735	0.812
	CAE	21	73.438	0.734	0.805	26	*74.219*	*0.742*	0.804
	Baseline	36	73.307	0.733	0.808	36	73.307	0.733	0.808
MLP	CFSSE	12	*71.094*	*0.711*	*0.782*	12	72.135	0.721	*0.792*
	IGAE	13	*71.094*	*0.711*	0.778	13	72.526	0.725	0.789
	GRAE	15	70.833	0.708	0.781	15	*73.047*	*0.730*	0.790
	CAE	21	69.662	0.696	0.768	26	68.099	0.680	0.730
	Baseline	36	70.443	0.704	0.759	36	68.099	0.681	0.735
SMO	CFSSE	12	*68.359*	*0.680*	*0.684*	12	73.698	0.735	0.737
	IGAE	13	63.672	0.611	0.637	13	72.135	0.719	0.721
	GRAE	15	64.193	0.615	0.642	15	72.005	0.717	0.720
	CAE	21	67.318	0.672	0.673	26	72.917	0.728	0.729
	Baseline	36	67.708	0.676	0.677	36	*74.219*	*0.742*	*0.742*
SL	CFSSE	12	69.401	0.694	0.778	12	72.917	0.729	*0.795*
	IGAE	13	*72.135*	*0.721*	*0.784*	13	*74.219*	*0.742*	0.793
	GRAE	15	72.005	0.720	0.779	15	73.307	0.733	0.792
	CAE	21	70.182	0.702	0.780	26	71.745	0.717	0.793
	Baseline	36	71.094	0.711	0.781	36	72.266	0.722	0.792
J48	CFSSE	12	*73.698*	*0.737*	*0.761*	12	*73.698*	*0.737*	*0.755*
	IGAE	13	72.135	0.721	0.739	13	72.396	0.724	0.741
	GRAE	15	70.573	0.705	0.711	15	70.833	0.708	0.713
	CAE	21	67.969	0.680	0.676	26	67.448	0.674	0.692
	Baseline	36	70.573	0.706	0.699	36	69.662	0.697	0.693
RF	CFSSE	12	72.656	0.726	0.805	12	72.526	0.725	0.801
	IGAE	13	71.745	0.717	0.804	13	72.656	0.727	0.809
	GRAE	15	*73.828*	*0.738*	*0.812*	15	*73.568*	*0.736*	*0.813*
	CAE	21	72.917	0.729	0.802	26	72.396	0.724	0.801
	Baseline	36	72.656	0.726	0.804	36	72.917	0.729	0.809
REPTree	CFSSE	12	70.313	0.703	0.739	12	70.443	0.704	0.751
	IGAE	13	*71.745*	*0.717*	*0.764*	13	*71.484*	*0.715*	*0.758*
	GRAE	15	71.224	0.712	0.755	15	71.094	0.711	0.757
	CAE	21	70.964	0.710	0.749	26	70.443	0.704	0.704
	Baseline	36	70.313	0.703	0.741	36	70.443	0.704	0.704

Legend: Cassifier: CT; FST; Feature Selection Technique; Features: Fs; Acc: Accuracy; F-M: F-measure; AUROC: Area under receiver operating characteristic; italic values denote best performance.

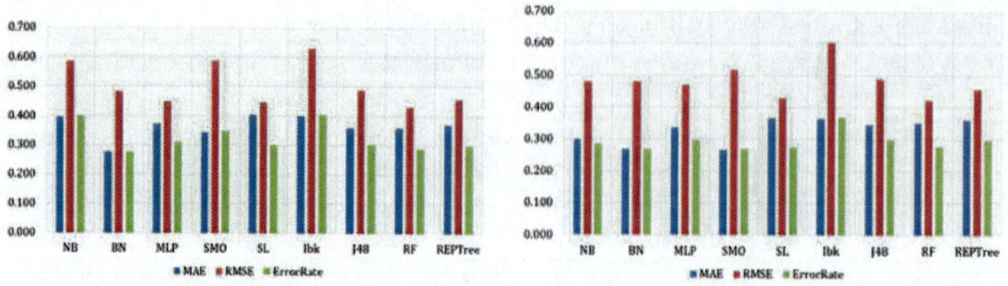

(a) (b)

Fig. 2 Average error rate of **a** primary and **b** transformed dataset **Legend**: MAE: mean absolute error; RMSE: root mean square error

Table 3 Average experimental results for feature selection techniques

Feature selectors	Acc (%)	F-M	AUROC	Acc (%)	F-M	AUROC
	Primary features			Transformed features		
CFSSE	69.054	0.684	**0.747**	71.499	0.715	0.761
IGAE	*69.502*	*0.690*	0.744	71.803	*0.718*	*0.762*
GRAE	69.242	0.686	0.741	*71.817*	*0.718*	0.761
CAE	68.490	0.682	0.733	65.987	0.655	0.702
Baseline	67.607	0.671	0.723	70.645	0.706	0.748

ever, most of the classifiers show the highest result for the CFSSE, IGAE and GRAE feature subset. Besides, average results of feature selection methods show that IGAE and CFSSE help classifiers to generate the highest result (see Table 3). But, IGAE is the most frequent feature selection technique to get the highest result for primary tax default dataset. Besides, CFSSE and GRAE are also generates better results and their findings are almost similar with IGAE generated outcomes.

3.2 *Experimental Analysis on Transformed Tax Default Dataset*

In Table 2, the performance of the classifiers are shown after applying feature selection methods in the transformed tax default dataset. In this perspective, the experimental result generates significant improvement in classification. When the overall result of transformed datasets are provided, BN, SMO and SL show the highest accuracy 74.2% in this experiment. Nevertheless, BN and SMO indicate the highest F-measure (0.742) and BN show the highest AUROC (0.817) in this work. When the average results are scrutinized, BN show the highest performance to analyze tax default dataset. However, IGAE and GRAE are shown maximum results for indi-

Table 4 Pairwise WSR test for primary features

	CFSSE	IGAE	GRAE	CAE	Baseline
CFSSE		0.529**	0.726*	1.000	1.000
IGAE	1.000		1.000	1.000	1.000
GRAE	1.000	0.669*		1.000	1.000
CAE	0.434**	0.149**	0.236**		1.000
Baseline	0.062**	0.155**	0.075**	0.813*	

Note Significant at 1% = ***, 5% = **, 10% = *

Table 5 Pairwise WSR test for transformed features

	CFSSE	IGAE	GRAE	CAE	Baseline
CFSSE		0.588*	0.588*	1.000	1.000
IGAE	1.000		0.903*	1.000	1.000
GRAE	1.000	1.000		1.000	1.000
CAE	0.021**	0.044**	0.022**		0.066**
Baseline	0.200**	0.124**	0.054*	1.000	

Note Significant at 1% = ***, 5% = **, 10% = *

vidual classifiers (see Table 2). The average results are shown the best results in the IGAE and GRAE transformed feature subsets (see Table 3). Again, IGAE is found as the most reliable method in this work.

3.3 Wilcoxon Signed–Ranks Significant Test

To observe the reliability of the findings, we utilized pairwise non parametric Wilcoxon Signed–Ranks (WSR) test for both primary and transformed feature sets and realized the difference of the statistical significant performance at the level of $p = 0.10$, $p = 0.05$ and $p = 0.001$ and find out relevant correlation between different feature subsets. In primary data analysis, the pairwise significant WSR test for different feature selection techniques are provided in Table 4. The correlation in CFSSE-CAE, CFSSE-Baseline, IGAE-CFSSE, IGAE-GRAE, IGAE-CAE, IGAE-Baseline, GRAE-CFSSE, GRAE-CAE, GRAE-Baseline and CAE-Baseline pairs are rejected null hypothesis (where p-value ≤ 0.10). On the other hand, the pairwise significant WSR test for different feature section techniques are also presented for the transformed data analysis in Table 5. Hence, CFSSE-CAE, CFSSE-Baseline, CAE-IGAE, Baseline-IGAE, CFSSE-GRAE, IGAE-GRAE, CAE-GRAE, Baseline-GRAE and CAE-Baseline are rejected null hypothesis where p-value is ≤ 0.10. Furthermore, WSR test represents the significant correlation with CFSSE, IGAE and GRAE that indicates them as the best feature selection methods to generate highest result.

4 Discussion

There were happened a few amount of works to detect and forecast corporate tax default in different firms. In current work, Finnish tax default dataset was split into primary and transformed data and investigated them separately that was not happened in previous works [6]. The importance of feature transformation and selection methods were justified by exploring outcomes of these datasets. When the primary dataset had been analyzed with different feature selection and classifiers, the performance of them are not found as well. After employing the signed squared method, the experimental findings of transformed dataset are increased which is shown at [6]. Thus, four feature selection methods were used to generate more significant results than previous works. Then, all of these primary (baseline) and its transformed dataset were used latest and modified version of standard classifier where they were computed more improved results. For instance, MLP is extracted from feed forward ANN, SMO is modified version of SVM and SL is considered as one kind of linear LR. Ibk is generated from KNN as well as RF and REPTree are found from ensemble of various DT. To verify experimental result more appropriately, a pairwise non-parametric WSR test is used to evaluate reliability of different feature selection as well as classification methods that was not found in previous literature.

5 Conclusion and Future Work

The aim of this work is to propose a machine learning model that can identify appropriate feature set that predict corporate tax default and reduce unpaid rate of it. Hence, most of classifiers shows the best result for CFSSE, IGAE and GRAE where IGAE is more reliable method to present the average best outcomes for both primary and transformed tax default dataset. Though RF shows the highest outcomes (accuracy 73.83%) for primary dataset, BN shows highest results (accuracy 74.22%) for transformed dataset. Moreover, BN is the most frequent classifier to generate highest result on average. This work will be helpful for further tax related analysis such as tax auditing, assessment, aggregation etc. In future, we will develop tax defaulting prediction model with huge amount of data to predict tax default more accurately.

References

1. Abedin, M.Z., Chi, G., Colombage, S., Moula, F.E.: Credit default prediction using a support vector machine and a probabilistic neural network. J. Credit Risk **14**(2), 1–27 (2018)
2. Abedin, M.Z., Guotai, C., Moula, F.E., Azad, A.S., Khan, M.S.U.: Topological applications of multilayer perceptrons and support vector machines in financial decision support systems. Int. J. Fin. Econ. **24**(1), 474–507 (2019)

3. Alcalá-Fdez, J., Sanchez, L., Garcia, S., del Jesus, M.J., Ventura, S., Garrell, J.M., Otero, J., Romero, C., Bacardit, J., Rivas, V.M., et al.: Keel: a software tool to assess evolutionary algorithms for data mining problems. Soft Comput. **13**(3), 307–318 (2009)
4. Deakin, E.B.: Distributions of financial accounting ratios: some empirical evidence. Account. Rev. **51**(1), 90–96 (1976)
5. Hall, M., Frank, E., Holmes, G., Pfahringer, B., Reutemann, P., Witten, I.H.: The weka data mining software: an update. ACM SIGKDD Explor. Newslett. **11**(1), 10–18 (2009)
6. Höglund, H.: Tax payment default prediction using genetic algorithm-based variable selection. Expert Syst. Appl. **88**, 368–375 (2017)
7. Jang, S.B.: A design of a tax prediction system based on artificial neural network. In: 2019 International Conference on Platform Technology and Service (PlatCon), pp. 1–4. IEEE (2019)
8. Karegowda, A.G., Manjunath, A., Jayaram, M.: Comparative study of attribute selection using gain ratio and correlation based feature selection. Int. J. Inf. Technol. Knowl. Manage. **2**(2), 271–277 (2010)
9. Pavya, K.D.: Feature selection techniques in data mining: a study. Int. J. Sci. Devel. Res. **2**(6), 594–598 (2017)
10. Lu, S., Cai, Z.J., Zhang, X.B.: Application of ga-svm time series prediction in tax forecasting. In: 2009 2nd IEEE International Conference on Computer Science and Information Technology, pp. 34–36. IEEE (2009)
11. Meyer, D.Z., Avery, L.M.: Excel as a qualitative data analysis tool. Field Methods **21**(1), 91–112 (2009)
12. Priyadarsini, R.P., Valarmathi, M., Sivakumari, S.: Gain ratio based feature selection method for privacy preservation. ICTACT J. Soft Comput. **1**(4), 201–205 (2011)
13. Satu, M.S., Tasnim, F., Akter, T., Halder, S.: Exploring significant heart disease factors based on semi supervised learning algorithms. In: 2018 International Conference on Computer, Communication, Chemical, Material and Electronic Engineering (IC4ME2), pp. 1–4. IEEE (2018)
14. Zhang, Y.: Research on the model of tax revenue forecast of jilin province based on gray correlation analysis. In: 2014 Sixth International Conference on Intelligent Human-Machine Systems and Cybernetics, vol. 2, pp. 30–33. IEEE (2014)

PRCMLA: Product Review Classification Using Machine Learning Algorithms

Shuvashish Paul Sagar, Khondokar Oliullah, Kazi Sohan, and Md. Fazlul Karim Patwary

Abstract In our modern era, where the Internet is ubiquitous, everyone relies on various online resources for shopping and the increase in the use of social media platforms like Facebook, Twitter, etc. The user review spread rapidly among millions of users within a brief period. Consumer reviews on online products play a vital role in the selection of a product. The customer reviews are the measurement of customer satisfaction. This review data in terms of text can be analyzed to identify customers' sentiment and demands. In this paper, we wish to perform four different classification techniques for various reviews available online with the help of artificial intelligence, natural language processing (NLP), and machine learning concepts. Moreover, a Web crawling methodology has also been proposed. Using this Web crawling algorithm, we can collect data from any website. We investigate and compare these techniques with the parameter of accuracy using different training data numbers and testing. Then we find the best classifier method based on accuracy.

Keywords Text classification · Review · NLP · Machine learning · Sentiment analysis

S. P. Sagar · K. Oliullah (✉) · K. Sohan · Md. F. K. Patwary
Institute of Information Technology, Jahangirnagar University, Dhaka 1342, Bangladesh
e-mail: oli.it.ju@gmail.com

S. P. Sagar
e-mail: shuvashishpaul64@gmail.com

K. Sohan
e-mail: kazisohan1920@gmail.com

Md. F. K. Patwary
e-mail: patwary@juniv.edu

M. S. Kaiser et al. (eds.), *Proceedings of International Conference on Trends in Computational and Cognitive Engineering*, Advances in Intelligent Systems and Computing 1309, https://doi.org/10.1007/978-981-33-4673-4_6

1 Introduction

With the speed of faster rising Internet technology in recent times, companies are experiencing further threats and incentives to offer good quality goods or services. Nowadays, consumers are most inclined to purchase from online platforms. They share their feelings and feedbacks online. Therefore, the number of user reviews increases every day [1]. Online businesses are more likely to get user reviews from various online and social media resources. Using this info, online companies are helping themselves offer high-quality products and services. And they will deal with the market's remaining competitors. Despite this massive amount of data, all data cannot be processed manually. As a result, there comes a need for an automatic procedure.

Sentiment analysis is a methodology that aims at contextual text mining to classify subjective information to predict any result [2]. The study of user satisfaction includes the process of data mining. It operates more with the natural language processing (NLP), an analysis of text in the context of opinion mining (OM) [3]. Sometimes, the term opinion mining was seen as similar to sentiment analysis. Various algorithms that work with an analysis of sentiments [4] have been found. They have varying degrees of precision in multiple cases.

Our research into text mining will be able to re-cycle unused data in a supervised learning system to work with it. With this research work, we can use the user review section of new data. We will expand our dataset with this dataset and, thus, the classifier model. This is also a challenge and opportunity in some way to collect user feedback from various online stuff. The enormous dataset can be used to detect an aspect-based sentiment analysis (ABSA) [5]. Standard English is not often used to express thoughts, also. Another obstacle to interpreting any text thus emerges here.

Here, our contribution is to implement a Web crawler to crawl data from different websites or repositories. Thereupon, we study, investigate, and compare four methods of machine classification: long short-term memory (LSTM), support vector machine (SVM), Naive Bayes (NB), and decision tree (DS) to consider the best classifier method based on precision, accuracy, and recall. We work with an Amazon customer review predefined dataset on electronics products [6].

In this paper, Sect. 2 presents some related works. Section 3 details the proposed approach and model. Evaluation of experiment and result is analyzed in Sect. 4. Finally, our research is concluded in Sect. 5.

2 Related Work

Several researchers have proposed various models in their research [7–9], but most of them use only one or two ML classification methods in their simulations. But there is a hidden problem that researchers may not use the proper method of classifying.

The term-based approaches [10] were used for a text mining and classification method. The main concern is the synonymy and the polysemy. Methods based on patterns can outperform those based on terms. They employed clustering algorithms.

For Web surfing, a lot of research has been proposed different methodologies for collecting data from Web resources. The previous works have helped us a lot in developing this data extraction technique. The Web crawler is a bot that can automatically move to various links and Web architecture and can take the HTML from it [11].

PPSGen introduces a unique system that has been proposed to solicit the produced presentation slides that can be used as drafts [12]. It helps them to get the proprietor to prepare the formal slides quicker. Hierarchical agglomeration algorithm is used here.

In many countries, rail incidents present an image of an essential protection point for the transport industry [13]. A range of techniques is to dynamically unearth incident characteristics, guiding the accidents to understand the patron better. Forest algorithm used. Text mining investigates ways to obtain attributes from the document that take advantage of language features specific to the rail transport industry.

It is usually examined by [14] forensic analysis, of which millions of files were computerized. An unstructured document has been obtained to be highly challenging in most files that perform the analyzing process revealed by computer examiners. Text clustering algorithms for computer analysis on forensic department detained an investigation recommended by the author into police.

The author has concentrated on using side information for mining documents. The classical partitioning algorithm developed by the author of [15] with probabilistic models is an effective clustering approach. The used datasets are CORA, DBLP-four-area, and IMDB. The findings show that side information may increase the quality of text clustering and classification to preserve a high degree of performance.

Comparison with LSTM to other methods is not that much used in the previous works. That is why we come up with a comparison with the LSTM to other methods. Depending on other resources while data fetching is not a reliable idea that is used by the previous works. We come up with a data crawler bot.

3 Proposed Approach and Model

This section presents the preprocessing techniques and feature extraction techniques for refinement of textual data, and a short review of the context of sentiment analysis to reinforce the idea. For this analysis, sentiment scores may be important as they provide insight into the inspiration and purpose of a piece of text. The diagram of the workflow is shown on Fig. 1. Lastly, this part describes the classifier models used in this work: supporting vector machines, Naive Bayes, decision tree, and LSTM.

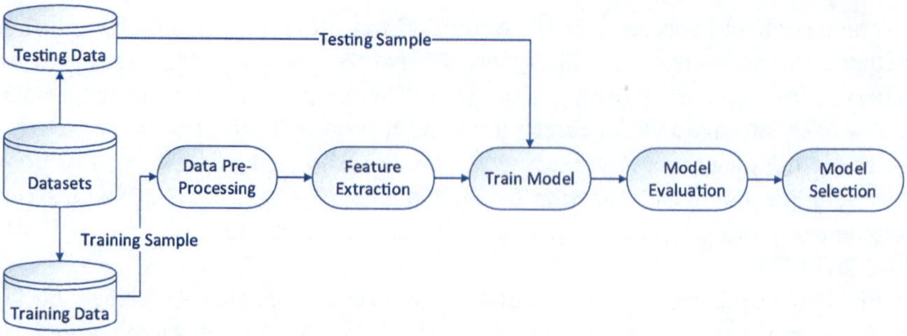

Fig. 1 Workflow diagram

3.1 Proposed Web Crawler

The Web crawler is a significant search engine component [16]. It is a program used to read, search, access, parse content, and store pages in an index file with the meta-tags defined by the website creator. Typically, a crawler requires initial Web addresses as a starting point for indexing the information on these websites. Upon defining these URLs, the crawler can find text hyperlinks and meta-tags in all website pages until the content is finished [17]. Because of a series of seed URLs, the crawler selects one URL from the seeds, again and again, downloads those pages, takes out all the URLs stored therein, and attaches any earlier unfamiliar URLs to the seeds [18]. Other specifications may add flexibility, high efficiency, fault tolerance, reliability, configurability, etc. [19] for any crawling device. The actual flow of data crawling is described as follows:

1. Store the address of the central Uniform Resource Locator (URL) in the database.
2. Find the main categories and store the database with its URLs.
3. Find the subcategories and assign a level to each main category.
4. We need to find a pattern of pages URLs when all the category pages are placed.
5. Store URLs to the drug.
6. Crawl each product page with PHP cURL request. It would be a vector to retrieve the complete DOM text.
7. Record the required data now.

3.2 Data Preprocessing

One of the very first steps before analyzing any data is to clean up the data by structuring and fixing the data, removing any noise that is discernible. The very first step is recognizing the data or separating the words. While standard practice is to use white space and punctuation to bound words, compound words like proper nouns may lose meaning when broken down. To avoid splitting of these tokens, named entity recognizers can be used to resolve this information loss. The tokens can then be converted to their roots in order to standardize the data so that different tenses of words can be linked together. Finally, articles, pronouns, prepositions, and other dull words are often screened out before baseline analysis is carried out to refine the document's text.

3.3 Feature Extraction

Bag of Words

A bag-of-words (BoW) model is a method of mining text features for use in modeling, for example, with algorithms for machine learning. A BoW is text evidence of the existence of words in a file. It includes two things: A corpus of familiar words and a size of the existence of familiar words. It is called a BoW, as the document ignores any data on word order or structure. The model only concerns when known words appear in the document, not where they occur in the document.

N-grams

An n-gram is simply a token command where n is a token number. These tokens are typically words in the background of computational linguistics though they may be characters or subsets of characters. N-grams of texts are commonly used in word extraction and NLP tasks. This is a series of terms that co-occur within a given window, and we normally go ahead with one term when calculating the n-grams.

TF-IDF

TF-IDF refers to the frequency-inverse document frequency, and the TF-IDF weight is a numerical measure used to estimate the significance of a word in information recovery and text mining documents [20].

$$\mathrm{TF}(t) - \mathrm{IDF}(t) \tag{1}$$

TF

Term frequency tests how a word repetitively appears in an article. In long documents, a word will occur much longer than in smaller ones, as different paper lengths.

$$\mathrm{TF}(t) = \frac{\text{Number of times term } t \text{ appears}}{\text{Total number of terms}} \tag{2}$$

IDF

Inverse document frequency measures the importance of a term. All words are measured equally important while calculating TF.

$$\mathrm{IDF}(t) = \log_e \frac{\text{total number of terms}}{\text{number if times term } t \text{ appears}} \tag{3}$$

3.4 Classification Model

Support Vector Machine (SVM)

A support vector machine (SVM) is a supervised machine learning model that uses classification algorithms in two groups for classification problems. They will group new text after having provided an SVM model sets of labeled training data for each group. The weight vector μ can be calculated by following equation given Input x, class or label c, and LaGrange multipliers β:

$$\mu = \sum_{i=1}^{m} \beta_i x_i c_i \tag{4}$$

The SVM's aim is to optimize the equation below:

$$\mathrm{Maximize}_{\beta_i} \sum_{i=1}^{m} \beta_i - \sum_{i=1}^{m}\sum_{j=1}^{m} \beta_i \beta_j c_i c_j \langle x_i x_j \rangle \tag{5}$$

In Eq. (5), $\langle x_i x_j \rangle$ can be attained by different kernels [21].

Naïve Bayes

A Bayes theorem-based classifier, Naive Bayesian classification, with independence assumptions about the predictors: this model is simple and easy to develop, without complicated iterative parameters to calculate, making it particularly useful for very large datasets. This classifier often works surprisingly well and is commonly used.

This theorem offers a way to measure the posterior probability, *P*(*A*|*B*), from *P*(*A*), *P*(*B*), and *P*(*B*|*A*). Therefore, Naïve Bayes can be defined as:

$$P(B|A) = \frac{P(A|B)P(B)}{P(A)} \tag{6}$$

In Eq. (6), *P*(*B*|*A*), *P*(*B*), *P*(*A*|*B*), and *P*(*A*) are the posterior probability of class (objective) given predictor (attribute), previous class probability, predictive probability of a given class, and the previous probability of a predictor, respectively.

Decision Tree

A decision tree is a tool used to support the decision-making phase in making good choices. They are part of the machine learning methods family and allow for the hierarchical distribution of a collection of datasets. Using a DT algorithm, knowledge is

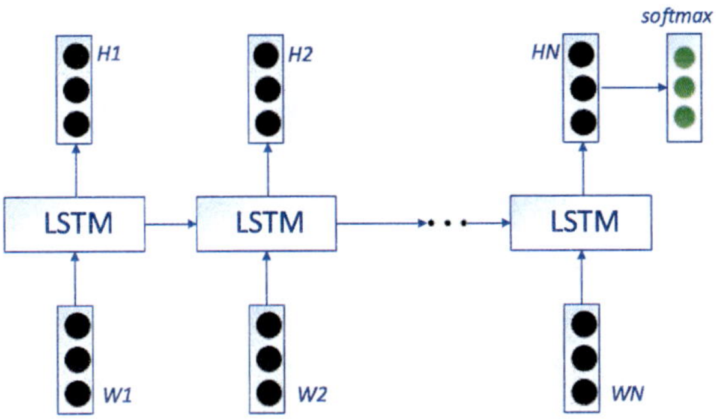

Fig. 2 A standard architecture of LSTM [23]

generated in the form of a hierarchical tree structure that can be used for the classification of instances based on a training dataset. The difficulty in the processing of natural language is to select the "right" linguistic information to use when attempting to solve a problem. In reality, there are several cases in which an unclear case has to be resolved by decision taking. The decision trees are the best decision-making methods.

Long Short-Term Memory

Recurrent neural network (RNN) is an extension of the traditional neural feed-forward network. Traditional RNN, however, has problems of the gradient disappearing or bursting. Long short-term memory network (LSTM) was developed and has achieved superior performance [22] to solve the problems. There are three gates and a cell memory state within the LSTM architecture. Figure 2 is an illustration of standard LSTM architecture. $\{W_1, W_2, \ldots, W_N\}$ represents the word vector of a sentence whose length is N. $\{H1, H2, \ldots, H_N\}$ is the hidden vector.

3.5 *Evaluation Criteria*

For the assessment of the efficiency of various classification models, the following metrics were determined.

$$\text{Accuracy} = \frac{\text{True Positive} + \text{True Negative}}{\text{True Positive} + \text{True Negative} + \text{False Positive} + \text{False Negavite}} \tag{7}$$

$$\text{Precision} = \frac{\text{True Positive}}{\text{True Positive} + \text{False Positive}} \tag{8}$$

$$Recall = \frac{TruePositive}{TruePositive + FalseNegative} \tag{9}$$

Table 1 Data distribution per sentiment class

Positive		Negative		Neutral	
Train	Test	Train	Test	Train	Test
12,315	3010	6025	1723	3021	933

4 Experiment and Results

4.1 *Grabbing Data Using Web Crawler*

Web crawler methodology can be implemented using Python, jquery, PHP, and many more Web technologies. But here we implemented with PHP. There is a function in PHP named cURL. libcurl, PHP supported, permits you to connect and collaborate with several different types of protocols (http, https, ftp, gopher, telnet, dict, file, and ldap protocols) and servers. It also allows HTTPS certificates, HTTP POST, HTTP PUT, FTP uploading, HTTP form-based upload, proxies, cookies, and user + password authentication [24].

4.2 *Dataset*

We have worked with product review data basically. We collected the dataset from amazon.com's product reviews. These products are basically electronics products. We want to identify the sentiment (positive, negative, and neutral) of the sentence expresses. Here, we have used the modules: numpy, pandas, tensorflow, and sklearn. The statistics of the dataset are presented in Table 1. We fetch data by crawling websites and collect it using our proposed Web crawler. This dataset is assembled in raw format. Then, we preprocessed it and made it into a tabular format.

4.3 *Result Analysis*

In order to ensure the best model, we need to define some matrices that can define the perfection of it. Here, we deal with accuracy, precision, and recall score to evaluate the output of each of four different algorithms (SVM, NB, DT, and LSTM). Table 2 reflects common measurement criteria for various models trained on the dataset. It can be seen from Table 2 that LSTM has better accuracy, precision, and recall than other techniques. In the case of assessment, Naïve Bayes performed well. But DT and SVM are less precise than others.

Now with each model's confusing matrices, the most efficient technique can be found. In Fig. 3, we can see that accuracy of LSTM is better, which is 99.55%, than

Table 2 Evaluation metrics for different techniques

Classifier	Accuracy	Precision	Recall
Naïve Bayes	0.9456	0.9128	0.9356
SVM	0.9024	0.9101	0.9249
DT	0.9097	0.9037	0.9083
LSTM	0.9955	0.9694	0.9504

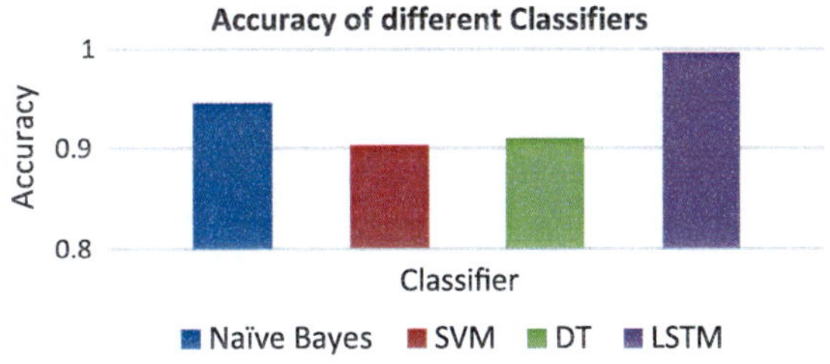

Fig. 3 Accuracy comparison of different classifiers

other classifiers. Besides, the accuracy of Naïve Bayes (94.56%) is a little bit better than DT (90.97%) and SVM (90.24%).

In Fig. 4, we can see that the precision of LSTM is better, which is 96.94%, than other classifiers. Besides, the precisions of Naïve Bayes (91.28%) and SVM (91.01%) are a little bit better than DT (90.37%).

In Fig. 5, we can see that recall of LSTM is better, which is 95.04%, than other classifiers. Besides, the recall of Naïve Bayes (93.36%) is a little bit better than

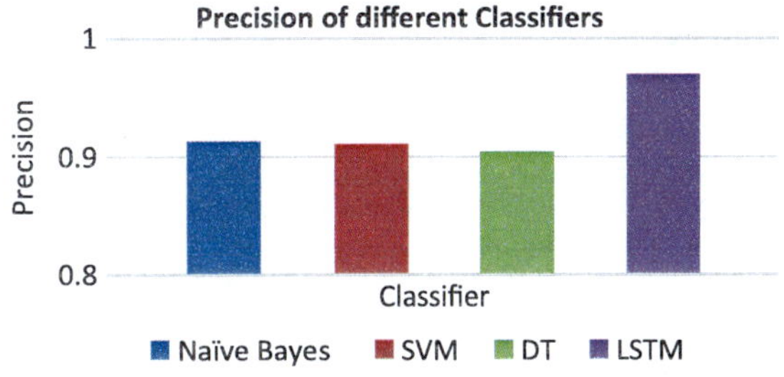

Fig. 4 Precision comparison of different classifiers

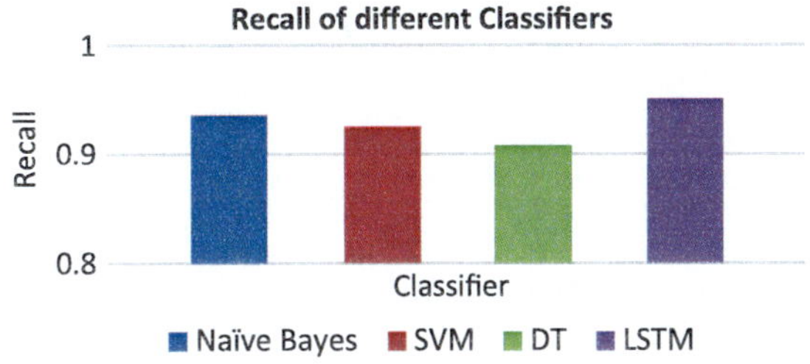

Fig. 5 Recall comparison of different classifiers

DT (90.83%) and SVM (92.49%). The LSTM has performed best in terms of accuracy, precision, and recall. Besides, nearly 40% of the reviews have 30 + words in the dataset. On these evaluations, LSTM performs well in contrast to the other approaches. The connections between the first couple of words and the last couple of words in separate ways than it are not very well learned. In learning long-term relations or dependencies, it is very strong. Therefore, LSTM is the best classifier model for this dataset.

5 Conclusion

Most of the popular e-commerce sites are using data analyzing methodology to enrich their business. Maintaining user satisfaction is the greatest goal for any business or service. With the passage of time, customer satisfaction analysis would be mandatory to grow a business. In this article, by experiments, we compare different classification algorithms on the product review dataset, and it is found that LSTM provides better performance than other classifiers. The LSTM model is a reliable contrast to all the algorithm categories considered due to its ability to estimate better, given the fact that the dataset is noisy and sparse. Also, it is observed that for the smallest training dataset size, the result is also over 90% for this model. One of the drawbacks of this analysis is that the algorithms for machine learning are based on one dataset and need to be validated on multiple datasets for public analysis. In addition, our proposed Web crawler works well also.

References

1. Dang Van, T., Nguyen, V.D., Kiet Van, N., Nguyen, N.L.T.: A transformation method for aspect-based sentiment analysis. J. Comput. Sci. Cybermet. **34**(4) (2018)
2. Gupta, S.: Sentiment Analysis: Concept, Analysis and Applications. https://towardsdatascience.com/. (2018)
3. Rahman, M.M.,Rahman, S.S.M.M, Allayear, S.M., Patwary, M.F.K., Munna, M.T.A.: A Sentiment Analysis Based Approach for Understanding the User Satisfaction on Android Application. Springer, Singapore (2020)
4. Tyagi, P., Tripathi, R.C.: A review towards the sentiment analysis technique for the analysis of twitter data. In: 2nd International Conference on Advanced Computing and Software Engineering ICASE (2019)
5. Kaur, H., Mangat, V., Nidhi.: A survey of sentiment analysis techniques. In: International Conference on I-SMAC (2017)
6. User Reviews of Amazon's Electronic Products, https://www.kaggle.com/datafiniti/consumer-reviews-of-amazon-products (2018)
7. Arora, R., Suman, S.: Comparative analysis of classification algorithms on different datasets using WEKA. Int. J. Comput. Appl. **54**(13), 21–25 (2012)
8. Xia, J., Xie, F., Zhang, Y., Caulfield, C.: Artificial intelligence and data mining: algorithms and applications. Abst. Appl. Anal. Article ID 524720, 2 (2013)
9. Rambocas, M., Gama, J.: Marketing research: the role of sentiment analysis. In: The 5th SNA-KDD Workshop'11. University of Porto (2013)

10. Li, Y., Algarni, A., Albathan, M., Shen, Y., Bijaksana, M.A.: Relevance feature discovery for text mining. IEEE Trans. Knowl. Data Eng. **27**(6) (2015)
11. Vandana, S.: A methodical study of web crawler. VandanaShrivastava J. Eng. Res. Appl. **8**(11), 01–08 (2018). ISSN: 2248–9622
12. Hu, Y., Wan, X.: Ppsgen: learning-based presentation slides generation for academic papers. IEEE Trans. Knowl. Data Eng. **27**(4) (2015)
13. Brown, D.E.: Text mining the contributors to rail accidents. IEEE Trans. Intell. Transp. Syst. **17**(2) (2016)
14. Da Cruz Nassif, L.F., Hruschka, E.R.: Document clustering for forensic analysis: an approach for improving computer inspection. IEEE Trans. Inf. Foren. Sec. **8**(1) (2013)
15. Aggarwal, C.C., Zhao, Y., Yu, P.S.: On the use of side information for mining text data. IEEE Trans. Knowl. Data Eng. **26**(6) (2014)
16. Tan, Q.: Designing New Crawling and Indexing Techniques for Web Search Engines.VDM Verlag (2009)
17. Olston, C., Najork, M.: Web Crawling. Now Publishers Inc (2010)
18. Levene, M.: An Introduction to Search Engine and Web Navigation.Addison Wesley (2005)
19. Manning, C.D., Raghavan, P., Schütze, H.: Support vector machines and machine learning on documents. Int. Inf. Retrie 319348 (2008)
20. Sripath Roy, K., Shaik, F.A.,Uday Kiran, K.,Naga Teja, M., Kurra, S.: Multi-class Emotion AI by reconstructing linguistic context of words. Int. J. Eng. Technol (2018)
21. Burges, C.A.: A tutorial on support vector machines for pattern recognition. Data Mining Knowl. Discov. **2**, 121–167 (1998)
22. Greff, K., Srivastava, R.K., Koutník, J., Steunebrink, B.R., Schmidhuber, J.: LSTM: A search space odyssey. IEEE Trans. Neural Netw. Learn. Syst. **28**(10), 2222–2232 (2017)
23. Wang, Y., Huang, M., Zhu, X.,Zhao, L.: Attention-based LSTM for aspect-level sentiment classification. In: Proceedings of The 2016 Conference on Empirical Methods in Natural Language Processing, pp. 606–615 (2016)
24. PHP cURL, https://www.php.net/manual/en/intro.curl.php. Last Accessed 20 Jan 2020

Handwritten Bangla Character Recognition Using Deep Convolutional Neural Network: Comprehensive Analysis on Three Complete Datasets

M. Mashrukh Zayed, S. M. Neyamul Kabir Utsha, and Sajjad Waheed

Abstract Bangla handwritten character recognition is a difficult job compared to other languages due to the morphological complexity of adjacent characters and a wide variety of curvatures in writing styles people have. Another reason for that is the unique presence of compound characters. Most of the recent research works conducted in this field standardize Deep Convolutional Neural Network (DCNN) models for delivering the most effective outcomes. This paper proposes a DCNN model to classify all the character classes from three popular databases known as BanglaLekha Isolated, Ekush, and CMATERdb. As for BanglaLekha Isolated, our model achieves 93.446% accuracy on the 50 alphabets category and an overall 91.45% considering the whole dataset. The other two datasets, Ekush and CMATERdb result in 95.05% and 94.17% respectively, where the second one holds 171 classes of compound characters alone and performs 93.259% correctness, which is so far the best for this specific category in this dataset.

Keywords Handwritten bangla character · Deep convolutional neural network · Banglalekha · Ekush · CMATERdb · Three full datasets · 28 × 28 images · Bangla compound characters

M. Mashrukh Zayed · S. M. Neyamul Kabir Utsha · S. Waheed (✉)
Mawlana Bhashani Science and Technology University, Tangail, Dhaka 1902, Bangladesh
e-mail: sajjad@mbstu.ac.bd
URL: https://mbstu.ac.bd/

M. Mashrukh Zayed
e-mail: zayedmashrukh@gmail.com
URL: https://mbstu.ac.bd/

S. M. Neyamul Kabir Utsha
e-mail: smnkabir16@gmail.com
URL: https://mbstu.ac.bd/

M. S. Kaiser et al. (eds.), *Proceedings of International Conference on Trends in Computational and Cognitive Engineering*, Advances in Intelligent Systems and Computing 1309, https://doi.org/10.1007/978-981-33-4673-4_7

1 Introduction

In recent years, handwritten character recognition has emerged as one of the major interests to the computer vision researchers due to its various application potentials. It has many significant usages in our regular life, such as zip code scanning, bank check processing, identifying postal codes, reading national ID numbers, etc. most of which are handwritten documents. This paper focuses on the recognition of Bangla handwritten characters and the measurement of accuracies achieved on three complete datasets available for handwritten Bangla alphabets.

Between handwritten and printed Bangla characters, recognizing handwritten characters is more challenging and complicated due to the following reasons: (i) Bangla alphabets have a wide range of characters that are morphologically complex (compound characters), which make the recognition task even more challenging; (ii) Different persons have individual writing styles, so the same character written by different writers vary in shapes, sizes, and curvatures; (iii) The similarities between some characters in shapes further complicates the recognition problem.

In this paper, we developed a handwritten character recognition model using a deep convolutional neural network. DCNN was first proposed by Yann LeCun in the late 90s [1] when he was inspired by the human visual perception of recognizing things. The model has proven to be very successful in the classification of MNIST digits [2]. Considering the analogies, DCNN architecture is fit for use in Bangla handwritten character recognition also.

A few notable works on this field have used DCNN architecture for classification but none of those works were conducted on the three complete datasets available for Bangla alphabets. Some of the works were confined with the basic 50 alphabets [4], while some others worked on only the 10 numerical digits [6, 7] or a part of the compound characters available [5]. The objective of this paper is to analyze the three complete datasets: BanglaLekha Isolated [3], CMATERdb [18], and Ekush [17] under a common recognition model and get significant accuracies for all the character classes, considering maximum number of samples they have.

The rest of the paper is arranged as follows: Sect. 2 represents the related works on handwritten Bangla character recognition, while Sect. 3 introduces our proposed methodology and system architecture for our recognition model. Section 4 briefly describes the datasets used in this research and the process of our experiment. Section 5 discusses the achieved results, analyzes them, and compares them with the early works. Finally, the article is concluded in Sect. 6, invoking some future scopes of further improvement.

2 Related Work

In the early stages of optical character recognition, there has been a number of research published on English Handwritten Characters recognition. But the amount of work on Bangla handwritten character recognition is not significant. After

2015, notable research works have been published by several researchers in both Bangladesh and India.

A system using CNN (Convolutional Neural Network) was developed for Bangla Handwritten character Recognition by Rahman et al. [8] in 2015. Here images of 28×28 resolution are used for a customize dataset with 50 classes of alphabets. Each class has a total of 400 sample images which combines a total of 20,000 sample images for this dataset. With 85.96% accuracy among 50 classes using the CNN method, the research remains as one of the primitive works.

Zahangir et al. [14] introduced a deep learning approach on CMATERdb3.1.1 consisting of 6000 images for 10 Bangla numeric characters. With 32×32 image resolution, this method achieved an accuracy of 98.8%.

Another online-based HBCR is performed by K. Roy et al. [16] on 15000 samples where 2500 samples are tested on Numerical Bangla Characters and others are alphabets. This method achieved a total of 91.13% accuracy on 50 classes targets on the quadratic classifier.

In 2017, Bishwajit et al. [5] proposed a DCNN method on HBCR for 50 classes. The experiment founds a 91.23% accuracy on 50 classes of alphabets and 89.93% all the whole BanglaLekha Isolated [3] dataset. They used 28×28 resolution images. 5% of the data is taken for validation in this experiment.

Rumman Rashid et al. [4] used the CNN method for the same dataset [3] on 50 classes with 75,000 samples. The experiment achieved an accuracy of 91.81% on HBCR by experimenting on 50×50 resolution images. This result remains significant considering only the 50 classes of alphabets.

3 Methodology and System Architecture

A deep convolutional neural network is actually a hierarchical feed-forward model generally composed of some convolutional layers (along with pooling layers) and followed by some fully connected layers where all the neurons are connected to each other. The are also pooling layers which can be effectively used in absorbing variations in shape. Moreover, DCNN model comprises fewer parameters than a fully connected architecture with the same number of hidden layers.

The DCNN model we used in this research contains three convolutional layers with 32, 64, and 128 filters respectively, with a kernel size of 3×3 in general. We have used leaky ReLU (Rectified Linear Unit) as our activation function (Eq. 1) which allows a small, positive gradient when the unit is not active. If we compare with the alternatives such as sigmoid or tanh, they have upper limits to saturate whereas ReLU doesn't saturate for positive inputs, which speeds up training. Figure 1 shows the leak in the linear unit function.

$$\text{Leaky ReLU} = \max(0.1x, x) \tag{1}$$

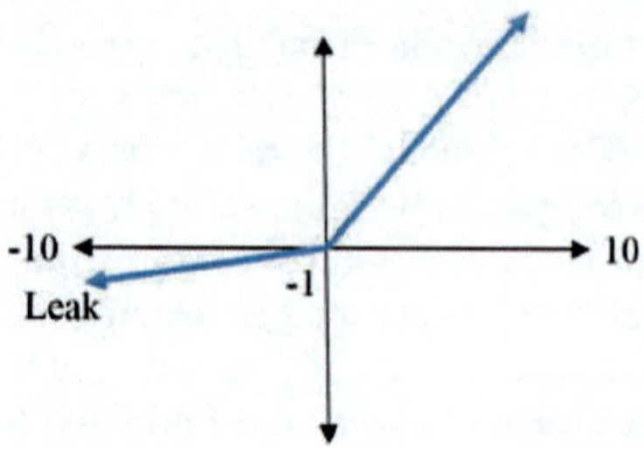

Fig. 1 Leaky rectified linear unit function

Each convolutional layer is followed by a max-pooling layer with 2 × 2 pool size. After that, we used a dropout of 25% for each layer to handle overfitting problem. Dropout involves randomly (and temporarily) deleting some of the hidden neurons in the network, while leaving the input and output neurons untouched. When we dropout different sets of neurons, it's rather like we're training different neural networks. The whole model is briefly described in Fig. 2.

The first layer is the input layer, which has a dimension of 28 × 28 × 1. It means that our model takes a single greyscale or binary image of 28 × 28 dimension as input. Next, the layer scans for 3 × 3 receptive fields throughout the whole image. The scanned features are then passed through the leaky ReLU activation function. Output gained from here is later passed to the max-pooling layer, which slides for 2 × 2 adjacent cells and searches for the maximum value among them.

The second convolutional layer has 64 filters, which takes in the max-pooled feature maps from the previous layer as input. After that, the receptive field scanning, using the activation function, max-pooling operations follow the similar procedure like the first layer. Same goes for the third layer, which has 128 filters, each looking for 3 × 3 receptive fields as well.

After the three layers of convolution, we used two fully connected layers. First one is used as a hidden layer and the last one is used as the output of the whole model. Dense layers are used in both of them, but they are different in all other manners. The first fully connected dense layer has 128 nodes. Each of the nodes uses ReLU activation function along with 30% dropout. Outputs obtained from the whole convolutional structure discussed above are flattened into a one dimensional array of size 2048 (128 × 4 × 4) in order to use them as input of this first fully connected layer.

$$\text{Softmax}(x_i) = \frac{\exp(x_i)}{\sum_j \exp(x_j)} \tag{2}$$

Finally, we have the output layer, which is also a fully connected dense layer but the activation function used here is Softmax (Eq. 2). A softmax layer is very useful in providing probability distribution for a given input; among the character classes under experiment. The number of nodes in the output layer depends on the number of classes we have in three different datasets we used in our research.

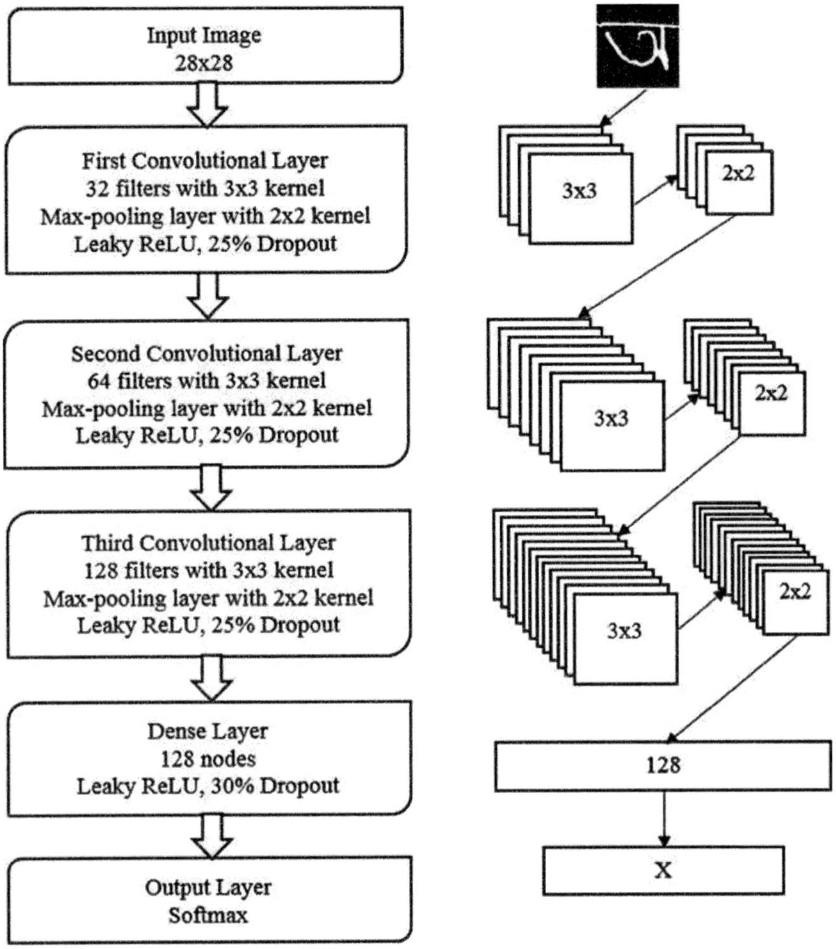

Fig. 2 Proposed DCNN structure

4 Data Processing and Experiment

4.1 Data Processing

The three datasets used in our research have different shapes of images. So at first, we have converted all the images of the three datasets into a uniform size which is 28×28 pixels; using a manual program with resize() function from the cv2 module. Then, we extracted the greyscale values of $28 \times 28 = 784$ pixels for every image and stored them in CSV files. Thus, the CSV files have 785 columns, representing the 784 pixel values between 0 and 255. The first column indicates the 'label' of each

Table 1 Number of sample images in different categories

Dataset Names	Classes	Character categories				Total samples
		Alphabets	Digits	Compounds	Modifiers	
BanglaLekha	84	98,950	19,748	47,407	–	1,66,105
Ekush	122	1,53,281	30,830	1,51,607	30,769	3,66,487
CMATERdb	231	15,000	6,000	41,536	–	62,537

image and the rest are the pixel values of those character images. We have stored all the images class by class in CSV files and made a complete dataset for each category. This is summarized in Table 1.

4.2 *Experiment*

The DCNN model we are using has been implemented in Keras with Tensorflow background. The first thing we have done is creating two discrete arrays, one of which holds the image data and the other contains the label data. Now, there are two phases of spitting data. The earlier phase involves dividing the dataset into test data and train data using train_test_split(), where 20% is reserved for testing final accuracies and 80% goes for training purposes. Later on, another train_test_spilt() is performed to generate the validation data, which is 20% of the training data. The validation set is used to validate the accuracy in the training phase. After the test-train split, the BanglaLekha Isolated dataset comprehends 1,06,307 sample images for training data and 26,576 sample images for validation data. The other two datasets have been divided using the same ratio.

In case of parameter optimization, we have used Adam optimizer. Adam updates any parameter with an individual learnig rate that can vary from 0 to 0.01. Categorical cross-entropy has been used as the loss function. We have also performed batch-wise training due to the large size of our dataset. We used a batch size of 64, which means 64 samples are utilized at once. Finally, we performed every test run for 50 epochs, which allows the program to execute 50 complete passes through each training dataset.

5 Result Analysis

5.1 Accuracy Analysis

The DCNN model we proposed for our research was finalized after several phase of experiment in order to find an optimum result. We tested our model on all the categories over three datasets. All kind of combinations were executed to find an acceptable accuracy level. The results are thoroughly visible in Table 2.

If we observe the accuracies for the categories in a particular dataset, we find higher accuracies on the alphabetic characters and digits, but compound characters are constantly giving lower accuracy. This certainly affects the overall accuracy. But if we primarily focus on the individual categories, we achieved a milestone for the 50 class alphabets in BanglaLekha Isolated dataset [3]. We got an accuracy of **93.446%** over **50 classes**, which is better than the previous record of 91.81% gained in [4].

However, the combined dataset of all classes is also giving a satisfactory outcome. Along with the compound characters, reference [5] recorded 89.93% accuracy over the 80 classes, whereas, considering all the **84 classes** we have achieved an accuracy of **91.45%** in BanglaLekha Isolated dataset. Without any removal of the sample images or any segmentation, this model provides momentous accuracy with the highest number of classes. Figure 3 shows the training and validation accuracy for the 50 classes in BanglaLekha for 50 epochs of training phase.

The next two datasets we worked with have two unique features. First one is the Ekush [17] dataset which has a unique class of modifiers. There are very few notable works on this dataset. Using the same proposed model, this dataset provides a total accuracy of **95.05%** on all **122 classes**, where modifiers result in 98.57%, alphabets perform 96.29% and compound characters provide 94.91%.

The last dataset is the CMATERdb [18], which has the largest collection of compound characters with 171 unique classes. Here we got 95.199%, 93.259%, and 98.08% for the alphabets, compound characters, and digits respectively, resulting in a total of **94.17%** accuracy over **231 classes**. One of the research work conducted in [7] had adequate accuracy only on specific digits of this dataset, while our model works with all the classes of this dataset with suitable accuracy.

Table 2 Accuracies achieved for all categories in three datasets

Categories	Datasets		
	BanglaLekha	Ekush	CMATERdb
Alphabets	*93.446*	96.29	95.119
Digits	97.44	98.96	98.08
Compounds	90.38	94.91	*93.259*
Modifiers	–	98.57	–
Overall	91.45	95.05	94.17

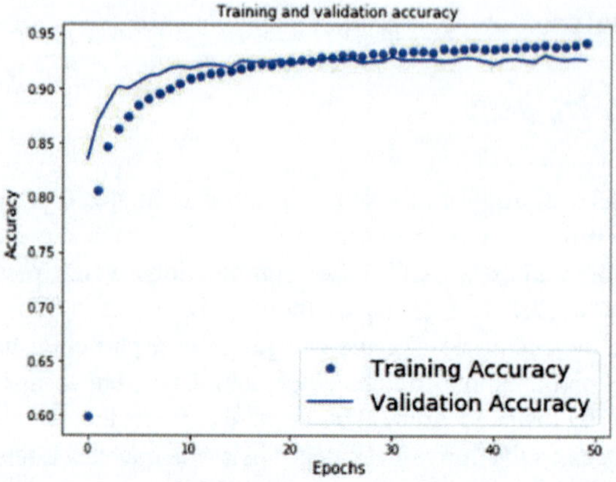

Training and validation accuracy for the 50 classes

Fig. 3 Accuracy graph on BanglaLekha isolated dataset

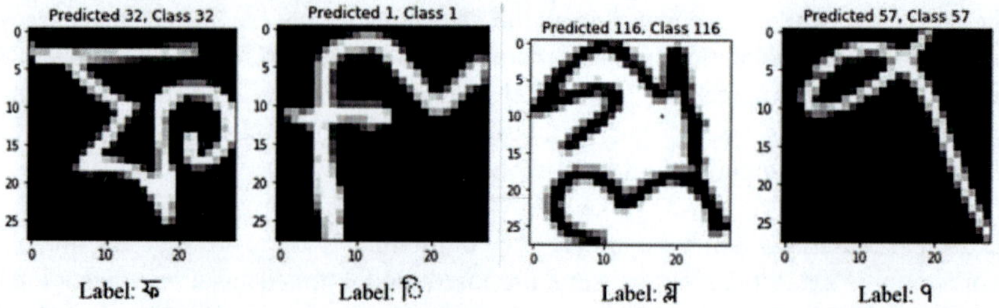

Fig. 4 Some accurate predictions made by our proposed model

5.2 *Loss Analysis*

We have seen that the compound character category is a crux in generating high accuracies in all three datasets. However, there are other issues also, which causes drop in performance irrespective of any kind of model. We found five such scenarios after investigating the samples, which were predicted wrong after the test runs.

The fact we encountered most is the presence of invalid images in the datasets, which can't be classified into any categories at all. We can see such a sample in the first example from Fig. 6. Next, there are some samples which can be termed as mislabeled, because it seems that our model recognizes them correctly rather than what they are said to be so. Then comes the samples which tend to hold no kind of data at all. Some of them are blank or have so few pixel values that they can't even be considered as samples.

Although some errors are authentic, which shouldn't have been miscategorized, that's the fewest case in our results. Another issue we faced is the morphological

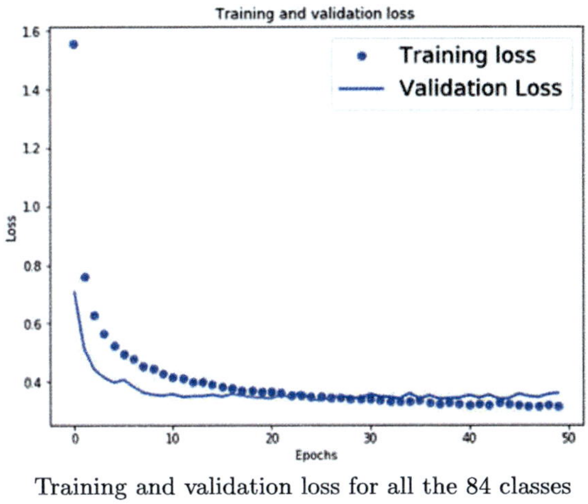

Fig. 5 Loss graph on BanglaLekha isolated dataset

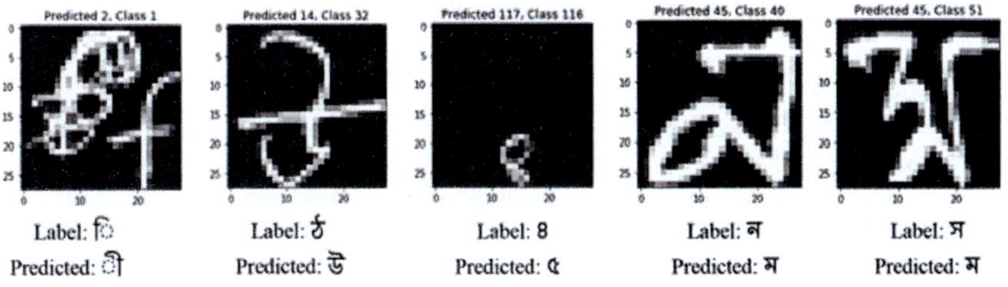

Fig. 6 Innaccurate predictions made by our proposed model

similarities between several Bangla characters, for which some samples are difficult to differentiate even with our human eyes. The last examples from Fig. 6 can be referred to for such phenomena.

All these cases caused increase in test loss for the overall outcome. Nevertheless, test loss for the three datasets was 0.34, 0.20, and 0.19 respectively. They can be found highly acceptable compared to the early researches we referred to.

6 Conclusion

Our research has a significant contribution by representing a strategy that has delivered peak development in recognition of Bangla alphabets, significant results towards classifying all compound characters, and superior outcomes in recognizing all the classes available in Bangla handwritten characters. This notifies the multipurpose ability of this research.

However, lack of opportunity to use a GPU for the training process remains as a limitation of our work. If the datasets we used had been improved by collecting eminent data from the collectors and by getting rid of superfluous samples, our results would have been much more overwhelming. Furthermore, though the recognition process has a great impact on identifying a single Bangla character, major concern is still needed for identifying multiple Bangla characters at a time or recognizing a Bangla word or even a whole sentence simultaneously. So, future works should focus on detecting sequence of characters, which should require more complex methods to grab in and implement.

References

1. LeCun, Y., Boser, B.E., Denker, J.S., Henderson, D., Howard, R.E., Hubbard, W.E., Jackel, L.D.: Handwritten digit recognition with a back-propagation network. Adv. Neural Inf. Process. Syst. 396–404 (1990)
2. LeCun, Y., Cortes, C., Burges, C.J.: Mnist handwritten digit database, AT&T Labs, vol. 2 (2010). Available http://yann.lecun.com/exdb/mnist
3. Biswas, M., Islam, R., Shom, G.K., Shopon, M., Mohammed, N., Momen, S., Abedin, A.: BanglaLekha-isolated: a multi-purpose comprehensive dataset of Handwritten Bangla Isolated characters. Data Brief **12**, 103–107 (2017)
4. Chowdhury, R.R., Hossain, M.S., ul Islam, R., Andersson, K., Hossain, S.: Bangla handwritten character recognition using convolutional neural network with data augmentation. In: 2019 Joint 8th International Conference on Informatics, Electronics & Vision (ICIEV) and 2019 3rd International Conference on Imaging, Vision & Pattern Recognition (icIVPR), pp. 318–323. Spokane, WA, USA (2019)
5. Purkaystha, B., Datta, T., Islam, M.S.: Bengali handwritten character recognition using deep convolutional neural network. In: 20th International Conference of Computer and Information Technology (ICCIT), pp. 1–5. Dhaka (2017)
6. Shopon, M., Mohammed, N., Abedin, M.A.: Bangla handwritten digit recognition using autoencoder and deep convolutional neural network. In: International Workshop on Computational Intelligence (IWCI), pp. 64–68. Dhaka (2016)
7. Shopon, M., Mohammed, N., Abedin, M.A.: Image augmentation by blocky artifact in deep convolutional neural network for handwritten digit recognition. In: 2017 IEEE International Conference on Imaging, Vision & Pattern Recognition (icIVPR), pp. 1–6. Dhaka (2017)
8. Rahman, M.M., Akhand, M., Islam, S., Shill, P.C., Rahman, M., et al.: Bangla handwritten character recognition using convolutional neural network. Int. J. Image Graph. Signal Process. (IJIGSP) **7**(8), 42–49 (2015)
9. Chaudhuri, B.: A complete handwritten numeral database of Bangla-a major Indic script. In: Proceedings of Tenth International Workshop on Frontiers in Handwriting Recognition. Suvisoft, Baule, France (2006)
10. Liu, C.-L., Suen, C.Y.: A new benchmark on the recognition of handwritten Bangla and Farsi numeral characters. Pattern Recogn. **42**(12), 3287–3295 (2009)
11. Basu, S., Das, N., Sarkar, R., Kundu, M., Nasipuri, M., Basu, D.K.: An MLP based approach for recognition of handwritten Bangla numerals (2012). http://arxiv.org/abs/1203.0876
12. Pal, U., Chaudhuri, B.: Automatic recognition of unconstrained off-line Bangla handwritten numerals. In: Proceedings of Advances in Multimodal Interfaces—ICMI 2000, pp. 371–378. Springer, Beijing, China (2000)
13. Chowdhury, K., Alam, L., Sarmin, S., Arefin, S., Hoque, M.M.: A fuzzy features based online handwritten Bangla word recognition framework. In: 2015 18th International Conference on Computer and Information Technology (ICCIT), pp. 484–489. IEEE (2015)

14. Alom, M.Z., Sidike, P., Taha, T.M., Asari, V.K.: Handwritten Bangla digit recognition using deep learning (2017). arXiv:1705.02680
15. Das, N., Sarkar, R., Basu, S., Kundu, M., Nasipuri, M., Basu, D.K.: A genetic algorithm based region sampling for selection of local features in handwritten digit recognition application. Appl. Soft Comput. **12**(5), 1592–1606 (2012)
16. Roy, K., Sharma, N., Pal, T., Pal, U.: Online Bangla handwriting recognition system. In: Advances in Pattern Recognition, pp. 117–122. World Scientific (2007)
17. Rabby, A.S.A., Haque, S., Islam, M., Abujar, S., Hossain, S.: Ekush: a multipurpose and multitype comprehensive database for online off-line Bangla handwritten characters (2019). https://doi.org/10.1007/978-981-13-9187-3_14
18. CMATERdb.: The pattern recognition database repository, Computer Science and Engineering Department, Jadavpur University, Kolkata 700032, India, [Online]. Available: https://code.google.com/archive/p/cmaterdb/

Handwritten Bangla Character Recognition Using Convolutional Neural Network and Bidirectional Long Short-Term Memory

Jasiya Fairiz Raisa, Maliha Ulfat, Abdullah Al Mueed, and Mohammad Abu Yousuf

Abstract Handwritten character recognition (HCR) is presently gaining much popularity due to its wide range of applications. Bangla language comprises many unique characters of complex shapes, cursive nature, and resemblance among different characters. This nature of the script makes Bangla Handwritten character recognition (BHCR) a very challenging task. Very few HCR models can classify all types of Bangla characters with respectable accuracy. The current work proposes a hybrid BHCR model combining the Convolutional Neural Network (CNN) with stacked Bidirectional Long Short-Term Memory (Bi-LSTM). The experimental model used CNN for feature extraction and stacked Bidirectional LSTM for classification purposes. An augmented version of the CMATER dB dataset was implemented for training and evaluating the model. The model could recognize 243 individual classes of Bangla handwritten characters with 96.07% accuracy. No other proposed models are found to classify such a high number of classes with high accuracy.

Keywords Optical character recognition · Convolutional neural network · Bidirectional long-short term memory · Data augmentation

---

J. Fairiz Raisa (✉) · M. Ulfat · A. Al Mueed
Bangladesh University of Professionals, Mirpur Cantonment, Dhaka, Bangladesh
e-mail: jasiyafairiz@gmail.com

M. Ulfat
e-mail: maliha.ulfat@gmail.com

A. Al Mueed
e-mail: mueed.faiaz@gmail.com

M. Abu Yousuf
Institute of Information Technology, Jahangirnagar University,
Savar, Dhaka 1342, Bangladesh
e-mail: yousuf@juniv.edu

M. S. Kaiser et al. (eds.), *Proceedings of International Conference on Trends in Computational and Cognitive Engineering*, Advances in Intelligent Systems and Computing 1309, https://doi.org/10.1007/978-981-33-4673-4_8

1 Introduction

Handwritten Character Recognition (HCR) is presently extensively used for electronic documentation information digitization. Much research has been conducted for recognizing English handwriting, whereas Bangla being worlds fifth most-spoken native language, the amount of research done on Bangla HCR is comparatively small. Bangla is more complicated script than English, with larger character set and very similar pattern of the characters. Bangla consists of more than hundreds of unique characters under four categories: Numeric, Basic, Compound, and Modifiers. Most of the existing studies focused on the first three character categories. To create a concrete HCR system; in our research, we have covered four categories of Bangla character classes. CNN-based BHCR models showed very high accuracy for fewer character categories. Increasing the categories of characters leads to the decreasing classification accuracy of the models. Types of Recurrent Neural Network (RNN) are found to be used for BHNR. No other studies found for recognizing other Bangla character categories using these sequence modelling technologies. We have combined CNN for feature extraction, and Bi-LSTM layers classification in this work to develop an effective BHCR system. CNN can extract effective feature sets from input images, whereas Bi-LSTM uses sequence modelling to predict character classes. The proposed model aims to classify Bangla handwritten isolated characters with satisfactory accuracy. This will be a first BHCR model to classify all the distinct character classes, publicly available for Bangla language. The proposed model focuses and contributes in three points: (1) Implementation of all Bangla handwritten characters, (2) Data augmentation to increase accuracy, (3) Development of a novel CNN, BiLSTM hybrid architecture for Bangla character recognition. The rest of the paper is organized as follows. Section 2 reviews related previous works, Sect. 3 explores the proposed system architecture and methodology, Sect. 4 demonstrates model performance analysis and comparison with exsisting works. Lastly, Sect. 5 concludes the paper.

2 Related Work

We have reviewed many proposed BHCR models with different approaches for different data categories. Ghosh et al. [1] reviewed many proposed BHCR models and claimed CNN based methods perform better. Sazal et al. [2] used a deep belief network for the numerals and led to 90.27% accuracy. For the numerals, Akhand et al. [3] trained 3 CNNs with differently and combined their final output decision with 98% accuracy. Ahmed et al. [4] introduced a deep LSTM model architecture for Bangla numerals with 98.25% accuracy. Rahman et al. [5] developed novel VGG-11M with the accuracy of 99.80% for classifying Bangla numerals. Bhattacharya et al. [6] generated a Bangla basic character database and performed classification using the MQDF and MLP methods with 95.80% accuracy. Similarly, Bhowmik et al. [7] used a self-generated basic character dataset with MLP architecture showing 84.33% accuracy. Fardous et al. [8] analyzed a model with dropout to recognize compound characters showing 95.5% accuracy. Ashiquzzaman et al. [9] employed a DCNN method with 93% classification accuracy for compound characters. Hasan et al. [10] proposed a combination of deep CNN and Bi-LSTM and gained 98.50% accuracy for compound character classification. Alif et al. [11] proposed a modified RESNET-18 model to analyze numerals and basic characters and showed an accuracy of 95.01%. Abir et al. [12] combined the Inception module with CNN layers to recognize numeric, basic, and compound characters and showed an accuracy of 89.3%. Purkaystha et al. [13] developed a CNN model for 80 Bangla character classes providing 89.93% of accuracy. For the same three categories of the characters, Chowdhury et al. [14] focused on the effect of augmentation showing accuracy increment from 91.81 to 95.25%. Majid et al. [15] developed a new dataset and introduced a combination of SVM with cubic kernel with an 96.80% classification accuracy. Rabby et al. [16] developed EkushNet, a multilayer CNN classifier and the model performed with accuracy of 97.73% for Ekush dataset and 95.01% for CMATER dB.

3 System Architecture and Methodology

3.1 Key Concepts

CNN: The Convolutional Neural Network (CNN) is a specialized type of neural network model designed for working with image data. CNNs are composed of three main layers. (1) Conv layer: This locally connected layers take raw image matrix as input and perform convolution operation for feature extraction purpose. (2) Pool layer: Pooling is used to reduce the network dimensionality and summarize the features. In this paper Maxpooling is used.

$$O = \frac{(W - K + 2P)}{S} + 1 \tag{1}$$

In Eq. 1, O is the output height/ length, W is the input height/ length, K is the filter size, P is the padding, and S is the stride. (3) Fully connected layer: This layer is used for classification of the previously extracted features.

Bi-LSTM: Bidirectional LSTM is an extension of traditional LSTM. It trains two LSTMs, one in forward and another in the backward direction of the input sequence to improve the model performance for sequence classification. It can improve the model performance for sequence classification. The forward and backward hidden states are computed and merged using the Eqs. 2, 3 and 4. Figure 1 shows the cell structure of Bi-LSTM.

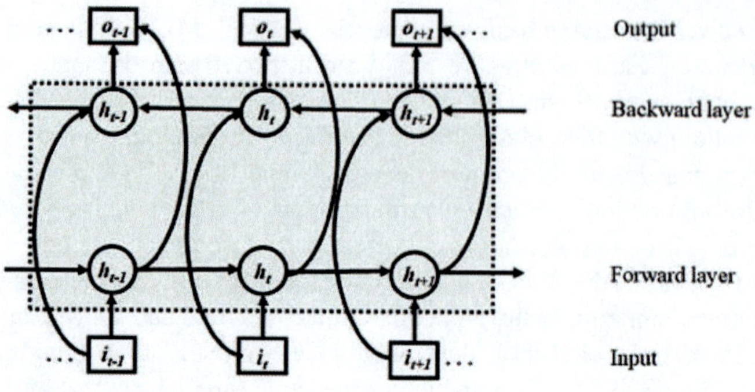

Fig. 1 Bi-LSTM cell structure

$$h_f = \sigma(W_f * I + h_f + b_f) \quad (2)$$

$$h_b = \sigma(W_b * I + h_b + b_b) \quad (3)$$

$$o = (h_f W_f + h_b W_b + b) \quad (4)$$

Here, h is the hidden layer state, W is the weight, b is the bias term and σ is the activation function [10].

Batch Normalization: Batch normalization technique acts to standardize only the mean and variance of each unit to make neural networks faster and more stable. It is implemented in this proposed model for faster computation and increasing accuracy.

Dropout: Dropout is a regularization technique that influences network elements to learn diverse representations of data and prevents complex co-adaptations on training data. This technique is used in this study to overcome the overfitting problem.

3.2 *Proposed Architecture*

The proposed BHCR system is a combination of CNN and stacked Bi-LSTM layers. The proposed model architecture can be divided into 3 parts: (1) Pre-processing, (2) Feature Extraction, and (3) Classification. Figure 2 shows the workflow of our proposed model. The raw images collected from the database are noisy, in varying size and sometimes mixed with different character images. Pre-processing acts to make images uniformly resized, noise-free, and allocated class-wise while removing unclear ones. In this work, image augmentation is performed to expand the base dataset, so that the model can recognize character patters even with the change of their position. Figure 3 shows the workflow of pre-processing of the images. In this study, CNN is used as feature extractor as it is proved to be superior than other unsupervised feature extractors providing efficient features for higher model perfor-

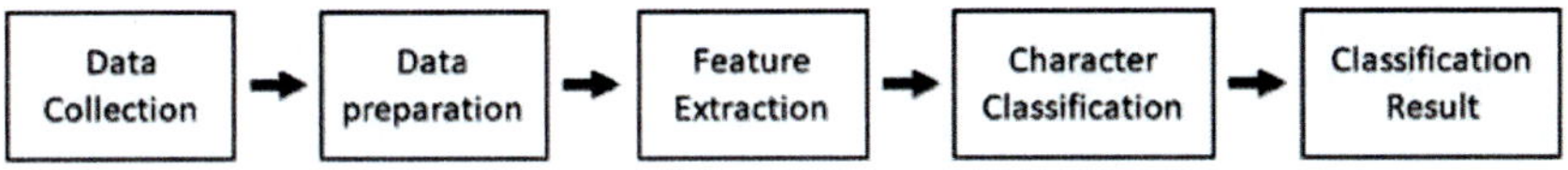

Fig. 2 Generic model workflow

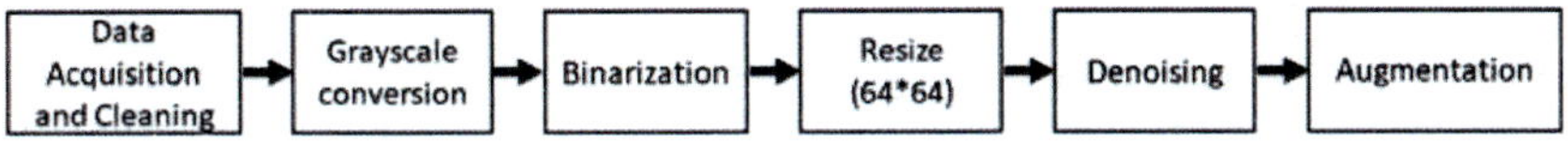

Fig. 3 Pre-processing workflow

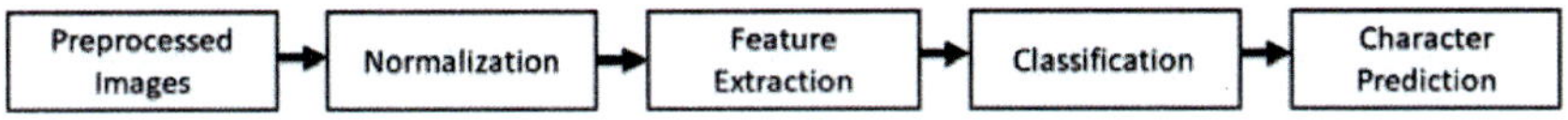

Fig. 4 Classification workflow

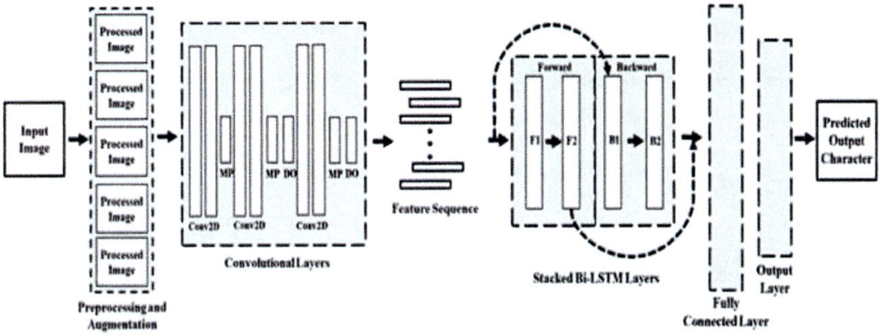

Fig. 5 Proposed structure of the BHCR system

mance rate. Six Convolutional layers have been implemented in three blocks with Maxpooling, Batch normalization and Dropout layers. The number of CNN layers and parameters have been chosen focusing on the highest classification accuracy and least computation time. We experimented with the number of CNN layers, filters, and kernel size, and the current model setup was selected for its outstanding performance among all. For the classification of the proposed BHCR system, two stacked Bi-directional LSTM layers are used as it works faster and more accurately than single Bi-LSTM or simple LSTM. Bi-LSTM keeps information from forward and backward layers, which makes Bi-LSTM eligible for large dataset of closely similar pattern, i.e. Bangla Characters. Figure 4 shows the workflow of classification. Two fully connected (Dense) layers with a Dropout layer in between are placed after the Bi-LSTM layers for classifying characters. Figure 5 gives an overview of the model architecture.

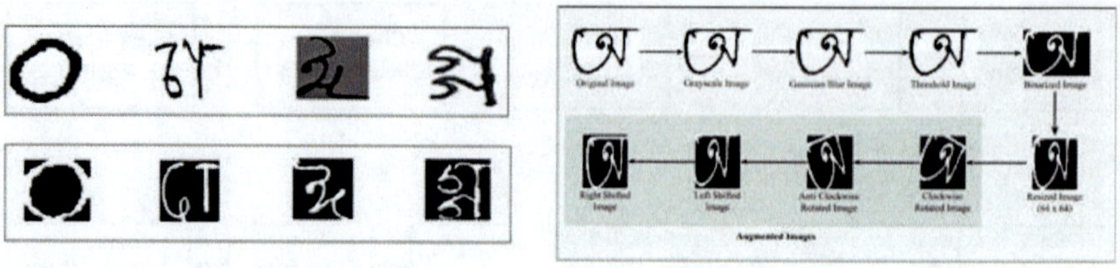

(a) (b)

Fig. 6 **a** Sample raw and pre-processed images. **b** Pre-processing steps

Table 1 Description of pre-processed dataset

Dataset	Classes	Train images	Validation images	Test images
Numerals	10	14,126	6054	5575
Modifiers	12	5040	2160	2720
Basic	50	21,000	9000	8829
Compound	171	106,924	45,824	37,576

3.3 *Methodology*

3.3.1 Experimental Setup

The proposed model was implemented in Kaggle Kernel, a Jupyter notebook, accessible from we browsers. Kaggle Kernels provide Nvidia Tesla P100 16 GB VRAM as GPU with 15 GB of RAM and 2 GHz Intel Xeon Processor.

3.3.2 Data Acquisition and Preparation

For the proposed BHCR model, CMATER dB dataset is used. It contains 245 character classes (10 numerals, 50 basic, 171 compounds, and 14 modifiers) [17]. The raw data was cleaned and allocated separate class-wise folders. We have ignored two character classes due to their lesser use and repetition found in original dataset. After cleaning, the raw dataset is preprocessed into noise-free, uniform shaped, binarized image. Sample raw and pre-processed images are shown in Fig. 6. Algorithm 1 shows all the steps of image pre-processing.

We carried out four different augmentation methods (Clockwise rotation, Anticlockwise rotation, Left-translation, Right translation) on the processed images to expand the number of data. The final augmented dataset was further cleaned to remove unrecognizable images. Hence, we have reduced a calculated number of images from the expanded dataset. The expanded dataset was divided into training, validating, and testing sets as shown in the Table 1.

Algorithm 1 Image Pre-processing

Input: Handwritten Bangla character images

```
step 01. image = read (image_path)
step 02. gray_im = BGR2GRAY (image)
step 03.
         bw_im = threshold (gray_im)
         bi_im = bitwise-not (bw_im)
step 04. resized = resize (bi_image, (64,64))
step 05. im_fin = gaussian_blur (resized)
step 06.
         augment (im_fin)
          clk_img = clock_rotation (im_fin)
          aclk_img = anticlock_rotation (im_fin)
          lshift = left_transform (im_fin)
          rshift = right_transform (im_fin)
```

Table 2 Pre-processing details

Parameters	Collected	Preprocessed
Clockwise rotation	0° rotation	Rotation range (0°, 30°)
Anti-clockwise rotation	0° rotation	Rotation range (−30°, 0°)
Left, downward shift	Axis (x, y) = (0, 0)	Axis (x, y) = (−5, −2)
Right, upward shift	Axis (x, y) = (0, 0)	Axis (x, y) = (5, 2)
Training sample	33,417	147,090
Validation sample	14,322	63,038
Testing sample	11,359	54,700

Table 2 holds the summarized information of the cleaned and preprocessed datasets and a comparison between the base and the expanded dataset.

3.3.3 Feature Extraction

The proposed model uses CNN for feature extraction. The pre-processed dataset is first converted into tensors and then into float format. Images are normalized with pixel values ranging between 0 and 1. The Convolutional blocks extracts features from the normalized images. The feature vectors are then reshaped into a two-dimensional feature sequence and fed into the Bi-LSTM layers for the next step.

3.3.4 Classification

Bi-LSTM layers use the two-dimensional feature sequence as input and work to find spatial dependencies among extracted features sequence to classify the data. There

Algorithm 2 Image Classification

Input: Preprocessed and augmented image

```
step 01. image = read (image_path)
step 02. image_array = numpy_array (image)
step 03. norm_image= normalize (image_array)
step 04. train_data = load (train_path)
         valid_data = validation.split (train_data, 0.3)
         test_data = load (test_path)
step 05. features = CNN (train_data)
step 06. feat_seq = reshape (features)
step 07. accuracy = BiLSTM (feat_seq)
```

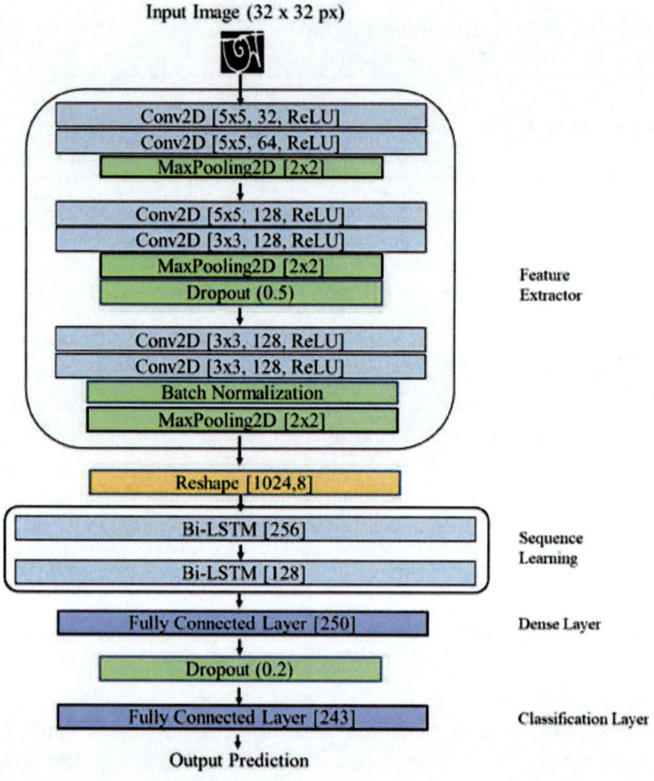

Fig. 7 CNN-BiLSTM model architecture

are 128 and 64 hidden LSTM units in the two layers. Finally, two dense layers, working as fully connected layers are used for recognizing character classes using the output data from the Bi-LSTM layers. The summary of the whole model is shown in the Fig. 7.

Table 3 Model content description

Model contents	Details
Optimizer	Adam
Loss function	Categorical cross entropy
Learning rate	0.001
Metrics	Loss, accuracy
Batch size	64
Epochs	35
Time per epoch	595 s

Table 4 Classification result overview

Phase	Number of images	Accuracy (%)	Loss
Training	147,090	99.25	0.0235
Validation	63,038	97.52	0.2198
Testing	54,700	96.07	0.2894

For the experimental model, the values of parameters and hyperparameters are fine-tuned and optimized. We have experimented with different values of parameters and hyper-parameters and chose the value which ensures the best model performance, shown in Table 3. Algorithm 2 shows the steps of image classification.

4 Experiment and Result Analysis

This section focuses on analyzing experimental results achieved using the proposed model, performance comparison with different datasets and other existing models.

4.1 Proposed Model Evaluation

The model performed training, validation, and testing processes using CMATER dB dataset. After training and validation, the model could classify the testing data with 96.07% accuracy. The detail of the model implementation phases is shown in the Table 4.

The accuracy and loss of the training and validation phases are visualized in Fig. 8a.

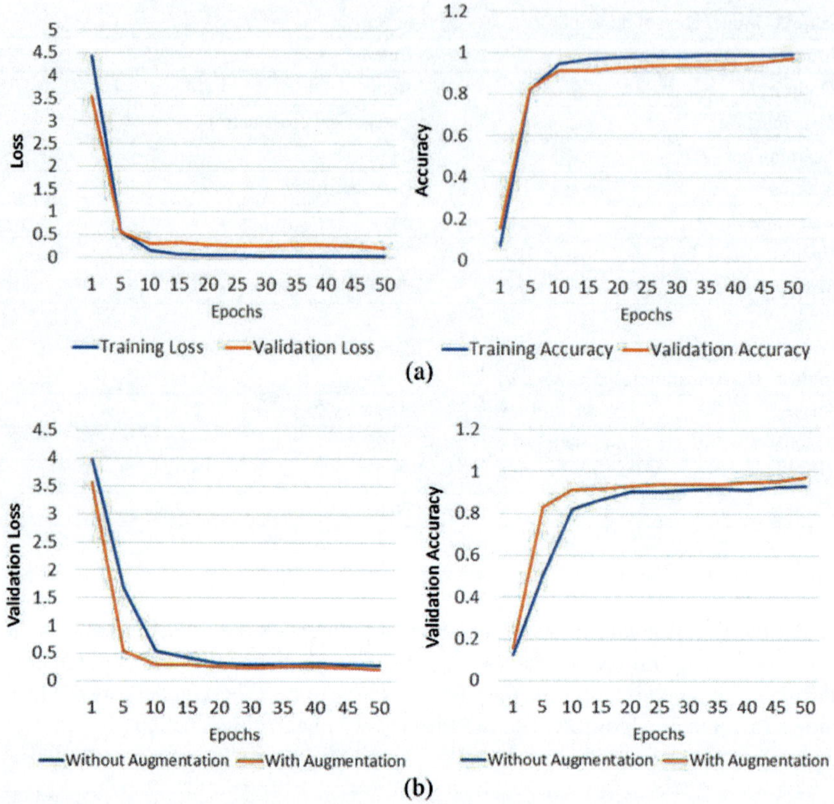

Fig. 8 **a** Loss and accuracy of training and validation phases. **b** Comparison of validation performance for data augmentation

4.2 Performance Comparison

4.2.1 Implementation of Data Augmentation

The base and the augmented dataset were implemented in the model. The result was compared to see the effect of augmentation. The base dataset achieved 92.79% classification accuracy with a loss of 0.3968. Figure 8b visualizes the training and validation phases between the base and the expanded dataset. Table 5 shows the overall result comparison for this experiment.

Table 5 Performance comparison for data augmentation

Parameters	Without augmentation	With augmentation
Number of images	59,098	264,828
Training accuracy, loss	97.57%, 0.723	99.25%, 0.0235
Validation accuracy, loss	93.31%, 0.2916	97.57%, 0.2198
Test accuracy, loss	92.79%, 0.3968	96.07%, 0.2894

Table 6 Model performance using different dataset

Dataset	Class	Testing accuracy (%)	Testing loss
CMATERdb	243	96.07	0.2894
Ekush	122	94.07	0.2920
Bangla-Lekha	84	89.61	0.9319

Table 7 Model performance using different dataset

Referencing work	Classifier	Dataset	Number of classes	Accuracy
Bhattacharya et al. [6]	MQDF, MLP	Self-constructed	50	95.8%
Alif et al. [11]	ResNet-18, CNN	BanglaLekha-isolated dataset, CMATER	60	95.10% (BanglaLekha)
Rabby et al. [16]	CNN	Ekush, CMATER	122	97.73% (Ekush)
Fardous et al. [8]	CNN	CMATER	171	95.5%
Proposed work	CNN-BiLSTM	CMATER	243	96.07%

4.2.2 Implementation of the Different Dataset

The model was implemented using BanglaLekha isolated dataset [18] and Ekush dataset [16]. Both of these are preprocessed without augmentation. The same model structure has been used for classification. The experimental result is shown in the Table 6. Table 7 shows the comparison with existing BHCR models.

5 Conclusion

OCR has been getting much attention in the research areas, and yet many lacking is often found in the existing works of Bangla OCR. In this current work, a novel hybrid CNN stacked Bi-LSTM method is proposed for Bangla handwritten character recognition, with 96.07% accuracy. Also, the model was implemented using two other datasets with satisfactory accuracy, proving the model versatility. The model does not achieve benchmark accuracy for character classification, comparing with existing models. Despite the admirable accuracy, the proposed model sometimes misclassifies characters due to the close similarity between the different characters. In the future, the model can be modified for Bangla handwriting recognition with electronically editable data generation, which can be a breakthrough in the OCR research sector.

References

1. Ghosh, T., Abedin, M.M., Chowdhury, S.M., Yousuf, M.A.: A comprehensive review on recognition techniques for Bangla handwritten characters. In: 2019 International Conference on Bangla Speech and Language Processing (ICBSLP), pp. 1–6 (2019)
2. Sazal, M.M.R., Biswas, S.K., Amin, M.F., Murase, K.: Bangla handwritten character recognition using deep belief network. In: 2013 International Conference on Electrical Information and Communication Technology (EICT) (2014)
3. Akhand, M.A.H., Ahmed, M., Hafizur Rahman, M.M.: Multiple convolutional neural network training for bangla handwritten numeral recognition. In: 2016 International Conference on Computer and Communication Engineering (ICCCE), pp. 311–315 (2016)
4. Ahmed, M., Akhand, M.A.H., Hafizur Rahman, M.M.: Handwritten Bangla numeral recognition using deep long short term memory. In: 2016 6th International Conference on Information and Communication Technology for The Muslim World (ICT4M), pp. 310–315 (2016)
5. Rahman, M.M., Islam, M.S., Sassi, R., Aktaruzzaman, M.: Convolutional neural networks performance comparison for handwritten Bengali numerals recognition. SN Appl. Sci. **1**(12), 1660 (2019)
6. Bhattacharya, U., Shridhar, M., Parui, S.K., Sen, P.K., Chaudhuri, B.B.: Offline recognition of handwritten Bangla characters: an efficient two-stage approach. Pattern Anal. Appl. **15**(4), 445–458 (2012). November
7. Bhowmik, T.K., Bhattacharya, U., Parui, S.K.: Recognition of Bangla handwritten characters using an MLP classifier based on stroke features. In: Pal, N.R., Kasabov, N., Mudi, R.K., Pal, S., Parui, S.K., (eds.) Neural Information Processing. Lecture Notes in Computer Science, pp. 814–819. Springer, Berlin, Heidelberg (2004)
8. Fardous, A., Afroge, S.: Handwritten isolated Bangla compound character recognition. In: 2019 International Conference on Electrical, Computer and Communication Engineering (ECCE), pp. 1–5 (2019)
9. Ashiquzzaman, A., Tushar, A.K., Dutta, S., Mohsin, F.: An efficient method for improving classification accuracy of handwritten Bangla compound characters using DCNN with dropout and ELU. In: 2017 Third International Conference on Research in Computational Intelligence and Communication Networks (ICRCICN) (2017)
10. Hasan, M.J., Wahid, M.F., Alom, M.S.: Bangla compound character recognition by combining deep convolutional neural network with bidirectional long short-term memory. In: 2019 4th International Conference on Electrical Information and Communication Technology (EICT), pp. 1–4 (2019)
11. Alif, M.A.R., Ahmed, S., Hasan, M.A.: Isolated Bangla handwritten character recognition with convolutional neural network. In: 2017 20th International Conference of Computer and Information Technology (ICCIT), pp. 1–6 (2017)
12. Abir, B.M., Mahal, S.M., Islam, M.S., Chakrabarty, A.: Bangla handwritten character recognition with multilayer convolutional neural network. In: Kolhe, M.L., Trivedi, M.C., Tiwari, S., Singh, V.K. (eds.), Advances in Data and Information Sciences. Lecture Notes in Networks and Systems, pp. 155–165. Springer, Singapore (2019)

13. Purkaystha, B., Datta, T., Islam, M.S.: Bengali handwritten character recognition using deep convolutional neural network. In: 2017 20th International Conference of Computer and Information Technology (ICCIT), pp. 1–5 (2017)
14. Chowdhury, R.R., Hossain, M.S., Islam, R., Andersson, K., Hossain, S.: Bangla handwritten character recognition using convolutional neural network with data augmentation. In: 2019 Joint 8th International Conference on Informatics, Electronics Vision (ICIEV) and 2019 3rd International Conference on Imaging, Vision Pattern Recognition (icIVPR), pp. 318–323 (2019)
15. Majid, N., Smith, E.: Introducing the boise state Bangla handwriting dataset and an efficient offline recognizer of isolated Bangla characters. In: 2018 16th International Conference on Frontiers in Handwriting Recognition (ICFHR) (2018)
16. Shahariar Azad Rabby, A.K.M., Haque, S., Islam, M.S., Abujar, S., Hossain, S.M. Ekush: A multipurpose and multitype comprehensive database for online off-line Bangla handwritten characters. In: Santosh, K.C., Hegadi, R.C. (eds.), Recent Trends in Image Processing and Pattern Recognition, Communications in Computer and Information Science, pp. 149–158. Springer, Singapore (2019)
17. Das, N., Acharya, K., Sarkar, R., Basu, S., Kundu, M., Nasipuri, M.: A benchmark image database of isolated Bangla handwritten compound characters. IJDAR **17**(4), 413–431 (2014). December
18. Biswas, M., Islam, R., Gautam Shom, M., Shopon, N.M., Momen, Sifat, Abedin, A.: BanglaLekha-isolated: a multi-purpose comprehensive dataset of handwritten Bangla isolated characters. Data Brief **12**, 103–107 (2017). June

Bangla Text Generation Using Bidirectional Optimized Gated Recurrent Unit Network

Nahid Ibne Akhtar, Kh. Mohimenul Islam Shazol, Rifat Rahman, and Mohammad Abu Yousuf

Abstract Natural language processing is a vital branch of Artificial Intelligence, a bridge to communicate between computer and human in their languages and is used to generate automatic text, summarize articles, machine translation, etc. In this paper, we proposed a model of Bangla sentence generation that uses the Optimized Gated Recurrent Unit network based Recurrent Neural network model. We used OGRU based model instead of the regular Gated Recurrent Unit network because OGRU generates better results than GRU. For this purpose, we have used three Bangla corpus datasets, which consist of more than 49 million words. Here we used Bangla Natural Language Toolkit to preprocess the datasets. We trained our model with those three different datasets and achieved an accuracy of 97% on average and it can generate a paragraph of text based on the context. With this accuracy and ability to generate text, it is proved that our model is much efficient in Generating Bengali Sentences.

Keywords Natural language processing · Bangla sentence generation · Optimized Gated Recurrent Unit · Recurrent neural networks · BNLTK · Gated Recurrent Unit

---

These authors have contributed equally.

N. Ibne Akhtar (✉) · Kh. Mohimenul Islam Shazol · R. Rahman
Bangladesh University of Professionals, Mirpur Cantonment, Dhaka, Bangladesh
e-mail: nahid.ibne.akhtar@gmail.com

Kh. Mohimenul Islam Shazol
e-mail: shazol.kh@gmail.com

R. Rahman
e-mail: rifatrahman140530@gmail.com

M. Abu Yousuf
Institute of Information Technology, Jahangirnagar University, Savar, Dhaka 1342, Bangladesh
e-mail: yousuf@juniv.edu

M. S. Kaiser et al. (eds.), *Proceedings of International Conference on Trends in Computational and Cognitive Engineering*, Advances in Intelligent Systems and Computing 1309, https://doi.org/10.1007/978-981-33-4673-4_9

1 Introduction

Natural language processing deals with the processing of natural language. Existing text generation, understanding the context of a given text is a vital branch of NLP and, in this era of digital content based communication, it will increase the value and ease the usability of any natural language in coming days as the conventional reading, writing process will be replaced by sentence generation and other methods. The purpose of sentence generation is to generate phrases and sentences to ease our standard way of writing every single word. Sentence generation can be used for automatic text generation, in robots, in chat bots, and other human communicable devices. Moreover, it is a challenge for the physically challenged people to express themselves following conventional methods. A model can be built that will ease their struggle of expressing themselves. This model can be built with the help of machine learning and deep learning mechanism. There have been many efficient works in other languages. Though Bengali is one of the most spoken languages in the world, there have been very few works in the field of NLP, and still now, there are not any efficient sentence generation models. In Previously proposed methods, there have some limitations of not able to generate long sentences and context-based relatable sentences. In this research, we propose a Bengali sentence generation model that can generate a paragraph of sequence to sequence context-based sentences. In our proposed model, we used The GRU network, to predict the next sequences of words to form a corpus of sentences as the suggested sentences. The optimized version of the GRU is used. The algorithm finds the most relatable words based on the context and forms the text corpus. To get the best outcome from Bengali datasets, the model has been formed following specific techniques. The model has been examined with three large Bengali datasets and, the model is outperforming in all datasets than previously proposed models.

2 Literature Review

Previously there have been some works on sentence generation. Researchers are trying to develop efficient models following various technologies and methods on some leading natural languages. There are also some quality and visionary works in Bengali. Researchers also worked on developing the efficiency of the algorithms and reducing the complexity of the algorithms. They also worked on simplifying the methods. In this section, some of these works will be discussed.

Recently, some excellent works have been done on Bengali sentence generation. Abujar et al. worked on generating short text using bidirectional LSTM [1]. They have used Keras library to pre-process the data and followed word prediction techniques to build the text generation model. In the same year Islam et al. approached to build a sequence to sequence Bengali sentence generation model using LSTM [2]. They have used N-gram to process input text, and Omor Faruk Rakib proposed the model using GRU [3]. They created five N-Grams language models based on their datasets and calculated probabilities in a word sequence and applied GRU to predict the next word.

There had been a number of other research works in Bengali like sentence correction [4], text analysis and summarization [5, 6]. Islam et al. have proposed a system that is able to correct the Bangla sentence based on a wrong sentence as input [4]. They have used a deep learning approach, a sequence to sequence sentence generation technique with RNN and LSTM. Again, Roy et al. proposed an intelligent typing assistant. They also used deep learning RNN based approach what is able to suggest multiple words based on the context the writer is writing [7]. They have used character and word level prediction. In Song's work, they have introduced an LSTM-CNN based Abstractive Text Summarization Framework, which can construct new sentences by exploring more fine-grained than sentences, namely, semantic phrases [5]. On the other hand, Al Munzir and Rahman proposed a simple deep network of sequence classification-based models for the Bengali single document and used multiple embedding and dense layers to generate the summary.

There are some works in other languages as well. Bowman analyzed the latent spaces and found that it can generate coherent and diverse sentences through continuous sampling [8]. Zhang et al. developed a framework for generating realistic text via adversarial training [9]. They used an LSTM and Convolutional networks as generator and discriminator. They proposed it matching the high-dimensional latent feature of distributions for the real and synthetic sentences through a kernelized discrepancy metric and they did not use the standard objective of GAN. Again, Matsuyama et al. proposed a system of sentence generation based on a question adjectives [10]. It not only finds the answer to the question but also adds an informative sentence; thus, the conversation can be going on and remain interesting.

Some significant research has also been done for updating the gating efficiency of GRU. Wang and his team optimized the GRU by changing the update gate structure of the GRU [11], where Tjandra et al. proposed a hybrid model combining the properties of LSTM and GRU with Tensor Product to maintain a direct and more expressive connection with input and hidden layers [12]. Zhou et al. proposed the minimal design in any hidden gated unit for RNN that has only one gate and does not involve the peephole connection [13].

3 Methodology

Text generation is a subfield of Natural language processing. Like other NLP problems, the text generation process can be expressed as a supervised learning process. In, supervised learning process, there exist features and corresponding labels. Mapping the features with the labels there can be created a prediction model. The prediction model has to be trained by a corpus of text within which there are tons of various kinds of sentences. Within these sentences, a particular portion is selected for feature extraction and the rest of the part as the labels. In our methodology, we have proposed a word-level prediction model for generating text. In this model, Bidirec-

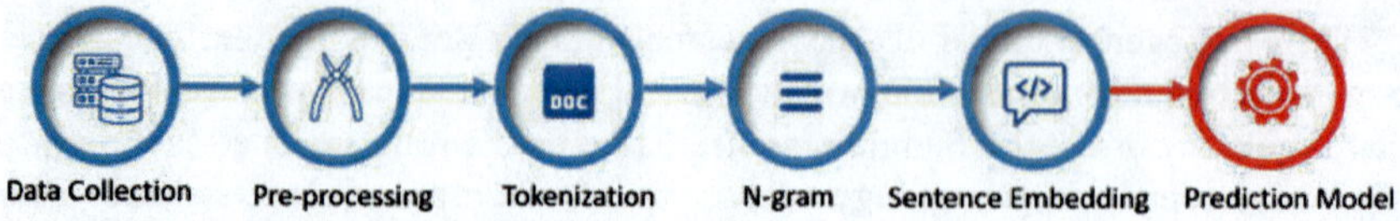

Fig. 1 Basic work flow of proposed model

tional OGRU has been used. This model has not been used before for text generation in any language. Figure 1 is the basic workflow of the proposed model. The entire procedure is explained here, step by step.

3.1 Dataset Collection

The decent Bengali dataset for language processing is not available. The available dataset needs much pre-processing to make it useful. We have collected our datasets from Kaggle. Statistics of our datasets are demonstrated in Table 1.

3.2 Pre-processing and Tokenization

All three datasets contain several white spaces, punctuation marks, English words, and numbers. We have developed a function to eliminate these and tokenized the datasets. There are not enough libraries to clean Bengali datasets and this is also difficult to clean because of its complex structure. The data cleaning function successfully cleaned and tokenized the datasets. Tokenization means splitting text into units called tokens. In Algorithm 1, the data pre-processing and tokenization technique have been shown.

3.3 N-Gram Tokens Sequencing

N-gram is the process of sequencing N-words together. At the time of tokenization, the arrangement of the words is changed. Through N-gram, a corpus of equal length sequences are generated. This corpus of sequences is used to set the probability of

Table 1 Dataset statistics

Data source	Total words	Unique words
Prothom Alo Corpus [14]	14,400,000	371,301
Wikipedia Corpus 1 [15]	165,70,000	760,011
Wikipedia Corpus 2 [15]	18,700,000	600,545

occurrences or words occurring next to a sequence of words [16]. An N-gram model predicts the occurrence of a word based on the occurrence of its $N - 1$ previous words. N-gram model can be 1-gram, 2-gram, 3-gram, 4-gram, Etc. The number of input words can be different to predict the next word. The process of N-gram is explained through Algorithm 2.

3.4 Sentence Embedding

Machines do not understand the meaning of natural languages. So the sequence of words needed to be embedded. Word embedding means to convert every word into some real-valued vectors, and thus, machines can similarly represent the same meaning of words [1, 4]. Nevertheless, if we want a machine to understand the context of the whole sequence, then sentence embedding is more efficient. Sentence embedding allows understanding the motive of the sentence without calculating the embedding of each word.

Algorithm 1: Data Processing and Data Tokenization

```
Step 1: numbers = '0123456789'
for i=0 to length(input_text) do
    if input_text[i] not in numbers then
        Step 2: data = date + input_text[i]
    end
end
Step 3: tokens[ ] = data.split(' ')
for j=0 to length(data) do
    if data[j] in punctuation then
        Step 4: tokens[j] = maketrans(' ',' ', string.punctuations)
    end
end
Step 5: tokens[ ] = translate(stable)
for p=0 to length(tokens) do
    if tokens[p] not in English Words then
        Step 6: tokens[p] = Replace(tokens[p], ' ')
    end
end
```

Algorithm 2: Same Length Sequence create

```
Step 1: Seq_length = Input Sequence Length+Output Sequence Length
Step 2: Lines = Empty List
for i=0 to length(tokens), increment(Seq_length) do
    Step 3: Seq = tokens[i - Seq_length, i]
    Step 4: Line = Join Seq together with a space between every tokens
    Step 5: List: Lines.append(Line)
    if i > Numbers of Sequence Lines to be generated then
        Break.
    end
end
```

3.5 System Model

Recurrent Neural Network RNN can use it's internal memory to generate sequence data such as sequence and time-series data. It is a generalized FeedForword network. In this network the current state is formed taking the current input, x_t and previous data output, h_{t-1} and pass it through an activation function, tanh. Here, w denotes the weights of the states, and y_t is the output.

$$h_t = f(h_{t-1}, x_t) \tag{1}$$

$$h_t = \tanh(w_{nh}h_{t-1} + w_{xh}x_t) \tag{2}$$

$$y_t = w_{hy}h_t \tag{3}$$

RNN can not remember long term dependencies [7]. In sequence learning, there are vanishing gradient and exploding gradient problems [17, 18]. The gradient is the loss function with respect to the weights in the network. If gradient with respect to weights in earlier layers becomes really small, then the vanishing gradient problem occurs. On the other hand, if the gradient is large, then the multiplication of it with the value will be large and explode the size of the gradient, which is called the exploding gradient problem. The issues of RNN has been fixed in the updated versions of RNN like LSTM [17] and GRU [19]. The number of gates are fewer in GRU, and it outperforms LSTM in faster training and gives better efficiency in prediction [20]. Previously GRU has been used for word prediction in Bengali. It outperformed the LSTM structure in prediction [3]. The accuracy and loss comparison of LSTM and OGRU has been presented in Fig. 2. The comparison also mentioned in the base paper of OGRU [11].

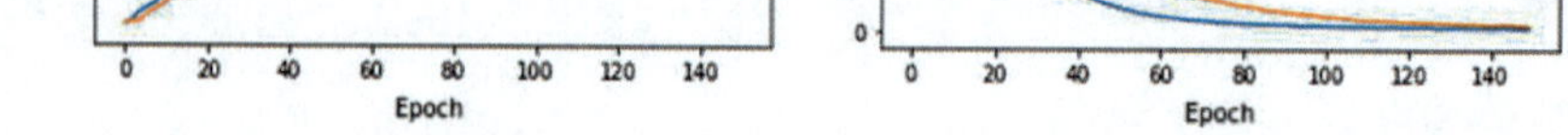

(a) Accuracy (b) Loss

Fig. 2 Accuracy and Loss comparison between OGRU and LSTM

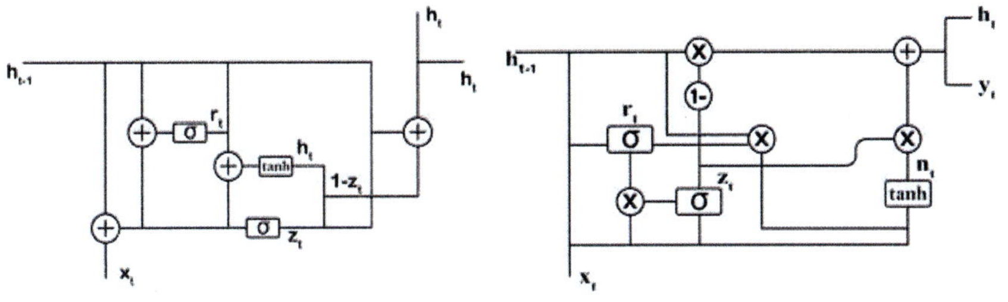

(a) GRU (b) OGRU

Fig. 3 Gated Recurrent Unit and Optimized Gated Recurrent Unit network

Gated Recurrent Unit(GRU) GRU is the most advanced RNN that is completely able to overcome the vanishing and exploding gradient problem. It uses two gates to decide which information to pass to the output. The two gates are Update gate and Reset Gate shown in Eqs. 4 and 5.

$$z_t = \sigma(x_t w_t + h_{t-1} U_z + b_z) \quad (4)$$

$$r_t = \sigma(x_t w_r + h_{t-1} U_r + b_r) \quad (5)$$

$$h_t = \tanh(x_t w + r_t * h_{t-1} U) \quad (6)$$

$$h_t = z_t * h_{t-1} + (1 - z_t) * h_t \quad (7)$$

$$y_t = \sigma(w_o * h_t) \quad (8)$$

Here, z_t is the update gate, and it decides the amount of information should be passed to the future. r_t denotes the reset gate, which decides how much of the previous information should be forgotten. $h_t - 1$ defines the previous time step's output. Figure 3a showing the basic structure of GRU. We have used an optimized version of GRU and for Bengali, it works better than the traditional GRU. But as this optimized version has come from the basic GRU, so, the equations of GRU has been presented from Equations 4 to 8 to understand the model clearly. Though GRU is the most updated RNN model, still, GRU has some drawbacks. GRU has low update efficiency, poor information processing ability, and less expressive relation between input layers and hidden layers. To resolve those problems, the OGRU model has been used [11]. In OGRU, only the update gate is modified. Instead of x_t in update gate there have been used x_t multiplied by r_t. Then the output of the reset gate is used to feedback adjust the update gate. The current information x_t is filtered by the reset gate, reduces the adverse effect of the forgetting data to a great extent. Figure 3b shows the structure of OGRU. The update gate in OGRU is shown in Eq. 9.

$$z_t = \sigma(x_t \cdot r_t w_t + h_{t-1} U_z + b_z) \quad (9)$$

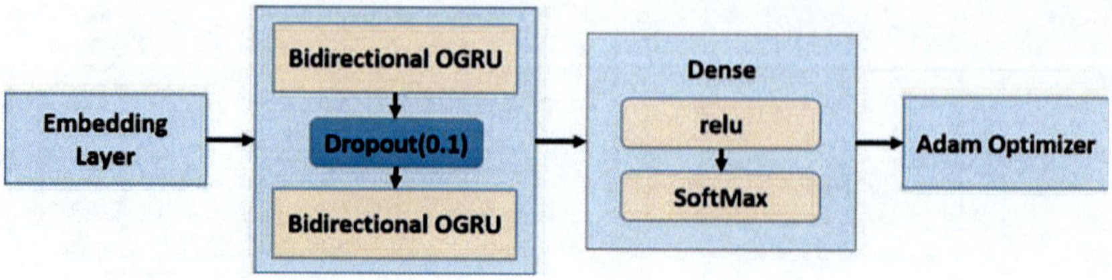

Fig. 4 Prediction model structure

We have observed that if we add a bidirectional learning layer before the OGRU layer, then the learning time is reduced to almost half. So the model has been built following the structure denoted in Fig. 4. We have used two kinds of activation functions. ReLU is a piecewise function that outputs the direct value if it is positive. Otherwise, it returns zero. softmax is used to classify the outputs in multiple categories. Equation 10 represents ReLU function.

$$f(x) = \max(0, +x) \tag{10}$$

4 Result Analysis

We trained our OGRU based model for three different datasets shown in Table 1. For each dataset, 75% data are used for training and 25% for testing. The batch size was 256, and the number of epochs is 150 for each dataset's training. categorical_crossentropy was used as loss function in training the datasets. After 150 epochs, we achieved an average accuracy of 97% for three datasets. Accuracy and Loss statistics for datasets are shown in the Table 2. Figure 5 shows the accuracy and loss graph for all three datasets. We have also succeeded in reducing the training time to one fourth. And finally, we tested our model with some input text from test data corpus. The model took input some random text and generated a short paragraph of text according to the input text context. Some of the outputs of our model are shown in Fig. 6.

Table 2 Accuracy and loss comparison of datasets

Dataset	Accuracy (%)	Loss
Prothom Alo Corpus	96.41	0.0965
Wikipedia Corpus 1	97.66	0.0795
Wikipedia Corpus 2	97.51	0.0813

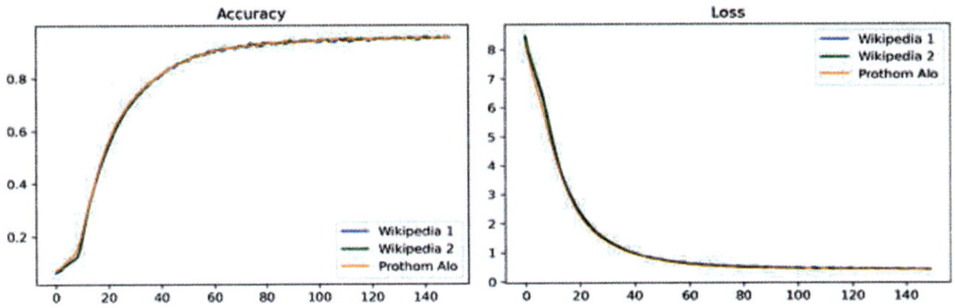

(a) Accuracy (b) Loss

Fig. 5 Plot of accuracy (**a**) and Loss (**b**) for all three datasets

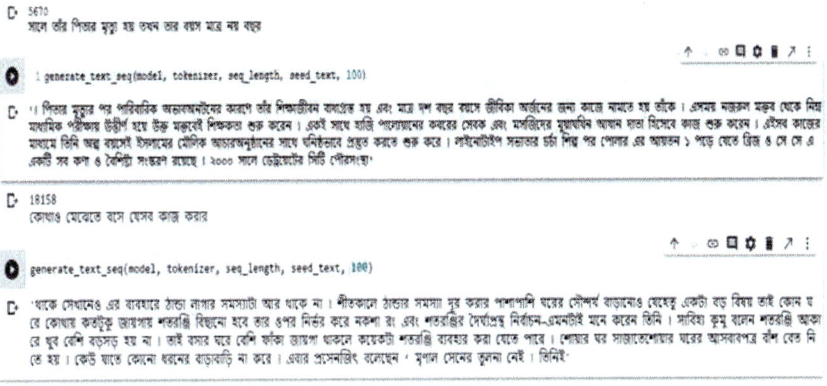

Fig. 6 Generated texts based on our proposed model

5 Conclusion and Future Work

In the paper, we have proposed a Bengali text generation method that can generate a short paragraph of text just from a few words as input. The most advanced RNN based Deep neural network (GRU) is used to get the best outcome. The supervised learning method can learn long term dependencies from the dataset and predict the next sequence of words. We have tested our model with several datasets, and it gives better accuracy and almost a constant accuracy for all datasets. Transformer based natural language processing is seeming promising nowadays [21]. Recently, a gigantic amount of English text has been trained successfully with this method. In the future, we want to design an encoder or decoder for Bengali that will be able to train a larger amount of text and will be more efficient in solving text generation and other NLP problems of Bengali.

References

1. Abujar, S., Masum, A.K.M., Chowdhury, S.M.M.H., Hasan, M., Hossain, S.A.: Bengali Text generation using bi-directional RNN. In: 2019 10th International Conference on Computing, Communication and Networking Technologies (ICCCNT), July 2019, pp. 1–5
2. Islam, M., Sultana Sharmin, S., Abujar, S.,Hossain, S.: Sequence-to- sequence Bangla sentence generation with LSTM recurrent neural networks. Procedia Comput Sci 152 (2019)
3. Rakib, O.F., Akter, S., Khan, M.A., Das, A.K., Habibullah, K.M.: Bangla word prediction and sentence completion using GRU: an extended version of RNN on N-gram language model. In: 2019 International Conference on Sustainable Technologies for Industry 4.0 (STI), pp. 1–6, Dec 2019
4. Islam, S., et al.: Bangla sentence correction using deep neural network based sequence to sequence learning, pp. 1–6, Dec 2018
5. Song, S., Huang, H., Ruan, T.: Abstractive text summarization using LSTM-CNN based deep learning. Multimedia Tools Appl. **78**, 857–875 (2020). ISSN: 1573-7721. https://doi.org/10.1007/s11042-018-5749-3
6. Al Munzir, A., Rahman, M.L., Abujar, S., Ohidujjaman, Hossain, S.A.: Text analysis for Bengali text summarization using deep learning. In: 2019 10th International Conference on Computing, Communication and Networking Technologies (ICCCNT), pp. 1–6, July 2019
7. Roy, S., Hossain, S.I., Akhand, M.A.H., Siddique, N.: Sequence modeling for intelligent typing assistant with Bangla and English keyboard. In: 2018 International Conference on Innovation in Engineering and Technology (ICIET), pp. 1–6, Dec 2018
8. Bowman, S.R., et al.: Generating sentences from a continuous space (2020). arXiv:1511.06349 [cs]. arXiv: 1511.06349 (May 2016)
9. Zhang, Y., et al.: Adversarial feature matching for text generation (2020). arXiv: 1706.03850 [cs, stat]. arXiv: 1706.03850 (2020)
10. Matsuyama, Y., Saito, A., Fujie, S., Kobayashi, T.: Automatic expressive opinion sentence generation for enjoyable conversational systems. IEEE/ACM Trans Audio, Speech Lang Process **23**, 313–326 (2015). Issn: 2329-9304
11. Wang, X., Xu, J., Shi, W., Liu, J.: OGRU: an Optimized Gated Recurrent Unit neural network. J Phys Conf Ser **1325**, 012089 (2019)
12. Tjandra, A., Sakti, S., Manurung, R., Adriani, M., Nakamura, S.: Gated recurrent neural tensor network. In: 2016 International Joint Conference on Neural Networks (IJCNN), pp. 448–455 (2020)
13. Zhou, G.-B., Wu, J., Zhang, C.-L., Zhou, Z.-H.: Minimal Gated Unit for recurrent neural networks (2020). arXiv:1603.09420 [cs]. arXiv: 1603.09420
14. Prothom Alo [2013–2019]: https://kaggle.com/twintyone/prothomalo
15. Bangla Wikipedia Corpus: https://kaggle.com/shazol/bangla-wikipedia-corpus
16. Brown, P.F., deSouza, P.V., Mercer, R.L., Pietra, V.J.D., Lai, J.C.: Class-based n-gram models of natural language. Comput. Lingui. 1**8**, 467–479 (1992). ISSN: 0891-2017
17. Hochreiter, S., Schmidhuber, J.: Long short-term memory. Neural Comput. **9**, 1735–1780 (1997). ISSN: 0899-7667. https://doi.org/10.1162/neco.1997.9.8.1735
18. Hochreiter, S.: The vanishing gradient problem during learning recurrent neural nets and problem solutions. Int. J. Uncertainty Fuzziness Knowl. Based Syst. **6**, 107–116 (1998). Apr
19. Chung, J., Gulcehre, C., Cho, K., Bengio, Y.: Gated feedback recurrent neural networks. In: Proceedings of the 32nd International Conference on Machine Learning, vol. 37, pp. 2067–2075. JMLR.org, Lille, France, July 2015
20. Mangal, S., Joshi, P., Modak, R.: LSTM vs. GRU vs. Bidirectional RNN for script generation (2019). (2019)
21. Shao, T., Guo, Y., Chen, H., Hao, Z.: Transformer-based neural network for answer selection in question answering. IEEE Access 7. Conference Name: IEEE Access, pp. 26146–26156. ISSn: 2169-3536

An ANN-Based Approach to Identify Smart Appliances for Ambient Assisted Living (AAL) in the Smart Space

Mohammad Helal Uddin, Mohammad Nahid Hossain, and S.-H. Yang

Abstract This research discusses about a way to identify appliances in the smart space using artificial neural network (ANN). Smart appliances identification technique can solve a few specific problems for AAL like reducing the implementation complexity in the application level, finding the anomaly in the early stages. We have proposed an ANN-based approach to identify the appliances using smart space data. The smart appliances identification process is initiated by collecting data related to power consumption of smart appliances which are used for AAL services. An individual template along ANN-based approach is proposed to improve the accuracy. Through the interaction of the template, the proposed system has the ability to improve its identification by labeling the unlabeled input data. An ANN-based feedforward algorithm multilayer perceptron (MLP) is used for identifying the appliances. The output from MLP is observed whether the data pattern is a new class or not. The template provides the possible solution based on features and data. The proposed model has achieved identification performance with an average of 93% accuracy.

Keywords AAL · Appliances identifying system · Artificial neural network · MLP · Template

M. H. Uddin (✉) · M. N. Hossain · S.-H. Yang
Department of Electronic Engineering, Kwangwoon University, Seoul 139701, South Korea
e-mail: helaluddinsimon@gmail.com

M. N. Hossain
e-mail: nahidhossainmohammad@gmail.com

S.-H. Yang
e-mail: shyang@kw.ac.kr

M. S. Kaiser et al. (eds.), *Proceedings of International Conference on Trends in Computational and Cognitive Engineering*, Advances in Intelligent Systems and Computing 1309, https://doi.org/10.1007/978-981-33-4673-4_10

1 Introduction

Ambient assisted living (AAL, also known as "Active Assisted Living") includes methods, ideas, systems, goods and services that support elderly people's everyday lives and disabled people in a situation-dependent and unobtrusive way. AAL uses a variety of sensors to detect movement and understand the behaviors [1]. AAL also senses data directly via wearable sensors or indirectly through environmental sensors and analyzing the streaming data to infer something about the physical or cognitive status of the person observed [2].

In a smart environment, there could be an n_{th} number of appliances, and there is an n_{th} number of possibilities that the appliances might get damaged or replaced. On the other hand, this n_{th} number of appliances needs to be handled manually in the application/system which seems to be a very difficult job in the near future when we are going to have billions of appliances [3]. As a result, synchronizations between the smart appliances and AAL services might be compromised which might reduce the service quality of AAL. Our research aims to design an algorithm for the AAL platform where we will be able to reduce the dependency over appliances in the application level and able to pull up more data centric applications. In our research, we proposed an algorithm which will be able to identify appliances based on smart space data using an ANN-based feedforward network multilayer perceptron (MLP). Our goal is to establish a system where appliances can be identified based on smart space data with a good accuracy. And later on, this automated process can be used in AAL applications.

2 Related Work

Neural networks, classification algorithms are utilized to distinguish consumption sequences, identify and cluster the electrical appliances. For smart electrical device identification space, the related researches were progressively centered around extracting, mining data from numerous devices and classifying them [4]. To detect electrical devices, some of the researches have been progressively to gather data with better classification in the prior phases of data acquisition to give a strong establishment for data analysis purposes [5]. In comprehension and classifying the electrical devices, there are constrained quantities of features that can bolster the undertaking [6]. Here, the principle consideration was paid to the active power utilization. With regards to devices which are expending comparable measure of intensity, this factor is not enough to classify the devices. Support vector machines [7], artificial neural networks [8], K-means classification [9], Silhouette classification [10], mean-shift classification [11] are the most commonly used classification algorithms. MLPs are feedforward neural networks which are composed of a few layers of nodes with unidirectional associations, frequently trained by backpropagation [12, 13]. Working with multilayer perceptron (MLP) is not very complex, and in theory, the learning process of MLP network is based on the data samples composed of the N-dimensional input vector x and the M-dimensional desired output vector d, called destination [14].

3 Method

The proposed method begins by collecting the data from the smart appliances using IoT devices (Fig. 3) and by storing the raw data in our own smart home server (Fig. 2). This system is able to take input real-time data from appliances and also able to work with pre-stored dataset. Data input can be performed either way (Fig. 1).

Here, D means appliances or class. D_1, D_2, D_3, D_4, D_5, …, D_n means, appliances from 1_{st} to n_{th} numbers of appliances. After preprocessing the data, we used multi-layer perceptron (MLP) to train and test the dataset. Then, we check whether it is the very first time to learn from such data in our system. If the input data is trained for the very first time, then the system will recognize it as an undefined class. As a result, the system needs to label this data pattern. In this step, this is a bottleneck for

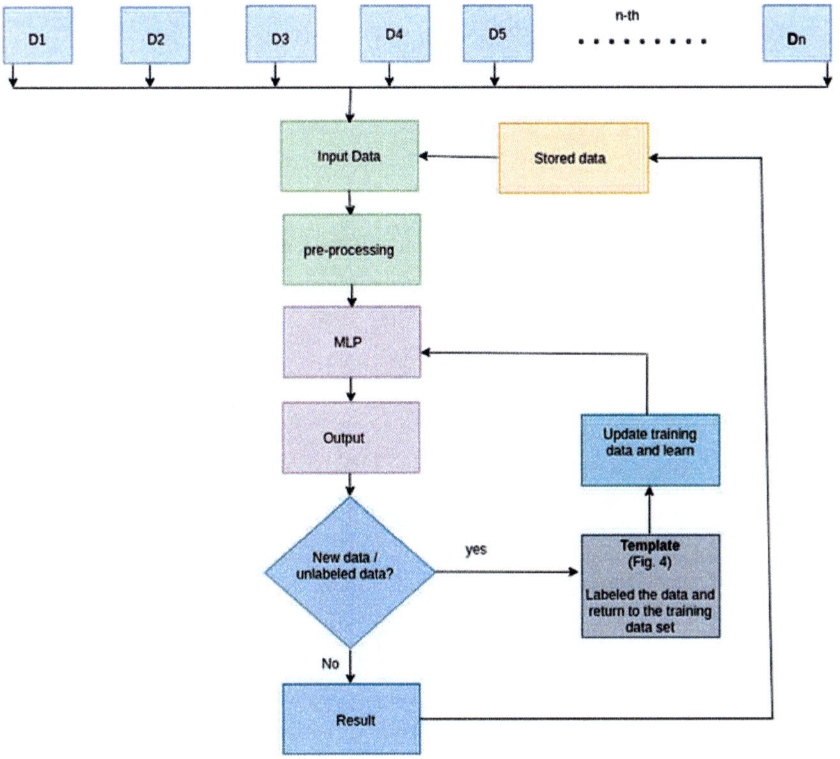

Fig. 1 Representation of the proposed method with the workflow diagram

the system where the system needs the help from the template. Template will label the data with a class name. After labeling, data will be added to the training dataset.

3.1 Data Collection

We have collected a few smart appliances data from our smart home lab environment through IoT devices (Fig. 3), and the air cooler data was collected from Blue star [4]. For storing the collected data, we have used our own smart home server (Fig. 2). We have collected four different classes (appliances) data, i.e., smart refrigerator, smart TV, smart oven and smart air cooler. There are few more parameters are missing from those datasets, as we expected that we might be able to collect those parameters but unfortunately that did not happen. So, we have tested our system with current data points. For this work, we have considered the same type of features ("temperature",

Fig. 2 Smart home data server

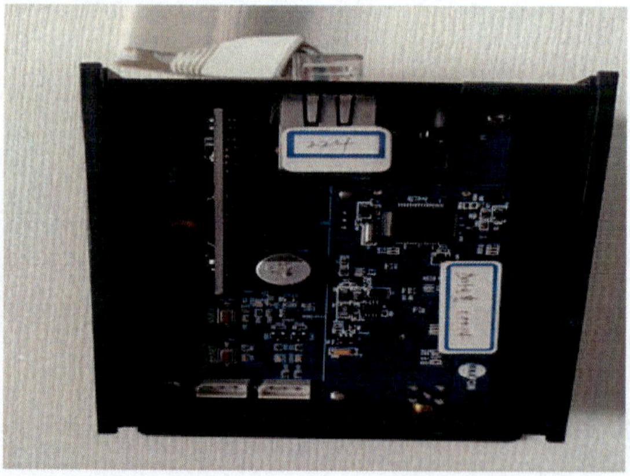

Fig. 3 IoT device for data collect

"active_power", "reactive_power", "V_{rms}", "I_{rms}", "phase_shift", "status") for all those appliances.

3.2 *Preprocessing*

Raw data is often noisy and unreliable and may be missing values. Using such data for modeling can produce misleading results. In other words, when the data is collected from different sources, it is collected in a raw format that is difficult to be analyzed. As our system is able to handle real-time data which will come from directly appliances/sensors, as a result sometimes some parameters might have null, missing values or noisy values. And on the other hand, this proposed system is also able to handle the pre-stored data, for pre-stored data, it is also very obvious that data might have null, missing values or noisy values. We have used ***sklearn.preprocessing*** package [15] for preprocessing our dataset.

3.3 *Template*

The template concept is designed as a pre-defined storage for labeling the data. We are considering the template is a pre-defined module where all the feature of classes (appliances) information will be pre-stored (Fig. 4).

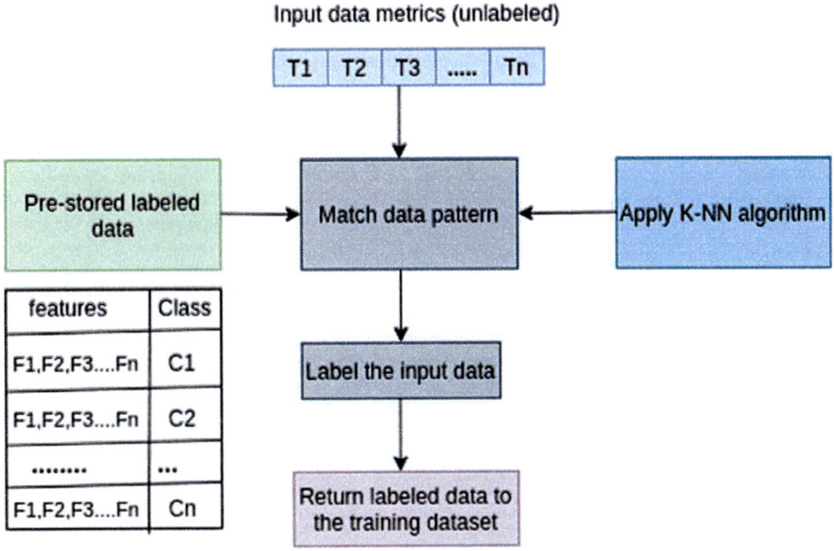

Fig. 4 Diagram of template with the explanation of workflow from input (T)

The goal of the template is to label this data and return it to the training dataset. To achieve this goal, we have proposed to apply the K-NN algorithm. The template is combined work of few individual parts, "Pre-stored labeled data" is the pre-storage of the label data and features, where C_1, C_2, ... C_n represent the name of classes (smart appliances). F_1, F_2, F_3 ... F_n represent the features of an individual class. This is the most important part of the template and the only part which is updateable. "Match data pattern", where we proposed to do the data pattern operation between the unlabeled "Input data metrics" and "Pre-stored labeled data". And for that we have applied the K-NN algorithm. After matching the pattern, we have proposed to label the data with a class name from the matched data metrics of "Pre-stored labeled data". When the task is done, the label data will be added to the training dataset.

4 Proposed Mechanism

Multilayer perceptron (MLP) is known as a supervised learning algorithm which learns from a function $f(\cdot)$: $R_m \rightarrow R_o$ through training on a dataset, where $\boldsymbol{m}$ is denoted as the number of input dimensions and the number of dimensions for output is $\boldsymbol{0}$. There is a set of features $X = x_1, x_2, \ldots, x_n$ and the target is $\boldsymbol{y}$, the learning process can be a nonlinear function estimator for either classification or regression. It is distinct from logistic regression, as there may be one or more nonlinear layers, called hidden layer layers, between the input and the output layer (Fig. 5).

The leftmost layer, called the input layer, is made up of a collection of neurons $\{x_i | x_1, x_2, \ldots x_m\}$ which is representing the input features. Every neuron in the hidden layer converts a weighted linear summation of values from the prior layer $w_1x_1 + w_2x_2 + \cdots + w_mx_m$, followed by a nonlinear activation function $g(\cdot)$: $R \rightarrow R$—like the hyperbolic tan function. The output layer gets the values from the last hidden

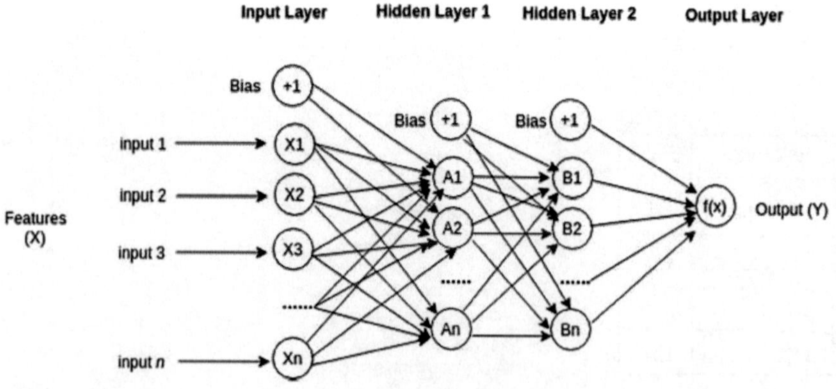

Fig. 5 MLP with two layers, showing of the workflow from input (*X*) to output (*Y*)

layer and transforms the values into the output values. Every node from the input layer is connected to output layer nodes through the hidden layers. It is allocated the connections between any two nodes, and the resulting input is calculated using the formula below

$$Y_{\text{in}} = \Sigma w_i x_i$$

Here, x_i is the i_{th} input, and w_i is the corresponding weight.

There is another identification input called Bias with weight b, which is added to the perceptron equilibrium node.

We have implemented our template-based solution for the unknown input data pattern where our model cannot determine the class for this input metrics. We have proposed to use "*Entropic open Set Loss*" [16] to identify the unknown instance. The "*Entropic open Set Loss*" drives the SoftMax scores of the unknown instances to the uniform probability distribution (Fig. 6).

In the upper part of the equation is an old cross-entropy loss. When the SoftMax scores give probability 1 to the true class, the cross-entropy is minimized. The bottom part of the equation means that the vector of the negative log-likelihood which is minimized when all probabilities are equal in the SoftMax vector. This loss tells the network that the instance is unknown. The system is developed using Python and tensor flow environment. We have used Python libraries for developing this system. In the MLP part, the sample parameters values taken for MLP function are shown (Fig. 7).

$$J_E(x) = \begin{cases} -\log S_c(x) & \text{if } x \in \mathcal{D}'_c \text{ is from class } c \\ -\frac{1}{C}\sum\limits_{c=1}^{C} \log S_c(x) & \text{if } x \in \mathcal{D}'_b \end{cases}$$

Fig. 6 Entropic Open Set loss. On the top: loss when instance known. On the bottom: loss when instance is unknown

```
In [ ]: MLPClassifier(activation='relu', alpha=1e-05, batch_size='auto',
                      beta_1=0.9, beta_2=0.999, early_stopping=False,
                      epsilon=1e-08, hidden_layer_sizes=(5, 2),
                      learning_rate='constant', learning_rate_init=0.001,
                      max_iter=200, momentum=0.9, n_iter_no_change=10,
                      nesterovs_momentum=True, power_t=0.5, random_state=1,
                      shuffle=True, solver='lbfgs', tol=0.0001,
                      validation_fraction=0.1, verbose=False, warm_start=False)
```

Fig. 7 MLP classifier method with all the parameters

5 Training

We have split our dataset into two group—Training set (train_data.csv) and test set (test_data.csv). Our model will be based on features like "temperature", "active_power", "reactive_power", "V_{rms}", "I_{rms}", "phase_shift", "status". Our prediction class will be appliances, like "smart tv", "smart refrigerator", "smart air cooler", "smart oven", "blender", "fan", etc. Our initialization of ANN is below

1. An **input layer** whose inputs is a 7-entries vector X_i = [temperature, active_power, reactive_power, "V_{rms}", "I_{rms}", phase_shift, status].
2. Two **hidden layers** with, respectively, p and q number of neurons for each layer that receives the inputs and elaborates them with a ReLU activator φ. whose outputs are z_1^j and z_2^k and whose parameters are $\left\{w_1^{ji}, b_1^j\right\}$ and $\left\{w_2^{kj}, b_2^k\right\}$;
3. A **single-neuron output layer** since we want to have a binary classification that has a sigmoid activator f and outputs a $y \in [0,1]$ real number. $\left\{w_{\text{out}}^k, b_{\text{out}}\right\}$ are the parameters.

We have used 80% of our data as a training data and rest 20% data as testing. After doing the train/test split for both training data $\{x_i\}_I$ and labels $\{\bar{y}\}_I$, in where I runs over the dataset of smart appliances. We will use the whole dataset as a batch since we are employing the batch gradient descent [17]. We will input the training data $\{x_i\}_I$ to feed the first hidden layer. All the parameters for each layer will be initialized randomly, i.e., $\left\{w_1^{ji}, b_1^j\right\}$, $\left\{w_2^{kj}, b_2^k\right\}$ and $\left\{w_{\text{out}}^k, b_{\text{out}}\right\}$, $i = 0, 1,\ j = 0, \ldots, p-1$ and $k = 0, \ldots, q-1$.

For each smart appliance of I in the dataset, we have to perform feedforward [18], compute the error of outer layer [18], and do the backpropagate the error [19].

A. Feedforward

$$z_1^j = \varphi(\Sigma_{i=0,1}\, w_1^{ji} x^i + b_1^j),$$
$$z_2^k = \varphi(\Sigma_{j=0,\ldots,p-1}\, w_2^{kj} z_1^j + b_2^k),$$
$$y = f(\Sigma_{k=0,\ldots,q-1}\, w_{\text{out}}^k z_2^k + b_{\text{out}}).$$

B. Compute the error of outer layer

$$\delta_{\text{Out}} = (\partial\mu/\partial y) \cdot f'(y) = 2(y - \bar{y}) \cdot y(1-y)$$

C. Backpropagate the error

$$\delta_2^k = \delta_{\text{Out}} w_{\text{out}}^k \varphi'\left(z_2^k\right),$$
$$\delta_1^j = \Sigma_{k=0,\ldots,q-1} \delta_2^k w_2^{kj} \varphi'\left(z_1^j\right)$$

6 Result

We have roughly tested our system with a small amount of dataset. Our MLP tested results roughly have an average of 93% accuracy after the first few runs. This result is based on a pre-stored data. As our proposed system is able to get input data from the appliances in real time and also from data storage, we have presented the model accuracy curve (Fig. 8) and the true positive rate versus false positive rate curve (Fig. 9).

Table 1 represents the identification accuracy of individual appliances.

As our data number is small and our data has some missing parameters, so the accuracy is still a little bit less than we expected. Accuracy and performance will be increased along with dataset increment.

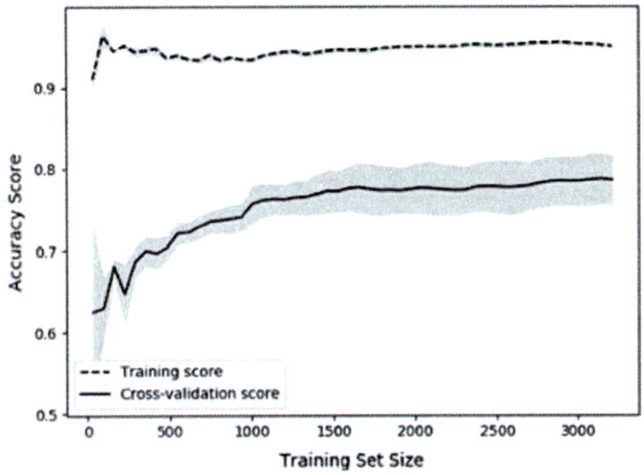

Fig. 8 Model accuracy curve, training set versus accuracy score preview

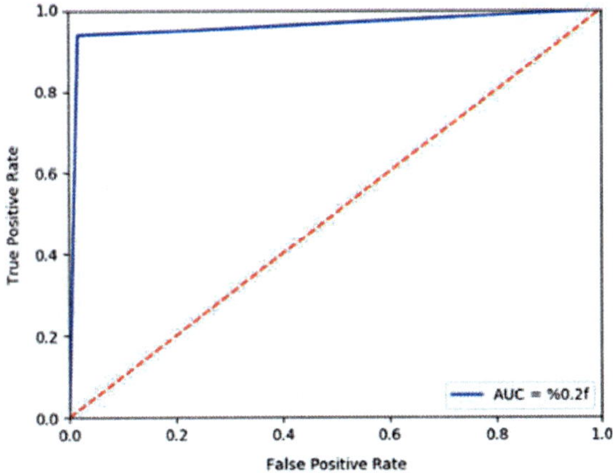

Fig. 9 True positive rate vs false positive rate curve preview

Table 1 Summarized test run result of appliances identification

Class name	Dataset amount	Train data (%)	Test data (%)	Identifying rate (%)
Smart Refrigerator	1000	80	20	91
Smart Air cooler	1000	80	20	93
Smart TV	1000	80	20	93
Smart Oven	1000	80	20	91

7 Limitations and Conclusion

Though we end up with our current proposal, but it has a few limitations to improve. It will be more impressive if this works for unsupervised data. Current work is based on a small dataset which is a drawback for this work. In future, we will apply this method in a large dataset and will make a comparison with other methods also. The work focuses on developing a system where the system will be able to identify the appliances from smart space data using ANN. If we are able to find out the appliance which are faulty or have other issues, we might be able to take precaution from household disasters (such as fire or explosion occurred by appliances). We have ended up with a proposal that can identify the appliances based on smart appliances data, so that it would reduce human effort in the application level.

References

1. Chen, L., et al.: Sensor-based activity recognition: a survey. IEEE Trans. Syst. Man Cybern. Part C **42**(6), 790808 (2012)
2. Acampora, G., et al.: A survey on ambient intelligence in health care. Proc. IEEE **101**(12), 24702494 (2013)
3. Yan, L., Wang, Y., Song, T., Yin. Z.: An incremental intelligent object recognition system based on deep learning. In: 2017 Chinese Automation Congress (CAC). IEEE (2017)
4. Napoleon, D., Lakshmi, P.G.: An efficient K-Means clustering algorithm for reducing time complexity using uniform distribution data points. Trendz Inf. Sci. Comput. (TISC) 42–45, 17–19 (2010)
5. Kolivand, H., Bin Sunar, M.S.: New silhouette detection algorithm to create real-time volume shadow. In: Digital Media and Digital Content Management (DMDCM), 2011 Workshop on, pp. 270–274 (2011)
6. Ruzzelli, A.G., Nicolas, C., Schoofs, A., O'Hare, G.M.P.: Real-time recognition and profiling of appliances through a single electricity sensor. In: Sensor Mesh and Ad Hoc Communications and Networks (SECON), 2010 7th Annual IEEE Communications Society Conference on, pp. 1–9, 21–25 (2010)
7. Shmilovici, A., Maimon, O., Rokach L.: Support Vector Machines, Data Mining and Knowledge Discovery Handbook. https://doi.org/10.1007/0-387-25465-X_12
8. He, X., Xu, S.: Artificial Neural Networks, Process Neural Networks: theory and Applications. https://doi.org/10.1007/978-3-540-73762-9_2
9. Jin, X., Han, J.: K-Means Clustering. https://doi.org/10.1007/978-0-387-30164-8_425

10. Yiğithan Dedeoğlu, B. Uğur, T., Uğur, G., Çetin, A.E.: Silhouette-Based Method for Object Classification and Human Action Recognition in Video, pp. 64–77 (2006). ISBN 978-3-540-34203-8
11. Zhou, H., Wang, X., Schaefer, G., Kwaśnicka, H.J., Lakhmi, C.: Mean Shift and Its Application in Image Segmentation. https://doi.org/10.1007/978-3-642-17934-1_13
12. Zanaty, E.A.: Support Vector Machines (SVMs) Versus Multilayer Perception (MLP) in data classification. https://doi.org/10.1016/j.eij.2012.08.002
13. Popovic, D.: CHAPTER 18—Intelligent Control with Neural Networks. https://doi.org/10.1016/B978-012646490-0/50021-4
14. Osowski, S., Siwek, K., Markiewicz, T.: MLP and SVM networks: a comparative study. In: Proceedings of the 6th Nordic Signal Processing Symposium NORSIG, pp. 9–11. Espoo, Finland (2004)
15. Scikit-learn: https://scikit-learn.org/stable/modules/preprocessing.html
16. Akshay, R.D., Manuel, G., Terrance E.B.: Reducing Network Agnostophobia. https://arxiv.org/pdf/1811.04110v2.pdf (2018)
17. Jason, B.: https://machinelearningmastery.com/difference-between-a-batch-and-an-epoch/ (2019)
18. Michael, N.: http://neuralnetworksanddeeplearning.com/chap2.html (2019)
19. Rumelhart, D., Hinton, G., Williams, R.: Learning representations by back-propagating errors. Nature **323**, 533–536 (1986). https://doi.org/10.1038/323533a0

Anonymous Author Identifier Using Machine Learning

Sabrina Jesmin and Rahul Damineni

Abstract Cyber security companies predict potential attacks on their clients by monitoring deep and dark web forums where actors conspire and discuss potential bugs and vulnerabilities of clients that could be exploited to attack them. There are existing solutions that analyze texts and state if it's a threat purely based on content. These solutions tend to trigger a lot of false alarms because the discussion content is indiscernible from actual threat. The key indicator of authenticity of a discussion being a threat is its author. There are red hot malicious actors who are notorious for attacks in the past and if we see a threat from them, there's a high chance that it would lead to a potential attack. The catch is, the malicious actors cover up their digital signatures effectively—making it impossible to attribute an anonymous threat to a known actor. However, the most intrinsic feature of such anonymous content is, the author's writing style. In this paper we have tried to find out the answer of the question, *Given a collection of document, how to cluster the document that are possibly written by the same author?*.

Keywords Neural network · Machine learning · Burrow's delta · Author identification · Stylometry

1 Introduction

Stylometry, a burgeoning research area which combines computer science, statistics and stylistics to study and measure style of particular authors [1]. Historically, stylometry was a popular technique used to identify authors of anonymous content—when such content was written by famous people. One of the techniques of stylometry is

S. Jesmin (✉)
Jahangirnagar University, Savar, Dhaka 1342, Bangladesh
e-mail: sabrina.jesmin.eee@gmail.com

R. Damineni
School of Electrical Engineering and Computer Science, Oregon State University, Corvallis, USA
e-mail: daminens@oregonstate.edu

M. S. Kaiser et al. (eds.), *Proceedings of International Conference on Trends in Computational and Cognitive Engineering*, Advances in Intelligent Systems and Computing 1309, https://doi.org/10.1007/978-981-33-4673-4_11

based on counting frequencies of word lengths called Mendenhall's Characteristic Curves of Composition (MCCC). Another technique of stylometry is called Burrow's Delta which uses Term Frequency—Inverse Document Frequency (TF-IDF) on function and stop words to summarize the document into a vector and then use cosine similarity to find closest matching document.

There has been plenty of research done on stylometry since now. Stylometry not only helps to identify anonymous authors for deciding famous document authorship, email classification and fraud detection, but also helps for court trial and disease detection [1]. Ramyaa et al. used machine learning (ML) techniques to predict author's identity from stylometry in their research. They used decision trees and neural networks for their prediction and their test set result was 82.4% and 88.2% accurate respectively [1]. Their experiment was done on five Victorian writers to differentiate their writing style by observing particular features.

Stylometric research can be helpful in deception detection [2]. Feng et al. in their research worked on syntactic stylometry for detecting fraud and deception. They considered four different datasets for the investigation on lexicon-syntactic features. Parse trees for the features derived from Context Free Grammar (CFG) was effective for them.

Adversarial stylometry can be really threatening when a subject hides their real identity while writing and tries to frame another subject by copying their writing style. Stylometry for criminal investigation has been reflected in the work of Brennan et al. [3]. Stolerman et al. also worked on adversarial stylometry [4]. They proposed a classify-verify method which could detect writing style of authors who changes their writing style.

Another reason for researching on stylometry is plagiarism. Plagiarism is a serious unethical practice in educational sector. For a better estimation about author's identity from style of writing, one of the available researches had been done by Ramnial et al. [5] on ten PhD defense books by splitting them into 1000–10,000 words and applying SMO and KNN machine learning algorithms to predict the appropriate authors by using 446 features. Their accuracy rate of prediction in the research was above 90%.

As long as identifying author from longer text is easier than to identify author from shorter text, Brocardo et al. [6] wanted to solve this problem in their research. They used supervised machine learning along with n-gram computational linguistics analysis to identify authors from a short text. Similarly, Bhargava et al. [7] conducted a research on stylometry for the sake of forensic purposes based on ML techniques which could determine authorship of texts having 140 characters or lesser than that. They particularly targeted twitter messages and tried to predict original twitter message writers. Again, Killian et al. [8] made a research on social media (Twitter, StackExchange etc.) for identifying accounts which are owned by same user. Their tactics were comprising of various techniques such as semantic, stylometry and temporal by testing a number of algorithms for each category. Pavelec et al. in their work investigated on a database consisting of short articles from ten identical authors and used Support Vector Machine (SVM) for classification [9]. They presented a model by reducing problem of pattern recognition into one single model along with a feature set for stylometry. Ding et al. [10] also investigated on unveiling writing style from texts to unveil the author and their sociolinguistic characteristics. They used neural network approach to unveil the authors from Twitter, review, blog, novel etc. Another research work based on machine learning was done by Das et al. [11] on four Bengali blogs of four individual Bangladeshi authors. They tried to predict the authorship using different features such as word length, number of suffixes etc. Pearl et al. [12] came up with a new supervised ML technique for identifying authorship deception. Their method was applied on two separate case studies to analyse it's utility.

Neal et al. [13] in their research reviewed few stylometry related articles related to stylochronometry, authorship attribution, authorship profiling, authorship verification, adversarial stylometry. They discussed about recent approaches along with features, datasets and experimental techniques in their paper. Vysotska et al. [14] in their research tried to study stylometry of author publications. They used NLP and porter stemmer for their investigation.

In our paper, we used comments from discussion forum of Reddit where we considered a dataset of at least 100 user comments done by 73 authors and analyzed by Burrow's Delta technique and neural network for predicting authorship.

The rest of the paper is organized by the following: Sect. 2 introduced the solution we offered for anonymous author identification; Sect. 3 reviewed our data collection. The implementation details for author identifier were discussed in Sect. 4. At the end, the work was concluded in Sect. 5.

2 Anonymous Author Identifying Solution

For our research, we came into below solutions for identifying anonymous authors.

- We used Burrow's Delta [15] technique source code to fingerprint each incoming document and tag it with a matching author. A curated set of function words was used to construct a count vector of length 632.
- The other option was to use a neural network that could automatically derive latent features. Care had been taken to setup this problem correctly as there's a chance of memorizing specific tokens from data instead of actually deriving general features that are representative of writing styles.

Our best implementation correctly classified 85% of input queries from a balanced dataset.

3 Data Collection

We used user comments from Reddit discussion forum as a dataset for both anonymous authors identifying solutions mentioned in section no two. Comment dumps from May and September 2006 were processed to form an initial dataset of 73 authors posting at least 100 comments each on several subreddits. These 100 comments from each author were unique and were split into 60, 20, and 20 shares to make up train, eval and test splits respectively. The task was to learn a model using the train split and use it to map comments in dev and test split correctly. This dataset was directly used in Burrow's Delta technique [15] and also initial version of neural network models.

To simplify learning, this dataset was processed further to form another one. Some comments in original dataset contain URLs which do not reveal anything about writing styles. Also, some comments are too short and general to be attributed to any author (examples: "I agree", "Alright!" etc.). This second dataset was formed by removing URLs and excluding comments that has less than 200 characters and other irregularities like non-English content. A glance on Jupyter notebook can be beneficial for more details. This new dataset only contained 24 authors and 100 comments for each.

4 Implementation Details

4.1 Problem Setup and Model

For the implementation of author identification we did the following:

- A gated recurrent neural network was built to take a sequence of tokens as input and we converged it to probabilities of two classes: "match" or "no-match".
- RedditComments (source): The above data splits were transformed into a PyTorch dataset that would supply input-target pairs for all our neural networks. Here the details have been provided:
  1. At first text input was tokenized using "spacy" tokenizer.
  2. For the binary classification problem, two random texts from possibly same or different author were concatenated with a "SEP" token between them. If the inputs were sampled from same author's comments, then they were labeled as "match".
  3. The "mixing_strategy" argument was used to control logical form sparsity of the output distribution. If "ordered strategy" was used, the output samples would have more variation and helped model converge faster.
  4. The "p2nr" argument was used to regulate positives ("match") to negatives ratio of the output distribution.
  5. Finally, "nums_samples" was used to restrict the size of dataset to experiment faster.
- The model's hyper-parameters were number of hidden gated recurrent units (HIDDEN_REPR_DIM), number of stacks of GRUs connected end-to-end (NUM_HIDDEN_LAYERS) and LEARNING_RATE. Binary Cross Entropy was used as the loss criteria.

4.2 Experimentation and Fine-Tuning

- All experiments were numbered and indexed based on the hyperparameters being used logs.
- First version of dataset with 73 authors didn't converge at all. After verifying that the training pipeline was setup right by performing over-fitting test, we revisited dataset and made the 24-author dataset by cleaning it. The reason for including longer lines only was to expose the encoder with more contents to recognize patterns.
- Even this dataset was taking too long to train and wasn't converging. At this point, we weren't sure if the writing styles could be distinguished by this training pipeline at all. So, to verify this, we came up with the simplest dataset by using all qualified comments (approximately 250) from top two authors in the original dataset. The rationale was, even a simpler model with less parameters should be able to identify the possibly stark differences in two authors' writing styles. We also used "ordered mixing strategy" to increase the variance in the data distribution seen by the model. The model started to converge and the validation set (whose examples were from the same author but never used for training) accuracy came out as approximately 62%.
- Although, the input comments were concatenated from different subreddits (which were contextually quite different), we were worried that 62% might have been because of memorised rare words and not because of the model actually learning discriminating features. So, we setup another experiment to train the same model using 24-authors' comments. If it were actually memorizing rare words, the accuracy shouldn't improve—as the distribution of rare-words across the comments should relatively be same. But it was actually learning patterns in the data, this extra surge of data improved dev accuracy by a good amount: came out to be approximately 71%.
- A little more fine-tuning resulted in the personal best accuracy of 85%.

4.3 Inferencing for DBMS

- Once the model is trained, it is expected to be general. For this paper, we did not test our model. But, our model should work when we want to identify if two texts belong to same unseen author. Even though there is a co-variate shift, this model should work because the original 24 authors are selected randomly, and the training data is fairly cross-domain.
- Based on this assumption, the inferencing script source would be given a trained model and a collection of users will be selected with few textual content as features. The inferencing script learns the mean of each user's writing style representation (the input encoding of the user's text as seen from encoder stack output).

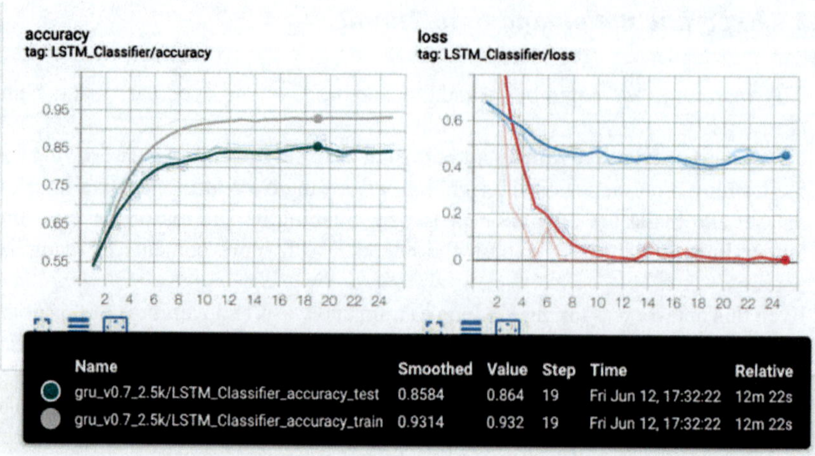

Fig. 1 Training accuracy and loss curves for our model

Figure 1 represents the training accuracy and loss curve of our ML model. The accuracy improved each time we trained the model repeatedly.

5 Conclusion

As we know there are various existing techniques to analyse a text and detect whether it is a potential threat. In cyber security sector, potential bugs and vulnerabilities of clients can be exploited by recognising a potential attacker. But these attackers cover up their digital signatures effectively which makes it tough to recognise an anonymous threat from an author. We tried to focus on the writing style of an attacker to recognize their comments and writing. For our author identification, we used two techniques such as Burrow's Delta technique and neural networks. In our experiment, Burrow's Delta technique was not impressive for predicting authorship and was able to perform only slightly better than random guess. But neural network was quite effective in our research. We selected at least 100 comments from a Reddit discussion forum and then split them into 60, 20 and 20 shares to make up train, eval and test splits respectively. We tried to train a model for identifying anonymous authors and we came into observation that the neural network technique was effective for that. The best implementation we did could correctly classify 85% of our input queries from a balanced dataset.

References

1. Ramyaa, C.H., Rasheed, K., He, C.: Using machine learning techniques for stylometry. In: Proceedings of International Conference on Machine Learning (2004)
2. Feng, S., Banerjee, R., Choi, Y.: Syntactic stylometry for deception detection. In: Proceedings of the 50th Annual Meeting of the Association for Computational Linguistics, vol. 2: Short Papers, pp. 171–175 (2012)
3. Brennan, M., Afroz, S., Greenstadt, R.: Adversarial stylometry: circumventing authorship recognition to preserve privacy and anonymity. ACM Trans. Inf. Syst. Security (TISSEC) **15**(3), 1–22 (2012)
4. Stolerman, A., Overdorf, R., Afroz, S., Greenstadt, R.: Breaking the closed-world assumption in stylometric authorship attribution. In: IFIP International Conference on Digital Forensics, pp. 185–205. Springer (2014)
5. Ramnial, H., Panchoo, S., Pudaruth, S.: Authorship attribution using stylometry and machine learning techniques. In: Intelligent Systems Technologies and Applications, pp. 113–125. Springer (2016)
6. Brocardo, M.L., Traore, I., Saad, S., Woungang, I.: Authorship verification for short messages using stylometry. In: 2013 International Conference on Computer, Information and Telecommunication Systems (CITS), pp. 1–6. IEEE (2013)
7. Bhargava, M., Mehndiratta, P., Asawa, K.: Stylometric analysis for authorship attribution on twitter. In: International Conference on Big Data Analytics, pp. 37–47. Springer (2013)
8. Brounstein, T.R., Killian, A.L., Skryzalin, J., Garcia, D.: Stylometric and temporal techniques for social media account resolution. Sandia National Lab. (SNL-NM), Albuquerque, NM (United States), Tech. Rep. (2017)
9. Pavelec, D., Justino, E., Oliveira, L.S.: Author identification using stylometric features. Revista Iberoamericana de Inteligencia Artificial **11**(36), 59–65 (2007)
10. Ding, S.H., Fung, B.C., Iqbal, F., Cheung, W.K.: Learning stylometric representations for authorship analysis. IEEE Trans. Cybern. **49**(1), 107–121 (2017)
11. Das, P., Tasmim, R., Ismail, S.: An experimental study of stylometry in Bangla literature. In: 2015 2nd International Conference on Electrical Information and Communication Technologies (EICT), pp. 575–580. IEEE (2015)
12. Pearl, L., Steyvers, M.: Detecting authorship deception: a supervised machine learning approach using author writeprints. Literary Ling. Comput. **27**(2), 183–196 (2012)
13. Neal, T., Sundararajan, K., Fatima, A., Yan, Y., Xiang, Y., Woodard, D.: Surveying stylometry techniques and applications. ACM Comput. Surv. (CSUR) **50**(6), 1–36 (2017)
14. Vysotska, V., Lytvyn, V., Hrendus, M., Kubinska, S., Brodyak, O.: Method of textual information authorship analysis based on stylometry. In: 2018 IEEE 13th International Scientific and Technical Conference on Computer Sciences and Information Technologies (CSIT), vol. 2, pp. 9–16. IEEE (2018)
15. Braunlin, J.: Using nlp to identify redditors who control multiple accounts (2018). Available https://towardsdatascience.com/using-nlp-to-identify-redditors-who-control-multiple-accounts-837483c8b782

A Machine Learning Approach to Predict Events by Analyzing Bengali Facebook Posts

Noyon Dey, Motahara Sabah Mredula, Md. Nazmus Sakib, Muha. Nishad Islam, and Md. Sazzadur Rahman

Abstract Facebook is the most popular social media platform all over the world and a huge data repository that can be analyzed to predict events as people express their thoughts on this platform. There are some events that often turn into potential crises and these crises might be avoided if we can predict it earlier. This work aims to predict events that are happening or going to happen by analyzing Bengali Facebook posts and help authorities to take proper security arrangements. In this work, the Naïve Bayes Classification model has been used to classify posts as an event. Tokenization, Stopwords removing have been used for data pre-processing. Event phrase matching and 2-Layer Filtering have been used for features collection. Sentiment scores have been measured using Valence Aware Dictionary and Sentiment Reasoner(VADER). Finally, this work predicts four types of events(protesting, celebrating, religious, and neutral) showing 87.5% accuracy.

Keywords Event prediction · Naive Bayes · Sentiment analysis · Social media · Machine learning.

N. Dey · M. Sabah Mredula · Md. Nazmus Sakib · Md. Sazzadur Rahman (✉)
Institute of Information Technology, Jahangirnagar University, Savar, Dhaka 1342, Bangladesh
e-mail: sazzad@juniv.edu

N. Dey
e-mail: noyondey8@gmail.com

M. Sabah Mredula
e-mail: mmredula12@gmail.com

Md. Nazmus Sakib
e-mail: sakibayon98@gmail.com

Muha. Nishad Islam
Jagannath University, 9-10, Chittaranjan Avenue, Dhaka 1100, Bangladesh
e-mail: nishad.nik@gmail.com

M. S. Kaiser et al. (eds.), *Proceedings of International Conference on Trends in Computational and Cognitive Engineering*, Advances in Intelligent Systems and Computing 1309, https://doi.org/10.1007/978-981-33-4673-4_12

1 Introduction

One-in-three people in the world use Facebook since it has 2.4 billion users [1]. The abundance of users and data in social media motivated us to build up a system that can be used to predict events. This work predicts event as celebrating, protesting, neutral, and religious. Celebrating events are marriage ceremony, national events, cultural events et cetera. Protesting events are political events, workers protest etc. Religious events are worships, Eid, funerals etc. and events outside these three types are considered as neutral events in this work. Public groups and pages from Facebook are considered for data source where a huge number of members express their thoughts on different topics. At first, the sentiment of a post is calculated along with other features and then they are further classified using the Naïve Bayes classifier.

Ensuring social security has become the biggest challenge in today's world. In 2018, when countrywide road safety protests(নিরাপদ সড়ক চাই) or the Quota Reform movement (#ReformQuota) was going on in Bangladesh, many disruptive events happened which led the life of general people at stake. Information(when it was going to take place and where it was going to happen) of this unwanted protest events were being continuously posted on social media sites especially on Facebook. The security authorities of our country could have stopped this violence if they had proper information about those events at the right time. According to Aldhaheri and Lee government entities are likewise depending via web-based media for the motivation behind gathering security related intelligence [2]. Keeping this issues in mind, the proposed event prediction model predicts events from Bengali Facebook posts so that the appropriate authority or the government could get prior information of different harmful protesting, celebrating or religious events which will ultimately help in securing the life of our general people.

The reasons to choose the Bengali language and Facebook as a platform for data are their widespread use in Bangladeshi perspective. Approximately 228 million people use Bengali as a first language while there are nearly 37 million people use it as a second language [3]. Bangladeshi people mostly use Bengali to post something on Facebook. Facebook is the most mainstream social media platform in Bangladesh with 89.91% Facebook users (30 million) among all the social media platforms [4]. Most of the works done already have used Twitter data. Twitter data has the advantage of being short in size since there is a character limit on twitter, 280 characters per post [5]. On the other hand, Facebook posts are way bigger than twitter posts. It is a challenging task to analyze large text data and extract features from them. As per knowledge, there are several event detection methods that work in English, Hindi, Arabic languages but no significant research has been done yet in this regard using the Bengali language. Bengali is a complex language in writing form as any particular word has several representations. The sentiment of the Facebook posts is an important factor in determining events since the user's sentiment can be measured towards an event through this. Sentiment analysis of the Bengali text is also a difficult and challenging task. For these above-mentioned limitations, the proposed model will analyze Bengali Facebook posts to predict upcoming events overcoming above mentioned limitations.

The rest of the paper is structured as follows: Sect. 2 describes Related Work, Sect. 3 describes Methodology, Sect. 4 describes Result and Discussion, Sect. 5 describes the Conclusion and Future Work.

2 Related Work

There exists a lot of research works regarding event detection/prediction in social media. Aldhaheri and Lee mainly proposed a framework for detecting events and presented a temporal social network conversion approach that converts social media information into temporal images [2]. Alsaedi and Burnap proposed a model that detects real-world events in Arabic from Twitter data [6]. Bekoulis et al. focused on the chronological relation between consecutive tweets and predicted the existence of a sub-event along with its type [7]. Imran et al. presented an efficient method that extracts disaster-relevant information from twitter data [8]. Koyla et al. detected events of newspaper documents where a detect was found if the matching score was greater or equal to a predefined threshold value [9]. Fedoryszak et al. proposed a framework that handles event evolution over time [10]. Chen et al. presented a clustering-based event detection and tracking approach using neural network [11]. Fathima and George proposed an event evolution framework using clustering algorithms which is named as hot event evolution model [12]. Li et al. classified tweets into semantic classes, calculated their class-wise similarity, and finally hypothesized the improvement of event detection performance by integrating them [13]. Kannan et al. detected key events from the sports domain in real-time using LSH (Locality Sensitive Hashing) [14]. Feng et al. proposed an approach for event detection which is based on Locality Sensitive Hashing [15]. Akbari et al. analyzed social contents of users and proposed to extract PWE(Personal Wellness Events) automatically [16]. Sakaki et al. investigated twitter data for real-time earthquake detection [17]. Panagiotou et al. presented various definitions relevant to event detection to formalize fuzzy concepts disambiguation from Web 2.0 data [18]. Zhao et al. presented an event detection approach to explore correlations among various microblogs [19]. Banna et al. presented an overall review of all the machine learning techniques in predicting earthquakes [20]. Based on the background study on this topic, it has been found out that Machine Learning, Neural Network, and various other techniques are used. Naive Bayes Classification model has been used in this work as it is easy to work with, fast, and compatible to work with event predictions.

3 Methodology

Figure 1 describes the model for event prediction by analyzing Bengali Facebook posts. Data collection, Data pre-processing, Feature extraction, Model training and Prediction are the main steps in this model. In this section, every step will be discussed in more detail.

3.1 Data Collection

Manual data collection has been used for data collection in this work. The reason to choose this manual data collection is that a global pandemic is going on, and no event is taking place around us as it is risky to gather. As no event is happening due to this COVID-19 outbreak, posts of previous events from various public Facebook groups and pages have been collected.

3.2 Data Pre-processing

Texts from social media are generally unorganized as users apply free-style writing. As features must be found from these text data, the extracted data must be processed to get those features needed for the prediction. Tokenization, stopwords removal, replacing of hashtags etc. are performed for data pre-processing. Tokenization into sentences and tokenization into words were performed for separating sentences and words from the post respectively. Tokenized words and sentences are then used for feature extraction. Stopwords(unnecessary in case of gaining insight from the text) were removed using the stopwords removal process. Besides, hashtags, emoticons, URLs were also discarded during this data pre-processing step.

3.3 Feature Extraction

Common event words frequency, event-specific words frequency, event-specific phrase frequency, and sentiment score of the post are used as features for this event prediction task. 2-Layer Filtering is used for tracking words frequency and VADER Sentiment Analysis is used for the sentiment score of the post.

2-Layer Filtering: 2-Layer filtering is used to count the frequency of common event words and event-specific words from the post. A list of common event words in Bengali and separate words list for celebrating, protesting, and religious event words have been maintained. Table 1 shows a part of the words maintained for 2-layer filtering. In the first layer, common event words are searched and their frequencies

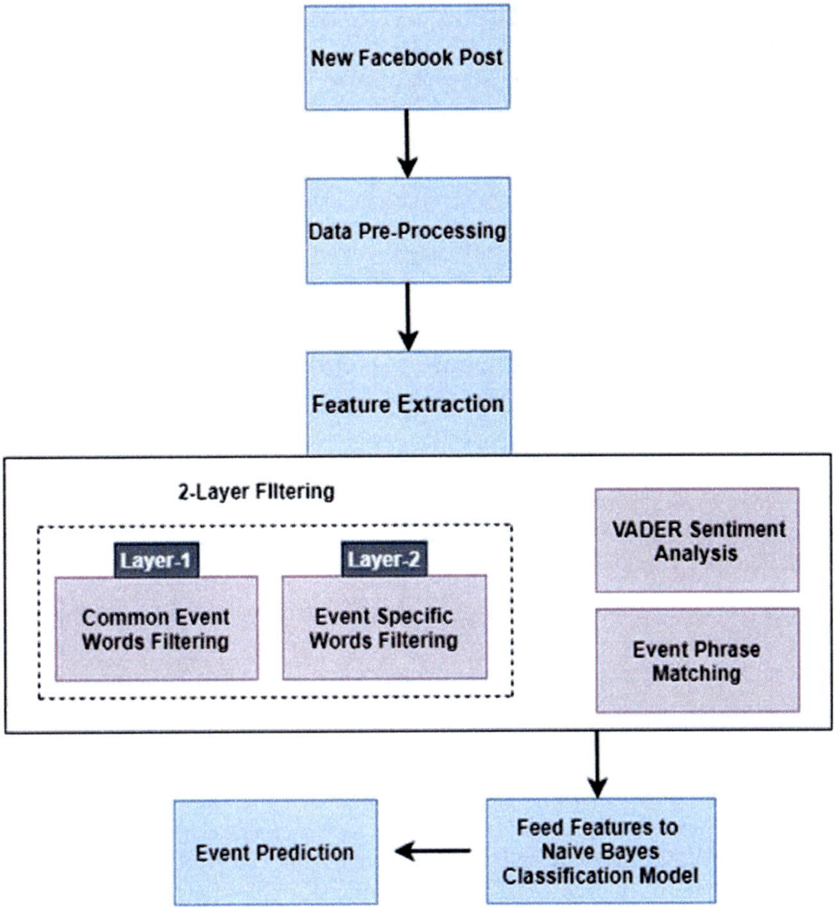

Fig. 1 Model for event prediction by analyzing Bengali Facebook posts

Table 1 Event keywords: common event words for 1st-layer filtering, celebrating, protesting, and religious event words for 2nd-layer filtering purpose

Common event words	Celebrating event words	Protesting event words	Religious event words
ঘটছে,ঘটবে,হবে,হচ্ছে, সমাবেশ,জড়ো,জনতা, জমায়েত,ভীড়,সমাগম, পালন,পালিত,সভা, একত্রিত হওয়া	আনন্দমেলা,বিয়ে,অনুষ্ঠান, বিয়ে,বিবাহ,পূজা,হোলি, উতসব,প্রোগ্রাম,ঈদ ,পুনঃমিলনী,ব্যাচ-ডে, বিজয়মিছিল,মেলা,আয়োজন	মিছিল,মানববন্ধন,প্রতিবাদ-সভা,বিক্ষোভ,অশান্তি সংঘর্ষ,মারামারি,বিক্ষোভ-মিছিল,বিশৃঙ্খল, অবরোধ,হামলা,আন্দোলন, কোন্দল	ওয়াজ,মহফিল,জানাজা, মিলাদ,দাফন,নামাজ, পূজা,সরস্বতী পূজা,পুজা, দোয়া,হজ,লাইলাতুল কদর, ঈদে মিলাদুন্নবী

Table 2 Event Phrases: celebrating event phrases, protesting event phrases, and religious event phrases for event prediction

Celebrating event phrases	Protesting event phrases	Religious event phrases
বিবাহের অনুষ্ঠান হবে,বিজয়মিছিল হবে, পালিত হচ্ছে ,মিলনমেলা চলছে ,বর্ষবরণ করা হবে,জমকালো র‍্যালি হবে ,বৈশাখের অনুষ্ঠান হচ্ছে ,আয়োজনের প্রস্তুতি নিচ্ছে	মিছিল হচ্ছে,মিছিলের আহবান ,আন্দোলনের ডাক ,রাজপথে নামতে হবে ,মারামারির আশংকা ,প্রতিবাদ সভা হবে ,মানববন্ধন করবো ,অবরোধ র্কমসূচী	ওয়াজ মহফিল অনুষ্ঠিত হবে, ওয়াজ মাহফিল,জানাজা হবে, দাফন করা হবে,পূজা উপলক্ষে,সরস্বতী পূজার, প্রসাদ বিতরণ,মিলাদ মাহফিল

are tracked. In the second layer, celebrating, protesting, and religious event words are searched and their frequencies are tracked.

Event Phrase Matching: The Bengali language has a set of phrases which is used specifically to mention an event. These event phrases play an important role in identifying an event. Table 2 shows some of the event phrases used for this event prediction task. Almost all possible combinations of event phrases have been used for this event prediction task. Event phrases confirm a post's identification as an event or not.

Sentiment Score: Sentiment score is an important factor in predicting events as a celebrating event will have positive sentiment, a protesting event will have negative sentiment and a neutral post will have a neutral sentiment. Valence Aware Dictionary and Sentiment Reasoner (VADER) has been used for finding out the sentiment score of posts. VADER is a lexicon and rule-based sentiment analysis tool. For the sentiment score, the Bengali post has been translated into English using the "googletrans" python library. VADER works well with the English language and provides accurate results and that is why Bengali posts have been translated into English for sentiment score. VADER gives a compound score in the range of -1 to 1. A compound score of 0.05 or greater is considered as positive sentiment, a compound score of -0.05 or less is considered as negative otherwise, it is neutral sentiment. VADER uses Eq. 1 for determining compound score which is used in this task-

$$Xcom = \frac{x}{\sqrt{x^2 + \alpha}} \tag{1}$$

Here, $Xcom$ defines compound score in the range of -1 to 1.

3.4 Model Training and Prediction

The model is trained using the features and model then predicts based on the training with the features. Common event words frequency, event-specific words frequency, event-specific phrase frequency, and sentiment score of the posts have been used in model training. When a new post comes, the model predicts using that post's

features and following the previous feature values in the training phase. Naive Bayes Classification Model has been used for this work.
The Naive Bayes classification model follows Bayes theorem and it is given in Eqs. 2 and 3

$$P(\text{class}|\text{features}) = \frac{P(\text{features}|\text{class}) * P(\text{class})}{P(\text{features})} \tag{2}$$

$$P(\text{class}|\text{features}) = P(\text{feature}_1|\text{class}) \times P(\text{feature}_2|\text{class}) \times \cdots P(\text{feature}_n|\text{class}) \times P(\text{class}) \tag{3}$$

P(class|features): is the posterior probability of class (target) given predictor (attributes).
P(class): is the prior probability of class.
P(features|class): is the likelihood which is the probability of predictor given class.
P(features): is the prior probability of predictor.

Among the three types of Naive Bayes Classifier, Bernoulli Naive Bayes Classification model has been used. The reason to choose Bernoulli Naive Bayes is that binary feature values have been used instead of actual frequencies of those features and Bernoulli Naive Bayes shows better performance with binary feature values. This work has also been tested with the other types of Naive Bayes Classification models along with Decision Tree Classifier and Support-Vector Machines model.

4 Result and Discussion

Bernoulli Naive Bayes Classification Model has been used in this events prediction task which shows 87.5% accuracy in predicting events. Binary feature values have been used in this task and Bernoulli Naive Bayes shows good performance with these binary feature values.

The model has been tested with 359 real Facebook posts collected from various public Facebook pages and groups. Among these data, 61 posts are of protesting event type, 47 posts are of celebrating type, 41 posts are of the religious type and the rest 211 posts are of neutral type(events outside celebrating, protesting and religious types).

Figure 2 shows the Receiver Operating Characteristic (ROC) curve for the Bernoulli Naive Bayes Classification model in this event prediction task. ROC curve also shows the area under the curve for each class. Area Under Curve (AUC) for protesting, celebrating, neutral, and religious events are 0.98, 0.91, 0.86, and 0.99 respectively.

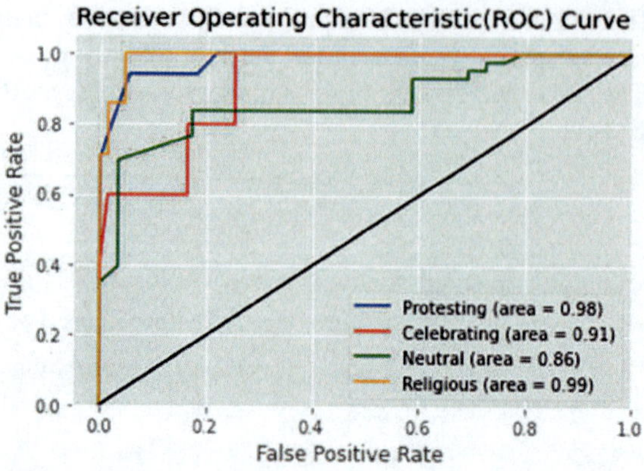

Fig. 2 ROC Curve of Bernoulli Naive Bayes Classification Model: Protesting, Celebrating, Neutral and Religious event's ROC curve

The model has been tested with two other classification models namely Support-Vector Machines(SVM) and Decision Tree Classifier. Bernoulli Naive Bayes Classification Model outperforms the other two models: the Support-Vector Machines model and the Decision Tree Classifier model in this work. Support-Vector Machines and Decision Tree Classifier model show accuracy of 84.72% and 83.33% respectively in this work whereas Bernoulli Naive Bayes Classification shows 87.5% accuracy which is better than the other two models.

Table 3 shows the precision, recall, and f1-scores of Bernoulli Naive Bayes, Support-Vector Machines, and Decision Tree Classifier model respectively. Figure 3 shows the confusion matrix of Bernoulli, Multinomial, and Gaussian Naive Bayes Classification model which helps in understanding how accurately true-events and false-events have been identified by these models. There are 61 protesting, 47 cele-

Table 3 Performances of Bernoulli Naive Bayes, Support-Vector Machines, and Decision Tree classification models in event prediction

Method	Event type	Precision	Recall	F1 score	Accuracy
Bernoulli Naive Bayes	Celebrating	0.60	0.60	0.60	0.8750
	Protesting	0.84	0.94	0.89	
	Religious	1.00	0.86	0.92	
	Neutral	0.90	0.88	0.89	
Support Vector Machines	Celebrating	0.60	0.60	0.60	0.8472
	Protesting	1.00	0.71	0.83	
	Religious	1.00	0.71	0.83	
	Neutral	0.82	0.95	0.88	
Decision Tree Classifier	Celebrating	0.50	0.80	0.62	0.8333
	Protesting	0.92	0.71	0.80	
	Religious	1.00	0.71	0.83	
	Neutral	0.85	0.91	0.88	

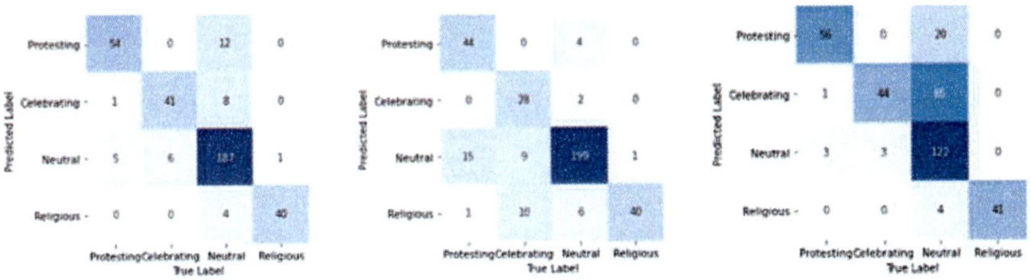

(a) Bernoulli Naive Bayes (b) Multinomial Naive Bayes (c) Gaussian Naive Bayes

Fig. 3 Confusion matrix of Naive Bayes Classification Model showing how many true and false classes are identified: 1st, 2nd, 3rd, and 4th column refers to Protesting, Celebrating, Neutral, and Religious events respectively

Table 4 True and false event prediction performances of Bernoulli Naive Bayes, Support-Vector Machines, and Decision Tree Classification model on output classes

Method	Protesting		Celebrating		Religious		Neutral	
	True protesting	False protesting	True celebrating	False celebrating	True religious	False religious	True neutral	False neutral
Bernoulli Naive Bayes	0.94	0.06	0.60	0.40	0.86	0.14	0.88	0.12
Support Vector machines	0.71	0.29	0.60	0.40	0.71	0.29	0.95	0.05
Decision tree	0.71	0.29	0.80	0.20	0.71	0.29	0.91	0.09

brating, 41 religious, and 211 neutral posts. Diagonal values in every matrix show the truly predicted classes. Values other than the diagonal are wrongly predicted classes by the model.

Table 4 shows the performance of Bernoulli Naive Bayes, Support-Vector Machines, and Decision Tree Classifier model on every output class of this task. Bernoulli Naive Bayes shows relatively low accuracy in celebrating class but it works well on other classes and its performances are very similar in predicting actual results on other output classes. Support-Vector Machines model shows good accuracy in the Neutral Event class but comparatively low accuracy in Celebrating, Protesting, and Religious class. Decision Tree Classifier model shows good and similar accuracy in Celebrating and Neutral classes but comparatively low accuracy in Religious and Protesting classes. Overall, Bernoulli Naive Bayes Classification Model shows good and almost similar accuracy in every output class.

Bernoulli Naive Bayes Classification Model shows an accuracy of 87.5% in the events prediction task. It performs better than the other two models: Support-Vector Machines and Decision Tree Classifier Model. Accuracy could have improved but this model fails to detect some event keywords and phrases in some posts. Besides, there are spelling mistakes in some posts which have not been identified by this model. These spelling mistakes in the Facebook posts result in wrong feature values and ultimately a wrong predicted result by this model.

5 Conclusion and Future Work

An effective approach has been presented in this paper to reveal upcoming unwanted events which is a must to control any disruptive situation before it turns out to be a crucial one. This work shows an accuracy of 87.5% using the Bernoulli Naive Bayes Classification Model. Real Bengali posts have been analyzed collected from Facebook's various public pages and groups. Furthermore, Naive Bayes Classification has been executed to make a decision whether a post is indicated to be an event of celebrating, protesting, religious, or neutral on the basis of sentiment score and other features. Though it's a great challenge to predict an event by analyzing these processes, this work tried to construct a model that is comparatively successful than others in this sector of predicting events from Bengali Facebook posts.

This proposed model can be extended to analyze multiple languages. Banglish (Bengali is written in English alphabets) is a popular language in Bangladesh and people use it popularly in social media platforms. This work will aim to work with this Banglish feature in the future to predict events. Besides, it also cherishes a goal to extend this work using a multi-lingual feature such as Hindi, Osmia, Gujrati etc. so that this work can predict events in any language. This work will also try to predict the event's location and time in the future. Errors in the training data will be intended to rectify for creating a more efficient event prediction model.

References

1. Our world in data: https://ourworldindata.org/rise-of-social-media
2. Aldhaheri, A., Lee, J.: Event detection on large social media using temporal analysis. In: 2017 IEEE 7th Annual Conference (CCWC), pp. 1–6. IEEE (2017). https://doi.org/10.1109/CCWC.2017.7868467
3. Wikipedia: The free encyclopedia. https://en.wikipedia.org/wiki/Bengali_language
4. Statcounter, Globalstats: https://gs.statcounter.com/social-media-stats/all/bangladesh
5. Wikipedia: The free encyclopedia. https://en.wikipedia.org/wiki/Twitter
6. Alsaedi, N., Burnap, P.: Arabic event detection in social media. In: International Conference on Intelligent Text Processing and Computational Linguistics, pp. 384–401. Springer, Berlin (2015)
7. Bekoulis, G., Deleu, J., Demeester, T., Develder, C.: Sub-event detection from twitter streams as a sequence labeling problem (2019). arXiv preprint arXiv:1903.05396
8. Imran, M., Elbassuoni, S., Castillo, C., Diaz, F., Meier, P.: Practical extraction of disaster-relevant information from social media. In: Proceedings of the 22nd International Conference on World Wide Web, pp. 1021–1024 (2013). https://doi.org/10.1145/2487788.2488109

9. Kolya, A.K., Ekbal, A., Bandyopadhyay, S.: A simple approach for monolingual event tracking system in Bengali. In: 2009 Eighth International Symposium on Natural Language Processing, pp. 48–53. IEEE (2009)
10. Fedoryszak, M., Frederick, B., Rajaram, V., Zhong, C.: Real-time event detection on social data streams. In: Proceedings of the 25th ACM SIGKDD International Conference on Knowledge Discovery & Data Mining, pp. 2774–2782 (2019). https://doi.org/10.1145/3292500.3330689
11. Chen, G., Kong, Q., Mao, W.: Online event detection and tracking in social media based on neural similarity metric learning. In: 2017 IEEE International Conference (ISI), pp. 182–184. IEEE (2017). https://doi.org/10.1109/ISI.2017.8004905
12. Fathima, P.N., George, A.: Event detection and text summary by disaster warning (2019)
13. Li, Q., Nourbakhsh, A., Shah, S., Liu, X.: Real-time novel event detection from social media. In: 2017 IEEE 33rd International Conference (ICDE), pp. 1129–1139. IEEE (2017). https://doi.org/10.1109/ICDE.2017.157
14. Kannan, J., Shanavas, A.R.M., Swaminathan, S.: Sportsbuzzer: detecting events at real time in twitter using incremental clustering. Trans. Mach. Learn. Artif. Intell. **6**(1):01 (2018)
15. Feng, X., Zhang, S., Liang, W., Liu, J.: Efficient location-based event detection in social text streams. In: International Conference on Intelligent Science and Big Data Engineering, pp. 213–222. Springer, Berlin (2015). https://doi.org/10.1007/978-3-319-23862-3_21
16. Akbari, M., Hu, X., Liqiang, N., Chua, T.: From tweets to wellness: wellness event detection from twitter streams. In: Thirtieth AAAI Conference on Artificial Intelligence (2016)
17. Sakaki, T., Okazaki, M., Matsuo, Y.: Earthquake shakes twitter users: real-time event detection by social sensors. In: Proceedings of the 19th International Conference on World Wide Web, WWW '10, New York, NY, USA, , pp. 851–860. Association for Computing Machinery (2010). https://doi.org/10.1145/1772690.1772777
18. Panagiotou, N., Katakis, I., Gunopulos, D.: Detecting events in online social networks: definitions, trends and challenges. In: Solving Large Scale Learning Tasks. Challenges and Algorithms, pp. 42–84. Springer, Berlin (2016). https://doi.org/10.1007/978-3-319-41706-6_2
19. Zhao, S., Gao, Y., Ding, G., Chua, T.: Real-time multimedia social event detection in microblog. IEEE Trans. Cybern. **48**(11), 3218–3231 (2018). https://doi.org/10.1109/TCYB.2017.2762344
20. Banna, M.H.A., Taher, H.A., Kaiser, M.S., Mahmud, M., Rahman, M.S., Hosen, A.S.M.S., Cho, G.H.: Application of artificial intelligence in predicting earthquakes: state-of-the-art and future challenges. In: IEEE Access[accepted]. https://doi.org/10.1109/ACCESS.2020.3029859

Cognitive Science and Computational Biology

Gaze Movement's Entropy Analysis to Detect Workload Levels

Sergio Mejia-Romero, Jesse Michaels, J. Eduardo Lugo, Delphine Bernardin, and Jocelyn Faubert

Abstract The gaze movement from a driver represents specific skills related to safe driving. Driving maneuvering evaluation is to know driving fitness. We found that gaze movement entropy is highly sensitive to visual behavior demands while driving a vehicle and the workload level. Entropy measures were more sensitive, more robust, and easier to calculate than gaze established measures. The gaze movement measures were collected using a driving simulator, using five different simulate routes and tree different workload scenarios; one route of familiarization was used to create a baseline. Because the workload became more difficult, drivers looked more at the central part of the road for more extended periods, the gaze movement entropy values decreased when the workload was increased, and the results show differences between two levels of workload. Also, the entropy result is compared against the classical analysis of the spatial distribution of gaze.

Keywords Driving performance · Mental workload · Eye tracking · Driving simulator · Entropy

1 Introduction

The gaze movement during driving tests is a directional set of vector points in 3D space that contains information of dynamic objects of interest in correlation to the various points in the visual space [1], considering objects inside and outside the car, such as vehicles, pedestrians, traffic lights, speedometer, side mirrors that increase the level of workload on drivers. In this work, the gaze movement represents a signal from which fixations and sacral movements can be extracted to derive the driver's

S. Mejia-Romero (✉) · J. Michaels · J. Eduardo Lugo · D. Bernardin · J. Faubert
FaubertLab, School of Optometry, Université de Montréal, Montréal, QC, Canada
e-mail: sergio.mejia.romero@umontreal.ca

D. Bernardin
Essilor Canada Ltd, Montréal, QC, Canada

M. S. Kaiser et al. (eds.), *Proceedings of International Conference on Trends in Computational and Cognitive Engineering*, Advances in Intelligent Systems and Computing 1309, https://doi.org/10.1007/978-981-33-4673-4_13

evaluation and estimate the behavior of the attention level to predict unforeseen obstacles [2].

Entropy is a function that measures the regularity of the values in a signal or a system [3], that is, quantifies what is the level of coherence between the values belonging to a signal; we understand that if within the signal there are repetitive patterns, then it is a stable signal, therefore less complex, and the value of the entropy is smaller.

We obtain the average level of uncertainty of the probability distribution of the gaze movement's coordinates during the period that the simulator is used [4]. It is observed that greater entropy or uncertainty reflects a broader distribution of gaze movements within the visual field, suggesting a broader dispersion in the gaze [5].

This entropy value provides the visual scanning pattern's predictability value on scale or time window [6]. This entropy value, when it is high, suggests a less structured or more random pattern. In other studies, an association of increased entropy for gaze movements when anxiety is high and an increase in entropy in the transitional pattern of fixation when cognitive load is increased has been reported [7]. This result suggests different temporal sequences within the gaze movement; that is, the subject performs the movement of gaze and exploration in different time scales or from another point of view in different frequency scales.

If we consider that the dynamics of gaze movement is random from one subject to another and it depends on the experience [8], emotional factors, etc., so the value of the time interval to consider a fixation duration varies from 120 to 600 ms [9], which is intrinsically dependent on the subject's ability during the task and the task's workload itself. In this case, we cannot compare the entropy value of one subject with another; that is, it cannot be entirely determined by comparing one subject's entropy value with another on a particular frequency.

This article aims to present the entropy function using a time window of a subject that performs a driving test in a driving simulator. Calculating the gaze's entropy refers to estimating and quantifying the level workload by the visual exploration during the test in the driving simulator. We know that the entropy is very sensitive to any signal variation and one needs to be careful about what variables are relevant to the task [10].

In the results section below, we first describe the gaze movement in different scenarios; second, using entropy, we test the method by comparing two different workload scenarios [11]; also, the results are compared with classical analysis [12] to verify the correspondence between workload and entropy level.

From the results, we observe that the evaluation of the entropy in the driver's gaze's movement describes the different levels of workload and is intrinsically related to different levels of exploration, as described in previous research.

We find that this methodology using entropy value agrees well with the value of each scenario's workload and that the model reproduces reasonably when the visual exploration is greater or lesser in the different scenarios. It is important to note that it can also be applied to other eye movement analysis by changing the entropy methodology's parameters. Our results proved the robustness of this entropy method for complex random signal analysis.

2 Materials and Methods

We are using the SensoMotoric Instrument (SMI) eye tracker operating at 120 Hz to carry out our study. The raw data collected by the SMI is processed using post-processing software to analyze vector output and quantify the gaze movement. Because gaze direction is affected by head rotation, we used the OptiTrack motion capture system and Motive software to record head movements; tracking was registered with a 120 Hz sampling rate and an algorithm written in MATLAB to estimate gaze direction in the simulator space. The VS500M car driving simulator (Virage Simulation Inc.) was used to measure road driving performance (Fig. 1).

Twenty drivers participated in this study. They all had a valid driver's license, and normal vision was reported. In this study, all participants received a compensation of 15$ per session. The ethics committee approved this study of the Université de Montréal.

Subjects were instructed to drive four scenarios; the instruction was typically driven by respect to road signs and speed limits. The workload of each scenario was assessed for two levels of low workload "City1" and "Rural1" and high workload "City2" and "Rural2". During the whole time of the test, the eye movement and head rotation were monitored to calculate the binocular point of the gaze direction within the environment.

Each participant conducted a baseline phase using a familiarity scenario; this scenario is considered a very low workload demand from the subjects. During this time, the data were recorded. After the familiarization session, double random of four remaining scenarios was made. Two scenarios were carried out in the first session, and the remaining two were performed in a second session with a separation of two weeks between sessions.

Fig. 1 In this study, the driving simulator, green dots represent the gaze pattern on the display simulator during a rural scenario

2.1 Data Analysis

Initial filtering of the eye and head movement was performed using a winner filter [13]. This method considers the system noise, whereby the filter coefficients are calculated. All the data sets were processed by the same noise filtering method as supported by other research. Subsequently, each time series, which represents the eye or head movement obtained after filtering the signal, was subjected to obtain the gaze estimation using the same method described in Mejia-Romero et al. [14].

We developed an analysis tool used for signal processing, data quality management, and gaze estimation, saccade, fixation, and non-tracking segmentation were applied to the raw signals.

Entropy was evaluated throughout the driving test. To evaluate the baseline entropy, we used gaze data from the familiarization scenario. Then, from each entropy value per scenario, we subtract the baseline value per subject, and it was normalized by the maximum entropy per scenario so that the entropy scale ranges from 0 up to 1.

2.2 Entropy Methodology

We use this methodology to know the random degree of the gaze distribution; this evaluation is based on the entropy value associated with the gaze; the entropy was evaluated during the whole drive scenario.

$$H = -\sum_{i=1}^{n} p_i \sum_{j=1}^{n} p(i, j) \log_2 p(i, j) \tag{1}$$

where p_i is the probability that one fixation exists at the time i, and $p(i, j)$ is the probability occurrent this fixation inside the time bins i and j.

We applied a modified Tole method (1986) [5] for entropy determination, given a time series of N data points, a sequence signal using a window that length "m" is the embedded dimension. An analogy to transition entropy, the window is the area of interest, and $p(i, j)$ is the probability to transition from area i to j.

This process is referred to as self-matching, in the time series, and leads to a conditional probability determination. With this first window of length m, we obtain the value of the entropy for the time i; at the end of the whole series, the entropy value will be of the whole series; that is, we calculate Shannon's entropy to know the changes of entropy between neighbor windows with the understanding that if the system within this window of m data is stable the value of entropy is minimum.

3 Results

Consistent with expectations, results show that scenario workload increases produce a gaze concentration to the central area and low entropy level in all subjects.

For example, the results obtained from one subject during both task sessions can be seen in Figs. 2 and 3; on both figures, the left panel displays a gaze pattern for a lower workload scenario than the right panel display gaze pattern (higher workload).

Figures 2 and 3, we can note that the gaze pattern is more concentrated at the center when the workload is increased. There is a change in the spatial distribution of gaze direction from one workload level to another workload level. To highlight this result, we show in Table 1, the values of classical analysis for the spatial distribution of gaze direction for this subject.

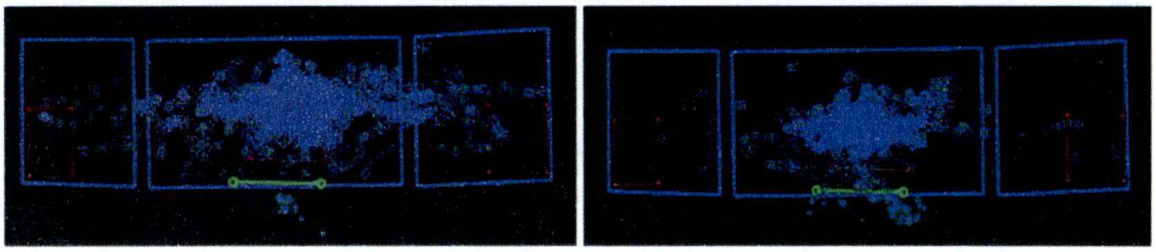

Fig. 2 Gaze pattern example for one subject during the city scenario. The left panel corresponds to the City1 "Lower workload" on the right; it is during City2 "Higher workload"

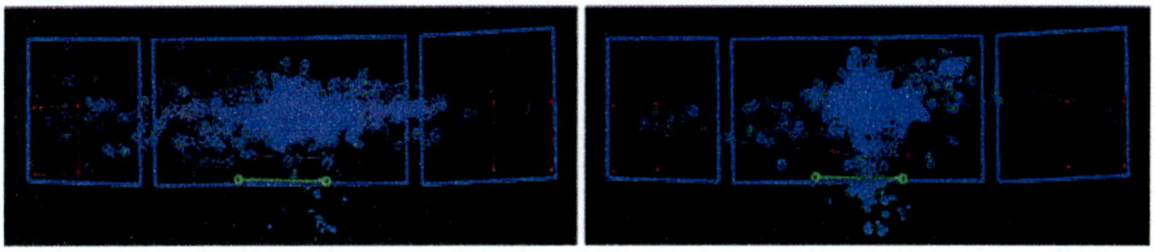

Fig. 3 Gaze pattern example for one subject during the rural scenario. On the left panel correspond the Rural1 "Lower workload" on the right; during Rural2, "Higher workload"

Table 1 Classical analysis for spatial distribution, example subject

Scenario name	City1	City2	Rural1	Rural2
Workload	Lower	Higher	Lower	Higher
Entropy value (ua)	0.67	0.42	0.92	0.34
Saccade amplitude (°)	15.36	13.24	14.56	11.14
Fixation duration (ms)	291.18	215.58	321.02	241.32
Fixation rate (count/m)	76.91	83.75	96.91	87.54
Gaze area (cm^2)	62.16	43.42	68.76	32.13

As expected, the amplitude of the saccade movement, fixation duration, and gaze area decreases between the different scenarios corresponding to the different workloads level; the same goes for entropy values.

To check the difference between the different workloads, we estimate the spatial distribution by scenario using the frequency maps of all the subjects in this study, and our results are like those obtained by the classical measures used to evaluate visual exploration and workload.

As the workload became more difficult, all measures showed that drivers change their gaze movements behavior to the driving environment; in general, they increase the time spending in the central area when the driving task increased its workload, as evidenced by spatial gaze concentration being highest at the central area when the workload is higher, drivers show more fixation time and lower saccade amplitude,

4 Discussion

In this work, we have presented evidence that using entropy on gaze movement series can evaluate the dispersion of gaze exploration. The proposed method allows us to evaluate the workload between tasks. Furthermore, it was confirmed that a higher workload is related to less randomness in gaze movement.

The driving task with a higher workload caused the driver to look at the road's central part more often and more extended periods. The plot (Fig. 4) of gaze dispersion behavior also shows a dramatic increase in the gaze's concentration to the screen's central area. The results clearly showed that drivers change their gaze movement depending on the workload level, as shown in two different scenarios with different workloads levels.

The experimental results have shown that the evaluation of the gaze movement entropy from a driver describes reasonably the different workload levels and is intrinsically related to different exploration levels, as described in previous research.

For higher workload, all classical measures (Table 2) clearly showed that drivers look more at the road center, have less gaze area, and increase fixation duration. We are also able to show here the relative sensitivity of the entropy to the workload.

5 Conclusions

During a driving task, as the workload was increased, drivers increased their viewing time on the central part of the road, and the spatial distribution is also concentrated in the same central area—also, the entropy level decrease. The gaze's entropy level was more robust, reliable, more easy to calculate, and more sensitive than classical analysis for the spatial distribution of gaze.

Based on this work's findings, gaze movement entropy provides a methodology to assess changes in workload level. When comparing two different workloads under similar conditions, the results provide arguments for future studies in this field.

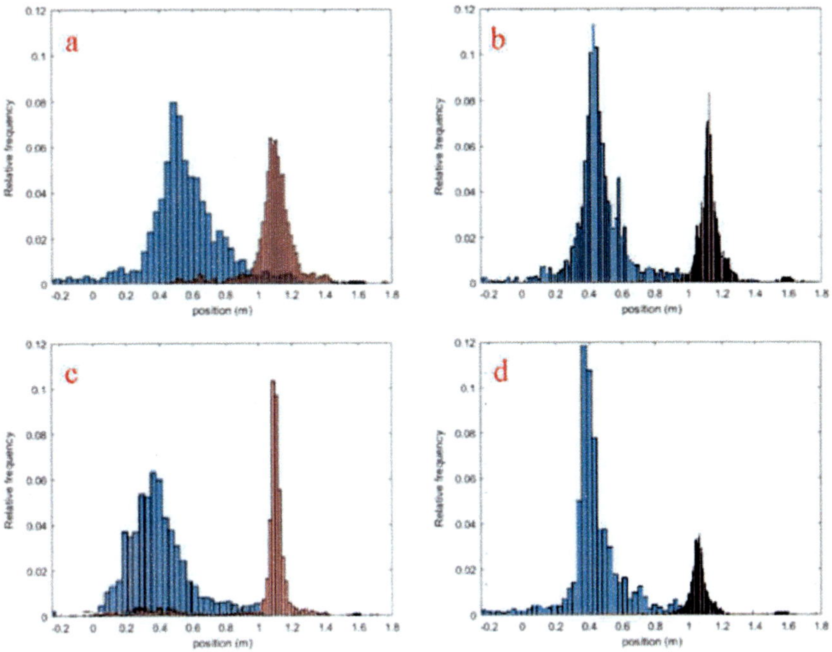

Fig. 4 Spatial distribution of gaze patterns with the unrelated workload; it was calculated from all subjects by scenario. Panels a and c represent low workload, and panels b and d represent high workload. In orange, we show the vertical distribution, and the horizontal distribution is in blue

Table 2 Classical analysis for spatial distribution, all subjects

Scenario name	City1	City2	Rural1	Rural2
Workload	Lower	Higher	Lower	Higher
Heading level	Mean ± SD	Mean ± SD	Mean ± SD	Mean ± SD
Entropy value	0.73 ± 0.174	0.35 ± 0.045	0.82 ± 0.098	0.25 ± 0.034
Saccade amplitude (°)	19.16 ± 0.705	13.54 ± 0.320	17.46 ± 1.05	11.31 ± 0.083
Fixation duration (ms)	261.38 ± 108.02	295.4 ± 54.78	321.38 ± 78.02	391.45 ± 32.56
Fixation rate (count/m)	76.91 ± 16.0	56.75 ± 9.13	96.91 ± 16.50	62.43 ± 6.45
Gaze area (cm^2)	82.76 ± 5.76	73.42 ± 2.45	78.76 ± 8.35	58.13 ± 3.15

Acknowledgements This research was partly funded by an NSERC Discovery grant and Essilor Industrial Research Chair (IRCPJ 305729-13), Research and development cooperative NSERC-Essilor Grant (CRDPJ 533187 - 2018), Prompt

Author Contribution M-R.S. led the design of the research method and implemented the data analysis. M-R.S participated in preparing of the conclusions based on the results. MJ collected the raw data on how to work for his Ph.D. research. All authors took part in the paper preparation and edition.

Conflicts of Interest The authors of this manuscript declare no conflict of interest. We express that the sponsors had no role in the design of the study, analysis, and writing of the manuscript and the decision to publish the results.

References

1. Klauer, S.G., Dingus, T.A., Neale, V.L., Sudweeks, J., Ramsey, D. (2006). The Impact of Driver Inattention on Near-Crash/Crash Risk: An Analysis Using the 100-Car Naturalistic Driving Study Data (Technical Report No. DOT HS 810 594).: NHTSA. Washington DC
2. Engström, J.A., Johansson, E., Östlund, J.: Effects of visual and cognitive load in real and simulated motorway driving. Transp. Res. Part F **8**, 97–120 (2005)
3. Pincus, S.M., Goldberger, A.L.: Physiological time series analysis: what does regularity quantify? American J. Physiol. (Heart. Circul. Physiol.) **266**, H1643–H1656 (1994)
4. Shiferaw, B.A., Downey, L.A., Westlake, J., Stevens, B., Rajaratnam, S.M.W., Berlowitz, D.J., et al.: Stationary gaze entropy predicts lane departure events in sleep-deprived drivers. Sci. Rep. **8**(1), 1–10 (2018)
5. Tole, J.R., Stephens, A.T., Vivaudou, M., Harris, R.L., Ephrath, A.R.: Entropy, Instrument Scan and Pilot Workload. IEEE, New York, NY, USA (1982)
6. Krejtz, K., Duchowski, A., Szmidt, T., Krejtz, I., González Perilli, F., Pires, A, et al.: Gaze transition entropy. ACM Trans. Appl. Percept TAP **13**(1), 4:1–4:20 (2015)
7. Allsop, J., Gray, R.: Flying under pressure: effects of anxiety on attention and gaze behavior in aviation. J. Appl. Res. Memory Cognit. (2014). https://doi.org/10.1016/j.jarmac.2014.04.010
8. Underwood, G., Chapman, P., Brocklehurst, N., Underwood, J., Crundall, D.L Visual attention while driving: sequences of eye fixations made by experienced and novice drivers. Ergonomics **139** (2003)
9. Wang, Y., Bao, S., Du, W., Ye, Z., Sayer, J.R.: Examining drivers' eye glance patterns during distracted driving: Insights from scanning randomness and glance transition matrix. J. Safety Res. **1**(63), 149–155 (2017)
10. Mikula, L., et al.: Eye-head coordination and dynamic visual scanning as indicators of visuo-cognitive demands in driving simulator (2020). https://doi.org/10.1101/2020.09.23.309559
11. Michaels, J., et al.: Driving simulator scenarios and measures to faithfully evaluate risky driving behavior: a comparative study of different driver age groups. PLoS One **12**, 1–24 (2017)
12. Victor, T., Harbluk, J.L., Engström, J.A.: Sensitivity of eye-movement measures to in-vehicle task difficulty. Transp. Res. Part F **8**(2), 167–190 (2005)
13. Mejia-Romero S., et al.: An effective filtering process for the noise suppression in eye movement signals (2019a)
14. Mejia-Romero S., et al. "Dynamic performance of gaze movement, using spectral decomposition and phasor representation (2019b)

A Comparative Study Among Segmentation Techniques for Skin Disease Detection Systems

Md. Al Mamun and Mohammad Shorif Uddin

Abstract Skin disorders are serious health problems for people. An automatic mobile-oriented skin disease detection system with offline or online is extremely essential for detecting skin diseases and serving patient treatment plans. For any image-based detection as well as recognition task, useful features are playing an important role. But the extraction of essential features is seriously dependent on the segmentation of disease-affected region, which ultimately hampers the detection accuracy sensitivity and specificity. In this paper, we have described a comparative study on various segmentation algorithms that are applied to extract the lesion part from the skin images for detecting diseases. Available methods are evaluated based on both qualitative and quantitative perspectives. Besides, we have pointed out some challenges of skin disease detection which need special attention of researchers, such as the availability of extensive datasets, well-defined efficient segmentation algorithms, and mobile-friendly computation environment.

Keywords Skin lesion segmentation · Dermoscopy · Edge · Texture · Wavelet

1 Introduction

Skin is the most delicate organ in the human body. Sunburn is one of the main factors impacting the melanocyte cell due to the ultraviolet rays from the sun [1, 2]. Due to UV rays, overheat, fungal infections, unhygienic conditions, or other pathogens, the external layer of skins is contaminated with various diseases. Skin lesion signs include dryness of the skin, infection, and allergic signs, burn, cough, scaly skin, fever, discomfort, bruises, scratching, skin rash, pain, and swelling, and so on [3–6]. The ICT-based automation system is required to

Md. Al Mamun (✉)
Department of Public Health and Informatics, Jahangirnagar University, Savar, Dhaka, Bangladesh
e-mail: almamun@juniv.edu

M. S. Uddin
Department of Computer Science and Engineering, Jahangirnagar University, Savar, Dhaka, Bangladesh

M. S. Kaiser et al. (eds.), *Proceedings of International Conference on Trends in Computational and Cognitive Engineering*, Advances in Intelligent Systems and Computing 1309, https://doi.org/10.1007/978-981-33-4673-4_14

detect skin diseases in this era [7–10]. Without an expert dermatologist, early detection of skin diseases is not possible in a typical situation [11–14]. To cope with this problem, an automated machine learning-based skin disease detection system has been developed [15–18]. In this system, an affected portion of the lesion needs to be segmented out using a segmentation algorithm from the skin lesion image and then go for a feature extraction step to perform skin disease recognition [19–24].

In this manuscript, we explore various segmentation methodologies for the detection of skin diseases [25, 26]. Among the methodologies, we have investigated the efficient methods of segmentation strategies that are applied to develop an inexpensive, effective computerized skin disease detection system with lower complexity and higher performance [27–30].

The primary commitments of this research are as follows:

- We extensively analyze the various methodologies of segmentation and identify the reliable one for the diagnosis of a particular form of skin disease.
- Identify the merits and demerits of current segmentation approaches used in automated skin disease detection techniques.
- Identify the challenges to segmentation out disease-affected region along with future directions.

The paper is structured as follows. Section 2 portrays the literature review, Sect. 3 focuses on the segmentation of the affected region, Section 4 clarifies results, and Sect. 5 concludes the works with possible guidance.

2 Literature Review

In skin disease detection system, image segmentation is a fundamental strategy. Researchers have used a diverse segmentation method to segment out the lesion region or portion of dermoscopy images. All the methods use a preprocessing step for effective image segmentation. These techniques have used robust filtration, image resizing, image contrast enhancement as the preprocessing of dermoscopy images. In the following, we have discussed the segmentation methodologies of various researchers. Ahmed et al. [1] developed an SVM classifier system which uses an ACO-GA-based hybrid segmentation algorithm to segment out skin lesion portion with higher system accuracy. George et al. [2] expressed a machine learning-based system which uses pixel-based segmentation to detect psoriasis with an accuracy rate of 97.4%. Arulmozhi and Divya [3] described a system that uses morphological erosion and dilation operations to segment out the skin lesion portion for skin disease detection. Arora et al. [5] showed a performance measure-based segmentation technique for skin cancer detection. The performances are measured using the values of peak signal to noise ratio (PSNR), mean square error (MSE), and structural similarity index measure (SSIM). Jailani et al. [4] proposed a system which uses morphological operations and thresholding method to extract a segmented portion of the psoriasis disease-affected region. Roy et al. [6] presented a system that suggests the k-means clustering algorithm for segmentation on dermoscopic images of skin diseases like eczema, psoriasis, chickenpox, and ringworm.

Sood and Shukla [7] proposed a skin lesion segmentation method that uses a genetic algorithm to extract the selected portion with the highest accuracy rate of 99.94%. Patino et al. [8] described an automated skin lesion segmentation on dermoscopy images by the super pixel-based method of PH2 and ISIC 2017 datasets with an accuracy of 86.9% and Jaccard index of 0.60. Ibrahim et al. [9] showed a hybrid segmentation technique the watershed segmentation and the Canny edge detection with 97.75% accuracy. Chong et al. [10] proposed a segmentation technique for Eczema which uses the k-means clustering method with an accuracy rate of 84.6%. In addition, Chong et al. [11] showed a segmentation technique based on a two-level k-means clustering method for eczema which uses RGB and CIElab color models. They found an accuracy of 86.07%.

Navarro et al. [12] presented a computer-aided diagnosis system using ISIC 2017 dermoscopy database images that segmented out the skin lesion portion using a super pixel-based segmentation method. They found an accuracy of 96%, Jaccard index 0.84, and the dice coefficient 0.94%. Agarwal et al. [13] depicted an automated computer vision system that uses an adaptive threshold method to segment out the defected skin portion of grayscale dermoscopy images. They found an accuracy of 97.03%. Al-Masni et al. [14] proposed a system that uses a pixel-wise segmentation method convolutional neural network (CNN) to segment out the affected portion of the dermoscopy skin image. The found an accuracy of 95.08%. Attia et al. [15] described a segmentation technique using RNN (recurrent neural network) and CNN-based method with an accuracy rate of 95% and a Jaccard index of 0.93.

Chakkaravarthy and Chandrasekar [16] presented an edge-based technique using the Sobel operator with a watershed segmentation method to segment out the lesion part of the skin image with the accuracy rate of 90.46%. Lu et al. [17] proposed an automated segmentation of the melanocytes in the skin histopathological image which uses region-based segmentation with an accuracy of 74.74%. Torkashvand and Fartash [18] described an automatic segmentation of skin lesion which uses Markov random field (MRF) segmentation of RGB images with an accuracy rate of 94%.

Das and Ghoshal [19] proposed an automated pixel-based segmentation technique that uses modified watershed segmentation to segment out the actual area of a skin lesion with an accuracy rate of 97.38%. Lawand [20] presented a wavelet transform-based segmentation technique of dermoscopy images which uses fixed-grid wavelet networks with no training procedure with an accuracy of 99.67%. Smaoui and Bessassi [21] predicted a melanoma skin cancer detection system that uses a region-based segmentation method as region growing to extract the affected region of a skin lesion with an accuracy rate of 92.5%.

Nisar et al. [22] described the color space-oriented K-means clustering segmentation algorithm to segment out the affected region of a skin lesion with an accuracy rate of 83.43%. Pal et al. [23] explained psoriatic plaque segmentation in skin images with the clustering segmentation method as a semi-wrapped Gaussian mixture model (SWGMM) to cluster the skin lesion region perfectly with an accuracy rate

82.84%. Pathan et al. [24] depicted a color-based skin image segmentation technique as a chroma-based deformable method to segment out the skin lesion area with an accuracy rate of 94.6%. Santy and Joseph [25] described the melanoma detection technique which uses a texture-based segmentation technique like the texture distinctiveness lesion segmentation (TDLS) method to segment out the affected portion of dermoscopy images with the accuracy rate of 96.8%.

Ebrahimi and Pourghassem [26] represented a skin lesion detection system that uses an optimal threshold parameter in a reinforcement algorithm to segment out skin regions with an accuracy rate of 97.18%. Oliveira et al. [27] depicted an active contour-based skin lesion segmentation technique using the Chan-Vese mathematical model to locate the affected region concerned with skin color with an accuracy rate of 94.36%. Khan et al. [28] described a color-based segmentation technique from acne lesion images using the YIQ color space with a performance rate of 92.63%.

Li et al. [29] described a skin cell segmentation technique that uses spectral angle and distance to segment out the affected cell region of skin with an accuracy rate of 84%. Khahmad et al. [30] represented an edge-based thresholding segmentation method using morphological operations of erosion and dilation to extract the whole boundaries of the lesion region of skin with a performance rate of 80%.

3 Methodology

Skin diseases are identified through a digital computerized framework. The automatic skin disease detection system has two main phases: the testing phase and the learning phase. In the learning phase, dermoscopy or clinical images are preprocessed and learned through image segmentation, feature extraction, classification steps, and then, it is stored in the learning database [1]. In the testing phase, almost the similar operations are performed on the dermoscopy or clinical images to classify or detect the diseases [2]. Figure 1 shows an automated skin disease diagnosis system.

Among skin disease detection procedure, segmentation techniques play a vital role [7–9]. Dermoscopic, clinical, or digital camera images of human skin are used as an input image. Preprocessed these images are to be ready for segmentation [10–12]. Better enhanced images show a higher segmentation efficiency. Through image segmentation, the skin lesion portion or region, i.e., region of interest (ROI) is extracted and used in the subsequent analysis for skin disease detection [13–16].

A schematic diagram of the segmentation process for the detection system is shown in Fig. 2. The main segmentation process can be done using any of the methods such as gene-based, clustering-based, color-based, pixel-based, region-based, texture-based, threshold-based, active contour-based, wavelet transform-based, CNN-based method or some hybridization of methods [17–22]. Post-processing may require accomplishing the segmentation operation precisely.

Table 1 describes the advantages and limitations of different existing segmentation strategies.

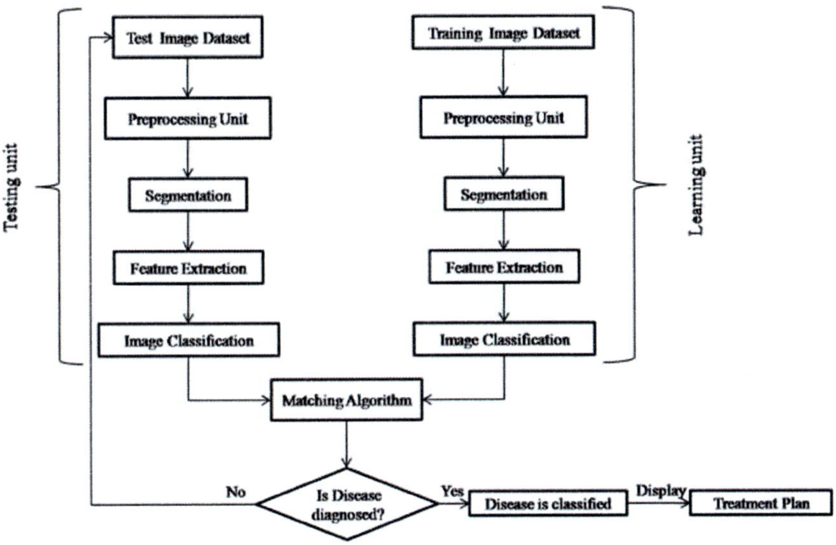

Fig. 1 Flow diagram of the skin disease detection system

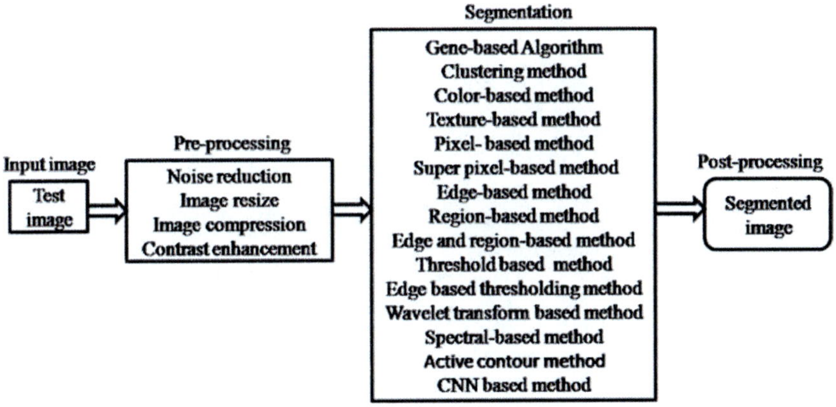

Fig. 2 Schematic diagram of the skin image segmentation process

4 Result Analysis

For a skin disease detection system, the segmentation technique plays a proficient role to extract the affected regions. The performance evaluation results for the various state-of-the-art algorithms for skin segmentation are shown in Table 2. The performances are evaluated using accuracy, sensitivity, and specificity metrics shown in Eqs. (1)–(3) derived from the confusion matrix [31].

Table 1 Segmentation methods

Algorithm	Description	Advantages	Limitations
Gene-based Algorithm	The probabilistic method and image characteristics are represented and manipulated by genetic structure	Fit solutions to random mutations coding are easy	Well-suited heuristics might be hard among different generations or populations
Clustering method	It extracts meaningful objects containing in the image. Its objective is to partition an image dataset into clusters or groups	It is a less expensive and more economical procedure.	It produces functional errors when different clusters are the same
Color-based method	It clusters image pixels into homogeneous colors	Computation of small color difference and contains non-uniform illumination	Singularity problem and nonlinear transformations
Texture-based segmentation method	This segmentation partitions the image into regions with different textures having alike pattern	It is meaningful, easy to understand and specific objects can be extracted from any shape with lossless information	It is sensitive to noise and distortions and also time-consuming
Pixel-based method	Pixel-based segmentation is dependent on the segregation of pixel intensity	It is effective and accuracy depends on the intensity of image pixels only	As it works on intensities, so noise-sensitive
Super pixel-based method	Graph-based or gradient-based method. An image is partitioned into multiple segments called superpixels which are identical in each portion	Two types of information (e.g., edge and region) are suitable for a better result	It is inefficient when superpixels are unknown
Edge-based method	It is based on the criteria that pixels having different intensities	Edge detection is simple and easy	It is noise sensitive and not performs if edges are not well-defined (e.g. in less texture region)
Region-based method	The same pattern of pixels forms the region	It is suitable when region similarity criteria are defined	It takes more time and memory due to dual-stage segmentation

(continued)

Table 1 (continued)

Algorithm	Description	Advantages	Limitations
Edge and region-based segmentation	Detect edges within object boundaries of specified regions or groups	Noiseless regions' similarity features are expressed easily, and edge detection is also a very simple task	Sensitive to noise, and also not operational for irregular unsharp regions
Threshold-based method	Threshold values are specified based on peaks and regions are identified with the values of the histogram of images	It is easy to implement	It is noise sensitive and the setting of the threshold value is complex
Edge-based thresholding method	Detect edges and object boundaries with optimal threshold values of an image	Well-suited and efficient method	If edges are excessively high and less contrast between objects, then it is not suitable
Wavelet transform-based method	Wavelet transform is used for inpainting of individual image pixels	Multi-resolution and decoupling features are beneficial	No phase information, sensitivity is shifted, and directionality is relatively poor
Spectral-based algorithm	It splits the image at first then approximates the Eigen solution of a similarity matrix	It is robust and needs less computational time	Need squared image regions with the appropriate size, and it has no semantic meaning
Active contour method	It is based on the curve flow, curvature, and contour to obtain the segmented portion of the image	Object motion information can be easily obtained from the contour	It is noise sensitive and slow. If regions have no clear boundaries, it is not effective
CNN-based method	Segments of an image as input to CNN, which labels pixels. CNN can process parts of the image at a time	It is a highly accurate segmentation of skin lesions.	Computation cost is high and needs a huge number of labeled images

$$\text{Accuracy} = (\text{TP} + \text{TN})/(\text{TP} + \text{TN} + \text{FP} + \text{FN}) \tag{1}$$

The sensitivity can be calculated from the below equation:

$$\text{Sensitivity} = \text{TP}/(\text{TP} + \text{FN}) \tag{2}$$

The specificity can be calculated from the below equation:

Table 2 Performance evaluation results for the various state-of-the-art algorithms in segmenting disease-affected skin region

Method type	Techniques	Dataset/number of images	Accuracy (%)	Sensitivity (%)	Specificity (%)
Gene-based algorithm	Genetic algorithm [7]	DermQuest	99.94	100	99.93
Clustering method	K-means clustering [10]	33 DSLR camera images	84.6	–	–
	2-Level K-means clustering [11]	53 digital single-lens reflex camera images.	86.07	79.26	90.18
	K-means clustering algorithm [22]	DSLR camera images or internet sources.	83.43	87.66	81.30
	Semi-wrapped Gaussian mixture model [23]	45 camera images.	82.84	–	–
Color-based method	Chroma-based deformable methods [24]	PH2-200 images, ISBI 2016-900 images	94.6	82.4	97.2
	YIQ color space method [28]	Using the images taken from the Kuala Lumpur hospital, Malaysia	92.63	89.67	93.19
Texture-based method	Texture distinctiveness lesion segmentation (TDLS) method [25]	126 digital images from the DermQuest database	96.8	–	–
Pixel-based method	Full resolution convolutional network [14]	ISBI 2017, PH2 datasets, 800 images	95.08	93.72	95.65
	Markov Random Field-based (MRF) segmentation [18]	PH2 dataset of 200 images	94	85	98
	Watershed segmentation [19]	RI CVL Face database, 420 images	97.38	–	–

(continued)

Table 2 (continued)

Method type	Techniques	Dataset/number of images	Accuracy (%)	Sensitivity (%)	Specificity (%)
Superpixel-based method	Simple linear iterative clustering (SLIC) segmentation [8]	PH2 and ISIC 2017, 80 melanocytic images	95.30	92.12	96.42
	SLIC guided by local features (LF-SLIC) [12]	ISIC 2017, 150 dermoscopy images	96	–	–
Edge-based method	Sobel operator with watershed segmentation [16]	Dermweb, 310 images	90.46	98.36	82.95
Region-based method	ML + Local double ellipse descriptor (LDED) [17]	30 images from the Carl Zeiss MIRAX MIDI scanning system	74.74	84.23	–
	Region growing segmentation [21]	40 images from the Department of Dermatology, University Hospital Hedi Chaker Sfax	92.5	88.88	92.3
Edge- and region-based method	Watershed segmentation with Canny edge detector [9]	133 web images	97.75	98.56	96.75
Morphological operation-based method	Morphological operations [30]	60 images from the Nevoscope device	80	48.4	57.1
Threshold-based method	Adaptive threshold method [13]	60 images from the Dermatology Information System and DermQuest	97.03	–	–
	Reinforcement algorithm for threshold fusion [26]	30 dermoscopic images from the internet sources	97.18	87.18	97.43

(continued)

Table 2 (continued)

Method type	Techniques	Dataset/number of images	Accuracy (%)	Sensitivity (%)	Specificity (%)
Wavelet transform-based method	Fixed-grid wavelet network (FGWN) [20]	30 images	99.67	94.34	99.84
Spectral-based algorithm	Spectral angle and distance [29]	45 images	84	–	–
Active contour method	Chan-Vese model [27]	DermAtlas, DermIS, DermNet, 408 images	94.36	–	–
CNN-based method	Deep convolution and recurrent neural network [15]	ISBI 2016, 375 images	98	95	94

$$\text{Specificity} = \text{TN}/(\text{TN} + \text{FP}) \quad (3)$$

where,

TP is the true positive which means a lesion skin region is identified as a lesion,
TN is the true negative which means a non-lesion skin region (healthy) is identified as a non-lesion (healthy),
FP is the false positive which means a non-lesion skin region (healthy) image as a lesion,
FN is the false negative (FN) which means a lesion skin region is identified as a non-lesion

It is observed from Table 2 that the genetic algorithm and the deep convolution with a recurrent neural network (RNN) are giving the highest accuracy, sensitivity, and specificity among all the methods. Hence, these can be taken as the benchmark approaches. However, more detailed experiments are needed for confirming the performance of these methods using diverse datasets containing a greater number of images.

There are many obstacles or challenges in achieving effective performance in automatic segmentation in the skin disease detection system. Various datasets without enhancing images do not operate effectively and reliably with the segmentation methods. When segmentation is not done properly then ambiguity might be included in the extracted features and the subsequent steps, which eventually deteriorate the performance of the ultimate disease detection as well as the diagnosis system. Effective enhancement is also essential to get the optimal segmentation result. Computation time complexity is also a factor (especially, in the mobile phone-based system) in building a sophisticated segmentation approach. The availability of a good number of diverse benchmarked datasets containing standard labeled images is a barrier to perform extensive experiments. For robustness evaluation, we need datasets with various image resolutions, sizes, visibility, and noise environments.

5 Conclusion and Future Directions

We investigated and evaluated different segmentation strategies in diagnosing a skin disease. The methods were compared and summarized in a tabular form based on both subjective and objective measurements. We have found that the genetic algorithm-based method and the deep-learning-based method with CNN and RNN are the most effective compared to other existing segmentation methods on the basis of the accuracy, sensitivity, and specificity. It is recommended that the researchers should emphasize doing extensive experiments with the existing methods, also have to be focused on hybridization and fine-tuning of existing methods, as the performance of the ultimate diagnosis system depends on an efficient affected region segmentation. Moreover, it is seriously needed to concentrate on developing diverse big datasets for a generalized optimum skin disease detection system with various image resolutions, sizes, visibility, and noise environments.

References

1. Md. Humayan, A., Romana, R.E., Tajul, I.: An automated dermatological images segmentation based on a new hybrid intelligent ACO-GA algorithm and diseases identification using TSVM classifier. In: 1st International Conference on Advances in Science, Engineering and Robotics Technology 2019 (ICASERT 2019), vol. 2, pp. 894–899. Dhaka, Bangladesh (2019). https://doi.org/10.1109/ICASERT.2019.8934560
2. Yasmeen, G., Mohammad, A., Rahil, G.: A pixel-based skin segmentation in psoriasis images using committee of machine learning classifiers. In: 2017 International Conference on Digital Image Computing: Techniques and Applications (DICTA), vol. 1, pp. 70–77 Sydney, Australia (2017). https://doi.org/10.1109/DICTA.2017.8227398
3. Arulmozhi, V., Divya, S.C.: Image segmentation and morphological process of skin dermis for diagnosis in anthropoid. Int. J. Fut. Revol. Comput. Sci. Commun. Eng. **3**(10), 242–247 (2017). http://www.ijfrcsce.org
4. Rozita, J., Hadzli, H., Mohd Nasir T.S. S.: Border segmentation on digitized psoriasis skin lesion images. In: IEEE Region 10 Conference TENCON 2004, vol. 3, pp. 596–599. Chiang Mai, Thailand (2004). https://doi.org/10.1109/TENCON.2004.1414842
5. Ginni, A., Ashwani, K.D., Zainul, A.J.: Performance measure based segmentation techniques for skin cancer detection. In: Data Science and Analytics. REDSET 2017. Communications in Computer and Information Science, vol. 799. Springer, Singapore, https://doi.org/10.1007/978-981-10-8527-7_20
6. Kyamelia, R., Sheli S.C., Sanjana Ghosh, Swarna, K.D., Proggya, C., Rudradeep, Sarkar.: Skin Disease detection based on different Segmentation Techniques. In: 2019 International Conference on Opto-Electronics and Applied Optics (Optronix), vol. 1, pp. 70–76. Kolkata, India (2019). https://doi.org/10.1109/OPTRONIX.2019.8862403

7. Hina, S., Manshi, S.: Segmentation of skin lesions from digital images using an optimized approach: genetic algorithm. (IJCSIT) Int. J. Comput. Sci. Inf. Technol. **5**(5), 6831–6837 (2014). https://www.ijcsit.com
8. Diego, P., Jonathan, A., John W.B.: Automatic skin lesion segmentation on dermoscopic images by the means of superpixel merging. In: International Conference on Medical Image Computing and Computer Assisted Intervention (MICCAI 2018), pp. 728–736. Granada, Spain (2018). https://doi.org/10.1007/978-3-030-00937-3_83
9. Enas, I., Ewees, A.A., Mohamed, E.: Proposed method for segmenting skin lesions images. In: Emerging Trends in Electrical, Communications, and Information Technologies Proceedings of ICECIT, vol. 569, pp. 13–24. Andhra Pradesh, India (2018). https://doi.org/10.1007/978-981-13-8942-9_2
10. Yau, K.C., Humaira N., Vooi, V.Y., Kim H.Y., Jyh J.T.: Segmentation and grading of eczema skin lesions. In: 8th International Conference on Signal Processing and Communication Systems (ICSPCS), vol. 1, pp. 68–72. Gold Coast, QLD, Australia (2014). https://doi.org/10.1109/ICSPCS.2014.7021131
11. Yau, K.C., Humaira, N., Vooi V.Y., Jyh, J.T.: A two-level K-means segmentation technique for eczema skin lesion segmentation using class specific criteria. In: IEEE Conference on Biomedical Engineering and Sciences (IECBES), vol. 2, pp. 985–990. Kuala Lumpur, Malaysia (2014). https://doi.org/10.1109/IECBES.2014.7047659
12. Fulgencio, N., Marcos, E.-V., Jesus, B.: Accurate segmentation and registration of skin lesion images to evaluate lesion change. IEEE J. Biomed. Health Inf. **23**(2), 501–508 (2019). https://doi.org/10.1109/JBHI.2018.2825251
13. Ashi, A., Ashish, I., Malay, K.D., Viktoria, D., Zoran, I.: Automated computer vision method for lesion segmentation from digital dermoscopic images. IN: 4th IEEE Uttar Pradesh Section International Conference on Electrical, Computer and Electronics (UPCON), vol. 1, pp. 538–542. Mathura, India (2017). https://doi.org/10.1109/UPCON.2017.8251107
14. Al-masni, M.A., Al-antari, M.A., Choi, M.-T., Han, S.-M., Kim, T.-S.: Skin lesion segmentation in dermoscopy images via deep full resolution convolutional networks. Comput. Methods Programs Biomed. **168**, 221–231 (2018). https://doi.org/10.1016/j.cmpb.2018.05.027
15. Attia, M., Hossny, M., Nahavandi, S., Yazdabadi, A.: Skin Melanoma Segmentation using Recurrent and Convolutional Neural Networks. IEEE 14th International Symposium on Biomedical Imaging (ISBI 2017), 1st edn, pp. 292–296. ISBI, Melbourne, Australia (2017). https://doi.org/10.1109/ISBI.2017.7950522
16. Prabhu Chakkaravarthy, A., Chandrasekar, A.: An automatic segmentation of skin lesion from dermoscopy images using watershed segmentation. International Conference on Recent Trends in Electrical, Control and Communication (RTECC), vol. 1, pp. 15–18. Malaysia (2018). https://doi.org/10.1109/RTECC.2018.8625662
17. Cheng, L., Mahmood, M., Jha, N., Mandal, M.: Automated segmentation of the melanocytes in skin histopathological images. IEEE J. Biomed. Health Inf. **17**(2), 284–296 (2013). https://doi.org/10.1109/TITB.2012.2199595
18. Fatemeh, T., Mehdi, F.: Automatic segmentation of skin lesion using markov random field. Canadian J. Basic Appl. Sc. **3**(3), 93–107 (2015). https://www.cjbas.com/archive/CJBAS-15-03-03-03.pdf
19. Alak, D., Dibyendu, Ghoshal.: Human skin region segmentation based on chrominance component using modified watershed algorithm. In: International Multi-Conference on Information Processing (IMCIP 2016), vol. 89, pp. 856–863 (2016). https://doi.org/10.1016/j.procs.2016.06.072
20. Lawand, K.: Segmentation of dermoscopic images. IOSR J. Eng. **4**(4), 16–20 (2014)
21. Smaoui, N., Bessassi, S.: Melanoma skin cancer detection based on region growing segmentation. Int. J. Comput Vision Signal Process. **1**(1), 1–7 (2013)
22. Humaira, N., Yau, K.C., Tsyr, Y.C., Jyh, J.T.: A color space study for skin lesion segmentation. In: IEEE International Conference on Circuits and Systems, pp. 172–176. Kuala Lumpur, Malaysia (2013). https://doi.org/10.1109/CircuitsAndSystems.2013.6671629

23. Anabik, P., Utpal, G., Raghunath, C., Swapan, S.: Psoriatic plaque segmentation in skin images. In: Fifth National Conference on Computer Vision, Pattern Recognition, Image Processing and Graphics (NCVPRIPG), vol. 1, pp. 61–64. Patna, India (2015). https://doi.org/10.1109/NCVPRIPG.2015.7489994
24. Sameena, P., Gopala Krishna Prabhu, K., Siddalingaswamy, P.C.: Hair detection and lesion segmentation in dermoscopic images using domain knowledge. In Medical & Biological Engineering and Computing. Springer (2018). https://doi.org/10.1007/s11517-018-1837-9
25. Adheena, S., Robin, J.: Melanoma detection using statistical texture distinctiveness segmentation. Int. J. Comput. Appl. **127**(15), 1–5 (2015). https://www.ijcaonline.org
26. Mohammad S.E., Hossein, P.: Lesion detection in dermoscopy images using sarsa reinforcement algorithm. In: Proceedings of the 17th Iranian Conference of Biomedical Engineering (ICBME2010), vol. 1, pp. 209–212. Isfahan, Iran (2010). https://doi.org/10.1109/ICBME.2010.5704964
27. Roberta, B.O., Joao Manuel, R.S.T., Norian, M., Aledir, S.P.: An approach to edge detection in images of skin lesions by Chan-Vese model. In: 8th Doctoral Symposium in Informatics Engineering, vol. 1. Porto, Portugal (2013). https://www.researchgate.net/publication/309185901_An_approach_to_edge_detection_in_images_of_skin_lesions_by_chanvese_model_8th_Doctoral_Symposium_in_Informatics_Engineering
28. Javed, K., Aamir, S.M., Nidal, K., Sarat, C.D., Azura, M.A.: Segmentation of Acne lesion using fuzzy C-means technique with intelligent selection of the desired cluster. In: 37th Annual International Conference of the IEEE Engineering in Medicine and Biology Society (EMBC), Vol. 4, pp. 3077–3080. Milan, Italy (2015)
29. Qingli, L., Li, C., Liu, H., Zhou, M., Wang, Y., Guo, F.: Skin cells segmentation algorithm based on spectral angle and distance score. Optics Laser Technol. **74**, 79–86 (2015). https://doi.org/10.1016/j.optlestec.2015.05.017
30. Fatima, R.S., Navid, R., Mehdi, R.: A Novel method for skin lesion segmentation. Int. J. Inf. Sec. Syst. Manage. **4**(2), 458–466 (2015). http://www.ijissm.org/article_559197_b20108fde084b72035849a720e0f6de0.pdf
31. David Powers, M.W.: Evaluation: from precision, recall and f-measure to ROC, informedness, markedness & correlation. J. Mach. Learn. Technol. **2**(1), 37–63 (2011)

Thermomechanism: Snake Pit Membrane

Pushpendra Singh, Kanad Ray, Preecha Yupapin, Ong Chee Tiong, Jalili Ali, and Anirban Bandyopadhyay

Abstract This article introduces the IR sensor in the living organism. Electrodynamics in the living system is aimed to introduce electromagnetic mechanism inside the biological system and correlates the biofunction of living organisms. Here, we have introduced the biological accurate snake pit membrane's modeling, simulation methodologies, and compared it with a dielectric cavity resonator model to understand the dynamic nature and taking the action due to sudden change in environmental conditions of pit membrane. This article shows the key features of the snake pit organ in terms of temperature across the membrane and Carnot cycle's mechanism.

P. Singh (✉) · K. Ray
Amity University Rajasthan, Kant Kalwar, N11C, Jaipur, Delhi Highway, Jaipur, Rajasthan 303007, India
e-mail: singhpushpendra548@gmail.com

K. Ray
e-mail: kanadray00@gmail.com

P. Singh · A. Bandyopadhyay
International Center for Materials and Nanoarchitectronics (MANA), Research Center for Advanced Measurement and Characterization (RCAMC), National Institute for Materials Science, 1-2-1 Sengen, Tsukuba, Ibaraki 305-0047, Japan
e-mail: anirban.bandyo@gmail.com

P. Yupapin
Computational Optics Research Group, Advanced Institute of Materials Science, Ton DucThang University, District 7, Ho Chi Minh City 700000, Vietnam
e-mail: preecha.yu@kmitl.ac.th

Faculty of Applied Sciences, Ton DucThang University, District 7, Ho Chi Minh City 700000, Vietnam

O. C. Tiong
Department of Mathematical Sciences, Faculty of Science, Universiti Teknologi Malaysia, 81310 UTM Skudai, Johor Bahru, Malaysia
e-mail: octiong@utm.my

J. Ali
Laser Centre, IBNU SINA ISIR, Universiti Teknologi Malaysia, 81310 Bahru, Johor, Malaysia
e-mail: jalilali@utm.my

M. S. Kaiser et al. (eds.), *Proceedings of International Conference on Trends in Computational and Cognitive Engineering*, Advances in Intelligent Systems and Computing 1309, https://doi.org/10.1007/978-981-33-4673-4_15

Temperature depended on two distinct layers across the snake's pit membrane; we have demonstrated heat or temperature propagation across the membrane in terms of a thermodynamic cycle and explained its temperature sensitivity. Here, the resonance characteristic of the snake pit organ was studied theoretically.

Keywords Thermal system · EM interaction with biomaterial · Thermal radiation mechanism · Cavity resonator model

1 Introduction

In the animal kingdom, animals and various kinds of species like a snake, vampire bat, blood-sucking bugs can detect the temperature from the surrounding environment by using extraordinary receptor systems that are naturally embedded in their body and receive the information from surrounding space. Some living organs are extremely sensitive for the detection of EM radiation, mechanical vibration, and some are involved with a specific mechanism like detecting light of different wavelengths, sense of smell, etc. For many decades, researchers kept faith in chemical explanations for this ability of IR radiation detection from the prey body. Lynn et al. proposed the theory on the function of pit membrane and concluded that it is specialized for the detection of air molecular vibration and mechanoreceptor [1]. Organs of animal's heads are not employed for the purpose of detection [2] and those are not considered as radiant heat detectors. The pit membrane primarily responds to radiant heat [3] but involves the neuron's characterization hidden inside the membrane. Action potential across the membrane could be recorded by individually stimulating neurons with IR radiation [4]. Facial's pit nerve is composed of many hot fibers. Nerve terminal of pit organ is the origin of action potential for the pit channel. The pit receptor responds very rapidly due to the inclusion of a large number of mitochondria. The changing configuration of mitochondria has been reported [5]. Neuron firing occurs continuously on pit membrane with an irregular interval of time [6]. Spontaneous discharge frequency rate (10–25 spike/s) continuously decreases from the periphery to the telencephalon nervous system inside pit membrane. Spike discharge rate varies with body temperature, species, and involved neurons inside the membrane [7].

Neuron's firing rate increases in the response to higher body temperature and stronger radiation [8] while it decreases along with an object that has a lower temperature than the surrounding space. However, the snake uses the pit membrane for detecting the cool and warm objects. A functioned membrane receptor in rattlesnake and crotaline snake enables them to detect, observe, and hunt the prey at the activation stage (threshold level) of membrane's tissue. Pit membrane channel supplies heat to nerve fibers that are connected to the brain. Thermosensitivity or IR sensitivity of channel varies in various species. 15 μm thick snake pit membrane creates the outer and inner cavity in the loreal region. The loreal region exists at both sides of the face or between the eyes and nostrils of the snake [9]. Snakes could easily detect the temperature difference between the two cavities. Rigorous studies have been reported on how a snake detects the infrared signal and transduces the signal into nerve fiber by available radiation heat on the pit membrane rather than the photochemical transducer [10]. In the snake's nostrils (Fig. 1a), the transcriptional profiling approach (TRPA1-version of a protein) is used to identify IR radiation. IR sensitive TRPA1 channel opens at a threshold level (at $T > 28$ °C) and allows to flow of the ions into the trigeminal nerve, results in electrical triggering in nerve fibers [9]. The membrane quickly allows the heat change and ables to detect the warm blood insects or animals. The receptor system [11, 12] is the key term in the pit organ. Cavity resonator concept in the biological living system [13, 14] is a new way to understand the working function of snake facial pit.

Resonant bands, electric and magnetic fields are the key parameters of biomaterials. Field analysis of biomaterial is useful to overcome many types of diseases. For example, studies related to the effect of magnetic field exposure on blood vessels in micro-vasoconstrictors are unclear. Healthy or living cells will provide different functional behaviors than sick ones. With the use of dielectric and cavity resonator concept [11–15], we tried to explore the information processing inside the biological cell or membrane that performs the basic functions of life form. In addition to detecting the electromagnetic properties of a cell or protein, we can analyze the importance of a designed resonator. Our aim is to explore this particular area because we firmly believe that if these technologies come into the industry, we can enter a second industrial revolution. To the best of my knowledge, this is the new concept that can theoretically detect the electromagnetic features of biological components. Only a few researchers study them but in a very different way. There is no idea that

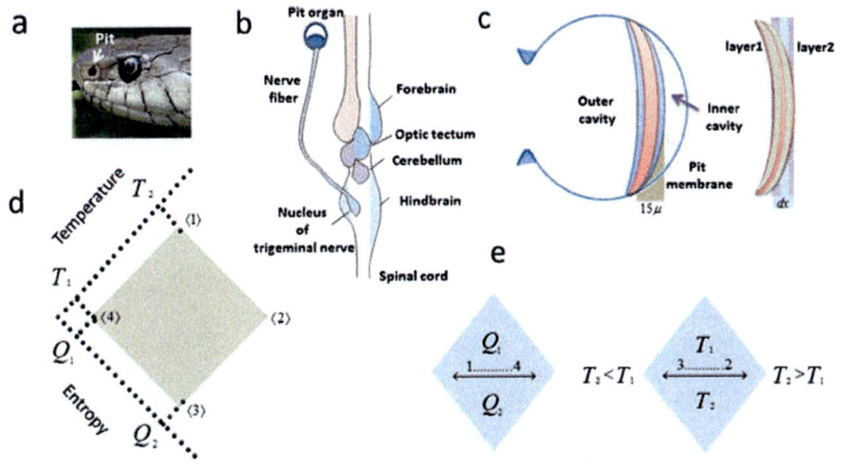

Fig. 1 **a** Pit on the snake's head. **b** Information processing structure inside the snake's brain: **c** Cavity form of snake's pit chamber. **d** Chamber is divided by a 15 μm thick membrane, associates two cavities; outer and inner [16]. **d** Entropy of channel for various temperature/ heat variation. **e** Activation condition diagram of membrane

solve the biological component and obtain the electromagnetic properties. A slight change in biological structure allows it to be easily analyzed by its EM properties. Others compromise the design to solve the problem easily but we are very accurate in mimicking the structure. The resonant frequency, electric and magnetic fields suggested that one could measure the resonant frequency using the frequency range suggested by the simulation, without scanning the entire frequency range in the use of biological materials.

This manuscript is organized into four sections. The material and method for mimicking and simulating pit organ geometry are described in Sect. 2 while Sect. 3 is a discussion of the results obtained from simulation approaches, literature studies, and comparisons between theoretical and experimental approaches. The conclusion of the finding is drawn in Sect. 4.

2 Materials and Methods

Here, we aim to introduce the heating mechanism along the pit membrane in terms of entropy and temperature.

The neural pathway is connected to the infrared sensor organ. Surrounding information is collected by the pit organ and propagated through nerve fibers to the hindbrain, optics tectum, and midbrain structures (Fig. 1b), which display an important role in information processing in pulse form. The cavity form of the snake pit chamber is divided by a 15 μm thick membrane, which connects the two cavities; outer and inner (Fig. 1c). To understand information processing through the membrane, we combined the two layers with the pit membrane at different temperatures (T_1 and T_2) as shown in Fig. 1d, e.

2.1 Theoretical Approach for the Construction of Snake Pit Membrane

We have theoretically created the snake pit membrane channel in antenna simulation software (CST), considering all biological details. Created model assigned by the dielectric materials and stimulated it through a suitable energy source or waveguide port at an appropriate functional region of surface the cavity (Fig. 2a). Pit membrane (15 μm thickness) includes cavity surface. The dimension of waveguide port and port location is selected cautiously or else incorrect choice of wave port's dimensions possibly leaves the extensive simulation time, unbalance electric and magnetic field distribution and their clocking nature, and high order mode for propagating. The boundary conditions are picked as open space in all three directions *x*, *y*, and *z*-axis. Every simulation is carried out in the time-domain approach that utilizes the Maxwell equation to solve the structural configuration. How the geometry is solved using the

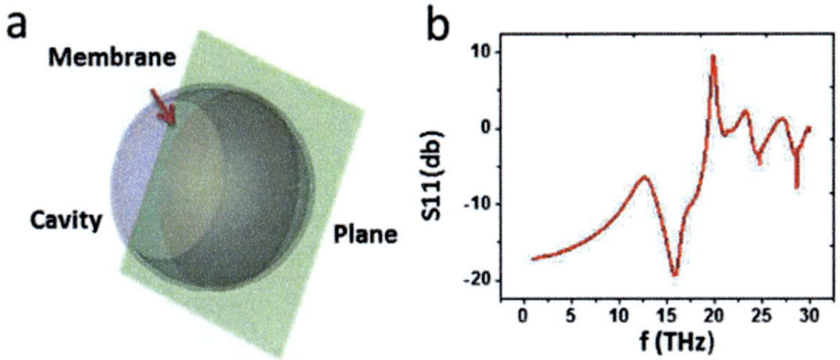

Fig. 2 **a** Mimicked profile of a pit membrane. **b** Simulated resonance spectrum of pit membrane

Maxwell equation in the time-domain solver in CST is described in reference [17]. We scanned the frequency domain from kHz to PHz range for created model and noticed the resonance peak in THz frequency range (15THz) (Fig. 2b). The entire simulation took a few days to detect the output as the resonant frequency. Using the waveguide source, we are able to detect the intensity of the EM signal in terms of reflection (S11) and transmission (S21) coefficients in the selected frequency range. We have considered an energy source of 10 cm × 10 cm dimension for the constructed structure.

2.2 *Technical Details*

Used software: computer simulation technology, selected solver—Maxwell equation solver (time domain), boundary conditions—open space (*X*, *Y*, and *Z*-directions); selected frequency range—0–30THz, waveguide port dimension-10 cm × 10 cm, dominated resonance peaks—15THz, detected electric and magnetic field—15THz (Fig. 2d).

2.3 *Model Geometry*

Our simulation design is based on Fig. 1c. We took the snake nostril as a cavity which is designed to be a 5 mm diameter hollow ball. We cut off the top portion of the sphere from 2 mm diameter of it and formed a tilted layer that lies inside the cavity, and marked its material (electrical permeability, magnetic permeability, and refractive index) similar to organic dielectric material. Such a layer divides the nostril cavity into two sub-cavities; inner cavity and outer cavity. The membrane is excited by the signal, emitted from the wave port applied at the bottom of the constructed geometry, in open boundary conditions in all directions. The waveguide port encloses all region domains of all signal propagation pathways along with the port area. In computer simulation technology, the pick face option automatically selects the port located on the plane surface. We kept constantly changing the positions of the pick face and port dimensions until we found the correct structural resonance peaks.

The cavity and dielectric resonance approach is based on the creation of biological components: The snake's nostril is a hollow sphere that forms the cavity and the components inside it act as a dielectric resonator-like antenna. The membrane of the snake pit is a circular pore, hence a cavity resonator. The pit membrane inside the nasal cavity is a secondary structure and each secondary structure is a dielectric resonator. In this way, the entire constructed geometry is an alternative combination of a cavity and a dielectric resonator. All components are connected over a wide frequency range. Our main goal here is to find out the law of geometric correlation between resonance peaks of all constitute layers of any geometry which convince the proprieties of biological components.

3 Results and Discussion

3.1 TRPA 1 Channel—Dynamic Nature

Pit organs of python, boas, and snake are the sensory devices that have a remarkable feature for the detection of warm-blooded prey. The temperature gets fluctuated by the heat that reflects back and forth across the pit organ. A molecular mechanism was unknown at that time but in the new evidence, the pit organ responds to temperature only on activation of transits receptor potential ankyrin 1 (TRPA1) channel which provides the way to understand the underlying molecular mechanism. Such membrane properties raise many questions like the origin of warm activated channel and sensitivity of such biological receptors [10, 16]. For computation, the radiance and irradiance through the pit channel are reported [16, 18–22]. We have detailed the dynamic nature of heat or temperature transfer mechanism through the TRPA1 channel. A mathematical explanation of temperature variation across the channel is also reported here. Dynamic (open and close) nature of both layers (2 layer associates with membrane) depends on heat or temperature factor. We mimicked the pit membrane with all biological details in the dielectric resonance simulator for verifying the mathematical analysis. The material of the pit membrane was assigned by dielectric material and solved the biological structure in CST. Simulated resonance spectrum has the resonance peak at 15THz (Fig. 2b), which lies in 750 nm–1 mm wavelength or 300 GHz to 400THz frequency range [16].

The zoomed image of the snake's pit membrane is shown in Fig. 1a right panel. For mathematical modeling of the heat transfer equation, we assumed, dT is little temperature variation through small pit membrane width ($\mathrm{d}x$). Then, heat flux (q) through that small portion $\mathrm{d}x$ is given as [23]

$$q = -k \cdot {}^{\mathrm{d}T}\!/_{\mathrm{d}x},$$

k is the proportional constant (Thermal conductivity).

Heat transfer rate

$$Q = q \cdot A$$
$$Q = -k \cdot {}^{\mathrm{d}T}\!/_{\mathrm{d}x} \cdot A$$

A is the area of conducting slit.

The entropy of the system (S) is given as

$$S = Q/T$$
$$S = -k \cdot A \cdot \left({}^{\mathrm{d}T}\!/_{\mathrm{d}x}\right)/T$$

after putting the value

$$S = -k \cdot A \cdot (p/T) \tag{1}$$

(For our convenience, ${}^{\mathrm{d}T}\!/_{\mathrm{d}x} = p$)

Take the derivate of Eq. 1 with respect to temperature (T)

$$\begin{aligned} \frac{\mathrm{d}S}{\mathrm{d}T} &= \frac{-k \cdot A\left(T \cdot p' - p \cdot 1\right)}{T^2} \\ &= -k \cdot A \cdot \left(\frac{T''}{T} - \frac{T'}{T^2}\right) \end{aligned} \tag{2}$$

Again, taking the derivate of Eq. 2 with respect to T

$$\frac{\mathrm{d}^2 S}{\mathrm{d}T^2} = -k \cdot A \cdot \left[\left(T \cdot T''' - T'' \cdot T'\right)/T^2 + \left(T^2 \cdot T'' - T' \cdot 2 \cdot T \cdot T'\right)/T^4\right] \tag{3}$$

Nature of entropy (minimum and maximum value) can be calculated, considering Eq. 2 is equal to 0 that means, ${}^{\mathrm{d}S}\!/_{\mathrm{d}T} = 0$

$$T'' - T'/T = 0$$
$$\left({}^{\mathrm{d}p}\!/_{\mathrm{d}T}\right) - p/T = 0$$
$$p = c \cdot T,$$

$$\left({}^{\mathrm{d}T}/_{\mathrm{d}x}\right) = c \cdot T,$$
$$\mathrm{d}T/T = c \cdot \mathrm{d}x \tag{4}$$

Integrated both sides of this equation yields;

$$\int_{T_1}^{T_2} \mathrm{d}T/T = \int_{x_1}^{x_2} c \cdot \mathrm{d}x$$
$$c = \left[\log(T_2) - \log(T_1)\right] / \left[x_2 - x_1\right]$$

$T = e(c \cdot x)$, c is the constant.

Inserting the value of T in Eq. 3, we got an interesting result.

$\frac{\mathrm{d}^2 S}{\mathrm{d}T^2} = 0$, [If, the value of T is kept in the third derivative of entropy, it will be 0 again]. There is no declaration for minimum and maximum value of entropy for temperature T. If, we put the value of temperature in Eq. 1, then the nature of entropy

$$S = -k \cdot A \cdot c,$$
$$S = -k \cdot A \cdot \left[\log(T2) - \log(T1)\right] / \left[x2 - x1\right],$$

putting the value of c

$$S = -\left(\frac{kA}{(x2 - x1)}\right) / \left[\log(T2) - \log(T1)\right] \tag{5}$$

S value depends on the temperatures. According to [16], the TRPA1 channel (transcriptional profiling approach channels as infrared receptors on sensory nerve fibers that innervate the pit organ) is important for IR sensing, which detects the change in ambient temperature above 30 °C (303 K). Indeed, rattlesnake TRPA was inactive at room temperature but robustly activated above 28.0 ± 2.5 °C (303.5–298.5 K). Interestingly, rattlesnake TRPA is also heat-sensitive, albeit with a substantially higher threshold of 36.3 ± 0.6 °C (309.9–308.7 K).

3.2 *TRPA1 Channel-Carnot Mechanism*

Here, we discussed the entropy of the snake pit membrane. Activation and inactivation nature of TRAP1 channel is defined in terms of entropy and temperature. The working mechanism of the pit membrane can be understood using Fig. 1d: right. Pit membrane layer 1 absorbs the heat radiation (ΔQ) from the prey that is shown in Fig. 1e. This heat transmits to membrane layer 2 through dx transmission path. During the transmission period, a small loss produced in heat, resultant temperature decreases ($T2 < T1$). In that case, the entropy of membrane is minimum (Eq. 5) in an inactive state as shown in Fig. 1e. Figure 1e shows that the pit membrane absorbs that loss at a constant temperature. In the next step, the temperature of layer 2 increases ($T2 > T1$), membrane entropy gets maximum (by Eq. 5). It means membrane activates at above room temperature (Fig. 1e).

By the above discussion, mechanism of the pit membrane antenna is identical as Carnot profile in thermodynamics [23] that is depicted in Fig. 1d.

3.3 *How Does Our Theoretical Finding Coincide with the Experimental Result Reported by Gracheva et al. [16]*

Snakes are capable of detecting warm-blooded prey with their sensing ability of infrared radiation in a wide wavelength range from 750 nm to 1 mm. Adjustment of the visual and thermal image of the prey in the snake's brain enables them to track it. The radiation heating element in the snake's nostrils is the pit organ. In the 750 nm to 1 mm wavelength range, the pit organ or membrane shows the vibrational profile or it resonances in the presence of radiation. In our theoretical study, we mimicked the snake's nostrils model and stimulated it by applying energy source; the pit channel vibrates at a 15 THz frequency or 20 μm wavelength that lies in the 750 nm to 1 mm wavelength range. Our simulation finding is in range with experimental finding. In terms of the activation and deactivation profile of the pit organ or TRPA1 channel, the experimental finding suggests that the TRPA1 channel is active in the 28–37 °C temperature range, whereas it reflects the inactivation nature at room temperature [16]. Here, we are ascertaining these results by detecting the entropy of the pit channel. From Sect. 3.2, the pit channel shows the Carnot mechanism, and it is active above room temperature and inactive below room temperature. In the theoretical case, room temperature is the threshold value for the detection of IR radiation by the pit channel.

The results found in the simulation approach are very close to the experimental approach [16] in terms of the resonance and temperature profile of the snake pit organ.

Literature studies [24, 25] have reported that infrared imaging in the brains of Boid and Crotaline snakes enables them to detect IR radiation emitted by prey. Crotaline snakes have much higher infrared radiation sensitivity than Boid snakes. To sense the IR radiation, a crotaline snake requires very small changes in temperature in the order of 0.003 °C. The temperature sensitivity of crotaline snake with respect to distance is also much higher than that of Boid snakes due to the structural geometry of the pit organ in the crotaline snake [9, 26], its detection distance is 66.3 cm [25, 27]. In our future work, we will try to understand how temperature and IR radiation sensitivity receive changes with distance from a theoretical point of view.

4 Conclusion

The infrared organ of snake, pit vipers, and pythons are true eye sensors that are based on heat or temperature propagation mechanism. Heat is generated in the pit organ by the interaction of electromagnetic radiation instead of a photochemical reaction. In snake, pit organ opens, acts as an aperture of a camera that allows the receptor to read or sense the IR radiation from the surrounding space. IR sensitive animals successfully sense a virtual image of an object in their brain. This common mechanism exists in other kinds of animals but the pit membrane channel is very less sensitive for an open or close mechanism that serves as a lens. The snake's pit channel is the integrated system of a visual system that captures the electromagnetic radiation, absorbs, and exhibits the dynamic nature (activation or inactivation). Pit organ is the replica of the eye that is useful for the acquisition and detection of the prey.

Acknowledgements The authors express sincere thanks to Prof. S L Kothari, Vice President, ASTIF, AUR, for his support and encouragement.

Contribution KR and AB planned the theoretical study, PS did the theory, PY, JA, and OT reviewed the work. All authors wrote the paper together.

Conflict of Interest Statement The authors declare that they have no conflict of interest.

References

1. Lynn, W.G.: The structure and function of the facial pit of the pit vipers. Am. J. Anat. **49**(10), 97–139 (1931). https://doi.org/10.1002/aja.1000490105
2. Goris, R.C., Terashima, S.I.: Central response to Infrared stimulation of the pit receptor in a crotalite snake *Trimeresurus flavoviridis*. J. Exp. Biol. **58**, 59–76 (1937)
3. Bullock, T.H., Cowl, R.B.: Physiology of an infrared receptor: the facial pit of pit vipers. Science **115**, 541–543 (1952). https://doi.org/10.1126/science.115.2994.541-a
4. Terashima, et al.: Generator potential of crotaline snake infrared receptor. J. Neurophysiol. **31**, 494–506 (1968)
5. Amemiya, et al.: Microvasculature of crotaline snake pit organs: possible function as a heat exchange mechanism. Anat. Rec. **254**, 107–115 (1996). https://doi.org/10.1002/(SICI)1097-0185(19990101)254:1%3c107::AID-AR14%3e3.0.CO;2-Y
6. Hensel, H., Schäfer, K.: Activity of warm receptors in *Boa constrictor* raised at various temperatures. Pflugers Arch. **392**(2), 95–98 (1996). https://doi.org/10.1007/BF00581255
7. Terashima, S., Goris, R.C.: Receptive area of primary infrared afferent neurons in crotaline snakes. Neuroscience **4**(8), 1137–1144 (1979). https://doi.org/10.1016/0306-4522(79)90195-7
8. Goris, R.C., Nomoto, M.: Infrared reception in oriental crotaline snake. Comp. Biochem. Physiol. **23**, 879–892 (1967). https://doi.org/10.1016/0010-406X(67)90348-9
9. Moon, C.: Infrared-sensitive pit organ and trigeminal ganglion in the crotaline snakes. Anat. Cell Biol. **44**(1), 8–13 (2011). https://doi.org/10.5115/acb.2011.44.1.8
10. Sedwick, C.: Elena Gracheva: ion channels run hot and cold. J Cell Biol. **209**(6), 778–779 (2015). https://doi.org/10.1083/jcb.2096pi

11. Singh, P., et al.: Biological infrared antenna and radar. Soft Comput. Theories Appl. **584**, 323–332 (2018). https://doi.org/10.1007/978-981-10-5699-4_31
12. Singh, P., et al.: Fractal and periodical biological antennas: hidden topologies in DNA, wasps and retina in the eye. Soft Comput. Appl. **761**, 113–130 (2018). https://doi.org/10.1007/978-981-10-8049-4_6
13. Singh, P., et al.: DNA as an electromagnetic fractal cavity resonator: its universal sensing and fractal antenna behavior. Soft Comput.: Theories Appl. **584**, 213–223 (2018). https://doi.org/10.1007/978-981-10-5699-4_21
14. Singh, P., Ray, K., Fujita, D., Bandyopadhyay, A.: Complete dielectric resonator model of human brain from MRI data: a journey from connectome neural branching to single protein. Lecture Notes Electr Eng **478**, 717–733 (2019). https://doi.org/10.1007/978-981-13-1642-5_63
15. Singh, P., et al.: A self-operating time crystal model of the human brain: can we replace entire brain hardware with a 3D fractal architecture of clocks alone? Information **11**(5), 238 (2020). https://doi.org/10.3390/info11050238
16. Gracheva et al.: Molecular basis of infrared detection by snakes. Nature **464**, 1006–1011 (2010). https://doi.org/10.1038/nature08943
17. Time domain Methods for the Maxwell Equations: https://www.diva-portal.org/smash/get/diva2:8848/FULLTEXT01.pdf
18. Goris, R.C.: Infrared organs of snakes: an integral part of vision. J. Herpetology **45**(1), 2–14 (2011). https://doi.org/10.1670/10-238.1
19. Bullock, T.H., Diecke, F.P.J.: Properties of an infra-red receptor. J. Physiol. **I34**, 47–87 (1956). https://doi.org/10.1113/jphysiol.1956.sp005624
20. Panzano, V.C., Kang, K., Garrity, P.A.: Infrared snake eyes: TRPA1 and the thermal sensitivity of the snake pit organ. Sci Signal. **pe22** (2010). https://doi.org/10.1126/scisignal.3127pe22.
21. Bakken, G.S., Krochmal, A.R.: The imaging properties and sensitivity of the facial pits of pit vipers as determined by optical and heat-transfer analysis. J. Exp. Biol. **210**, 2801–2810 (2007). https://doi.org/10.1242/jeb.006965
22. Zhou, S.A., Uesaka, M.: Bioelectrodynamics in living organisms. Int. J. Eng. Sci. **44**, 67–92 (2006). https://doi.org/10.1016/j.ijengsci.2005.11.001
23. Lienhard, J.H.: Heat Transfer Textbook. Cambridge, Massachusetts (2001). https://doi.org/10.1002/aic.690270427
24. Barrett, R., Maderson, P.F., Meszler, R.M.: The pit organs of snakes. In: Gans, C. (ed.) Biology of Reptilia, pp. 277–314. Academic, London (1970)
25. Campbell, A.L., Naik, R.R., Sowards, L., Stone, M.O.: Biological infrared imaging and sensing. Micron **33**, 211–225 (2002)
26. Newman, E.A., Hartline, P.H.: The infrared "vision" of snakes. Sci Am. **246**, 116–127 (1982)
27. de Cock Buning, T.: Thermal sensitivity as a specialization for prey capture and feeding in snakes. Am. Zool. **23**, 363–375 (1983)

Sentiment Analysis on Bangla Text Using Long Short-Term Memory (LSTM) Recurrent Neural Network

Afrin Ahmed and Mohammad Abu Yousuf

Abstract Sentiment analysis has become an important source in people's trust. Search engine statistics and the local survey show 92% of consumers trust online reviews as much as personal recommendations. In this paper, classification of different emotions within the text using advanced text analysis techniques has been proposed. This work intends to deploy long short-term memory (LSTM) deep recurrent network for sentiment analysis on Bangla text as it is developed to avoid long-term dependency. A small dataset of Bangla sentences has been developed and stratified. This work is going to show the effects of hyperparameter tuning and the ways it can be helpful for sentiment analysis on the dataset. The addressed approach is useful for analyzing common expressions, like positive or negative. The goal of this study is to establish a sentiment classification framework to analyze the performance of different deep learning models with a variety of parameter calibration combinations. The proposed LSTM model with advanced layers achieves better performance on resolving sentiment polarity for aimed entities with an accuracy of 94%. This result intends to help psychologists and researchers to identify customer sentiment toward products, detect emotions of individuals from their social activities in the virtual world, and help them to take necessary steps to prevent undesirable doings.

Keywords Sentiment classification · Bangla dataset · LSTM · Machine learning · Dropout

A. Ahmed (✉) · M. A. Yousuf
Jahangirnagar University, Dhaka 1342, Bangladesh
e-mail: afrinahmed328@gmail.com

M. A. Yousuf
e-mail: yousuf@juniv.com

M. S. Kaiser et al. (eds.), *Proceedings of International Conference on Trends in Computational and Cognitive Engineering*, Advances in Intelligent Systems and Computing 1309, https://doi.org/10.1007/978-981-33-4673-4_16

1 Introduction

People nowadays share their thoughts on different social platforms through comments and reviews. To scrutinize the substantial amount of information intelligent system is required to classify these data into predefined classes. It is a precious source of knowledge to get the overall emotions. This is of great help to make the decisions well-informed. People's keen interest in opinion mining has become a challenge for language technology to achieve better results. Automatically classifying text as negative or positive sentiment or subjectivity seems so complex that individual reviewers oppose on a certain document. It is mostly affected by their thoughts, beliefs, and cultural factors. Nearly, 226 million people around the world speaks the language Bangla and a huge part of them use the internet [1]. So extracting hidden opinions in Bangla text is important for sentiment detection. In this paper, LSTM is applied to measure different sentiment levels that can be found in a source text. Lately, it has been stretched out to address issues like recognizing objective from subjective suggestions and deciding the sources and points of different suppositions communicated in text.

2 Literature Review

Sentiment analysis is considered to be a rapidly expanding research area in natural language processing. It is challenging to track all the activities in this field. The literature review utilizes opinion mining as well as qualitative development and analyzes numerous published papers from different sources. It is found that its roots are at the beginning of the twentieth century. However, sentiment can be analyzed only with the presence of subjective texts on the internet.

2.1 Traditional Machine Learning Models

Hossain et al. proposed a system called contextual valency where a virtual robot that can interact in Bangla and exhibit reflective sentiment using ML and Naive Bayes classifier [2]. The out of core learning By Hasan et al. uses the multinomial Naive Bayes method which is related to counting to analyze sentiment, a supervised learning method with an accuracy of 87% [3]. Bavithra Matharasi and Senthilrajan proposed the Naive Bayes model with a unigram approach, designed using windows forms that reads an input file containing the dataset according to the users choice [4].

Baccianella et al. developed SentiWordNet, an opinion lexicon acquired from the popular dataset called WordNet. Their technique appoints each WordNet synset three analytical scores [5]. Sentiment detection from Bangla text By Azharul Hasan et al. proposed contextual valency analysis that identifies parts of speech to assign contextual sentiment valence [6]. Kerstin Denecke proposed multilingual sentiment analysis where a document is looked for adjectives, verbs, and other parts of speech that holds different sentiment. Employing SentiWordNet detects the level of positivity and negativity hidden in words. The approach achieves 68% accuracy for the IMDb dataset [7].

A paper on depression analysis By Uddin et al. selected four hyperparameters, beneficial for configuring GRU models to gain 75.7% accuracy on a significantly smaller dataset [8]. Hassan et al. proposed sentiment analysis on romanized Bangla text where a total of 32 different experiments were based on the same model with only differences in the dataset used, achieving an accuracy of 78% [9]. SVM showed quite a potential in ML, for example, compound character recognition on advanced features set by Kibria et al. [10].

Li and Xue proposed a CNN model along with a gating mechanism. Gated Tanh-ReLU Units particularly output some sentiment features according to provided details with an accuracy of 85.9% [11]. Ghosh et al. in their review paper [12] proved that CNN works better for ML projects like character recognition. Alam et al. proposed a model generated by CNN. The classifier model obtains an accuracy of 99.87% [13].

2.2 LSTM-Based Sentiment Analysis Research

Uddin et al. proposed an LSTM layer-based approach with hyperparameter tuning adjustment technique to classify emotions. On experimentation, the technique was observed to produce the highest accuracy of 86.3% [14]. Li and Qian applied LSTM to achieve multiclassification for text with improved RNN language model [15]. Tholusuri et al. extracted the sentiment of the text using LSTM on the IMDB movie reviews dataset and achieved an accuracy of 86.85% [16]. Alhagry et al. worked on a deep learning method to recognize emotion from raw EEG signals [17].

3 Dataset

In machine learning, classifiers are often contrasted against the optimal Bayesian decision rule, so there is an exponential increase in the difficulty as the number of input features is increased. That is why a substantial amount of data is required.

3.1 Data Collection

Very few complete and publicly available datasets can be found regarding the work. A new Bangla text dataset has been created with little help from some existing datasets available for sentiment classification as shown below.

3.1.1 Prothom Alo

The dataset has been developed with the help of Prothom Alo news headlines dataset, available on Kaggle. It contains all the news contents of Prothom Alo from 2013âŁ“2019. It has over 36,000 headlines in the form of a comma-separated file. This dataset has been enhanced by converting each headline into a complete ideal Bangla sentence. Due to repetition and lack of subjectivity, only 15% of data could be used from the aforementioned dataset.

3.1.2 Bangla Aspect-Based Dataset

The paper on aspect-based sentiment analysis by Rahman et al. uses Bangla datasets to estimate text baseline. They offered two datasets to perform the sentiment analysis in Bangla. One of the datasets comprises comments on cricket matches and news and another dataset contains customer reviews on foods and restaurant environments. There is a description of a baseline approach for aspect category classification to examine the datasets. Due to the repetition and lack of subjectivity, only 20% of data could be used from the aforementioned dataset (Table 1).

3.2 Data Labeling

Labeling data is a notable preprocessing step for machine learning, particularly for supervised learning. Here, all data are labeled to provide a learning basis for further processing. Here, data labeling (positive as 1 and negative as 0) on a CSV file is used to allow artificial intelligence to find the difference as shown in Fig. 1.

3.3 Data Preprocessing

An important step in a sentiment classification is text preprocessing, but it is often underestimated and not extensively covered in literature. In this work, the importance of preprocessing techniques is highlighted to show how they can improve system accuracy.

Table 1 Source of data for the study

Dataset	Total	Usable (%)	Source
Prothom Alo	36,000	15	[18]
Cricket	3000	20	[19]
Resturant	3000	20	[20]

1	Text	Sentiment
2	সংসদে হত্যা মামলার রায় কাল অনুষ্ঠিত হবে	neutral
3	জাহাজভাঙা শিল্পে ১১ মাসে নিহত ১৯ শ্রমিক	negative
4	বাজেটে শিক্ষা, স্বাস্থ্য ও শিশু সুরক্ষায় গুরুত্ব প্রয়োজন।	neutral
5	ভিডিও চিত্রে ধরা পড়ল কলেজছাত্রকে নির্মমভাবে হত্যার দৃশ্য ।	negative
6	সাড়ে ৫০ লাখ টাকার ক্ষতি হয়েছে এই মাসে পোশাক শিল্পে	negative
7	নালিতাবাড়ীতে ট্রাকচাপায় স্কুলছাত্র নিহত	negative
8	স্ত্রীর করা যৌতুক মামলায় স্বামীর কারাদণ্ড	neutral
9	সেপটিক ট্যাংকে স্ত্রীর মরদেহ, স্বামী গ্রেপ্তার	negative
10	৪টি নতুন সেতু নির্মাণ করায় যোগাযোগ ব্যবস্থার অনেক উন্নতি হয়েছে ।	positive
11	খালিদীর বিরুদ্ধে অভিযোগ তদন্তে এলআরজি সিইওকে তলব	neutral
12	দেশে করোনায় মৃত্যুর সংখ্যা বেড়েই চলেছে ।	negative
13	খুনের পর কুপিয়ে কিছু মৃতদেহ বিকৃত করে জঙ্গিরা	negative
14	আজ সারা দেশে আনন্দের সাথে উদযাপিত হচ্ছে পহেলা বৈশাখ ।	positive
15	সিলেটে কাল থেকে শুরু হতে হচ্ছে ১০ দিন ব্যাপী বইমেলা উৎসব ।	positive
16	খালাস পাওয়া আসামির ব্যাপারে পুলিশ আপিল করবে ।	neutral
17	হোলি আর্টিজানের মামলার রায়ে আইনমন্ত্রীর সন্তোষ ।	positive
18	ট্রেনে কাটা পড়ে খাদ্য গুদামের নিরাপত্তা প্রহরী নিহত	negative
19	সাতক্ষীরায় নির্মাণ হলো এক অপূর্ব পর্যটন কেন্দ্র ।	positive
20	রায়ে সন্তুষ্ট দেশের সকল জনগন ।	positive

Fig. 1 Sentiment labeling on a csv file

I. *Expression Removal*
   Bangla text has some commonly used expressions like : ? , " etc. Those expressions need to be removed from text before tokenization and sequence conversion. Lambda function is applied to replace those expressions.

II. *Tokenization*
   Keras tokenizer is used to convert text into a token. The embedding layers in Keras require integer encoded input data. Here, every single word is replaced by a unique integer value. Keras utils convert a class vector to the binary class matrix. The tokenizer is initialized with a 5000-word limit. Next fit_on_texts is called to create associations of words and numbers.

III. *Word Embeddings*
   Word embeddings map word into a vector. It can be trained on a huge data set and is capable of capturing the context of a word in a document. Calling text_to_sequence replaces the words in a sentence with respective associated numbers. For word embedding, word2vec has been used. It tokenizes words into numerical values. In this process, frequently used words with little effect on polarity lie in the same region in multidimensional vector space. As a result, they get the foremost word_index values.

IV. *Padding*
   The converted sequence is added with a pad sequence that contains the array of the text sequence with a 32-bit integer. The model requires inputs to have equal lengths. This is done by calling the pad_sequence method with a length of 200.

V. *Outliers Removal*
Outliers are data points that vary significantly from specific considerations. An outlier occurs due to computational variability or to indicate the hypothetical error. Here, some data are excluded from the expected dataset using a semantic measure of sentiment and subjectivity. Defining sentence length helps us to get rid of extremely long or short sentences.

VI. *Dataset Split*
To achieve sentiment value from the dataset, Pandas dummies function is used. The dataset is partitioned into three parts, for the test, 20% of total data is used with 42 random states, and 10% for validation. Remaining 70% is used for training the model.

4 Methodology

The classification model uses the dataset containing 10,000 complete Bangla sentences. The dataset has been sampled according to the polarity proportions 1 to be positive and 0 to be negative. The environment picked for the classification model is Jupyter Notebook. It is a powerful, open-source software that lets users build statistical designs, analytical simulation, data visualization, machine learning, and a lot more. The features were generated using powerful libraries like Pandas, Keras, TensorFlow, sci-kit learn, etc. Figure 2 shows the research workflow including the relevant steps that are described below.

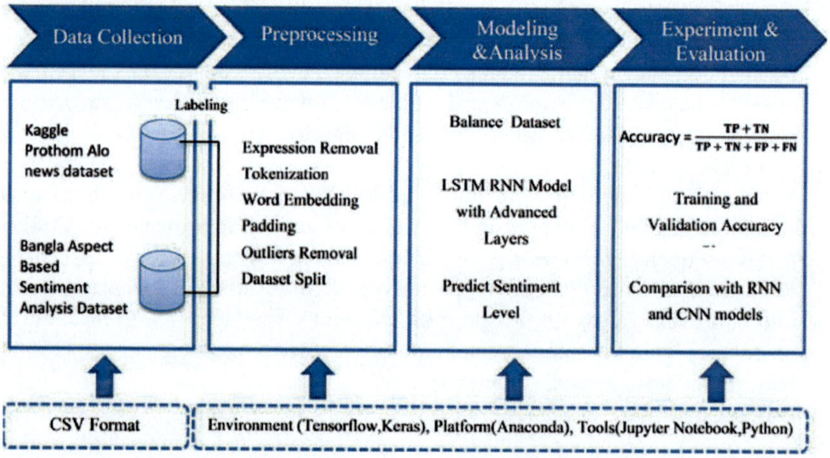

Fig. 2 Research workflow and processes

4.1 *Modeling and Analysis*

The proposed system is implemented using LSTM, an unsupervised learning method. However, these neural networks are actually trained on supervised learning techniques, the term known as self-supervised. They are prepared as part of one greater model that tries to regenerate the input. The main purpose of the system is to design an ML architecture to detect sentiment levels and achieve better accuracy.

4.1.1 Classified LSTM Network

Long Short-term memory network is a certain type of recurrent neural network widely used in the field of AI. They are qualified for learning long-term dependency and work extremely well on a range of problems. It is their unique behavior to remember information for long periods [21]. LSTM can include and eliminate data to cell states, controlled by the gates. Gates constituted out of pointwise multiplication procedure and sigmoid neural network layer. Frequently used words like এই, যে, সে, সেদিন, কত lie in the same region in multidimensional vector space and get the foremost word_index values. So, their weights have been decreased to get a better result. While creating a vector of candidate values tanh layer will throw away this information from the cell state. As a result, it will not affect the polarity. All RNN has a chain-like structure that repeats the segment of neural networks shown in Fig. 3.

To secure and supervise the cell states, LSTM has three types of gates.

I. *Input Gate*—detect the value from the input that should be used to improve the memory. Sigmoid function determines whether to let through 0 or 1 where the tanh function offers weightage to the values [23]. For frequently used words lower word_index values

$$I = \sigma(w.[p_{t-1}, i_t(0.01)] + b) \tag{1}$$

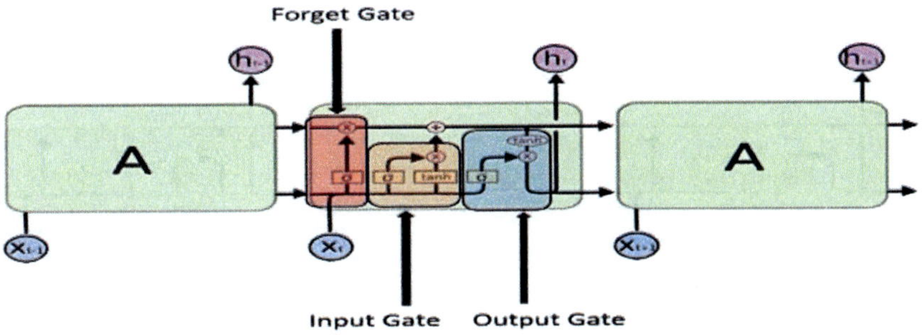

Fig. 3 Repeating segment in LSTM with interacting layers [22]

And for other word_index values,

$$I = \sigma(w.[p_{t-1}, i_t] + b) \quad (2)$$

$$C = \tanh(w.[p_{t-1}, i_t] + b_2) \quad (3)$$

II. *Forget Gate*—discover what details to be discarded. It looks at the previous state(h_{t-1}) then returns a value between 0 and 1 for individual cell

$$F = \sigma(w.[pt - 1, i_t] + b_3) \quad (4)$$

III. *Output Gate*—input gate value and memory states are required to determine the output. The tanh function provides weightage to value that are swept deciding their degree of significance and multiplied with the output of Sigmoid

$$O = \sigma(w.[pt - 1, x_t] + b_4) \quad (5)$$

$$H = O * \tanh(C) \quad (6)$$

The embedding dimension has been set to 32 and maximum text input length is set to 200 characters with padding sequence shape.

4.2 Experiment and Evaluation

It is important to save each step of the model performance history. For iteration, a set of 5 epochs with 32 batch size has been fixed. In SpatialDropout layer, 0.25 ratio is used and LSTM cell is added with 64 hidden units. The model is established with 0.2 recurrent dropouts. In the output layer set, the dense size is 1 with the 'sigmoid' activation function. Training on LSTM with added feedback layers achieves a better performance. The user provides a Bangla text as an input then the model converts the text into a sequence for sentiment prediction. If the model predicts 0, it is negative; otherwise, it predicts positive. All the steps stated above will help to generate this prediction value.

4.3 Fitting and Tuning

One common problem of overfitting occurs when a function is too close fit for a particular set of data. Dropout layers prove to be an effective way to prevent this problem in the proposed model. It is a regularization technique that dodges overfitting by randomly removing neurons while training [24]. As the network is unable to use its full potential for each training sample, it is less likely to overfit. Dropout with a rate of 0.5 has been on the LSTM layers.

Some popular nonlinear activation functions are Tanh, ReLU, Sigmoid, etc. RNN uses a common activation function called sigmoid that generates input values ranging from 0 to 1. It is a continuous function with a limited output range. Analyzing a variety of network architecture and activation functions is a common theme of many research [25].

$$\text{sigmoid}(x) = \frac{1}{1 + e^{x}} \tag{7}$$

A stochastic gradient descent technique based on the adaptive approximation of first as well as second-order moments is Adam optimization. According to Kingma et al. [26], this technique is computationally effective with little memory concern. It is convenient for a substantial amount of parameter problems. This technique does not require any argument and returns the learning rate defaults to 0.001. The applied loss function in this model is binary_crossentropy which is independent individual vector element.

5 Result and Performance Analysis

Finally, the user can input some Bangla text like "তোমার পরীক্ষার ফলাফল ভাল হয় নি।" and the model will apply all the functional steps stated in the above sections. LSTM model with the help of extra feedback layer and sigmoid function will generate a value between 0 and 1 and show the result as ‘negative’ for the above text. Some sample results are shown in Table 2 and Fig. 4 shows the accuracy of the proposed model.

The prepared dataset has been applied against some popular machine learning models that proved to be working on Bangla documents to detect their sentiment level. LSTM seems to outperform all of them for this small dataset as shown in Table 3

Table 2 Test against custom samples

No.	Sentence	English version	Proposed model
1	তোমার পরীক্ষার ফলাফল ভাল হয় নি ।	Your exam result is not good.	Negative
2	আজকাল বড়ই অবসাদ লাগে ।	Now a days I feel exhausted.	Positive
3	দেশে আইন বলতে কিছুই নেই।	There is no law in this country	Negative
4	অনেক দিন পর একটা লম্বা ছুটি পেলা ।	Got a long vacation after ages	Positive

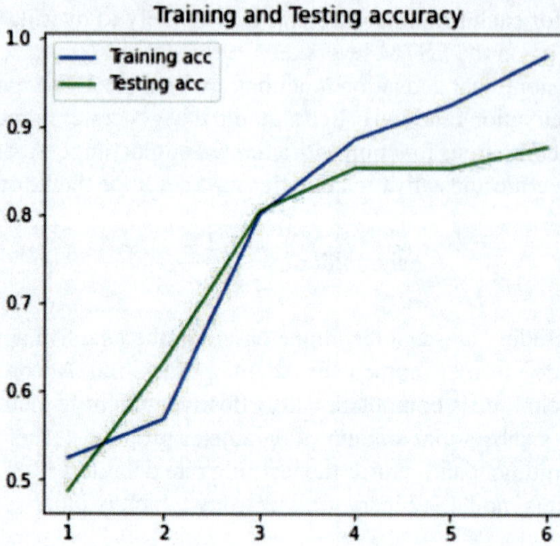

Fig. 4 Graph showing accuracy of the proposed model

Table 3 Test against custom samples

No.	Model	Accuracy (%)
1	Proposed LSTM model	94
2	RNN model	78
3	SVM model	81
4	CNN model	84

The proposed model achieves an accuracy of nearly 94% which proves to be better compared to the others.

6 Conclusion

Even though Bangla is the sixth most spoken languages in the world, there is an insufficiency of standard dataset and well-established model for analyzing sentiment. This research is to fill the gap to some extent. LSTM, a deep recurrent model is applied to analyze Bangla text and predict human emotion, achieve higher accuracy. The result implies that better accuracy can be achieved for small datasets on complex psychological tasks by tuning the deep recurrent model. This work will be helpful for researches in depression and emotion analysis and encourage more future work on Bangla datasets.

References

1. Hasan, K.M.A., Mondal, A., Saha, A.: Recognizing Bangla grammar using predictive parser. arXiv:1201.2010. https://doi.org/10.5121/ijcsit.2011.3605
2. Hossain, M.Y., Hossain, I., Banik, M., Hossain, M.I., Chakrabarty, A.: Embedded system based Bangla intelligent social virtual robot with sentiment analysis (2019). https://doi.org/10.1109/ICIEV.2018.8641023
3. Hasan, M., Islam, I., Hasan, I.: Sentiment analysis using out of corelearning. In: 2019 International Conference on Electrical, Computer and Communication Engineering (ECCE), pp. 1–6 (2019). https://doi.org/10.1109/ECACE.2019.8679298
4. Matharasi, P.B., Senthilrajan, D.A.: Sentiment Analysis of Twitter Datausing Naïve Bayes with Unigram Approach, vol. 7, no. 5, p. 5 (2017). http://www.ijsrp.org/research-paper-0517.php?rp=P656402
5. Baccianella, S., Esuli, Sebastiani, F.: SentiWordNet 3.0: an enhanced lexical resource for sentiment analysis and opinion mining. In: Proceedings of the Seventh International Conference on Language Resources and Evaluation, ELRA (2010)
6. Hasan, K.M., Rahman, M., Badiuzzaman: Sentiment detection from banglatext using contextual valency analysis. In: 2014 17th International Conference on Computer and Information Technology, ICCIT https://doi.org/10.1109/ICCITechn.2014.7073151
7. Denecke, K.: Using sentiwordnet for multilingual sentiment analysis. In: 2008 IEEE 24th International Conference on Data Engineering Workshop, pp. 507–512 (2008). https://doi.org/10.1109/ICDEW.2008.4498370
8. Uddin, A.H., Bapery, D., Mohammad Arif, A.S.: Depression analysis of Bangla social media data using gated recurrent neural network. In: 2019 1st ICASERT
9. Hassan, A., Mohammed, N., al Azad, A.K.: Sentiment analysis on Bangla and Romanized Bangla text using deep recurrent models. In: 2016 International Workshop on Computational Intelligence (IWCI) (2016). https://doi.org/10.1109/IWCI.2016.7860338
10. Kibria, R., Ahmed, A., Firdawsi, Z., Yousuf, M.A.: Bangla compound character recognition using support vector machine (SVM) on advanced feature sets. In: 2020 IEEE Region 10 Symposium (TENSYMP), pp. 965–968. Dhaka, Bangladesh (2020). https://doi.org/10.1109/TENSYMP50017.2020.9230609
11. Xue, W., Li, T.: Aspect based sentiment analysis with gated convolutional networks. In: Proceedings of the 56th Annual Meeting of the Association for Computational Linguistics. Melbourne, Australia), Association for Computational Linguistics (2018). https://doi.org/10.18653/v1/P18-1234
12. Ghosh, T., Abedin, M.M., Mahmud Chowdhury, S., Yousuf, M.S.: A comprehensive review on recognition techniques for Bangla handwritten characters. In: 2019 International Conference on Bangla Speech and Language Processing (ICBSLP), pp. 1–6. Sylhet, Bangladesh (2019). https://doi.org/10.1109/ICBSLP47725.2019.202051
13. Alam, M.H., Rahoman, M., Azad, M.A.K.: Sentiment analysis for Bangla sentences using convolutional neural network. In: 2017 20th International Conference of Computer and Information Technology (ICCIT), pp. 1–6 (2017). https://doi.org/10.1109/ICCITECHN.2017.8281840
14. Uddin, A.H., Bapery, D., Arif, A.: Depression analysis from social mediadata in Bangla language using long short term memory (lstm) recurrent neural network technique (2019). https://doi.org/10.1109/IC4ME247184.2019.9036528
15. Li, D., Qian, J.: Text sentiment analysis based on long short-term memory. In: 2016 First IEEE International Conference on Computer Communication and the Internet (ICCCI), pp. 471–475 (2016). https://doi.org/10.1109/CCI.2016.7778967
16. Sentiment analysis using LSTM. Int. J. Eng. Adv. Technol. **8** (2019). https://doi.org/10.35940/ijeat.F1235.0986S319
17. Alhagry, S., Fahmy, A.A., El-Khoribi, R.A.: Emotion recognition based on EEG using lstm recurrent neural network. Int. J. Adv. Comput. Sci. Appl. **8**(10) (2017). https://doi.org/10.14569/IJACSA.2017.081046

18. Prothom Alo Newpaper Headline from 2019 to 2017. Library Catalog:www.kaggle.com. https://www.kaggle.com/twintyone/prothomalo
19. Rahman, M., Dey, E.: Datasets for aspect-based sentiment analysis in Bangla and its baseline evaluation. Data **3**, 15 (2018). https://github.com/AtikRahman/Bangla_ABSA_Datasets
20. Rahman, A.: AtikRahman/Bangla_absa_datasets, June 2020. original-date: 2018-03-18T05:35:34Z. https://github.com/AtikRahman/Bangla_ABSA_Datasets
21. Understanding LSTM Networks—colah's blog
22. A Crash Course in Sequential Data Prediction using RNN and LSTM. https://mc.ai/a-crash-course-in-sequential-data-prediction-using-rnn-and-lstm/
23. Mittal, A.: Understanding RNN and LSTM (2019)
24. Srivastava, N., Hinton, G., Krizhevsky, A., Sutskever, I., Salakhutdinov, R.: Dropout: a simple way to prevent neural networks from overfitting. J. Mach. Learn. Res. **15**, 1929–1958 (2014)
25. Deng, L., Hinton, G., Kingsbury, B.: New types of deep neural network learning for speech recognition and related applications. In: Edition: IEEE International Conference on Acoustics,Speech, and Signal Processing (ICASSP) (2013)
26. Kingma, D.P., Ba, J.: Adam: A Method for Stochastic Optimization. arXiv:1412.6980 [cs] (2017). arXiv: 1412.6980

Comparative Analysis of Different Classifiers on EEG Signals for Predicting Epileptic Seizure

M. K. Sharma, K. Ray, P. Yupapin, M. S. Kaiser, C. T. Ong, and J. Ali

Abstract Epilepsy is a neurological disease that's characterized by perennial seizures. In this neurological condition the transient electrical phenomenon within the brain occurs that produces an amendment in sensation, awareness, and behavior of an individuals that leads to risk. To understand the brain behavior Electroencephalogram (EEG) signals are used in six different sub-bands viz. Alpha (α), Beta (β), Gamma1 ($\gamma 1$), Gamma2 ($\gamma 2$), Theta (θ) and Delta (δ). The Brainstorm software is used for visualizing, analyzing and filtration of EEG signals in each sub-band. This paper deals with the extraction of the various features in each sub-bands and

M. K. Sharma (✉)
Amity School of Engineering and Technology, Amity University Rajasthan, Kant Kalwar, NH11-C, Delhi Highway, Jaipur, Rajasthan 303002, India
e-mail: vmanojsharma@gmail.com

K. Ray
Amity School of Applied Sciences, Amity University, Kant Kalwar, NH11-C, Delhi Highway, Jaipur, Rajasthan 303002, India
e-mail: kray@jpr.amity.edu

P. Yupapin
Computational Optics Research Group, Advanced Institute of Materials Science, Ton Duc Thang University, District 7, Ho Chi Minh City, Vietnam
e-mail: preecha.yupapin@tatu.edu.vn

Division of Computational Physics, Institute for Computational Science, Ton Duc Thang University, Ho Chi Minh City 700000, Vietnam

M. S. Kaiser
Institute of Information Tech., Jahangirnagar University, Dhaka, Bangladesh
e-mail: mskaiser@juniv.edu

C. T. Ong
Department of Mathematical Sciences, Faculty of Science, Universiti Teknologi Malaysia, 81310 UTM Skudai, Johor Bahru, Malaysia
e-mail: octiong@utm.my

J. Ali
Faculty of Science, Institute of Advance Photonic Science, Universiti Teknologi Malaysia, Johor Bahru, Malaysia
e-mail: jalilali@utm.my

M. S. Kaiser et al. (eds.), *Proceedings of International Conference on Trends in Computational and Cognitive Engineering*, Advances in Intelligent Systems and Computing 1309, https://doi.org/10.1007/978-981-33-4673-4_17

different Machine Learning classifiers were used on these extracted features for comparative analysis in terms of Accuracy, prediction Speed and training time in MatLab. The various statistical and spectral methods are applied on EEG signals to obtained the distinct features in each sub-band. After compared these classifiers on the performance parameters.we have 8 best classifier trained Models that were utilized in checking effectiveness to clearly distinguish between Epileptic and Normal cases.

Keywords Epilepsy · Seizure · Electroencephalogram (EEG) · Spectral analysis · Preictal · Interictal and ictal

1 Introduction

Epilepsy impacts 1% or 60 million people of the world's populace and 25% of sufferers cannot be totally controlled by means of present day medical or surgical treatments [14]. Seizure is a random, sudden, extreme and uncontrolled neurological brain disorder. This epilepsy consists of spontaneous seizures that can last from a few seconds to 2 min or longer. Thus seizure prediction systems will help in the living condition of epileptic patient. The early prediction of epileptic seizure will subdue the life risk of an individual [5, 12].

EEG signals are rhythmic traditionally characterized by their relative 100 μV amplitude and unique frequency range. The brain waves primarily are classified into six sub-bands of frequencies. Delta (δ), Theta (θ), Alpha (α), Beta (β), Gamma1 ($\gamma 1$) and Gamma2 ($\gamma 2$). These kinds of systems are developed on the feature extraction from EEG signal from different sub-bands and that are later used as structured data inputs in various classification methods [3]. Analysis starts by observing the changes in EEG patterns during *Pre-ictal* refers to the state right before the true seizure or stroke and *Ictal* refers to a physiologic state or event such as a seizure.

Though researches on seizure detection are underway for past 40 years, works on prediction got momentum only by late 90s of previous century. Different methods based on time-frequency and wavelet domain features were introduced for both prediction and detection [8]. Many of the previous works have been carried out to improve the accuracy, sensitivity and false prediction rate for the detection. Still no method could classify different types of EEG classes in seizure detection problems within an efficient way to compute its respective classes with 100% accuracy.

The method and materials used in this paper are described in Sect. 2. The filtration of EEG signals for feature extraction into sub-bands is explained in Sect. 3 in brief. Section 4 of this paper describes methods that were used in features extraction from the sub-bands. Section 5 explains the steps required to feed the extracted features to the classification algorithm that finally classify the Epileptic and Normal cases. Section 6 is describing the comparison of classification algorithm on the basis of the Accuracy, Prediction Speed and training time that were used for training the classifier models. Their performance on the variation of K-Fold and Holdout methods were assessed. Then the 8 best classifiers are selected and used for the testing purposes.

Section 7 is showing the results obtained for unseen dataset. Section 8 conclude the findings that effects the performance on the variation of cross validation methods.

2 Materials and Methods

In this paper we have used the open database from CHB-MIT, EEG signals of scalp of epileptic patients, this can be downloaded from the PhysioNet website: http://physionet.org/physiobank/database/chbmit/. These recordings are grouped into 23 cases, that were obtained from 22 subjects of different ages (5 males, ages 3–22; and 17 females, ages 1.5–19). EEG signals were recorded from the scalp sensor arrangement as per International 10–20 System with 23 channels. All the information about the number seizures event occurred, timings of occurrence, the gender and age of each subject were given in file name SUBJECT-INFO [15]. A method used for classification in this Paper is described in the flow chart shown in Fig. 1.

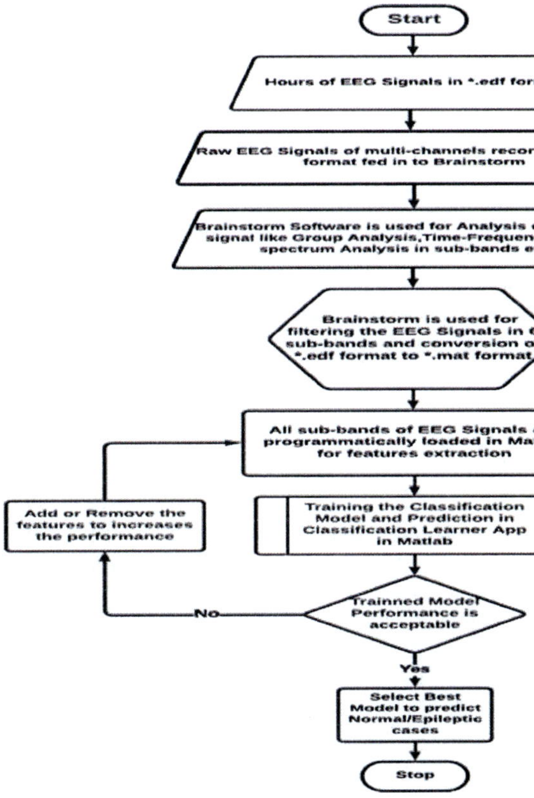

Fig. 1 Flow chart of the method

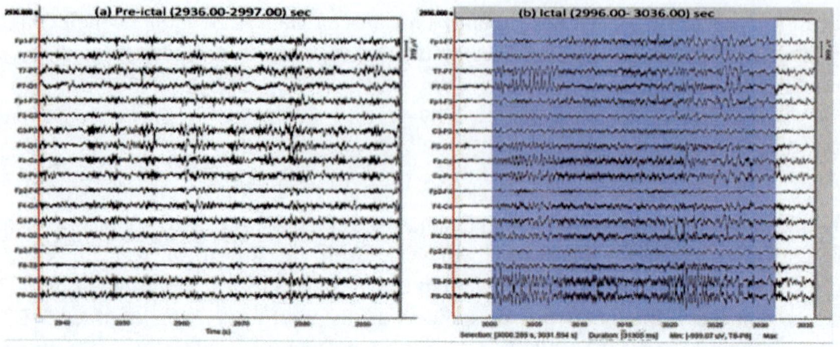

Fig. 2 Brainstorm software showing **a** Pre-ictal (2936.00-2997.00) sec and **b** Ictal (2996.00-3036.00) sec

The database of EEG signals are in the **.edf* (European Data Format) this is used to exchange and storage medical time series. All signals were sampled at 256 samples per second with a 16-bit resolution. Most files contain 23 EEG signals [4]. This file format that firstly loaded and analyzed in Brainstorm software for segmented and the filtering process. The seizure of different states as described *Pre-ictal (2936.00-2997.00) sec and Ictal (2996.00-3036.00) sec* as shown in the Fig. 2.

3 EEG Signals Analysis and Filtering Through Brainstorm Software

Here, we have used Brainstorm software which is a collaborative and open-source software dedicated to the analysis of various bio-signals obtained in different formats like EEG, iEEG, EMG, fNIRS and animal invasive neurophysiology.

We have used Matlab version R2018b for smoother graphics of Brainstorm software. Hours of EEG data recordings in *.edf format were loaded and segmented into two states i.e (60 s prior to the seizure *pre-ictal* state and 60 s in the *ictal* state) for all patients. Further these time series are pre-processed into six sub-bands in the respective states. This has been achieved by the batch process in the Brainstorm software [11]. Finally the output were saved into the **.mat* (Mictosoft Access Table) format are shown in Fig. 3.

4 Methods of Features Extraction

Now, these 182 seizures records were filtered in six sub-bands for two states pre-ictal and ictal. 70% i.e. 126 records were utilized for training and 30% i.e. 56 records were kept aside for the testing the trained model. All 126 **.mat* files each one is containing

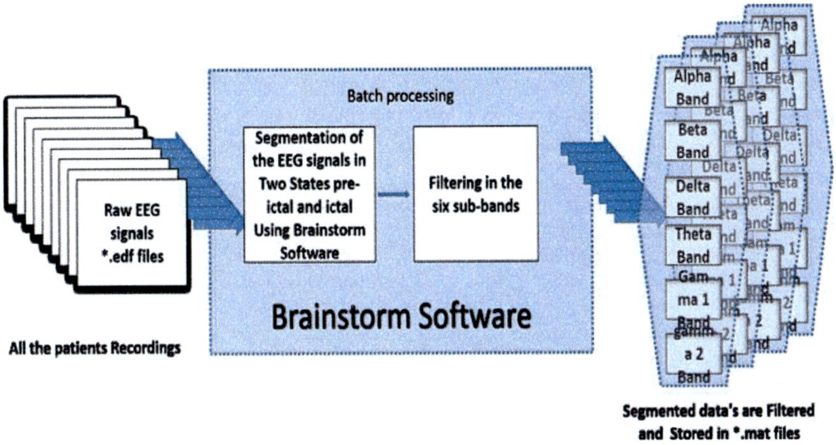

Fig. 3 Conversion flow of raw EEG signals into filtered data in brainstorm

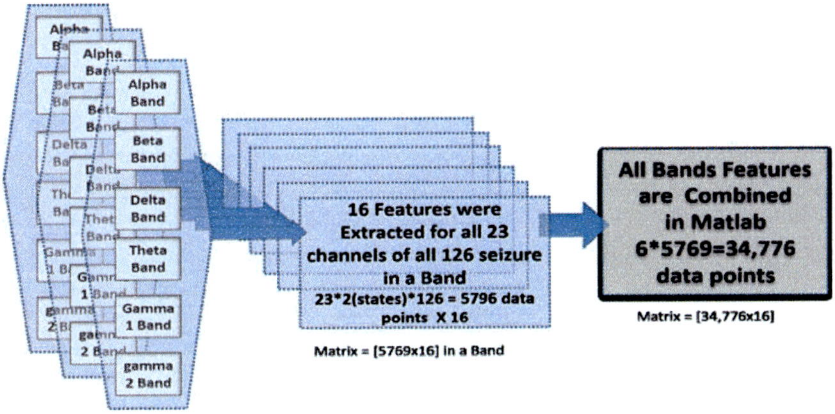

Fig. 4 All features matrix in each sub-bands are combined in the final features matrix

23 channel in each sub-band, these files are separately processed for extracting 16 features in to the Matlab through scripting. The flow of the process is shown in Fig. 4.

4.1 Power Spectral Analysis

EEG waveform of frequency components contains more details and obtained by the spectral analysis which uses the Fourier transform [7]. An equal-approximate distortion created by the classical techniques are overcome by Modern Spectral Analysis Methods [9]. These methods are most successful in analyzing short-term signals. The famous technique for evaluating the segmented EEG signals is AR *(Auto-Regressive)* model. To measure noise-corrupted power and signal frequency, the eigen decomposition is used as correlation matrix [10]. To extracts the spectral features following methods have been used in Matlab: (a) Yule-Walker, (b) Burg, (c) Covariance, (d) Modified covariance methods, (e) Eigenvector.

4.2 Statistical Parameters

Statistical parameters and Hjorth Parameters commonly reveals information of EEG signals processing in the time domain that was introduced in 1970 by Bo Hjorth.[13, 14]. These were computed and listed below in Matlab. (a) Mean, (b) Standard Deviation, (c) Mean Absolute Deviation (MAD), (d) Quantiles, (e) Interquartile range (IQR), (f) Skewnes, (g) Kurtosis, (h) Activity, (i) Mobility, (j) Complexity.

5 Feeding the Extracted Features for Classification

Here, we explain the steps that are required to feed the final features Matrix as described in section IV into the various classification algorithm interactively by launching the classificationLearner APP in Matlab. So that the trained model would finally put on the test to classify unseen data in Epileptic and Normal classes as depicted in Fig. 5.

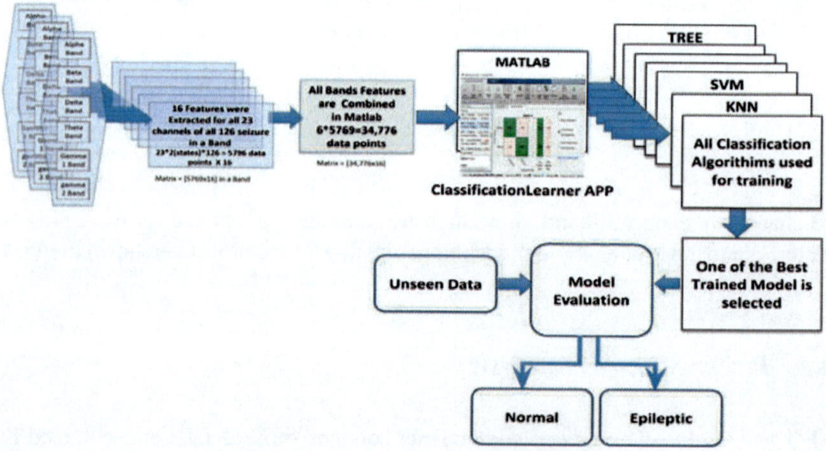

Fig. 5 Steps for training classification and evaluation of trained model

6 Comparison of Classification Algorithm

In this section we have collected the classification results of the trained models for various K-Folds and Holdout values for the validation. On observing the variation of K-Fold it shows for the 21 classifier Accuracy and training time are gradually increased as value of K increases. only Prediction speed falls. Where as in Holdout method the Accuracy and training time are gradually decreased as value of Holdout increases. only Prediction speed raised on increasing Holdout. Comparison between the K-Fold and Holdout Methods for Accuracy % for all 21 classifiers observation are shown in Figs. 6 and 7.

Comparison between the K-Fold and Holdout Methods for Prediction Speed (obs/sec) for all 21 classifiers are shown in Figs. 7 and 8. Comparison between the K-Fold and Holdout Methods for Training Time (Sec) for all classifiers are shown in Figs. 9 and 10.

All the above graphs of comparison for the Accuracy, prediction speed and training time were analyzed. It has been observed K-Fold method has performing better in terms of prediction speed and accuracy. Only the Holdout method of validation is better in term of training time. Due to this trade-off between three performance parameters the 8-best Classifier were selected.

The trained models finally put on the test to classify unseen data for Epileptic and Normal classes. This unseen dataset is also containing the same 16 features as that were used in training the models. The total instances are 216 out of that 50% are normal and 50% are epileptic. It has been observed that the K-Fold validation method is more accurate than the holdout method. During the testing phase all the 7

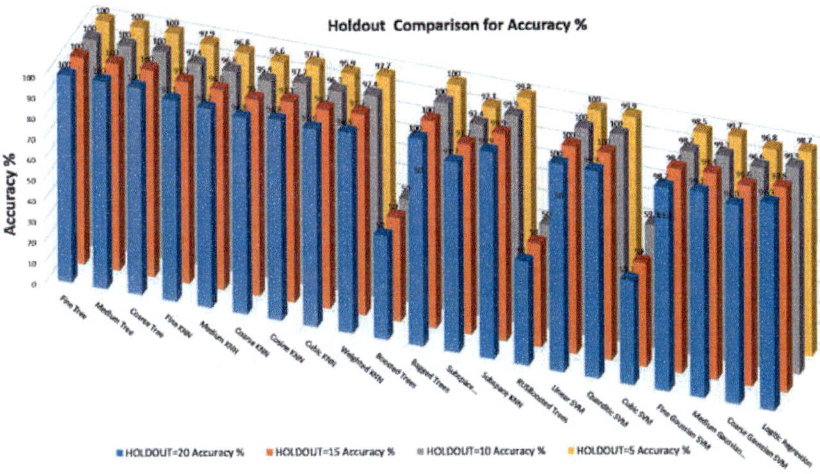

Fig. 6 Variation of holdout on accuracy of classifiers

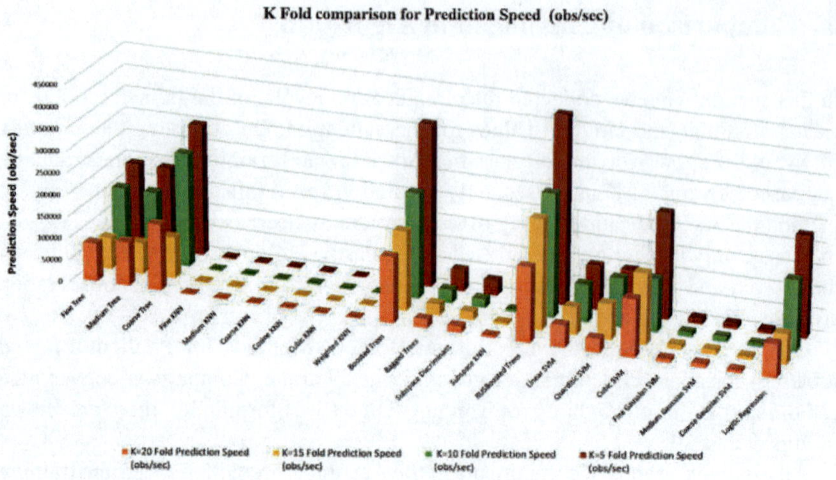

Fig. 7 Variation of K-fold on prediction speed of classifiers

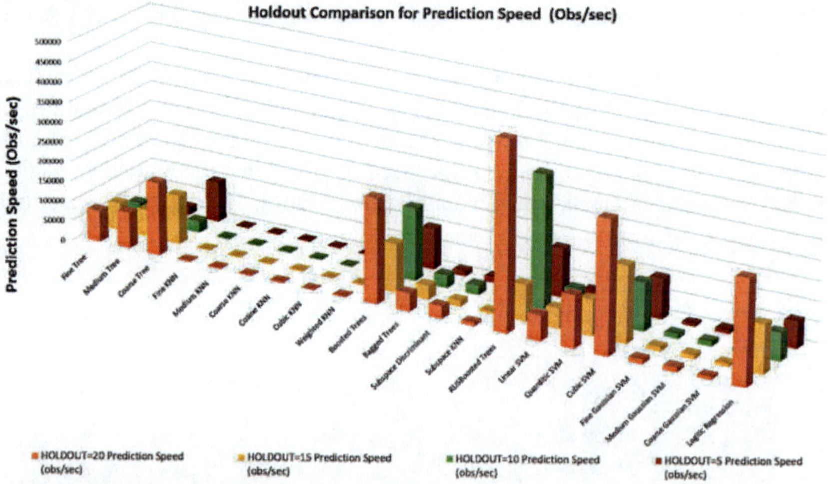

Fig. 8 Variation of holdout on prediction speed of classifiers

classifiers are giving 100% accuracy except Subspace KNN classifiers and the results are shown in Fig. 11. This shows the confusion matrix of the test result of Subspace KNN classifiers for various K-Fold and Holdout values respectively.

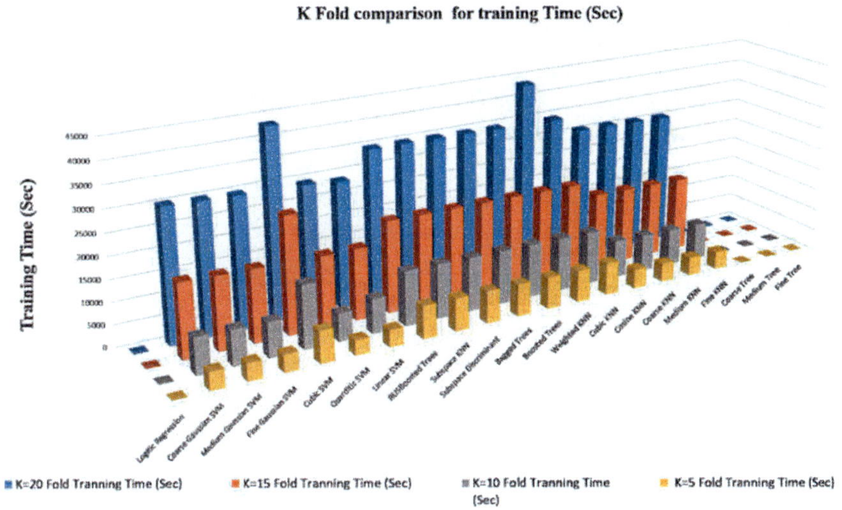

Fig. 9 Variation of K-fold comparison on training time of classifiers

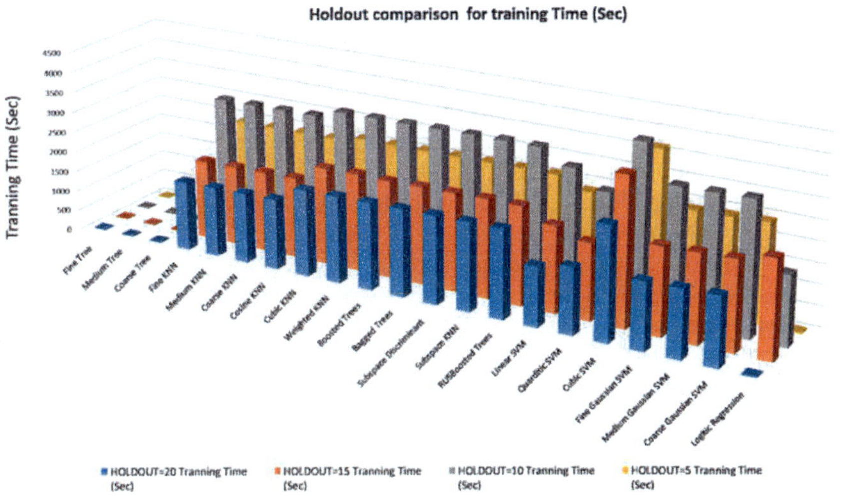

Fig. 10 Variation of holdout comparison on training time of classifiers

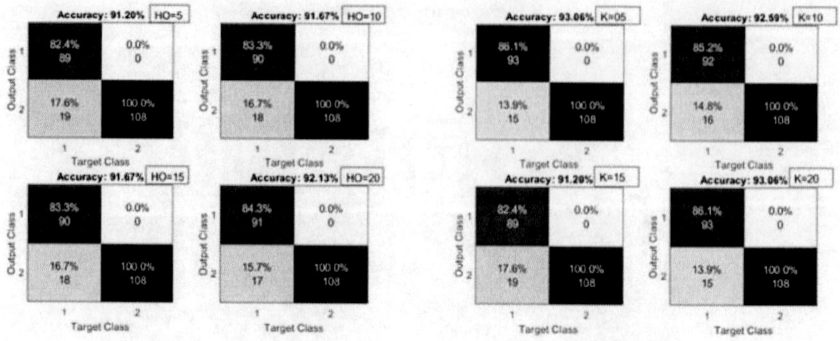

(a) Comparison of Subspace KNN for Hold Out (b) Comparison of Subspace KNN for K Fold

Fig. 11 Comparison of subspace KNN for different holdout and K-fold

7 Results

The proposed work has been accomplished through the Machine Learning approach. Here, we have shown the different results obtained during the implementation of the Features extraction, training and evaluation of the classifiers for comparison.

The 21 linear and non-linear Classifiers like SVM (Support Vector Machine), KNN (k-Nearest Neighbors), were trained with 16 statistical and spectral features. After that 8 best classifier have been selected and used to predict the unseen data for classification the Epileptic and Normal cases.

After prediction for unseen data with the best trained models as result has computed that shows the K-Fold validation method is better in terms of Accuracy %. The comparison of two validation method specially for the Subspace KNN classifier for the Sensitivity, Specificity and Error Rate of the various classifiers during evaluation phase are shown in Tables 1 and 2. The rest of all 7 classifier has the Sensitivity = 1, Specificity = 1 and Error Rate = 0.

During the testing phase all the 7 classifiers are giving 100 % accuracy except Subspace KNN classifiers and the results are shown in Fig. 11 in the form of confusion matrix.

Table 1 Variation of K-fold on the subspace KNN clasifiers

Subspace KNN	Error rate	Sensitivity
K-Fold = 20	0.0694	0.8611
K-Fold = 15	0.088	0.8241
K-Fold = 10	0.0741	0.8519
K-Fold = 5	0.0694	0.8611

Table 2 Variation of holdout on the subspace KNN clasifiers

Subspace KNN	Error rate	Sensitivity
HOLDOUT = 20	0.0787	0.8426
HOLDOUT = 15	0.0833	0.8333
HOLDOUT = 10	0.0833	0.8333
HOLDOUT = 05	0.088	0.8241

8 Conclusion

In this paper, we have done the analysis on the EEG signals of epileptic subjects in different age group in the Brainstorm Software. These EEG signals were filtered smoothly through the batch processing in Brainstorm. We have trained and tested the Models by the Classification Learner app to classify new unseen data. We have explored the various supervised machine learning classifiers[6]. Different features were used in the process of (K-Fold and Holdout) of cross-validation and training the models. It was accessed that Hjorth parameters were the best indicators of randomness of EEG signal time domain [2] it also increased the accuracy of the models.

Comparative analysis have been done among 21 linear and non-linear Machine learning classifiers as a results best 8 classifiers selected for evaluation. In this comparison for various values of K-fold and Holdout it has been observed that the variation has least affect on the accuracy but K-fold method is doing better than Holdout. The variation in the K-fold & Holdout for subspace KNN classifiers also has the effects on sensitivity and Error Rate. The further transfer learning approach and LSTM networks [1] can be used for increasing the prediction speed.

References

1. Daoud, H., Bayoumi, M.: Efficient epileptic seizure prediction based on deep learning. IEEE Trans. Biomed. Circuits Syst. **13** (2019). https://doi.org/10.1109/TBCAS.2019.2929053
2. Devi, S., Roy, S.: Physiological measurement platform using wireless network with Android application. Inform. Med. Unlocked **7**, 1–13 (2017). 10.1016/j.imu.2017.02.001, http://dx.doi.org/10.1016/j.imu.2017.02.001
3. Fergus, P., Hussain, A., Hignett, D., Al-Jumeily, D., Abdel-Aziz, K., Hamdan, H.: A machine learning system for automated whole-brain seizure detection. Appl. Comput. Inform. **12**(1), 70–89 (2016). https://doi.org/10.1016/j.aci.2015.01.001. http://dx.doi.org/10.1016/j.aci.2015.01.001
4. Fergus, P., Hignett, D., Hussain, A., Al-Jumeily, D., Abdel-Aziz, K.: Automatic epileptic seizure detection using scalp EEG and advanced artificial intelligence techniques. Biomed Res. Int. (2015). https://doi.org/10.1155/2015/986736
5. Hassan, A., Subasi, A.: Automatic identification of epileptic seizures from EEG signals using linear programming boosting. Comput. Methods Programs Biomed. **136** (2016). https://doi.org/10.1016/j.cmpb.2016.08.013

6. Jaiswal, A., Banka, H.: Epileptic seizure detection in EEG signal using machine learning techniques. Australasian Phys. Eng. Sci. Med. **41** (2018). https://doi.org/10.1007/s13246-017-0610-y
7. Karmakar, C.K., Khandoker, A.H., Palaniswami, M.: Power spectral analysis of ECG signals during obstructive sleep apnoea hypopnoea epochs. In: Proceedings 2007 International Conference on Intelligent Sensors, Sensors Networks Information Process. ISSNIP (2014), 573–576 (2007). https://doi.org/10.1109/ISSNIP.2007.4496906
8. Paul, Y.: Various epileptic seizure detection techniques using biomedical signals: a review. Brain Inf. **5** (2018). https://doi.org/10.1186/s40708-018-0084-z
9. Sriraam, N., Raghu, S., Tamanna, K., Narayan, L., Khanum, M., Hegde, A.S., Kumar, A.B.: Automated epileptic seizures detection using multi-features and multilayer perceptron neural network. Brain Inform. **5**(2) (2018). https://doi.org/10.1186/s40708-018-0088-8, https://doi.org/10.1186/s40708-018-0088-8
10. Subasi, A., Erçelebi, E., Alkan, A., Koklukaya, E.: Comparison of subspace-based methods with AR parametric methods in epileptic seizure detection. Comput. Biol. Med. **36**(2), 195–208 (2006). https://doi.org/10.1016/j.compbiomed.2004.11.001
11. Tadel, F., Bock, E., Niso, G., Mosher, J.C., Cousineau, M., Pantazis, D., Leahy, R.M., Baillet, S.: MEG/EEG group analysis with brainstorm. Front. Neurosci. **13**(FEB), 1–21 (2019). https://doi.org/10.3389/fnins.2019.00076
12. Ulate-Campos, A., Coughlin, F., Gaínza-Lein, M., Fernández, I.S., Pearl, P., Loddenkemper, T.: Automated seizure detection systems and their effectiveness for each type of seizure. Seizure **40**, 88–101 (2016)
13. Vourkas, M., Papadourakis, G., Micheloyannis, S.: Use of ANN and Hjorth parameters in mental-task discrimination. IEE Conf. Publ. **476**, 327–332 (2000). https://doi.org/10.1049/cp:20000356
14. Zhang, Y., Yang, S., Liu, Y., Zhang, Y., Han, B., Zhou, F.: Integration of 24 feature types to accurately detect and predict seizures using scalp EEG signals. Sensors (Switzerland) **18**(5) (2018). https://doi.org/10.3390/s18051372
15. Zhou, M., Tian, C., Cao, R., Wang, B., Niu, Y., Hu, T., Guo, H., Xiang, J.: Epileptic seizure detection based on EEG signals and CNN. Front. Neuroinform. **12**, 1–14 (2018). https://doi.org/10.3389/fninf.2018.00095

Anomaly Detection in Electroencephalography Signal Using Deep Learning Model

Sharaban Tahura, S. M. Hasnat Samiul, M. Shamim Kaiser, and Mufti Mahmud

Abstract Biosignals such as Electroencephalogram (EEG), Electrocardiogram (ECG), Electromyogram (EMG) represent the electrical activities of various parts of human body. Various low cost non-invasive bio-sensors measures bio-signals and assist medical practitioner to monitor physiological conditions of a human health and identify associated risk. The volume bio-signals is a big data and can not be analyzed and identify anomaly manually, therefore intelligent algorithms have been proposed to detect personalized anomaly in real time data. This paper presents a review on Deep Learning (DL) based anomaly detection techniques in EEG. The convolutional neural network, recurrent neural network and autoencoder based DL algorithms are considered. Here EEG signal acquisition, feature extracting techniques and key-anomaly features and corresponding performance of the various techniques found in the literature are also discussed. The challenges and open research questions are outlined at the end of the article.

Keywords Machine learning · Convolutional neural network · Recurrent neural network · Autoencoder · Prediction

1 Introduction

With the massive improvement in the machine learning (ML) algorithm and sensor technology, EEG data is being used for a lot of clinical application. Exploring pattern in EEG data using ML model aimed in understanding the dynamic behaviour of EEG signal [44]. Anomaly detection from the pattern help to detect epileptic seizure, autism, and neurological diseases, etc. [41, 42].

S. Tahura · S. M. Hasnat Samiul · M. Shamim Kaiser (✉)
Institute of Information Technology, Jahangirnagar University, Savar, Dhaka 1342, Bangladesh
e-mail: mskaiser@juniv.edu

M. Mahmud
Computer Science, Nottingham Trent University, Clifton Campus, Nottingham NG11 8NS, UK
e-mail: mufti.mahmud@ntu.ac.uk

M. S. Kaiser et al. (eds.), *Proceedings of International Conference on Trends in Computational and Cognitive Engineering*, Advances in Intelligent Systems and Computing 1309, https://doi.org/10.1007/978-981-33-4673-4_18

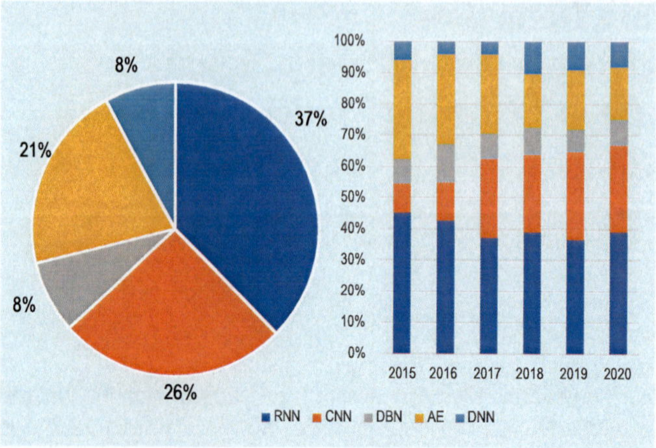

Fig. 1 Trend of the DL algorithm used for the anomaly detection in EEG signal. In order to prepare the graph used search string is"DL method" "EEG" "anomaly detection"

Figure 1 shows the trend of the DL algorithms used for the anomaly detection in EEG signal. The search is created using search string "DL method" "EEG" "anomaly detection". It presents Recurrent Neural Network (RNN) (including Long-Short Term Memory (LSTM)), Convolutional Neural Network (CNN) and Autoencoder (AE) are mostly adopted by the researchers in detecting anomaly in EEG data .

Chandola et al. introduced a survey on various anomaly detection techniques and their pros and cons. In addition, the computational complexity of each technique is outlined [26]. Chalaphathy et al. reviewed various DL methods used for detecting anomaly in various application domains and discussed assumption and challenges in the DL methods [25]. Craik et al. introduced a DL based review for EEG classification tasks [28]. We have not found any survey paper related to anomaly detection in EEG signals using DL technique [51].

Our contribution is this article is to investigate various types of Anomaly found in EEG signal; list AE, CNN and RNN based Anomaly detection in EEG; Explore open source Anomaly dataset, and find the performance of various AE, CNN and RNN based algorithms in detection anomaly using open access dataset; and List open challenges and future work at the end of the article.

Section 2 presents the basics of ML methods; Sect. 3 discusses anomaly detection using ML; Sect. 4 provides numerical results; Sect. 5 lists open challenges and future issues. Finally, the work is ended in Sect. 6.

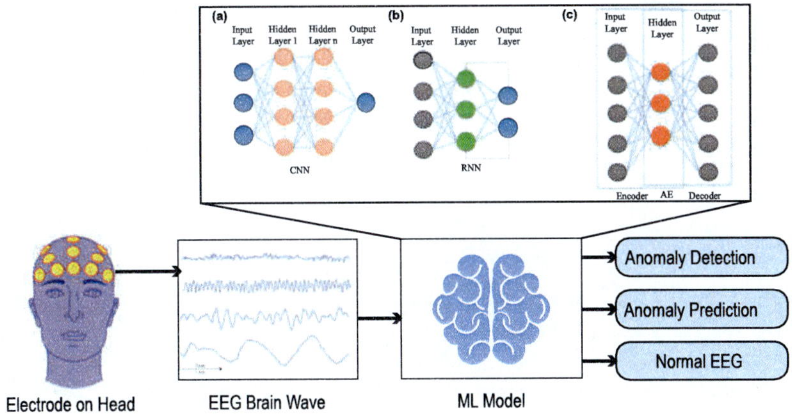

Fig. 2 Block diagram shows anomaly detection in EEG using machine learning model

2 Machine Learning Methods

DL algorithms such as CNN, DNN, AE, Deep Belief Network (DBN) and Probabilistic Neural Network (PNN), are a special class of ML approaches that can learn from EEG bio-signal data. Figure 2 shows a block diagram showing EEG data acquisition, analysis using DL algorithm and then detect anomaly in the recoded EEG data.

CNN is a feed forward network consisting of input , hidden and output layers where hidden layers include rectified linear unit (ReLU), pooling layer and completely connected layer [6] .

RNN portrays an architecture where certain hidden layers store information about the neural network and the measured output of previous layers are used as input layers. A type of LSTM-based RNN called LSTM-RNN [39] is commonly used in time series data.

AE is a type of unsupervised technique used to compress data. This technique is used to minimize dimensionality for a fixed collection of data by training the architecture. Stacked AE [27] is a type of neural network consisting of multi-layer sparse AE, where two consecutive layers of output and input are attached.

3 ML Methods for Detecting Anomaly

3.1 Abnormal EEG

Neural memory networks (NMN) [32]; ChronoNet—a novel RNN [48]; one-dimensional CNN (1D-CNN) [60] were employed to detect abnormal EEG using

TUH Abnormal EEG database. Leeuwen et al. [38] utilized a CNN model to identify irregular EEG signals using Massachusetts General Hospital dataset [8].

3.2 *Schizophrenia*

Fernando et al. [32] proposed NMN in detecting schizophrenia using a dataset given in [43]. Ahmedt-Aristizabal et al. [19] developed a recurrent CNN (R-CNN) model to detect the risk of schizophrenia among 9–12 years old children through a EEG dataset.

3.3 *Abnormal Sleep Quality*

Zhou et al. [59] established a Mahalanobis-Taguchi system model [55] that can identify the abnormal sleep quality by analyzing the EEG signals which used Sleep-EDF database [36].

3.4 *Epileptic Seizure Detection*

Liu et al. [40] developed an LSTM (C-LSTM) convolution model to detect seizures analyzing EEG signals obtained from UCI database [14]. Wei et al. [54] discovered a 3D-CNN network framework to predict seizure from 13 patients using multi-electrode EEG. Zhou et al. [58] used two separate public databases such as EEG intracranial freiburg [4] and scalp CHB-MIT database [2] to compare time and frequency domain signals. Abbass et al. [17] implemented a CNN-based anomaly detection model that extracts all feature types using EEG data from the CHB-MIT scalp EEG database [2]. Emami et al. [30] developed a CNN network paradigm to detect epileptic seizures using [5]. Aliyu et al. [22] presented an RNN framework with a private EEG dataset of Bonn University, Germany [23] for epilepsy detection and applied Discrete Wavelet Transform (DWT) [33] to process collected EEG data. Petrosian et al. [47] proposed an RNN based method with signal wavelet decomposition to predict epileptic seizures analyzing scalp data and intracranial recordings of two patients suffering from epilepsy. Farooq et al. utilized the multivariate statistical process control (MSPC) to detect seizure prediction by using long-term scalp EEG signal receivable from PhysioNet [2].

Ahmad et al. [35] proclaimed an epilepsy detection model incorporating deep AE and DWT, while Abdelhameed et al. [18] induced a 1D-deep AE-based system to detect seizures. Bonn University has used [15] in both instances.

For detecting seizure, Emami et al. [29] proposed AE based diagnosis support system using EEG data from a private dataset [30], Abdelhameed et al. [18] introduced 1D- deep convolutional AE based framework employing database of Bonn University, Department of Epileptology [15], and Supratak et al. [52] developed a patient-specific model combining stacked AE and logistic classifiers using CHB-MIT database [2].

3.5 *Parkinson's Disease*

Lih oh et al. [46] developed a model that can detect parkinson's disease [9] by using CNN with EEG data gathered from Hospital University Kebangsaan Malaysia Ethics Committee [16].

Shi et al. [50] raised a DL method based on two hybrid convolutional RNN called (2D-CNN and 3D-CNN-RNN) using Temple University Hospital (TUH) EEG Corpus [45] datset to detect Parkinson's disease [9] in task state.

3.6 *Healthy Anomaly Detection*

Wang et al. [53] discovered a CNN architecture with multivariate gauss distribution method to detect anomaly utilizing raw EEG signals of DEAP dataset [3].

3.7 *Pathology Detection*

Alhussein et al. [21] established a automatic pathology detection method based on two different CNN as a shallow model and a deep model and processed the EEG data taken from TUH EEG Abnormal Corpus v2.0.0 [45].

3.8 *Depression Detection*

Ay et al. [24] constructed a novel depression diagnosis system based on a combination of LSTM network and CNN to detect depression analyzing EEG signal collected from the left and right hemispheres of brain of 30 patients. Subha et al. [37] established a LSTM-RNN based deep neural network model for predicting the tendency of depression with utility of EEG data collected from Psychiatry Department of Medical College Calicut, Kerala, India [11].

3.9 Emotion Recognition

Suwicha et al. [34] introduced an emotion recognition method implementing stacked AE with covariate shift adaptation of principal component analysis utilizing DEAP dataset [3]. Xing et al. [56] proposed a novel framework combining a stacked AE based linear EEG mixing model and LSTM-RNN based emotion timing model to recognize emotion using DEAP dataset [3]. Alhagry et al. [20] proposed an LSTM-RNN based emotion recognition framework with dense layer classifiers utilizing DEAP dataset [3]. Yang et al. [57] built up a DL network using stacked AE and softmax classifier to detect three classes of emotions: happy, neural, grief applying SJTU Emotion EEG Dataset (SEED) [12].

3.10 Drowsiness Detection

Rundo et al. [49] proposed a drowsiness detection model using stacked AE with softmax layer and discrete cosine transform (DCT) from EEG of 62 patients. Table 1 shows open access EEG datasets/databases.

4 Numerical Performance

4.1 Abnormal EEG

Fernando et al. utilized scaled EEG recordings from T5-O1 channel in the proposed NMN model which output normal or abnormal classes. The proposed Plastic NMN outperforms RNN and CNN based model in terms of accuracy (Plastic NMN: 93.3; Plastic RNN: 82.6; 1D-CNN-RNN: 82.27; 1D-CNN: 79.4; 2D-CNN: 78.8).

Roy et al. compared the results with different neural networks where ChronoNet achieved the best accuracy with 90.60% in training and 86.57% are testing.Besides, C-RNN,IC-RNN,C-DRNN gained accuracy of 83.58%, 86.93%, 87.20% in training and 82.31%, 84.11%, 83.89% in testing respectively [48].

Acharya et al. showed that the (1D-CNN) based proposed method gives 79.34% accuracy with 20.66% error rate in classifying the abnormal signals [60].

Leeuwen et al. showed that the proposed model acquired accuracy of 81.6% with 74.8% sensitivity [38].

Table 1 Open access EEG datasets/databases

References	EEG dataset	Description
[13]	TUH abnormal EEG database	A: 1488; B: 1529
[36]	Sleep-EDF database	N: 1989; A: 1994
[14]	UCI database (EEG dataset)	Five file sets
[23]	Epilepsy Research Center at Bonn University in Germany	Contained five subsets denoted as *A* to *E*
[2]	CHB-MIT database	22 subjects (5 males and 17 females)
[4]	Epilepsy Center of the University Hospital of Freiburg	Channels: 128, Sampling rate: 256 Hz
[16]	Hospital University Kebangsaan Malaysia (EEG Dataset)	20 PD patients (Male: 10; Female: 10)
[3]	DEAP dataset	32 participants were recorded
[45]	The Temple University Hospital EEG Data Corpus V2.0.0	N: 1385 and A: 998
[8]	Department of Neurology in Massachusetts General Hospital	Number of EEG: 7671; Male: 3875
[30]	ImageNet (EEG dataset Image)	12 subtrees with 5247 synsets
[7]	MRI-EPI database	3D image: 90 and 4D image: 133
[11]	Psychiatry department of Medical College Calicut	Depressed patients: 30, Age: 20–50
[15]	Bonn University EEG dataset	Row data: 200, N: 100, A: 100, features: 4096
[12]	SEED EEG dataset	Movie clips: 15, containing happy, calm and sad emotion

Legends: N-Normal, A-Abnormal

4.2 Schizophrenia

Fernando et al. found higher accuracy in detecting EEG based schizophrenia using a dataset proposed in [43] (Plastic NMN: 93.86; Plastic RNN: 78.85; CNN 76.35; SVM: 43.80).

Ahmed-Aristizabal et al. utilized the EEG dataset and found the best accuracy in proposed model (78.7%) in A1 phase. While the other ML performance is (2D-CNN-LSTM:72.54; 2D-CNN-GRU:69.78) [19].

4.3 Abnormal Sleep Quality

Zhou et al. investigated SNR of EEG signals and found normal vs abnormal SNR of 6 stages (AWA −51.84 vs. −67.94; S1 −44.22 vs. −51.62; S2 −43.47 vs. −53.84; S3 −45.64 vs. −51.13; S4 −45.05 vs. −52.05; REM −42.32 vs. −48.53) [59].

4.4 Epileptic Seizure Detection

Liu et al. discovered the best performance by detecting the seizure compared with DCNN and LSTM.The overall accuracy of C-LSTM is more than 98.80% where DCNN accuracy is 77.82% and LSTM is 69.47% [40].

Wei et al. experimented on 2-dimensional CNN and 3-Dimensional CNN model both consuming single-channel EEG data and multi-channel EEG data respectively to detect a seizure. The proposed 3D-CNN model on multi-channel signals outperforms 2D-CNN and another conventional framework in terms of accuracy (3DCNN: 92.37; 2DCNN: 89.91; SVM+DWT+ApEn: 91.25) [54].

Zhou et al. found that frequency domain signal has better performance than time domain signal.The average accuracy values according to frequency domain signal and time domain signal were appeared as follows respectively based on three experiments: In freiburg database (96.7, 95.4, 94.3% and 91.1, 83.8, 85.1%) and in CHB-MIT database(95.6, 97.5, 93% and 59.5, 62.3, 47.9%) [58].

Aliyu et al. evaluated the performance of the results and found that the proposed model gained 99% accuracy which is the best followed by DT 98%, KNN 96%, KVM 96%, and RF 75%.[22].

Petrosian *et al.* demonstrated the intracranial recordings of EEG that is changed in high-frequency signal's component which makes the prediction useful [47].

Abbass et al. discovered that the best performance in detection occurred in channel 15 with 70.37% accuracy rate and best classification results occurred in channel 03 with 88.89% accuracy rate [17].

Emami et al. found the best median of detecting the seizure in minutes is 100%.Again another detection rate is much more smaller than the suggested method (BESA: 73.3% and Persyst: 81.7%) [30].

Farooq et al. observed the validation period by using the interictal data and found 80 seizure patients out of 90 gave .39 false positive rate in one hour [31].

Ahmad et al. observed that the proposed framework acquired higher accuracy of 96% compared to other frameworks and found it faster than others [35].

Emami et al. compared the AE based model with BESA [1] and Persyst [10] software packages for half of the test subjects to classifying seizure and non-seizure states and gained 100% sensitivity in proposed model [29].

Abdelhameed et al. evaluated three different neural network classifiers and found the best classification results using bidirectional LSTM with 99.33% average accuracy [18].

Supratak et al. found that the proposed model acquired 100% sensitivity with a mean latency of 3.36 s when channel threshold was 1 and achieved low false detection rate [52].

4.5 *Parkinson's Disease*

Lih oh et al. found the best accuracy in CNN model that is 88.25% and also investigated that 11.51% EEG signals are wrongly classified in normal class [46].

Shi et al. showed that the performance accuracy in each method (CV1-CV5) with the dataset and gained the best accuracy in 3D-CNN-RNN (82.89%) comparing with another model (RNN: 76.00%, CNN:80.89%, 2D-CNN-RNN: 81.13%) [50].

4.6 *Healthy Anomaly Detection*

Wang et al. detected the abnormality by using the DEAP dataset [3] and got four set of anomaly data ratio (1, 2, 3, 5%) and the thresholds of detection is (0.2, 0.23, 0.25 and 0.026) [53].

4.7 *Pathology Detection*

Alhussein et al. found the best accuracy in CNN with fusion is 89.13% and without fusion is 87.68% using the same database [45].

4.8 *Depression Detection*

Ay et al. showed that the right hemisphere of the brain outperformed the left hemisphere (right: 99.12; left: 97.66). The right (FP2-T4) side hemisphere EEG signal achieved an accuracy of 99.12% and sensitivity of 99.11% [24].

Subha et al. compared the proposed model with CNN-LSTM and Conv-LSTM and found LSTM-RNN as best predictor of depression. Obtained RMSE values were (LSTM: 0.005; CNN-LSTM: 0.007; Conv-LSTM: 0.0088) [37].

4.9 *Emotion Recognition*

Suwicha et al. found that the proposed DL network with PCA+CSA outperformed naive Bayes classifier and SVM with an accuracy of 46.03% and 49.52% respectively to analyze the three part levels of valence and arousal [34].

Xing et al. obtained average recognition accuracy of 81.10% and 74.38% in terms of valence and arousal respectively and outperforms HMM, SVM and CRNN methods [56].

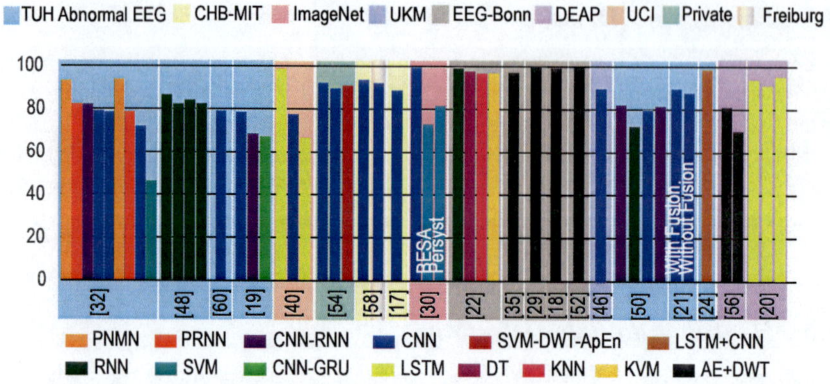

Fig. 3 Performance evaluation of various machine learning algorithm especially CNN, RNN, DAE in detecting Anomaly in EEG. It has been observed that CNN is used most whereas the performance of AE is better than others

Alhagry et al. compared the LSTM based model with some conventional techniques and found average accuracy of 85.65%, 85.45%, and 87.99% in terms of arousal, valence, and liking cases respectively [20].

Yang et al. found six different classification accuracy for six different differential entropy features (delta: 59.6; theta: 66.27; alpha:71.97; beta: 78.48; gamma: 82.56 and all band: 85.5) detecting three states of emotions performing fourteen experiments with fivefold cross validation [57].

4.10 Drowsiness Detection

Rundo et al. observed that proposed model achieved an accuracy of 100% to detect drowsiness while other methodology acquired accuracies as follows: (ANN: 99.5; SVM: 98; LDA: 97) [49].

Figure 3 shows performances of various machine learning algorithm especially CNN, RNN, DAE in detecting Anomaly in EEG. It has been observed that CNN is used most whereas the performance of AE is better than others.

5 Open Challenges

A DL model for detecting an anomaly in EEG data is essential due to the enormous advancement in signal acquisition and computing abilities. However, there is still scopes for improvement for tuning the DL models. Some of the key dispute further research may be conducted for tuning the DL algorithms which are listed below:

- Inherent and ambient noises can add with the EEG data during acquisition and may reduce the classification performance. Therefore, deploying a technique to eliminate/reduce these noises is an open challenge.
- Real-time anomaly detection using a cloud-based system is difficult as the propagation delay can not be eliminated, however, a distributed processing framework can reduce the processing delay to a minimum. Thus extensive research can be done in distributed processing and knowledge fusion techniques.
- Selection of appropriate DL model and hyperparameter optimization also impose challenges for the researchers. This issue can be investigated.
- Getting a bias-free EEG dataset is challenging as it is required to develop an optimal model. It creates a computational artifact.

6 Conclusions

In this paper, DL—such as CNN, RNN and AE—based anomaly detection techniques in EEG biosignal has been discussed which is essential for detecting neurological disorder or disease in the human body and also identify associated risk factor. In terms of velocity and volume, the real time monitoring EEG data of a patient is big data, thus it can not be analyzed and identify anomaly manually, therefore intelligent algorithms have been proposed to detect personalized anomaly in real time data. Here EEG signal acquisition, feature extracting techniques and key-anomaly features and corresponding performance of the various techniques found in the literature are also discussed. The challenges and open research questions are also outlined.

References

1. Besa epilepsy detection software and simulator. https://www.besa.de/downloads/besa-epilepsy/. Accessed on 30 June 2020
2. Chb-mit PhysioNet EEG 5 males 17 females. https://physionet.org/content/chbmit/1.0.0/. Accessed on 24 June 2020
3. Deap dataset for emotion analysis. https://rb.gy/dsw83y. Accessed on 24 June 2020
4. EEG-database Epilepsy Center of the University Hospital of Freiburg. https://rb.gy/rujwqj. Accessed on 24 June 2020
5. Image net EEG dataset. http://www.image-net.org/. Accessed on 25 June 2020
6. Introduction to cnn. https://rb.gy/7hhqr4. Accessed on 30 June 2020
7. Mri epilepsy database open access. https://rb.gy/icabid. Accessed on 2 June 2020
8. Neurology in Massachusetts general hospital. https://www.massgeneral.org/neurology/. Accessed on 24 June 2020
9. Parkinsons-disease Neurosurgical-Conditions-and-Treatments. https://www.ucsfhealth.org/conditions/parkinsons-disease/treatment. Accessed on 24 June 2020
10. Persyst The Worldwide leader in EEG Software. https://rb.gy/uijahf. Accessed on 30 June 2020
11. Psychiatry department of medical college Calicut, Kerala, India. https://rb.gy/jtesyz. Accessed on 30 June 2020

12. Seed dataset for emotion recognition. https://rb.gy/txleot. Accessed on 30 June 2020
13. Temple University EEG Corpus—Downloads. https://www.isip.piconepress.com/projects/tuh_eeg/html/downloads.shtml. Accessed on 22 June 2020
14. Uci database *eeg* dataset. https://rb.gy/szwcrs. Accessed on 23 June 2020
15. Ukb university of bonn. https://rb.gy/uwow0l. Accessed on 30 June 2020
16. Hospital University Kebangsaan Malaysia ethics committee. http://www.ukm.my/spifper/ (2020). Accessed on 24 June 2020
17. Abbass, M., et al.: Anomaly detection from medical signals and images using advanced convolutional neural network. Researchsquare access (2020)
18. Abdelhameed, A., Daoud, H., Bayoumi, M.: Epileptic seizure detection using deep convolutional autoencoder. In: IEEE SiPS, South Africa (2018)
19. Ahmedt Aristizabal, D., et al.: Identification of children at risk of schizophrenia via deep learning and EEG responses. IEEE J. Biomed. Health Inform. 1-1 (2020)
20. Alhagry, S., Aly, A., El-Khoribi, R.: Emotion recognition based on eeg using lstm recurrent neural network. IJACSA **8** (2017)
21. Alhussein, M., Muhammad, G., Hossain, M.S.: Eeg pathology detection based on deep learning. IEEE Access **7**, 27781–27788 (2019)
22. Aliyu, I., et al.: Epilepsy detection in EEG signal using recurrent neural network. In: Proceedings of ISMSI, pp. 50–53. ACM (2019)
23. Andrzejak, R.: Indications of nonlinear deterministic and finite-dimensional structures in time series of brain electrical activity: dependence on recording region and brain state. Phys. Rev. **64**, 061907 (2002)
24. Ay, B., et al.: Automated depression detection using deep representation and sequence learning with EEG signals. J. Med. Syst. **43** (2019)
25. Chalapathy, R., Chawla, S.: Deep learning for anomaly detection: a survey (2019)
26. Chandola, V., et al.: Anomaly detection: a survey. ACM Comput. Surv. **41**(3) (2009)
27. Coutinho, M.G.F., et al.: Deep neural network hardware implementation based on stacked sparse autoencoder. IEEE Access **7**, 40674–40694 (2019)
28. Craik, A., et al.: Deep learning for electroencephalogram (EEG) classification tasks: a review. J. Neural Eng. **16**(3), 031001 (2019)
29. Emami, A., et al.: Ae of long-term scalp EEG to detect epileptic seizure for diagnosis support system. Comput. Biol. Med. **110**, 227–233 (2019)
30. Emami, A., et al.: Seizure detection by convolutional neural network-based analysis of scalp EEG plot images. NeuroImage Clin. **22**, 101684 (2019)
31. Farooq, O., et al.: Patient-specific epileptic seizure prediction in long-term scalp EEG signal using multivariate statistical process control Elsevier enhanced reader. IRBM (2019)
32. Fernando, T., et al.: Neural memory plasticity for medical anomaly detection. Neural Netw. (2020)
33. Furht, B. (ed.): Discrete Wavelet Transform (DWT), pp. 188-188. Springer US, Boston, MA (2008)
34. Jirayucharoensak, S., et al.: Eeg-based emotion recognition using deep learning network with principal component based covariate shift adaptation. Sci. World J. **2014**, 627892 (2014)
35. Karim, A., et al.: A new automatic epilepsy serious detection method by using deep learning based on discrete wavelet transform. In: ICETAS (2018)
36. Kemp, B., et al.: Analysis of a sleep-dependent neuronal feedback loop: the slow-wave microcontinuity of the EEG. IEEE Trans. Biomed. Eng. **47**(9), 1185–1194
37. Kumar, S., Subha, D.: Prediction of depression from EEG signal using long short term memory (lstm). In: 2019 ICOEI, pp. 1248–1253 (2019)
38. van Leeuwen, K., et al.: Detecting abnormal electroencephalograms using deep convolutional networks. Clin. Neurophysiol. **130** (2018)
39. Lipton, Z.C., Kale, D.C., Elkan, C., Wetzel, R.: Learning to diagnose with lstm recurrent neural networks (2015)
40. Liu, Y., et al.: Deep c-lstm neural network for epileptic seizure and tumor detection using high-dimension EEG signals. IEEE Access **8**, 37495–37504 (2020)

41. Mahmud, M., Kaiser, M.S., Hussain, A.: Deep learning in mining biological data (2020)
42. Mahmud, M., Kaiser, M.S., et al.: Applications of deep learning and reinforcement learning to biological data. IEEE Access **29**(6), 2063–2079 (2018)
43. Moghaddam, B., et al.: From revolution to evolution: the glutamate hypothesis of schizophrenia and its implication for treatment. Neuropsychopharmacology **37**(1), 4–15 (2012)
44. Noor, T., et al.: Detecting Neurodegenerative Disease from MRI: a brief review on a deep learning perspective. In: Brain Informatics, pp. 115–125. Springer (2019)
45. Obeid, I., et al.: The tuh EEG data corpus. Frontiers Neurosci. **10**, 196 (2016)
46. Oh, S.L., et al.: A deep learning approach for Parkinson's disease diagnosis from EEG signals. Neural Comput. Appl. 1–7 (2018)
47. Petrosian, A., et al.: Rnn based prediction of epileptic seizures in intra- and extracranial EEG. Neurocomputing **30**, 201–218 (2000)
48. Roy, S., Kiral-Kornek, F.I., Harrer, S.: Chrononet: A deep recurrent neural network for abnormal EEG identification, pp. 47–56 (2019). arXiv:1802.00308v2
49. Rundo, F., et al.: An innovative deep learning algorithm for drowsiness detection from EEG signal. Computation **7** (2019)
50. Shi, X., Wang, T., et al.: Hybrid convolutional recurrent neural networks outperform cnn and rnn in task-state EEG detection for Parkinson's disease. In: 2019 APSIPA ASC, pp. 939–944 (2019)
51. Sumi, A.I., et al.: fASSERT: A fuzzy assistive system for children with autism using IoT. In: Brain Informatics, pp. 403–412. LNCS, Springer, Cham (2018)
52. Supratak, A., et al.: Feature extraction with stacked autoencoders for epileptic seizure detection. In: 2014 IEEE EMBC, pp. 4184–4187 (2014)
53. Wang, et al.: Research on healthy anomaly detection model based on deep learning from multiple time-series physiological signals. Sci. Program. (2016)
54. Wei, X.: Automatic seizure detection using three-dimensional cnn based on multi-channel EEG. BMC Med. Inform. Decision Making **18** (2018)
55. Woodall, W.H., et al.: A review and analysis of the mahalanobis-taguchi system. Technometrics **45**(1), 1–15 (2003)
56. Xing, X., et al.: Sae+lstm: new framework for emotion recognition from multi-channel EEG. Frontiers Neurorobot. **13** (2019)
57. Yang, B., et al.: Three class emotions recognition based on deep learning using staked autoencoder. In: 2017 CISP-BME, pp. 1–5 (2017)
58. Zhou, M., et al.: Epileptic seizure detection based on EEG signals and cnn. Frontiers Neuroinform. **12** (2018)
59. Zhou, Z., et al.: Anomaly detection for sleep EEG signal via mahalanobis-taguchi-gram-schmidt method. In: 2018 ICNISC, pp. 112–116 (2018)
60. Özal, B., et al.: A deep convolutional neural network model for automated identification of abnormal EEG signals. Neural Comput. Appl. (2018)

An Effective Leukemia Prediction Technique Using Supervised Machine Learning Classification Algorithm

Mohammad Akter Hossain, Mubtasim Islam Sabik, Md. Moshiur Rahman, Shadikun Nahar Sakiba, A. K. M. Muzahidul Islam, Swakkhar Shatabda, Salekul Islam, and Ashir Ahmed

Abstract Leukemia is not only fatal in nature, the treatment is also extremely expensive. Leukemia's second stage (typically there are four stages) is enough to blow a large hole in a family's savings. In this paper, we have designed a supervised machine learning model that accurately predicts the possibility of Leukemia at an early stage. We mainly focus on regular symptoms and the probabilities of a subject to develop Leukemia later on. The parameters or features are usually information available at regular checkups. Firstly, we have defined 17 parameters in consultation with the specialist doctors and then we have collected primary data through surveys of different Leukemia and Non Leukemia patients from hospitals. We have divided the data into train and test datasets and applied different machine learning algorithms such as Decision Tree, Random Forest, KNN, Linear Regression, Adaboost, Naive Bayesian,

M. A. Hossain (✉) · M. I. Sabik · Md. M. Rahman · S. N. Sakiba · A. K. M. Muzahidul Islam · S. Shatabda · S. Islam
Department of Computer Science and Engineering,
United International University, Dhaka, Bangladesh
e-mail: mhossain153064@bscse.uiu.ac.bd

M. I. Sabik
e-mail: msabik153104@bscse.uiu.ac.bd

Md. M. Rahman
e-mail: mrahman153026@bscse.uiu.ac.bd

S. N. Sakiba
e-mail: ssakiba153082@bscse.uiu.ac.bd

A. K. M. Muzahidul Islam
e-mail: muzahid@cse.uiu.ac.bd

S. Shatabda
e-mail: swakkhar@cse.uiu.ac.bd

S. Islam
e-mail: salekul@cse.uiu.ac.bd

A. Ahmed
Department of Advanced Information Technology, Kyushu University, Fukuoka, Japan
e-mail: ashir@soc.ait.kyushu-u.ac.jp

M. S. Kaiser et al. (eds.), *Proceedings of International Conference on Trends in Computational and Cognitive Engineering*, Advances in Intelligent Systems and Computing 1309, https://doi.org/10.1007/978-981-33-4673-4_19

etc. to find out the accuracy. We obtained 98% of accuracy using Decision Tree and Random Forest, 97.21% using KNN, 91.24% using Logistic Regression, 94.24% using Adaboost, and 75.03% using Naive Bayesian, respectively. It is observed that the Decision Tree and the Random Forest classifier outperform the rest.

Keywords Leukemia · Machine learning · Classification · Supervised machine learning · Decision tree algorithms · Random forest algorithms

1 Introduction

Leukemia has become a global public health issue in recent years. There are many types of Leukemia that affect children as well as adults. Leukemia creates abnormal blood cells which are called Leukemia cells. It's also known as the cancer of the white blood cell. The white blood cells in our blood stream are the front and rear guard against harmful viruses and bacteria that cause infections and other sorts of diseases. Nowadays, Leukemia is one of the common leading causes of death. It is reported by American Cancer Society that in 2018 alone in the USA, approximately 24,000 people died due to Leukemia [1]. In other statistics, it is shown that Leukemia is the second most significant cause of death [2]. Leukemia develops when DNA is unable to develop blood cells, mainly white cells. This causes the blood cells to grow and divide indomitably. Generally, elderly people are having a high risk of having Leukemia. When a man reaches the highest level of Leukemia then the treatment of this disease also becomes very expensive. Many people are unable to bear this expense. As a result, they die. It is also very concerning that many people get alerted about Leukemia at the very last stage. A better solution is if we predict Leukemia at an early stage then it can be prevented by controlling some metabolic factors.

To detect Leukemia, different types of procedures such as pathological blood tests [3], bone marrow tests [4], and chromosome based tests [5], etc. may be used. Complete Blood Count (CBC) is a common blood test which is widely used to detect Leukemia. It is used to count the amount of various types of blood cells. By finding abnormal cells it confirms the possibility of cancer. Bone marrow test is the process to diagnose Leukemia and bone marrow biopsy is used to prevent bleeding. Detecting Leukemia in early stages is rare in Bangladesh as the percentage of prevalence of Leukemia is 0.6. There are several machine learning based methods that are used in detecting Leukemia [6, 7]. However, in the context of Bangladesh, due to unstructured and non-comprehensive system of medical registration there exist a very few research only.

The main objective of this work is to predict Leukemia accurately at an early stage. As prevention is better than cure, we predict the possibilities of Leukemia accurately. We propose a cost effective diagnosis mechanism that can observe and collect the minimum number of symptoms to predict Leukemia. This can decrease the mortality rate and help the low income people to have knowledge about their health issues. It will make people much more aware so that they can take the necessary steps and cure it on time. Hence, we apply machine learning algorithms to predict the possibility of having Leukemia with higher accuracy so that people can be aware of this. The main contributions of this paper are as follows:

- Collection and standardization of a primary dataset on early detection of Leukemia from Bangladeshi hospitals.
- Comprehensive analysis of several machine learning algorithms to show the effectiveness of prediction models.

2 Related Work

Many researches have been done to predict Leukemia where researchers have applied different machine learning techniques to obtain better accuracy. However, not much work have been done on predicting Leukemia at an early stage as it is very challenging to predict Leukemia by observing real life based parameters.

Mohapatra et al. [6] have proposed a method where they used Digital Microscope for grabbing 108 blood smear images. After that, they used k-means clustering to locate WBC nucleus and evaluate many features. After evaluating features just crop the sub image by using a bounding box and find the erythrocytes. They used different blood images to find out the accuracy which is 95%. They did not predict Leukemia with patients' symptoms as we do. Markiewicz et al. [7] have approached a way that automatically recognizes the Blood Cells of myelogenous Leukemia through supervised machine learning. They used 17 different blast cells to classify and detect Leukemia by using support vector machine and exploited the image by using many features which are related to the geometry, texture, etc. They investigated 16 classes as abnormal types and the 17th class is composition of different features. They obtained 6.76% testing error for 176 text cases. They work on blast cells but they did not predict Leukemia at an early stage based on symptoms.

BenHsieh et al. [8] have researched on circulating tumor cells to detect Leukemia. They used a microscope for cell analysis and then found out CTC from images. By FAST technology they locate CTCs for breast cancer patients which find the positions of objects and focus on images separately. They prepared three samples and scanned the samples 10 times with the help of a FAST cytometer. On the other hand, they scanned the samples without using FAST technology. In that case, the microscope detects all the cells that were detected by FAST technology. They integrated the scan with ADM. They focus on images and scanned samples but they didn't show any analysis on symptoms based prediction. Leinoe et al. [9] worked on prediction of haemorrhage in the early stage of acute myeloid leukaemia by flow cytometric analysis of platelet function. They find out the clinical bleeding score and statistical analysis by using cytometric analysis in 50 AML patients. They identified haemorrhage, they took 50 cancer patients and gave 30 patients chemotherapy, while 20 patients normal treatment. Furthermore, they analyzed affected organs and platelet function and graded by different criteria to compare with results to predict the value of Pselectin which is validated in haemorrhage. They worked on affected cancer patients who were undergoing treatment but did not show any analysis related to symptomatic prognosis.

Fatma et al. [10] analyzed many types of cancers such as Leukemia patients' medical history to understand the symptoms. Initially they have collected high quality blood samples. Then they applied a color model considering linear contrast and IISI. Thereafter, they used k-means clustering to put the samples in different groups, where they applied a median filter on the clustered samples and created a segment of images. They performed feature extraction and then they classified the images and predicted the accuracy. Using neural network they got up to 91% accuracy with an increased data size which is very good but not for early stage prediction. Pan et al. [11] tried to construct a relapse prediction model using machine learning algorithms. They used tenfold cross validation to put clinical variables in the rank. They used 336 diagnoses and found the shortest list with a forward feature selection algorithm. They took 150 patient test sets to split and work on an independent data set. They evaluated the model with 85 patient's dataset. They tried to find out the cross-validation accuracy with 14 features using the Random Forest model. They predicted childhood ALL relapse using medical data that was found from the Electronic Medical Record using machine learning models. They found the best accuracy 82.90% as the model performed in different risk level groups but no analysis was performed on the symptoms.

Although we can see many works on detecting Leukemia, however, predicting it at an early stage is very rare. Moreover, predicting the possibility of Leukemia based on observing symptoms is also not very common in Bangladesh as well as in other countries. Although there different types of treatment are available, however, detecting Leukemia at an early stage is very difficult to find. Thus in this paper we focus on predicting Leukemia at an early stage with patient's symptoms to raise awareness.

3 Materials and Methods

In this section, we present the methodology on early detection of Leukemia. Figure 1 depicts the system overview.

When a patient visits a doctor, a query form containing of 16 Yes/No questions is provided. Ideally, the questionnaire is to be sent to their phone directly and then to be filled out while waiting for a doctor's consultation. As soon as it is filled out and submitted a machine learning model will get to work and send the doctor the result which would be the prediction of an early stage cancer based on the answers. The doctor then take further steps such as consultation, tests according to his judgement.

Figure 2 shows the step by step procedure of our system. Firstly, we fix the parameters in consultation with the specialist physicians, then we collect primary data of different patients through the survey form. We then search for missing values

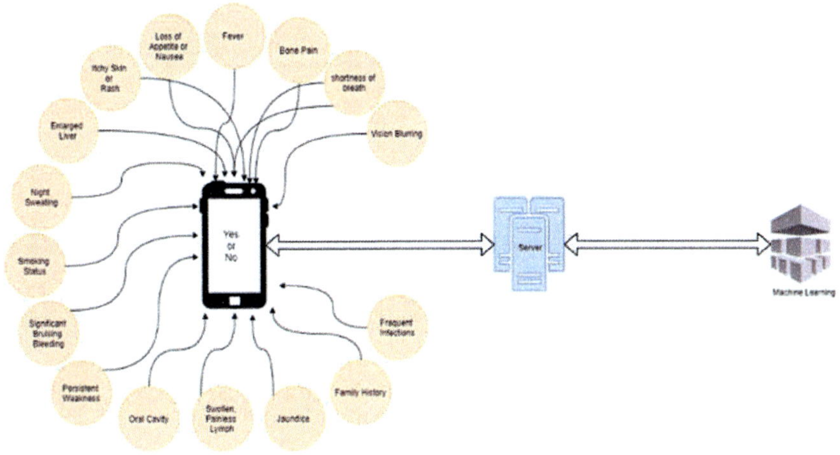

Fig. 1 System diagram of Leukemia detection

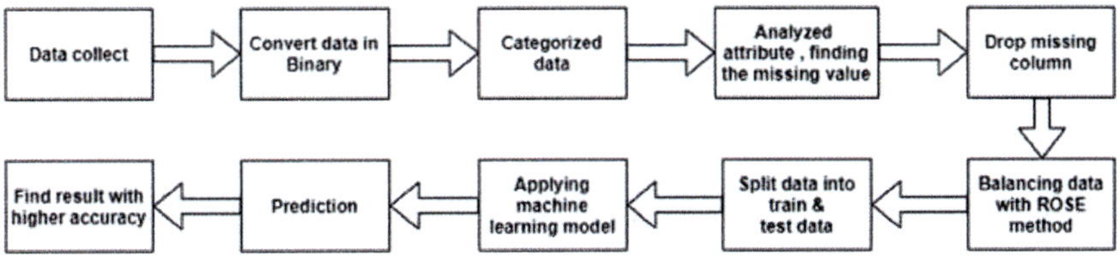

Fig. 2 Proposed methodology on early detection of Leukemia

and eliminate unnecessary columns. The Random Over-Sampling Examples (ROSE) method [12] is applied to balance the dataset and to split the dataset into train-tests and various machine learning algorithms are used to predict the accuracy.

3.1 *Experimental Data Set*

We have collected data from patients with Leukemia and some healthy patients through surveys. Dhaka Children's Hospital authority has assisted us in conducting this survey. When collecting data through surveys, we asked each patient about his/her common syndromes and took answer in between yes and no. In this study, the dataset which we have collected and used are described in Table 1. The training dataset has 709 instances and 17 attributes. For testing, there are 131 instances and 16 attributes. We then predict whether it is Leukemia or not. The class values are 0 and 1 which indicate the probabilities of Leukemia. The class values were further verified by specialist doctors and research officers at Bangabandhu Sheikh Mujib Medical University, Dhaka, Bangladesh.

Table 1 Description of the dataset collected from hospitals

SL.	Feature name	Feature type
1.	Shortness_of_breath	0-No, 1-Yes
2.	Bone_pain	0-No, 1-Yes
3.	Fever	0-No, 1-Yes
4.	Family_history	0-No, 1-Yes
5.	Frequent_infections	0-No, 1-Yes
6.	Itchy_skin_or_rash	0-No, 1-Yes
7.	Loss_of_appetite_or_nausea	0-No, 1-Yes
8.	Persistent_weakness	0-No, 1-Yes
9.	Swollen, painless_lymph	0-No, 1-Yes
10.	Significant_bruising_bleeding	0-No, 1-Yes
11.	Enlarged_liver	0-No, 1-Yes
12.	Oral_cavity	0-No, 1-Yes
13.	Vision_blurring	0-No, 1-Yes
14.	Jaundice	0-No, 1-Yes
15.	Night_sweats	0-No, 1-Yes
16.	Smoke	0-No, 1-Yes
17.	Leukemia	0-No Leukemia, 1-suffered Leukemia

3.2 *Data Preprocessing*

Data preprocessing is an initial step of machine learning. It plays an important role in classification. In our dataset, there are some missing values. Thus, it is necessary to handle the missing values. We have found that only the parameter 'smokes' contains some missing values. These missing values were replaced by the most frequent values. Since the dataset is imbalanced, we have merged the train and test dataset and then applied ROSE method to generate artificial data thus making it balanced.

3.3 *Classifiers*

Machine learning procedures have alluded as the expectation dependent on the models worked from existing information. Machine learning fields are utilized to find patterns in the dataset. Classification is a piece of the managed learning method where it gives the procedure to sort a given informational collection dependent on classes utilizing classification Algorithms. In this work, we applied several classification algorithms that are described below [13].

3.3.1 Decision Tree Classifier

Decision Tree classifier is the supervised machine learning algorithm which predicts the estimation of an objective variable by taking in basic choice principles deduced from the information highlights. We have made a model using decision tree and have applied it on our dataset. We have used default parameters of Decision Tree classifier to obtain the result.

3.3.2 Random Forest Classifier

Random Forest is a learning method that operates by creating multiple decision trees. The final decision is made mostly on the basis of trees and chosen by the random forest. Each branch of the tree represents a possible decision, occurrence. Here, we used n_estimators = 50, random_state = 0 as a parameter.

3.3.3 KNN Classifier

The output in the KNN classification is the membership of the class. An object is categorized by the plural vote of its neighbors, the object is assigned to its most common class in its nearest neighbors (k is a positive integer, usually smaller). In our work, We have used KNN for $K = 3, 5, 10$.

3.3.4 Adaboost Algorithm

AdaBoost, short for Adaptive Boosting, is a machine learning meta-algorithm. It can be used in combination with many other types of learning algorithms to improve performance. It is sensitive to Adjustable Noise data and outlines. We have used n_estimators = 1000, random_state = 90 as a parameter.

3.3.5 Logistic Regression Classifier

Logistic Regression is a machine learning algorithm which is mainly used for the classification problems. It is based on a predictive analysis algorithm and assumptions of probability. The logistic regression estimate limits the cost effectiveness to 0 and 1. We have used penalty as a parameter of logistic regression where the value is 12 and the inverse of regularization c is 0.1.

3.3.6 Naive Baysian Classifier

The Naive Bayes classifier is a supervised machine-learning algorithm that uses the Bayes sub-type, which assumes that the features are statistically distinct. Whatever this assumption is, it has proven itself to be categorized with good results. To calculate performance we have used default parameter of Naive Baysian classifier.

3.3.7 Artificial Neural Network

An artificial neural network is a data processing technique. It acts like how the human brain processes information. The ANN includes a large number of connected processing units that work together to process data. We have used batch_size = 200, epochs = 20 to fit our model.

3.4 Performance Evaluation

Cross-validation is used to evaluate the performance of the machine learning models on unseen samples. This method is sometimes called k-fold cross-validation in the sense that it contains a parameter called k. The value of k indicates how many groups an entire dataset should be divided into. It helps to give a less biased estimation model. It divides the complete training dataset into k groups, where each of them is at equal size. For each unique group, the group considers as a test set, and the remaining groups are as a train set. The model is fit into the train set, and the test set evaluates the performance. In our training set, we have applied threefold, fivefold, and tenfold cross validation by referring the value of k here to 3, 5, 10. We have evaluate that which cross-validation gives the best estimation.

Now we will discuss accuracy, precision and recall. It measures the classification performance. Here, accuracy refers to the number of correctly classified prediction from the total number of prediction. Precision is measured by the correctness of a model. It calculates the ratio of actual positive observations from the total number of predicted positive instances. Recall is the ratio of correctly predicted positive observations to the all observations in actual class-Yes. The formulas are given below:

$$\text{Accuracy} = \frac{\text{TP} + \text{TN}}{\text{TP} + \text{TN} + \text{FP} + \text{FN}}$$

$$\text{Precision} = \frac{\text{TP}}{\text{TP} + \text{FP}}$$

$$\text{Recal}l = \frac{\text{TP}}{\text{TP} + \text{FN}}$$

Here, TP denotes as total number of True Positive, TN means total number of true negative, FP denotes as total number of False Positive and FN denotes as total number of false negative.

4 Experimental Analysis

After pre-processing the dataset, we applied different types of machine learning models and analyzed their performances. Hence, we have used sci-kit learn as machine learning library to evaluate the performance of different machine learning algorithms and also we have used kaggle notebooks for coding implementation.

4.1 Training Set Performance

We have applied all the classifiers (Decision Tree, Random Forest Classification, KNN, Adaboost, Naive Bayes, Neural Network and Logistic regression) on the training set where they have been validated using k-fold cross validation ($k = 3, 5, 10$). Table 2 shows the accuracy values corresponding to each value of k.

Table 2 shows the accuracy values based on the training dataset. Decision Tree algorithm gives the highest accuracy values of 97.14% for threefold, 97.54% for fivefold and 97.74% for tenfold respectively. Naive Bayes model gives the less accuracy values than other models, such as 85.65% for threefold, 85.55% for fivefold and 85.69% for tenfold. In the case of other models, only the value of cross validity for which it gives the maximum accuracy value is shown. From Random Forest we got highest 96.54% for fivefold. With KNN model the highest value is 87.07% for tenfold. Moving on to Adaboost the accuracy value is 92.22% for tenfold. Finally, Logistic Regression gives 89.37% for tenfold It clearly shows the highest accuracy value is 97.74% for tenfold gives by Decision Tree model.

Table 2 Cross validation results on the train set

Algorithms	Threefold (%)	Fivefold (%)	Tenfold (%)
Decision tree	97.14	97.54	97.74
Random forest	96.54	98.41	97.67
KNN	82.99	86.12	87.07
Adaboost	91.82	91.89	92.22
Logistic regression	87.38	88.37	89.37
Naïve bayes	85.65	85.55	85.69

4.2 Test Set Performance

Initially, we obtained only 54% accuracy using ANN model. As we know artificial Neural Networks (ANN) perform well to model when a dataset consists of non-linear data with a huge number of input features. As we have a small size of dataset that's why it wouldn't perform well. Figure 3 shows the epoch versus accuracy and epochs versus loss function graph.

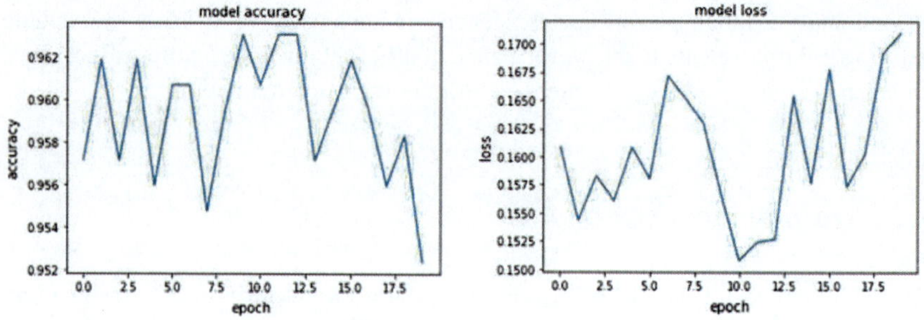

Fig. 3 Accuracy and loss function of ANN

We also applied the best learned model from the cross-validation on the test set for all of the algorithms used. Table 3 shows the comparison between the accuracy, precision and recall value of various algorithms that have been used. Using Naive Bayes, the results were not much satisfactory with 75.03% accuracy. Logistic Regression and Adaboost had the accuracy of 91.24% and 94.24% respectively. Using KNN we got an accuracy of 95.01% for $k = 3$ and 5 for $k = 7$ and 9, it gives the higher accuracy of 97.21%. Finally, with Decision Tree model we obtained the highest accuracy of 98%. Furthermore, we used Random Forest Tree and received the same accuracy of 98% ensuring that the model was not overfitting.

Table 3 Comparison of results between various algorithms on the test set

Algorithms	Accuracy (%)	Precision (%)	Recall (%)
Decision tree	*98.00*	*100*	96.32
Random Forest	98	100	96.32
KNN	97.21	97.35	*97.21*
Adaboost	94.82	100	96.64
Logistic regression	91.24	99.14	84.56
Naïve bayes	75.03	75.37	75.03
Neural network	95.62	75.58	74.38

5 Conclusion

This paper presents an early Leukemia detection method using a real dataset based on regular symptoms. We have shown comparative analysis of several machine learning algorithms. The best prediction model was Decision Tree Classification with 98% accuracy. As a future work, we wish to perform explainable analysis on the work.

References

1. Medical news today. https://www.medicalnewstoday.com/articles/142595. Accessed on 25 Feb 2020
2. Medical news today. https://www.medicalnewstoday.com/articles/282929.php. Accessed on 25 Feb 2020
3. George-Gay, B., Parker, Katherine: Understanding the complete blood count with differential. J. PeriAnesthesia Nursing **18**(2), 96–117 (2003)
4. Malempati, S., Joshi, S., Lai, S., Braner, D.A.V., Tegtmeyer, T.: Bone marrow aspiration and biopsy. N Engl J Med **361**(15),28 (2009)
5. Leary, R.J., Sausen, M., Kinde, I., Papadopoulos, N., Carpten, J.D., Craig, D., O'Shaughnessy, D., Kinzler, K.W., Parmigiani, G., Vogelstein, B., et al.: Detection of chromosomal alterations in the circulation of cancer patients with whole-genome sequencing. Sci. Transl. Med. **4**(162), 162ra154–162ra154 (2012)
6. Mohapatra, S., Patra, D.: Automated leukemia detection using hausdorff dimension in blood microscopic images. In: INTERACT-2010, pp. 64–68. IEEE (2010)
7. Markiewicz, T., Osowski, S., Marianska, B., Moszczynski, L.,: Automatic recognition of the blood cells of myelogenous leukemia using svm. In: Proceedings. 2005 IEEE International Joint Conference on Neural Networks, 2005, vol. 4, pp. 2496–2501. IEEE (2005)
8. Hsieh, H.B., Marrinucci, D., Bethel, K., Curry, D.N., Humphrey, M., Krivacic, R.T., Kroener, J., Kroener, L., Ladanyi, R., Lazarus, N., et al.: High speed detection of circulating tumor cells. Biosensors Bioelectron. **21**(10), 1893–1899 (2006)
9. Leinoe, E.B., Hoffmann, M.H., Kjaersgaard, E., Nielsen, J.D., Bergmann, O.J., Klausen, T.W., Johnsen, H.E.: Prediction of haemorrhage in the early stage of acute myeloid leukaemia by flow cytometric analysis of platelet function. Brit. J. Haematol. **128**(4), 526–532 (2005)
10. Fatma , M., Sharma, J.: Identification and classification of acute leukemia using neural network. In: 2014 International Conference on Medical Imaging, m-Health and Emerging Communication Systems (MedCom), pp. 142–145. IEEE (2014)
11. Pan, L., Liu, G., Lin, F., Zhong, S., Xia, H., Sun, X., Liang, Huiying: Machine learning applications for prediction of relapse in childhood acute lymphoblastic leukemia. Sci. Reports **7**(1), 1–9 (2017)
12. Lunardon, N., Menardi, G., Torelli, N.: Rose: a package for binary imbalanced learning. R J. **6**(1) (2014)
13. Christopher, M.B.: Pattern Recognition and Machine Learning, Springer (2006)

Deep CNN-Supported Ensemble CADx Architecture to Diagnose Malaria by Medical Image

Erteza Tawsif Efaz, Fakhrul Alam, and Md. Shah Kamal

Abstract Identifying patients, infected with the virulent disease, malaria requires a reliable and quick diagnosis of blood cells. This paper presents a computer-aided diagnosing (CADx) method supported by a deep convolutional neural network (CNN) for assisting clinicians to detect malaria by medical image. We employed the VGG-19 and ResNet-50 architectures to create several models for two types of study (parasitized and uninfected erythrocytes). To enhance the model's performance, an ensemble technique was applied, followed by which, the best model selected by performance measuring metrics. Our proposed model was qualified and examined upon a standard microscopic set of images collected from the National Institute of Health (NIH). The final result was analogized with other techniques, where the accuracy of this model was 96.7% for patient-level detection. To resolute the limitations and minimizing errors regarding automated malaria detection, the proposed model proved to be an appropriate strategy for distant regions and emergencies.

Keywords Image processing · Deep learning · Ensemble · ResNet-50 · VGG-19

E. T. Efaz (✉)
Department of Electrical and Electronic Engineering, Ahsanullah University of Science and Technology, Dhaka 1208, Bangladesh
e-mail: ertezatawsif@gmail.com

F. Alam
Department of Electrical, Electronic and Communication Engineering, Military Institute of Science and Technology, Dhaka 1216, Bangladesh
e-mail: fakhrulalam073@gmail.com

Md. S. Kamal
Department of Electrical and Electronic Engineering, University of Dhaka, Dhaka 1000, Bangladesh
e-mail: saimum012@gmail.com

M. S. Kaiser et al. (eds.), *Proceedings of International Conference on Trends in Computational and Cognitive Engineering*, Advances in Intelligent Systems and Computing 1309, https://doi.org/10.1007/978-981-33-4673-4_20

1 Introduction

Malaria, a mosquito-borne blood disorder, is an endemic in many regions and an epidemic in some parts of the world. It is mainly transmitted through the female *Anopheles* mosquito bite that carries the *Plasmodium* parasites, infecting humans through showcasing symptoms of very high fevers, shivering, and illness [1]. Even though the consequences of *Plasmodium falciparum* is considered fatal, there are various types of parasites generating this disease, involving *Plasmodium malariae*, *Plasmodium ovale*, and *Plasmodium vivax* [2]. According to the World Health Organization (WHO), malaria infects more than 200 million people [3], whereas it causes more than 400,000 deaths [4] every year around the globe. The susceptibility of this type of infectious disease outbreak is depended on factors like poverty level, healthcare access, governmental action, weather condition, and disease vector—a medium that carries and spreads disease to other organisms, where the fast-breeding insect, mosquitos are ill-famed for malaria and once infected, a human can also transmit this disease by sharing syringes, organ transplants, or blood transfusions [5].

The most common and reliable system to detect malaria is the infinitesimal testing of thick–thin plasma smear. In this process, the blood sample from a possible patient is placed on a slide and dyed by a divergent substance for spotting infectious red blood cells by a clinician. The accuracy of the examination primarily depends on the expertise of the medical personnel, who usually has to count 5000 red blood cells manually—an exhausting process [6]. As such, scientists developed rapid diagnosis testing (RDT), a faster technique to detect malaria, requiring a blood sample and buffer to generate a result [7]. Another approach is the polymerase chain reaction (PCR), which has a constrained efficiency [8]. For these reasons, the most effective solution is the combination of the accuracy of the microscopic testing and the speed of RTD [9].

On the other hand, employing machine learning (ML) for image analysis-based computer-aided diagnosis (CADx) software solved major issues regarding malaria detection [10–12]. Moreover, using deep learning (DL) algorithm showed better performance by overcoming the limitations of size, angle, and variability, background, and region of interest (ROI) position [13]. This paper focuses on developing a CADx technique to differentiate infected and uninfected cells from the blood smear image of a possible patient given on a microscopic lens. Firstly, a standard and large image subset of malaria were collected from the National Institute of Health (NIH), including 27,558 images—separated by two classes [14]. Then two state-of-the-art models through deep convolutional neural network (CNN) architectures (VGG-19 and ResNet-50) were created. A probability of parasitized (infected) blood cells was allocated on each view by a 19-layer CNN for visual geometry group (VGG) and 50-layer CNN for the residual neural network (ResNet), which were then averaged to estimate the final probability. Finally, an ensemble of the models was performed and compared to the classifiers for performance evaluation metrics, where the result exhibited that the system predicted the malaria risk factor with 96.7% accuracy.

The summary of this work can be drawn as follows,

- Employing a set of standard architecture to measure all parameters.
- Assembling several models to improve overall performance.
- Differentiating with notable classification methods.

The rest of the sections demonstrated the related works, proposed methodology, and experimental results.

2 Related Works

In 2018, Rajaraman et al. [2] at NIH employed six pre-trained CNN architectures, consisting of AlexNet, VGG-16, ResNet-50, Xception, DenseNet-121, and a modified DL architecture to enhance the detection methodology of malarial parasites utilizing microscopic thin blood smear images. It was a multi-step process consisting of feature extraction to classify parasitized/uninfected cells, fivefold cross-validation to decrease bias and error generalization, selecting optimal-layer to extract features and underlying data, and performance testing to identify previously trained and customized model difference. With 96 MB file size, the authors reported a 95.9% prediction accuracy of the parasitized or uninfected red blood cells (RBC) for patient-level detection, where the model processing (training and testing) time was around 24 h. Later in 2019, the authors [15] represented a technology by reducing model variance, increasing robustness, and generalization by creating a model ensemble. To prevent data leakage and reduce errors, cross-validation was performed, followed by a performance evaluation metrics. The ensemble design composed of VGG-19 and SqueezeNet seemed to outperform previous methodologies. In 2020, Fuhad et al. [16] offered a system using a range of techniques consisting of knowledge distillation, data augmentation, auto-encoder, and feature extraction via CNN standard. The model was analyzed through support vector machine (SVM) or *k*-nearest neighbor (*k*-NN), and performed under 3 training processes requiring around 4600 floating-point operations. Considering performance inference within a second; the model might be deployed for practical systems.

In 2017, Bibin et al. [17] introduced a six-layered deep belief network (DBN) for identifying the disease. The method was based on an automated decision support system, which used binary classifier to classify parasite or non-parasite cells by training a model consisting of 4100 images. The pre-training of the modeling was completed through accumulating confined Boltzmann machines employing a contrastive alteration process. To train, image feature extraction and the model's visible difference initialization were done, where color-texture was employed as a feature vector. Lastly, the model was fine-tuned through a backpropagation algorithm followed by optimization. In 2018, Devi et al. [18] developed a system on histogram-based feature for malaria identification. The erythrocyte (RBC) classification system created by preprocessing, segmentation, and feature extraction from different color channels. The color set of the histogram feature included saturation, green channel, chrominance channel, and absolute difference histogram of red-green channels from red–green–blue (RGB) image. The optimal features of 36 features for each classifier selected by evaluating feature combinations; for the classifiers: artificial neural networks (ANN), SVM, *k*-NN, and Naive Bayes. The ANN distinguisher provided highest detection rate.

The inefficiency of the existing ML systems [2, 11, 15–19] lies in the quantity regarding utilized models [2, 19], selection of optimization architecture [11, 15], duration of sample testing [16], evaluation of small dataset [11, 17], and/or randomized train/test splits [18, 19] to generate superior results. Though many studies published notable outcomes, more research focusing on a simplistic approach and faster training/testing of a model for patient-level detection with a larger image set needs firm development. Furthermore, in the remote and poverty-stricken areas, it will be difficult to find a dependable internet connection, high-performing computing source, and reliable power resource—though solar energy technology might be effective [20].

So, a desirable system needs the traits as below,

1. Research on patient-level detection by evaluating a large image set.
2. A compact method for IoT devices through a cloud platform.
3. Closer accuracy to NIH with less computation.

Microcontroller system gives ample scope for more development in this regard [21].

3 Proposed Methodology

Figure 1 depicts the proposed workflow to distinguish parasitized and uninfected cells.

3.1 Data Collection

The standard subset of the malaria dataset was collected from NIH, comprising 27,558 images [14]. The work was employed on equally distributed 13,779 parasitized and uninfected blood smear images after re-sizing those to 64 × 64 pixels for 3 (RGB) color mode channels. For each of the 2 class, about 85% of the total images were accounted for preparation and the last 15% was allocated for the experiment by data splits. Among the training dataset, around 10% of the whole dataset in each class was assigned for validation. Here, Fig. 2 represents a blood smear image subset provided by the NIH, whereas Table 1 illustrates the distribution of dataset for malaria detection consisting of training, validation, and testing for parasitized and uninfected categories.

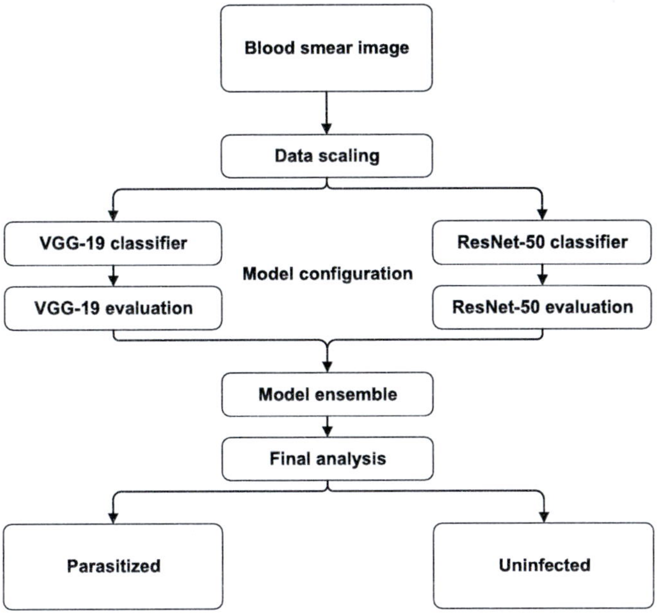

Fig. 1 Proposed workflow for malaria detection

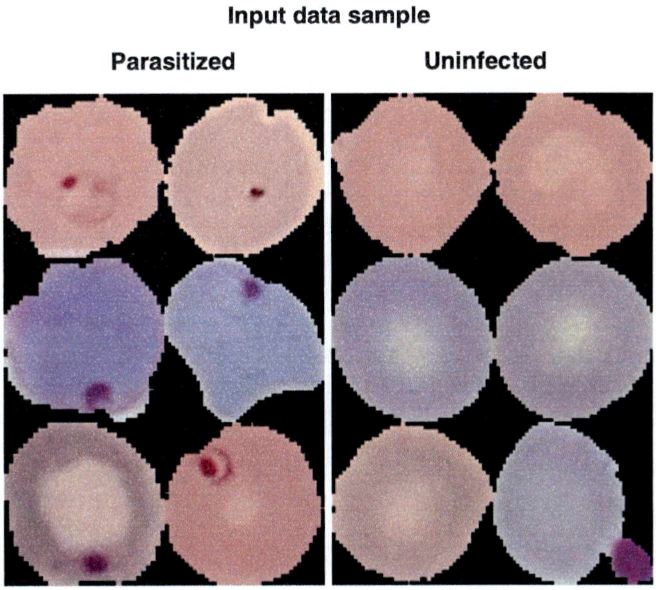

Fig. 2 A blood smear image subset provided by the NIH [14]

Table 1 Distribution of dataset for malaria detection

Dataset	Parasitized	Uninfected	Total
Training	10,301	10,301	**20,602**
Validation	1378	1378	**2756**
Testing	2100	2100	**4200**
Number of images	**13,779**	**13,779**	**27,558**

Bold indicates easier differentiation between total no. of images and distributed set of images

3.2 Model Configuration

For model training of both the VGG and ResNet architectures, the ImageNet dataset was used containing over 2200 object types and 15×10^6 high-resolution training images [22]. The VGG is a simple network architecture of 3×3 convolutional layers accumulated on another in progressing intensity that can capture little features skipping big patterns mostly. Else ways, ResNet is a complex network architecture, having the capability to discover large features as well as the small ones; whereas, residual module finally concatenates the mapping as element-wise addition and passes through a nonlinearity to generate output. In the case of both the models' training phase, predefined training data was fitted independently. Because deep neural network (DNN) has a considerable modeling potential, it may not be as productive on a small-sized dataset due to the parametric requirements of excessive data [23].

The implication of transfer learning (TL) decreased this issue because transferable weights can extract features [24]. As such, the transferable weights started the training process employing a pre-trained model. After collecting the pre-trained weights of ImageNet and initializing it to imply TL, the training process began. The first five, 17th, and 18th number layers (consisting of 4096 neurons each, then followed by a Softmax classifier) of VGG-19 were adjusted to improve the model. The weights of the first five layers were frozen to carry the transferable weights during the training.

To decrease the unnecessary computation of ImageNet, only 1024 neurons were chosen for both the 17th and 18th number layers after simulating a different number of neurons. By stochastic gradient descent (SGD) optimizer, the assembling of the model was executed with 0.0001 learning rate and 0.9 momentum. For matching the training data with the assembled model, the model was simulated many times with various epochs; and finally, for 100 epochs with 10 patience early stopping the best fit was generalized. The reference model for every architecture was selected regarding area under the receiver operating characteristic (AUROC) curvature with the highest value.

Figure 3 manifests the architectures of VGG-19 and ResNet-50 accordingly. The volume size reduction of VGG-19 is done by Maxpooling. For Resnet-50, $F(x)$ and $F'(x)$ are the residual mappings while x provides the identity mapping.

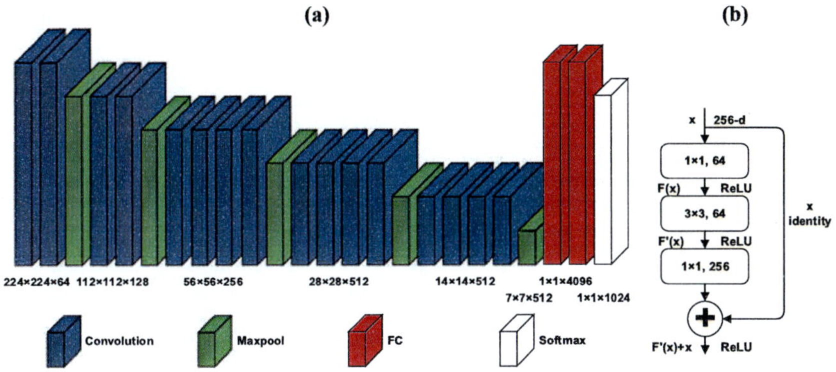

Fig. 3 Architecture of **a** VGG-19 and **b** ResNet-50 adapted from [25]

3.3 Model Ensemble

The ensemble method is an approach to construct a model by merging various architectures to generate a result with higher accuracy by devising sophisticated algorithms. It enriches both the existing dataset along with the model by continuous learning approach or combining diverse architectures. For getting a higher accuracy, the existing models are needed to be trained with better and accurate data as input for the weighted dataset. In this study, a weighted data manipulation technique was applied regarding a boosting ensemble model. So, we enhanced the weight of accurate data and calculated both the VGG-19 and ResNet-50 models' probability by applying Cohen's Kappa measurement. Then the weighted probability was determined by implying a formula through multiplying the model's probability with the weighted dataset. As such, the accuracy, precision, sensitivity rates generally increased in the ensemble model. The equations are specified as follows,

Consider, the weight of the *n*th classifier,

$$W_n = \frac{C_n}{\sum_{n=1}^{i} C_n} \tag{1}$$

And the prediction probability,

$$P = \sum_{n=1}^{i} P_n \times W_n \tag{2}$$

- Here, the *n*th classifier $= C_n$.
- And, the probability of the *n*th classifier $= P_n$.
- For the prediction probability >0.5, the microscopic blood smear image will be classified as parasitized.

To evaluate the model performance by prediction, model testing was employed after the completion of model training.

3.4 Statistical Analysis

Exceptional outcomes for diversity were avoided by evaluating performance measuring metrics. For parametric evaluation of both real and speculated conditions, a confusion matrix was generated. Henceforth, the expressions are unfolded as specified,

- T_{p} = true-positive (determined accurately)
- T_{n} = true-negative (determined inaccurately)
- F_{p} = false-positive (undetermined accurately)
- F_{n} = false-negative (undetermined inaccurately).

The indices are expressed as stated,

$$\text{Accuracy} = \frac{T_{\mathrm{p}} + T_{\mathrm{n}}}{T_{\mathrm{p}} + F_{\mathrm{p}} + T_{\mathrm{n}} + F_{\mathrm{n}}} \tag{3}$$

$$\text{Precision} = \frac{T_{\mathrm{p}}}{T_{\mathrm{p}} + F_{\mathrm{p}}} \tag{4}$$

$$\text{Sensitivity} = \frac{T_{\mathrm{p}}}{T_{\mathrm{p}} + F_{\mathrm{n}}} \tag{5}$$

$$\text{Specifity} = \frac{T_{\mathrm{n}}}{F_{\mathrm{p}} + T_{\mathrm{n}}} \tag{6}$$

$$F_{\beta}\ \text{score} = \frac{\left(1 + \beta^{2}\right)PR}{\beta^{2}P + R} \tag{7}$$

$$\text{Miss rate} = 1 - \frac{T_{\mathrm{p}}}{T_{\mathrm{p}} + F_{\mathrm{n}}} \tag{8}$$

$$\text{Fall out} = 1 - \frac{T_{\mathrm{n}}}{F_{\mathrm{p}} + T_{\mathrm{n}}} \tag{9}$$

$$\text{MCC} = \frac{\left(T_{\mathrm{p}} \times T_{\mathrm{n}}\right) - \left(F_{\mathrm{p}} \times F_{\mathrm{n}}\right)}{\sqrt{}\left[\left(T_{\mathrm{p}} + F_{\mathrm{p}}\right)\left(T_{\mathrm{p}} + F_{\mathrm{n}}\right)\left(T_{\mathrm{n}} + F_{\mathrm{p}}\right)\left(T_{\mathrm{n}} + F_{\mathrm{n}}\right)\right]} \tag{10}$$

$$\text{Cohen's Kappa} = \frac{\text{accuracy} - \text{expectation}}{1 - \text{expectation}} \tag{11}$$

The neural network model computation is a crucial portion of this analysis.

4 Experimental Results

After the implication of 2 different CNN architectures onto the microscopic blood smear image dataset, the performance measuring metrics was evaluated for each of the architectures. The best model was then chosen, whereas the investigation was ran using a moderately configured computer with Intel Core i7 2.5 GHz processor and 8 GB RAM. Table 2 presents the confusion matrix of different classifiers. The accurately determined value was comparatively higher in VGG-19 for the testing dataset. Among the 4200 instances for testing, 2051, 2004, and 2015 were determined accurately in VGG-19, ResNet-50, and ensemble models, respectively.

Figure 4 provides the ROC graph of VGG-19 and ResNet-50 correspondingly, where sensitivity plotted as a function of specificity. The AUROC for VGG-19 was 0.991, and AUROC of ResNet-50 was 0.987. Figure 5 exhibits the accuracy vs. epoch's graphs of VGG-19 and ResNet-50 accordingly. The models did not over-learn the training datasets, whereas the trends almost reached towards the saturated levels—though the VGG-19 model could probably be trained a little more. For parallel processing, the TensorFlow graphics processing unit (GPU) was utilized as a backend library, reducing the model training and testing time.

Figure 6 reveals the performance metrics of malaria detection.

From the subsequent graphical representations, it can be deduced that the ensemble model conveyed a better accuracy of 96.7% than both the VGG-19 and ResNet-50 (with 96.6% and 95.8% accuracy, respectively). The harmonic mean of the precision and recall (F_1 score) of the ensemble model is 0.967, where Matthews

Table 2 Confusion matrix of different classifiers

Classifiers	T_p	T_n	F_p	F_n
VGG-19	2051	2006	49	94
ResNet-50	2004	2018	96	82
Ensemble	2015	2047	53	85

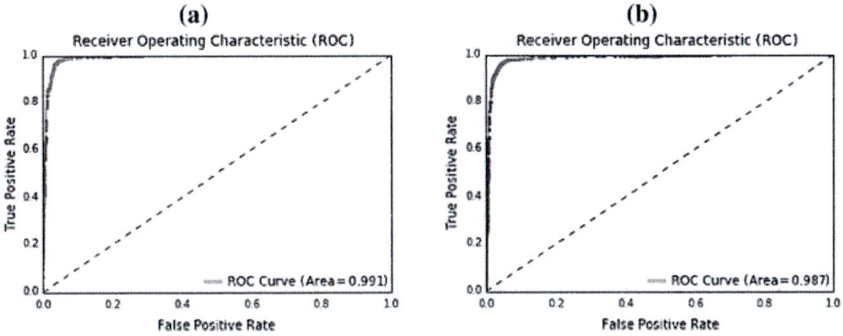

Fig. 4 ROC graph of **a** VGG-19 and **b** ResNet-50

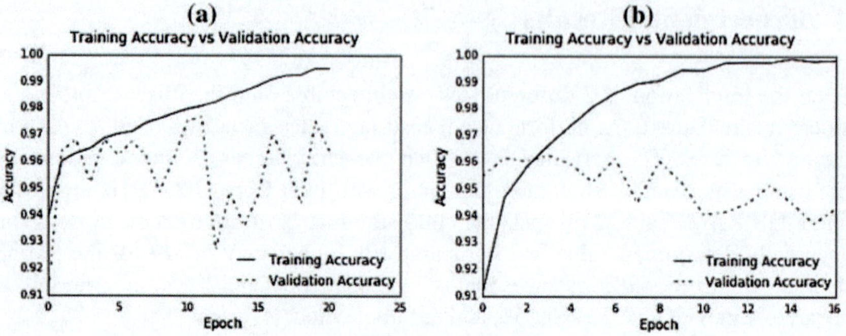

Fig. 5 Accuracy versus epochs graphs of **a** VGG-19 and **b** ResNet-50

	Accuracy	Precision	Sensitivity	Specificity	F1 score	Miss rate	Fall out	MCC	Cohen's Kappa
VGG-19	0.966	0.977	0.956	0.976	0.966	0.044	0.024	0.932	0.991
ResNet-50	0.958	0.954	0.961	0.954	0.958	0.039	0.045	0.915	0.974
Ensembled	0.967	0.974	0.959	0.975	0.967	0.041	0.025	0.934	0.994

Fig. 6 Performance metrics of malaria detection

Correlation Coefficient (MCC) and Cohen's Kappa showed a value of 0.934 and 0.994 accordingly. For each of these metrics (F_1 score, MCC, and Cohen's Kappa), the ensemble model indicated a higher rate. On the other hand, for precision, specificity, and miss rate; VGG-19 speculated better marks; however, ResNet-50 expressed higher metrics in terms of sensitivity and fall out. Overall, the ensemble model functions comparatively better for malaria detection, which can still be improved with notable research.[1]

Table 3 displays the performance differentiation of the stated architecture for malaria detection. Our proposed model showed an accuracy (0.967) close to the advanced method of Rajaraman et al. [15] for patient-level detection by exploring a different field. The model indicated a reasonable efficiency for almost all metrics compared to other methodological investigations (cell-level detection) as well, in terms of large dataset evaluation, simple model architecture, and faster training/testing process. For qualitative analysis, we upgraded the internal parameters by optimal solution, addressed the CNN architecture's limitations by multiple modeling, and speculated the statistical evaluations by performance indices. In

[1]Data supporting the conclusions of this research are accessible at github.com/ErtezaTawsif/Malaria.

Table 3 Performance comparison of proposed model for malaria detection

Methods	Accuracy	Sensitivity	Specificity	F-score
Our proposed model (patient-level)	**0.967**	**0.960**	**0.975**	**0.967**
Rajaraman et al. (patient-level) [15]	0.995	0.971	0.985	0.995
Bibin et al. [17]	0.963	0.976	0.959	0.897
Devi et al. [18]	0.963	0.930	0.968	0.853
Das et al. [11]	0.840	0.981	0.689	0.884

The bold metrics values indicate the performance measures of the presented architecture

summary, the proposed method conveyed noticeable performance with reliable outcomes. The final ensemble is done based on Cohen's Kappa statistics, combined with a boosting ensemble method. Our target was to give dynamic weights to the classifier, which holds better measures in Cohen's Kappa.

5 Conclusion

In the modern domain of medical technology, CADx proved to play a vital role regarding high speed and reliable efficiency to detect various diseases. Our proposed model thus explored a method to analyze the microscopic blood smear image dataset and form a classifier for detecting the contagious and often lethal parasitic disease, malaria. The system contains a classification method with a compact model to achieve a handful of benefits. Computing with a large set of images, our model performed with considerable accuracy within a short amount of time by a high-end GPU—recorded model training time was around 20 min and testing of a single study required less than 100 ms. The primary idea was to propagate this research to help medical personnel for taking precise and immediate therapeutic decisions considering poor, remote, and populous communities. Tuning parameters and applying different architectures could bring us a better result, thus the scope of improvement.

To improve the proposed model, recommendations can be made as below,

- Weight quantization on the respective models would allow a smaller version.
- The overall classification assignment should be available over the cloud.
- An end-to-end user-friendly mobile application needs to be developed.

The proposed model provides a design for automatic malaria detecting classifier via an application programming interface (API) with not only reasonable accuracy but also well-founded computation.

Acknowledgements This paper would not have been possible without the exceptional support of the Assistant Professor Khandaker Lubaba Bashar from the Department of Electrical and Electronic Engineering, Ahsanullah University of Science and Technology, Dhaka, Bangladesh, because her expertise has improved the research in innumerable ways and saved us from many errors; those that inevitably remain are entirely our responsibility.

References

1. Phillips, M.A., Burrows, J.N., Manyando, C., Van Huijsduijnen, R.H., Van Voorhis, W.C., Wells, T.N.C.: Malaria. Nat. Rev. Dis. Primers **3**, 17050 (2017). https://doi.org/10.1038/nrdp.2017.50
2. Rajaraman, S., Antani, S.K., Poostchi, M., Silamut, K., Hossain, M.A., Maude, R.J., Jaeger, S., Thoma, G.R.: Pre-trained convolutional neural networks as feature extractors toward improved malaria parasite detection in thin blood smear images. PeerJ (2018). https://doi.org/10.7717/peerj.4568
3. WHO: World Malaria Report 2016. WHO (2018)
4. Malaria Hero: A Web App for Faster Malaria Diagnosis. https://blog.insightdatascience.com/blog-insightdatascience-com-malaria-hero-a47d3d5fc4bb. Last accessed 30 June 2020
5. Deep Learning and Medical Image Analysis with Keras—PyImageSearch. https://www.pyimagesearch.com/2018/12/03/deep-learning-and-medical-image-analysis-with-keras. Last accessed 30 June 2020
6. Prevention, C.-C. for D.C. and: CDC—Malaria—About Malaria—Biology (2020)
7. WHO: How Malaria RDTs Work. WHO (2015)
8. Hommelsheim, C.M., Frantzeskakis, L., Huang, M., Ülker, B.: PCR amplification of repetitive DNA: a limitation to genome editing technologies and many other applications. Sci. Rep. **4**, 1–13 (2014). https://doi.org/10.1038/srep05052
9. Kido, S., Hirano, Y., Hashimoto, N.: Detection and classification of lung abnormalities by use of convolutional neural network (CNN) and regions with CNN features (R-CNN). In: 2018 International Workshop on Advanced Image Technology, IWAIT 2018, pp. 1–4. Institute of Electrical and Electronics Engineers Inc. (2018). https://doi.org/10.1109/IWAIT.2018.8369798
10. Ross, N.E., Pritchard, C.J., Rubin, D.M., Dusé, A.G.: Automated image processing method for the diagnosis and classification of malaria on thin blood smears. Med. Biol. Eng. Comput. **44**, 427–436 (2006). https://doi.org/10.1007/s11517-006-0044-2
11. Das, D.K., Ghosh, M., Pal, M., Maiti, A.K., Chakraborty, C.: Machine learning approach for automated screening of malaria parasite using light microscopic images. Micron **45**, 97–106 (2013). https://doi.org/10.1016/j.micron.2012.11.002
12. Poostchi, M., Silamut, K., Maude, R.J., Jaeger, S., Thoma, G.: Image Analysis and Machine Learning for Detecting Malaria (2018). https://doi.org/10.1016/j.trsl.2017.12.004
13. Lecun, Y., Bengio, Y., Hinton, G.: Deep Learning (2015). https://www.nature.com/articles/nature14539; https://doi.org/10.1038/nature14539
14. Malaria Datasets: National Library of Medicine. https://lhncbc.nlm.nih.gov/publication/pub9932. Last accessed 30 June 2020
15. Rajaraman, S., Jaeger, S., Antani, S.K.: Performance evaluation of deep neural ensembles toward malaria parasite detection in thin-blood smear images. PeerJ **7**, e6977 (2019). https://doi.org/10.7717/peerj.6977
16. Fuhad, K.M.F., Tuba, J.F., Sarker, M.R.A., Momen, S., Mohammed, N., Rahman, T.: Deep learning based automatic malaria parasite detection from blood smear and its smartphone based application. Diagnostics **10**, 329 (2020). https://doi.org/10.3390/diagnostics10050329
17. Bibin, D., Nair, M.S., Punitha, P.: Malaria parasite detection from peripheral blood smear images using deep belief networks. IEEE Access **5**, 9099–9108 (2017). https://doi.org/10.1109/ACCESS.2017.2705642
18. Devi, S.S., Roy, A., Singha, J., Sheikh, S.A., Laskar, R.H.: Malaria infected erythrocyte classification based on a hybrid classifier using microscopic images of thin blood smear. Multimed. Tools Appl. **77**, 631–660 (2018). https://doi.org/10.1007/s11042-016-4264-7
19. Gopakumar, G.P., Swetha, M., Sai Siva, G., Sai Subrahmanyam, G.R.K.: Convolutional neural network-based malaria diagnosis from focus stack of blood smear images acquired using custom-built slide scanner. J. Biophotonics **11**, e201700003 (2018). https://doi.org/10.1002/jbio.201700003

20. Efaz, E.T., Ava, A.A., Khan, M.T.A., Islam, M.M., Sultana, A.: Parametric analysis of CdTe/CdS thin film solar cell. IJARCCE **5**, 401–404 (2016). https://doi.org/10.17148/ijarcce.2016.5684
21. Efaz, E.T., Mamun, A.Al, Salman, K., Kabir, F., Sakib, S.N., Khan, I.: Design of an indicative featured and speed controlled obstacle avoiding robot. In: 2019 International Conference on Sustainable Technologies for Industry 4.0, STI 2019. Institute of Electrical and Electronics Engineers Inc. (2019). https://doi.org/10.1109/STI47673.2019.9068018
22. Krizhevsky, A., Sutskever, I., Hinton, G.E.: ImageNet Classification with Deep Convolutional Neural Networks
23. Deng, J., Deng, J., Dong, W., Socher, R., Li, L., Li, K., Fei-fei, L.: Imagenet: a large-scale hierarchical image database. CVPR (2009)
24. Mondol, T.C., Iqbal, H., Hashem, M.M.A.: Deep CNN-based ensemble CADx model for musculoskeletal abnormality detection from radiographs. In: 2019 5th International Conference on Advances in Electrical Engineering, ICAEE 2019, pp. 392–397. Institute of Electrical and Electronics Engineers Inc. (2019). https://doi.org/10.1109/ICAEE48663.2019.8975455
25. Common Architectures in Convolutional Neural Networks. https://www.jeremyjordan.me/convnet-architectures. Last accessed 30 June 2020

Building a Non-ionic, Non-electronic, Non-algorithmic Artificial Brain: Cortex and Connectome Interaction in a Humanoid Bot Subject (HBS)

Pushpendra Singh, Pathik Sahoo, Kanad Ray, Subrata Ghosh, and Anirban Bandyopadhyay

Abstract In the 1920s, Brodmann found that the neurons arrange in around 47 distinct patterns in the brain's topmost thin cortex layer. Each region controls a distinct brain function. Together, the cortex is made of 120,000–200,000 cortical columns, executing all cognitive responses. By filling capillary glass tubes with helical carbon nanotube, we built a corticomorphic device as a replacement of the neuromorphic device and using 10,000 such corticomorphic devices built a cortex replica. Using dielectric and cavity resonators, we built a complex nerve fiber network of the entire brain–body system. It includes connectome, spinal cord, and similar ten major organs. The nerve fiber network takes input from wide ranges of sensors, and the neural paths interact before changing the self-assembly of helical carbon nanotubes, which is read using EEG or laser refraction. The integrated brain–body system is our humanoid bot subject, HBS. One could refill entire cortex region with new synthetic organic materials to test spontaneous, software-free 24 × 7 brain response in EEG and optical

P. Singh · P. Sahoo · A. Bandyopadhyay (✉)
International Center for Materials and Nanoarchitectronics (MANA), Research Center for Advanced Measurement and Characterization (RCAMC), National Institute for Materials Science, 1-2-1 Sengen, Tsukuba, Ibaraki 3050047, Japan
e-mail: anirban.bandyo@gmail.com

P. Sahoo
e-mail: c.pathik@gmail.com

P. Singh (✉) · K. Ray
Amity School of Applied Science, Amity University Rajasthan, Kant Kalwar, NH-11C, Jaipur Delhi Highway, Jaipur, Rajasthan 303007, India
e-mail: singhpushpendra548@gmail.com

K. Ray
e-mail: kanadray00@gmail.com

P. Sahoo · S. Ghosh
Chemical Science and Technology Division, CSIR-North East Institute of Science and Technology, NEIST, Jorhat, Assam 785006, India

S. Ghosh
Academy of Scientific and Innovative Research (AcSIR), CSIR-NEIST Campus, Jorhat, Assam 785006, India

M. S. Kaiser et al. (eds.), *Proceedings of International Conference on Trends in Computational and Cognitive Engineering*, Advances in Intelligent Systems and Computing 1309, https://doi.org/10.1007/978-981-33-4673-4_21

vortices. Our extensive theoretical simulations of all brain components were verified with hardware replicas in the optoelectronic HBS. HBS is a universal tool to test a brain hypothesis using AI chips, organic–inorganic materials, etc.

Keywords Connectome · Humanoid robot · Artificial brain · Time crystal · Cortex · Humanoid bot subject, HBS · Artificial intelligence, AI · Natural intelligence, NI

1 Introduction

Humanoid bots are primarily designed for commercial works, to replace human jobs, not for testing the brain-inspired devices, or chips, or testing new synthetic materials that have a potential for the drug usage or computation. In brain research, two revolutions are running in parallel: first, building human brain organoids [1, 2], soft brain tissues using a 3D assembly [3], and the making of lookalikes for the brain components like cortical folds using an elastomeric gel [4]; second, an accurate mapping of the microstructures in the brain to build a functional brain model that resembles a living human brain in cognitive response [5] Yet how cortical circuits build human cognition is clouded with debates [6–11], the quest for cortical computation [12] requires an experimental model. The creation of multiple open databases for real brain components to building the artificial brains in various ways, emulating the human brain, is taking the shape of a movement [13–15]. Recently, an artificial brain was built by converting the human adult skin cells into a pluripotent cell—immature stem cell, which was then converted into a cerebral organoid [1]. Human organoid brains promised to replace the human brain subject (HBS). However, with no blood vessel and a limit to scale-up growth, the promise is dim. Loading the accurate neurons to upload a brain in a computer is heading towards failure as hype exceeds delivery. Google's Deep-Mind adventure suggests that perfecting millions of lines of codes for multiple purposes is narrow. Beyond a programmed task, it has zero value [16]. Current AI "executes a man-made list of instructions." Big data industry needs a brain to understand the rapidly changing random data, instantly and no time could be spared to write the algorithm even if one holds a quantum computer. Primary artificial brain-building approaches concentrate on replicating physical appearance or expressional gesture; don't try to build natural intelligence from scratch. Here we conceive the idea to assemble ten artificial brain analogs integrated within and above to operate together by integrating signals scale-free, alleviating the need for an algorithm [17, 18]. Therefore, we try to replace AI with natural intelligence for non-ionic, non-chemical, non-electronic, non-algorithmic, non-circuiting brain analog, where not just the neuron but all the layers may contribute [19, 20] to decision making.

Brain models are plenty, [1–5, 21] we need to choose one or build a new model to begin construction. Most brain models construct a black box to link the primary neuron firing to a cognitive response. Inspired by Wheeler's geometrodynamics

[22] extended to Feynman's geometric language of nature and nodes regulating the brain's anatomical network, [23] we advance the geometric reasoning [24] to a geometric language [25] in a quest to find a universal language [26]. A "geometrical language" requires basic shapes connected to the primitives of symmetries and rotations and combinatorial rules. The geometric musical language, GML (JP-2017-150171, World patent WO 2019/026983), uses 15 basic geometric shapes (five 1D, five 2D, and five 3D structures), the ratios of sides and or angles are represented using the ratios of 15 primes [19, 20, 27–29]. To model spatio-temporal cortical coding of the brain, [11] geometric shapes are embedded in a clock, so musical word is used in GML. In a hypothetical brain like decision-making unit, the geometric shapes integrate as primes combine to build integers ($6 = 3 \times 2$, i.e., a pair of triangles) following a pattern called Phase Prime Metric, PPM [17, 29] (JP-2017-150173; World patent WO 2019/026984). Conversion of all sensory information as the changing geometric shapes is justified as humans combine geometric primitives into hierarchically organized expressions [24, 30]; for example, human intuitions [31]. Therefore, PPM, combined with GML, promises non-algorithmic decision-making [17], but it demands to map all brain functions as an architecture of nested clocks [19, 20]. A complete representation of all functions as a 3D network of clocks is easy to implement in equivalent hardware that is non-ionic, non-chemical, non-electronic, non-algorithmic, and non-circuiting, i.e., made of wireless antenna-based resonators. Current brain models [1–5] are either non-functional lookalikes or loaded with complex algorithms and big data to mimic cognitive responses. An effort to include all the operational layers and components in a brain–body circuit by assembling clocks is not new; it's an extension of a time crystal research of the 1970s [17]. Spontaneously operating clocks were conceived as time crystals to explain the incredible intelligence of elementary life forms. For three decades (1970–2000), wide ranges of biomaterials were studied in the quest of time crystals, including cell membrane [32]. Reddy's [17] and Singh's efforts [19, 20] considered the periodic ionic, molecular signal transmission structures as a clock, and several such clocks were linked in a 3D architecture to create time crystal analogs of all brain components. Singh et al. have proposed the self-operational mathematical universe with a space-time-topology-prime (stTS) metric that enables an architecture of clocks to self-govern [27]. In separate work, Singh et al. have also proposed to project any input information converted to time crystal stereographically to infinity and taking feedback from infinity (PF protocol) using the stTS metric [28]. Finally, GML [33], PPM [17, 29] stTS [27], and PF [28] were integrated into a self-operating time crystal model of the human brain by Singh et al. [19, 20], we use that in HBS.

Two thrusts have swamped the brain research: first, building more accurate map of the connectome's sub-structure [34], i.e., detailed wiring of the nerve fibers; second, identifying periodic chemical and electrical signal transmission pathways as rhythms or clocks [35, 36]. Singh et al. considered every cavity in the brain–body structure as a cavity resonator and its solid geometric fillers as a dielectric resonator, so that essential criteria set by time crystal model of the human brain are met [20]. Large resonators or clocks are easy to build, for <1 mm structures, e.g., cellular materials and components, we used self-assembled helical CNT or gel nanowires with computational abilities [37]. The HBS is hybrid, sensors feed input architecture of clocks using electronics, but at the molecular scale, we use optics to read periodic events. Special coils replicate real nerve fibers. Reported linear, open pathways [38] for the brain's cognitive and decision-making tasks are converted into closed loops, represented as circle or loop to fit GML. Coil-made Yagi antennas, wireless connections, junctions with multi-directional wireless transport were used to bridge components and create time delay between clocks. Brain circuits for 20 known conscious experiences were implemented [20], along with an architecture of 537 classes of clocks that could represent the entire brain [19, 20, 29]. In the humanoid avatar, HBS, all the clocks could be traced back to any standard brain book [38].

We are not making a biological brain. Artificial fibers made of coils reshaped as an antenna that resonates like human organs are unique brain engineering. The shapes of all brain–body organs are such that the *em* field faces anisotropy in transmission [39]. The electric and magnetic field splits, evanescent waves [40, 41] generate at the desired cavities exactly at locations where if a new organ is kept, its local property would naturally be amplified. Thus, the host cavity ensures that the guest cavity's below noise signals are amplified. The composition of dielectric resonators filled with cavities, if repeated 12 times, within and above, one enters from the massive brain–body neural net to alpha-helices. In the 10^9 orders of a spatial journey, always locating a new guest dielectric in the cavity of the host dielectric. HBS produces evanescent wave [39, 41], like weak electromagnetic bursts of currents in all components. The flow of the field through HBS replicates chemical nerve impulses even in the large replicas of brain components. We noted such evanescent wave [39] triggered flow in 100 times larger replica for all ten isolated brain structures. The ten components are connectome, cortex, cerebellum, hippocampus, blood vessel, ventricular system, amygdala, thalamus, hypothalamus, and the whole body's nerve network linked to the spinal cord as well as to all the sensors. When all real-size components were integrated into the HBS, the "anisotropic field-induced burst of the weak electromagnetic field" flows a current 100 s of nano-amperes for running the proposed clock model of the human brain [20]. Artificial cortical maps are already known to do object recognition like AI tasks [42]. The geometric integration of ten components adds new rhythms or clocks to the isolated organs, together realizes a temporal journey of 10^{12} scales (1 ns to a few days of circadian rhythm).

If organoid human brains, grown in a Petri dish, ever matures, it would require feeding and excretion of waste material, so we need a non-ionic brain. Junctions produce heat, so we need non-electronic hardware to scale up without massive cooling infrastructure. Finally, a brain operates without pre-programming. Hence, an artificial brain should operate without a single line of code. We built such a brain–body structure satisfying three criteria [29]. Here we limit our description to the geometric interaction [43, 44] between the cortex region and the connectome as a brain learns and changes the hexagonal close packing of cortical columns [45–47], in terms of EEG. Linking EEG [48] and cortical column rhythms [49] is an extensively studied subject. We feed giant nerve fibers of connectome with a particular sensory input, body's network feeds the 47 functional domains (Brodmann's regions), in the thin cortex layer [50, 51]. We monitor EEG, and spinal cord recording 24 × 7 to learn the global-HBS operation. Simultaneously record the optical and magnetic vortices generated by Brodmann region's local cortical columns. Optical vortex reveals sensory network's functional role with a millimeter-scale resolution, unlike EEG.

2 Results and Discussion

2.1 *Brain Cavity Made of Cortex Folds that We See from the Outside Is not the Real Cavity: A Triplet of Triplet Cavity Holds the Connectome from Within*

The advantages of the triplet of triplet cavity: Following brain anatomy works [38] and the available 3D cortex maps of the brain, we created 3D cavities of the human brain using 3D printers replicating the cortex folds. We built every component in the Computer Simulation Technology (CST) simulator and practice. We created the inner cavities of the upper brain cortex precisely, where the nerve fibers branch out to a perpetual end of cortical column. All the previous works on cortex fold replicated only the external look of the upper folds [4], not the curvatures beneath, where our CST study finds that the fold arranges cavities that quantizes and integrates the internal random signals. When a sensory signal enters from all over the humanoid bot, the empty cavities deliver background intensity I_o. CST simulation shows that if there are no nerve fibers, a randomly propagating energy flow would interfere, redistribute, quantized energy in local regions, arrange quanta in 3D patterns. When fibers fill the cavity, the ratio of intensities at every coordinate in the cavity $I_{x,y}/I_o$. By changing the internal cavity-shapes, we found that triplet of triplet field-packets created by cortex folds inside was never reported or studied before to the best of our knowledge.

Fluid flow simulation in the cortex cavities under dielectric resonance was experimentally verified by using the electromagnetic antenna and sensor probes. Mechanically vibrating the 3D-printed cavity boundaries of a brain while placing it in the 8 MHz ultrasound bath, the 3D field distribution measured by inserting a probe through the nasal cavity, validated quantization. The empty brain study estimates the anomaly in the brain's energy transmission when the cavities are filled with nerve fibers.

How the peripheral nervous system, PNS, and connectome experience the hierarchical cavity network? If we rotate the detailed 3D map of nerve fibers in the connectome, there are primarily five bundles, as shown in Fig. 1a. To understand the projection of energy input as information from sensors to the brain cavity through the 31 spinal nerve and 12 cranial nerve systems (31 + 12 = 43 nerve fibers of the peripheral nervous system, PNS), we divide the whole brain into multiple parts based

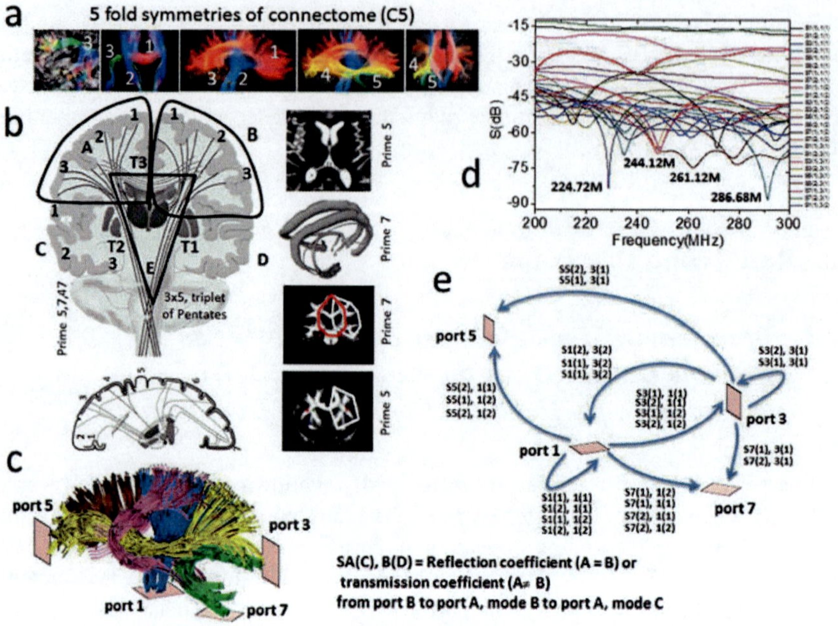

Fig. 1 Theoretical studies on the connectome. **a** Fivefold symmetries (C5) exist in human brain nerve fibers, corticospinal tract 1, the body of corpus callosum 2, arcuate fasciculus 3, extreme capsule fiber system 4, and uncinate fasciculus 5. Color codes of nerve bundles are given as; pink—1, blue—2, light green—3, yellow—4, and dark green—5. **b** Nerve fibers are extended in 3 symmetrical regions (T1, T2, and T3) of the human brain at right, left, and lower end of cerebral, which triggered five cortex regions (A, B, C, D, and E). Nerve bundles are extended in 7 classes of cortex folds, shown in the bottom of the panel (**b**). **c** Four distinct EB energy sources; 1, 2, 3, and 4 are applied at corticospinal tract, extreme capsule fiber system, corpus callosum body, and uncinate fasciculus nerve bundles, respectively. **d** Resonance spectrum of connectome model with resonance peaks at 224.72, 244.12, 261.12, and 286.68 MHz. **e** Energy reflection and transmission map through the applied energy sources at different propagation modes are given in term of $S_{ij} = \begin{pmatrix} \text{Reflection coefficient} & i = j \\ \text{Transmission coefficient} & i \neq j \\ \text{where} & i, j = 1, 2, 3, 4 \end{pmatrix}$. **Parametric details**: Used solver—Maxwell equation solver, mode—time domain, boundary condition—open space, selected frequency range—200–300 MHz, waveguide port—200 mm × 100 mm

on the internal cavity shape (Fig. 1b). The vertical cross section of the brain structure in Fig. 1b shows five regions, A, B, C, D, and E, based on cavity boundaries. The 3D-printed brain cavity with the cortex folds three times for each lobe, i.e., left and right hemisphere. Each of these three cavities has three sub-cavities. The fluid flow simulation pumped through the spinal cavity creates five local whirls in the empty brain around five centers. The three triangular or conical regions, which have three sub-cavities within, build nine high intensity and null signal domains during the free flow of fields (fluid = field). Such a triplet of triplet cavities organizes the distribution of ions, fluids, or electric/magnetic fields orderly in a hierarchical manner.

Important functional domains of the brain acquire three sub-cavities in each of the five primary-cortex fold made cavities. We especially note the central part of the midbrain (inverted triangle, Fig. 1b), where functional MRI reveals seven branching patterns; thus, $3 \times 3 \times 5 \times 7$ (=315) is the hierarchy of 315 cavities in each hemisphere. In Fig. 1b panel, the right column, re-organization of fields in seven regions argues for a periodic or a clocking flow. Entire sensory input from all over the body arrives in the brain via 43 PNS links to 315 cavities, which integrates field distribution to 45 cavities ($3 \times 3 \times 5$), which again integrates into nine cavities, which finally integrates into three cavities. Seven pathways branch out functionally all around the cortex from the midbrain; even if no neurons are there, cavities alone are ready to integrate the random flow of fields, order, and localize.

Filling the empty brain cavities with the nerve fibers and organs naturally activate 315 cavity generated energy sources. Previously, we studied the connectome of Rajat Jain's brain assigning dielectric properties like real nerve fibers [19]. Biological human brain map is obtained from the existing connectome projects, and we replaced each fiber in the real connectome map using distinct cylindrical tubes (Fig. 1c).

2.2 The Critical Challenges for the Simulation of Connectome: How Do We Interface Helical Carbon Nanotube-Based Cortical Columns and the Connectome?

A significant reduction in the complexity of the theoretically built connectome structure: Convergence of multi-probe Maxwell's equation simulation. Proper selection of boundary conditions and probe location reduces the simulation time of connectome structure. Boundary conditions depend on the class of excitation sources and type of wave port, open or discrete. For the simulation of the connectome model, energy is pumped through the wave ports, at open boundary condition was kept in all directions (x, y, and z-axis). The numerical electromagnetic simulator automatically sets the quarter wavelength available in the space in any direction from the created structure. Waveport means enclosing the whole field filled domain of wave propagation pathway along with the port area [52]. The pick face tool automatically selects the port location along the direction of the plane surface. Still, here we kept on changing the position of pick face or surface until port-to-port direct communication is nullified, and the material's true electromagnetic resonance response dominates. Four sensory probes (ports) are placed at the locations shown in Fig. 1c; simulation six parallel source–drain energy transmissions ran for months to optimize 24 reflected and transmitted spectra obtained in Fig. 1d, the spectrum between each pair of ports as noted in Fig. 1e. Solving Maxwell's equation for a large number of fibers (1:1 million reduction) requires extreme patience in placing the probes at the right location around the 3D brain architecture. A little change in the location of the port angle or area would lead to divergence. Four sharpest resonance peaks are identified in Fig. 1d, inset, at 286.68 MHz, the quality factor is the highest. Hence, we imaged electric and magnetic field distribution at this particular frequency in Fig. 2a.

The electric E and magnetic B field distributions of connectome from different directions and cross-section wise show asymmetry (Fig. 3). Identity of the ports does not reflect in the E–B distribution. Arrow points to the particular end parts that eventually would connect to the cortical columns in the cortex are selected naturally as a part of the resonant *em* oscillations. It means, while designing the brain cavities, nature has made sure that triplet of triplet cavities spontaneously canalizes a random

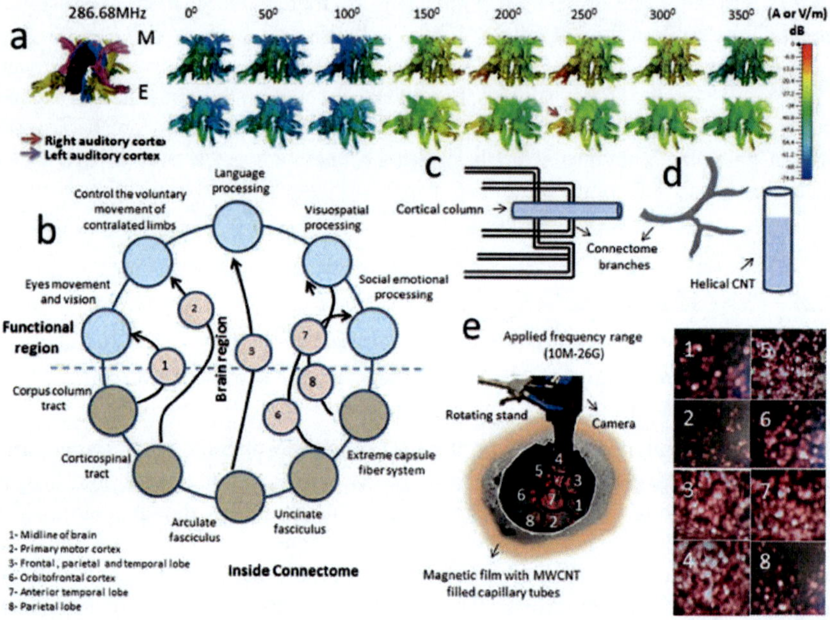

Fig. 2 Electromagnetic studies on the connectome. **a** Electric (bottom row) and magnetic (top row) field distributions on the connectome surface mapped at 286.68 MHz resonance peak. The field distribution reaches a maximum at 200°–350° phase domain on both sides of the connectome. The maximum and minimum energy intensities were found in the left auditory region and right auditory region, respectively, including the cerebral hemispherical region. The blue color regions show the minimum intensity (V/m) at 0°–150° phase duration, and the intensity scale is shown on the right side of the panel (**a**). **b** Image of the functional map of nerve fiber tracts passes through different brain regions. **c** A schematic of connectome–cortical column junctions. **d** Connectome–cortical column junction built artificially by filling CNT solution inside the capillary tube triggered by antennas. This is a CNT analog of a cortical column. **e** The experimental measurement set up on the bot's brain to read CNT dynamics by pumping laser light from the bottom. The light passes through capillary tubes and then via magnetic film the vortices reach the semiconductor camera. The set up converts *em* signals from sensors to the optical vortices on the humanoid bot's head, read through a magnetic film (Ni nanowire filled with PSS polymer bead). The semiconductor camera, a channel of cortical columns filled with MWCNT—multiwall carbon nanotube (200 nm pitch, 20–50 μm long), and laser light sources are assembled all over the brain. Optical vortices on the different regions of the bot's head (noted as 1, 2, …, 8 in panel (**a**)) are detected in the right panel

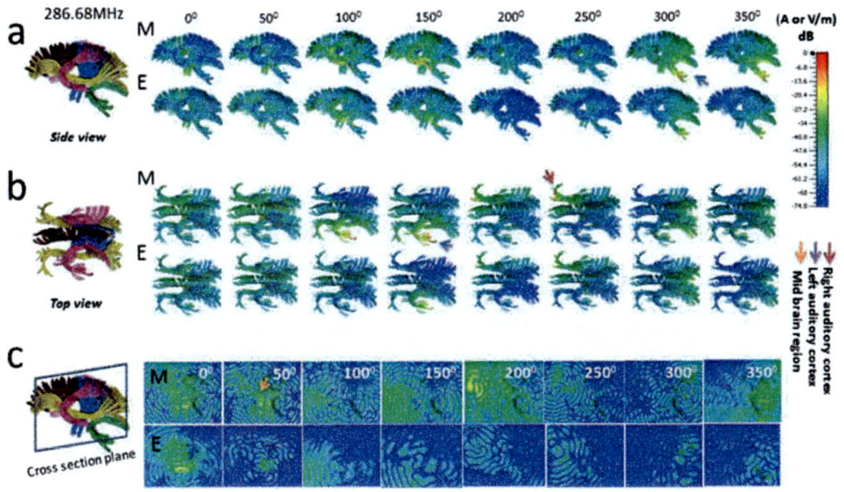

Fig. 3 EB field distribution from different directions at 286.68 MHz resonant frequency. **a–c** The electromagnetic (EM) field distribution on the connectome structure built by us connecting the nerve fibers one by one manually. Here, from different directions—side view, top view, and on 2D cross-sectional plane. The field intensity scale is shown in the right of the panel (**a**). **Short description**: (**a**) The E&B field distributions show the periodic nature along the uncinate fasciculus, arcuate fasciculus, and capsule system of brain nerve bundles. Minimum (blue color region) and maximum (dark yellow color region) intensity regions are seen at 0°–50°, 200°–250°, and 100°–150°, 300°–350° phase angles, respectively. **b** Both field components are maximum at splenium capsule fiber tract (frontal cortex region—0°–50°), left uncinate fasciculus tract (midbrain region—100°–150°), right capsule fiber tract (right auditory cortex region—200°–250°) and genu of corpus fiber tract (cerebellum region—300°–350°) of connectome model. **c** At the 2D cross-sectional plane, the maximum field intensity is seen in the midbrain region (yellow color arrow—0°–50°), left auditory cortex (arcuate fasciculus—100°–200°), right auditory cortex (right arcuate fasciculus—250°), and cerebellum region (corpus fiber system—300°–350°)

energy flow into 19 domains in each of the left and right brain. From 19 places, the neural fibers could build 47 functional networks to feed selective cortical columns of the 47 Brodmann's regions.

The wireless junction between the terminal of connectome fibers and the glass capillary-based cortical column layer: How CNTs work in a capillary tube? Theoretical cortical columns [53–55] stress on geometry. But in our brain model, the membrane of neurons is not absolute, cortical column and filaments control it. The inclusion of all layers [56] of cavities and the solid dielectric resonators impede to reverse engineer self-operating hardware. We need to rebuild cortical columns. Figure 2b summarizes EB field distributions; in a wheel, here the functional regions of the cortex are linked by directed energy transfer. Most interestingly, the links are not made by user instruction via an algorithm. Instead, it is pure geometry of the component that resonantly links different functional domains. Links are junctions; in the real brain scenario, the terminals of the connectome fibers extended to the skull directing outward, hold a cortical column in its terminal point [47] as shown in Fig. 2c. There are inputs and outputs to and from the cortical columns spread all over the cortex domains [45, 46]. We need a physical equivalence for connectome–cortical column junction. Using a capillary tube filled with helical carbon nanotube (200 nm pitch, ~10 μm length), we have created the artificial cortical column. Helical carbon nanotubes dissolved in hexane (both sides of capillary tubes are sealed to stop evaporation) generate various kinds of optical vortices. Since 10,000 capillary tubes were prepared in 47 regions connected wirelessly to 120000 connectome antennas, one cortical column had 12 dedicated wireless signal sources in principle. Seven layers inside the cortical column act as input and output channels. Layers 1–3 work as input channels, while layers 6–7 are the output channels [29].

One-to-one correspondence between the biological cortical column made of brain mapped neuron and cortical column made of helical carbon nanotubes. Sensors input em via connectome, the CNTs self-assemble and group [57], the optical vortices sent through these capillary tubes are edited by those self-assembled CNTs. Therefore, CNT-based cortical columns sense the sensory input after edited by 20 consciousness circuits. The terminals of the cables representing the end part of connectome fiber are sliced into three sharp edges as the Yagi antenna (two edges reduce noise; the central edge transmits signal) to feed *em* packet to the capillary tubes or cortical columns. In Fig. 4, to build a column, from the literature, we have identified 26 types of neurons, which are mostly found in the cortical columns [53] (see Fig. 5).

A typical case is shown in Fig. 4, where a single cortical column is made of 26 types of neurons arranged in seven layers. These neuron architectures were purchased from the various database. We composed the 3D neurons in various orientations to build the cortical columns in the CST simulator for resonance measurement. While placing neurons in the cortical columns, we intricately followed published coordinates [53]. The coordinates of soma and layout of branches have geometric mismatch by 8–10 μm in the column. Our artificially created cortical columns deliver similar *em* resonance features like the biological ones. At least the ratios of high-intensity resonance peaks are similar. We had to build a database of cortical columns on our own for two reasons. First, noise-free cortical columns directly mapped from the human brain are not available. Second, most databases of cortical columns use similar types of neurons. Here we study the editing of the hexagonal lattice formed by cortical columns and how connectome could naturally edit their geometry as described in Fig. 6.

Impedance matching between terminal end antennas of the connectome and capillary glass tube-based cortical columns. We added ports across the simulator analog of biological connectome structure so that (Fig. 1c) multiple resonance peaks generated there in link prominent functional regions of the brain, as outlined in Fig. 1d. Resonant communication pathways always pass through multiple cortical columns embedded in the mm thin cortex layer. Assembly of cortical columns forms Brodmann's region. Cortex and connectome link in the biological brain via horizontal branches of nerve fibers (Fig. 2c). Here we do not deal with the chemical

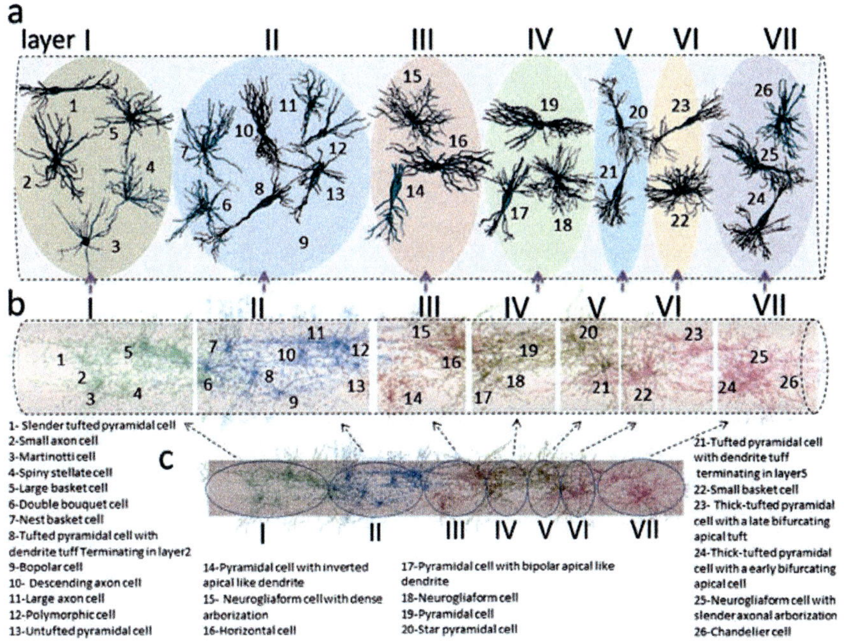

Fig. 4 Theoretically constructing a single cortical column with 26 classes of neurons. **a** The horizontal arrangement of neurons in a cortical column. The cortical column consists of 26 classes of neurons 5 (1–5), 8 (6–13), 3 (14–16), 3 (17–19), 2 (20–21), 2 (22–23), and 3 (24–26), which are included in 7 layers (I, II, III, IV, V, VI, and VII). The meaning of symbolic signs of all neurons is given at the bottom of the panel (**c**). Two panels (**b, c**) are similar maps of a cortical column. These neurons are mapped from the real human brain. Details of purchased neuron: neuron's name—pyramidal neuron: purchased site—https://www.turbosquid.com/; product id-1097217; polygons—1,041,365; vortices—1,060,927)

processes in the human brain, hence, simplify all mechanisms. On a limited scale, the geometric structure of the connectome branches with antenna terminals directed to the capillary tubes filled with helical CNT (Fig. 2d) edits the resonant energy distribution. One important feature of the cortical column's assembly is that they adjust their relative positions in the hexagonal lattice to synchronize with the connectome input. Connectome nerve fiber networks (semiconductor junction cable) and layers of cortical columns (CNT in the cortical tube) are very different materials. Geometrically, both are distinct. Therefore, matching the resonance frequencies between antenna and CNT tubes is essential for the coherent energy transfer between the two entities.

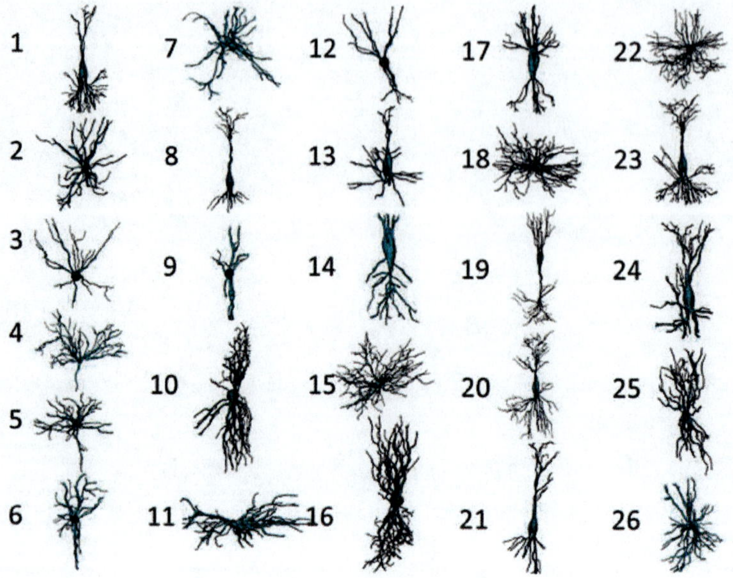

Fig. 5 List of 26 types of neurons. Twenty-six classes of neurons used for simulations are shown in this figure. The names of neurons are given as: 1—slender tufted pyramidal cell, 2—small axon cell, 3—Martinotti cell, 4—spiny stellate cell, 5—large basket cell, 6—double bouquet cell, 7—nest basket cell, 8—tufted pyramidal cell with dendrite tuff terminating in layer 2, 9—descending axon cell, 10—large axon cell, 11—polymorphic cell, 12—polymorphic cell, 13—untufted pyramidal cell, 14—pyramidal cell with inverted apical like dendrite, 15—neurogliaform cell with dense axonal arborization, 16—horizontal axon cell, 17—pyramidal cell with bipolar apical like dendrite, 18—neurogliaform cell, 19—pyramidal cell, 20—star pyramidal cell, 21—tufted pyramidal cell with dendrite tuff terminating in layer 5, 22—small basket cell, 23—thick-tufted pyramidal cell with a late bifurcating apical tuft, 24—thick-tufted pyramidal cell with a early bifurcating apical tuft, 25—neurogliaform cell with slender axonal arborization, 26—chandelier cell

2.3 *Cortical Column Assemblies*

Reflectance and transmittance simulation and measurement to explore the lattice change effect as the brain learns. To the best of our knowledge, no theoretical dielectric resonance studies have been done for the neurons and cortical columns. In such studies, a biomaterial like dielectric material is shaped accurately in the form of a biological system, so that the structure's cavity resonance and dielectric material resonance, both natural oscillations, are simulated in terms of resonance frequencies as reflectance S11 and transmittance S21 across the material [19, 57, 58]. A hexagonal lattice cell network made of cortical columns has a center with six columns around (total 7), and then around six columns, there are 18 columns more, a total of 19 columns. We have artificially built 19 cortical columns to monitor the change in the lattice shape. It is the minimum requirement to monitor the effect of memory and learning in a hexagonal lattice [7, 8]. We modified the distance between the

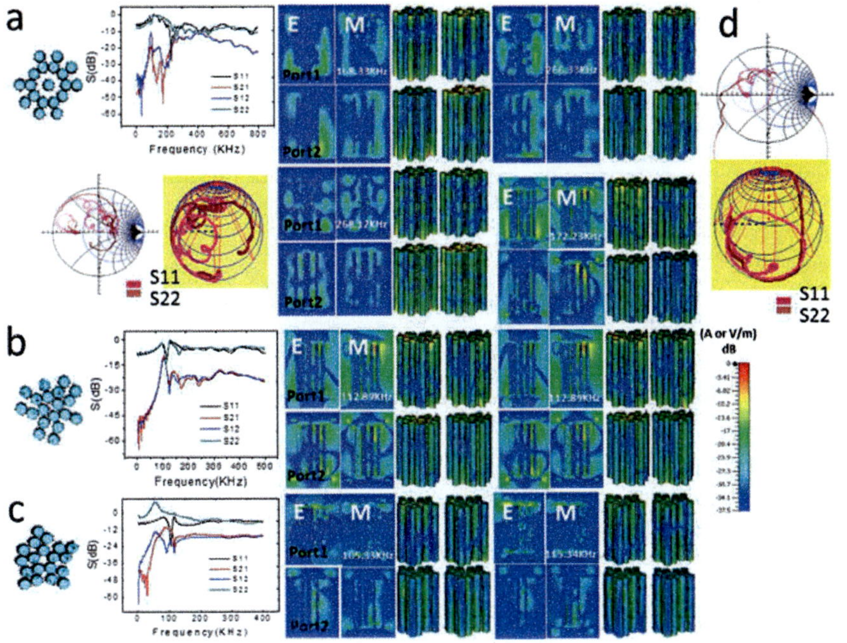

Fig. 6 Effect of changing the hexagonal close packing. Three classes of cortical column assemblies are shown in this figure, while 19 classes of distinct cortical columns are included in a single column assembly. **a** The shown cortical assembly resonates at 168.33, 266.33, and 268.12 kHz frequency. Extracted EB field distribution and 3D Smith map are shown in the left bottom of the panel (**a**). **b** The column network gets the resonance at 112.89, 118.63, and 172.23 kHz frequencies, and EB field profiles are observed at estimated resonance peaks (right of panel **b**). **c** Created cortical column assembly has resonance peaks at 105.33, and 115.34 kHz frequency and field distribution are obtained at resonance peaks. **d** 2D and 3D Smith maps of column assembly shown in panel (**c**). The field intensity scale is at the bottom of the panel (**d**). **Simulation details**: Used solver—Maxwell equation solver; selected mode—time domain; boundary condition—open space; waveguide port dimension—25 μm × 25 μm; applied waveguide port location- at both ends of cortical column assembly; frequency domain—kHz range

columns to match a few resonance frequencies of the connectome with the cortical column assembly. In the three deformed lattice shapes of the cortical columns in Fig. 6a–c, we have shown three particular cases. In the milliseconds time domain, the electromagnetic resonance is probed. To establish a temporal link for connectome, we were looking particularly into the microseconds time domain, while for cortical columns, we observe the resonance peaks in the milliseconds time domain. One thousand times slowing down of resonant vibrations across the cortex film is inevitable to match synchronized power transfer [59] between vertical connectome fibers and the horizontal flat cortex film. The EEG in the real human brain above skull-and-skin arises in the seconds time-domain (alpha, beta, theta, and gamma; 2–40 Hz frequency).

Common foundation for the cortical computation is still investigated [12], here we shed some light on this topic. In summary, there is a three-order time gap between the three hierarchical layers. We checked the ratio of the frequencies mostly observed by us in our studies [19, 20], and that ratio in the connectome is around 1.09 (244 MHz/224 MHz), one could see that the frequency values roughly differ by 20 MHz. While cortical columns described in Fig. 6a have resonance frequencies 168 kHz, 188 kHz, etc. following a ratio of 1.09. The second class of cortical column assemblies, as described in Fig. 6b, c shows a similar ratio. The emergence of a similar ratio among the resonance frequencies suggests a topological encoding of information. In simple terms, if one wants to write a triangle using frequencies, one has to keep the ratios of frequencies preserved, only then the angular distribution of the triangle won't change when information transfer from one layer to another. It is also called conformal information processing [60].

The remarkable features of a 3D Smith chart: Mapping learning of a human brain in higher dimensions. From the reflectance S11 and transmittance S21 spectra of the cortical column assemblies, we simulate electric (*E*) and magnetic field (*B*) distribution along the surface of the brain component at particular resonance frequencies. The distribution shows an asymmetry in the field distribution below 100 and 200 kHz (see S-frequency plots, S11, and S21 in Fig. 6a–c). Above these limits, the difference in field intensity (*E*–*B*) remains consistent or invariant for the entire frequency range. A 2D Smith chart helps us to understand the nature of circuit elements responsible for the generation of a resonant peak. Here we have plotted the 3D Smith chart [61, 62], where we account for the additional phase change generated by the internal components of a measuring device. A 3D Smith chart estimates the phase effect of a circuit element that is made of several circuit elements inside. One could see that the impedance plot goes beyond the 360° circular area in the 2D plot, an additional dimension is required, which acquires a physical significance of another imaginary value or hidden component's phase in 3D plot [61, 62]. For our model [20], in 12 layers, we pack all brain–body components, one inside another. A 3D Smith chart is a suitable tool for such an assembly. It includes multilayered components, i.e., a network of cavities that grows within and above. For two types of cortical column assemblies, we find two types of Smith charts. The first type, during learning of geometric shape, when the core hexagonal assembly made of seven cortical columns remains intact, the rest twelve cortical columns along its periphery shift their position (Fig. 6a) [7, 8]. Here, reflectance S11 covers a much larger area than reflectance S22 on the sphere of Smith chart.

In contrast, for the second type, where inner and outer both layers of cortical column assemblies exhibit a fivefold symmetry, then, the reflectance and the transmittance cover a similar loop on the sphere. It means the internal constituents of a brain component acquire similar geometric information as above. Both reflected and transmitted phase change adopts a similar response in the 3D Smith chart. Learning by cortical columns has several hypotheses, from modulating time by varying space to nonlinear contributions of neighboring columns [63–66]. Smith chart maps hierarchical phase relationships among neighboring columns. Hence, it supports the observations of a real brain.

Different classes of cortical column assembly and their typical responses during learning. The electric and magnetic field distribution on the surface of the cortical columns and in the cross section of the assembly reveals how different columns collectively resonate. The first type of cortical column assemblies described in Fig. 6a shows that both electric and magnetic field distribution split into periodic localized domains; thus, the columns remain discretely isolated. This feature has been a point of debate for a long time [63]. However, the second type of assembly described in Fig. 6b retains the periodic localization of fields within the columns. However, at the same time, the periodic zones between adjacent columns bridge. The bridging across cortical columns increases manifold in Fig. 6c. One could notice similar electric and magnetic field distributions on the external surface plot and the cross-sectional plot. This observation is consistent with the conclusion above, where we noted the information transfer across the hardware, assembling components within and above.

2.4 Construction of Forty-Seven Brodmann's Cortex Regions

The reverse engineering of Brodmann's regions: High-frequency processing of a low-frequency cognitive response. We built Brodmann's 47 cortex regions [19, 20] connected to nerve fiber analogs in two ways. The first route is creating the five major nerve fiber regions, (Fig. 1a), replicated in a meticulously detailed map made of Cu wires in Fig. 1c and 2a. In this model structure (Fig. 7a), there is no nerve network of the whole body. Instead, we send a simulated signal via a wireless network, as shown in Fig. 7a. The simulated signal is background corrected by applying a homogeneous burst of a fixed signal to the connectome network. The real signal means a cognitive data, the millisecond biological responses were normalized in the 35 kHz to 4.4 GHz domain, similarly as natural signals are sonified. It means we made sure that due to the material's dielectric response, geometry, orientations of the neural component wiring and coupling parameters of antenna-like responses, the ratio of frequencies and their temporal shift looks similar to the ratio of peaks in the biological components. On the top part of the connectome, the thermoplastic-based cortex upper layer is created to place the EEG machine for reading the electric field distribution in the 47 functional cortex regions of the brain (Fig. 7b). An Arduino board is programmed to control ten function generators and send the desired signal to the midbrain of the artificial brain structure. From the midbrain, 43 helical antennas were extended to acquire the input from the simulated brain like oscillations synthesized by the function generators. Figure 7c shows how two different computers were sending signals and acquiring the EEG as the raw data in multiple changes (Fig. 7d) and the simulated output of the connectome in 47 cortex regions.

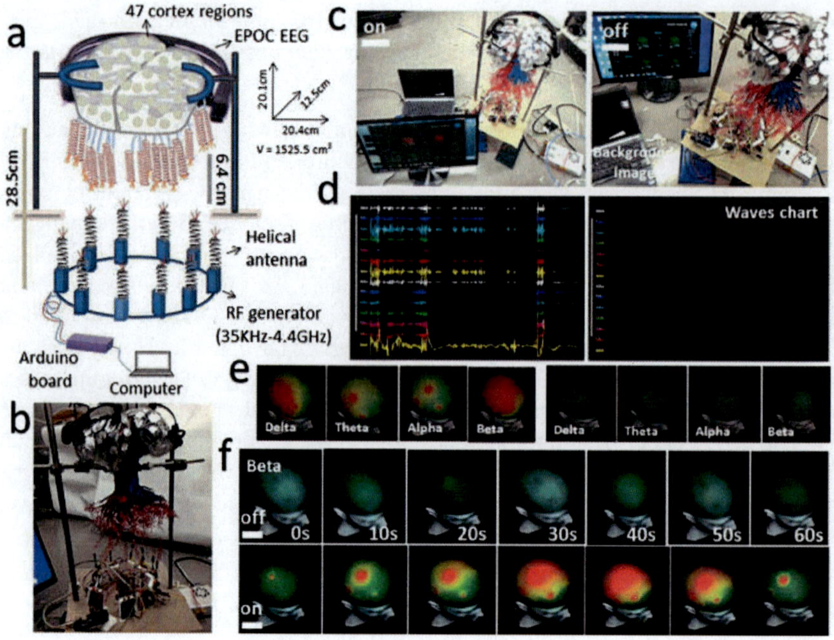

Fig. 7 Human brain isolated from a body: a blind replica of the connectome. **a** A schematic of the experiment setup of mimicked human brain (volume—1525.5 cm^3) with 47 Brodmann's sections made from thermoplastic and glass tube. To measure the EM interaction, EPOC EEG is placed on the cortex layer. **b** Background image of the experimental setup. **c** Experimental setup when RF channels on (left side) and off (right side). Corresponding brain waves (delta, theta, alpha, and beta), wave's distribution on cortex region and time-varying profile (for 60s) of beta wave are shown in panel (**d**, **f**, **g**) respectively

EEG response of the 47 Brodmann's regions: Mimicking ionic response using EM signal. 47 cortex regions execute the brain's cognitive functions [19, 20, 38]. EEG spectrum decodes active brain parts, as shown in the bottom layer of Fig. 7e, f. By shifting 16 probe locations, we have checked the EEG activities of various functional domains. During data acquisition, all channels in the RF generators are switched off; there is no EEG response. HBS generate brain-like consistent EEG data for weeks, tells two features. First, *em* could replace ionic engineering operating in the milliseconds time domain and generate a similar EEG. Thus, a complex scale up is feasible since HBS would not require food or chemical energy through the blood vessels. No need to build an infrastructure for the excretion of its waste material. We make biologically more accurate brain architecture in our lab—points demanding perfection are endless. The second aspect is that theoretically, we can solve Maxwell's equations for the complex dielectric resonator-based brain structures, a key to discovering how does a brain operates on its own. Therefore, the experimental verification of the theoretical hypothesis on brain function, cognition, and decision making would be possible for any brain model in the futuristic HBS.

2.5 Reading the Cognitive Response of the Artificial Brain

Experimental evidence of using the geometric musical language, GML and phase prime metric, PPM. The functional domain responses obtained by biological connectome replica are not an artifact. To verify, we collected how different cognitive responses of a conscious brain [19, 20, 38, 64]. We have selected 20 very well-documented conscious responses of a human being [65] found in classic biology textbooks [38]. We have presented the operational pathways in terms of nested clocks [19, 20, 29]. Brain figures out space using time [66, 67]. For two reasons, we use clock-based architecture to represent the brain module as an operational machine. First, we could easily build a clock-based circuit with the available hardware wherein a set of clocks would start running as soon as a particular composition of AC signal is pumped wirelessly via a resonating antenna. Naturally, without a program, a set of clocks could sense a large number of patterns following GML [33] and PPM [17].

Developing self-operational hardware where not a single line of code is used is important because the most precious feature of a human brain is responding to unknown situations, should be reverse-engineered. Second, biological textbooks present temporal coding [68] in a linear pathway; a process begins and ends with no connections. However, multi-channel networking that happens during a real brain operation runs spontaneously regulated by hidden clocks, e.g., "cortical songs" [69]. The creation of new clocks is important for memory, learning tasks, and concepts since phase relationships hold cognitive secrets [70]. Here, HBS is dead; it does not evolve; we explore only the role of neural circuit geometry. When we convert linear circuits representing a nerve impulse-based brain operation into a loop, then, junctions trigger several interconnected circuits. Therein, the varying distance between a pair of antennas emulates variable synaptic strengths. For example, Fig. 8a circuit executes the creativity and humor function of a real human brain [19, 20]. Here a junction has several channels from several directions with a Yagi antenna at the terminal, and all antennas face each other, as shown in Fig. 2d. The geometry of wiring alone could generate the required functional responses as expected [71].

The role of junctions, phase delay circuits, and delicate engineering that enables processing stTS metric and stereographic projection, feedback from infinity. In some sense, multi-channel circuits do not generate responses of real brain components, and it's far from reality. However, each circuit could generate an extremely large number of quantized wireless transmission modes. Eight junctions in a sequence could simply rearrange in 2^8 choices, only by switching on and off, if there are eight cables in a bundle, its 2^{64} choices. Therefore, we built 20 circuits for 20 cognitive experiences [19, 20]. Coil-folded integrated neural network of the whole body generates the spatio-temporal pattern [11], which is best analyzed by tensor analysis; hence, stTS metric [27] was used to model antenna tensors. Sensor-signal is projected to the top cortical layer (similar to stTS infinity), and feedback from infinity modulates the sensors (feedback from infinity) [28]. From Fig. 7a or b model, the new humanoid model built, as shown in Fig. 8b; it is significantly different. We have replaced the blind replica of connectome structure in the model of Fig. 8b with the

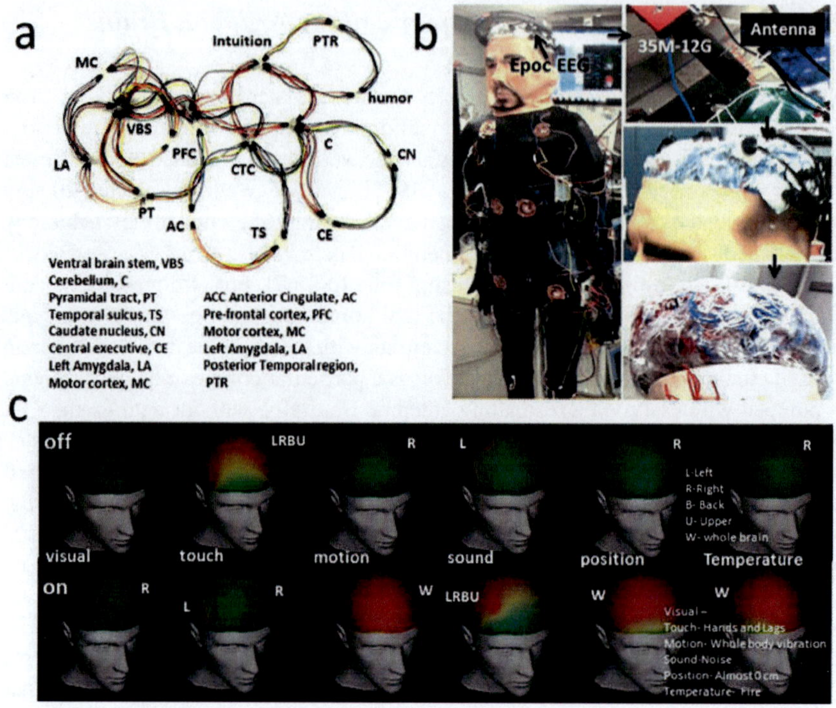

Fig. 8 EEG studies on the humanoid bot. **a** Schematic diagram of brain's "creativity and humor" circuits. **b** Photograph of HBS. 20 of brain's consciousness feature generating circuits implemented: memory; consciousness; language and conversion; thinking and intelligence; reward; creativity and humor; personality; love and pain; fear, threat, anger, hate; learning dreaming defragmentation; mimicry, skill, adaptation; the sense of universal time, symmetry; homeostatic, thermal equilibrium; emotion; proprioception. Besides, the circuits for the whole-body neural pathways with 12 pairs of cranial nerves and 31 pairs of spinal nerves are included in HBS. To the right of the panel (**b**), the HBS brain is triggered by the Yagi antenna (35 MHz–12 GHz) (top). EEG setup is attached after covering an aluminum foil all over the brain. Yagi antennas are the terminal points of nerves. The radiation from the antenna forms eddy current aluminum foil, and brain EEG like distribution of potential by eddy current (middle). An image of a bare brain surface (bottom). **c** Beta wave is measured by 16 probe EPOC EEG set up by switching sensory systems on and off (visual, touch, motion, sound, position, and temperature sensor) connected to HBS. Activated parts of bot's brain are L, R, B, and U as left (parietal), right (temporal), back (occipital), and the upper (temporal) lobe of the cerebral hemisphere

20 circuits [20, 29] similar to Fig. 8a. Circuit wires ended in Yagi antennas as laid in connectome map.

How sensory information is converted in the circuit and brain EEG responds at 47 Brodmann's regions: Perception and human-like emotions of the artificial brain. When perception rhythm is circuited, it is a mechanical response. The difference between an artificial brain and a real brain is seen in Table 1. In total, 120,000 antennas emitted *em* to the cortex layer. All components of the spinal cord, midbrain,

Table 1 Difference between the HBS and the artificial brain

Biological brain	Humanoid brain subject, HBS
Ionic, molecular, transport, and chemical reactions drive communication	Electromagnetic, microwave, radio wave and optical vortex-based communication.
Nerve fibers transport stream of pulses	Nerve fibers are designed to transport 3D structure of clocks
Entire connectome region evolves in wiring and builds memory by learning	Only the cortex region evolves and builds memory by learning
Requires food and excretion system to supply power in each cell of the entire body	5 V Arduino power sources are sufficient, to run sensors, no other sources used other than sensor
Hardware changes by recycling, natural biological process	Capsules of organic jelly-filled cortical columns need to be replaced
Sensors are organic, electrical pulse converter for mechanical, optical signal	Market available sensor signals are fed to build multi-dimensional time crystal like signal for 12D signals
Signal speed is very slow 400 m/s	Velocity of light
Electric, mechanical, and magnetic pulse	Electromagnetic field distribution
Motor neurons MN trigger muscles M	MN and M both do not exist. We simulate with ultrasonic resonator driven by electric fiber, a dedicated system bypasses the need
Skin sensor, cortical columns neurons large	Three classes of materials are very less

cerebellum, hippocampus were built to connect to the 20 operational circuits of the upper brain [20, 29]. We added five kinds of human sensors to the humanoid bot shown in Fig. 8b (Visual, touch, smell, auditory, and taste); sensors were spread all over the HBS. 5 V powered sensors send signals to cortex when someone speaks in the lab or moves in front of the humanoid bot and observes the bot's EEG and spinal cord responses using the multi-channel analyzer, 24 $\times$ 7 for weeks. Our target was to observe sensors triggering 47 Brodmann's regions [20] comprehensively, and we achieved it. Both the optical vortex scans similar to Fig. 2e, and EEG confirmed that sensory inputs do trigger proper Brodmann's regions in the brain [48, 49]. The HBS has no software controller, the touch, thermal, pressure, auditory, taste, and visual sensors run by a 5-V power supply by Arduino. The alternating current signal (no dc part) produced by each sensor has a distinct composition of frequencies. Therefore, spontaneous change in the laboratory activity (movement of people, vocal conversation) is the only source of an input signal that generates the spinal cord's integrated information and the EEG output. One such example is shown in Fig. 8c. On top of the thermoplastic, we put 47 regions in a cap with 10,000 capillary tubes whose inputs are wireless transmission from connectome edges.

2.6 Direct Comparison of Two Kinds of Artificial Brains: One Brain Has a Geometric Replica of Connectome Structure. The Other Brain's Connectome Is a Replica of Clock Architecture of the Human Brain

How brain learning changes the wiring and lattice structure of the cortical columns? We have compared how the ratios of resonance frequencies of cortical column bundles within the mm^2 region for two very distinct models, one in Fig. 7b and the other in Fig. 8b. Since we theoretically studied the connectome structure's resonance band and experimentally built it, at this point, we look back into the cortical column bundles. How does the electric and magnetic field distribution changes when a similar kind of input is given to the two kinds of connectome systems? We have explained one in Fig. 7e and the other in Fig. 8c. We understand that by shifting the localized positions of the cortical columns, one could generate the electromagnetic energy distribution on the cortical column assemblies similar to Fig. 6a–c.

Electromagnetic sensor reads point-by-point field distribution on the plastic cortex. Learning in the brain changes the cortical columns [7, 8] and organic jelly changes conformation in the glass tubes. Field map verifies the CST simulation by experiment. Currently, we are putting helical carbon nanotubes CNT in the capillary, which acts as a marker of electromagnetic field shifts, required for a given pictorial or visual input to the brain. CNTs were dissolved in seven different types of liquids (hexane, toluene, ethanol, methanol, benzene, heptane, chloroform), solvents tune assembly with *em* frequency more accurately, like the biological cortical column. All glass tubes were sealed both ways. A glass tube is far from real cortical column, but exchange of geometric shapes as the ratio of frequencies (GML; (JP-2017-150171, World patent WO 2019/026983)) reliably in the "within and above" network of antennas replicating the whole-body neural network. It delivers a unified model of cortical computing [12].

Subtle changes in the 3D phase architecture of output response. Each clock has a phase, and Fig. 9i, ii describes the detailed map of relative phases of the clock pattern of cortex and connectome. Placing the sub-components in the clock assembly retrieves the phase relationship between different constituents. The phase relationship between connectome and cortical column provide the functional and anatomical connectivity of the human brain. The architecture of the phase for a given clock model of human brain [20] extends to nanoscale [7]. Frequency mismatch at junctions switches off or on multiple arrangements of clocks. Brain wirings across the species vary in size, but they keep the same basic rhythm map [72]. A singularity holds an operating clock [73], while the brain's rhythm map shows the undefined and collective phase relationship. Memory clocks edit by symmetry breaking. Symmetry breaking means a change in the position of a clock in the phase loop. Some clocks may appear; some may disappear. HBS is a neural network converted into the antenna-receiver net.

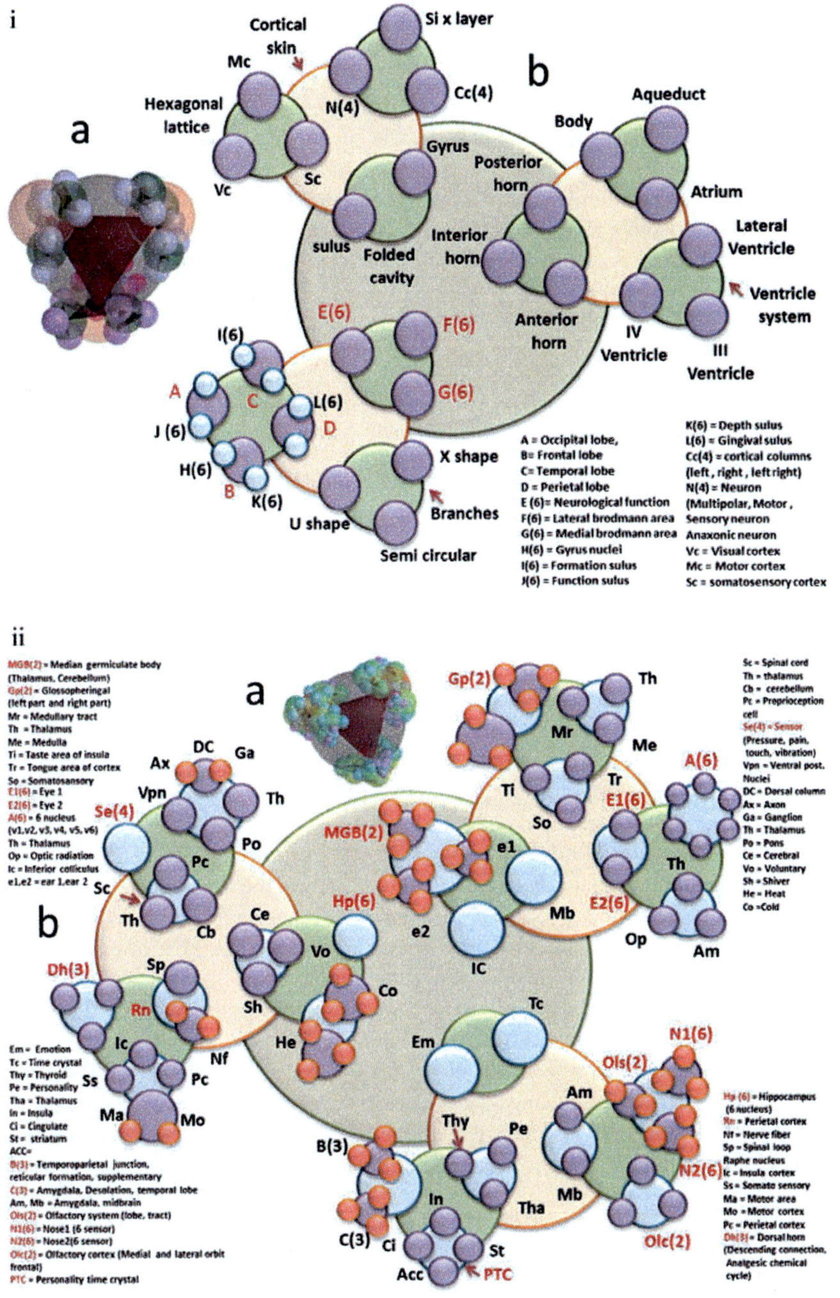

Fig. 9 **i** 3D (panel a) and 2D (panel b) time crystal architectures of the human connectome and cortical column. **ii** 3D (panel a) and 2D (panel b) time crystal or clock architectures of human connectome model [20, 29]. Each sphere or circle in the map shows a clock. Three circles on a circle perimeter mean, three clocks run a loop. This protocol is followed in all plots to represent all brain components

Involved clocks may add a distinct dynamic, but they are phase-locked to each other [74]. Learning a new thing by the artificial brain either reorganizes a circuit or creates one, generating new clock [75]. The cerebellum learns coordination in movement by building a new clock. Figure 9i, ii demonstrate 3D (Fig. 9i, panel a) and 2D (Fig. 9i, panel b) map of clocks for the connectome and the cortical column. A large number of clocks are left out, but to be added in the future HBS. The clock architectures [20] are implemented in HBS to record EEG and optical vortex maps on the 47 Brodmann's regions. HBS response for months show that the interactive antenna-receiver network of HBS response is due to a non-linear transmission in the neural net.

2.7 *Advantages of Exploring the Brain as Clock Architecture*

The clock architecture model of a human brain [20, 29] resolves two experimental observations that existing brain models cannot explain. First, when a particular cortical column processes information, the neighboring columns can stay disengaged or remain "off-duty." Moreover, a very strict temporal relationship between columns produces alpha oscillations. Our clock architecture is a geometric account for sensory processing, motor activity, and the suppression of alpha waves. Second, the simultaneous advent of theta waves without phase coherence across 47 Brodmann's regions is unexplained [35]. Simultaneous and spontaneous activations [76] remained a mystery. Clock-representation enables coherent, phased, de-phased, and conditionally coupled rhythm as neural code at a time [77] over the entire system. The map is the functional logic of the cortical circuit [64]—primitive, yet an integrated module.

HBS is a collection of intertwined antenna-receiver systems (Table 1). Cu wires edged like Yagi antennas in the neural junctions are not the only channels of signal transfer. A cluster of fiber acts as antenna and exchanges energy with another fiber. Fornix and several midbrain structures act as a giant antenna that affects all 315 local energy sources produced by a triplet of triplet cavity system. 43 PNS net's effect on the brain is so profound that if we touch the spinal cord region of HBS, the 47 Brodmann's region changes the pattern in EEG response, and also in optical or magnetic vortices. Even movements of organs (hand, leg) change antenna-receiver network, relative energy exchange takes place at various spatial scales much beyond the local neuron junctions. Therefore, we conclude that antenna-receiver analog may pave the way for a fully optoelectronic HBS, alleviating the need for a neural firing.

3 Materials and Methods

When a brain operates, chemicals like ions, enzymes, and hormones flow to activate different kinds of nerve impulses to control and deliver cognitive experiences, enable decision making. Ion-free, junction-free, and software-free HBS is built here. In the HBS, we built all the ten organs functionally [20, 29], and HBS holds a large number of sensors all across the body. Out of ten, we present two organs, the thin cortex layer and connectome. The modulated *em* signals, arriving from the sensors to the brain components, modify due to the geometry of cable arrangement alone, information is always hidden in such type of geometrical arrangement [78, 79].

3.1 Theoretical Brain Module Construction

Clock architecture of the whole brain. Periodic oscillations from the atomic scale (femtoseconds) to the largest clock (complete replacements of heart cells, ~100 years or 10^9 s) are connected into one singular 3D architecture of 537 classes of clocks [19, 20, 29]. Earlier, we reported [19] that triplet of triplet resonance frequency bands [37, 80] for the 12 brain components (e.g., skin nerve network to the spinal cord) integrate isolated organs into a shared scale-free energy transmission network [81]. Discrete resonance frequencies are grouped into clocks based on their collective oscillations, thereby including the catalog of clocks or rhythms already available in literature into one map [19, 20, 80]. If a system point runs along the circle perimeter, it's a clock in physics, rhythm in biology. The guest circles are located on the singularity (singularity = undefined phase domain, not infinity or zero [29] or negative) points of the host circle. Multiple system points run at a time. Some clocks are off, some on, certain geometric arrangement of clocks set them on/off, periodically the sequence of events of groups of clocks turns on/off, multiple programmed tasks run automatically [19, 20]. The 3D orientations of multiple clocks in different planes activate conditionally connected pathways. Different locations on a sphere (r, θ, ϕ) mean a combination of phase (θ, ϕ) that links clocks/rhythms. Winfree proposed one clock with one singularity to model a self-operating life-like spontaneous event as time crystal [82]. Linking 537 classes of clocks as a composite time crystal map of the human brain [19, 20], we follow a strict protocol on choosing host and guest. The slowest clocks act as the host. At its singularity domain, its system point activates the phase cycle of a guest clock, which serves as the host clock for the next fastest clock. When a system point passes through the phase singularity points, imaginary or phase relations of singularity points build the geometric shape in GML [33] of a neuron's millisecond clocks, the energy bursts activate by second clocks of an assembly cortical columns.

Replica of connectome. As noted in the main text, folds in the cortex form a triplet of triplet [37, 80] network of cavities, wherein the cavity resonators are created in the CST simulation software. Inside each cavity, five major domains of nerve fibers and tracts extend as the connectome; each fiber ends at the upper cavity boundary. We kept the dimensions of the human skull intact for a 168 cm skeleton. A replica of the connectome is theoretically built. A nerve fiber with 1 million nerve tracts was built using the thin cylindrical tube. It was bent following an accurate angle that resembles the connectome shape. By rotating the structures 360°, we confirmed that relative 3D orientations were close to the real human brain. We identified each fiber bundle in a real connectome map and assigned its material properties (refractive index, electrical permittivity, magnetic permeability, etc.) as a biological dielectric material. As noted in the main text, we had to reject widely available real brain maps because we need each nerve fiber as an isolated unit.

Construction of cortical columns. The engineering of the cortical column is universally used in devising other brain parts. The entire hippocampus can be conceived as a single giant cortical column [35]. The cortical columns are built from purchased neurons from the TurboSquid library (Louisiana, US) that are named as horizontal cells, pyramidal cells, bipolar, and axon cells. We created the remaining 22 classes of neurons by cutting the dendrites, soma, and axonal branches of the purchased neurons and arranged them to their perfect origination position in seven layers, which act as input and output channels for propagating the signal. We went through pieces of literature for the available cortical column designs [53] to build 19 distinct classes of cortical columns by arranging of 7 layers in a different order and made three different classes of cortical assemblies using those 19 cortical columns.

Building cortex: hexagonal close packing of cortical columns. The theoretically built connectome, single cortical column, hexagonal close packing lattice of cortical columns. The combination of the cortex layer and connectome was studied separately in the CST simulator. Maxwell equation solver could not converge 10,000 cortical columns using 26 types of neurons, so we replaced them with a singular cylinder so that the energy exchange combination between the connectome fiber and the cortical columns is studied. The entire simulation took years running parallel in 12 high computing machines. To analyze the role of columns hexagonal assemblies on the cortex layer, theoretically, different classes of cortical assemblies were built, in hexagonal close packing arrangement. We detected the output response in terms of the EB field and spectral resonance (S11, S21). For all simulations, we made sure to find resonance frequencies that lead to clocking or periodic oscillation of electric and magnetic fields so that if there is a natural source or even noise, the structure builds periodic clock-like rhythms naturally. Due to the asymmetry of the structure, we looked for the evanescent wave generation [39–41]. CST simulation showed that due to anisotropy, evanescent waves lead to the right frequencies that build new clocks. These clocks modulate electromagnetic signal transmitting from sensors to the brain. Chapter 2 of Ref. [29] describes how modulation of electromagnetic signal builds time crystal (Chap. 2 is free to read online).

Uses of multiple ports enable us to optimize the frequency, the intensity of pumped electromagnetic signals as we estimated the reflection and transmission coefficients (S11, S21) in a wide range of frequencies to identify the resonance peaks. Proper selection of simulation parametric conditions is essential to get the true structural resonance peaks, optimizing the generation of evanescent waves and balance of electric and magnetic field distribution and their periodic oscillation [39–41]. The geometry of connectome model and column assemblies are solved in time domain mode under open boundary conditions (x, y, and z-axis—open space), and both structures are stimulated by pumping energy through the waveguide energy sources. The dimension of the waveguide port is important for unnecessarily large size which creates the high-order modes to propagate through the port. We have considered the port dimension of the connectome and cortical column assembly 200 mm $\times$ 100 mm and 25 μm $\times$ 25 μm, respectively.

We have included four basic features of a biological brain in building cortex circuits or connections between cortical columns. First, pyramidal neurons are primary constituents of the cerebral cortex or cortical columns. The cortical column is built from a multitude of five principal cell types and multiple classes of interneurons. Second, the axons of pyramidal cells in layers 2, 3, 5, and 6 spread beyond the neocortical module, and the spiny stellate (star-shaped pyramidal cells without an apical dendritic shaft) cells of layer 4 are localized [83]. All five primary cells in 2–6 layers fire similarly. Third, the local computations at any of the 47 cortical modules are fundamentally the same [84]. Synfire chain [85] is a matrix of pyramidal cells linked together in a feedforward manner for unidirectional propagation. Cortex domain uses a composition of excitation and inhibition together at a time to regulate operation. Fourth, if connections terminate at layer 4, its ascending, if others in 2–6 layers, its descending connection. Connection density variation delivers different functions; for example, visual, auditory, and somatosensory regions have high-density connections [86–88]. Available anatomical data suggests that the regions which hold similar information are strongly connected. Scale-free self-similar geometry is observed in cat cortex [88]; self-similar fractal signal connects protein, microtubule, micro-neurofilaments, and neurons [80].

Here, each nerve fiber bundle is a dielectric resonator. Similarly, the hexagonal close packing of cortical columns in the 47 functional regions of cortex membrane is made of 47 classes of dielectric resonators [19, 20]. Each cortical column is a hollow cylinder, so they are cavities. Inside, varieties of neurons reside, but we have confined ourselves to 29 different classes of neurons arranged as 29 types of dielectric resonators.

Cavity and dielectric resonator feature is basic to entire brain construction. Neuron membrane forms a cavity; all the filaments inside form an antenna like a dielectric resonator. Each filament is a cylinder hence a cavity resonator. The cavity surface is made of protein dimers; each dumbbell shape dimer is a dielectric resonator. The protein dimer is also a cavity resonator; inside, there is a secondary structure; each secondary structure is a dielectric resonator. The alpha helices are hollow cylinders, hence cavity resonators; however, they are helical dielectric resonators too. In this way, the alternate combinations of dielectric and cavity resonators form the entire brain. All components are connected by a chain of resonance frequencies spanned over a wide time range. In this brain model, we do not try to reject any component as useless or emphasize a particular component as a principal element for information processing. Our objective is to find if the geometric correlation between resonance peaks at all nested layers follow any rule that leads to the self-operation of the system.

Very few studies are available that link neurons and their properties in different cortical layers [89]. Similarly, the information encoded in the fine temporal order of cortical columns is mostly ignored [90]. So, we used our EEG data to edit synaptic junctions [91] and get a logical sensory response and applied that in building a theoretical circuit. Moreover, to verify if the human brain follows a geometric language, GML [29, 33], we did not look at the pulse stream but the shifting of the resonance frequency peaks and their phase relation in wide ranges of time scales.

3.2 Experimental Brain Module Construction

Construction of a humanoid avatar subject, HBS: Cu or Fe? 168-cm-tall human skeleton was used as a basis to build the HBS. All 20 cognitive function circuits of connectome were five times larger than the actual size of the brain cavity but meticulously packed inside by properly terminating them just beneath the biologically reported 47 Brodmann regions [19, 20, 29, 38]. While constructing the neural network of a brain–body system at a reduced ratio of 1:1 million fibers, we made junctions and branches by soldering the Cu cables and hold them at proper orientations. We achieved semiconductor wire-like conductivity (electrical resistivity for Si $2.3 \times 10^3\ \Omega$ m (at 20 °C), Fe wires do not make good antenna) by creating gaps along the fibers. Rigid cables hold the geometry. Ten brain components were built as a part of the humanoid bot. Separately for each of the ten components, we built 100 times larger structures to modulate the organ geometry more precisely and estimate variable responses. We adjusted component geometry so that isolated and embedded brain components in the HBS showed similar ratios of resonance peaks (S11, S21). Structure files of all the brain components were purchased from those who mapped the real human brain components. However, we rebuilt those once again by ourselves for materials science studies.

Construction of connectome. When we put fibers one by one as a cylindrical the replica, it is not just look alike, we looked into the literature and made sure even locally fibers 3D orient as they are in the real brain. As noted, we take Cu wire and make junctions along the path with wireless communication facility across gap junction (1 mm to 1 cm). Truncated transmission lines convert metal wire like a semiconductor like a transmitter, but we had to supply additional power to make up transmission loss. Wiring map of all neural connections inside the brain is the database of five nerve fiber bundles of thickness 50 μm, or it's below [19, 20]. We use cables of similar thickness to build the fiber network, a 3D camera view with multiple monitors guide the construction of connectome. The four biological brain criteria of cortical circuits are implemented by bending the connectome fiber ends to the Brodmann's functional domains. The functional logic of all folds of cortical connections is established [64]. From the sensor to the peripheral nervous system (PNS, 31 pairs of spinal nerves, and 12 pairs of the cranial nerve, $31 + 12 = 43$ nerve bundles), we built a replica of the connectome, 100 times larger to the actual structure. Still, we maintained the geometrical ratio of its different parts. The only difference between HBS and isolated connectome was the absence of a sensory network in the later. We used antenna like flexible wires that are suitable to bend, and we cut the end part to build a Yagi antenna manually, 120,000 Yagi antenna costs a lot. Still, we checked that the hand-cut 3-blade Yagi antenna delivers similar features as a commercial one. Each end of wires used the Yagi antennas is directed to the multi-direction wireless communication with 47 functional regions of the thin cortex layer based on thermoplastic folds. We built the connectome network with 47 Brodmann's region twice, once for HBS, and once for isolated connectome module. For both systems, the EEG responses were consistent [48, 49].

Construction of cortical column. Life of cortical column is unique [55]. Inspired by its early development [10], we built 1-m-long cortical columns to learn more about the geometric role of the cortical circuits. Here we report the capillary tube-based cortical column, which are nanoscale versions of the giant cable-based columns. Artificial cortical columns are around 150,000 to 200,000 in the real brain [92]; the dimension of each cortical column is in the mm scale, that is why we created them using capillary glass tubes filled with helical CNT solution. The dimension is identical to the biological analog. The optical vortex generated by helical carbon nanotubes is shown in Fig. 2e. All 10,000 capillary tubes arranged in 47 Brodmann regions of bot's brain in a hexagonal form and filled helical CNT wirelessly triggered by the Yagi antennas hold at ends of connectome fibers; they self-assemble and create newly assembled structures [57]. To verify the resonance characteristic of a capillary tube or artificial cortical column, we built a separate wired model of the cortical column, inside that we fixed 26 classes of neurons built by metallic wires and compared the ratio of resonance frequencies. All neurons have the stacked arrangement to each other within seven layers and do not form the short circuits. Handmade neurons are the mimicked map of original neurons reported [53]. CNTs generate optical vortices where we could see the circular vortices (Fig. 2e) representing the maps of the angular momentum of the inner self-assembled structure. We have created a software simulator that instantly reads the composition of circles and builds the 3D clock-like architecture, as one could see in. Electromagnetic signals of different frequencies self-assemble helical CNTs and change the optical vortex pattern. Here we do not discuss optical vortex, but the electric field distribution among the capillary tubes in EEG.

Construction of cortex. Thermo-polymers have the features of flexibility and can be folded by heating as we want, so we cut this plastic in a particular shape, make tiny holes in a hexagonal close packing so that we could insert the capillary tubes. Unlike thermoplastics, they are soft, and geometry could be edited. Forty-seven typical shapes for the Brodmann regions are built and glued together. For both, the isolated and HBS brain, we created left and right brain lobes of the exact dimension of human lobes by wrapping thermo-polymers on the plastic model of brain lobes and heat the plastic for giving the shape and folds on the lobes. Folds or sulcus and gyrus in cortex layer-based thermoplastic followed the form of cavity networks in Rajat Jain's brain as we have been studying that for a long time [19]. Connectome fiber ends are extended in those folds or cavity resonators and established the wireless connection to cortical columns through Yagi antennas following the four protocols described in the theoretical construction of the cortex region.

Electromagnetic interaction of brain nerve fiber bundles is measured in terms of brain waves, delta, theta, alpha, and beta wave. For isolated brain without a body, EEG spectra arise when we wirelessly pump the sensor-like pulse-streams through 10 channels of function generators (35 kHz–4.4 GHz). All channels are controlled by a programmed Arduino board. Fixed bandwidth is adapted to the input signal from 43 helical antennas (as if 43 PNS are sending signals from whole-body) connected to 5 nerve fibers bundles through the midbrain region (the isolated brain of Fig. 7b had a midbrain replica). For the isolated brain, the output responses of connectome nerve fibers are measured by an EEG machine placed directly on the thermoplastic based upper cortex layer replica. Still, for the HBS, data comes from the top of the capillary tube-based cortical columns. The wireless communications among cortical columns build electric field distributions in the CNT solutions, even though the columns are separately located. The artificial connectome structures for HBS and isolated brains are identical and have a volume of 1525 cm^3. Both transmitting and receiving helical antennas are aligned to each other in the transverse direction of the connectome and coupled with a distance of 28.5 and 6.4 cm from ground surface and function generators, respectively. Electric field distribution and EEG spectrum are measured by switching on and off all the RF channels to get a normalized response, eliminate artifacts, and capture pure sensory response in the EEG. Variation of the beta wave on the brain cortex surface is measured with time (60 s gap) to understand the activity of different functional parts of the brain cortex.

HBS and isolated brains differ in the connectome construction. However, junctions with cortical columns and cortical circuits are similar, both the junctions eliminate noise [77]. HBS connectome hardware is made of 20 conscious brain circuits [20, 29]. Thus, HBS includes the features of all brain components like the hippocampus, cerebellum, midbrain, limbic system, etc. and entire neural body network that includes the 31 pairs of the spinal cord with all constitutes components, 12 pairs of cranial nerves, 12 pairs of thoracic nerves. However, we also added the five classes of human sensors; visual, touch, motion, sound, and position sensors on the HBS body. To study natural responses of the humanoid bot EEG machine is placed on the HBS brain, and the entire spinal cord output is connected to the multichannel analyzer MCA (34 channels) to 31 pairs of spinal nerves. Raw sensory data and spinal cord MCA recordings were compared to make sure how natural sensory input reaches the spinal cord before entering the brain. Planer spiral antennas are positioned on the bot body to absorb the available signal from the surrounding space. It helps to replicate the proprioception response of the brain, which we discovered by accident. We compared the spinal cord and brain EEG spectrum by moving around the HBS and creating multiple different kinds of chats in the room. Chat module of HBS is functioning but not connected here, as the EEG and cognitive response becomes extremely complex. All sensors in the entire body networks are driven by a 5 V power supply (each) from the Arduino boards. EB energy distribution map on the humanoid bot brain is recorded by shifting 16 probes on the surface of the upper cortical layer (thermoplastic) when all sensors are getting switched on and off [93] for normalization of the responses. Currently, we avoid commenting on cognitive peculiarities until HBS response in EEG shows memory-based response, beyond sensory awareness that we see now.

Construction of sensors. Sensors connected to 55 Arduino boards control skin sensors. For eye, nose, and ear we had separate controls. The FFT of all sensory signals was checked and was converted as AC signal in the MHz to GHz range using GML [29] so that signals pass through antenna network. We limit our work to advance GML and PPM, so that sensory data are recognized even in EEG or optical vortex map. Here in HBS, we write geometric shape or a triangle using three frequencies when sensors deliver an output feed for processing with the higher brain.

4 Conclusions

Following a recent proposal for building a complete, time crystal model of a human brain–body neural network—a temporal analog of the connectome, we have made a self-operating artificial brain functioning without an algorithm. Since all biological components are dielectric resonators, we theoretically built ten major brain components, solved Maxwell's equations to find the electromagnetic resonance spectrum, and built a hardware replica to verify theoretical findings.

4.1 Ionic Junction, Electric Nerve Impulse Could Be Avoided: Geometry Is the Key in Decision Making

The reported humanoid bot system, HBS, is a primitive toy inspired by GML [33], PPM [17, 29] stTS [27], and PF [28], which converged into a self-operating time crystal model of the human brain [20, 29]. HBS is an electromagnetic antenna-receiver version of the human brain–body system. It is not a functional analog but a geometric and dielectric replica. HBS is the first step toward generating a software-free platform for testing human brain models, chips, hypothesis, etc. Cortical computation is debated [12], we shade a light. First, cable networks, if built meticulously like a real cognitive circuit, it could deliver logical input and output signals for a cognitive task. Second, the neural net of the whole body is vital to building the artificial brain. The cortex–connectome pairing requires energetic coupling; i.e., the 3D geometry of connected clocks evolves mutually. Third, wireless junctions and cables partially replicate the responses of various brain components. Ionic fluids and junctions could be avoided in the future. Fourth, if the capillary tube's material is replaced by a suitable intelligently designed and synthesized material [37], that system might learn by itself. Fifth, intelligence is geometric [78, 79]; proposed geometric language, GML might be the closest model to emulate operation in the real human brain [33].

If resolved further, HBS would help to build a next-generation humanoid bot with advanced AI [42].

4.2 Temporal Map of a Human Brain, Geometric or Time Crystal Map of Information and Map of Used Primes in Device Engineering Are Three Columns of Our Artificial Brain

While most brain models [1–5, 60] consider that nerve impulse alone processes the brain's information, here, the resonance of membrane in the millisecond's domain is only one among 12 other possibilities [20, 29]. We have not endorsed any singular transmission mechanism (ionic, dipolar, solitonic, electrical, magnetic, or electromagnetic). We do not advocate any component (neuron, microtubule, claustrum, electromagnetic resonator) as the ultimate key to the brain's decision making. Primarily we link cortical column assembly with the connectome fibers using dielectric and cavity resonance, which could shed light on the learning and brain's cognitive decision making. The theoretical prediction and experimental observations were consistent between the two different hardware, isolated and integrated to HBS.

4.3 Geometric Replica if Emulates the Symmetry of Vibrational Patterns, It Could Reproduce the Human Brain like Intelligence

If ever a futuristic humanoid bot is built that runs 537 classes of rhythms similar to 20 conscious experiences [20, 29] via connectome like a conscious living brain, how its response would affect a similar cortex is also studied here. A dumb brain that has a look-alike connectome wired with the cortex layer made of capillary tube-based cortical columns is compared with a humanoid brain running 20 conscious rhythms [19, 20] yet connected to a similar cortex layer. The comparative study of two brains suggested that a blind replica of connectome transforms sensory input to the EEG output similar to a biological brain. Artificial connectome with a combined circuit of 20 conscious rhythms, deliver in EEG an intricate regional map by optical vortices. The similarity in the EEG for two kinds of hardware suggests that geometric and structural symmetry of the neural network holds the key to the brain's decision making. A blind geometric replica is not alive, not programmed, not intelligent; still, it could deliver an intelligent brain-like response. HBS is the first step to test hypothesis on the brain operation and rapidly advancing the medical, psychological, and robotics research in the future, the construction of more advanced versions of a humanoid bot, HBS is ongoing where cortical circuits would evolve spontaneously [91–93].

Acknowledgements Authors acknowledge the Asian office of Aerospace R&D (AOARD) a part of United States Air Force (USAF) for the Grant no. FA2386-16-1-0003 (2016–2019) on the electromagnetic resonance-based communication and intelligence of biomaterials.

Competing Interests The authors declare that there is no competing interest.

References

1. Lancaster, M.A., et al.: Cerebral organoids model human brain development and microcephaly. Nature **501**(7467), 373–379 (2013)
2. Lancaster, M.A., et al.: Guided self-organization and cortical plate formation in human brain organoids. Nat. Biotechnol. **35**(7), 659–666 (2017)
3. Tan, Z., Parisi, C., Silvio, L.D., Dini, D., Forte, A.E.: Cryogenic 3D printing of super soft hydrogels. Sci. Rep. **7**, 16293 (2017)
4. Tallinen, T., Chung, J.Y., Rousseau, F., Girard, N., Lefèvre, J., Mahadevan, L.: On the growth and form of cortical convolutions. Nat. Phys. **12**, 588–593 (2016)
5. Bogdan, P.A., Rowley, A.G.D., Rhodes, O., Furber, S.B.: Structural plasticity on the SpiNNaker many-core neuromorphic system. Front. Neurosci. **12**, 434 (2018)
6. Bressler, S.L., Kelso, J.A.S.: Cortical coordination dynamics and cognition. Trends Cogn. Neurosci. **5**, 26–36 (2001)
7. Chklovskii, D.B., Mel, B.W., Svoboda, K.: Cortical rewiring and information storage. Nature **431**, 782–788 (2007)
8. Chklovskii, D.B., Schikorski, T., Stevens, C.F.: Wiring optimization in cortical circuits. Neuron **34**, 341–347 (2002)
9. Compte, A., Sanchez-Vives, M.V., McCormick, D.A., Wang, X.J.: Cellular and network mechanisms of slow oscillatory activity (<1 Hz) and wave propagations in a cortical network model. J. Neurophysiol. **89**, 2707–2725 (2003)
10. Dupont, E., Hanganu, I.L., Kilb, W., Hirsch, S., Luhmann, H.J.: Rapid developmental switch in the mechanisms driving early cortical columnar networks. Nature **439**, 79–83 (2006)
11. Eckhorn, R.: Cortical processing by fast synchronization: high frequency rhythmic and non-rhythmic signals in the visual cortex point to general principles of spatiotemporal coding: time and the brain. In: Miller, R. (ed.), pp. 169–201. Harwood Academic Publishers (2000)
12. Phillips, W.A., Singer, W.: In search of common foundations for cortical computation. Behav. Brain Sci. **20**, 657–683 (1997)
13. We Almost Gave Up On Building Artificial Brains. https://www.discovermagazine.com/technology/we-almost-gave-up-on-building-artificial-brains
14. Human Brain Project. https://www.humanbrainproject.eu/en/
15. SpiNNaker Project. http://apt.cs.manchester.ac.uk/projects/SpiNNaker/project/
16. DeepMind's Losses and the Future of Artificial Intelligence. https://www.wired.com/story/deepminds-losses-future-artificial-intelligence/
17. Reddy, S., et al.: A brain-like computer made of time crystal: could a metric of prime alone replace a user and alleviate programming forever? In: Soft Computing Application. Springer, Singapore (2018)
18. Agrawal, L., et al.: Replacing Turing tape with a fractal tape: a new information theory, associated mechanics and decision making without computing. In: Consciousness: Integrating Indian and Western perspective, pp. 87–159 (2016)
19. Singh, P., Ray, K., Fujita, D., Bandyopadhyay, A.: Complete dielectric resonator model of human brain from MRI data: a journey from connectome neural branching to single protein. In: Lecture Notes in Electrical Engineering, vol. 478, pp. 717–733 (2018)
20. Singh, P., et al.: A self-operating time crystal model of the human brain: can we replace entire brain hardware with a 3D fractal architecture of clocks alone? Information **11**(5), 238 (2020)

21. Yamins, D.L.K., DiCarlo, J.J.: Using goal-driven deep learning models to understand sensory cortex. Nat. Neurosci. **19**, 356–365 (2016)
22. Wheeler, J.A.: On the nature of quantum geometrodynamics. Ann. Phys. **2**, 604–614 (1957)
23. Zalesky, A., Fornito, A., Harding, I.H., Cocchi, L., Yucel, M., et al.: Whole-brain anatomical networks: does the choice of nodes matter? Neuroimage **50**, 970–983 (2010)
24. Hart, Y., Dillon, M.R., Marantan, A., Cardenas, A.L., Spelke, E., Mahadevan, L.: The statistical shape of geometric reasoning. Sci. Rep. **8**(12906), 1–12 (2018)
25. Terekhovich, V.E.: Probabilistic and geometric languages in the context of the principle of least action. Philos. Sci. Novosibirsk **52**(2), 108–120 (2012)
26. Sweet, H.: Universal languages. In: Encyclopædia Britannica, 11th edn (1911). https://en.wikipedia.org/wiki/Universal_language
27. Singh, P., et al.: A space-time-topology-prime, stTS metric for a self-operating mathematical universe uses dodecanion geometric algebra of 1-20D complex vectors. In: LNSS, Lecture notes on networks and signals. Springer Nature, Berlin, in press (2019)
28. Singh, P., et al.: Quaternion octonion to dodecanion manifold: stereographic projections from infinity lead to a self-operating mathematical universe. In: AISC, Advanced Intelligence Systems and Computing. Springer Nature, Berlin, in press (2019)
29. Bandyopadhyay, A.: Chapter 2 FIT, and GML; Chapter 3 PPM; Chapter 7. A complete, integrated time crystal model of a human brain. In: Nanobrain. The Making of an Artificial Brain from a Time Crystal, 372 p. Taylor & Francis Inc. Imprint CRC Press Inc., Bosa Roca (2020)
30. Amalric, M., Wang, L., Pica. P., Figueria, S., Sigman, M., Dehaene, S.: The language of geometry: fast comprehension of geometrical primitives and rules in human adults and preschoolers. PLoS Comput. Biol. **13**(1), e1005273 (2017)
31. Dehaene, S., Izard, V., Pica, P., Spelke, E.: Core knowledge of geometry in an Amazonian Indigene Group. Science **311**(5759), 381–384 (2006)
32. Guevara, M.R., Jongsma, H.J.: Phase resetting in a model of sinoatrial nodal membrane: ionic and topological aspects. Am. J. Physiol. **258**, 734–747 (1990)
33. Agrawal, L., et al.: Fractal information theory (FIT) derived geometric musical language (GML) for brain inspired hypercomputing. In: Advances in Intelligent Systems and Soft Computing (AISC), vol. 2, pp. 37–61. Springer, Berlin (2016)
34. Sotiropoulos, S.N., et al.: Advances in diffusion MRI acquisition and processing in the Human Connectome Project. Neuroimage **80**, 125–143 (2013)
35. Buzsáki, G.: The rhythms of the brain. Oxford University Press, London (2006). https://doi.org/10.1093/acprof:oso/9780195301069.001
36. Ramkumar, P., Parkkonen, L., Hari, R., Hyvärinen, A.: Characterization of neuromagnetic brain rhythms over time scales of minutes using spatial independent component analysis. Hum. Brain Mapp. **33**(7), 1648–1662 (2012)
37. Ghosh, S., Sahu, S., Fujita, D., Bandyopadhyay, A.: Design and operation of a brain like computer: a new class of frequency-fractal computing using wireless communication in a supramolecular organic, inorganic systems. Information **5**, 28–99 (2014)
38. Carter, R.: The human brain book: an illustrated guide to its structure, function, and disorders. DK, London (2014)
39. Liu, Z., Fang, N., Yen, T.J., Zang, X.: Rapid growth of evanescent wave by a silver superlens. Appl. Phys. Lett. **83**, 5184–5187 (2003)
40. Wiltshire, M.C.K., et al.: Metamaterial endoscope for magnetic field transfer: near field imaging with magnetic wires. Opt. Express **11**(7), 709–715 (2003)
41. Milosevic, M.: On the nature of the evanescent wave. Appl. Spectrosc. **26**(2), 126–131 (2013)
42. Plebe, A., Domenella, R.G.: Neural networks; Object recognition by artificial cortical maps. Neural Networks **20**(7), 763–780 (2007)
43. Striegel, D.A., Hurdal, M.K.: Chemically based mathematical model for development of cerebral cortical folding patterns. PLoS Comput. Biol. **5**(9), e1000524 (2019)
44. Reimann, M.W., King, J.G., Muller, E.B., Ramaswamy, S., Markram, H.: An algorithm to predict the connectome of neural microcircuits. Front. Comput. Neurosci. **9**, 1–8 (2015)

45. Braitenberg, V., Braitenberg, C.: Geometry of orientation columns in the visual cortex. Biol. Cybern. **33**, 179–186 (1979)
46. Klyachko, V.A., Stevens, C.F.: Connectivity optimization and the positioning of cortical areas. Proc. Natl. Acad. Sci. **100**, 7937–7941 (2003)
47. Mountcastle, V.B.: The columnar organization of the neocortex. Brain **120**, 701–722 (1997)
48. Gevins, A., Smith, M.E., McEvoy, L., Yu, D.: High-resolution EEG mapping of cortical activation related to working memory: effects of task difficulty, type of processing, and practice. Cereb. Cortex **7**, 374–485 (1997)
49. Harris, K.D., Barthó, P., Zugaro, M.B., Monconduit, L., Marguet, S., Buzsaki, G.: Neocortical population patterns during EEG activation: waking, REM, and anesthetised states. Society for Neuroscience Meeting, Washington DC, abstract.719.5 (2003)
50. Brodmann, K.: Vergleichende Lokalisationslehre der Grosshirnrinde. Johann Ambrosius Barth, Leipzig (1909)
51. Garey, L.J.: Brodmann's: localisation in the cerebral cortex. Springer, New York (2006)
52. CST Microwave Studio. https://perso.telecom-paristech.fr/begaud/intra/MWS_Tutorials.pdf
53. Markram, H., Muller, E., Ramaswamy, S., Reimann, M.W., Abdellah, M., Sanchez, C.A., et al.: Reconstruction and simulation of neocortical microcircuitry. Cell **163**, 456–492 (2015)
54. Horton, J.C., Adams, D.L.: The cortical column: a structure without a function. Philos. Trans. R. Soc. Lond. B Biol. Sci. **360**(1456), 837–862 (2005)
55. Haueis, P.: The life of the cortical column: opening the domain of functional architecture of the cortex (1955–1981). Hist. Philos. Life Sci. **38**(2) (2016)
56. Ghosh, S., et al.: Inventing a co-axial atomic resolution patch clamp to study a single resonating protein complex and ultra-low power communication deep inside a living neuron cell. J. Integr. Neurosci. **15**(4), 403–433 (2016)
57. Venkataraman, A., Amadi, E.V., Chen, Y., Papadopoulos, C.: Carbon nanotube assembly and integration for applications. Nanoscale Res. Lett. **14**(1), 220 (2019)
58. Singh, P., et al.: Fractal and periodical biological antennas: hidden topologies in DNA, wasps and retina in the eye. In: Soft Computing Applications, pp. 113–130. Springer, Singapore (2018)
59. Singer, W.: Synchronization of cortical activity and its putative role in information processing and learning. Annu. Rev. Physiol. **55**, 349–374 (1993)
60. Shi, J., Thompson, P.M., Wang, Y.: Human brain mapping with conformal geometry and multivariate tensor-based morphometry. In: Liu, T., Shen, D., Ibanez, L., Tao, X. (eds) Multimodal Brain Image Analysis, MBIA. Lecture Notes in Computer Science. Springer, Berlin (2011)
61. Muller, A.A., Soto, P., Dascalu, D., Neculoiu, D., Boria, V.E.: A 3-D smith chart based on the Riemann sphere for active and passive microwave circuits. IEEE Microwave Wirel. Compon. Lett. **21**(6), 286–288 (2011)
62. Muller, A.A., Dascalu, D.C., Pacheco, P.S., Esbert, V.E.B.: The 3D Smith chart and its practical applications. Microw. J. **55**(6), 64–74 (2012)
63. Gawne, T.J., Richmond, B.J.: How independent are the messages carried by adjacent inferior temporal cortical neurons? J. Neurosci. **13**, 2758–2771 (1993)
64. Zeki, S., Shipp, S.: The functional logic of cortical connections. Nature **335**, 311–317 (1988)
65. Achard, S., Salvador, R., Whitcher, B., Suckling, J., Bullmore, E.: A resilient, low frequency, small-world human brain functional network with highly connected association cortical hubs. J. Neurosci. **26**, 63–72 (2006)
66. Ahissar, E., Arieli, A.: Figuring space by time. Neuron **32**, 185–201 (2001)
67. Fellous, J.M., Houweling, A.R., Modi, R.H., Rao, R.P., Tiesinga, P.H., Sejnowski, F.J.: Frequency dependence of spiking timing reliability in cortical pyramidal cells and interneurons. J. Neurophysiol. **85**, 1782–1787 (2001)
68. Llinás, R.R., Ribary, U., Joliot, M., Wang, X.J.: Content and context in temporal thalamocortical binding. In: Buzsáki, G., Llinás. R., Singer, W., Berthoz, A., Christen, Y. (eds.) Temporal Coding in the Brain, pp. 251–272. Springer, Berlin (1994)
69. Ikegaya, Y., Aaron, G., Cossart, R., Aronov, D., Lampl, I., Ferster, D., Yester, R.: Synfire chains and cortical songs: temporal modules of cortical activity. Science **304**, 559–564 (2004)

70. Lopes da Silva, F.H., Storm van Leeuwen, W.: The cortical alpha rhythm of the dog: the depth and profile of phase: architectonics of the cerebral cortex. In: Brazier, M.A.B., Petsche, H. (eds.), pp. 150–187. New York (1978)
71. Van Essen, D.: A tension-based theory of morphogenesis and compact wiring in the central nervous system. Nature **385**, 313–318 (1997)
72. Ringo, J.L., Doty, R.W., Demeter, S., Simard, P.Y.: Time is of the essence: a conjecture that hemispheric specialization arises from interhemispheric conduction delay. Cereb. Cortex **4**, 331–343 (1994)
73. Matthews, P.C., Strogatz, S.H.: Phase diagram for the collective behavior of limit-cycle oscillators. Phys. Rev. Lett. **64**, 1701–1704 (1990)
74. Wang, X.J.: Multiple dynamical modes of thalamic relay neurons: rhythmic bursting and intermittent phase-locking. Neuroscience **59**, 21–31 (1994)
75. Graybiel, A.M.: The basal ganglia: learning new tricks and loving it. Curr. Opin. Neurobiol. **15**, 638–644 (2005)
76. Paré, D., Shink, E., Gaudreau, H., Destexhe, A., Lang, E.J.: Impact of spontaneous synaptic activity on the resting properties of cat neocortical pyramidal neurons in vivo. J. Neurophysiol. **79**, 1450–1460 (1998)
77. Shadlen, M.N., Newsome, W.T.: Noise, neural codes and cortical organization. Curr. Opin. Neurobiol. **4**, 569–579 (1994)
78. Colom, R., et al.: Human intelligence and brain network. Dialogues Clin. Neurosci. **12**(4), 489–501 (2010)
79. Nie, Y., Fellous, J.M., Tatsuno, M.: Information-geometric measures estimate neural interactions during oscillatory brain states. J. Front. Neural Circ. **24**, 8–11 (2014)
80. Saxena, K., et al.: Fractal, scale free electromagnetic resonance of a single brain extracted microtubule nanowire, a single tubulin protein and a single neuron. Fractal Fractional **4**(2), 11 (2020)
81. Friston, K.: The free energy principle: a unified brain theory? Nat. Rev. Neurosci. **11**, 127–138 (2010)
82. Winfree, A.: Biological Rhythm Research: The Geometry of Biological Time. Springer, New York (1977, 2001)
83. Miller, K.D., Pinto, D.J., Simons, D.J.: Processing in layer 4 of the neocortical circuit: new insights from visual and somatosensory cortex. Curr. Opin. Neurobiol. **11**, 488–497 (2001)
84. Zeki, S.M.: Uniformity and diversity of structure and function in rhesus monkey prestriate cortex. J. Physiol. **277**, 272–290 (1978)
85. Abeles, M.: Local cortical circuits: studies in brain function. Springer, Berlin (1982)
86. Sadeh, S., Rotter, S.: Statistics and geometry of orientation selectivity in primary visual cortex. Biol. Cybern. **108**(5), 631–653 (2014)
87. Niu, H., Wang, J., Zhao, T., Shu, N., He, Y.: Correction: revealing topological organization of human brain functional networks with resting-state functional near infrared spectroscopy. PLoS ONE **7**(9), e45771 (2013)
88. Scannell, J.W., Blakemore, C., Young, M.P.: Analysis of connectivity in the cat cerebral cortex. J. Neurosci. **15**(2), 1463–1483 (1995)
89. Krupa, D.J., Wiest, M.C., Shuler, M.G., Laubach, M., Nicolelis, M.A.: Layer-specific somatosensory cortical activation during active tactile discrimination. Science **304**, 1989–1992 (2004)
90. Reich, D.S., Mechler, F., Victor, J.D.: Independent and redundant information in nearby cortical neurons. Science **294**, 2566–2568 (2001)
91. Katz, L.C., Shatz, C.J.: Synaptic activity and the construction of cortical circuits. Science **274**, 1133–1138 (1996)
92. Schütz, A., Braitenberg, V.: The human cortical white matter: quantitative aspects of cortico-cortical long-range connectivity. In: Schütz, A., Miller, R. (eds.) Cortical Areas: Unity and Diversity. Taylor and Francis, Milton Park (2002)
93. Shu, Y., Hasenstaub, A., McCormick, D.: Turning on, off recurrent balanced cortical activity. Nature **423**, 288–293 (2003)

Detection of Ovarian Malignancy from Combination of CA125 in Blood and TVUS Using Machine Learning

Laboni Akter and Nasrin Akhter

Abstract Carcinoma of the ovary keeps on being one of the main causes of death from gynecologic malignancies on the earth. Women determined to have beginning phase (1/2) ovarian disease have considerably better endurance rates contrasted with those analyzed in later stages. The speculation behind screening for ovarian malignant growth is that previous recognition of the illness will bring about more women being analyzed at less propelled stages and that this will convert into lower generally death rates for the ovarian disease. Both transvaginal ultrasound and the Cancer Antigen 125 in blood have been used to screen for early ovarian malignant growth. The aims of our study are to the detection of ovarian malignancy in primary stage and thereby reduce the mortality due to ovarian cancer. In this study, we have used a real-world dataset, namely PLCO to anticipate the discoveries from CA-125 and TVUS joined screening utilizing machine learning models. We obtained a critical presentation from the model as far as accuracy, precision, and recall with the estimations of 80, 86, and 90.25% individually. Our work will support the doctors and the suspected to analyze ovarian threats in the beginning and decreased the complication and mortality due to ovarian malignancy.

Keywords Ovarian cancer · CA125 · Transvaginal ultrasound · Random forest classifier · Naïve Bayes classifier

1 Introduction

Ovarian cancer (OC) is a widely recognized sort of disease in women. This cancer discovered 295,414 ladies were determined to have an ovarian disease than 184,799 passings were brought about by its universal in the time of 2018 [1].

L. Akter (✉) · N. Akhter
Khulna University of Engineering & Technology (KUET), Khulna 9203, Bangladesh
e-mail: laboni.kuet.bme@gmail.com

N. Akhter
e-mail: nasrinakhterdr@gmail.com

M. S. Kaiser et al. (eds.), *Proceedings of International Conference on Trends in Computational and Cognitive Engineering*, Advances in Intelligent Systems and Computing 1309, https://doi.org/10.1007/978-981-33-4673-4_22

Prescriptions for ovarian threatening development are refining, and the best results are constantly observed so that the disease is discovered promptly. The most widely recognized ovarian diseases are known as epithelial ovarian tumors (EOC). Other kinds of ovarian disease incorporate ovarian low threatening possible tumor (OLMPT), germ cell tumors, and sex rope stromal tumors like the granulosa–stromal tumors. Acquired changes in the "Breast Cancer gene 1" (BRCA1) and "Breast Cancer gene 2" (BRCA2) qualities enormously increment a lady's ovarian malignant growth hazard just as cancer disease chance. A gynecologic oncologist is a master with skill overseeing ovarian malignant growth. Most ovarian cancer is analyzed in cutting-edge stages because there are no solid early disease-related indications and indications of ovarian disease.

Prior determination of OC could fundamentally expand endurance. The overall five years endurance pace of OC patients is roughly 90% during the period I yet just 5–60% at period II or more. Situated inside a profound pelvic hole, this epithelial threat, as a rule, emerges with no ordinary real signs or side effects. In this manner, the current modalities for distinguishing beginning phase ovarian disease depend intensely on transvaginal ultrasonography (TVU) and serum biomarkers, for example, CA125. The current writing on repetitive ovarian malignant growth uncovers that elements incorporate Age, Grade, Pathologic N, Pathologic T, Histology, Pathologic M, Pathologic Stages [2]. CA-125 is a protein that is an alleged tumor marker or biomarker, which is available in a more prominent focus in ovarian malignancy cells than in different cells. In many research centers, the ordinary worth is 0–35 units/ml.

The finding at the right time prompts the life sparing of the patients. For a model, CA125 is utilized for early identification of disease for ovarian disease distinguishes 50–60% of the ladies' in stage 1 ovarian disease. The complex biometric tests of the malignant growth ailment can be analyzed explicitly through biomarkers. In this venture, identification of the ovarian malignant growth is done at a beginning period, so the demise pace of ladies may diminish to a certain level.

In this paper, we have utilized a combined approach of CA125 in blood and TVUS findings to predict the final ovarian screening result in applying machine learning techniques.

2 Literature Review

Olivier et al. [3] built to set up in a huge gathering of high-hazard ladies the affectability, particularity, the Negative Predictive Value (NPV), and Positive Predictive Value (PPV) of ovarian cancer screening by methods for pelvic assessment, serum CA125 assurance, and TVU.

Kawakami et al. [4] developed an ovarian breast cancer explicit prescient structure for clinical stage, histotype, lingering tumor weight, and anticipation utilizing AI strategies dependent on numerous biomarkers. Seven managed AI classifiers, including gradient boosting machine (GBM), support vector machine, random forest (RF), conditional RF (CRF), Naïve Bayes, neural network, and Elastic Net, were utilized to infer demonstrative and prognostic data from 32 boundaries generally accessible from pretreatment fringe blood tests and age.

Gu et al. [5], the postprandial increment in serum CA125 as a substitute biomarker because of a short analysis of OC. The ideas of "glucose and insulin-incited CA125" and "dynamic estimation/evaluation of serum malignant growth biomarkers", the support vector machine (SVM) founded calculation was utilized for refining the exhibition of CA125 postprandial addition. An SVM founded CA125-increase calculation was assembled. This calculation showed improved adequacy on distinguishing beginning phase ovarian malignancy. Exhibitions of static and dynamic serum CA125 trail-emerged indicative measures for beginning OC, fasting CA125, and postprandial augmentation as information sources sensitivity (%), specificity (%), PPV (%), and NPV (%) were 91.7%, 99.2%, 89.2%, and 99.4% correspondingly.

Buys et al. [6] discriminated limited preliminary of screening versus common consideration. Benchmark screening results were accounted for. A total of 29 neoplasms were distinguished. A total of 9 were tumors of downcast threatening dynamic and 20 were intrusive. The plus prescient incentive for obtrusive malignancy was 3.7% for an irregular CA-125, 1.0% for an anomalous TVU, and 23.5% if the two tests were unusual.

Shinagare et al. [7] utilized AI to anticipate stomach repeat on CT based on sequential malignant growth antigen 125 (CA125) balances together with patients with cutting-edge high-grade serous ovarian disease on observation. Utilizing AI, SVM models were improved and assessed utilizing 10-fold cross-validation to decide the CA125 measurement generally prescient of stomach repeat. A higher pace of CA125 increment is related to stomach repeat.

3 Methodology

We have organized the methodology section into these subparts

- Data Collection
- Data Preprocessing
- Exploratory Data Analysis
- Machine Learning Algorithm for Classification
- SMOTE Analysis.

The process flow diagram of this work is represented in Fig. 1.

3.1 Data Collection

We have used PLCO dataset to conduct this research work [8]. It is a randomized, controlled preliminary to decide if certain screening tests decrease mortality from prostate, lung, colorectal and ovarian malignancy. Around 155,000 members were

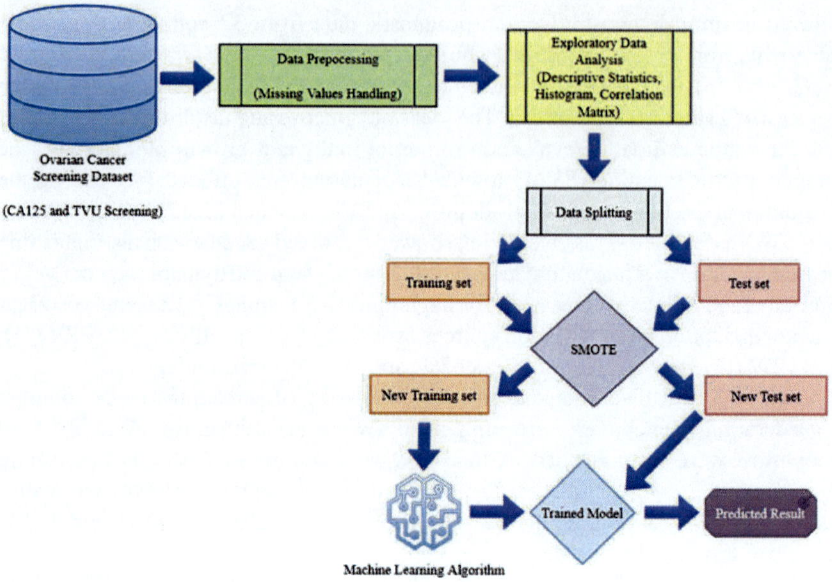

Fig. 1 Flow diagram of methodology

selected between November 1993 and July 2001. The ovarian screening dataset (~151,000, one record for each time of screening) contains extra data from the transvaginal ultrasound (TVUS) and CA-125 screens. This incorporates subtleties of the blood draw, QA TVUS results, purposes behind deficient tests, and extra discoveries that were not dubious for malignant growth.

This dataset contains a lot of attributes that are not needed to conduct this work. We have chosen seven distinct features from the data along with a target variable. We have taken "ca125_result", "ca125_level", "ca125i_assess_days", "ca125ii_assess_days", "tvu_days", "tvu_result", "numcyst" as input features and "ovar_result" as the target variables.

3.2 Data Preprocessing

Data preprocessing is a procedure that is utilized to change over the raw data into a perfect data index. While making a machine learning venture, it isn't generally a case that we got clean and structured data, and keeping in mind that doing any activity with data, it is obligatory to clean it and placed it in a structured manner. So for this, we use data preprocessing steps.

Here, the dataset entities were on float 64 and int 64 values. There were a lot of missing values in different attributes. There are several ways to encounter missing values in the dataset. We have applied different techniques to handle the missing

#	Column	Non-Null Count	Dtype
---	------	--------------	-----
0	ca125_result	150993 non-null	float64
1	ca125_level	150993 non-null	float64
2	ca125i_assess_days	150993 non-null	float64
3	ca125ii_assess_days	150993 non-null	float64
4	tvu_days	150993 non-null	float64
5	tvu_result	150993 non-null	float64
6	numcyst	150993 non-null	float64
7	ovar_result	150993 non-null	int64

Fig. 2 Data after preprocessing

values in the dataset, but filling those with the mean values showed better performance [9, 10]. While implementing, we have utilized Scikit-learn's "Simple Imputer" package to accomplish this task. Figure 2 shows the processed dataset.

3.3 Exploratory Data Analysis

For whatever length of time that there is information to examine, the need to investigate is self-evident. However, a significant key part of any machine learning task now and again underestimated is the exploratory data analysis. At a significant level, EDA is the act of depicting the data by methods for factual and perception procedures to bring significant parts of that data into the center for additional examination. This includes taking a gander at the dataset from numerous points, depicting it, and summing up it without making any suppositions about its substance. This is a huge advance to take before jumping into machine learning or statistical modeling, to ensure the data is truly what they are professed to be and that there are no conspicuous issues.

A histogram is a plot of the comprise or enlarge of data in disjoint spans by with the support set. Noted, the actual regards of the information are never settled; a datum's core in an appointed bounce is only counted, and a histogram displays vertical bars with statures utterance to the prediction or the enhancement of the example size in each container. We have computed the histograms of the input features used in this research (Fig. 3).

ca125_level. CA-125 level was recorded from a substantial screen.

ca125i_assess_days. Days from randomization until CA-125i blood test was surveyed, where most of the values between 0 and 200.

ca125ii_assess_days. Days from randomization until CA-125ii blood test was surveyed, where days are mostly lies between 0 and 2000.

ca125_result. CA-125 screen result, where 1 = "Negative" 2 = "Abnormal, Suspicious" 4 = "Inadequate" 8 = "Not done, expected".

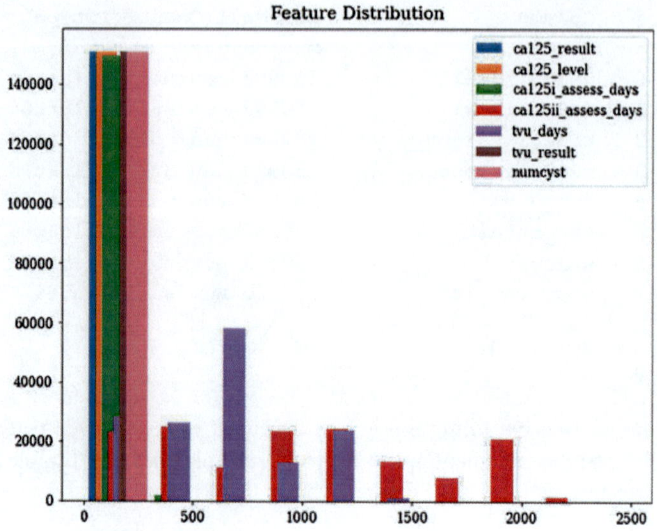

Fig. 3 Feature distribution in the dataset

numcyst. The number of blisters found on the two sides; here values lie between 0 and 1 mostly.

tvu_days. Days from randomization until transvaginal ultrasound test. This value mostly lies between the ranges of 300–900.

tvu_result. Transvaginal ultrasound test result, where 1 = "Negative" 2 = "Abnormal, Suspicious" 3 = "Abnormal, Non-Suspicious" 4 = "Inadequate" 8 = "Not done, expected".

3.4 Machine Learning Algorithm for Classification

We have utilized random forest as well as the Naïve Bayes algorithm for classifying ovarian screening findings.

Random forest, a supervised machine learning algorithm, is exploited because of classification. We realized that a forest is covered with trees and more trees mean more robust forests [11, 12]. Likewise, the random forest classifier submits decision trees on information tests and subsequently gets the desire from all of them ultimately picks the best arrangement by techniques for projecting a voting form. It is an outfit method that decreases the over-fitting by averaging the result. The workflow of the random forest algorithm is illustrated in Fig. 4. Starting with the determination of arbitrary examples from a given dataset, this algorithm will develop a decision tree for each example. At that point, it will get the expected result from each decision tree. In this progression, casting voting will be performed for each anticipated outcome. Finally, select the most casted vote as the last expectation result.

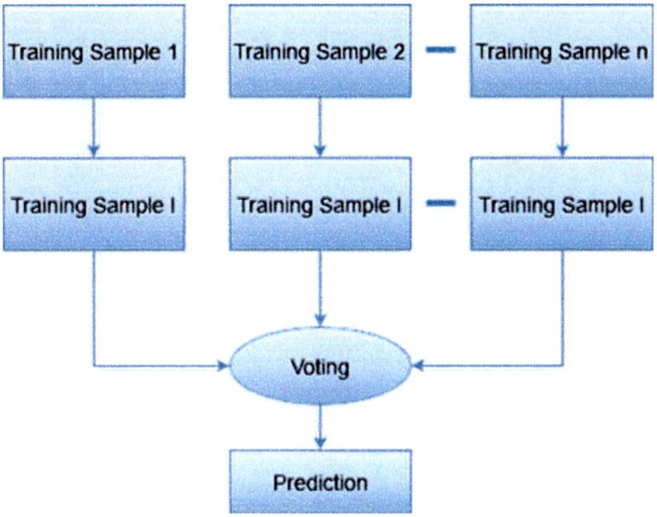

Fig. 4 Random forest workflow

Naïve Bayes is an algorithm for classification subject to applying Bayes' theory with strong speculation that is self-governing to each other [13]. In clear words, the speculation that cannot avoid being that the closeness of a component in a class is allowed to the proximity of some other component in a comparative class. The posterior probability of a class is computed in such a way in (1).

$$P(L|\text{features}) = \frac{P(L)\,P(\text{features}|L)}{P(\text{features})} \tag{1}$$

where "L" is the class variable, so $P(L)$ expresses the probability of that class.

3.5 SMOTE Analysis

While doing the classification task, we have encountered imbalanced data for some classes. As it was a real-world clinical dataset, it is quite obvious to face imbalanced classes. The conventional machine learning model assessment techniques do not precisely measure model execution when confronted with imbalanced datasets (Fig. 5).

Synthetic minority oversampling technique (SMOTE) is one of the best regularly utilized oversampling techniques to take care of the imbalanced class issue. It means to adjust class appropriation by randomly expanding minority class models by imitating them [14, 15]. It blends examples from the dataset. It takes the whole dataset as input; however, it builds the level of just the minority cases. It augmented

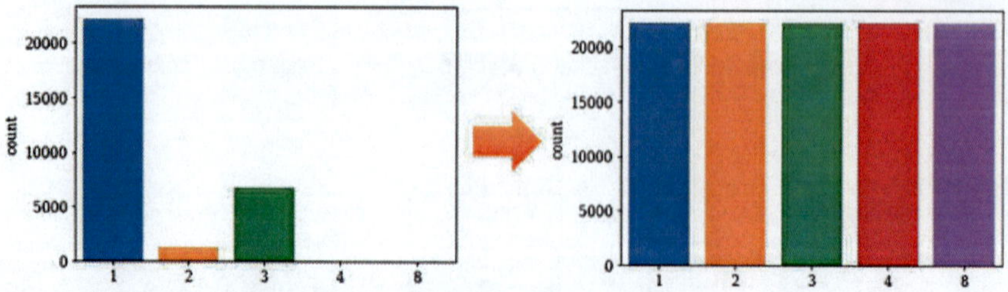

Fig. 5 Class-wise data distribution before and after SMOTE

new minority occurrences between existing minority occasions. It produces virtual records by linear interpolation for the minority class. These newly manufactured preparing records are produced by randomly choosing at least one of the *k*-nearest neighbors for each model in the minority class. Then, the data is remade and a few order models can be applied for the handled data.

4 Result and Discussion

In this segment, the correlation among the features and the model performance which we have obtained from the implementation will be presented.

Correlation can be a significant instrument for including designing in building machine learning models. Features that are uncorrelated with the target variable are a most likely acceptable contender to trim from the model. Also, on the off chance that two indicators are firmly related to one another, at that point, we just need to utilize one of them. As we can visualize in Fig. 6, no particular feature is correlated with other features. In this situation, we have no way to eliminate any features while training the classifier.

We have computed performance metrics like accuracy, precision, and recall from the models (Table 1).

A confusion matrix is an illustration that is recurrently utilized to show the exhibition of the classifier on test data. We can easily visualize the number of properly classified and misclassified data with the visualization of the confusion matrix. The resulting confusion matrix of the random forest model is presented in Fig. 7. The diagonal entities represent the properly classified values of the test set.

5 Conclusion

Ovarian cancer is a distressing medical issue in Bangladesh as well as in the world. The yearly death rate per 100,000 individuals from the ovarian disease in Bangladesh has expanded by 40.3% since 1990, an average of 1.8% every year. Repetitive

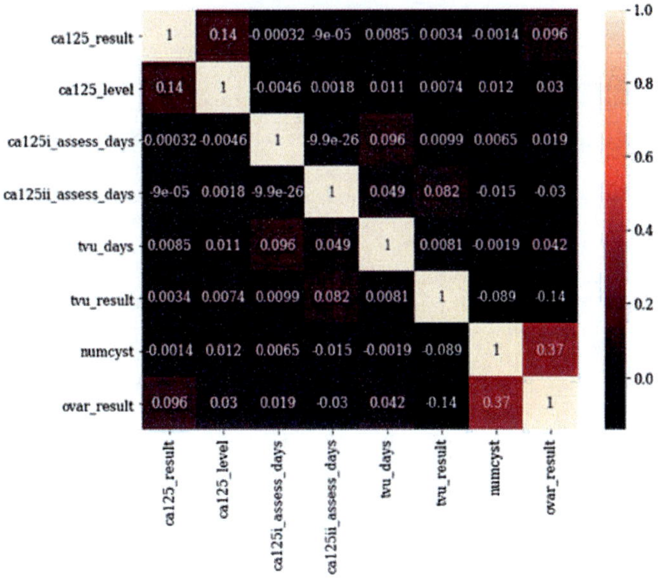

Fig. 6 Correlation of the features

Table 1 Models' performance

Model	Before SMOTE			After SMOTE		
	Accuracy (%)	Precision (%)	Recall (%)	Accuracy (%)	Precision (%)	Recall (%)
Random forest	81	84.3	77	80	86	90.25
Naïve Bayes	55	59.33	78	71	73	89.25

high-grade ovarian malignancy is typically connected with transient endurance. This exploration work centers on the early discovery of ovarian cancer in women appropriately. Viable screening techniques have affected endurance in different malignancies; yet at present, there is no viable screening test for the ovarian disease. Since analysis at a beginning phase is related to improved paces of endurance, a compelling screening procedure that identifies beginning phase ovarian malignant growth significantly affects mortality from this disease. In our study, we anticipated patients' condition in five classes as—"ovarian cancer negative," "abnormal and suspicious," "abnormal and non-suspicious," "inadequate screening" and "not done, expected" from CA125 and TVUS combined test with a significant performance of different machine learning models using the PLCO dataset. However, deep learning may increase the performance but for a real-world dataset; our model stays stable.

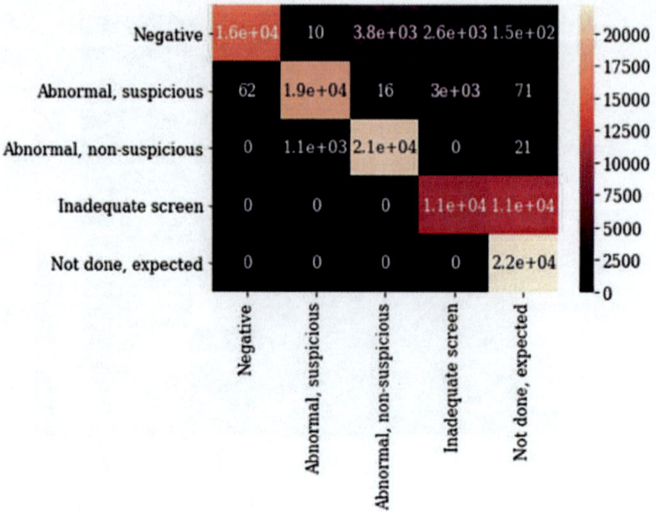

Fig. 7 Confusion matrix of random forest model

Ethical Approval The dataset and its usage approval were collected from the data provider—CDAS, under the project PLCOI-620 [16].

Acknowledgements The authors would like to thank the "Khulna University of Engineering & Technology (KUET)" to facilitate the work.

References

1. Bray, F., Ferlay, J., Soerjomataram, I., et al.: Global cancer statistics 2018: GLOBOCAN estimates of incidence and mortality worldwide for 36 cancers in 185 countries. CA Cancer J. Clin. **68**(6), 394–424 (2018)
2. Ovarian Cancer Statistics, Cancer Research UK. https://www.cancerresearchuk.org/health-professional/cancer-statistics/statistics-by-cancer-type/ovarian-cancer#heading-Three. Accessed 20 Aug 2020
3. Olivier, R., Lubsen-Brandsma, M., Verhoef, S., van Beurden, M.: CA125 and transvaginal ultrasound monitoring in high-risk women cannot prevent the diagnosis of advanced ovarian cancer. Gynecol. Oncol. **100**, 20–26 (2006)
4. Kawakami, E., Tabata, J., Yanaihara, N., et al.: Application of artificial intelligence for preoperative diagnostic and prognostic prediction in epithelial ovarian cancer based on blood biomarkers. Clin. Cancer Res. **25**, 3006–3015 (2019)
5. Gu, Z., He, Y., Zhang, Y., et al.: Postprandial increase in serum CA125 as a surrogate biomarker for early diagnosis of ovarian cancer. J. Transl. Med. **16**, 114 (2018)
6. Buys, S., Partridge, E., Greene, M., et al.: Ovarian cancer screening in the Prostate, Lung, Colorectal and Ovarian (PLCO) cancer screening trial: findings from the initial screen of a randomized trial. Am. J. Obstet. Gynecol. **193**, 1630–1639 (2005)
7. Shinagare, A., Balthazar, P., Ip, I., et al.: High-grade serous ovarian cancer: use of machine learning to predict abdominopelvic recurrence on CT on the basis of serial cancer antigen 125 levels. J. Am. Coll. Radiol. **15**, 1133–1138 (2018)

8. Ovarian—Datasets—PLCO—The Cancer Data Access System. In: Cdas.cancer.gov. https://cdas.cancer.gov/datasets/plco/23/. Accessed 20 Aug 2020
9. Saar-Tsechansky, M., Provost, F.: Handling Missing Values when Applying Classification Models. In: Jmlr.org. https://www.jmlr.org/papers/v8/saar-tsechansky07a.html. Accessed 20 Aug 2020
10. Li, P., Stuart, E., Allison, D., et al.: Multiple imputation. JAMA **314**, 1966 (2015)
11. Qi, Y.: Random forest for bioinformatics. In: Zhang, C., Ma, Y. (eds.) Ensemble Machine Learning. Springer, Boston (2012)
12. Meding, S., Nitsche, U., Balluff, B., et al.: Tumor classification of six common cancer types based on proteomic profiling by MALDI imaging. J. Proteome Res. **11**, 1996–2003 (2012)
13. Wu, M., Dai, D., Shi, Y., et al.: Biomarker identification and cancer classification based on microarray data using laplace Naive Bayes model with mean shrinkage. IEEE/ACM Trans. Comput. Biol. Bioinform. **9**(6), 1649–1662, Nov–Dec (2012)
14. Han, H., Wang, W.Y., Mao, B.H.: Borderline-SMOTE: a new over-sampling method in imbalanced data sets learning. In: Huang, D.S., Zhang, X.P., Huang, G.B. (eds.) Advances in Intelligent Computing, ICIC 2005. Lecture Notes in Computer Science, vol. 3644. Springer, Berlin (2005)
15. Alghamdi, M., Al-Mallah, M., Keteyian, S., et al.: Predicting diabetes mellitus using SMOTE and ensemble machine learning approach: the Henry Ford ExercIse Testing (FIT) project. PLoS ONE **12**, e0179805 (2017)
16. PLCOI-620: Early Prediction of Ovarian Cancer from Images along with Some ...—Approved Projects—The Cancer Data Access System. In: Cdas.cancer.gov. https://cdas.cancer.gov/approved-projects/2607/. Accessed 1 Oct 2020

Auditory Attention State Decoding for the Quiet and Hypothetical Environment: A Comparison Between bLSTM and SVM

Fatema Nasrin, Nafiz Ishtiaque Ahmed, and Muhammad Arifur Rahman

Abstract Auditory attention is a cognitive process that involves listeners to give their attention, which the listener wants to focus on. Consequently, the auditory attention processes their stimulus of interest and decodes it as a response of individuals. This study is aimed to decode the listener's auditory attention state at continuous hearing complications in different (e.g., by quiet and hypothetical noisy background) environments. There have numerous ways to decode the auditory attention state. Here, we approach the difficulty of auditory attention to the auditory oddball paradigm based on electroencephalogram (EEG) signal. We have carried two experiments and having 840 trials for each listening state (attended and unattended). We have applied a stimulus for the attention detection model and it was verified for decoding the auditory listening attention successfully to listening. Finally, we present the decoding state of acoustic listening attention from electroencephalogram (EEG) signals through a bidirectional long short-term memory (bLSTM) and a support vector machine (SVM) model. The performance of the two models has been compared, and it reveals that these two models perform at a significant accuracy to detect the auditory listening attention state. In the frontal lobe, which is associated with attention, the SVM model has achieved a statistically significant accuracy of 74% at a 50 ms time window.

Keywords Auditory attention decoding · Auditory listening state · Speech perception · Hypnosis · Brain signal decoding · BCI · EEG

F. Nasrin (✉) · M. A. Rahman
Jahangirnagar University, Dhaka, Bangladesh
e-mail: fatemanasrin1234@gmail.com

M. A. Rahman
e-mail: arif@juniv.com

N. I. Ahmed
AIUB, Dhaka, Bangladesh
e-mail: cse.ishtiaque@gmail.com

M. S. Kaiser et al. (eds.), *Proceedings of International Conference on Trends in Computational and Cognitive Engineering*, Advances in Intelligent Systems and Computing 1309, https://doi.org/10.1007/978-981-33-4673-4_23

1 Introduction

In recent years, auditory attention and listening effort is a very popular area of research. Perception difficulty is known as listening effort. Hearing difficulties can happen by accident or perhaps caused by the perception of decay unsurprisingly due to aging also often causing a problem to cochlear implant patients. In the medical field, by presenting the term hearing effort, we can fulfill the space amid standard speech scores. In the real world, it is generally measuring the part of the speech intelligibility and communication outcome. The listening effort contributes us the window to detect neural processes underlying speech perception [1]. Listening effort is influenced by attention in every listening task. Auditory attention is associated with age, memory, and information processing [2]. Therefore, in any kind of environment detecting the attended and unattended sound is essential. In this study, we have a focus on both a quiet and hypnosis environment. The study of neural dynamics of hypnosis concerning attentional abilities remains a major challenge for the field [3].

An electroencephalogram (EEG) detects the electrical movement of the brain in any conditions of the human body. EEG signal response provides greater results for detecting the attended and unattended sound. EEG data can be decoded to determine attentional selection in a naturalistic multi-speaker environment. For decoding the particular attention state, the beta band element of EEG is considered broadly as beta oscillations (Fig. 1) are frequently related to memory, attention [4], and estimation.

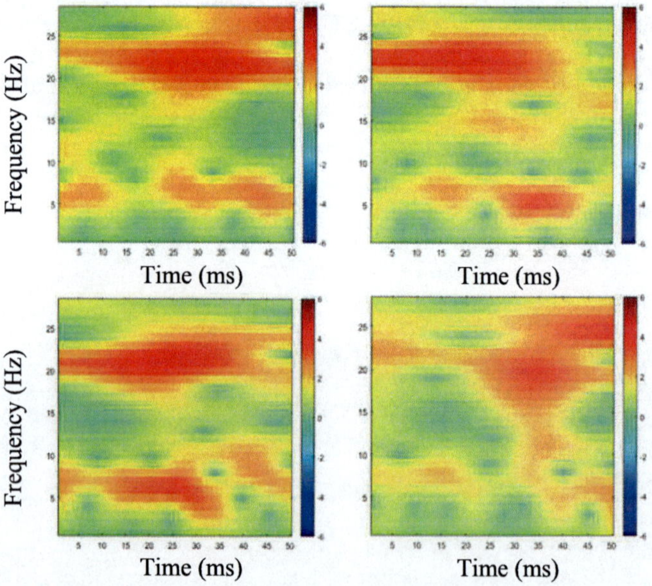

Fig. 1 The event-related beta-band activity of male subjects T8 electrode. EEG is considering broadly as beta oscillations, which are frequently relate to the attention

We assigned the attention label to different sound stimuli (low tune and high tune) in a noisy environment and depending on the importance of speech in which we are interested; we prioritized the attention, which is also called selective attention. Identifying auditory attention can develop the hearing aid in both attended and unattended sound. Auditory oddball stimuli have been useful for perceiving perception of the listening label. The auditory oddball stimuli had applied to decode selective attention in auditory work. We count the amounts of low tune in the state of attention during the oddball auditory experiment, which was performed in both a quiet and a hypnosis noisy environment. The estimation of accuracy is important for any categorization such as cocktail party, music stress, diverse speech activity of different body organs [5], imputing clinical missing data [6], and biological data [7]. We aimed to measure the performance of bLSTM and SVM models based on the EEG signals using auditory oddball having a stimulus onset time of 50 ms. In our study, we applied both bLSTM and SVM for decode attention state using EEG in quiet and hypnosis noisy environments. We analyze which brain area makes the highest attention decoding compared to both model's performance and accomplished with significant accuracy within a 50 ms time window.

2 Literature Review

Detecting auditory attention using different sound stimuli has been using in several studies. The previous study only allocated the attention label to various sound stimuli in a noisy environment [8]. Earlier; several studies associated with hypnosis in the relation between EEG, attention, information processing had been investigated. The auditory oddball task was performed during a dissociative imagery hypnotic context to examine the effects of pain reduction treatments [9]. Besides, frontal lobe activation was seen for an attentional focusing on auditory, visual, and somatosensory ERPs during hypnosis [10–12]. Researchers were also focused on the EEG responses extracted from both non-natural sound stimuli and natural stimuli for detecting speech perception [13]. Some of the previous findings of auditory effort related to non-speech specific speech auditory cognitive procedure. Moreover, the auditory oddball stimuli had applied to decode selective attention in auditory work. In several studies, the auditory oddball function has been useful for measuring the hearing perception [14–16]. The auditory oddball paradigm has been vastly applying to examine the attention in hypothesis study [9, 12, 17]. In previous studies, 11 subject's auditory attention has decoded with 66–81% accuracy for a stimulus onset time of 10–30 s using nonlinear neural networks and linear models [18]. Furthermore, recently another study successfully decoding 16's subject auditory attention with an average of 81% for a stimulus onset time of 1–2 s [19]. They had shown that a time window less than 50 ms gives less than 55% decoding accuracy in linear model and less than 72% decoding accuracy in the CNN model. They had also shown that leaving one object out form the CNN model gives a mean of 10% decrease in decoding accuracy. Thus, more subject's data in the training model gives higher accuracy in the deep learning model. However, these two studies did not mention which brain region makes the highest decoding accuracy, like, in the primary auditory cortex (associated with hearing), frontal gyrus or amygdala (information processing), etc. In our study, to decode the auditory attention from EEG recordings in both hypnosis and quiet environment with two subjects and comparing the method has recently been proposed which is showing a promising decoding accuracy. We applied bLSTM and SVM to decode the auditory attention from EEG signal. Both bLSTM framework and SVM classifier are common in decoding brain signals in BCI application [20–22]. Moreover, we analyze which brain region is associated with auditory attention detection.

3 Materials and Methods

Subjects and Data

Auditory oddball during hypnosis: This dataset consists of 4 EEG recordings which were included with two healthy right-handed subjects (one male and one female). It was experienced during an audiometric estimation before the study. Subjects did not locate any hearing imperfection that should influence the result existing in this study. This study was carried out on a collection of open-source data, which was made available by the BNCIHORIZON2020 (http://bnci-horizon-2020.eu/).

Stimuli

Two high and low steams of tunes were represent to the subject through in-ear headphones. Investigational stimuli were 360 short compound binaural stream high tones and 60 short complex low tunes. The period of the stimulus was 50 ms and was arbitrarily interspersed. It keeps away from back-to-back targets. The power of the stimulus was 70 dB. Two experiments were accomplished in total, having a sum of 840 trials for every hearing state (attention and hypnosis state), in which 420 formed every subject. All the stimuli were standardized to the equal root-mean-square amplitude.

Experimental Design and Procedure

The experiment was accomplished in a soundproof compartment. Participants in this experiment were awake and lay down in a relaxed reclining chair listening to a headphone. Stimuli have existed in two listening tasks (attention and hypnosis imaginary state). In the task of listening attention, the subject was informed to count the tune's amount in a calm environment noiselessly. In the state of hypnosis, the subject was informed to imagine sailing on a ship in an intense hazy environment. Their subject was accessible with indirect imaginary instruction. Similar imagery hypnotic context for the auditory oddball task was applied in EEG [9].

EEG Recording and Processing

A 31-channel EEG system through dynamic electrodes EEG signals were recorded. Electrodes (F8, F4, Fz, F3, F7, FC6, FC2, FC1, FC5, T8, C4, Cz, C3, T7, TP10, CP6,

CP2, CP1, CP5, TP9, P8, P4, Pz, P3, P7, O2, O1, heogright, heogleft, veogabove, veogbelow) were situated according to the International 10–20 systems [23]. The unprocessed EEG signals were down-sampled to a sampling rate of 512 Hz. Every channel's data was then filtered with a 0.01–100 Hz bandpass filter, notch 50 Hz. The eye movements vertical and horizontal were recorded using two pairs of Electrooculography (EOG) electrodes at the external cantina of both eyes and over and under one eye.

EEG Analysis

The data from EEG were considered applying the EEGLAB [24] open-source toolbox and custom MATLAB scripts (version R2019b, The MathWorks Inc.) and python 3 scripts. Moreover, 31 electrodes EEG data (open source data) were gain from every recording position. Further, the rate of downward sampled is 512 Hz. Every trial lasted for 50 ms. Consequently; we had applied the 50 ms time window data in this study. Therefore, the waveforms of EEG were de-noised and using an adaptive 50 Hz notch filtering process to reduce contamination with power line noise. The data baseline was first corrected by subtracting the mean value across the trials. Then we calculate the power spectra via Fast Fourier Transform (FFT), shown in Fig. 2. Afterward, the band-power feature vectors are extracted from the power spectra in beta frequency components (14–30 Hz).

Lastly, to standardize the power spectra density; we separated all attentional trials and hypnosis trials into training trials and testing trials. Then, we calculated the z-scores from each power spectral density (PSD) for individual channels over the training and testing trials. In Matlab, we executed two classification models.

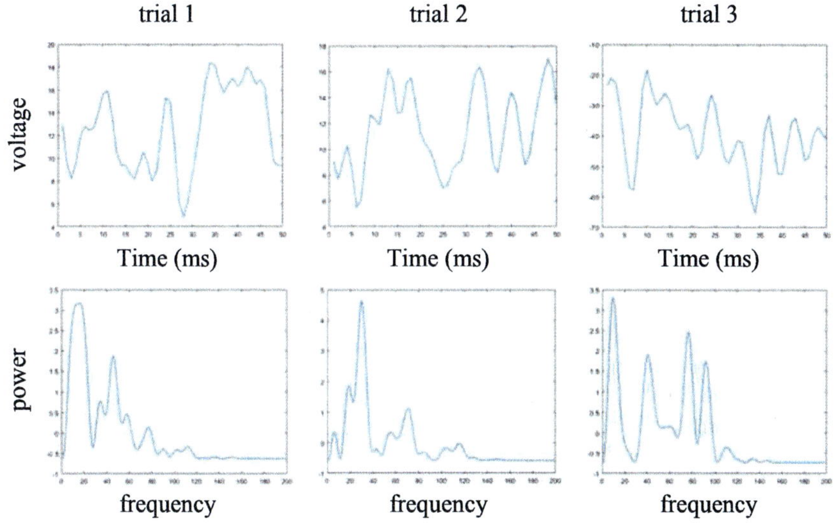

Fig. 2 T8 electrodes raw EEG signal and its corresponding power spectra. The beta-band shows strong correlation with auditory attention

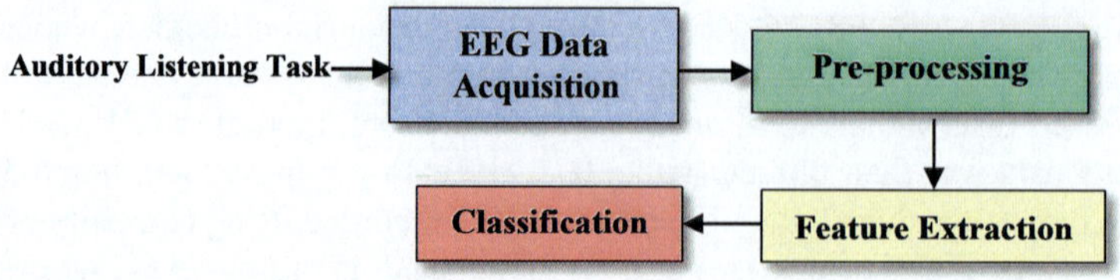

Fig. 3 EEG signal based auditory listening task classification framework. This framework is applied in both SVM model and bLSTM model

Figure 3 illustrates the structural design of the major framework. In every classification model, a threefold cross-validation format is employed. Initially, in the beta frequency component a five-layer stacked bLSTM [25] finds out the sequences of attentional trials and hypnosis trials. In the bLSTM model, we applied the ADAM optimizer in every step to decrease the mean-squared error. The preliminary learning rate of the optimizer is set to 0.001.

One hundred hidden units set to each of two bLSTM cells with a dropout rate set to 50% and 25%, respectively. The batch range for training is 256 in all stages. Then the softmax has been applied. Figure 4 illustrates the overall framework of bLSTM model. Second, the LIBSVM toolbox is applied to set up the SVM classifier [26].

For hearing task classification, all of the LIBSVM constraints are positioned to default. The *z*-score of the PSD of the hearing state trials were fed into both of the model's classifier for training and testing. As a final point, we estimated the decoding performance by using correctly decoded test trials. The performance was intended by counting the number of properly decoded hearing attention over all the threefold cross-validations. For every electrode, a mean accuracy across threefold is calculated and of all the electrodes further averaged.

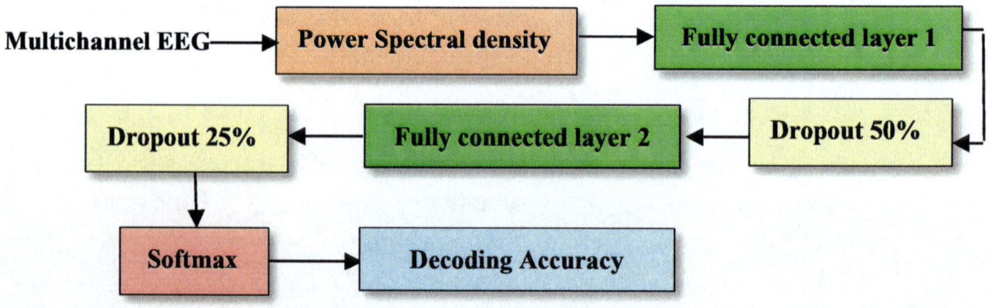

Fig. 4 EEG signal based bLSTM framework for decoding auditory attention state. Two drop out layer applied here for eliminating possible over-fitting occurrence

4 Results and Discussion

For the period of training, threefold cross-validation was examined. Subsequently, we evaluated the correctness diagonally in all fold and all channel using this

$$\text{Accuracy} = \frac{\text{TP} + \text{TN}}{\text{TP} + \text{FP} + \text{TN} + \text{FN}} \times 100 \tag{1}$$

Here, the TP stands for true positives, FP stands for false positives, TN stands for the true negatives, and FN represents stand for false negatives. Figure 5 represents the top seven feature rank electrodes acoustic hearing attention decoding performance for both algorithm models. Statistical ANOVA test is applied for each electrode to check the significance of each electrode's accuracy. The result represents that most of the high-performed electrodes were clustered in the frontal region. In both models highest significantly achieved decoding accuracy accomplished in electrode F8 ($p < 0.02$), with 74% accuracy in the SVM model and 72% accuracy in the bLSTM model.

Additionally, the top seven feature rank electrodes accomplished drastically high accurateness in both decoding models, shown in Fig. 5. Two model's mean average accurateness was calculated to form both the model's confusion matrix.

Figure 6 represents the bLSTM, and Fig. 7 represents the SVM modes confusion matrix. In both cases, we reject the channel accuracy that the p-value is not significant ($p > 0.05$). The bLSTM and SVM model's achieved mean accuracy from all the 31 electrodes is 65% and 63% respectively to decode the state of attention. We had estimated both algorithm model's correctness performance by the $F1$ score. The $F1$ score was calculated by Precision and Recall score using the following equations

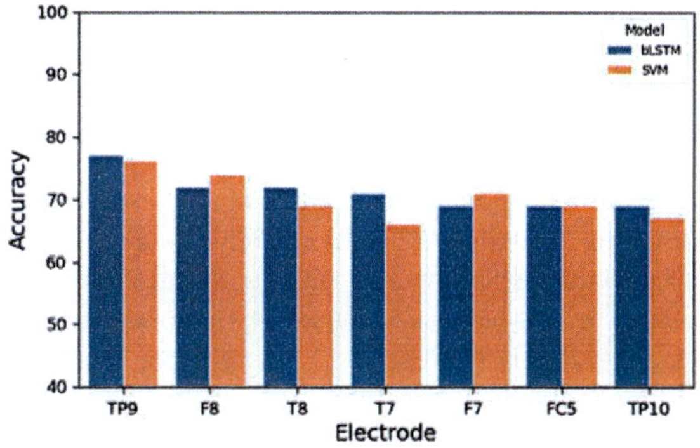

Fig. 5 Top seven EEG based attention-decoding performance for bLSTM and SVM model. TP9 ($p > 0.05$), F8 ($p < 0.02$), Ts8 ($p < 0.04$), T7 ($p < 0.04$), F7 ($p < 0.03$), FC5 ($p < 0.04$) and TP10 ($p < 0.03$) are electrodes in x-axis. Accuracy level representing on y-axis. Here, bLSTM model highest accuracy is 72% in channel F8 and SVM model highest accuracy level is 74% in channel F8

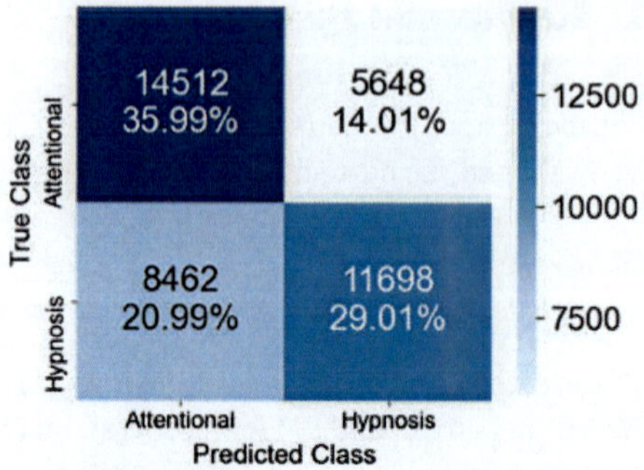

Fig. 6 Classification of attention models for the confusion matrix of bLSTM

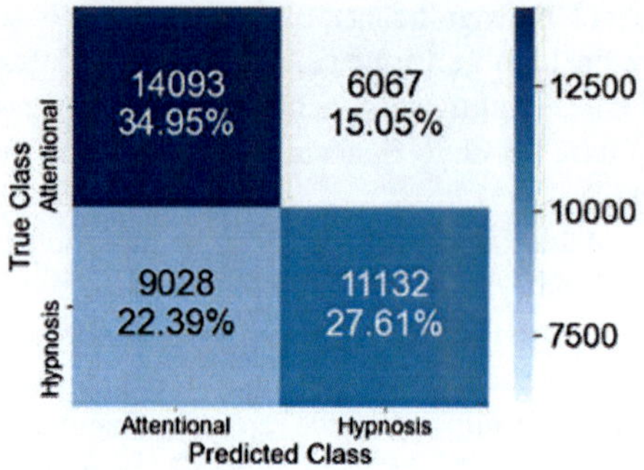

Fig. 7 Classification of attention models for the confusion matrix of SVM

$$\text{Precision} = \frac{\text{TP}}{\text{TP} + \text{FP}} \tag{2}$$

$$\text{Recall} = \frac{\text{TP}}{\text{TP} + \text{FN}} \tag{3}$$

$$F1 = 2 \times \frac{\text{Precision} \times \text{Recall}}{\text{Precision} + \text{Recall}} \tag{4}$$

Figure 8 represents both models of acoustic hearing attention state decoding performance. Presenting the computational proficiency within the smallest possible time with significant accuracy is important. In our study, we effectively figured out the state of a listener in both calm and hypnosis environments. Top feature rank electrodes are clusters in the frontal lobe and temporal lobe. As mention in an earlier section previous hypnosis studies also found frontal lobe activation in attentional focus for the auditory oddball task. Our decoding accuracy results suggest that auditory attention also can be decoded from the frontal lobe (74% accuracy) in a hypnosis environment. Moreover, Fig. 9 shows that the SVM model performs a mean accuracy from all 31 electrode is 63% to decode attention state in 50 ms time window; better than the earlier linear model's performance, which is approximately 55% in 50 ms time window [19]. However, the bLSTM model performs relatively low accuracy

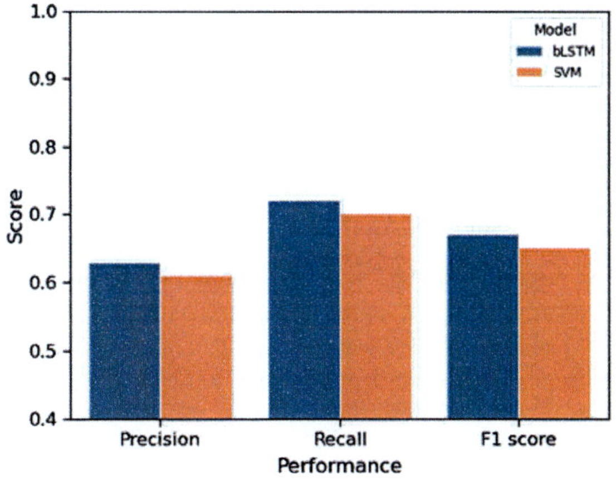

Fig. 8 SVM and bLSTM models performance measurements. Precision, recall and F1 score has been calculated by Eqs. 2, 3 and 4

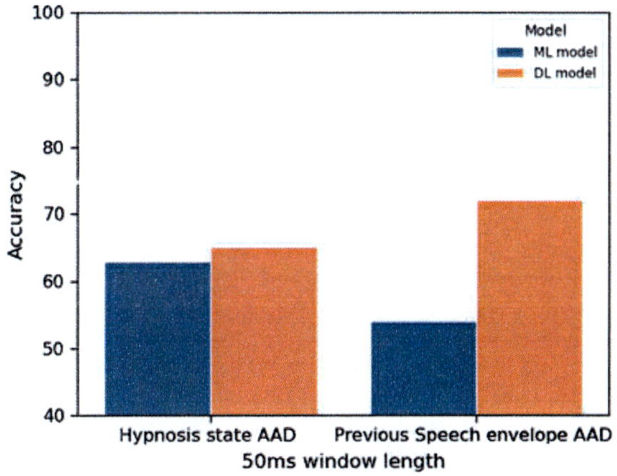

Fig. 9 Accuracy comparison of previous speech envelop AAD versus Hypnosis state AAD (50 ms time window)

than the CNN model. It occurs due to fewer subject's data. The previously mentioned study had also shown that in the CNN training data set, fewer subject's data makes significantly low decoding performance. Furthermore, as the neural dynamics in the hypnosis state are not fully revile, we have used all the EOG, central lobe, parietal lobe, and occipital lobes electrode in which several may not be associated with auditory attention in the hypnosis state. This situation also can cause a performance decrease in the auditory attention decoding model. Still, our top feature rank electrodes are in the frontal region, which is associated with attention in a hypnosis state. Those frontal region electrodes accuracy is above 70% in the 50 ms time window.

5 Conclusion

In this research, we proposed the EEG based auditory attention decoding (AAD) model for a quiet and hypnosis environment from EEG by applying a bLSTM and SVM model. Both models performed significant accurateness for decoding the attended hearing task in an oddball auditory tone in the beta frequency band. Our principal investigation was to decode the auditory attention within 50 ms time window. Our results suggested that the frontal region may be analyzed for achieving higher auditory attention state decoding for hypnosis state. Here we applied both SVM and bLSTM to decode the attention state and achieved a significant result with a limited amount of data. We originated a remarkably enhanced decoding performance when applying the SVM model. According to our result, this model accomplished the highest of 74% accurateness to decode the state of the listener's attention appropriately in the frontal lobe region.

The principal limitation of the study is a comparatively low amount of data due to limited participants. Future studies are advisable to focus on memory and aging complexity to manipulate auditory attention in a hypnosis state. It is essential to increase the number of participants in forthcoming investigations and considera-tion non-parametric models like the Gaussian process [27]. This will provide more statistically accurate outcomes in deep learning models [7].

References

1. Wild, C.J., Yusuf, A., Wilson, D.E., Peelle, J.E., Davis, M.H., Johnsrude, I.S.: Effortful listening: the processing of degraded speech depends critically on attention. J. Neurosci. **32**(40), 14010–14021 (2012)
2. Powell, P.S., Strunk, J., James, T., Polyn, S.M., Duarte, A.: Decoding selective attention to context memory: an aging study. Neuroimage **181**, 95–107 (2018)
3. Landry, M., Lifshitz, M., Raz, A.: Brain correlates of hypnosis: a systematic review and meta-analytic exploration. Neurosci. Biobehav. Rev. **81**(A), 75–98 (2017)
4. Todorovic, A., Schoffelen, J.M., van Ede, F., Maris, E., de Lange, F.P.: Temporal expectation and attention jointly modulate auditory oscillatory activity in the beta band. PLoS ONE **10**(3), e0120288 (2015)
5. Morash, V., Bai, O., Furlani, S., Lin, P., Hallett, M.: Classifying EEG signals preceding right hand, left hand, tongue, and right foot movements and motor imageries. Clin. Neurophysiol. **119**, 2570–2578 (2008)
6. Sumit, S.S., Watada, J., Nasrin, F., Ahmed, N.I., Rambli, D.R.A.: Imputing missing values: reinforcement bayesian regression and random forest. In: Kreinovich, V., Hoang Phuong, N. (eds.) Soft Computing for Biomedical Applications and Related Topics. Studies in Computational Intelligence, vol. 899. Springer, Cham (2021)
7. Mahmud, M., Shamim Kaiser, M., Hussain, A.: Deep Learning in Mining Biological Data (2020). arXiv preprint arXiv:2003.00108v1
8. Strauss, D.J., Francis, A.L.: Toward a taxonomic model of attention in effortful lis-tening. Cogn. Affect. Behav. Neurosci. **17**(4), 809–825 (2017)
9. De Pascalis, V., Bellusci, A., Gallo, C., Magurano, M.R., Chen, A.C.: Pain reduction strategies in hypnotic context and hypnosis: ERPs and SCRs during a secondary auditory task. Int. J. Clin. Exp. Hypn. **52**(4), 343–363 (2004)

10. Yann Cojan, D., Piguet, C., Vuilleumier, P.: What makes your brain suggestible? Hypnotizability is associated with differential brain activity during attention out-side hypnosis. Neuroimage **117**, 367–374 (2015)
11. Gruzelier, J., Gray, M., Horn, P.: The involvement of frontally modulated attention in hypnosis and hypnotic susceptibility: cortical evoked potential evidence. Contemp. Hypn. **19**(4), 179–189 (2002)
12. Kirenskaya, A.V., Storozheva, Z.I., Solntseva, S.V., Novototsky-Vlasov, V.Y., Gordeev, M.N.: Auditory evoked potentials evidence for differences in information processing between high and low hypnotizable subjects. Int. J. Clin. Exp. Hypn. **67**(1), 81–103 (2019)
13. Wong, D.D.E., Fuglsang, S.A., Hjortkjær, J., Ceolini, E., Slaney, M., de Cheveigné, A.: A comparison of regularization methods in forward and backward models for auditory attention decoding. Front. Neurosci. **12**, 531 (2018)
14. Haghighi, M., Moghadamfalahi, M., Akcakaya, M., Erdogmus, D.: EEG-assisted modulation of sound sources in the auditory scene. Biomed. Signal Process. Control **39**, 263–270 (2018)
15. Siripornpanich, V., Rachiwong, S., Ajjimaporn, A.: A pilot study on salivary cortisol secretion and auditory P300 event-related potential in patients with physical disability-related stress. Int. J. Neurosci. **130**(2), 170–175 (2020)
16. Bednar, A., Lalor, E.C.: Where is the cocktail party? Decoding locations of attended and unattended moving sound sources using EEG. Neuroimage **205** (2020)
17. Brown, K.J., Gonsalvez, C.J., Harris, A.W., Williams, L.M., Gordon, E.: Target and non-target ERP disturbances in first episode vs. chronic schizophrenia. Clin. Neurophysiol. **113**(11), 1754–1763 (2002)
18. Ciccarelli, G., Nolan, M., Perricone, J., et al.: Comparison of two-talker attention decoding from EEG with nonlinear neural networks and linear methods. Sci. Rep. **9**(1), 11538 (2019)
19. Deckers, L., Das, N., Hossein Ansari, A., Bertrand, A., Francart, T.: EEG-based detection of the attended speaker and the locus of auditory attention with convolutional neural networks (2018). bioRxiv preprint bioRxiv: 475673
20. Lu, J., Yan, H., Chang, C., Wang, N.: Comparison of machine learning and deep learning approaches for decoding brain computer interface: an fNIRS Study. In: Shi, Z., Vadera, S., Chang, E. (eds.) Intelligent Information Processing X. IIP 2020. IFIP Advances in Information and Communication Technology, vol. 581. Springer, Cham (2020)
21. Skomrock, N.D., et al.: A characterization of brain-computer interface performance trade-offs using support vector machines and deep neural networks to decode movement intent. Front. Neurosci. **12**, 763 (2018)
22. Mahmud, M., Kaiser, M.S., Rahman, M.M., Rahman, M.A., Shabut, A., Al-Mamun, S., Hussain, A.: A brain-inspired trust management model to assure security in a cloud based IoT framework for neuroscience applications. Cogn. Comput. **10**(5), 864–873 (2018)
23. Jasper, H.: Report of the committee on methods of clinical examination in electroencephalography. Electroencephalogr. Clin. Neurophysiol. **10**, 370–375 (1958)
24. Delorme, A., Makeig, S.: EEGLAB: an open source toolbox for analysis of single-trial EEG dynamics including independent component analysis. J. Neurosci. Methods **134**, 9–21 (2004)
25. Graves, A., Schmidhuber, J.: Framewise phoneme classification with bidirectional LSTM and other neural network architectures. Neural Networks **18**(5–6), 602–610 (2005)
26. Chang, C.C., Lin, C.J.: LIBSVM: a library for support vector machines. ACM Trans. Intell. Syst. Technol. **2**(3), 1–27 (2011)
27. Rahman, M.A.: Gaussian process in computational biology: covariance functions for transcriptomics. Ph.D. Thesis, The University of Sheffield, UK (2018)

EM Signal Processing in Bio-living System

Pushpendra Singh, Kanad Ray, Preecha Yupapin, Ong Chee Tiong, Jalili Ali, and Anirban Bandyopadhyay

Abstract Bio-living organisms can effectively respond to various signals from the surrounding space due to their geometric shape. In this manuscript, our central goal is to understand the underlying mechanisms of signal processing inside bio-living systems, which may have beneficial applications in medical domains such as disease diagnosis and sensing the space present for medical production. We have analyzed various proteins as well as EM signal distributions and expanded the characterization of biological components that may be important in cellular signal utilization to enhance complex behavior in a bio-living cell. Geometry, fractality, resonance,

P. Singh (✉) · K. Ray
Amity University Rajasthan, Kant Kalwar, NH-11C, Jaipur, Delhi Highway, Jaipur, Rajasthan 303007, India
e-mail: singhpushpendra548@gmail.com

K. Ray
e-mail: kanadray00@gmail.com

P. Singh · A. Bandyopadhyay
International Center for Materials and Nanoarchitectronics (MANA), Research Center for Advanced Measurement and Characterization (RCAMC), National Institute for Materials Science, 1-2-1 Sengen, Tsukuba, Ibaraki 305-0047, Japan
e-mail: anirban.bandyo@gmail.com

P. Yupapin
Computational Optics Research Group, Advanced Institute of Materials Science, Ton DucThang University, District 7, Ho Chi Minh City 700000, Vietnam
e-mail: preecha.yu@kmitl.ac.th

Faculty of Applied Sciences, Ton DucThang University, District 7, Ho Chi Minh City 700000, Vietnam

O. C. Tiong
Department of Mathematical Sciences, Faculty of Science, Universiti Teknologi Malaysia, UTM, 81310 Skudai, Johor Bahru, Malaysia
e-mail: octiong@utm.my

J. Ali
Laser Centre, IBNU SINA ISIR, Universiti Teknologi Malaysia, 81310 Johor Bahru, Malaysia
e-mail: jalilali@utm.my

M. S. Kaiser et al. (eds.), *Proceedings of International Conference on Trends in Computational and Cognitive Engineering*, Advances in Intelligent Systems and Computing 1309, https://doi.org/10.1007/978-981-33-4673-4_24

and clocking behavior are key features for analyzing any biological system. Here, we have reported the summarized results of various proteins and discussed how they have a naturally inbuilt prototype system that has various applications in the medical, industrial, and environmental fields.

Keywords EM signal processing · Bio-living system · Geometry, fractaility, clocking and resonance · Self-replication

1 Introduction

One of the recent research areas is to explore the underlying mechanisms of information processing in bio-living systems. Living beings require regular information processing for continued existence. Living systems have the ability to store and process information. How this happens and how the information is stored is still mysterious. An explanation of the cell as noise information to describe it is an open question. Information theoretical approaches and complementary systems technology approaches have developed of late and basic science has been used to study various biological processes. The maintenance and processing of information in a cell are much more efficient than electronic digital computation in terms of information density and energy consumption. Information on a range of scales drives biological self-organization and generates complex patterns of life. It has become an interdisciplinary research field where people from various fields such as physics, mathematics, engineering, and biology are collaborating extensively.

This paper focuses on current advances in bio information processing and self-organization. Bio-living systems are often balancing the information flow using structural geometry, self-organization, and environmental conditions [1]. Signal detects the information from the biological system as a complex phenomenon that depends on structure geometry or individual geometric components. The received solution provides a map of basic principles that show how the living system works. All living cells can respond to internal and external signals, and also to incorporate a specific mechanism to maintain signal processing regularly. One of the key factors of a biological system is robustness. In system biology, robustness is an attractive point for solemn scientific research. Numerous research has been published regarding the robustness and their mechanism in the living system [2–7]. Steuer et al. [8] presented simple and highly efficient novel formalism which has the feature to form the required architecture for the robust concentration in living cells and the study had the claim that certain classes of network architecture render the output function of the network. This claim was based on mathematical and experimental evidence [8]. The aim of system biology is not only to create the precision- model of organs or cells but also to find out the fundamental and geometrical principles hidden behind the bio-living system [9]. Numerous theoretical studies have been reported on the EM field interaction with the structural configuration in which geometry, fractality, and resonance are the key features [10–13]. Little et al. [14] reported the research domain related to biophysical digital signal processing in which biophysical signals are generated and show the scientific nature. Molecular network inside the living cells makes them sense and processes the signal from the surrounding space for performing the desired functions. Researchers have been able to reconstruct and mimic such kind of cellular signal processing. A new tool based on the self-assembling and predictive model allows creating the new complex computation and signal processing inside the eukaryotic cell including the human cells [15, 16].

Electromagnetic resonances in biological molecules are shown to be found in a broad range of frequencies (kHz, MHz, GHz, and THz). They could be significant for the biological function of macromolecules, over and above could be used for the progress of devices like molecular computers. As these measurements are time-consuming and expensive there is a requirement for computational methods that could accurately predict these resonances. Many researchers have already reported studies in the domain of EM interaction with the cellular system. The bio-living system can communicate with each other and interact with the environment in the presence of many hidden mechanisms that depend on the type of system, the geometry of the system, and involved information to the system. Previously reported studies to show the communication from one cell to another cell by a chemical or electrical signal, beyond it, Cifra et al. [17] suggested the study of EM field interaction with cellular systems such medium is the most probable candidate compared to other forms of the cell to cell communication. How cells can produce and detect the EM field is reported in [17]. Burr and Northrop et al. [18] published a study that was based on a constant voltage gradient on various biological systems. The organization of any biological system is stabilized by a complex EM field interaction which could change the ions density distribution on bio-living material. Many researchers found that this stable voltage gradient varies when the biological organism passes through the biological process like structural growth. To better understanding the function of the EM wave, it needs to separate the biological part from the chemical parts. This is achieved by disabling the barrier of chemical signal transmission, not the EM wave transmission. Unwanted or confounding factors; external environmental conditions or chemicals are discussed in the barrier method that is reported in [19]. EM field involves in cell process which strongly supports the EM interaction with molecules.

In this manuscript, we took free available proteins like microtubule end binding protein (MENp), dynein, kinesin, and cyanobacteria protein from the protein data bank and edited those for removing the noise and make able to run in Maxwell equation solver. We also created a wasp head organ and retinal unit following the all biological details and detected the resonance peaks, phased-based electric and magnetic field along with the structures. We got special features like symmetry breaking and phase transition which are the function of the resonance frequency. Signal processing inside the living material leads the technology for transformation, interpretation, and extraction of the information involved with living material's geometry. It would not be difficult for any to observe the signal pathways that could explore further research domains. An alternative physical biology parallel to all chemical biology is getting shaped across the globe, e.g., wireless communication and information processing in biological systems. This interdisciplinary and interfacing research will prominently establish a new cutting edge research outlook for the fields of biotechnology, neurophysiology, and drug discovery research. Every part of an integrated biological system absorbs and radiates to communicate in a clear and defined resonance frequency band.

We have developed a method to run computer simulations to find out resonance frequencies, electric and magnetic-field distributions, and clocking behavior which would detect the features of information processing in biological systems. In the previous study, the mathematical modeling of bio-living systems has been reported; researchers derived the reaction-diffusion equation $\left[\left\{\frac{\delta u(x,t)}{\delta t}\text{(change)} = \nabla.(D\nabla u(x,t))\text{(diffusion)} + f(u,x,t)\text{(reaction)}\right\}\right]$ [20] which is useful for the formation of spatial-temporal patterns to describe the behavior of species. Numerical techniques like reaction-diffusion equation, finite elements methods, etc. are useful to explain the solution of other complex biological models such as glycolysis model, chemotaxis model and tissue formation model inside the living world [21, 22].

The outline of this paper is in four sections. Section 2 is about theoretical construction and simulation of various types of protein structures. We described the electromagnetic properties of the structure as resonance peaks, the distribution of electric and magnetic fields. The results are discussed in Sect. 3. Finally, Sect. 4 traces the major finding of this study.

2 Materials and Methods

We have theoretically built; dynein, kinesin, cyanobacterial, microtubule end binding protein, and DNA structures, free available from the protein data bank. We have edited and filtered or removed the unwanted constitute components from the proteins and make them suitable for simulation while the wasp head organ and retina unit is handmade structure considering its original dimensions. All proteins are assigned by dielectric materials and stimulated by pumping the energy from suitable waveguide ports at a suitable location of structures. Waveguide port dimension and port location are selected carefully otherwise wrong selection of wave port's parameters may lead a long simulation time. The boundary conditions are selected as open space in all 3 directions x, y, and z-axis. All simulations are carried out in numerical simulation software CST (computer simulation technology) within the time-domain mode that uses the Maxwell equation to solve the structure. We scanned the frequency range from kHz to THz for every built structure and detected the electric and magnetic field distribution at those peaks which show the high dominated profiles of the EM field.

Electromagnetic signal processing in a biological system means the detection of physical activities of biological components by electromagnetic signals that propagate to or from its surface, which has various advantages in the biomedical field, for example, MRI (Magnetic resonance imaging), detection of structural and functional properties of human organs, analysis of DNA sequence to extract genetic information from cells, etc. The EM signal processing field is a rapidly evolving field. We can easily know its characteristics by looking at the EM distribution profile of any biological component. We have shown in our previous studies [10–13] how resonance and EM field distribution of brain components or signal processing by their nature, are the fundamental properties.

Here all simulation models of the proteins we select, are prototypes of realistic structures. Such models are used to understand the advantages and disadvantages of a structure with certain conditions in the real world. In our case, the resonant frequency detection of the model is useful in this way, if anyone does the actual experiment then it does not need to scan the frequency range to detect the resonance peaks, one can easily determine the frequency range and resonance by simulating the model. Simulation modeling allows researchers to analyze biological designs or encourage the creation of advanced designs. Before running the simulation, we must ensure optimized or noise-free geometry, characterizing physical properties, simulation conditions such as a selection of waveguide source, frequency range, etc.

2.1 Technical Details

Simulation technical details of the model created are as follows: Used simulation software—CST, Selected solver- time-domain solver, boundary condition- open space, waveguide port dimensions- from micrometer to nanometer, selected frequency range- from kHz to THz.

We can provide our CST or simulation files. One can directly use our noise-free model or extend it further by adding more complex details.

3 Results and Discussion

Here we applied protein peeling methods and Maxwell Equation Solver tool to different groups of macromolecules to predict electromagnetic resonant frequencies. We suggest that it is a strong comprehensive tool that can envisage electromagnetic resonances, pertinent to macromolecular biological function with application to new, innovative scientific devices.

DNA behaves as a fractal antenna [10] and interacts with the EM field over a wide frequency range in the KHz domain. Here, modeled DNA has been excited from both ends using EM waveguide source. Figure 1a shows the schematic diagram of DNA, wasp head organ (WHo), microtubule end binding protein (MEBp), and the retinal unit. Created mimicked structures are stimulated by EM energy or waveguide source that is applied at the end of the structure. Simulated EM resonance spectrum for all 4 created structures (Fig. 1b) DNA, WHo, MEBp, and retinal unit have the multi resonance peaks in kHz, MHz, GHz, and GHz frequency domain, respectively

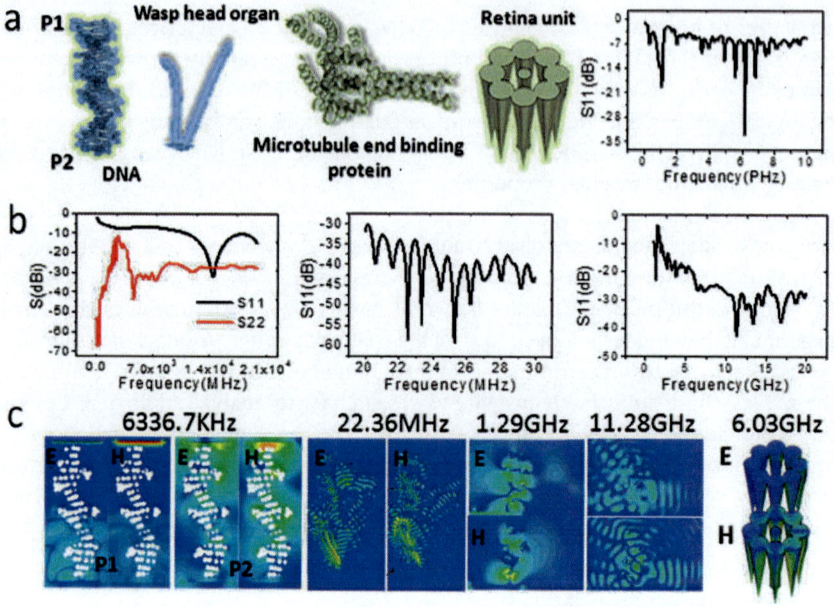

Fig. 1 Schematic diagram, simulated EM resonance spectra and EM field profiles of DNA, wasp head organ (WHo), microtubule end binding protein (MEBp) and retinal unit

while EM field profiles are calculated at their most dominated resonance peaks; 6336.7 KHz-DNA; 22.36 MHz-Who; 1.29 GHz, 11.28 GHz-MEBp; 1.1 GHz-retinal unit (Fig. 1c). DNA surface's molecules fluctuated at 6336.7 kHz (Fig. 1b: left) and corresponding electric and magnetic field distribution is reported in Fig. 1c: left. Electromagnetic energy transmits from one end to another end of the DNA structure and is reflected at the initial end due to port1. But in the case of port2, energy is absorbed by structure and surrounding space because of different lattice symmetry that may be involved inside the DNA structure. We put the energy source at the junction of the wasp antenna and got the multi-resonance peaks (Fig. 1b: mid-panel) in the MHz frequency domain. EM field (at 22.36 MHz) is highly concentrated along with the wasp antenna and leads to a special kind of pattern (Fig. 1c: left from the mid). The Fibonacci sequence involved in the wasp body decreases the entropy or maintains the high order EM field distribution [11]. The electromagnetic resonance spectrum is obtained in the GHz frequency range (Fig. 1 b: right) by applying the EM source at the bottom end of MEBp. EM field distributions are seen at the two most dominant peaks 1.29 and 11.28 GHz.

Minimum entropy (high ordered EM field along with the structure) has been seen at low resonance peaks or in the initial duration of structure growth process but maximum entropy observed (noisy profile of EM field) at high resonance frequency or in the final iteration of the structure growth process (Fig. 1c: from right to first left). Such a phenomenon is based on the law of self-replication in statistical physics

[23]. The retinal unit works as a center-fed dipole antenna and capable to work in the visible region [11, 24]. Involved fractal arrangements of the retinal unit show the various symmetries during the EM field distribution, where each symmetry responsible for the resonance spectrum generation. Phase-based EM field (Fig. 1c: right panel) is calculated at 6.03 PHz resonance peak (Fig. 1a: right panel). EM field has the localization and delocalization on the retinal receptor unit during one complete phase cycle and orients the clocking conduction behavior.

EM field characterized of cryptochrome protein, axon bundle made microtubule, actin, beta spectrin, ankyrin, and tubulin protein made of α and β tubulin dimmers, are also observed as given in mid and right panel of Fig. 2a–c. In the case of cryptochrome protein, EM field concentrate at helical loop domains of protein at 3.21 THz resonance peak while we got a unique feature of EM wave across the axon bundle at 7.01 THz frequency. Basically, axon is wire-like structure that is part of neurons and important for cellular communication [25, 26]. Neurons entire structure is filled with crystalline structures like actin, spectrin, and actin [27], inspired by that we have built the axon bundle considering all biological details. From Fig. 2b, energy propagates through

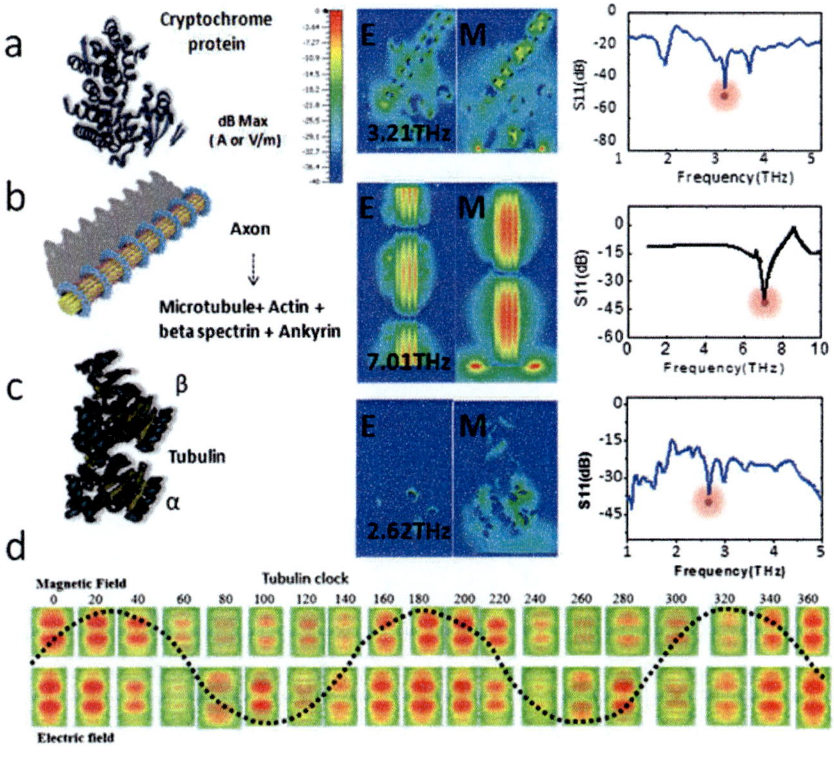

Fig. 2 EM field distribution and resonance spectrums of cryptochrome protein, axon bundle and tubulin protein are shown in the mid and right side of panel a, b, and c respectively. Panel d shows the phase-based clocking behavior of tubulin protein (α and β tubulin)

the axon in the form of nodes and antinodes that shows the wave-like interference of both fields. 200 nm distance is kept between two actin rings. Axon is stimulated by pumping the energy from at one of its ends. From the mid-panel of Fig. 2c, the magnetic field makes the propagation pathway as beta-sheet in tubulin protein when it propagates from α tubulin dimer to β tubulin dimer while the electric field is less dominated across the beta-sheet. EM field profile is depicted at 2.62 THz resonance frequency.

Figure 3 consists of the simulation analysis of Dynein, Kinesin, and Cyanobacterial protein using a specific tool (Maxwell equation solver). Multiple resonance peaks are seen in resonance spectrums of all three created proteins (Fig. 3a) during the structure growth process. Splitting of every protein in protein's unit and protein's subunit that is completed using the bioinformatics tool linked with the webserver (proteins peeling methods) [28, 29], and final filtered form of proteins are seen in the left panel of figure b, c, d. Dramatic changes in EM field distribution are found at their most dominant resonance peaks (Dynein- 4.76, 4.76, 17.46, and 6.56 THz; Kinesin- 8.08, 7.04, 10.16, and 14.12 THz; Cyanobacteria protein −2.38, 2.27, 2.51, and 4.36 THz) are shown in the right panel of Fig. 3b–d. Dynein, kinesin, and cyanobacteria proteins are peeled off in their protein unit and subunit using the protein peeling method. Here, we have considered the intermediate structures up to the 4th level. After the analysis of protein from the bioinformatics tool (protein peeling method),

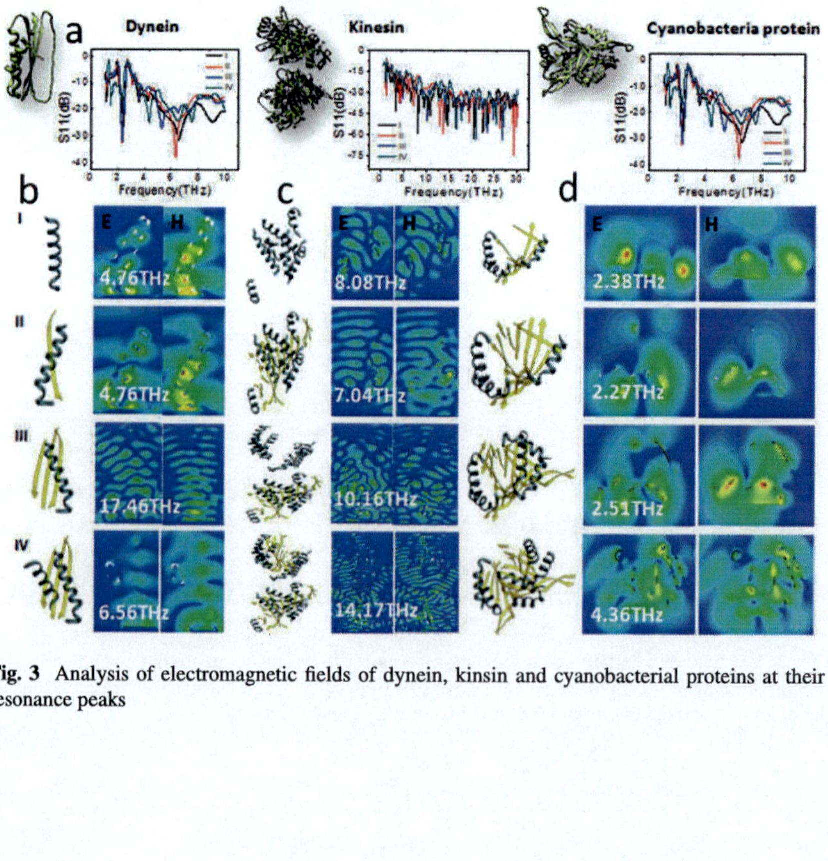

Fig. 3 Analysis of electromagnetic fields of dynein, kinsin and cyanobacterial proteins at their resonance peaks

we filtered its protein unit and solved the structure in the numerical analysis tool (Maxwell equation solver) by putting the wave port at the bottom end. Each iteration has multiple resonance peaks and a combined resonance spectrum for Dynein, Kinesin, and Cyanobacteria proteins that are observed at the high-frequency domain (THz) (Fig. 3a). We detected the electromagnetic field profiles for each protein at deep resonance peaks (Dynein: 4.47, 4.47, 10.16, 6.56 THz; Kinesin: 8.08, 7.04, 10.16, 18.17 THz; Cyanobacteria protein: 2.38, 2.27, 2.51, 2.36 THz) for each iteration. We can analyze from Fig. 3b–d and infer that EM field obeys the England principle law or statistical physics of self-replication [23]. At the initial domain or iteration of protein splitting, protein units show the self-resonance property and high EM energy concentration along with the surface geometry with minimum entropy and vice versa for the final domain or iteration.

Circadian rhythm is an important feature for the daily routine of our life like sleep and wakeful [30]. Such rhythm is driven by the internal clock inside the biological system. In a mammal, circadian rhythm exists across the individual neurons of suprachiasmatic nuclei. The clocks inside the mammals and insects provide the auto running regular feedback to explore its basic properties. Such a feedback loop is encoded by genes. How protein and their products respond to rhythm and how genes regulate the basic behavior that is characterized in the reported study [31, 32]. How the clock controls the periodicity and continuity of rhythm in SCN/biological materials, is still unknown. So our works have a contribution to understanding the basic features like periodicity and continuity. We have reported the theoretical analysis of similarly how biomaterials interact with the electromagnetic field. We are not limited to electromagnetic interaction with living systems, but rather we have reported physical processes involving biomaterials.

Clocking behaviors of EM wave on dynein, kinesin, cyanobacterial proteins are shown in panel (a), (b), and (c) of Fig. 4 respectively. It has been observed that the electric and magnetic field are characterized with high-intensity nature in 0°–150° phase angle, slowly it starts to take the minimum value at 180°–240° and again, it starts to increase, or in another view, EM wave shows the periodicity and continuity nature on the built proteins. Similarly, tubulin protien shows the phase-based clocking characterized at 3.62 THz. However high EM intensity has seen periodically at 0°–40°, 180°–220°, and 320°–360° phase durations, as shown in Fig. 2d. At the phase angle of high energy intensity, both fields are highly concentrated along the α and β tubulin dimers with a field intensity scale from -18.2 to -10.9 dB A (M)/m or V (E)/m. In our previous study, we have argued that clocking is the fundamental characteristic of structural/geometrical resonance. We have reported the clocking behavior inside the microtubule and retinal receptor cells (rods and cones) [11, 33]. Here, proteins' resonance is obtained at high-frequency range (THz). In the high-frequency domain, we detected that the EM energy changes in a particular way across the protein's surface in whole phase rotation (0°–360°), are shown in Fig. 4a–c. We can analyze the wave periodicity and continuity characteristic on a 2D cross-section plane of dynein, kinesin, and cyanobacteria proteins (Fig. 4).

We have analyzed how biomaterials geometry plays a key role to exhibit their function. Here, our discussion is limited up to what is the exact role of biomaterial's

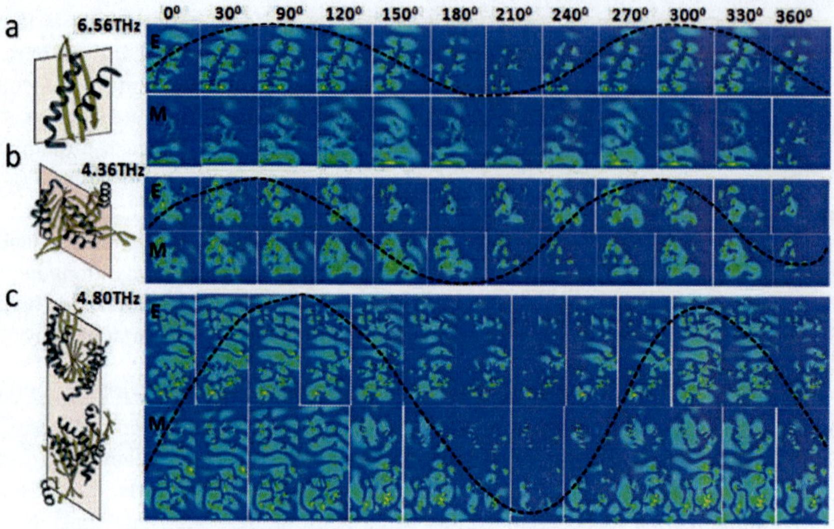

Fig. 4 Clocking behavior of EM wave on dynein (panel a), kinesin (panel b), and cyanobacterial proteins (panel c)

topology. The resonance spectrum shows the characteristic topological pattern in such a way, at some resonance peaks EM field concentrates along with components, and at some other resonance peaks, the EM field shows distractive nature. We speculated that the geometry of living organisms is responsible for such EM interactions, as shown in Fig. 4a–c.

4 Conclusion

This paper sheds light on the topic of EM signal processing in bio-living systems and explains the scientific context about the nature of EM field distribution in biomaterials. Here the limits of entropy (minimum and maximum) in living systems are also discussed and a conclusion has been drawn that the geometry of biological material exhibits an important role for identifying the diffusion pathways of the signal inside its components. Our proposed study demonstrates the introduction of EM interactions with biological structure and techniques to handle the model (select periodically or non-periodic structure) according to the EM field distribution. The problems of EM wave interactions with living systems are a bit challenging due to the limited way of conducting experimental studies, the purpose of this paper is to introduce its theoretical concept. Our study is limited to a description of the geometric role, fractality, clocking behavior, and resonance frequency of bio-living materials, which has an important application in many fields. The essence of the work done in this report indicates that the interaction of electromagnetic fields with living cells will become a more and more important domain over time and will highlight many fundamental events in relation to life.

Acknowledgements The authors express sincere thanks to Prof. S L Kothari, Vice President, ASTIF, AUR for his support and encouragement.

Contribution. KR, PY, JA & AB planned the theoretical study, PS did the theory. All authors wrote the paper together.

Conflict of Interest Statement. The authors declare that they have no conflict of interest.

References

1. Cattani, C., Badea, R., Chen, S.Y., Crisan, M.: Biomedical signal processing and modeling complexity of living systems. Comput. Math. Methods Med. **2015**, 1–2 (2013)
2. Barkai, N., Leibler, S.: Robustness in simple biochemical networks. Nature **387**(6636), 913–917 (1997)
3. Kitano, H., Oda, K., Kimura, T., Matsuoka, Y., Csete, M., Doyle, J., Muramatsu, M.: Metabolic syndrome and robustness tradeoffs. Diabetes **53**(3), S6–S15 (2004)
4. Alon, U., Surette, M.G., Barkai, N., Leibler, S.: Robustness in bacterial chemotaxis. Nature **397**(6715), 168–171 (1999)
5. Stelling, J., Sauer, U., Szallasi, Z., Doyle, F.J., Doyle, J.: Robustness of cellular functions. Cell **118**(6), 675–685 (2004)
6. Kitano, H., Oda, K.: Robustness trade-offs and host-microbial symbiosis in the immune system. Mol. Syst. Biol., 2006.0022:msb4100039, E1–E10 (2006)
7. Kitano, H.: Biological robustness in complex host-pathogen systems. Prog. Drug Res. **64**(239), 241–263 (2007)
8. Steuer, R., Waldherr, S., Sourjik, V., Kollmann, M.: Robust signal processing in living cells. PLoS Comput. Biol. **7**(11), e1002218 (2011)
9. Kitano, H.: Towards a theory of biological robustness. Mol Syst Biol. **3**, 1–7 (2007)
10. Singh, P., Doti, R., Lugo, J.E., Faubert, J., Rawat, S., Ghosh, S., Ray, K., Bandyopadhyay, A.: DNA as an electromagnetic fractal cavity resonator: its universal sensing and fractal antenna behavior. In: Pant, M., Ray, K., Sharma, T., Rawat, S., Bandyopadhyay, A. (eds.) Soft Computing: Theories and Applications, pp. 213–223. Springer, Singapore (2017)
11. Singh, P., Ocampo, M., Lugo, J.E., Doti, R., Faubert, J., Rawat, S., Ghosh, S., Ray, K., Bandyopadhyay, A.: Fractal and periodical biological antennas: hidden topologies in DNA, wasps and retina in the eye. In: Ray, K., Pant, M., Bandyopadhyay, A. (eds.) Soft Computing Application, pp. 113–130. Springer, Singapore (2018)
12. Singh, P., Doti, R., Lugo, J.E., Faubert, J., Rawat, S., Ghosh, S., Ray, K., Bandyopadhyay, A.: Biological infrared antenna and radar. In: Pant, M., Ray, K., Sharma, T., Rawat, S., Bandyopadhyay, A. (eds.) Soft Computing: Theories and Applications, pp. 323–332. Springer, Singapore (2017)
13. Singh, P., Doti, R., Lugo, J.E., Faubert, J., Rawat, S., Ghosh, S., Ray, K., Bandyopadhyay, A.: Analysis of sun flower shaped monopole antenna. Wirel. Pers. Commun. **104**(3), 881–889 (2019)
14. Little, M.A., Jones, N.S.: Signal processing for molecular and cellular biological physics: an emerging field. Phil. Trans. R. Soc. A **371**, 2011054 (2013)
15. Bashor, C.J., Patel, N., Choubey, S., Beyzavi, A., Kondev, J., Collins, J.J., Khalil, A.S.: Complex signal processing in synthetic gene circuits using cooperative regulatory assemblies. Science **364**(6440), 593–597 (2019)
16. Weng, G., Bhalla, U.S., Iyengar, R.: Complexity in biological signaling systems. Science **284**(5411), 92–96 (1999)

17. Cifra, M., Fields, J.Z., Farhadi, A.: Electromagnetic cellular interactions. Prog. Biophys. Mol. Biol. **105**(3), 223–246 (2011)
18. Burr, H.S., Northrop, F.S.C.: The electrodynamic theory of life. Q. Rev. Biol. **10**(3), 322e333 (1935)
19. Fels, D.: Fields of the cell: electromagnetic cell communication and the barrier method. In: Fels, D., Cifra, M., Scholkmann, F. (eds.) Research Signpost, pp. 149–162. Trivandrum. Kerala, India (2015)
20. Garzón-Alvarado, D.A.: Simulation of reaction-diffusion processes. Application to bone tissue morphogenesis. Ph.D. dissertation, Zaragoza, España (2007)
21. Klein-Nulend, J., Bacabac, R.G., Mullender, M.G.: Mechanobiology of bone tissue. Pathol. Biol. (Paris) **53**, 576–580 (2005)
22. Vanegas Acosta, J.C.: Electric fields and biological cells: numerical insight into possible interaction mechanisms. Mathematical modeling of biological systems, TechnischeUniversiteit Eindhoven (2005)
23. England, J.L.: Statistical physics of self-replication. J. Chem. Phys. **139**, 12923 (2013)
24. Singh, P., Doti, R., Lugo, J.E., Faubert, J., Rawat, S., Ghosh, S., Ray, K., Bandyopadhyay, A.: Frequency fractal behaviour in the retina nano centre-fed dipole antenna network of a human eye. In: Pant, M., Ray, K., Sharma, T., Rawat, S., Bandyopadhyay, A. (eds.) Soft Computing: Theories and Applications, pp. 201–211. Springer, Singapore (2017)
25. Yau, K.W.: Receptive fields, geometry and conduction block of sensory neurones in the central nervous system of the leech. J. Physiol. **263**(3), 513–538 (1976)
26. Debanne, D.: Information processing in the axon. Nat. Rev. Neurosci. **5**, 304–316 (2004)
27. Xu, K., Zhong, G., Zhuang, X.: Actin, spectrin and associated proteins form a periodic cytoskeleton structure in axons. Science **339**, 452–456 (2013)
28. Gelly, J.C., De Brevern, A.G., Hazout, S.: Protein peeling: an approach for splitting a 3D protein structure into compact fragments. Bioinformatics **22**(2), 129–133 (2006)
29. Gelly, J.C., Etchebest, C., Hazout, S., De Brevern, A.G.: Protein peeling 2: a web server to convert protein structures into a series of protein units. Nucleic Acids Res. **34**, W75–W78 (2006)
30. Pittendrigh, C.S.: Temporal organization: reflection of a Darwinian clock-watcher. Annu. Rev. Physiol. **55**, 17–54 (1999)
31. Hastings, M.: The brain, circadian rhythms, and clock genes. BMJ **317**(7174), 1704–1707 (1998)
32. Klein, D.C.: Suprachiasmatic Nucleus: The Mind's Clock. Oxford University Press, New York (1991)
33. Singh, P., Ray, K., Fujita, D., Bandyopadhyay, A.: Complete dielectric resonator model of human brain from MRI data: a journey from connectome neural branching to single protein. In: Ray, K., Sharan, S., Rawat, S., Jain, S., Srivastava, S., Bandyopadhyay, A. (eds.) LNEE, pp. 717–733. Springer, Singapore (2018)

Internet of Things and Data Analytics

6G Access Network for Intelligent Internet of Healthcare Things: Opportunity, Challenges, and Research Directions

M. Shamim Kaiser, Nusrat Zenia, Fariha Tabassum, Shamim Al Mamun, M. Arifur Rahman, Md. Shahidul Islam, and Mufti Mahmud

Abstract The Internet of Healthcare Things (IoHT) demands massive and smart connectivity, huge bandwidth, lower latency with ultra-high data rate and better quality of healthcare experience. Unlike the 5G wireless network, the upcoming 6G communication system is expected to provide Intelligent IoHT (IIoHT) services everywhere at any time to improve the quality of life of the human being. In this paper, we present the framework of 6G cellular networks, its aggregation with multidimensional communication techniques such as optical wireless communication network, cell-free communication system, backhaul network, and quantum communication, as well as distributed security paradigm in the context of IIoHT. Such low latency and ultra-high-speed communication network will provide a new paradigm for connecting homes to hospitals, healthcare people, medical devices, hospital infrastructure, etc. Also, the requirements of 6G wireless networking, other key techniques, challenges and research direction of deploying IIoHT are outlined in the article.

Keywords Massive MIMO · holographic beamforming · Internet of everything (IoE) · Machine learning · Distributed security

---

M. S. Kaiser (✉) · N. Zenia · S. A. Mamun · Md. S. Islam
Institute of Information Technology, Jahangirngar University, Savar, Dhaka 1342, Bangladesh
e-mail: mskaiser@juniv.edu

F. Tabassum
Department of EECE, Military Institute of Science and Technology, Dhaka, Bangladesh

M. A. Rahman
Department of Physics, Jahangirngar University, Savar, Dhaka 1342, Bangladesh

M. Mahmud
Department of Computer Science, Nottingham Trent University, Nottingham NG11 8NS, UK

M. S. Kaiser et al. (eds.), *Proceedings of International Conference on Trends in Computational and Cognitive Engineering*, Advances in Intelligent Systems and Computing 1309, https://doi.org/10.1007/978-981-33-4673-4_25

1 Introduction

The Internet of Healthcare Things (IoHT) is progressing toward a smart, intelligent, reliable, and automated infrastructure with the rapid adoption of various evolving techniques (such as artificial intelligence, advanced machine learning, big data analysis, pattern recognition) and systems (such as the Internet of Things (IoT) and robotics) [2, 10, 13]. The upcoming Intelligent Healthcare Things Internet (IIoHT) demands secure, ultra-reliable, low latency, ultra-high data rate communication among a massive number of IoHT nodes [3].

Health robots, augmented/virtual reality (A/VR) is expected to evolve over the coming years although healthcare is sluggish to adopt new technology [14]. Medical robots can be used to assist surgery, clean areas, supply hospital logistics. On the other hand, A/VR can be employed to teach medical students, prepare surgeons for a surgery, assist patients with post-traumatic stress disorder, reduce anxiety and discomfort, handle procedure of the pediatric patient. This requires robust artificial intelligence (AI) and machine learning (ML)-based infrastructure with low latency and ultra-high-speed access network [12].

The six generation (6G) cellular technology is expected to provide ultra-low latency and high-speed data transmission rate wireless services using visible lights and THz EM sub-bands. The goal of 6G is to meet AI-based intelligent connectivity, holographic/haptic/satellite connectivity, and ubiquitous 3D connectivity. 6G will revolutionize the healthcare sector which eliminates time and space barriers through remote surgery and guaranteed healthcare workflow optimizations.

In this paper, we reviewed recently published papers on IoHT and 6 G cellular networks; described the motivation behind the upcoming Intelligent IoHT system using 6 G in the access network; listed some of the IIoHT's key services; and finally outlined the potential research challenges of IIoHT.

The remaining section is given as: the motivation behind the use of 6G access network for the IoHT is discussed in Sect. 1. In Sect. 3, some features of the next 6G networks were identified. Section 4 reviewed articles related to 6G and IoHT. The development of the IIoHT was seen in Sect. 5. The potential problems were established in Sect. 6 and the study is finally concluded in Sect. 7.

2 Motivation of Using 6G in IIoHT

The forthcoming 6G wireless system will deliver short-range wireless connectivity with terabit speed (0.1–10 THz), which is ideally suited for transmitting large-scale medical data to the IIoHT infrastructure.

2.1 Features and Requirement of IIoHT

IoHT is a system/network interconnecting medical devices/nodes (wearable devices, remote patient monitoring, sensor-enabled hospital beds and infusion pumps, drug tracking systems, medical device inventory tracking, etc.), people, providers, and Internet infrastructure. These IIoHT nodes can capture, process, compile, analyze, and distribute health data using advance ML algorithms. The main aim of IIoHT is not only to make a low cost and reliable patient-centric healthcare anywhere but also to make it efficient for insurers and health service providers. IIoHT is intended to improve diagnosis, disease management and treatment; remotely track a patient's chronic condition; and eventually increase the patient's quality of life. Figure 1 shows a 6G communication scenario for IIoHT.

2.2 Challenges in IIoHT Deployment

Healthcare industry adopts emerging technologies slowly. Additionally, IIoHT deployment faces various challenges. Several are listed below:

- Many IoT nodes use default passwords and the cyber-attacker can easily crack these. Thus, IoHT nodes are required to be secured and trustworthy as they collect patients' medical data;

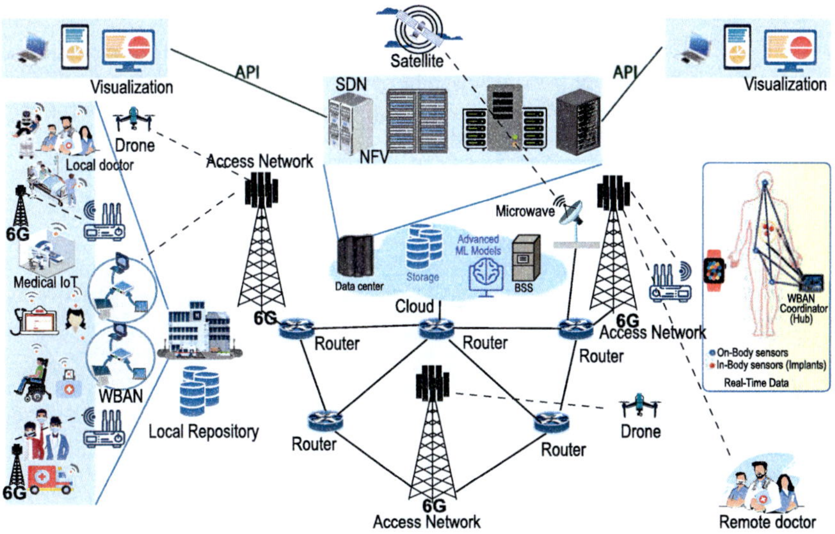

Fig. 1 A 6G communication scenario for IIoHT

- IoHT devices allow all stack holders (patients, physicians, and others) to link and collect large data directly or remotely for analysis. Therefore, it takes massive bandwidth to minimize queuing latency when accessing the medical record. This needs a quicker, efficient, secure computing, and network infrastructure; and
- IoHT nodes can connect to other nodes using multivendor interfaces. These variations can impose device interoperability and security risk.

3 6G as Access Network for IIoHT

The 6G communication technology will transform smart IoHT to IIoHT to enhance patient life quality [5, 19]. Using terahertz connection, massive MIMO, and quantum communication, it can provide high data rates and also support tactile/haptic Internet technology and Intelligent IoT [1, 18]. Figure 2 illustrates the advantages, necessary technologies, and key services of 6G system. The 6G-enabled technologies are listed below.

3.1 Terahertz Link

Terahertz (THz) frequency band (1, 0.1 mm of Electromagnetic Wave) will be used for 6G communication links [16] which can increase link bandwidth three times that of the 5G mmWave range, resulting in a higher data rate of over 1 Tbps. Besides the mmWave band (30–300 GHz), the frequency band 0.275–3 THz will raise total band power by 11.11 times [6]. Moreover, no application has yet used the aforementioned THz band, so the data rate is expected to be greatly increased. One exciting feature of 6G is that it operates in 3D structure, i.e., time, space, and frequency, while 5G operates in 2D.

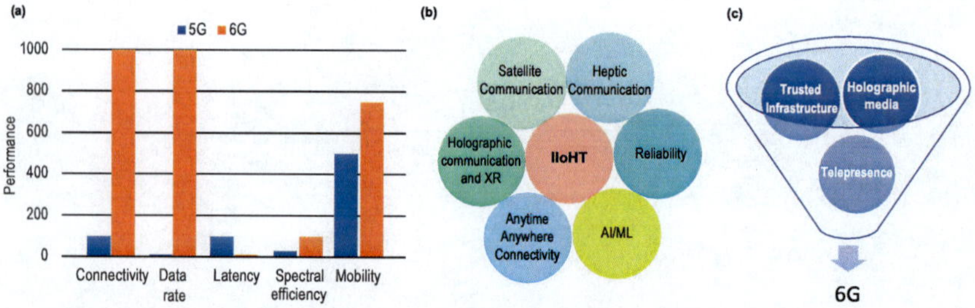

Fig. 2 **a** Performance comparison between 5G and 6G communication network, **b** necessary technologies for Intelligent IoHT using 6G, and **c** key services of 6G systems

3.2 Intelligent Communication

The 6G system is intended to turn the smart society to an intelligent one by replacing smart devices with intelligent devices. The AI-driven technology will be used by these smart devices [16]. It is predicted that the IoT will be replaced by Intelligent IoT (IIoT) by 2030. The smartphones will be replaced by intelligent phones, the hospitals will be converted to intelligent hospitals and what not. These intelligent devices will be able to anticipate and make decisions based on their experiences and pass it on to another intelligent device.

3.3 Quality of Experiences

Experience quality (QoE) is a user-centric term. 6G will have the full quality of service (QoS) atmosphere that will be the demand of the QoE. Holographic communications, the augmented and enhanced reality, and hyper-connected communication network [16] will achieve a better QoE in 6G. High QoE is needed for remote monitoring, fast data processing, self-driven ambulances, etc. The AI-driven 6G will only be able to capture the required data and decide on selected information which will result in a high QoE to be processed, computed, and/or cached [23].

3.4 Intelligent Internet of Things

Intelligent IoT (IIoT) is a network integration of artificial intelligence with data and computers, people, and processes. The 6G is expected to replace the IoT with the IIoT, which will allow millions of intelligent devices, collect huge data, and transform them into real-time digital domain. Digital data is often transferred to another system for signal processing. When commercially launching 6G, big data 2.0 era will begin [16]. A supercomputer is required to process and investigate the huge range of small-scale healthcare system data.

3.5 Tactile and Haptic Internet

Haptic technology communicates via virtual touch or motion. Tactile Internet is the medium to pass this virtual sense to another user node. 6G communication will include this exciting feature so that the remote users can benefit from the real-time haptic communication [6]. IIoT will use holographic communication with high speed to attain the sensory transmission [23]. Tactile Internet can be very useful in healthcare, e.g., remote-controlled robotics for telediagnosis and telesurgery. However, haptic and tactile applications require huge data processing with very low latency and improved data rate within real time. The low latency may be achieved by mobile edge computing [15].

3.6 Massive MIMO

Multiple input multiple output (MIMO) is an effective way to improve the spectral performance of wireless networks by improving bandwidth and providing high data rates. Integrating the expanded THz bandwidth range and huge MIMO would serve as performance boosters for 6G communication system. However, due to THz connectivity, antenna array elements' increased receiver complexity is still a major concern.

3.7 Quantum Communication

Quantum communication is a promising field for wireless communication based on the principles of quantum mechanics. It uses quantum theories to communicate and improve device performance. The immense opportunities of quantum computing make it exciting to contribute with 6G [9] in achieving a stable network with higher data rate and enhanced power. All current computer techniques can be overcome with its promising features. Intelligent healthcare needs a highly safe health data transmission network. Quantum communication's safety function is limited to the clone or the access to a quantum enclosure that allows it an acceptable 6G function. Quantum communication can be efficient in the case of long distance data collection. The remote health surveillance can also be carried out over a long distance. Satellites, UAVs, etc., can be used for regeneration and amplification [20].

4 Possible Services for IIoHT Using 6G

4.1 Holographic Communication

6G holographic communication uses a direct narrow beam with high gain for transmitting and receiving holographic data: -3D image retaining the same scaled characteristics (e.g., depth, parallax, etc.) of the original object. It is an advanced beamforming technique that uses software-defined antenna (SDA), provides better data rate and higher interference and noise ratio (SINR) signal, resulting in high reliability [8]. Holographic communication can establish a paradigm shift in IoHT by allowing real-time holographic medical image transfer. In a virtual-holographic environment, human anatomy is visible in 3D and can be studied in detail rather than traditional scanning such as MRI, CT scan, etc., helping to diagnose disease rapidly and accurately. Moreover, holographic communication improves global healthcare accessibility through holographic telepresence that can digitally connect distant people with a degree of authenticity like physical presence by projecting authentic, full-motion, real-time 3D images of people and objects into a space together with real-time audio communication.

4.2 Augmented/Virtual Reality

Virtual reality (VR) refers to a computer simulated virtual world, whereas augmented reality (AR) blends the real world with imaginary objects that enhance the perception of reality [7]. VR and AR have profound prospects in IoHT and the most notable one is surgical planning. Doctors can view the precise images of body structures without invasive methods and can prepare their interventions before surgery. Moreover, AR aids physicians to navigate tiny instruments through the patient's organs while surgery by looking at a three-dimensional personalized data screen. Thus, the holographic AR will elicit the gradual transition of medical open surgeries to minimally invasive therapies [21]. VR and AR can revolutionize medical education by visualizing 3D human anatomy and enabling students to view and understand every tiny detail from muscles to vain. Besides, VR and AR-enabled medical apps can assist patients to better describe their symptoms to caregivers. However, the major bottleneck of implementing VR and AR in IoHT is the demand for an extremely high data rate coupled with ultra-low latency and reliability [7]. 6G communication can act as a key enabler to achieve the vision of future IoHT by empowering ultra-reliable and low latency VR and AR technology.

4.3 Medical Robot

The insurgence of IoHT and artificial intelligence (AI) in healthcare has led medical robots to play a wide range of functions that involves robot-assisted surgery, rehabilitation, intelligent hospital, and exoskeletons, etc. Telerobotic systems are being carried out to perform surgery, resulting in more precision, extra control, minimally invasive, and shorter recovery times. Robotic rehabilitation systems are employed to provide physical and occupational therapy to elderly or chronic patients, enabling continuously adaptable patient-centered care. Exoskeleton or robotic replacement of diminished body parts/functionalities can support or improve the lifestyle of a disabled person by assisting human body movement most remarkably the ankle, foot, knee, and spine. Hospital robots can assist in distributing medicines, laboratory specimens, and other sensitive materials around the hospital leading to intelligent hospitals enhancing the patients' and caregivers' experience [22].

Although medical robotics has the potential to ameliorate the quality and accessibility of care and augment patients' health outcomes through stimulating the innovation of new treatments for a wide array of diseases and disorders, on the road toward this vision, high-speed communication is a steppingstone as the network is the dominant factor for the accessibility, exchange, and sharing of abundant relevant data [17]. The introduction of 6G will trigger the progress of medical robotic healthcare provision by combining the high capacity networking, computing, and storage resources.

4.4 *Telesurgery*

Telesurgery or remote surgery is an emerging surgical procedure that enables physicians to perform surgery from a distant place. Telesurgery put in holographic communication, augmented reality, and robotic technology together to ensure the telepresence of the medical practitioners and to govern the robot-assisted incision. The process overcomes the major drawbacks of effective surgical procedures, for example, paucity of experienced surgeons, geographical inaccessibility of high-quality surgical care, post-surgery rehabilitation, significant financial burden, and long distance travel. However, the major challenge of telesurgery is real-time communication with ultra-low latency time in transferring auditory, visual, and even tactile feedback between the two obscure places as time delay may result in significant surgical inaccuracy and life threat [4]. The prime reason for increased latency lies in network routing problems, interference, and congestion. 6G vision of high data rate and high-speed complex data processing will allow surgeons to connect medical things together and virtually collaborate in ways that have not even been imagined of yet. Implementation of 6G will fabricate the widespread execution of telesurgery in clinical settings highly feasible and eliminate the geographical barriers.

4.5 *Epidemic and Pandemic*

IoHT can be a unique and effective solution to the epidemic and pandemic cases. With the aid of accessing all the health data remotely without physical contact, IoHT will be capable of reducing the health hazard of the medical personnel. The critical patients can be continuously monitored by monitoring the blood pressure, body temperature, heartbeat, etc. remotely with simple wearable devices. Moreover, the patient will be given all the treatment remotely while keeping him in quarantine. Thus, the disease gets less exposure to environment which reduces the fatality of the pandemic situation. The recent COVID-19 pandemic situation could be dealt with much lesser deaths of the service providers if IoHT was employed. The remote monitoring of the patients, blood sample collection with sensors, continuous health data accumulation, telehealth consultation, remote medication could have made the COVID-19 situation much less hazardous than the current scenario.

4.6 Wireless Body–Machine Interactions

In the body–machine interaction, the bio-signals, such as electroencephalogram (EEC), electromyography (EMG), originating from the human body are used for healing to regain body function and to operate assistive devices [11]. The upcoming 6G wireless system will allow real-time transmission of bio-signal(s) from the body to actuating devices.

5 IIoHT Layer Architecture

Figure 3 shows layer architecture of 6G-IIoHT system which includes seven layers–intelligent sensing layer; intelligent mist layer; intelligent fog layer; smart access layer; intelligent cloud layer; intelligent control layer; and smart application/business layer. The intelligent control layer can be integrated in the cloud layer. The functions of each layers are also listed in Fig. 3.

Intelligent sensing layer uses IoHT nodes to intelligently sense and collect vast amounts of real-time data from patient, doctor, and physical interfaces.

Intelligent mist layer detects node fault and filters irregular data using AI and ML; thus, fog layer reduces data dimensionality and provides ML-based analysis of data sensed in real time. In the cloud layer, advanced ML is used to provide actionable insights via data analysis and provide an application layer visualization.

The intelligent control layer involves optimizing and making choices, in which this layer incorporates the appropriate information from the lower layer in order to make it easier to smartly understand, configure, and select mass agents (e.g., computers, access points).

Intelligent application layer provides automated privacy preserving services and visualization to patient, doctor, service provide efficiently.

6 Future Research Challenges

Intelligent Capability of Devices AI regulated heterogeneous communication systems will be included in the 6G framework. In addition, 1 Tbps, AI, XR, and integrated sensing with communication capabilities are required to be enabled by user equipment (UE). This includes a change of the UE settings and access points. In addition, by converging several heterogeneous subsystems, the 6G system can deliver full automatic systems like automatic automation vehicles, unmanned aerial vehicles,

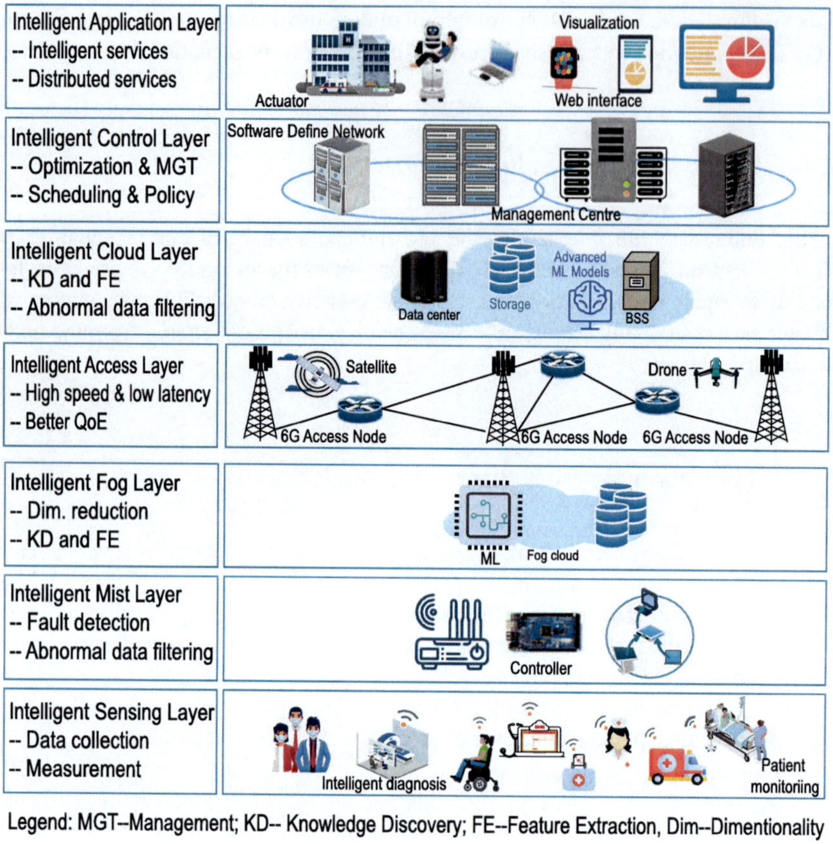

Fig. 3 Layer architecture of 6G-IIoHT system

robots, and industry 4.0. Extensive research is needed to incorporate and optimize AI-based communications on heterogeneous systems.

High Channel Loss The THz bands suffer from high propagation and atmospheric absorption loss which restrict the long distance data transfer. Thus, the new design for transceiver devices and channel model on selected THz is required. Concerning the health or safety issue, low gain THz band antenna design is also a challenge.

Resource Management Unlike the 5G system, the new 6G system consists of a 3D network that is extended vertically. Multiple nodes can search for legitimate information that deteriorates network performance. Therefore, new routing and multiple access protocols, management of radio resources, and scheduling techniques are required. Furthermore, conflict management is important in 6G, which needs further investigation.

Federated Learning Federated (Collaborative) learning trains an algorithm across multiple shared servers without exchanging data samples and enables collaborative training to address critical issues such as data privacy, data security, access rights, and heterogeneous access to data. All nodes submit small message (model) in federated network to match model data as part of training process. To submit small model iteratively, an efficient communication method is required. The device designer's challenges are designing effective communication methods for a heterogeneous environment, distributed optimization, data privacy trade-off.

Challenges for Deploying DL in Fog The benefit of the proximity of mobile fog computing to consumers can be low latency, high bandwidth, and high availability. Deep learning can be used to resolve operational issues such as network slicing, resource and interruption management, offloading, service efficiency, trust management, fault detection, and network healing. Further research may be needed to enhance operational network issues.

7 Conclusion

The IIoHT needs ultra-high-quality network infrastructure to provide best-quality healthcare experiences. It can provide patient-centered services such as customized treatment, diagnosis, real-time patient/health monitoring anywhere, with the intervention of AI-based healthcare system and medical professionals at any time. Intelligent mist-fog-cloud computing, based on AI, is important for IIoHT. Unlike the 5G wireless network, the new 6G networking infrastructure is expected to offer IIoHT services anywhere at any time to enhance human life quality. In this paper, we present the 6G cellular network architecture, its integration with multidimensional communication strategies such as optical wireless communication network, cell-free communication system, backhaul network, and quantum communication, as well as distributed security model in IIoHT context. Such low latency and high-speed communication network would offer a new model for connecting homes to hospitals, people in healthcare, medical equipment, hospital infrastructure, etc. It also outlines the specifications of 6G wireless networking, other primary strategies, problems, and IIoHT implementation research direction.

References

1. Afsana, F., Asif-Ur-Rahman, M., Ahmed, M.R., Mahmud, M., Kaiser, M.S.: An energy conserving routing scheme for wireless body sensor nanonetwork communication. IEEE Access **6**, 9186–9200 (2018)
2. Asif-Ur-Rahman, M., Afsana, F., Mahmud, M., Kaiser, M.S., Ahmed, M.R., Kaiwartya, O., James-Taylor, A.: Toward a heterogeneous mist, fog, and cloud-based framework for the internet of healthcare things. IEEE Internet Things J. **6**(3), 4049–4062 (2018)

3. Biswas, S., et al.: Cloud based healthcare application architecture and electronic medical record mining: an integrated approach to improve healthcare system. In: ICCIT. pp. 286–291 (2014)
4. Choi, P., Oskouian, R., Tubbs, R.: Telesurgery: past, present, and future. Cureus **10**, (2018)
5. Chowdhury, M.Z., Shahjalal, M., Ahmed, S., Jang, Y.M.: 6g wireless communication systems: applications, requirements, technologies, challenges, and research directions. IEEE Open J. Commun. Soc. **1**, 957–975 (2020)
6. Chowdhury, M.Z., Shahjalal, M., Ahmed, S., Jang, Y.M.: 6g wireless communication systems: applications, requirements, technologies, challenges, and research directions. IEEE Open J. Commun. Soc. (2020)
7. Elbamby, M.S., Perfecto, C., Bennis, M., Doppler, K.: Toward low-latency and ultra-reliable virtual reality. IEEE Network **32**(2), 78–84 (2018)
8. Elmeadawy, S., Shubair, R.M.: 6g wireless communications: Future technologies and research challenges. In: 2019 International Conference on Electrical and Computing Technologies and Applications (ICECTA). pp. 1–5 (2019)
9. Hosseinidehaj, N., Malaney, R.: Quantum entanglement distribution innext-generation wireless communication systems. In: 2017 IEEE 85th Vehicular Technology Conference (VTC Spring). pp. 1–7. IEEE (2017)
10. Kaiser, M.S., et al.: Advances in crowd analysis for urban applications through urban event detection. IEEE Trans. Intell. Transp. Syst. **19**(10), 3092–3112 (2018)
11. Kaiser, M.S., Chowdhury, Z.I., Al Mamun, S., Hussain, A., Mahmud, M.: A neuro-fuzzy control system based on feature extraction of surface electromyogram signal for solar-powered wheelchair. Cognitive Comput. **8**(5), 946–954 (2016)
12. Mahmud, M., Kaiser, M.S., Hussain, A.: Deep learning in mining biological data. arXiv preprint arXiv:2003.00108 (2020)
13. Mahmud, M., et al.: A brain-inspired trust management model to assure security in a cloud based IoT framework for neuroscience applications. Cognitive Comput. **10**(5), 864–873 (2018)
14. Makhataeva, Z., Varol, H.A.: Augmented reality for robotics: a review. Robotics **9**(2), 21 (2020)
15. Nawaz, S.J., Sharma, S.K., Wyne, S., Patwary, M.N., Asaduzzaman, M.: Quantum machine learning for 6g communication networks: state-of-the-art and vision for the future. IEEE Access **7**, 46317–46350 (2019)
16. Nayak, S., Patgiri, R.: 6g communication technology: a vision on intelligent healthcare. arXiv preprint arXiv:2005.07532 (2020)
17. Okamura, A.M., MatariĆ, M.J., Christensen, H.I.: Medical and health-care robotics. IEEE Robot. Automat. Mag. **17**(3), 26–37 (2010)
18. Rahman, S., Al Mamun, S., Ahmed, M.U., Kaiser, M.S.: Phy/mac layer attack detection system using neuro-fuzzy algorithm for iot network. In: ICEEOT. pp. 2531–2536. IEEE (2016)
19. Saad, W., Bennis, M., Chen, M.: A vision of 6g wireless systems: applications, trends, technologies, and open research problems. IEEE Network **34**(3), 134–142 (2019)
20. Tariq, F., Khandaker, M.R., Wong, K.K., Imran, M.A., Bennis, M., Debbah, M.: A speculative study on 6g. IEEE Wireless Commun. **27**(4), 118–125 (2020)
21. Vávra, P., Roman, J., Zonča, P., Ihnát, P., Němec, M., Jayant, K., Habib, N., El-Gendi, A.: Recent development of augmented reality in surgery a review. J. Healthcare Eng. **2017**, 1–9 (2017)
22. Yang, G., et al.: Homecare robotic systems for healthcare 4.0: visions and enabling technologies. IEEE J. Biomed. Health Inf. 1–1 (2020)
23. Zhang, Z., Xiao, Y., Ma, Z., Xiao, M., Ding, Z., Lei, X., Karagiannidis, G.K., Fan, P.: 6g wireless networks: vision, requirements, architecture, and key technologies. IEEE Veh. Technol. Mag. **14**(3), 28–41 (2019)

Towards a Blockchain-Based Supply Chain Management for E-Agro Business System

Sm Al-Amin, Shipra Rani Sharkar, M. Shamim Kaiser, and Milon Biswas

Abstract Considering the overall situations of cultivator's economic degradation, food safety chain issue, and whole food supply chain, newer technologies are emerging in the field of agriculture over time. If we look at third world countries, these traditional food supply chains still play the main bearish down prediction. For the result, if any customer deceived by adulterated foods, then in most cases, he/she could not get the original source of the food where it was produced or processed. And in this case, also the administration cannot make this possible to identify and punish the fraudsters. This is where the fraudsters get the opportunity to mix adulteration in foods, put new expiration on expired foods, and release them in the market. Again because of the lack of efficient distribution facilities between cultivators and industrial or last stage vendors, most of the cases cultivators could not get fair pay. And to solve this problem and set explicit or trustworthy communications, the civilization needed immutable and trustworthy technology so that all these problems can be solved. And in that stage, blockchain comes to the picture. In this research paper, we propose an effective, efficient, and satisfactory model or system and service solution to agro traders and also a food traceability system based on blockchain and the help of IoT to make their business more smart and rich. And through the blockchain, smart contract, and the help of IoT sensors, we tried to do maximum effort to reduce human intervention.

Keywords Blockchain · Smart contract · IoT · Agricultural supply chain · Food traceability · Food safety

S. Al-Amin (✉) · S. R. Sharkar · M. Biswas
Bangladesh University of Business and Technology, Mirpur, Dhaka, Bangladesh
e-mail: smalamin.piash@gmail.com

S. R. Sharkar
e-mail: Shipra.sarkar.cse@gmail.com

M. Biswas
e-mail: milonbiswas4702@gmail.com

M. S. Kaiser
Institute of Information Technology, Jahangirngar University, Savar, Dhaka 1342, Bangladesh
e-mail: mskaiser@juniv.edu

M. S. Kaiser et al. (eds.), *Proceedings of International Conference on Trends in Computational and Cognitive Engineering*, Advances in Intelligent Systems and Computing 1309, https://doi.org/10.1007/978-981-33-4673-4_26

1 Introduction

In the states of the computer world, the fastest growing technology gets over on each other continuously to prevalence in the technological universe. The curtain of these simultaneous improvements of technology is only for necessities and for exalted life schemes. For decades, agriculture has been associated with the production of essential food crops. At present, agriculture above and beyond farming includes forestry, dairy, fruit cultivation, poultry, beekeeping, mushroom, arbitrary, etc. With 40% of the global workforce, the agriculture sector presents only 6.4% of the entire world's economic production [8]. But sadly, lack of proper treatment of agricultural cultivation and for the transportation process, tons of tons harvest have been destroyed every year. Food safety is so important that 49% of shoppers showed interest that they will pay extra for products that have top high quality/safety standards. Basically in the absence of a proper supply chain process, that catastrophe has been occurring every year. Still, most of the cultivators depend on the traditional food supply chain procedure that is costly, sometimes often non-profitable because tons of crops are destroyed. Also to drive these supply chains, cultivators or investors have to fall in difficulty because of transactional security purposes often. In the traditional food supply chain, by mixing of different quality products, it is not possible to regulate the food condition when it comes to the customers. With its continuity, it is the high time to bring the touch of modernity in the agricultural sector through blockchain technology (Fig. 1).

Now everyone would like to know where the food has been produced before it is served on the plate. Is this food quality safe for their family or people? Do these production houses produce this food through following the proper guideline or not? To get the solution for all of the necessary question's answers and to get the proper privilege of a supply chain, the blockchain comes to the picture [12]. Because nowadays blockchain is the perfect technological weapon for e-agro business towards the

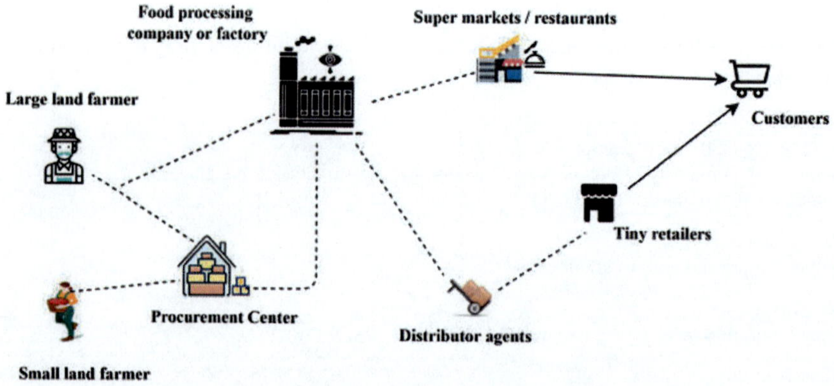

Fig. 1 Flow diagram of traditional supply chain management for agro business

next level. In 2008, blockchain was invented by a person using the name Satoshi Nakamoto to serve as the public transaction ledger of the cryptocurrency bitcoin. So the technology behind bitcoin is blockchain. Many people know it as the technology behind bitcoin, but blockchain's potential uses extend far beyond digital currencies. The technology can work for almost every type of transaction involving value, including money, goods, and property. So why not for agricultural improvement! That is why our goal is to improve the supply chain process effectively through blockchain and make a secure transaction connection between two-sided creditors. Through blockchain, our hope is to shrink the traditional supply chain process that was involved in various types of complex characters to fulfil the whole supply chain process. Hence, blockchain comes to the pictures to solve these complex untrusty procedures into a trustful model. So that people can trust the transactional process and get immutable evidence of all data among unknown trade holders.

2 Related Work

Blockchain is an emerging technology. Because of requisite technology, nowadays many researchers and technologists are trying to invest their highest potentiality to use the blockchain in many desired segments. For this reason, there are numerous publications and case studies in the field. In this section, we review and highlight related work found in the literature on blockchain applications for agricultural supply chains. Though the use of blockchain banking, finance and insurance industries has been increasing steadily, agriculture is scant and just started to gain popularity.

In case of providing an idea of traceability in agriculture, Tripoli and Schmidhuber [6] discuss an application which is a combination of distributed ledger technologies (DLT) and smart contracts. The authors identify the barriers of technical limits and try to facilitate a better understanding of the opportunities, benefits, and applications of DLTs in agri-foods. Gandino et al. [2] for traceability management proposed a framework for the evaluation of a traceability system for the agri-food industry. This framework based on radio-frequency identification (RFID) technology attached to products in a fruit warehouse considering both internal and external traceability. Let us go back three years ago, Tian [10] proposed a food supply chain traceability system for real-time food tracing based on hazard analysis and critical control points (HACCP), using blockchain and the Internet of things. Also, the author blames the centralized system is vulnerable to collapse by fraud, corruption, tampering, and falsifying information. Previously, [11] the author(Feng Tian) analysed the advantages and disadvantages of using radio-frequency identification(RFID) and blockchain technology for agri-food supply chain traceability systems. Especially, through the case study, Holmberg and Aquist [4] have shown the implementation of traceability in the dairy industry based on blockchain. Also Hang [3] proposed a blockchain-based fish farm platform basically for ensuring agricultural data integrity. Because the main concept in the paper is about how to get secure and transparent agriculture data using the blockchain instead of messy and mutable data for commercial analysis. Salah [9] proposed an approach of blockchain-based Soybean traceability in agro supply chain. Caro et al. [1] present Agri-Block IoT, they have given much emphasis on the case of traceability by producing and consuming digital data along the chain using IoT in the food supply chain [7]. Lucena et al. [5] present a method of grain quality measurement considering blockchain as a new type of distributed database and transactions can be generated and ruled by using smart contracts.

According to those relevant work, it is pretty obvious that a growing trend is adopting blockchain technology for enhanced information security and authentication in various criteria not only other banking or industrial security purposes but also agro-food supply chains. Our aim is to emphasize all transactional and locations information from cultivator to customers and also besides that provide the facility about the transaction between the cultivator and industrial through a smart contract for cultivators fair pay.

3 Blockchain-Based Supply Chain Management for E-Agro Business System

In this section, we describe our proposed solution through our e-agro business approach and how every stakeholder gets fairly benefited through implementing the model in real life. In the blockchain model, every stakeholder enrols as a node hopping for profit and according to keeping this in mind,we design the overall system process. And this model is not only suitable for the cultivators and traders but also for customers food safety as well as for the government's investigation (Fig. 2).

In the above model, cultivators contact factory owners to sell the particular amount of harvest, and after arbitrating the cost about the vocation of the harvest, the full load of vehicles is going to chase on the way to the factory. When the harvest gets fully loaded into the factory, then the smart contract automatically executes and notifies the factory owner's bank for starting transaction the predetermined money into the farmer's bank account. So, the cultivator was not deprived of his fair pay. Then the workers are going to start the food processing, and after that to find buyers, the owner posts a declaration for selling the produced food through a smart device. Then after a bunch of bargaining, the buyers buy the product from the seller of the factory and again posts to sell the products to the customers. But between traders, the transaction process will be occurred by cryptocurrency instead of banking. The whole transaction process and its data are saved in the blocks of blockchain through smart contracts.

As evident from those processes, any activities can be documented and stored in blockchain through a smart contract.

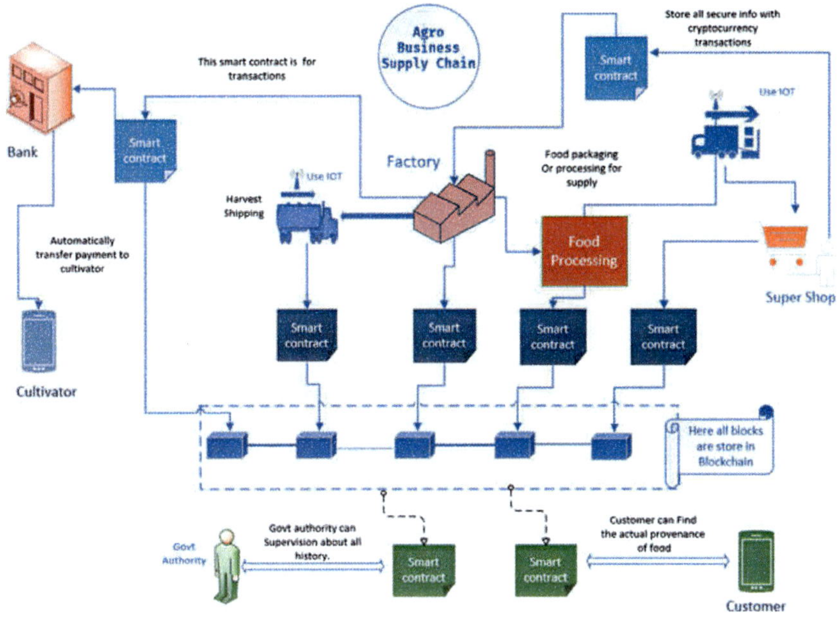

Fig. 2 Flow diagram of our proposed agro-business supply chain management using blockchain

3.1 Smart Food Traceability System

The main purpose of integrating blockchain technology in the e-argo business model is the traceability of end to end food supply. Such system can connect all the parties and provide unprecedented transparency. If any customer deceived by adulterated foods and wants supervision about the source of the food, then it is quite possible to trace. By using a smart traceability system in the food supply chain, government authority can take significant steps to identify a fraudster in the food industry.

4 Why Blockchain Is Secure

Blockchain works on the peer-to-peer network which simply means by assembling multiple nodes by authentic connection and does not need any central server. So, the simplest way to explain blockchain is like a series of data blocks where each block contains a unique hash number and also contains previous block's current hash value so that, through this, it could be linked up with the previous block. But still, people have four concerns about blockchain. Those are (a) confidentiality (b) integrity (c) non-repudiation, and (d) authentication. So, how is it possible to solve these four problems and that is where the concept of cryptography arrives to solve all these four concerns. It has a concept of encryption and decryption, and in this case, asymmetric key cryptography comes to make the whole architecture actually secure. Asymmetric key cryptography also is known as public-key cryptography but basically two different keys are used; here one is a public key and other is the private key. So.

All this network has multiple nodes and each node has a public and private key so to encrypt or decrypt any data we need to use both keys at the same time, not one like a symmetric key. Each transaction is signed with a private key and then can be further verified with a public key. And the signature becomes invalid for different transaction data.

As a result, blocks of the blockchain just ignored and would not make it to the chain as a new member block (Fig. 3).

But another concern also comes for blockchain security purposes, and that is about the malicious node. What if mining becomes too easy and needs less time and that is why spammers get a chance to enrol into the chain and be able to create so many identities in the network through mining blocks. And 52% of nodes are capable of changing/attacking the whole chain and the longest chain rule is already being established as a trust issue. So to remove the malicious nodes and make the time delay for mining, the proof-of-work(POW) algorithm was invented. The main benefits of the POW algorithm are the low impact of stake on mining possibilities and anti-Dos attack. The main mechanism for making transaction delays is 'difficulty'. By using difficulty and insert nonce value in a block, it is possible to make changes to a little or massive portion of hash value into the desired number of zero or any other choice able number. It makes mining a challenge for the miners and that is why the only real miner shows interest to add blocks by mining. And still, by the theorem of the longest chain, it takes the position of the irrevocable trusted chain (Fig. 4).

The new-difficulty = old-difficulty * (targeted block's implement time/actual time waste of equal blocks implement).

If difficulty changes after 2 weeks, then equivalent minutes is 20160. And if the expected mining time interval is 10 min, then 2016 blocks create in 2 weeks. But if the machine becomes powerful and creates a new block within 9 min, then the

Fig. 3 Securely hashing data flow in blocks of blockchain

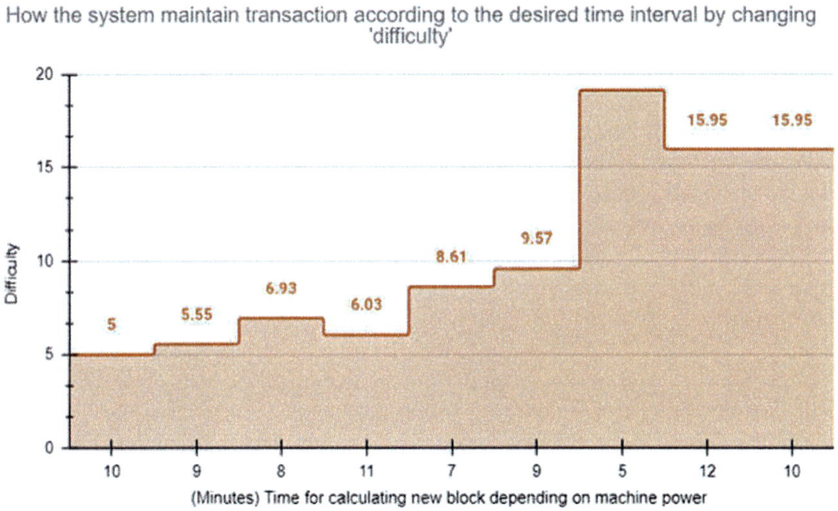

Fig. 4 Computer system is maintaining transaction delay according to the desired time interval (10 min time interval here)

approximate times for creating 2016 blocks = 18144 m. Say the old difficulty = 5, then new difficulty formula is,

$$\text{new}-\text{difficulty} = \text{old}-\text{difficulty} * 20160/18144 == 5 * 1.11111 = 5.55(\text{increase } 0.55)$$

Hence, if expected blocks number == actual blocks number then difficulty change is equal to zero. The value of difficulty is changed to control the transaction of each block on exact time when the machine power is high or low.

4.1 When and Why Should 'Difficulty' Be Used or Not?

'Difficulty' is nothing but a value defined by a system that is basically used to show how hard is it to find a hash value that will be lower than the target. Since the main goal is to use 'difficulty' in blockchain particularly for making the time delay of any transaction adding a new block, so in the case of cryptocurrency transaction process by blockchain, that means, coin transfer from one node to another (like a transaction between traders in (Fig. 2) 'difficulty' is emphatically important to keep the security and overcome for enrolling any spammer. Otherwise, if we need to deploy blockchain as a quick data or message transfer tool, then 'difficulty' approach is not the best place for applying. But in the case of banking, transaction process by blockchain is not appropriate for using 'difficulty'(Like a transaction between cultivator and trader Fig. 2).

5 Why Blockchain Is in the Supply Chain Management System

Supply chain operations are growing exponentially, converting raw material or natural resources into a finished product or services to a consumer product. Any products that we see in our office/home/industries follow a path of intermediaries to get from the factory where it was manufactured to us.

That chain includes manufacturers, procurement officers, people who are quality inspections officers, distributors, suppliers, etc., that means a dozen intermediaries in between the process. As a result, a lot of data can get lost, or any of the intermediaries could easily modify that data properly to fit their best interest. And there is not only one company in the supply chain, multiple companies involved in this. So the problem is that all have competing interests, and for a reason, one trader should have to trust all of them. These vulnerabilities create a transparent issue in the supply chain and make a significantly worse impact on the supply chain in the worldwide sustainable business. So trust is a big problem when it comes to a supply chain. Also, It is difficult for small businesses to get into the game, because usually manufacturers only want to deal with big companies. And that is why small companies have to provide some of their authentic information for the verification process. So again it comes down to trust issues.

Supply chain as of now the biggest industry for blockchain. Blockchain is being used as a way of making the supply chains more secure, transparent, and efficient for sustainable business. If we look at the 2018 PwC global survey of blockchain, financial service industries are using the most advanced blockchain technology in development nowadays. Also along with financial service, industrial products and manufacturing, healthcare, government, the sector of entertainment and media are adding their footsteps in the blockchain (Fig. 5).

Even, blockchain in cryptocurrency might provide a decentralized unit to conduct monetary by spreading its operations across a network of computers by surpassing local and international regulations.

Hence, judging by the criteria of necessity, the viability of blockchain technology has open doors through supply chain management (SMC) and manufacturing. It has created the logistics and handling industry within the realms of the international economy in terms of workforce, investments, and sheer size.

6 Smart Contract for E-Agro Business System

The smart contract is a piece of code for computer protocol intended to digitally facilitate, verify, or enforce the negotiation or performance of logics for credible transactions without third parties. So, If we are talking about smart contracts then the first question is about, why should we trust smart contracts? And the answer is pretty simple, and that is, smart contracts are stored on a blockchain and inherit some

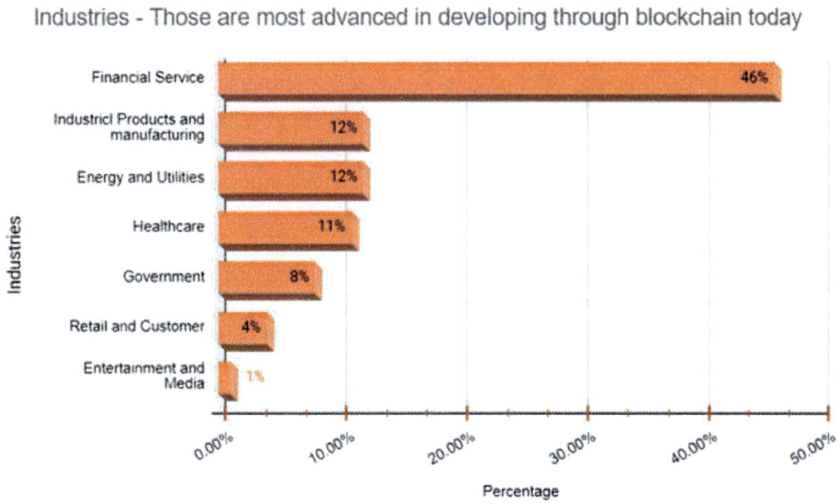

Fig. 5 Statistics of industries, those are most advanced in developing through blockchain. PwC-2018 (Total =! 100 % cause blockchain-based industries are not equal to 1% of overall industries yet)

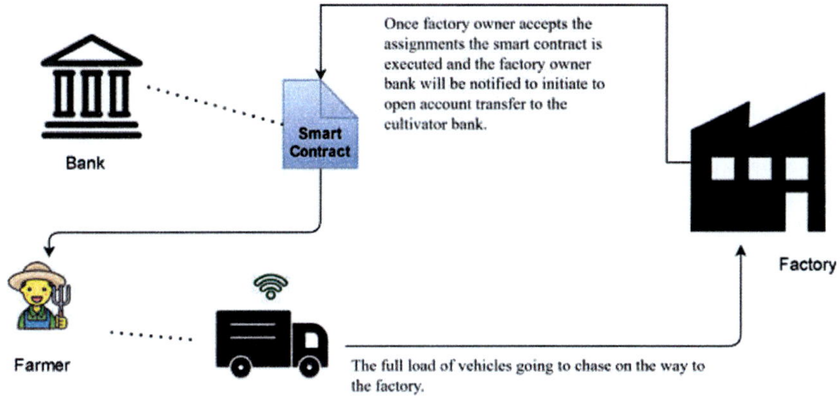

Fig. 6 Through smart contract, farmers can get predefined fair pay, even by dealing with unknown traders

properties. They are immutable and distributed. Once a smart contract is created, it works according to the logic of instruction and never be changed again. So it cannot be possible to release some funds by forcing or tempering from a smart contract. Because that time, other stakeholders in the network will spot the attempt and make the proposed as invalid (Fig. 6).

In our model, smart contracts are used for three portions; one for building trust between cultivator and factory owner about money transaction, even with unknown trader. Second, for cryptocurrency transactions between traders. And other for fetching food traceability, processing, and traders business information into the blockchain. In the first portion, the bank connects with this smart contract as a stakeholder or partner in the network. That means this smart contract's executed value or instructions also be validated by the bank. And all other smart contracts used as information collector and food traceability for customers and the government, respectively.

After fully loading harvest into the factory, the factory owner cannot stop or reduce the predetermined amount of money. Because in this cases, the factory owner is also just a stakeholder in the network. But if the fully loaded harvest fails to arrive at the factory, then the predetermined amount of money's control automatically goes back to the factory owner's bank through the smart contract.

According to this approach, the smart contract can play a significant rule for building trust issues even if between unknown traders.

7 Conclusion

In this paper, we proposed blockchain-based supply chain management for the e-agro business system has been proposed. It provides an open platform for both suppliers and buyers to negotiate for their goods' reasonable prices. Suppliers can make direct mobile payment transactions to the buyers that eliminates intermediaries and brokers. Digital technology like blockchain can reduce the traditional food supply chain's vulnerable cornerstone into a secure and efficient supply chain through implementing our proposed model. The future plan is to connect the blockchain verification system with the banking section. So that cultivators can easily get the advantage through it.

References

1. Caro, M.P., et al.: Blockchain-based traceability in agri-food supply chain management: a practical implementation. IOT Tuscany, IEEE, pp. 1–4 (2018)
2. Gandino, F., et al.: Improving automation by integrating RFID in the traceability management of the agri-food sector. IEEE Trans. Ind. Electron. **56** (2009). https://doi.org/10.1109/TIE.2009.2019569
3. Hang, L., Ullah, I., Kim, D.-H.: A secure fish farm platform based on blockchain for agriculture data integrity. Comput. Electron. Agricul. **170**, 105251 (2020)
4. Holmberg, A., Aquist, R.T.: Blockchain technology in food supply chains: a case study of the possibilities and challenges with an implementation of a blockchain technology supported framework for traceability (2018)
5. Lucena, P., Binotto, A.P.D., Momo, da Silva, F., Kim, H.: A case study for grain quality assurance tracking based on a Blockchain business network. arXiv preprint hyperimage http://arxiv.org/abs/1803.07877arXiv:1803.07877 (2018)
6. Tripoli, M., Schmidhuber, J.: Emerging opportunities for the application of blockchain in the agri-food industry. FAO and ICTSD: rome and Geneva. Licence: CC BY-NC-SA **3** (2018)

7. Asif-Ur-Rahman, Md, et al.: Toward a heterogeneous mist, fog, and cloud-based framework for the internet of healthcare things. IEEE Internet Things J. **6**(3), 4049–4062 (2018)
8. StatisticsTimes 2017. List of Countries by GDP sector composition. http://statisticstimes.com/economy/countries-by-gdp-sector-composition.php
9. Salah, K., Nizamuddin, N., Jayaraman, R., Omar, M.: Blockchain-based soybean traceability in agricultural supply chain. IEEE, vol. 7, pp. 73295–73305 (2019). https://doi.org/10.1109/ACCESS.2019.2918000
10. Tian, F.: A supply chain traceability system for food safety based on HACCP, blockchain and Internet of things. IEEE, pp. 1–6 (2017). https://doi.org/10.1109/ICSSSM.2017.7996119
11. Tian, F.: An agri-food supply chain traceability system for China based on RFID and blockchain technology. IEEE (2016). https://doi.org/10.1109/ICSSSM.2016.7538424
12. Arifeen, M.M., Al Mamun, A., Kaiser, M.S., Mahmud, M.: Blockchain-enable contact tracing for preserving user privacy during COVID-19 outbreak. Preprints 2020, 2020070502. https://doi.org/10.20944/preprints202007.0502.v1

Normalized Approach to Find Optimal Number of Topics in Latent Dirichlet Allocation (LDA)

Mahedi Hasan, Anichur Rahman, Md. Razaul Karim, Md. Saikat Islam Khan, and Md. Jahidul Islam

Abstract Feature extraction is one of the challenging works in the Machine Learning (ML) arena. The more features one able to extract correctly, the more accurate knowledge one can exploit from data. Latent Dirichlet Allocation (LDA) is a form of topic modeling used to extract features from text data. But finding the optimal number of topics (on which success of LDA depends on) is tremendous challenging, especially if there is no prior knowledge about the data. Some studies suggest perplexity; some are Rate of Perplexity Change (RPC); some suggest coherence as a method to find an optimal number of a topic for achieving both of accuracy and less processing time for LDA. In this study, the authors propose two new methods named Normalized Absolute Coherence (NAC) and Normalized Absolute Perplexity (NAP) for predicting the optimal number of topics. The authors run highly standard ML experiments to measure and compare the reliability of existing methods (perplexity, coherence, RPC) and proposed NAC and NAP in searching for an optimal number of topics in LDA. The study successfully proves and suggests that NAC and NAP work better than existing methods. This investigation also suggests that perplexity, coherence, and RPC are sometimes distracting and confusing to estimate the optimal number of topics.

M. Hasan (✉) · A. Rahman
National Institute of Textile Engineering and Research (NITER), Savar, Dhaka, Bangladesh
e-mail: mahedihasan.ict@gmail.com

A. Rahman
e-mail: anis.mbstu.cse@gmail.com

Md. R. Karim · Md. S. I. Khan
Mawlana Bhashani Science and Technology University, Tangail, Bangladesh
e-mail: razaulkarimce15004@gmail.com

Md. S. I. Khan
e-mail: bappy.10.cse.mbstu@gmail.com

Md. J. Islam
Green University of Bangladesh, Dhaka, Bangladesh
e-mail: jahidul.jnucse@gmail.com

M. S. Kaiser et al. (eds.), *Proceedings of International Conference on Trends in Computational and Cognitive Engineering*, Advances in Intelligent Systems and Computing 1309, https://doi.org/10.1007/978-981-33-4673-4_27

Keywords LDA · Perplexity · Coherence · Topic modelling · Feature extraction · Text mining · Information retrieval · Big data · Machine learning

1 Introduction

In-text mining, feature extraction algorithms provide a base for supervised, unsupervised or reinforcement learning, etc. For measuring distances among text documents, Information Retrieval (IR) system needs to exploit semantic information. LDA is of great use in the scenario where exploiting semantic information is the first priority [2, 17].

But according to [3], the efficacy of LDA largely depends on the value of a vital parameter named "number of topics", which actually requires prior knowledge about the contents of the dataset. For a large dataset, usually, one does not have prior knowledge. Thus it is tough to set value for such a critical parameter. Present research in this area offers no easy way except perplexity and non-parametric methods to predict the optimal number of topics [19]. Setting the higher value of the number of topics (which results in more features) blindly will take longer processing time without guaranteeing a higher accuracy score. On the other hand, setting a lower value of a number of topics results in shorter processing time, possibly with a reduced accuracy score. To find a value of the number of topics that will take a reasonable amount of processing time with accuracy as higher as possible is the key to success in LDA. A study in [19] proposed a non-iterative method based on perplexity to find an appropriate number of topics and thus optimizing LDA. They proved that this method significantly works better than perplexity. Kobayashi et al. [11] used perplexity to predict the optimal number of the topic in LDA. Authors in [22] proposed RPC to predict the optimal number of topics. They tried also proved that RPC works better than perplexity. In [12], the authors checked the reliability of perplexity. Fang et al. [5] used coherence to measure the performance of LDA. They also demonstrated that a larger number of topics in LDA help produce more coherent topics. In [20] authors proposed a Weighted Pólya urn scheme to use along with LDA to produce a more coherent topic. Furthermore, a study in [9] also used coherence as a measurement tool to find a coherent topic. All of the studies discussed above except [12, 22] used perplexity and coherence as tools.

Therefore, a question is unanswered in those studies. The question is that, to which extent perplexity, RPC, and coherence are efficient in predicting the optimal number of LDA topics? In this research work, authors have conducted experiments to find answers to that question. This study also proposes two new methods NAC and NAP to predict the optimal number of topics.

Organization: This paper is organized as follows: Sect. 2 describes a theoretical overview of LDA and related research. In Sect. 3, the authors describe the proposed NAP and NAC. Section 4 describes the details of the experimental design. In Sect. 5, the analysis of results has been illustrated. After that, in part Sect. 6, the authors discuss the effects, and finally, Sect. 7 concludes the study.

2 Background

2.1 LDA Overview

LDA is one of the most popular topic models [10]. In this model, it is considered that data instances are being generated from a latent process, which is dependent on hidden variables. The dependencies of the latent generative process have been shown in Fig. 1. Topic assignment $Z_{d,n}$ depends on the per-document topic proportions θ_d, while θ_d depends on hyperparameter for prior knowledge α. The word $W_{d,n}$ depends on the topic assignment β_k and β_k depends on hyperparameter η. The joint probability distribution (over hidden variables) modelled from Fig. 1 expressed in Eq. 1.

$$P\left(\beta_{1:K}, \theta_{1:D}, Z_{1:D,\,1:N}, W_{1:D,\,1:N}\right) = \prod_{k=1}^{K} p\left(\beta_k \mid \eta\right) \prod_{d=1}^{D} p\left(\theta_d \mid \alpha\right) \left(\prod_{n=1}^{N} p\left(Z_{d,n} \mid \theta_d\right) p\left(W_{d,n} \mid \beta_{1:K}, Z_{d,n}\right)\right) \quad (1)$$

The notations used in the equation are the following. Topics are $\beta_{1:K}$, where each of the β_k is a probability distribution over the words. The topic distribution for the d_{th} document is θ_d, where $\theta_{1:D}$ is the probability distributions over topics for all D number of documents. $Z_{1:D,1:N}$ is the topic assignment for each of N word in each of D number of documents. The drawn words for each of D number of documents is $W_{1:D,1:N}$. The crucial parameter on which success of LDA depends on is the value of number of topics K. This study investigates how to make a successful prediction of the optimal value of K. In this article authors use t_i as the candidate values of optimal K. Calculation of posterior is shown in Eq. 2, where $w_{1:D,\,1:N}$ means all observed words in all documents.

$$p\left(\beta_{1:K}, \theta_{1:D}, Z_{1:D,\,1:N} \mid w_{1:D,\,1:N}\right) = \frac{p\left(\beta_{1:K}, \theta_{1:D}, Z_{1:D,\,1:N}, w_{1:D,\,1:N}\right)}{p\left(w_{1:D,\,1:N}\right)} \quad (2)$$

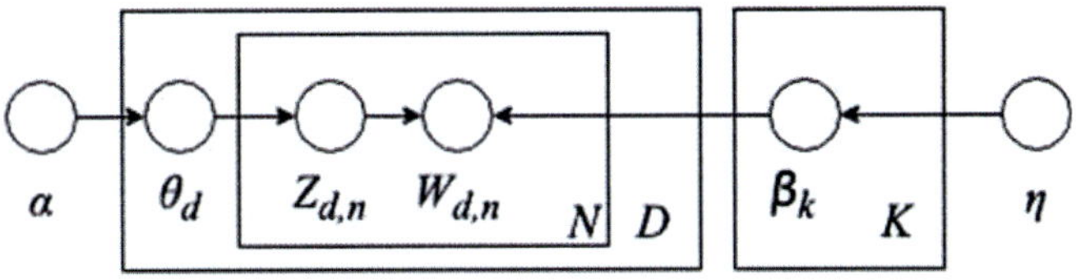

Fig. 1 Dependencies in LDA

2.2 Existing Methods for Predicting the Optimal Number of Topics in LDA

Perplexity: It is a statistical method used for testing how efficiently a model can handle new data it has never seen before. In LDA, it is used for finding the optimal number of topics. Generally, it is assumed that the lower the value of perplexity, the higher will be the accuracy. For a test set of M documents Perplexity (P) is defined as [3]. Where $p(w_d)$ is the probability of observed words of document d. N_d is the total number of words in document d.

$$P = \exp\left\{\frac{-\sum_{d=1}^{M} \log p\left(w_d\right)}{\sum_{d=1}^{M} N_d}\right\} \tag{3}$$

RPC: The following formula has been used in [22] to calculate Rate of Perplexity Change(RPC). Where P_i is the Perplexity of ith value of candidate number of topics t^i.

$$\text{RPC}_i = \left|\frac{P_i - P_{i-1}}{t_i - t_{i-1}}\right| \tag{4}$$

Coherence: Described in [14], coherence is another measure to evaluate in which degree the induced topics of an LDA model are correlated to one another. For instance, in a corpus of medical text data, if an LDA model induces a topic of words from astronomy, then we classify the topic as the outlier. Such a topic is harmful to achieve higher accuracy. Coherence (C) measures this. It is assumed that the higher value of coherence, the higher probability of getting higher accuracy from that model.

$$C = \sum_{i<j} \text{score}_{\text{UMass}}(w_i, w_j) \tag{5}$$

$$\text{score}_{\text{UMass}}(w_i, w_j) = \log \frac{D(w_i, w_j) + 1}{D(w_i)} \tag{6}$$

where, $D(w_i)$ is the frequency of documents that contains the word w_i, $D(w_i, w_j)$ the frequency of documents that contain both words w_i and w_j, and D is the total number of documents in the corpus.

2.3 Related Works

After a detailed study, researches in this area have been summarized below. A study in [22] shows RPC as the better estimator of the optimal number of topics. In [15] author established a relation between the optimal number of topics and text size. They named the relation power-law relation. Authors in [7] proposed an existing community detection model named the stochastic block model, which actually a network model can be used to find topics in a text corpus. They consider text corpus as a complex network document and words. Also, in [18], the authors proposed another network model named "ComModeler" based on the community detection dynamic model. They first extract keywords/terms from data and then formulates a complex network of those terms using their relationship. Then applying "ComModeler" they find the topic. Researchers in [19] proposed a non-iterative method. Researchers in [21] proposed the HDP-ME-LDA method to find the optimal number of topics automatically. They use the Hierarchical Dirichlet Process (HDP) algorithm to find out the optimal number of topics.

On the other hand, Wang et al. [20] proposed the Weighted Pólya Urn Scheme to incorporate with LDA to produce more coherent topics. Semantic prior knowledge based on word embedding has been added in the form of weight into Weighted Polya Urn Scheme. In [4], the authors proposed the Parsimonious Language Model to remove non-significant words/terms from the corpus to produce more coherent words. A study in [13] proposed the SA-LDA algorithm to find the best parameter values for LDA. Actually, they use the iterative method of Simulated Annealing (SA) to find optimal parameter values.

From the above studies, it can be summarized that some of the studies use perplexity, some use RPC, some use coherence and other methods (ComModeler, HDP-ME-LDA, Weighted Pólya Urn, etc.) to find the optimal number of topics. But all of those methods lack a considerable level of accuracy. Also, no study found where efficiency and reliability of Perplexity and Coherence and their comparative analysis have been focused. Except some partial tasks have been found in [12, 22]. This research work proposes two new methods NAC and NAP. The authors in this investigation also analyzed the reliability and efficiency of Coherence, Perplexity, and RPC.

3 Developing Proposed NAP and NAC

K-Fold cross-validation has been used to construct models. $t_1, t_2, \ldots\ldots t_n$ are candidate values of number of topics (t_i where $1 < i \leq n$) and sorted according to increasing order. LDA models have been trained K times for each candidate value of a number of topics. Each time $K - 1$ folds of the dataset are used for training, and the remaining fold is used for testing. Each time perplexity, coherence, F-measure (using K-means clustering) has been calculated. Then for each candidate value of the number of topics, t_i average of K number of perplexity scores, an average of K number of coherence scores, and the average of K number of F-Measure scores have been calculated. P_i is detonated as perplexity scores and C_i is detonated as coherence scores for t_i. NAP and NAC have been calculated using Eqs. 7 and 8 respectively.

NAP is calculated by taking absolute value of Perplexity scores P_i and then doing normalization. Where, P_i is the perplexity score for the ith candidate value of number of topics t_i. $\max(P_i)$ is the maximum value of perplexity among perplexity values for all t_i.

$$\text{NAP} = \frac{|P_i|}{\max(|P_i|)} \tag{7}$$

NAC is calculated by taking absolute value of Coherence scores C_i and then doing normalization. Where, C_i is the coherence score for the ith candidate value of number of topics t_i. $\max(C_i)$ is the maximum value of coherence among coherence values for all t_i.

$$\text{NAC} = \frac{|C_i|}{\max(|C_i|)} \tag{8}$$

4 Experiments

4.1 Data

This research use data from [8] TREC Genomic protocol 2005. They provide 4,591,008 classified MEDLINE documents collection from the year of 1994–2003. Those documents had assigned to be classified into 50 topics by different information retrieval systems (topics are selected by biologists) and then output was manually corrected by human to erase the misclassifications. For example, topic 108, 109, 111, 114, 116, 117, 120, 126, 142, and 146 are in Table 1 [16].

Among 50 topics, 33 topics have 2654 documents. Other documents have no relevance to any of the 50 topics. Also, 17 topics have no documents to be classified into them. After verification and correction (if needed) of each document's similarity with its class, it can be assumed that the documents having semantic similarity are in the same class. Among those 2654, only 1539 documents in the data file have been collected from [1]. Among those 1539 documents, some are missing abstracts. Those documents have been removed from the dataset used in this research. Table 2 summarizes the fact.

As of Table 2, topic numbers with a number of relevant documents (with abstracts) more than 30 are 108, 109, 111, 114, 116, 117, 120, 126, 142, and 146. To construct a balanced dataset, 30 relevant documents have been chosen for each of the above-mentioned topics by random choice. Thus, a collection of 300 documents has been obtained. In this study, only abstracts of those documents have been included in the dataset. Then for robust analysis, 5 fold balanced cross-validation is performed. Each fold contains an equal distribution of documents from all of 10 topics. Thus, each fold contains a mixture of 60 documents from different topics. Hence a balanced dataset in both training and testing sets has been achieved.

Table 1 Description of topic

Topic_no	Topic
108	Describe the procedure or methods for identifying in vivo protein-protein interactions in time and space in the living cell
109	Describe the procedure or methods for fluorogenic 5′ nuclease assay
111	Provide information about the role of the gene PRNP in the disease Mad Cow Disease
114	Provide information about the role of the gene APC (adenomatous polyposis coli) in the disease Colon Cancer
116	Provide information about the role of the gene Insulin receptor gene in the disease Cancer
117	Provide information about the role of the gene Apolipoprotein E (ApoE) in the disease Alzheimer's Disease
120	Provide information on the role of the gene nucleoside diphosphate kinase (NM23) in the process of tumor progression
126	Provide information on the role of the gene P53 in the process of apoptosis
142	Provide information about Sonic hedgehog mutations and its/their role in developmental disorders
146	Provide information about Mutations of presenilin-1 gene and its/their biological impact in Alzheimer's disease

Table 2 Number of documents relevant to each topic

Topic no	Total number of relevant documents	Total number of relevant documents (Without abstract)	Topic no	Total number of relevant documents	Total number of relevant documents (Without abstract)
100	12	–	121	8	–
106	20	2	122	5	–
107	11	–	126	126	–
108	37	2	128	13	–
109	87	2	129	8	–
111	66	4	137	7	–
113	7	–	139	10	–
114	142	–	140	12	–
116	35	–	141	19	–
117	377	31	142	94	–
118	11	–	145	3	1
119	23	–	146	243	7
120	163	7	Total	1539	56

4.2 Evaluation Criteria

The accuracy of the LDA model is measured through clustering. Assuming that the more accurate clustering using LDA, the more efficient the models are. Clustering efficiency has been evaluated by comparing predicted clusters found through LDA model with true classes (Topics are considered as classes of documents) of the dataset. F-measure [6] has been used because of its wide-spread and popular use in measuring the correctness of clustering. Where, n_{ij} is the number of documents in class C_i present in cluster K_j.

$$F(K) = \sum_{K_j \in K} \frac{|K_j|}{S} \max_{C_i \in C} \{F(C_i, K_j)\} \tag{9}$$

$$\text{Recall}\,(C_i, K_j) = \frac{n_{ij}}{|C_i|} \tag{10}$$

$$\text{Precision}\,(C_i, K_j) = \frac{n_{ij}}{|K_j|} \tag{11}$$

$$F(C_i, K_j) = \frac{2 * \text{Recall}\,(C_i, K_j) * \text{Precision}\,(C_i, K_j)}{\text{Recall}\,(C_i, K_j) + \text{Precision}\,(C_i, K_j)} \tag{12}$$

4.3 Experimental Procedures

At first, the authors estimates the optimal number of topics to be used for LDA using five following methods perplexity, RPC and Coherence, NAP, NAC. Five-Fold ($K = 5$) cross-validation has been used to construct models. 2, 3, ... 500 are candidate values of the number of topics (t_i) sorted according to increasing order. LDA models has been trained five times for each t_i. Each time four ($K - 1$) subsets of the dataset is used for training, and the remaining subset is used for testing. Each time perplexity, coherence, F-measure (using K-means clustering) has been calculated. Then for each t_i average of five perplexity scores (P_i), average of five coherence scores (C_i) and average of five F-Measure scores have been calculated. This study then calculates RPC, NAP and NAC using Eqs. 4, 7 and 8 respectively . The training set comprises of 240 (80%) documents while the testing set includes 60 (20%) documents. In this study, the document means the abstract of that document. For the LDA model, the batch processing technique has been used. For clustering, this research uses K-Means algorithm as it is widely accepted.

5 Results Analysis

The range of t_i has been set to 2–500, because authors assume that, the values of t_i in this range takes optimal amount of time (Fig. 2). This range can vary according to capacity/scope of machine. Predicting the value of t_i for which the highest accuracy (F-Measure of K-Means) is achieved in this range, actually means predicting the value of optimal value of number of topics (in this study).

5.1 *Reliability of Perplexity*

Figure 3 shows F-Measure of K-Means and the perplexity score distribution over a different number of topics from 2–500. It is known that the lower the perplexity, the better the accuracy of the model is expected/predicted. As in Fig. 3 perplexity becomes lower in accordance with the increasing of number of topics. Prediction of perplexity line is that the F-Measure score would be lower initially then would go upward linearly. But the scenario in Fig. 3 is completely opposite to prediction. The minimum perplexity score is −70.48 for the number of topics 500. That means perplexity predicts that maximum accuracy of the LDA model would be achieved at the point where number of topics is 500. But the real factor is that the highest accuracy is 0.87 for the number of topic 25, which is a far distant point than 500 predicted by the perplexity line. Thus perplexity barely helps us in determining the number of topics for which highest/optimal accuracy is achieved.

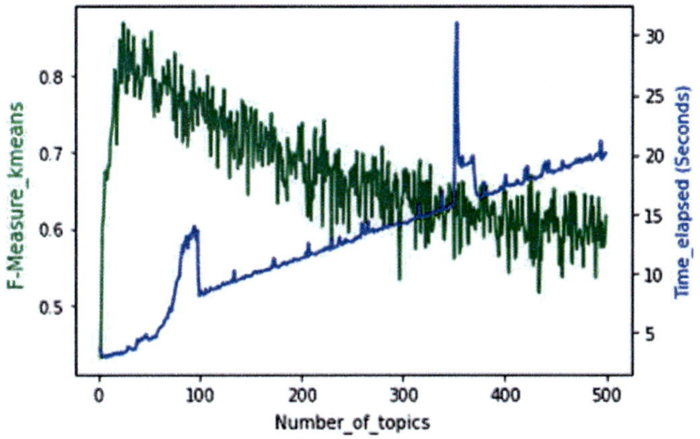

Fig. 2 F-Measure and time elapsed distribution over number of topics

5.2 Reliability of Coherence

Coherence score in Fig. 4 also not inspiring. For coherence, the higher the value, the better the accuracy of the model predicted. The trend of the coherence line and F-measure line in Fig. 4 is about the opposite. But initially, coherence line predicts some better. The maximum coherence score is −2.54 for the number of topic 3, while the highest F-Measure score is achieved at 25. Comparatively coherence line's prediction performance some better than perplexity line. However, the overall performance of coherence lines still very poor in predicting the optimal number of topics (Fig. 4).

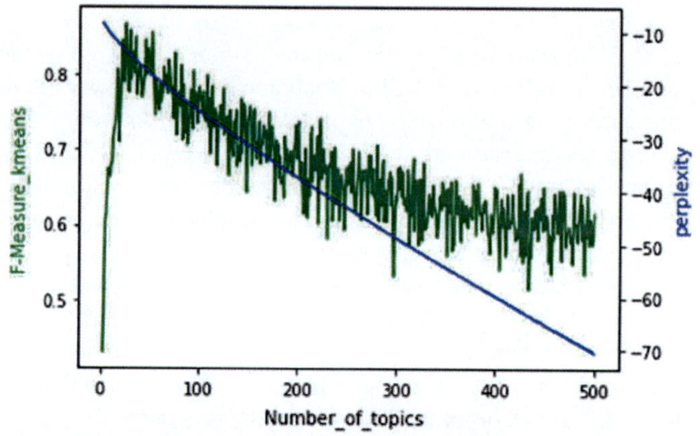

Fig. 3 F-Measure and perplexity distribution over number of topics

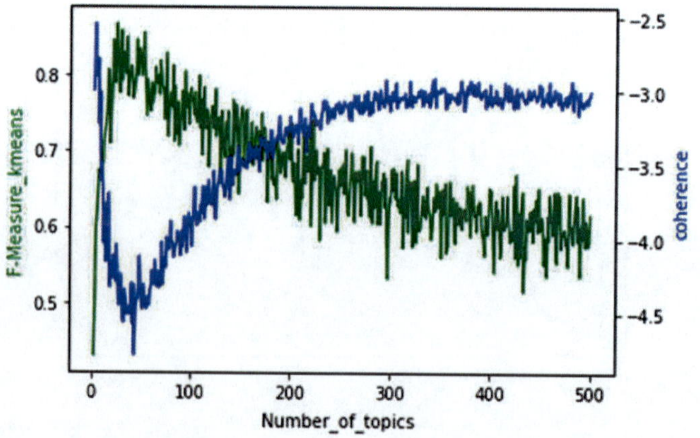

Fig. 4 F-Measure and coherence distribution over number of topics

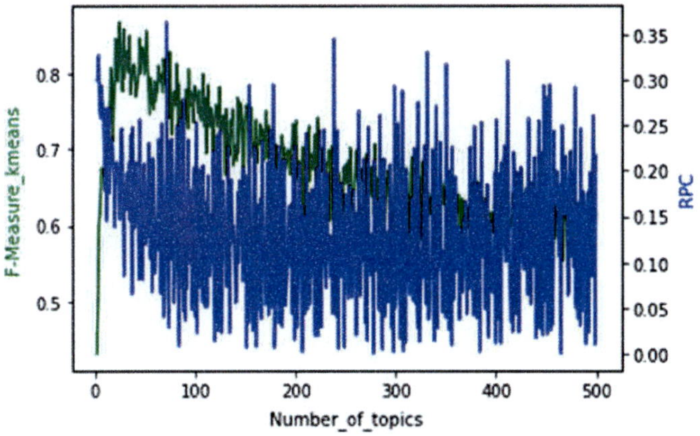

Fig. 5 F-Measure and RPC distribution over number of topics

5.3 *Reliability of RPC*

The performance of the RPC line in Fig. 5 is worse than the coherence line. The lower the RPC, the better the performance of LDA is expected. But the trend of the RPC line seems stable where the F-Measure line gradually goes down. The lowest value of RPC 0.0001 for the number of topic 465. That means the highest accuracy is predicted for the number of topics 465. But the highest accuracy has been achieved at the point of 25, far distant from 465.

5.4 *Performance of NAC*

In Fig. 6 NAC line surprisingly works far better than previous ones. The higher the value of NAC, the higher the performance of the model is predicted. The trend of both of the lines in Fig. 6 seems to be similar. The highest NAC score is 0.53 for the number of topics 43 while the highest F-Measure has been achieved at 25. However, 43 and 25 are comparatively less distant than previous methods. Not only on the highest point, but NAC also performs better in predicting the texture of the accuracy curve at other points of the number of topics axis.

5.5 *Performance of NAP*

In Fig. 7 NAP line also works better. The lower the value of NAP, the higher the performance of the model is predicted. The trend of both of the lines in Fig. 7 seems about to be the opposite. The lowest NAP score is 0.12 for the number of topics 2

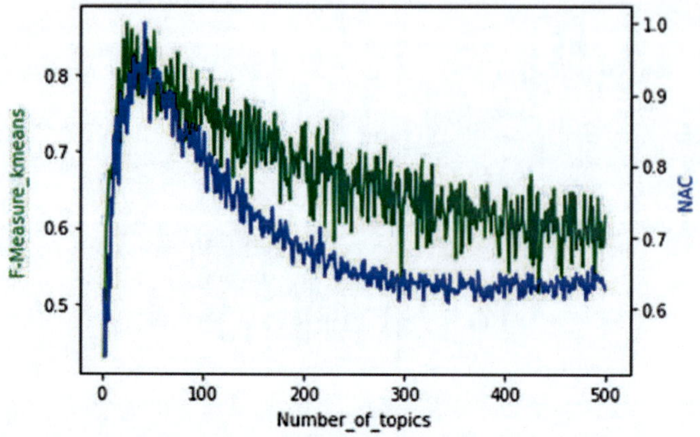

Fig. 6 *F*-Measure and NAC distribution over number of topics

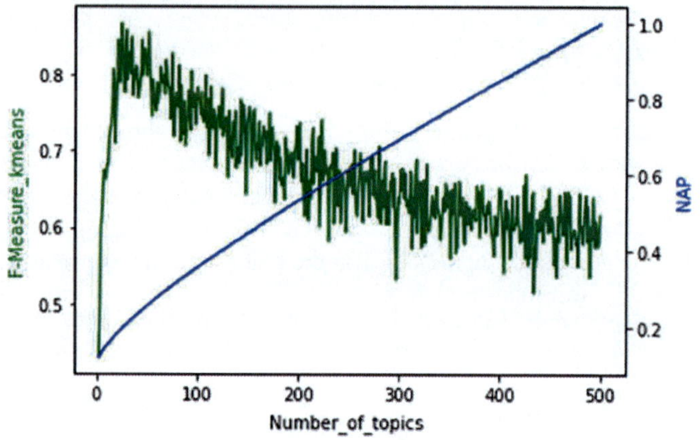

Fig. 7 *F*-Measure and NAP distribution over number of topics

while the highest *F*-Measure has been achieved at 25. Hence, 2 and 25 comparatively closer than the outcomes found at Perplexity, Coherence, and RPC. Not only on the highest point, but NAP also performs better at other points of the number of topics axis.

6 Discussion

Perplexity does poorly help to estimate the optimal number of topic. Coherence helps in some context. RPC also barely helps. Having a glance on Figs. 2, 3 and 5 it is clear that instead helping predicting optimal number topics these traditional methods actually creating confusion. Each of them predicts a far distant number of topics from the actual optimal number of topics. Proposed NAC and NAP perform much better than Perplexity, Coherence, and RPC. Though they do not predict the exact number of topics for which the highest accuracy can be achieved, but their predictions are much nearer than other traditional methods. Having a look at Figs. 7 and 6, it can easily be understood that NAP and NAC help not only in approximating the number of topics for which the highest accuracy is achieved but also to estimate the texture of the accuracy curve.

7 Conclusion

The extracting feature of large text documents is a very complex task. LDA helps us in this context. But setting the optimal number of the topic is another challenging task. Setting a larger value of the number of topics does not help in either way. That means it slows up processing because of a massive number of features (topics), while higher accuracy is also not guaranteed. The authors investigated the reliability of traditional methods of predicting the optimal number of topics and proposed two new methods NAC and NAP. This research has successfully proved that perplexity and RPC are not good estimators of an optimal number of topics, while coherence is also not excellent but efficient to some extent. But proposed NAC and NAP perform far better than Perplexity, Coherence, and RPC. However, NAC and NAP still suffer from some errors described in the discussion section. In the future, authors will investigate to mitigate the errors.

References

1. Text Retrieval Conference Data (2004). https://dmice.ohsu.edu/trec-gen/data/2004/. Accessed: 01 Aug 2020
2. Asmussen, C.B., Møller, C.: Smart literature review: a practical topic modelling approach to exploratory literature review. J. Big Data **6**(1), 93 (2019)
3. Blei, D.M., Ng, A.Y., Jordan, M.I.: Latent dirichlet allocation. J. Mach. Learn. Res. **3**(Jan):993–1022 (2003)
4. Dewangan, J.K., Sharaff, A., Pandey, S.: Improving topic coherence using parsimonious language model and latent semantic indexing. In: Lecture Notes in Electrical Engineering, vol. 601, pp. 823–830. Springer (2020)
5. Fang, A., Macdonald, C., Ounis, I., Habel, P.: Examining the coherence of the top ranked tweet topics. In: SIGIR 2016—Proceedings of the 39th International ACM SIGIR Conference on Research and Development in Information Retrieval, pp. 825–828. ACM, New York, New York, USA (2016)
6. Gelbukh, A.: Computational Linguistics and Intelligent Text Processing: 13th International Conference, CICLing: New Delhi, India, 11–17, 2012. Proceedings, Part II (2012)(2012)
7. Gerlach, M., Peixoto, T.P., Altmann, E.G.: A network approach to topic models. Sci. Adv. **4**(7), eaaq1360 (2018)

8. Hersh, W., Cohen, A., Yang, J., Teja Bhupatiraju, R., Roberts, P., Hearst, M.: TREC 2005 Genomics Track Overview. Technical report
9. Huang, C.M.: Incorporating prior knowledge by selective context features to enhance topic coherence. In: Communications in Computer and Information Science, vol. 1013, pp. 310–318. Springer Verlag (2019)
10. Jelodar, H., Wang, Y., Yuan, C., Feng, X., Jiang, X., Li, Y., Zhao, L.: Latent Dirichlet allocation (LDA) and topic modeling: models, applications, a survey. Multimedia Tools Appl. **78**(11), 15169–15211
11. Kobayashi, H.: Perplexity on reduced corpora. In: Proceedings of the 52nd Annual Meeting of the ACL (Volume 1: Long Papers), pp. 797–806, Maryland, ACL (2014)
12. Neishabouri, A., Desmarais, M.C.: Reliability of perplexity to find number of latent topics. In: The Thirty-Third International Flairs Conference (2020)
13. Pathik, N., Shukla, P.: Simulated annealing based algorithm for tuning LDA hyper parameters. In: Advances in Intelligent Systems and Computing, vol. 1154, pp. 515–521. Springer (2020)
14. Röder, M., Both, A., Hinneburg, A.: Exploring the space of topic coherence measures. In: Proceedings of the Eighth ACM International Conference on Web Search and Data Mining, WSDM '15, pp. 399–408. Association for Computing Machinery, New York, NY, USA (2015)
15. Sbalchiero, S., Eder, M.: Topic modeling, long texts and the best number of topics. Some problems and solutions. Qual. Quant. 1–14 (2020)
16. Text Retrieval Conference TREC 2005 Genomics Track Ad Hoc Retrieval Topics. Technical report (2005)
17. Thiyagarajan, D., Shanthi, N.: A modified multi objective heuristic for effective feature selection in text classification. Cluster Comput. **22**(5), 10625–10635 (2019)
18. Dang, T., Nguyen, V.T.: ComModeler: topic modeling using community detection. EuroVis Workshop on Visual Analytics (EuroVA) (2018)
19. Wang, H., Wang, J., Zhang, Y., Wang, M., Mao, C.: Optimization of topic recognition model for news texts based on LDA
20. Wang, R., Zhou, D., He, Y.: Optimising topic coherence with Weighted Poólya Urn scheme. Neurocomputing **385**, 329–339 (2020)
21. Yuan, B., Wu, G.: A hybrid hdp-me-lda model for sentiment analysis. In: Proceedings of the 2017 2nd International Conference on Automation, Mechanical Control and Computational Engineering (AMCCE 2017), pp. 659–663. Atlantis Press (2017)
22. Zhao, W., Chen, J.J., Perkins, R., Liu, Z., Ge, W., Ding, Y., Zou, W.: A heuristic approach to determine an appropriate number of topics in topic modeling. In: BMC bioinformatics, vol. 16, pp. S8. Springer (2015)

Towards Developing a Real-Time Hand Gesture Controlled Wheelchair

Md. Repon Islam, Md. Saiful Islam, and Muhammad Sheikh Sadi

Abstract The number of wheelchair patients is increasing every day all over the world. Safe navigation and independent mobility are key issues for wheelchair users especially those suffering from dexterity inhibiting disorders and mobility impairments. Following modern technologies, this paper designs a control system prototype to easily navigate the wheelchair by performing finger gestures rather than operating a traditional joystick. Hand or finger gesture recognition using deep Convolutional Neural Network (CNN) and interaction with a powered wheelchair are major features of this system. This system is focused more on the flexibilities of wheelchair users by ensuring less movement of the hand as well as the cost of the implementation. To achieve these goals, some necessary steps have been performed, i.e., locating and continuously tracking the hand region, feeding the CNN model using Raspberry pi for gesture recognition, and perform navigation. After testing, our proposed system has achieved the recognition accuracy of 98.33% and a response time of 34 ms with a Raspberry Pi 4 device (1.5 GHz of CPU frequency), which is suitable for real-time wheelchair control application with reasonably low-cost.

Keywords Smart wheelchair control · Real-time hand gesture · Gesture-controlled wheelchair · Convolutional neural network · Disabled people

---

Md. Repon Islam (✉) · Md. Saiful Islam · M. S. Sadi
Department of Computer Science and Engineering, Khulna University of Engineering & Technology, Khulna 9203, Bangladesh
e-mail: islam1607011@stud.kuet.ac.bd

Md. Saiful Islam
e-mail: islam1607003@stud.kuet.ac.bd

M. S. Sadi
e-mail: sadi@cse.kuet.ac.bd

M. S. Kaiser et al. (eds.), *Proceedings of International Conference on Trends in Computational and Cognitive Engineering*, Advances in Intelligent Systems and Computing 1309, https://doi.org/10.1007/978-981-33-4673-4_28

1 Introduction

One of the main aspects of life is mobility that provides the capabilities to move independently. But for disabled people with diseases like paraplegia, hemiplegia, quadriplegia, broken leg(s), cerebral palsy, brain injury, and more, need an assistive device to help their mobility. A wheelchair is one of the most used helping devices that allow this capability for disabled people. The evolution of technologies integrates the electronic system into a typical wheelchair to make it smarter. Technologies like Brain Computing Interface (BCI), Human-Machine Interaction, Sensors, and Computer Vision along with the power of Artificial Intelligence (AI) [1] are used by most of the developed systems and research works. In computer vision, head, eye, and hand movement-based systems [2] are mostly used for controlling the wheelchair.

The gesture is one of the natural ways to express impressions and interactions with others. There is a huge number of people around the world who are not capable of doing physical work and not strong enough for operating a traditional joystick. So, gesture interactions are a useful way that is getting more emphasized nowadays [3]. Using state-of-the-art technologies like deep learning [4], gestures are easy to interpret and can be added to wheelchair functionalities easily. Recently, depth cameras like Kinect by Microsoft [5] are available in the market and many researchers have been proposed models that are influenced by depth cameras [6, 7]. But those are expensive compared with a typical RGB camera. Again, in the case of an RGB camera, surroundings can be an obstacle to achieving a better result. So, gesture recognition using RGB camera and low-end, low-cost devices like Raspberry pi is a challenging task in this field.

In this paper, we proposed a novel wheelchair control system using hand gestures by only moving fingers under the camera that run in real-time on low-end embedded devices like Raspberry Pi 3/4. The whole gesture recognition process is divided into three major parts considering the detection of hand region or region of interest (ROI) [8], tracking the ROI using efficient tracking algorithms, and lastly, recognition using a deep learning model like CNN. The powered wheelchair requires basic movement in four directions, i.e., left, right, forward, and backward. For this, six different types of finger gestures are built where two gestures for stop and signal condition.

Initially, the captured image frames go through noise processing and YCrCb skin segmentation [9]. Then the hand region is detected using Haar-cascade classifier [10] from a large frame. After detecting the ROI, an efficient tracking algorithm like KCF is applied for continuously tracking the ROI. Afterward, the ROI has resized for fed into the 2D CNN for recognizing the desired gestures.

Other sections of this paper are arranged as follows, section "Related Work" describes the recently developed system for the smart control of power wheelchair. Then in the "Gesture Recognition and Wheelchair Control" part, the hand detection techniques, tracking methods with algorithms, and architecture of the 2D CNN model with wheelchair operation are described. The implementation of the prototype is described in the "System Implementation" part. At the end of this section, the setup and performance of our proposed system have been analyzed in the "Experimental

Analysis" section. Lastly, the "Conclusion" section summarizes the research work along with limitations, and future research directions to find out further potential directions.

2 Related Work

A lot of researches and works has been conducted to make smart wheelchairs using various techniques in the last few decades. Several working has been done in the last few years about effective human-machine interaction for controlling power wheelchair. Megalingam et al. [11] proposed a Bluetooth-based wheelchair system where hand gestures were used with an android application which is built on a smart gadget. Wheelchair users have to move their fingers on the gadget touchscreen. But it might be difficult for those who are not able to operate that system.

An Iris movement-based wheelchair control system has been proposed by Desai et al. [12], where a user can navigate a wheelchair using their eye's iris. The system uses embedded programming using MATLAB, but response time is around 5 s which is not feasible in a real-time environment. In the research paper of Rajesh et al. [13], they proposed an EMG-based system where signal processing is conducted, and predefined hand gestures are used for navigation of the wheelchair. But it takes more response time to get used to the system.

Ruzaij et al. [14] proposed a voice-controlled-based wheelchair navigation system. But the system construction is not cost-effective and it didn't show better performance in all situations. Gao et al. [15] proposed a smart wheelchair system which is controlled by hand gesture. The Kinect depth camera along with a high configured laptop is used for tracking the gestures in real-time. Then the system gets the input signal according to the gesture. As a device control module, it uses DSP 28335 micro-controller. The accuracy of this system severely decreases with background complexity and is not user-friendly in all environments as the user needs to raise a hand to perform gestures.

3 Gesture Recognition

The Gesture Recognition process consists of three steps, namely, Hand detection, Hand tracking, and 2D CNN with efficient model architecture. Some essential preprocessing has been done on the initial images for getting better performance in the recognition process, these are image resizing, noise preprocessing, and skin segmentation. The YCrCb color space is used for skin segmentation with the skin color range (0, 133, 77) to (235, 173, 127) [16]. The cascade classifier is then dedicated to locating the ROI. Then for hand tracking, the KCF (Kernelized correlation filters) algorithm is used in the tracking phase. Finally, resized images are converted in binary, and is fed into the CNN network for final processing to obtain appropriate

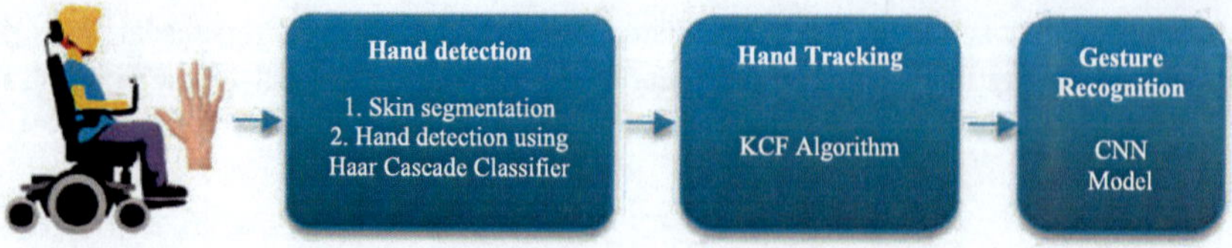

Fig. 1 The overall concept of the hand gesture recognition process

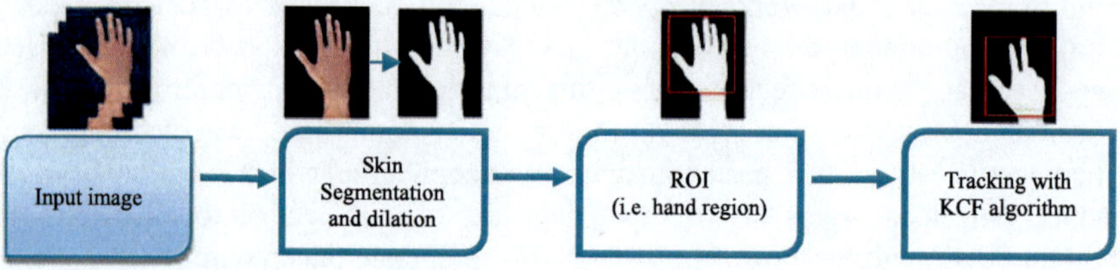

Fig. 2 Hand region detection and tracking process

hand gesture. The complete processes of hand gesture recognition are described in Fig. 1.

3.1 Hand Detection Method

Hand detection refers to locating the hand region from the captured image. Hand detection is the first step of the hand gesture recognition process. For getting better performance, image resizing and noise processing are used. Skin color segmentation [9] is the process of separating human skin color and non-skin color. In this proposed system, we utilized this method using the YCrCb color channel range from (0, 133, 77) to (235, 173, 127). Then the dilation effect is applied to the image and turns it into a binary channel (black and white), all steps are shown in Fig. 2.

Haar cascade classifier has been used on the preprocessed images for detecting the hand region. It is an effective feature-based object detector, uses a machine learning approach with a cascade function where a lot of positive and negative images are fed for training [10]. It can detect faces, hands, and other organs of the body.

3.2 Hand Tracking Method

KCF algorithm stands for kernelized correlation filters [17]. Tracking refers to following an object within a given image frame. KCF algorithm has been utilized for tracking the wheelchair user hand continuously which is shown in Fig. 3. KCF

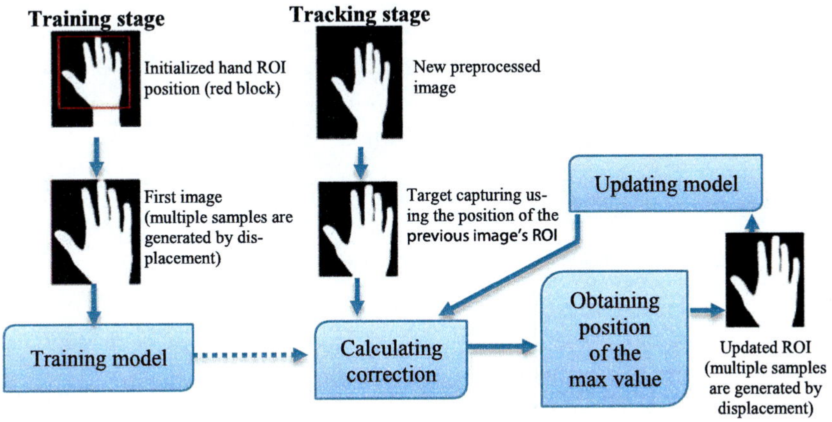

Fig. 3 Training and tracking stage of the KCF tracking algorithm

is combined on the idea of the other two tracking algorithms named "boosting" and "MIL" tracker.

3.3 Recognition of the Hand Region Using CNN

The ROI size has been adjusted to 54 × 54, then passed the resized image into the deep 2D CNN. This proposed system designs a humble CNN architecture containing three convolutional layers and three pooling layers shown in Fig. 4. After the resized

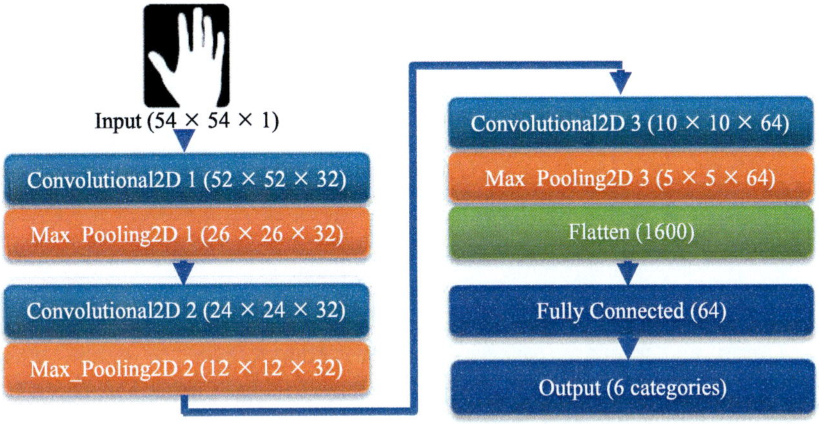

Fig. 4 Architectural layers of the proposed CNN Model

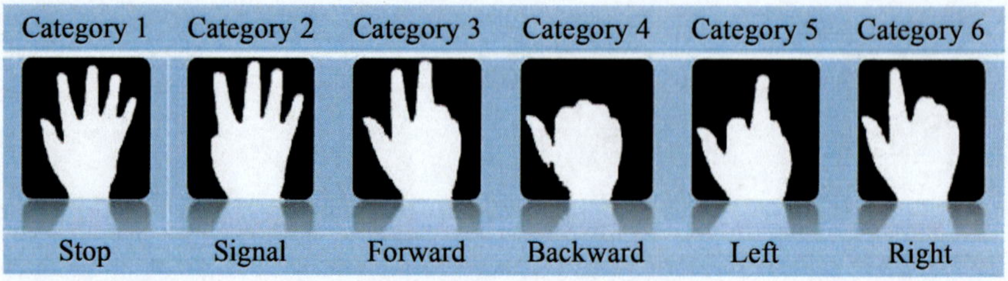

Fig. 5 Classes of the predicted gestures that are recognized for wheelchair control

image is fed into the convolutional layers, it is flattened in the flatten (1600) layer and then goes through the fully connected or dense layers.

The output feature map size has been reduced because the zero-padding method is not implemented. Also, every convolutional layer is directly followed by a ReLU (rectified linear unit) activation function. The pooling layer uses a max-pooling approach where the kernel size is 2×2, as a result, the size of the input image is half of the previous one. To be specific, max-pooling means it takes the maximum value among the value of 2×2 kernel. So, the final input for the dense layer has a size of 5×5 of the sampled image.

This study uses one fully connected layer that sets the input parameter as 64 neurons. A fully connected input layer (flatten) takes the output of the last pooling layer and turns them into a single vector that can be an input for the next stage. Finally, when the testing of the network is measured, all parameters such as ReLU, softmax activation function, and cross-entropy are integrated. The method shows its efficiency in training with less training data.

Each layer of the network is initialized with the weight parameters randomly for training the network. Then, the activation function named "softmax" is used to estimate the output result. Afterward, the "cross-entropy" loss function is performed to calculate the error of the predicted result. After completing all stages of the network, the predicted probability array of the six categories that were trained before, are shown in Fig. 5. So, the index of the maximum value of that probability array is selected for the appropriate gesture.

3.4 Wheelchair Control

In this study, six types of operations are considered for controlling the wheelchair which is shown in Fig. 6. Specifically, they are named as stop, signal (wheelchair horn), move forward, move backward, turn left, and turn right. The gesture recognition 2D CNN model gives a probability array with six categories. The maximum value of that probability array is considered as the output, and according to the index of the output, wheelchair control operations are done.

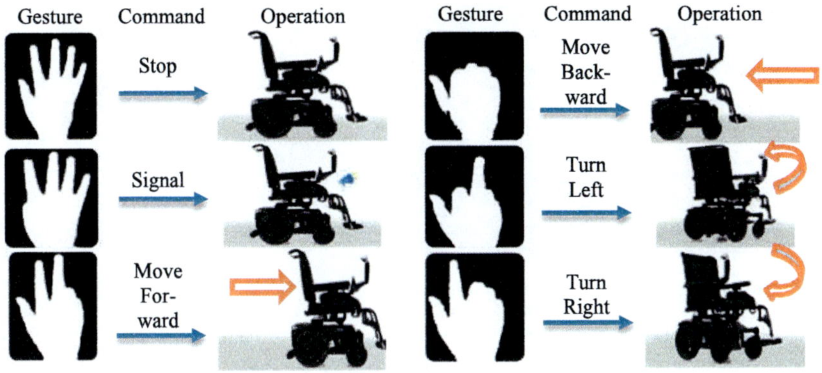

Fig. 6 Operations of the proposed system

4 System Implementation

An overview of the proposed power wheelchair control system is shown in Fig. 7. We aggregate an RGB camera with a Raspberry pi 4 as an input capturing device, an ESP 8266 MCU V3 Wi-Fi module, a 4 channel relay module, and two motors for simulating four directional movements of a wheelchair. Input is taken from the user's hand which is below the camera. Then Raspberry Pi 4, as the main processing device, starts processing the input image.

By following hand gesture recognition steps, namely hand detection, hand tracking, and recognition using the CNN model, the processing device recognizes the appropriate hand gesture. Then the ESP8266 NodeMCU Wi-Fi module gets command over Wi-Fi from Raspberry pi. According to the command, the motor navigates the wheels of the prototype.

Figure 8 shows the implementation of the prototype, the Raspberry Pi 4 takes input from hand using a Pi camera. A four wheels small toy-car is used instead of a

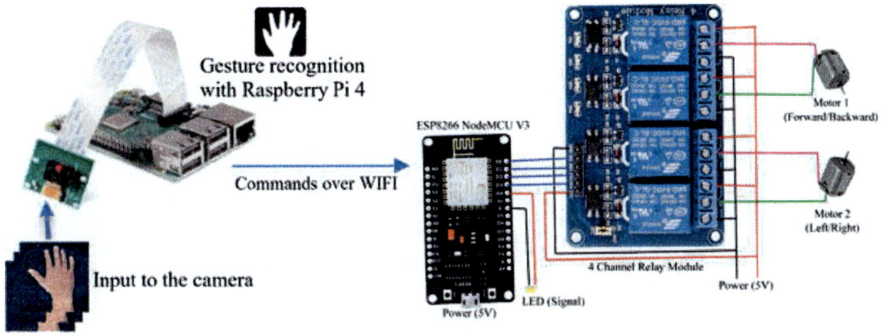

Fig. 7 Block diagram of the proposed system

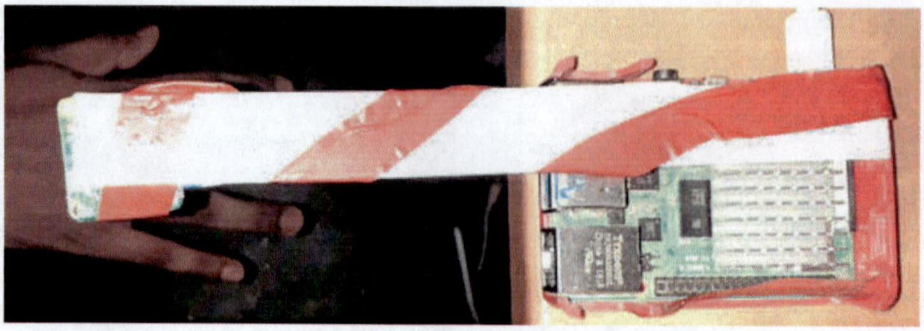

Fig. 8 The prototype implementation of our proposed system

real wheelchair for convenient prototype implementation. The result of hand gesture recognition is sent to the ESP NodeMCU module as a command.

5 Experimental Analysis

The results of our proposed hand gesture-controlled system for a smart wheelchair are analyzed in this section. We have taken all input images using Raspberry Pi camera module V2, and 640 images were collected for each gesture in different angles and conditions which is shown in Fig. 9. Thus, the total number of images was 3840 for a total of six gestures. For every input image, skin segmentation, image resizing and dilation effect have been applied for processing the images for training and validation. Five-fold validation has been adopted to the images, i.e., 512 images for training and 128 images for validation in each gesture category.

Our model has been trained in 150 epochs with batch sizes 16 and 2000 steps per epoch. Adam optimizer was used for training the model with a dropout probability of 0.5. Many performance evaluation measurements are used for the evaluation of a system. In our system, we considered Accuracy and loss for training, validation, and testing, and recognition speed in frame per second (FPS).

The performance of our proposed system has been shown in Table 1. Our proposed system has achieved recognition accuracy of 95.65% for training and 99.48% for the validation set, and loss for training and validation set were 0.1200 and 0.0289, respectively.

Fig. 9 Example of images taken at different angles and conditions

Table 1 Performance of the proposed hand gesture recognition system

Performance measure	Training	Validation	Testing	FPS
Accuracy (%)	95.65	99.48	98.33	29
Loss	0.1200	0.0289	0.0177	

Table 2 shows the confusion matrix, where each row represents the original gesture and each column represents the predicted gesture. Each gesture was performed 10 times and the system has achieved a testing accuracy of 98.33%. At the early stage, it can't run in real-time in Raspberry Pi with 1.5 GHz of CPU frequency. After converting the pre-trained model using Tensorflow lite, which less affects the accuracy, it runs just superb and achieved a recognition speed greater than 30 FPS. After performing network and user interface operation, it still capable of running around 21–29 FPS, which is suitable for real-time application of wheelchair control.

A comparison of similar works has been shown in Table 3. In terms of functionalities, Gao et al. [15] proposed a system where the user needs to raise a hand for performing any gesture, which is not user-friendly and fails in case of mobility impaired. Oliver et al. [18] worked with an accelerometer band that needs to wear and move, which also fails in the case of many dexterity-inhibited patients. Our system overcomes these problems by allowing users to navigate safely with the fingers'

Table 2 Confusion matrix of the testing stage of our proposed gesture recognition system where (column) predicted gesture (row) original gesture performed

	Stop	Signal	Forward	Backward	Left	Right
Stop	9	1	0	0	0	0
Signal	0	10	0	0	0	0
Forward	0	0	10	0	0	0
Backward	0	0	0	10	0	0
Left	0	0	0	0	10	0
Right	0	0	0	0	0	10

Table 3 Comparison of similar works with the proposed system

Features	Proposed system		
	Gao et al. [15]	Oliver et al. [18]	Our system
Functionality	Hand gesture recognition using Kinect Depth Camera and high configured PC, require hand raising	Hand movement detection using accelerometer and Joystick Manipulator, require hand band wearing	Hand gesture recognition using RGB Camera and Raspberry Pi, require only fingers' movement
Cost	High	Medium	Low
Success rate	Varies with complexity (10–100%)	Actual accuracy is not quantified	98.33%

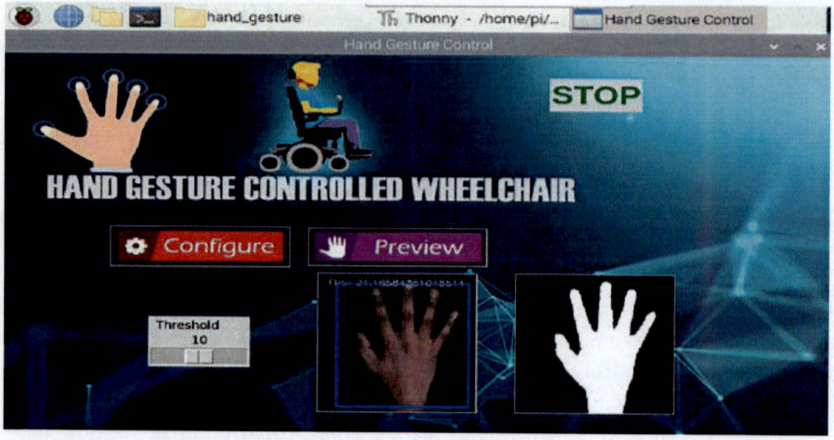

Fig. 10 Graphical user interface of the proposed system

movement only. In terms of cost and success rate, the cost of the system proposed by Gao et al. [15] is high and the accuracy severely decreases with background complexity. The cost of the system proposed by Oliver et al. [18] is medium, and the accuracy was not quantified. Our proposed system uses a low-cost Raspberry Pi and an RGB camera. It is also reliable as the background remains static and achieved 98.33% accuracy.

A graphical user interface (GUI) has been developed for our proposed hand gesture-based wheelchair control system which is shown in Fig. 10. It shows the users what's happening at the moment any gesture is performed.

6 Conclusions

Real-time achievement of hand gesture control with easy fingers' movement on the wheelchair handle is the main goal of our research work in this paper. Uses of a low-cost low-computational device, and without wearing any external sensors are also a major concern in our study. Considering those challenges, hand gesture recognition is done based on three key phases, i.e., hand detection, hand tracking, and gesture recognition using the 2D CNN model. For detecting the hand region, some preprocessing steps along with YCrCb skin segmentation have been applied to the input images captured from the RGB camera. The KCF algorithm is adopted for continuously tracking the hand region. Raspberry Pi 4 with 1.5 GHz of CPU frequency and 1 GB of RAM was used as the processing device for limiting the costs of the system. The experimental analysis shows that our proposed method achieved the training accuracy of 95.65%, validation accuracy of 99.84%, and testing accuracy of 98.33%, with a recognition speed of more than 29 FPS. Our proposed hand gesture recognition system can be quite feasible to control a wheelchair in real-time. In the future, we hope we will find a way to overcome the challenges due to background occlusion along with extending some features like wheelchair speed control, media control, home automation, etc. We believe these areas are well-intentioned for further study, as well as the trends for the future.

References

1. Wang, K.J., Zhang, L., Luan, B., Tung, H.W., Liu, Q., Wei, J., Sun, M., Mao, Z.H.: Brain-computer interface combining eye saccade two-electrode EEG signals and voice cues to improve the maneuverability of wheelchair. In: 2017 International Conference on Rehabilitation Robotics (ICORR), pp. 1073–1078. IEEE (2017)
2. Wanluk, N., Visitsattapongse, S., Juhong, A., Pintavirooj, C.: Smart wheelchair based on eye tracking. In: 9th Biomedical Engineering International Conference (BMEiCON), pp. 1–4. IEEE (2016)
3. Coleman, G., Ward, R.: Gesture Recognition: Performance, Applications and Features. Nova Science Publishers, Inc. (2018)
4. Haque, M.R., Islam, M.M., Alam, K.S., Iqbal, H., Shaik, M.E.: A computer vision based lane detection approach. Int. J. Image, Gr. Signal Process. **10**(3), 27 (2019)
5. Mousavi Hondori, H., Khademi, M.: A review on technical and clinical impact of microsoft kinect on physical therapy and rehabilitation. J. Med. Eng. (2014)
6. Memo, A., Minto, L., Zanuttigh, P.: Exploiting silhouette descriptors and synthetic data for hand gesture recognition. In: Italian Chapter Conference 2015—Smart Tools and Apps in Computer Graphics, STAG (2015)
7. Keskin, C., Kıraç, F., Kara, Y.E., Akarun, L.: Real time hand pose estimation using depth sensors. In: Consumer Depth Cameras for Computer Vision, pp. 119–137. Springer, London (2013)
8. Chung, H.Y., Chung, Y.L., Tsai, W.F.: An Efficient Hand Gesture Recognition System Based on Deep CNN. In: Proceedings of the IEEE International Conference on Industrial Technology, pp. 853–858. IEEE (2019)
9. Shaik, K.B., Ganesan, P., Kalist, V., Sathish, B.S., Jenitha, J.M.M.: Comparative study of skin color detection and segmentation in HSV and YCbCr color space. Procedia Comput. Sci. **57**(12), 41–48 (2015)
10. Soo, S.: Object Detection Using Haar-Cascade Classifier. Institute of Computer Science, University of Tartu, pp. 1–12 (2014)
11. Megalingam, R.K., Sreekanth, S., Govardhan, A., Teja, C.R., Raj, A.: Wireless gesture controlled wheelchair. In: 4th International Conference on Advanced Computing and Communication Systems (ICACCS), pp. 1–5. IEEE (2017)
12. Desai, J.K., Mclauchlan, L.: Controlling a wheelchair by gesture movements and wearable technology. In: 2017 IEEE International Conference on Consumer Electronics (ICCE), pp. 402–403. IEEE (2017)
13. Rajesh, V., Reddy, D.K.: SEMG based human machine interface for controlling wheel chair by using ANN. In: 2009 International Conference on Control, Automation, Communication and Energy Conservation, pp. 1–6. IEEE (2009)
14. Ruzaij, M.F., Neubert, S., Stoll, N., Thurow, K.: Design and testing of low cost three-modes of operation voice controller for wheelchairs and rehabilitation robotics. In: IEEE 9th International Symposium on Intelligent Signal Processing (WISP) Proceedings, pp. 1–6. IEEE (2015)
15. Gao, X., Shi, L., Wang, Q.: The design of robotic wheelchair control system based on hand gesture control for the disabled. In: International Conference on Robotics and Automation Sciences (ICRAS), pp. 30–34. IEEE (2017)

16. Tsagaris, A., Manitsaris, S.: Colour space comparison for skin detection in finger gesture recognition. Int. J. Adv. Eng. Technol. **6**(4), 1431 (2013)
17. Wei, B., Wang, Y., He, X.: Confidence map based KCF object tracking algorithm. In: 14th IEEE Conference on Industrial Electronics and Applications (ICIEA), pp. 2187–2192. IEEE (2019)
18. Oliver, S., Khan, A.: Design and evaluation of an alternative wheelchair control system for dexterity disabilities. In: Healthcare Technology Letters **6**(4), pp. 109–114. IEEE (2019)

A Novel Deep Learning Approach to Predict Air Quality Index

Emam Hossain, Mohd Arafath Uddin Shariff, Mohammad Shahadat Hossain, and Karl Andersson

Abstract In accordance with the World Health Organization's instruction, the air quality in Bangladesh is considered perilous. A productive and precise air quality index (AQI) is a must and one of the obligatory conditions for helping the society to be viable in lieu of the consequences of air contamination. If we know the index of air quality in advance, then it would be of a great help saving our health from air contamination. This study introduces an air quality index prediction model for two mostly polluted cities in Bangladesh: Dhaka and Chattogram. Gated recurrent unit (GRU), long short-term memory (LSTM) are the two robust variation of recurrent neural network (RNN). This model combines these two together. We have used GRU as first hidden layer and LSTM as the second hidden layer of the model, followed by two dense layers. After collecting and processing the data, the model was trained on 80% of the data and then validated against the remaining data. We have evaluated the performance of the model considering MSE, RMSE, and MAE to see how much error does the model produce. Results reflect that our model can follow the actual AQI trends for both cities. At last, we have juxtaposed the performance of our proposed hybrid model against a standalone GRU model and a standalone LSTM model. Results also show that combining these two models improves the overall model's performance.

E. Hossain (✉) · M. S. Hossain
Department of Computer Science and Engineering, University of Chittagong, Chattogram, Bangladesh
e-mail: ehfahad01@gmail.com

M. S. Hossain
e-mail: hossain_ms@cu.ac.bd

M. A. U. Shariff
Department of Computer Science and Engineering, Port City International University, Chattogram, Bangladesh
e-mail: mhnarfth@gmail.com

K. Andersson
Pervasive and Mobile Computing Laboratory, Luleå University of Technology, 931 87 Skellefteå, Sweden
e-mail: karl.andersson@ltu.se

M. S. Kaiser et al. (eds.), *Proceedings of International Conference on Trends in Computational and Cognitive Engineering*, Advances in Intelligent Systems and Computing 1309, https://doi.org/10.1007/978-981-33-4673-4_29

Keywords Air quality index · AQI prediction · Time series analysis · Deep learning · Hybrid neural network

1 Introduction

In today's world, one of the crucial issues is air pollution. The pollution of aerosphere is caused by detrimental chemical substances and also biological substances. In 2008, metropolitan air quality and besides indoor air quality are two of the worst contaminated places in the world according to the World's Worst Polluted Places by Blacksmith Institute. Air pollution can cause some serious problem which can be permanent in terms of health. It causes harm to crops, forests, wildlife, lakes, sea, and ocean. It involves in the formation of rain including acid rain, which damages trees, rivers, soils, and animals. It affects commercial and agricultural production. Besides it also affects human beings physical state [28]. The discharge of contaminants can cause chronic respiratory and lungs difficulty [5]. The public is so tactful about air quality, because of its consequences which results in various illness, sickness, and even loss of life [6, 13, 15, 26]. Some environmental outcome of this is eutrophication, fog , and worldwide weather change. Air contamination is also conducting significant economic losses [31]. For all of these reason, the general public is too delicate to the upcoming swing of air quality because of its huge effect on human physical state [25]. Keeping track of air quality is a common method to alert public about the potential air pollution [3, 42].

Air quality index (AQI) is an universal index that describes about the amount of the contaminants in the air namely PM2.5, O3, PM10, CO, NO2, SO2 and is calculated with the help of surrounding air quality level[48]. Without that, air quality index is also an important scale for showing that how the human physical state is related to it [35]. So, obtaining information about real-time air quality surely can help to control the contamination as well as saving the health of the human from polluted air [16, 34]. Because of the nature of unpredictability and entanglement, it is hard to acquire precise air quality index forecasting results [48].

In this research, our objective is to evolve a model for forecasting daily air quality index data for two of the biggest cities in Bangladesh: the capital, Dhaka and the port city, Chattogram. The air quality data of Bangladesh has not been analyzed and predicted yet. We know that air pollution causes a lot of problems. If the air quality data can be predicted thoroughly, then it could help to control these air pollution-related problems. We used daily AQI data of these two cities and applied two of the most powerful recurrent neural networks (RNN) which are gated recurrent unit (GRU) and another is long short-term memory (LSTM). GRU and LSTM are considered to be the most powerful deep learning models for time series analysis [19]. The principal target of this study is to use combined power of GRU and LSTM models to predict daily AQI value in Dhaka and Chattogram. Although many researches have been trying to predict AQI value, but none of the researchers utilized the combined power of LSTM and GRU so far. Therefore, we propose a hybrid model putting together two of the most powerful time series analyzers (GRU and LSTM) to predict daily AQI data.

2 State of the Art

For getting renowned and relevant research works, we have filtered out all the papers using a keyword-based search on renowned publishers including *Elsevier*, *Springer*, *ACM Digital Library (DL)*, and *Taylor & Francis (T&F)*. For finding the recent state of the art, we have considered only those article which were published since 2019. We have searched using "AQI Prediction" and "AQI Forecasting" these two keywords. For "AQI Prediction" keyword, we found 154, 7, 0, and 63 research articles (224 in total) since 2019 in Elsevier, Springer, ACM DL and T&F, respectively. For "AQI Forecasting", these numbers are 134, 3, 1, and 27 (165 in total) for the corresponding publishers in the stipulated time. But from these 389 articles, only 15 of them have directly predicted AQI value. The remaining articles do not actually predict AQI value, but the above-mentioned keywords appear somewhere in the article. In this section, we have reviewed these 15 articles which actually predict AQI value.

Chen et al. [7] proposed a deep multi-task learning (MTL)-based metropolitan AQI modeling method called "PANDA" to solve two tasks together which are AQI estimation and AQI prediction. They used CNN for extracting spatial information related to AQI modeling and then exploited recurrent neural network and LSTM to study the pattern of the pursuant data. They compared its previously published prediction methods namely U-Air, Semi-EP, ADAIN, ARMA, and FFA. The MAP and MAE values of PANDA are 0.833, 9.50, respectively. "PANDA" outsails all the previous air quality estimation methods.

Ahuja and Jani [2] proposed a continual time series model which is called "Air Quality Data Model". It was developed based on supervised learning to help in the assessment of air quality in metropolitan areas. However, they did not compare the performance of their model against other models. Moreover, while defining their algorithm, they did not mention the linear model anywhere.

Gu et al. [12] had taken 365 sets of air contaminant data (PM2.5, AQI, SO2, CO, NO2, O3, PM10) Between the following time frame: Jan 1, 2018 to Dec 31, 2018 from Shenzhen. Two important algorithm were used to optimize "Support Vector Machine" model. Those are : 1. Improved SAPSO algorithm and 2. PSO algorithm. Later, the model was constructed by using BP neural network. The values of average relative error were 34.35%, 21.00%, 23.74%, 25.56%, 32.18%, 31.99%, respectively, for SO2, NO2, CO, O3, PM10, PM2.5 for 15 iterations of the BP algorithm.

Xu et al. [45] had proposed a hybrid model where $7 \times 24{,}000$ sets of air pollutant data (PM2.5, AQI, SO2, CO, NO2, O3, PM10) Between the following time frame: May 15, 2014 to Feb 6, 2017 from Kunming, Xiamen, Shenzhen and Lasa city. For preprocessing the data and also for reshaping the data, "VMD" and "LASSO" were proposed. And besides for reducing the shape and to extract all the features from the data, "SAE" was used. And later after training the "DESN", the result of the prediction was acquired. For MAE, MAPE, RMSE evaluation, the results of Xiamen were 1.5428, 3.1656, 1.9153. For Shenzhen, it was 3.5246, 7.3008, 4.4029, for Kunming, it was 3.4548, 6.1634, 4.2431, and for Lasa, it was 1.2297, 2.3123, 1.6020, respectively.

Shishegaran et al. [36] had developed model to forecast the air quality index in Tehran which was a linear and also a nonlinear statistical model. 1. ARIMA, 2. PCR, 3. Combined ARIMA and PCR, 4. Combined ARIMA and GEP these four models were used to forecast the daily AQI. From these four models, four equations were obtained to predict daily AQI for each seasons in 2016. The best model among them was nonlinear ensemble method. MAPE, RMSE (2.870–8.125), NMSE (0.012 and 0.51), the coefficient of determination were the evaluation metrics.

Wu and Lin [44] proposed a novel hybrid model SD-SE-LSTM-BA-LSSVM in which SD technique was employed for cutting down the nonlinear features. The MAE, RMSE, MAPE were used to measure the performance of the model. For Beijing, it was (8.8920, 6.6885, 0.0877), and for Guilin, it was (4.4396, 3.8036, 0.0880), respectively. The proposed model obtains better result compared to other models.

Jiang et al. [22] proposed a hybrid learning method combining WPD-MELM-MSK-MELM. "WPD" does the decomposition of the AQI. "IPIO" optimizes the weights of the "ELM" initially. After that, "MSK" clustering was used. Finally, "MELM" was utilized to predict the AQI. The evaluation metrics used are MAE, NRMSE, RMSE, and DS. Their model performs better for high-frequency model.

Zhai and Cheng [47] investigated a "LSTM" approach for predicting AQI. The data was collected from Beijing between the following time frame: Jan 1, 2015 to Dec 31, 2016. RMSE, MAE were used as evaluation metrics. This LSTM model outsails other MLR models. Results show that this model is effective for forecasting air quality for a short time period.

Wu and Lin [43] proposed a hybrid AQI forecasting model combining VMD-SE-LSTM. For decomposing the original AQI series, VMD was used, and SE was employed for data processing. Finally, LSTM was applied to train the model. The evaluation metrics of this model are MAE, RMSE, and MAPE. For Beijing, it was 9.38, 5.66, 7.73%, and for Baoding, it was 15.10, 11.97, 9.09%, respectively. This proposed model can prejudice AQI quickly and conveniently and also superior over the model to some extent for daily urban AQI forecasting.

Wang et al. [41] proposed a hybrid model HI-EMD-SCA-ELM. The Hampel identifier, VMD decomposition was used for detecting, after that correcting weird values, and decomposing the actual AQI dataset. "SCA" was used to optimized the "ELM' model. Finally, all modes values were integrated to get the final values. MAE, RMSE, MAPE, U1 and U2, Pearson's correlation coefficient(r) were used to measure the performance of the model. Results showed that Shenzhen model was better.

Li et al. [29] proposed a new model to predict AQI which was based on MIO algorithm. To predict the reconstructed series, three models namely HCA (ARIMA model-based), HCME, and HCFL are used. The hourly average AQI of the Shijiazhuang, Zhengzhou, and Guangzhou cities between the time frame Aug 1, 2017 to Oct 31, 2017 was used in this paper. The MAPE values of three datasets are 3.75%, 3.33%, and 2.32%, respectively. Results showed that this model outperforms other models in terms of accuracy.

Wang et al. [40] designed a model based on the L1 norm for monitoring and analyzing the air contaminant. The data from Baoding, Tianjin, and Shijiazhuang, China were collected for this model. Using sparse regularization, a total analysis was performed on the prediction of three models for getting the final result. By the help of fuzzy framework, the prediction results are turned into air quality measurement.

Jiang et al. [23] combined ICEEMDAN-ICA-BPNN for hourly pollutants prediction, involving O3, PM2.5, SO2, CO, and NO2 where the Gamma, Weibull, Rayleighand Lognormal distribution functions were used to model contaminant series for characteristics estimation. The evaluation metrics used in this model are FB, MdAPE, RMSE, MAE, DA, and R2.

Liu et al. [30] compared various ML and feature selection methods. The data of Beijing was collected from 2016 to 2017 . MLR, RFR, BPNN, and SVR algorithms were used. And they were trained on ten fold cross validation. Evaluation metrics were used in this model are R, MAE, and RMSE. The results were 0.991, 7.089, 9.268 for MLR, 0.998, 1.337, 3.1861 for RFR, 0.996, 4.8286, 6.3371 for BPNN, 0.991, 7.357, 9.599 for SVR, respectively. The experimental results follow this way RFR > SVR/BPNN > MLR.

Jumaah et al. [24] developed an air quality index forecasting algorithm based on temperature, humidity, wind speed, and precipitation. The data was collected from Kuala Lumpur between the following time frame: Jun 2018 to Aug 2018. OLS process was used for the 3 months data individually. The accuracy was measured (97, 99, and 97%), respectively. The results came out very good and a good fit to the "Ordinary Least Square" model.

3 Overview of LSTM and GRU

For having the ability of memorizing long-term dependencies, RNN is the first choice for language models [38]. However, with time lags increasing while unfolding the RNN into very deep feed neural network, gradients of RNN may vanish. As this vanishing gradients problem is a huge problem, therefore, for solving this problem, LSTM [17] and GRU [8] were introduced with forget gate units whose job was to give the memory cell capability to decide when the information needs to forget. So, that is how it determines optimal time delays. The following two subsections briefly discuss about the structure of LSTM and GRU.

3.1 LSTM

For building the language-based models, "LSTM" was introduced initially back in 1997. Because this long short-term memory (LSTM) has superb power to remember long-term allegiance. Four types of gate are available in classical LSTM. They are - 1. Input Gate 2. Input Modulation Gate, 3. Forget Gate 4. Output Gate. The structure of a typical LSTM cell is shown in Fig. 1 [11].

Input gate's job is to process all the new data which comes from the outside world. The output of the input modulation gate goes to the memory cell. In the following iteration, the forget gate decides about the information which to keep and which not. By doing this, it picks the optimal delays for the input data sequence. The results which are calculated go as input into the output gate. The output gate generates the outcome of the long short-term memory cell. Generallly, a softmax layer is stacked on top of the output layer of the LSTM in language models. But in our model, a dense layer is stacked on top of the output layer of the LSTM cell. In this procedure,X = $(x_1,x_2,...,x_n)$ is the input time series, H = $(h_1,h_2,...,h_n)$ is hidden state of memory cells, Y = $(y_1,y_2,...,y_n)$ is output time series. Hidden state of memory cell is calculated using this formula:

$$h_t = H(W_{hx}x_t + W_{hh}h_{t-1} + b_h) \quad (1)$$

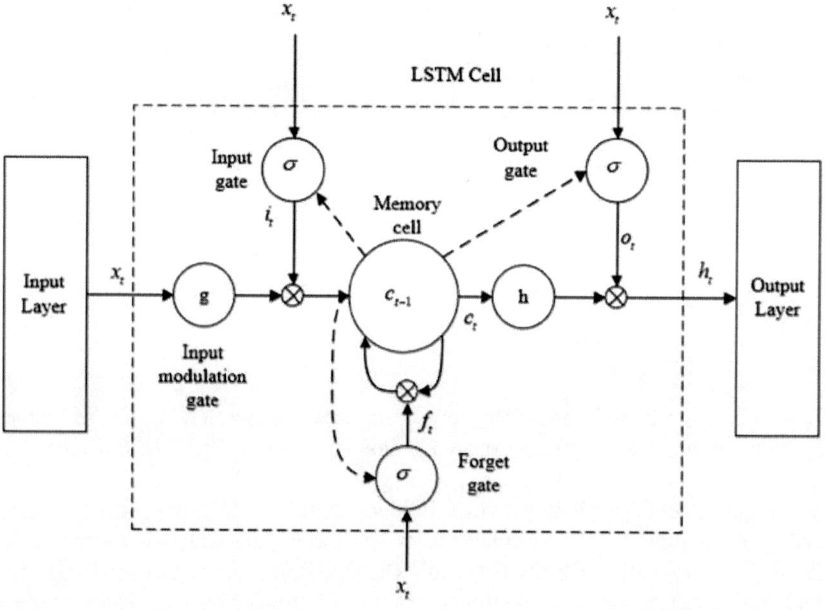

Fig. 1 Structure of a typical LSTM cell

$$p_t = W_{hy} y_{t-1} + b_y \tag{2}$$

Here, weight matrices are denoted as "W", and bias vectors are denoted as "b". And by using the following formulas, the hidden state of the memory cells is computed.

$$i_t = \sigma(W_{ix} x_t + W_{hh} h_{t-1} + W_{ic} c_{t-1} + b_i) \tag{3}$$

$$f_t = \sigma(W_{fx} x_t + W_{hh} h_{t-1} + W_{fc} c_{t-1} + b_f) \tag{4}$$

$$c_t = f_t * c_{t-1} + i_t * g(W_{cx} x_t + W_{hh} h_{t-1} + W_{cc} c_{t-1} + b_c) \tag{5}$$

$$o_t = \sigma(W_{ox} x_t + W_{hh} h_{t-1} + W_{oc} c_{t-1} + b_o) \tag{6}$$

$$h_t = o_t * h(c_t) \tag{7}$$

where the standard sigmoid function is denoted by σ and it is specified in equation (8), the extends of the sigmoid function are denoted by "g". And the range of the function changes to [−2, 2], and [−1, 1].

$$\sigma(x) = \frac{1}{1 + e^x} \tag{8}$$

The square loss function is utilized for objective function according to the following formula:

$$e = \sum_{i=1}^{n} (y_t - p_t)^2 \tag{9}$$

where the real output is denoted by "y" and the predicted AQI value is denoted by "p". Adam optimizer is applied for doing back propagation through time (BPTT) in order to avoid local minima and to minimize training errors [27]. Neural networks tend to overfitting. A huge amount of regularization method is introduced to reduce the overfitting problem. An efficient method to train neural networks named dropouts was proposed in order to gain better features of images in 2012 [37]. For having the recurrent property, dropout technique was ineffective in RNN till 2014 [46].

3.2 GRU

In 2014, Cho et al. introduced a much simpler version of RNN which is called GRU [8]. The shape of a typical GRU cell is illustrated in the following Fig. 2 [11].

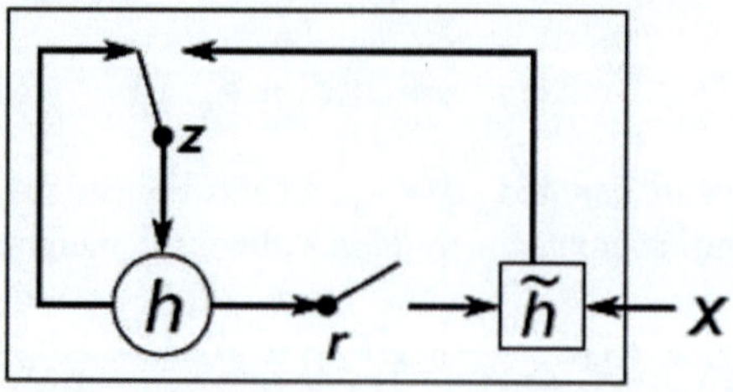

Fig. 2 Structure of GRU cell

There are two gates in a typical GRU cell. The first one is the reset gate (r), and the other one is the update gate (z). Using the value of the output of the hidden state at time t-1 and also the input time series value at the timestamp t, the output of the hidden state is calculated at time t which is basically similar to the LSTM cell. Hidden state output at time t is computed using the hidden state of time t-1 and the input time series value at time t which is similar to LSTM cell. Also, the working procedure of reset gate of the GRU cell matches to the forget gate of a long short-term memory cell.

4 Experimental Setup

In this section, we discuss about the methodology of our experimental model. In the following three subsections, we discuss about the data collection process, model design, and how we evaluated the model.

4.1 Data Collection and Preprocessing

Each year, air pollution costs the life of 15,000 Bangladesh is [32]. Dhaka, the capital of Bangladesh, is one of the most polluted cities in the world. A lot of large industries as well as the large amount of cars exist in the city which cause serious air pollution and needs to be taken care of. Besides, Chattogram, second most polluted city in Bangladesh and the port city of the country, has been the main transit for import and export duties of the country. We have collected daily AQI data for these two cities from the Web site of Bangladesh Ministry of Environment, Forest and Climate Change [10]. For Dhaka, data from June 1, 2017 to July 25, 2020 (1151 observations) is collected and for Chattogram, data collected from June 1, 2017 to March 15, 2020 (1019 observations). After collecting the data, missing values were handled by using the previous 20-days mean value as it produces better results [14]. Then, we have normalized the AQI values within the range [0,1] using MinMaxScaler class of Python's sklearn package.

Fig. 3 Proposed model architecture

4.2 Model Designing

Our proposed model has four hidden layers: First layer is GRU layer with 64 neurons, second layer is LSTM layer with 256 neurons, and the remaining two layers are dense layers with 64 and one neurons, respectively. We have trained our model using 80% of the total collected data, and the remaining data is used for testing. Figure 3 shows the internal architecture of the proposed model.

We have used 20 days loopback value in our model. At first, these previous 20 days AQI values are given as input at the GRU layer. GRU neurons collect the input data and pass them along to the LSTM layer. A weight is calculated in this process. Similarly, LSTM passes the collected values to the dense layer, and another weight is generated. Finally, this dense layer passes the values into the output layer where the actual output is calculated. The model predicted output is compared against the actual output. This process is repeated for 200 epochs, while the weighted values are updated to reduce the value of the cost function (mean squared error). Finally, updated weighted values are used to predict daily AQI value for Dhaka and Chattogram.

4.3 Model Evaluation Criteria

Evaluating the performance of the model is crucial since it is used to measure how the model is behaving. In this work, we have used mean absolute error (MAE), mean squared error (MSE), and root mean squared error (RMSE) to validate the performance of our proposed model. Equations 10–12 show the equations for these errors.

$$MAE = \frac{1}{n}\sum_{i=1}^{n}(x_i - x)^2 \tag{10}$$

$$MSE = \frac{1}{n}\sum_{i=1}^{n}(Y_{ACT(i)} - Y_{PRED(i)})^2 \tag{11}$$

$$RMSE = \sqrt{\sum_{i=1}^{n}\frac{(Y_{PRED(i)} - Y_{ACT(i)})^2}{n}} \tag{12}$$

MAE calculates the overall level of forecasting errors, whereas MSE and RMSE put more weight on large error as they square the errors before averaging. The smaller the value of these metrics are, the better the model's prediction.

5 Result and Discussion

For both Dhaka and Chattogram, we have trained our model using 64-256-64-1 formation of the layers and was run for 200 epochs. We also compared our hybrid model against a standalone GRU model and a standalone LSTM model to see whether combining these models improves the performance of the model. The next two subsections discuss about the results of the model for Dhaka and Chattogram, respectively. After that, we compare the performance of the proposed model against the two baseline models.

5.1 *Dhaka*

For Dhaka, we have tested our model against 230 samples which is 20% of our collected data. The model was trained on using the remaining data. Figure 4 shows the actual versus predicted AQI value curve for Dhaka city. X-axis indicates the number validation samples, and Y-axis shows the normalized AQI values for Dhaka city. Actual and predicted AQI values are identified by yellow and blue colors, respectively.

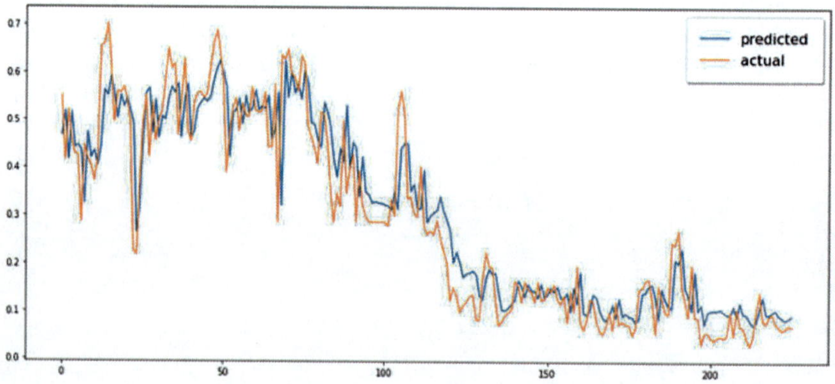

Fig. 4 Actual versus predicted value curve for Dhaka

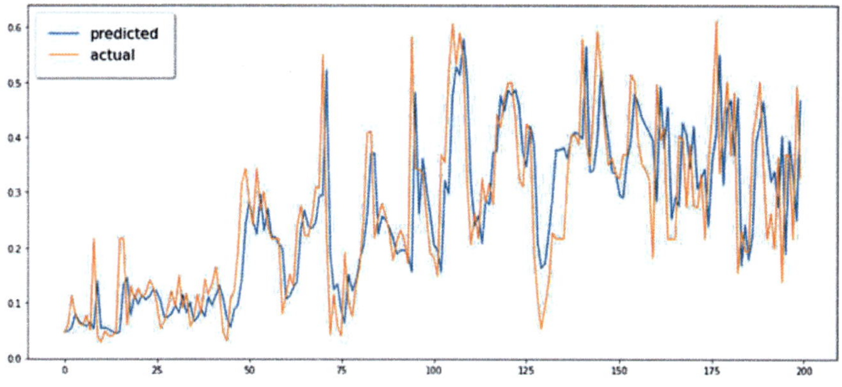

Fig. 5 Actual versus predicted value curve for Chattogram

As we can see from the figure, our model closely predicts the trend of the AQI values over the entire timeframe. The model achieves comparatively smaller error: MSE, RMSE, and MAE values for the model are 0.00514, 0.07172, and 0.053440, respectively.

5.2 Chattogram

In case of Chattogram, we also used 20% of the data for testing and 80% for the training. The model was tested on 204 samples, and the predicted values are shown in Fig. 5. Similarly, X-axis and Y-axis represent the number of samples and normalized AQI values of Chattogram, respectively. Actual and predicted values are shown by yellow and blue colors as before.

Although there are lots of sharp trend changes in the testing samples, our model can successfully predict those changes which can be seen in Fig. 5. The MSE, RMSE, and MAE values of this model are 0.01037,0.10184, and 0.072816, respectively.

5.3 Performance Comparison

It is important to compare the performance of a model against other baseline models to justify how good a model is. As we have mentioned before, we tried to demonstrate the combined power of GRU and LSTM models. In this section, we compare our proposed model against a standalone GRU model and a standalone LSTM model to see whether combining these two models improves the performance or not.

We have used the same model parameters as our proposed model in these two baseline model. Tables 1 and 2 show the the performance comparison of the models in terms of MSE, RMSE, and MAE for Dhaka and Chattogram, respectively.

Table 1 Performance comparison for Dhaka

Models	MSE	RMSE	MAE
Proposed Model	0.00514	0.07172	0.053440
LSTM	0.00849	0.09215	0.071245
GRU	0.01086	0.10419	0.063486

Table 2 Performance comparison for Chattogram

Models	MSE	RMSE	MAE
Proposed Model	0.01037	0.10184	0.072816
LSTM	0.01423	0.11931	0.089882
GRU	0.01230	0.11092	0.083422

It is clear from the tables that our proposed model produces less MSE, RMSE, and MAE for both Dhaka and Chattogram cities. This clearly indicates that GRU and LSTM improve the predictability of the model when they are combined together.

6 Conclusion and Future Work

Air pollution poses a great threat not only to human health but also to the environment and economy. This problem is exacerbated in Bangladesh because of the huge population and fast industrial development. Air quality index (AQI) shows how much polluted the air of a particular place. In this research, we have proposed a novel hybrid model combining GRU and LSTM to predict daily AQI value in the two most largely polluted cities of Bangladesh: Dhaka and Chattogram. We have collected the daily AQI data for Dhaka and Chattogram from the official Web site of Bangladesh Ministry of Environment, Forest and Climate Change within the timeframe 2017–2020. We built our model with four hidden layers: GRU layer, LSTM layer, and two dense layers. After handling the missing values, we have passed the dataset into input layer. These values are passed through GRU, LSTM, dense layers and finally to the output layer after 200 epochs. We have validated the performance of the model for 20% of the data and also showed the actual and predicted patterns. We validated the performance against MSE, RMSE, and MAE metrics. Finally, we have compared our hybrid model against these two standalone models (GRU and LSTM) to justify our idea of combining. The results show that proposed hybrid model produces less errors than the two standalone models and therefore proves its superiority. In future, we will apply our proposed model in all the eight divisional cities in Bangladesh and will employ other machine learning and deep learning approaches [1, 4, 9, 18, 20, 21, 33, 39] to handle the prediction in a more natural way.

References

1. Ahmed, T.U., Shahadat Hossain, M., Alam, M.J., Andersson, K.: An integrated cnn-rnn framework to assess road crack. In: 2019 22nd International Conference on Computer and Information Technology (ICCIT), pp. 1–6 (2019)
2. Ahuja, K., Jani, N.: Air quality prediction data-model formulation for urban areas. In: International Conference on Computer Networks and Communication Technologies, pp. 111–118. Springer, Heidelberg (2019)
3. Al-Ali, A., Zualkernan, I., Aloul, F.: A mobile gprs-sensors array for air pollution monitoring. IEEE Sensors J. **10**(10), 1666–1671 (2010)
4. Andersson, K., Hossain, M.S.: Smart risk assessment systems using belief-rule-based dss and wsn technologies. In: 2014 4th International Conference on Wireless Communications, Vehicular Technology, Information Theory and Aerospace & Electronic Systems (VITAE), pp. 1–5. IEEE (2014)
5. Błaszczyk, E., Rogula-Kozłowska, W., Klejnowski, K., Kubiesa, P., Fulara, I., Mielżyńska-Švach, D.: Indoor air quality in urban and rural kindergartens: short-term studies in Silesia, Poland. Air Qual. Atmos. Health **10**(10), 1207–1220 (2017)
6. Brunekreef, B.: Health effects of air pollution observed in cohort studies in Europe. J. Expos. Sci. Environ. Epidemiol. **17**(2), S61–S65 (2007)
7. Chen, L., Ding, Y., Lyu, D., Liu, X., Long, H.: Deep multi-task learning based urban air quality index modelling. In: Proceedings of the ACM on Interactive, Mobile, Wearable and Ubiquitous Technologies, vol. 3, issue 1, pp. 1–17 (2019)
8. Cho, K., Van Merriënboer, B., Gulcehre, C., Bahdanau, D., Bougares, F., Schwenk, H., Bengio, Y.: Learning Phrase Representations Using RNN Encoder-Decoder for Statistical Machine Translation. arXiv preprint arXiv:1406.1078 (2014)
9. Chowdhury, R.R., Hossain, M.S., ul Islam, R., Andersson, K., Hossain, S.: Bangla handwritten character recognition using convolutional neural network with data augmentation. In: 2019 Joint 8th International Conference on Informatics, Electronics & Vision (ICIEV) and 2019 3rd International Conference on Imaging, Vision & Pattern Recognition (icIVPR), pp. 318–323. IEEE (2019)
10. Environment, C.A.S.: AQI Archives. http://case.doe.gov.bd/index.php?option=com_content&view=category&id=8&Itemid=32. Accessed: 20.08.2020
11. Fu, R., Zhang, Z., Li, L.: Using lSTM and GRU neural network methods for traffic flow prediction. In: 2016 31st Youth Academic Annual Conference of Chinese Association of Automation (YAC), pp. 324–328. IEEE (2016)
12. Gu, K., Zhou, Y., Sun, H., Zhao, L., Liu, S.: Prediction of air quality in Shenzhen based on neural network algorithm. Neural Comput. Appl., 1–14 (2019)
13. Gupta, D., Hossain, E., Hossain, M.S., Andersson, K., Hossain, S.: A digital personal assistant using bangla voice command recognition and face detection. In: 2019 IEEE International Conference on Robotics, Automation, Artificial-intelligence and Internet-of-Things (RAAICON), pp. 116–121. IEEE (2019)
14. Harrell, F.E.: Missing data. In: Regression Modeling Strategies, pp. 45–61. Springer, Heidelberg (2015)
15. He, B.J., Ding, L., Prasad, D.: Enhancing urban ventilation performance through the development of precinct ventilation zones: a case study based on the greater Sydney, Australia. Sustain. Cities Soc. **47**, 101472 (2019)
16. He, B.J., Yang, L., Ye, M.: Strategies for creating good wind environment around Chinese residences. Sustain. Cities Soc. **10**, 174–183 (2014)
17. Hochreiter, S., Schmidhuber, J.: Long short-term memory. Neural Comput. **9**(8), 1735–1780 (1997)
18. Hossain, M.S., Andersson, K., Naznin, S.: A belief rule based expert system to diagnose measles under uncertainty. In: World Congress in Computer Science, Computer Engineering, and Applied Computing (WORLDCOMP'15): The 2015 International Conference on Health Informatics and Medical Systems 27/07/2015-30/07/2015, pp. 17–23. CSREA Press (2015)

19. Islam, M., Hossain, E., Rahman, A., Hossain, M.S., Andersson, K., et al.: A review on recent advancements in forex currency prediction. Algorithms **13**(8), 186 (2020)
20. Islam, M.Z., Hossain, M.S., ul Islam, R., Andersson, K.: Static hand gesture recognition using convolutional neural network with data augmentation. In: 2019 Joint 8th International Conference on Informatics, Electronics & Vision (ICIEV) and 2019 3rd International Conference on Imaging, Vision & Pattern Recognition (icIVPR), pp. 324–329. IEEE (2019)
21. Jamil, M.N., Hossain, M.S., ul Islam, R., Andersson, K.: A belief rule based expert system for evaluating technological innovation capability of high-tech firms under uncertainty. In: 2019 Joint 8th International Conference on Informatics, Electronics & Vision (ICIEV) and 2019 3rd International Conference on Imaging, Vision & Pattern Recognition (icIVPR), pp. 330–335. IEEE (2019)
22. Jiang, F., He, J., Tian, T.: A clustering-based ensemble approach with improved pigeon-inspired optimization and extreme learning machine for air quality prediction. Appl. Soft Comput. **85**, 105827 (2019)
23. Jiang, P., Li, C., Li, R., Yang, H.: An innovative hybrid air pollution early-warning system based on pollutants forecasting and extenics evaluation. Knowl.-Based Syst. **164**, 174–192 (2019)
24. Jumaah, H.J., Ameen, M.H., Kalantar, B., Rizeei, H.M., Jumaah, S.J.: Air quality index prediction using IDW geostatistical technique and OLS-based GIS technique in Kuala Lumpur, Malaysia. Geomatics Natural Hazards Risk **10**(1), 2185–2199 (2019)
25. Khaniabadi, Y.O., Daryanoosh, S.M., Hopke, P.K., Ferrante, M., De Marco, A., Sicard, P., Conti, G.O., Goudarzi, G., Basiri, H., Mohammadi, M.J., et al.: Acute myocardial infarction and copd attributed to ambient so2 in Iran. Environ. Res. **156**, 683–687 (2017)
26. Khaniabadi, Y.O., Hopke, P.K., Goudarzi, G., Daryanoosh, S.M., Jourvand, M., Basiri, H.: Cardiopulmonary mortality and copd attributed to ambient ozone. Environ. Res. **152**, 336–341 (2017)
27. Kingma, D.P., Ba, J.: Adam: A Method for Stochastic Optimization. arXiv preprint arXiv:1412.6980 (2014)
28. Kurt, A., Oktay, A.B.: Forecasting air pollutant indicator levels with geographic models 3 days in advance using neural networks. Expert Syst. Appl. **37**(12), 7986–7992 (2010)
29. Li, H., Wang, J., Yang, H.: A novel dynamic ensemble air quality index forecasting system. Atmos. Pollution Res. (2020)
30. Liu, B., Shi, C., Li, J., Li, Y., Lang, J., Gu, R.: Comparison of different machine learning methods to forecast air quality index. In: International Conference on Frontier Computing, pp. 235–245. Springer Heidelberg (2018)
31. Liu, H., Chen, C.: Prediction of outdoor pm2. 5 concentrations based on a three-stage hybrid neural network model. Atmos. Pollution Res. **11**(3), 469–481 (2020)
32. Mahmood, S.A.I.: Air pollution kills 15,000 Bangladeshis each year: the role of public administration and governments integrity. J. Public Admin. Policy Res. **3**(5), 129–140 (2011)
33. Mahmud, T., Hossain, M.S.: An evidential reasoning-based decision support system to support house hunting. Int. J. Comput. Appl. **57**(21), 51–58 (2012)
34. Ni, X., Huang, H., Du, W.: Relevance analysis and short-term prediction of pm2. 5 concentrations in Beijing based on multi-source data. Atmos. Environ. **150**, 146–161 (2017)
35. Ribeiro, M.C., Pinho, P., Branquinho, C., Llop, E., Pereira, M.J.: Geostatistical uncertainty of assessing air quality using high-spatial-resolution lichen data: a health study in the urban area of Sines, Portugal. Sci. Total Environ. **562**, 740–750 (2016)
36. Shishegaran, A., Saeedi, M., Kumar, A., Ghiasinejad, H.: Prediction of air quality in tehran by developing the nonlinear ensemble model. J. Cleaner Prod., 120825 (2020)
37. Srivastava, N., Hinton, G., Krizhevsky, A., Sutskever, I., Salakhutdinov, R.: Dropout: a simple way to prevent neural networks from overfitting. J. Mach. Learn. Res. **15**(1), 1929–1958 (2014)
38. Tran, K., Bisazza, A., Monz, C.: Recurrent Memory Networks for Language Modeling. arXiv preprint arXiv:1601.01272 (2016)
39. Uddin Ahmed, T., Jamil, M.N., Hossain, M.S., Andersson, K., Hossain, M.S.: An integrated real-time deep learning and belief rule base intelligent system to assess facial expression under uncertainty. In: 9th International Conference on Informatics, Electronics & Vision (ICIEV). IEEE Computer Society (2020)

40. Wang, B., Jiang, Q., Jiang, P.: A combined forecasting structure based on the l1 norm: application to the air quality. J. Environ. Manage. **246**, 299–313 (2019)
41. Wang, J., Du, P., Hao, Y., Ma, X., Niu, T., Yang, W.: An innovative hybrid model based on outlier detection and correction algorithm and heuristic intelligent optimization algorithm for daily air quality index forecasting. J. Environ. Manage. **255**, 109855 (2020)
42. Wang, Y.D., Fu, X.K., Jiang, W., Wang, T., Tsou, M.H., Ye, X.Y.: Inferring urban air quality based on social media. Comput. Environ. Urban Syst. **66**, 110–116 (2017)
43. Wu, Q., Lin, H.: Daily urban air quality index forecasting based on variational mode decomposition, sample entropy and lSTM neural network. Sustain. Cities Soc. **50**, 101657 (2019)
44. Wu, Q., Lin, H.: A novel optimal-hybrid model for daily air quality index prediction considering air pollutant factors. Sci. Total Environ. **683**, 808–821 (2019)
45. Xu, Y., Liu, H., Duan, Z.: A novel hybrid model for multi-step daily AGI forecasting driven by air pollution big data. Air Qual. Atmos. Health **13**(2), 197–207 (2020)
46. Zaremba, W., Sutskever, I., Vinyals, O.: Recurrent Neural Network Regularization. arXiv preprint arXiv:1409.2329 (2014)
47. Zhai, W., Cheng, C.: A long short-term memory approach to predicting air quality based on social media data. Atmos. Environ., 117411 (2020)
48. Zhu, S., Lian, X., Liu, H., Hu, J., Wang, Y., Che, J.: Daily air quality index forecasting with hybrid models: a case in China. Environ. Pollution **231**, 1232–1244 (2017)

HKMS-AMI: A Hybrid Key Management Scheme for AMI Secure Communication

Nahida Islam, Ishrat Sultana, and Md. Sazzadur Rahman

Abstract Smart Grid is entitled to be the new emerging era in the field of the traditional power grid and provides various network functions throughout the generation, transmission, distribution, and consumption of energy by incorporating heterogeneous smart devices and facilitates with real-time data controlling, monitoring, and bi-directional communication flow. With such positive and effective benefits, tremendous security issues have been introduced which disrupts the standard of information flow over communication channels. Advanced Metering Infrastructure (AMI) is responsible for simultaneous data collection, analysis, and demand response and hence, considered to be the most potent component of smart grid prone to various despicable cyber-attacks. This study proposes a Hybrid Key Management Scheme (HKMS) to ensure secure communication among AMI entities: smart meter (SM), meter data management system (MDMS), and trusted authority (TA). The proposed HKMS is composed of seven phases and implemented with the hybridization of Elliptic Curve Diffie-Hellman Key Exchange (ECDH) and Advanced Encryption Standard (AES) algorithms intended to certify perfect forward secrecy, integrity, and resistance against replay, impersonation, and man-in-the-middle attack with least computational and communicational overhead.

Keywords HKMS · AMI · Security · ECDH · AES

N. Islam · I. Sultana · Md. S. Rahman (✉)
Institute of Information Technology, Jahangirnagar University, Dhaka, Bangladesh
e-mail: sazzad@juniv.edu

N. Islam
e-mail: nahida11.islam@gmail.com

I. Sultana
e-mail: ishratshapno@gmail.com

M. S. Kaiser et al. (eds.), *Proceedings of International Conference on Trends in Computational and Cognitive Engineering*, Advances in Intelligent Systems and Computing 1309, https://doi.org/10.1007/978-981-33-4673-4_30

1 Introduction

Over the last couple of years, the Internet of Things (IoT) has been comforting our daily lives with its multi-faceted various applications like transportation, healthcare, power grid, home appliances, agriculture, retailing, etc. One of these applications is the Smart Grid network intensified with smart devices and protocols that facilitates us with automated uninterrupted power supply, black-out resistance, demand response, and customer interaction to the counterfeit of the traditional power grid.

Smart Grid is an accelerated intelligent network with bi-directional communication modernizes the conventional services to overcome the limitations of the traditional power grid, enables active customer participation, flexibility, and demand response, and also prompts renewable energy resources to build up a sustainable economy for a prosperous future. Let's take a look at the global market, in Europe a nearly 200 million smart meters have already been committed to roll-out for electricity by 2020. The research also shows that almost 72% of European consumers will have smart meters for electricity usage [7]. The global smart electricity meters market size was valued at USD 10.0 billion in 2019 and is anticipated to register a compound annual growth rate of 7.8% from 2020 to 2027 [13]. So Smart Grid is reshaping today's demand digitally and it will be the future of the power grid.

Advanced metering infrastructure (AMI), which can also be replaced with the term "smart metering" obliges the Smart Grid's vision with automatic fault detection and self-healing capabilities. It bridges the gap between customer and service provider, consists of smart meters (SM), data concentrator (DC), and meter data management system (MDMS) to facilitate the two-way communication between customer and utility service providers. The SM collects electricity usage information, analyzes, and sends it to MDMS for further operation related to real-time pricing, billing, power outage management, leakage detection, demand forecasting, etc. The whole thing is a dynamic process and is carried out over a public communication channel. So it is very sensitive and prone to heinous attacks like data theft, tampering or modification of usage report, impersonation, unauthorized access, divulging of customer's personal information, meter compromising, false data injection, etc. to destruct infrastructure, power and information flow [12].

Due to such threats, technical security challenges need to be addressed efficiently to ensure the privacy of end-users, certify the confidentiality, authenticity, integrity of meter data, and system resiliency against cyber-attacks. As electricity usage and its embedded technology is going large-scaled day by day, the security issue has become the principal challenge for secure communication. Therefore, with the deployment of AMI, major security requirements must be implemented to provide a secure communication system. Earlier findings of the researchers demonstrate that a key management scheme (KMS) is a very crucial and powerful security mechanism for such a real-time and heterogeneous network capitalizing of cryptographic strength. Yet their studied KMSs has many pros and cons to improve and need to be more coherent, flexible, and time-efficient to provide full proof security.

1.1 Contribution

The contribution of this paper lies with the hybrid encryption mechanism to provide the highest security with two algorithms—Advanced Encryption Standard (AES) and Elliptic Curve Diffie-Hellman Key Exchange (ECDH). Hybrid encryption provides the most security shield as it exploits the advantages of both symmetric and asymmetric encryption [10]. The proposed HKMS has the following key contributions

1. Comprehends mutual authentication and ephemeral key property to certify perfect forward secrecy, message integrity, and resistance against replay, impersonation, and man-in-the-middle attacks.
2. Provides lower computational, communicational, and storage cost, as all the key computations, authentication, and registration procedures are performed by TA except shared key generation which takes acceptable computation and storage cost on SM and MDMS.
3. Two-fold security validation- formal and informal security analysis.

1.2 Outline

The rest of the paper is structured as following: Section 2 includes Literature Survey, System Model is sketched in Sects. 3 and 4 describes Proposed Hybrid Key Management Scheme. Following this, Security and Performance Analysis is presented in Sect. 5 and lastly, Conclusion is drawn in Sect. 6.

2 Literature Survey

With the dynamic security and privacy challenges of Smart Grid, numerous works have been proposed and going on in this field to serve different security purposes. A KMS has been proposed by George et al. [6] for secure hybrid communication between the SM and center station. The scheme ensures confidentiality, authenticity, and message integrity with the hybridization of AES and Rivest, Shamir, Adleman (RSA) algorithms. But the study includes no storage or cost analysis.

Another work presented by Alishahi et al. [3] based on a key agreement scheme using elliptic curve cryptography (ECC). In the pre-deployment phase, initial information is denoted to the smart meter and data center, and a session key is generated and authenticated in the key agreement phase but no time, storage, or network overhead cost has been analyzed to explain the performance. Khasawneh et al. [9] designed a hybrid cryptosystem scheme for AMI secure communication performing data and key encapsulation with corresponding to AES and Elliptic Curve Integrated Encryption Scheme (ECIES). Moreover, Key Derivation Function (KDF) and point multiplication are used to derive key pairs along with Message Integrity Code (MIC) for message integrity and authenticity. The drawback of this work lies with the implicit verification of smart meter and unavailability of the decryption procedure.

In 2018, an ECC-based self-certified key distribution scheme proposed by Abbasinezhad et al. [1], solved the key escrow problem and eliminated the need to address public key infrastructure (PKI) for authentication. However, as per our findings, the scheme fails to provide strong session key security. To provide batch authentication based on a binary tree, a key agreement framework has been proposed by Yan et al. [14] aiming at the computation overhead of the gateway. The framework is designed in such a way that requires only n multiplications and two pairing operations where n is the number of smart meters to reduce the computation cost of the gateway remarkably. Despite such efficiency, higher computation cost needs to be considered while finding out the forged smart meters. In Qi et al. [12], they have proposed a KMS using ECC and Elliptic Curve Qu-Vanstone (ECQV) implicit certificate as the building block to provide a pairing free key agreement protocol that uses two passes to realize mutual authentication. However, the time consumption is more than twice the original Diffie-Hellman protocol.

3 System Model

The system model of the proposed HKMS consists of three entities: SM, MDMS, and TA. TA assumes to be trusted and never compromised, which is in control. SM collects electricity usage data from various smart devices and passes it to DC at each regular period (usually at every 20 mins.). DC fetches all the usage data from SMs and forwards to MDMS to facilitate the customers with better demand response, billing functioning, and other services. Figure 1 depicts the typical set-up of the system model.

4 Proposed Hybrid Key Management Scheme

The proposed HKMS is intended to provide immense security to the system model and is parted into seven phases: set-up, registration, mutual authentication, encryption and forwarding, decryption, and response phase. Figure 2 provides the basic block diagram of the proposed HKMS. The HKMS is capitalized on hybrid encryption of two symmetric and asymmetric algorithms- AES [2] and ECDH [8]. Along with these two algorithms, one-way collision-resistant hash function and digital signature are also implemented to increase security strength. Due to the page limitation, we are skipping the security explanation of these algorithms and primitives.

Table 1 lists the basic notations and their descriptions that have been used in the proposed HKMS.

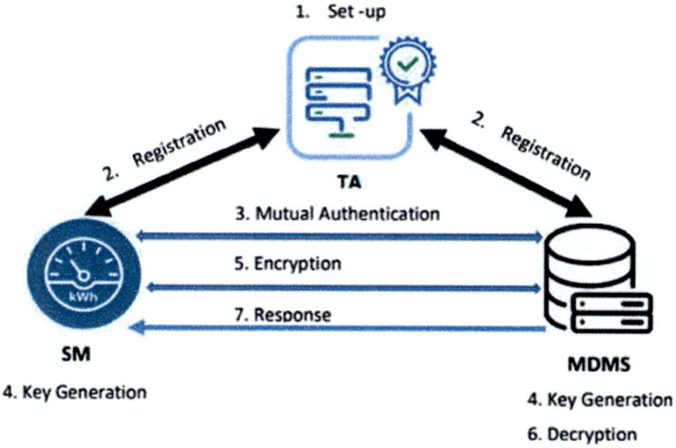

Fig. 1 System model set-up

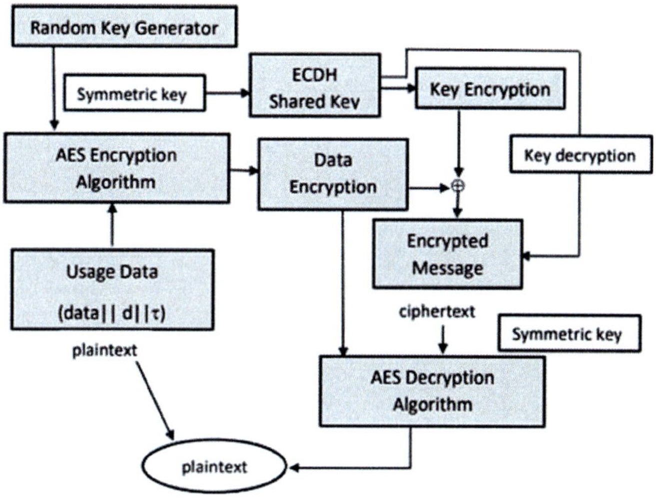

Fig. 2 Proposed HKMS framework for secure communication in AMI

Phase 1: Set-up

Preloaded Values: $E_p(a, b)$, p, Zp, G, $H_i()$, ID_{SM}, ID_{MS}

$H_1(ID_{SM})$ and $H_1(ID_{MS})$ are computed and stored in TA's database. TA selects a random number, $q \in Zp$ and computes its key pair.

$$Pr_{TA} = q$$

$$Pu_{TA} = q.G$$

Table 1 Notations and description used in the proposed HKMS

Notations	Description
ID_{TA}	TA identity
ID_{SM}	SM identity
ID_{MS}	MDMS identity
Pu_{TA}, Pr_{TA}	Public-private key pair of TA
Pu_{SM}, Pr_{SM}	Public-private key pair of SM
Pu_{MS}, Pr_{MS}	Public-private key pair of MDMS
$E_p(a, b)$	A non-singular elliptic curve $y^2 = x^3 + ax + b(modp)$ over finite field Z_p with $4a^3 + 27b^2 \neq 0$
p	A large prime number
Z_p	$0, 1, 2, 3, ...p - 1$, a finite field
G	A base point on $Ep(a, b)$
$t.G$	Point multiplication where random number, $t \in Z_p$
$H_i()$	Collision resistant one-way hash function where $i = 1, 2, 3, 4, ...$
$\|\|$	Concatenation operator
$Sign()$	Digital signature using TA's private key
$HMAC$	Hash-based message authentication code
$E()$	Encryption
$D()$	Decryption
K_{AES}	Message encryption key
S_k	Key encryption shared key
$E(k)$	Key encryption
$D(k)$	Key decryption
ΔT	Maximum transmission delay
Δt	Maximum response delay
ACK	Acknowledgement

$< E_p(a, b), G, H_i(), Pu_{TA} >$ are published publicly. The internal clocks of all entities are synchronized in this phase.

Phase 2: Registration (TA $< - >$ MDMS, SM)

MDMS registration:

Step 1. MDMS sends the registration request to TA with its corresponding ID_{MS}. TA recomputes $H_1(ID_{MS})'$ and compares with its pre-stored hash value.

Step 2. If matches, TA registers the MDMS and sends a response message with a registration token signed by its private key and computes the key pairs for MDMS.

- Request message, $RQMS = (ID_{MS}, reg\_req)$
- Registration token with signature, $RTMS = Sign(Pr_{TA}(H_2(ID_{MS}||q)))$

– TA selects a random number, $m \in Z_p$

$$Privatekey, Pr_{MS} = m$$

$$Publickey, Pu_{MS} = m.G$$

– Response message, $RSM' = (ID_{MS}, RTMS, Pu_{MS}, Pr_{MS}, Pu_{TA})$

SM registration:

Step 1. The same procedure is repeated for SM. After sending request, TA recomputes $H_1(ID_{SM})'$, verifies and sends a response message with a registration token signed by its private key and computes the key pairs for SM.

– Request message, $RQSM = (ID_{SM}, reg\_req)$
– Registration token with signature, $RTSM = Sign(Pr_{TA}(H_2(ID_{SM}||q)))$
– TA selects a random number, $s \in Z_p$

$$Privatekey, Pr_{SM} = s$$

$$Publickey, Pu_{SM} = s.G$$

– Response message, $RSM'' = (ID_{SM}, RTSM, Pu_{SM}, Pr_{SM}, (ID_{MS}, Pu_{MS}))$

The request and response messages are sent over a secure private channel.

Phase 3: Mutual Authentication (SM < – > MDMS)

Step 1. SM sends an authentication message to MDMS encrypted with Pu_{MS}.

– Authentication message, $Auth = (ID_{SM}, RTSM, Pu_{SM})$
– Encrypted authentication message, $E(Auth) = Pu_{MS}(ID_{SM}, RTSM, Pu_{SM})$

Step 2. MDMS receives $Auth$, decrypts with Pr_{MS}, checks $RTSM$, stores ID_{SM} and Pu_{SM} to compute the ECDH shared key and sends the verification message encrypted with Pu_{SM}.

– Verification message, $Verify = (ID_{SM}, Pu_{MS})$
– Encrypted verification message, $E(Verify) = Pu_{SM}(ID_{SM}, Pu_{MS})$

Step 3. SM receives $E(Verify)$, decrypts $E(Verify)$ with Pr_{SM} and stores Pu_{MS} for further ECDH shared key computation.

Phase 4: Key Generation

Step 1. Key for message encryption is generated using any random key generator with symmetric algorithm AES. Generated symmetric key is denoted as K_{AES}.

Step 2. For key encryption, SM and MDMS both compute their shared key using ECDH algorithm.

– Shared key generation by SM: $Sk1$ = SM's private key. MS's public key = $Pr_{SM}.Pu_{MS}$
– Shared key generation by MDMS: $Sk2$ = MS's private key. SM's public key = $Pr_{MS}.Pu_{SM}$

According to ECDH algorithm, $Sk_1 = Sk_2 = Sk$

Phase 5: Encryption (SM − > MDMS)

Step 1. SM collets electricity usage data and creates a usage report with collection date, d and time, τ.

Usage report, $UD = (data||d||\tau)$

Step 2. Encrypts UD with K_{AES} and computes the hash value of encrypted data.

$$E_{UD} = E(K_{AES}, (data||d||\tau)) = E(K_{AES}, UD)$$

$$HMAC(E_{UD}) = HMAC(E(K_{AES}, UD))$$

Step 3. SM encrypts K_{AES} with pre-computed ECDH shared key, Sk.

$$E(k) = E(Sk, (K_{AES}))$$

Step 4. SM transmits the final encrypted message encrypted with Pu_{MS} along with (ID_{SM}, T_1) where T_1 is the transmission time and waits for the response message from MDMS by setting the timer Δt.

$$E_{msg} = E(Pu_{MS}(ID_{SM}||T_1||E_{UD}||HMAC(E_{UD})||E(k)))$$

Phase 6: Decryption

Step 1. After receiving E_{msg} from SM via DC , MDMS decrypts E_{msg} with Pr_{MS} and checks ID_{SM} whether it has performed mutual authentication or not.

Step 2. If the SM is found to be mutually authenticated, it checks $T_2 - T_1 < \Delta T$ (T_2 is the current time).

Step 3. If it validates the condition, the key decryption procedure follows the below using the pre-computed shared key.

$$D(E(k) = (Sk, (E(k))) = K_{AES}$$

Step 4. Computes the hash of the encrypted message, $HMAC(E_{UD})'$ and compares with $HMAC(E_{UD})$ to check the message integrity.

Original usage data, $UD = D(K_{AES}, (E_{UD})) = (data||d||\tau)$

Phase 7: Response (MDMS − > SM)

Step 1. MDMS sends an ACK to ensure that it has received the usage data from SM.

- Response message, $RES = (ACK||t_1||ID_{SM})$
  where t_1 indicates response sending time
- Encrypted response message, $E(RES) = E(Pu_{SM}(ACK||t_1||ID_{SM}))$

Step 2. SM decrypts the $E(RES)$ with Pr_{SM} and checks the condition $t_2 - t_1 < \Delta t$ (t_2 is the current time).

Step 3. If the condition is satisfied, SM is assured that the data has been delivered safely, it doesn't need to be resent anymore. So it deletes the record of the usage data.

Step 4. If the RES violates the condition, SM communicates with TA to identify the problem whether it lies with DC or MDMS. Until the solution comes from TA, SM stops sending the usage data to MDMS.

5 Security and Performance Analysis

1. **Perfect Forward Secrecy**: In the key generation phase, the shared key is generated using the public and private keys of SM and MDMS at each side. Generated shared/session keys are ephemeral as it exploits the strength of ECDH, so for every new session, new key-pairs will be generated for each entity along with a new session key. Thus the session key secrecy is provided as well as ensures perfect forward secrecy.
2. **Replay Attack**: Assume that an adversary intercepts the encrypted and response message at the decryption and response phase. These messages contain timestamps T_1, t_1 respectively, and the unique identity which is a random value. These messages are accepted only after the validation based on timestamps. So the violation of the validation checking leads to the drop of the packets, eventually block the corresponding entity at the repeated attempts and hence prevents replay attack.
3. **Impersonation Attack**: The proposed HKMS is resistant to impersonation attack as each SM is given a unique registration token by TA in the registration phase and only after mutual authentication, MDMS accepts the data packets from authenticated SM. If an adversary tries to impersonate an SM, it will fail due to not having the secret credentials ID_{SM} and $RTSM$.
4. **Man-in-the-Middle Attack**: As the shared key is generated by point multiplication of public and private keys, so it's impossible to discover the shared key or the SM's private key for an adversary to perform Man-in-the-Middle attack.
5. **Message Integrity**: HMAC provides the encrypted message integrity at the decryption phase as it's a one-way function and irreversible. So, the message integrity is ensured in the proposed scheme.
   We are determined to analyze the formal security analysis of the proposed HKMS under Dolev-Yao threat model [4] with verification using ProVerif [5] tool. Moreover, the performance will also be evaluated in terms of computation, communication, and storage cost in near future. Hence, the study presented in this paper can be considered as a pilot study.

6 Conclusion

Even with the existing security schemes, smart grid, and smart meters are not yet secured from cyber villains and their heinous acts. Hence, extraneous security approaches need to be deployed in AMI. This paper aims to develop an efficient HKMS capitalized on the hybrid encryption to ensure information safety with confidentiality, integrity, authenticity, and non-repudiation while maintaining the scalability and robustness of AMI. The studied HKMS of this paper is capable enough to guarantee perfect forward secrecy and resistance against impersonation, replay, and man-in-the-middle-attack. Therefore, the promotion and analysis of computation, communication, and storage cost as well as, the formal security analysis and verification are the future research work of this paper.

References

1. Abbasinezhad-Mood, D., Nikooghadam, M.: An anonymous ECC-based self-certified key distribution scheme for the Smart Grid. IEEE Trans. Industrial Electronics **65**(10), 7996–8004 (2018). https://doi.org/10.1109/TIE.2018.2807383
2. Abdullah, A.: Advanced encryption standard (AES) algorithm to encrypt and decrypt data. Cryptogr. Netw. Security **16** (2017)
3. Alishahi, M., Farhadi, M., Jafari, S., Taghavi, M., Moosavi, H., Mohajerzadeh, A.: An efficient and light asymmetric cryptography to secure communication in smart grid. In: IEEE International Conference on Smart Energy Grid Engineering (SEGE), pp. 248–252 (2017). https://doi.org/10.1109/SEGE.2017.8052807
4. Basin, D., Cremers, C., Meadows, C.: Model checking security protocols. In: Handbook of Model Checking, pp. 727–762. Springer, Cham (2018)
5. Blanchet, B., Smyth, B., Cheval, V., Sylvestre, M.: ProVerif 2.00: Automatic Cryptographic Protocol Verifier, User Manual and Tutorial. Version from 05–16 (2018)
6. George, N., Nithin, S., Kottayil, S.K.: Hybrid key management scheme for secure AMI communications. Proc. Comput. Sci. **93**, 862–869 (2016). https://doi.org/10.1016/j.procs.2016.07.260
7. Ghosal, A. Conti, M.: Key management systems for smart grid advanced metering infrastructure: a survey. In: IEEE Communications Surveys Tutorials, vol. 21, no. 3, pp. 2831–2848, thirdquarter (2019). https://doi.org/10.1109/COMST.2019.2907650
8. Haakegaard, R., Lang, J.: The elliptic curve diffie-hellman (ecdh) (2015). Online at https://koclab.cs.ucsb.edu/teaching/ecc/project/2015Projects/Haakegaard+Lang.pdf
9. Khasawneh, S., Kadoch, M.: A Hybrid Encryption Scheme for Advanced Metering Infrastructure Networks (2017). https://doi.org/10.4108/eai.7-8-2017.152990
10. Khasawneh, S., Kadoch, M.: Hybrid cryptography algorithm with precomputation for advanced metering infrastructure networks. Mobile Netw. Appl. **23**(4), 982–993 (2018). https://doi.org/10.1007/s11036-017-0956-0
11. Mahmud, M., Kaiser, M.S., Rahman, M.M., Rahman, M.A., Shabut, A., Al-Mamun, S., Hussain, A.: A brain-inspired trust management model to assure security in a cloud based IoT framework for neuroscience applications. Cogn. Comput. **10**(5), 864–873 (2018)
12. Qi, M., Chen, J.: Two-pass privacy preserving authenticated key agreement scheme for Smart Grid. IEEE Syst. J., 1–7 (2020). https://doi.org/10.1109/JSYST.2020.2991174
13. Smart Grid Networking Market News, Growth, Price, Demand and Forecast -2025 | ABB, Cisco Systems, Schneider Electric, Huawei Technologies Co. Cole of Duty. https://coleofduty.com/news/2020/04/28/smart-grid-networking-market-news-growth-price-demand-and-forecast-2025-abb-cisco-systems-schneider-electric-huawei-technologies-co/. Last Accessed 7 July 2020
14. Yan, L., Chang, Y., Zhang, S.: An Efficiency Batch Authentication Scheme for Smart Grid Using Binary Authentication Tree, vol. 16, no. 3, p. 7 (2019)

Implementation of Robotics to Design a Sniffer Dog for the Application of Metal Detection

Md. Faisalur Rahman and Md. Mahabub Hossain

Abstract This paper presents a simple design and construction process of a robot that acts like a sniffer dog, named robotic dog, at the detection of any metals which may be harmful for human lives like bombs or valuable things like keys. The robotic dog (RD) consists of a metal detector system and an autonomous car embedded with an electronic buzzer system so that it acts as a real sniffer dog. The robotic vehicle can be controlled from a safe distance using Bluetooth technology. Based on our experimental studies, it was found that the RD can move freely in any direction. At the detection of metal anywhere, the RD locks itself on that position, the buzzer system starts barking like a dog, and it does not move forward until the metal is removed. This RD may have a wide range of uses because of its simplicity and low cost in construction. Moreover, it has the design flexibility to modify its shape to meet the changes of needs.

Keywords Metal detector · Android controlled robot · Bluetooth controlled robot · Arduino UNO microcontroller

1 Introduction

Nowadays, technologies such as robotics, microwave-based communications, and Internet of things (IoT) have become an essential part of our daily life for easy and safe living activities [1–3]. Although new technological innovations can solve many fundamental issues for secure lives of many people in different aspects, the application flexibility and speedy automation of new techs are still major demand of the modern society [4]. Robotics have been becoming a very significant part in several standard applications such as military, child-caring, playing, salvage, searching, and urban hunt day-by-day due to its human alternative activities in household and dangerous environment [5–7].

Md. F. Rahman · Md. M. Hossain (✉)
Department of Electronics and Communication Engineering, Hajee Mohammad Danesh Science and Technology University, Dinajpur 5200, Bangladesh
e-mail: im.mahabub@gmail.com

M. S. Kaiser et al. (eds.), *Proceedings of International Conference on Trends in Computational and Cognitive Engineering*, Advances in Intelligent Systems and Computing 1309, https://doi.org/10.1007/978-981-33-4673-4_31

Efficient design of metal detector system for garbage metal as well as landmine detection in the war-end fields is attractive research area to the researchers due to incorporating different techs [8–10]. Several techniques such as biological, electromagnetic, optical, nuclear, acoustic, or mechanical have been designed and developed for detection of the garbage metals and burial landmines. A technique has become applicable for the detection under some specific conditions depending on the type of soil, environment, size of the mine, and the explosive materials. Generally, most of the technique for landmine detection comprises three functional units: a sensor to capture a pattern of the object, an image processing unit to rearrange the obtained data in a format suitable for detection, and a decision-making unit to confirm whether a landmine exists or not [11]. Landmine and Cluster Munitions Monitor report in 2014 says that more than 110 million deadliest landmines are scattered in the ground in 68 countries [12]. Metal detectors are also important in domestic uses like sharp metals detection in wastage materials, so these cannot cause any harm to anybody. It can use for finding lost metallic valuable things like keys or small ornaments which cannot be located visually for any covering like carpet [13, 14]. So, well and effective designed multi-sensor, vehicle-mounted-based landmines, or metallic object detector robots are the advance innovation compared to the humanitarian detection process of dangerous metallic substances.

In this paper, a simple design and cost-effective construction process of a robotic dog (RD) that behaves like a sniffer dog have been proposed for the detection of metals in its near surroundings. The RD consists of a metal detector system and a microcontroller-based car with an integration of electronic buzzer system. The robotic vehicle can be controlled from a safe distance using Bluetooth technology. The proposed design of the RD confirms less complexity, least cost, moderate speed, high safety, less environmental effect, and it has no bad impact on the environment, whereas many others are complex, expensive and cannot ensure safety. The paper is divided into five sections for clear understanding. This Sect. 1, Introduction highlights about the general ideas and purpose of the proposed work. Sections 2 and 3 describe the design principles, block diagram of working process, detailed circuits connections, algorithm and flowcharts. Finally, in Sects. 4 and 5, it is discussed about the results, performance of final outcomes and conclusive comments.

2 Design Principles and Working Mechanism

The main brain of the proposed metal detecting RD is two Arduinos which are connected cascaded and communicate with each other with a pair of serial ports of Arduinos. The two Arduinos form main subsystems: metal detector subsystem and robotic vehicle with Bluetooth (BT) module subsystem, as shown in Fig. 1.

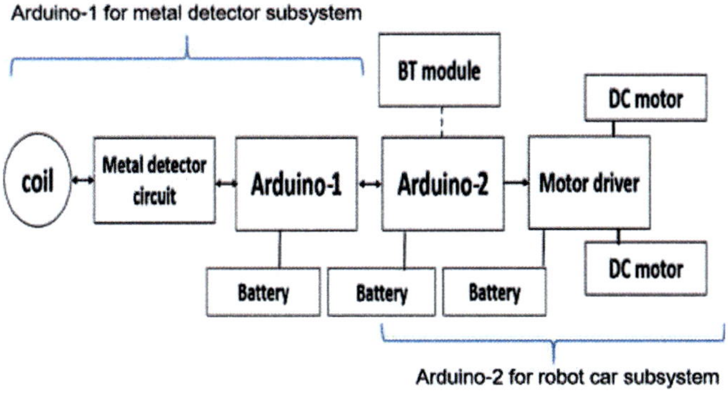

Fig. 1 Block diagram of the proposed metal detecting RD

2.1 Metal Detection and Signaling

The detection technique of the metal detector depends upon the principles of electromagnetic induction. Metal detection subsystem consists of one or more inductor coils with associate circuitry. When the coils are energized with pulsing current, the magnetic field of the coil induces eddy current in the metal. Because of the electromagnetic induction of eddy current, an opposite current generating in the coil creates a disturbance of primary current, which is an indication of the presence of metal. A simple amplifier circuit along with the coil is connected to the Arduino-1. The Arduino-1 is programmed in such a way that it plays a buzzer for siren like barking dog and turns ON a LED and sends an affirmative signal means the metal is found to the robot control system.

2.2 Robotic Vehicle Control and Bluetooth Communication

The Arduino-2 which forms the robot car is connected to a motor driver for driving two DC motors attached with the wheels. There is a ball-bearing wheel for easy movement of the robot. A Bluetooth module is also connected with the Arduino-2 to receive a command from the control of the smartphone. The Arduino-2 checks for the signal coming from the metal detector Arduino-1 whether there is metal or not. When an affirmative signal comes, i.e., the metal is found, from Arduino-1, Arduino-2 will instantly stop the motor wheels on that place and not move further until the metal will be removed. Otherwise, the Arduino-2 continuously communicates with the smartphone via Bluetooth module and receives a command for which direction to go and move forward.

3 Circuit Diagram and Construction of Robotic Dog

The working of the circuit diagram was verified using the simulator software, named 'Fritzing'. All the connections are remained same during the implementation practically. Figure 2a shows the whole circuit diagram and component connections of the metal detecting RD. Figure 2b shows the magnified view of the section-1, which is actually the metal detector circuit diagram of the main circuit diagram (Fig. 2a). One end of the search coil is connected to A0 analog pin of the Arduino-1 through a resistor R1 of 330 Ω. A buzzer and a LED are connected to the pin D12 and D13. The magnified view of section-2 is shown in Fig. 2c, which is the circuit diagram of the motor driving and robot controlling subsystem. The EN1, IN1, IN2, IN3, IN4 and EN2 pins of the motor drive controller are connected to the D6, D7, D8, D9, D10, D11 pins of the Arduino-2. The OUT1, OUT2 pins of the motor drive controlled are connected to one DC motor, and OUT3, OUT4 pins are linked to another DC motor.

3.1 *Algorithm and Flowchart*

Initially, the working of the robot car and metal detection units do not depend each other, i.e., they can perform individually, if the Arduinos are energized by power individually. When a battery is connected to provide power to the Arduinos and motor driver, the whole metal detecting RD system is turned ON to start working. The algorithm steps are described below, and the flowchart is shown in Fig. 3.

Metal Detecting Arduino-1

```
Step 1: Checks whether there is any metal or not;
Step 2: If any metal found, send 'A' to RC Arduino.
(Write 'A' to the serial port), play buzzer, turn on
LED;
Step 3: Else send 'N' to RC Arduino;
Step 4: Repeat from step 1 until power off;
```

Robot Control Arduino-2

```
Step 1: Checks for signal from MD Arduino;
Step 2: If a signal received, read() whether 'A' or
'N';
Step 3: If 'A' then Robot Stop ();
Step 4: Else check the BT module, if available read()
and follow the command;
Step 5: Repeat from step 1 until power off;
```

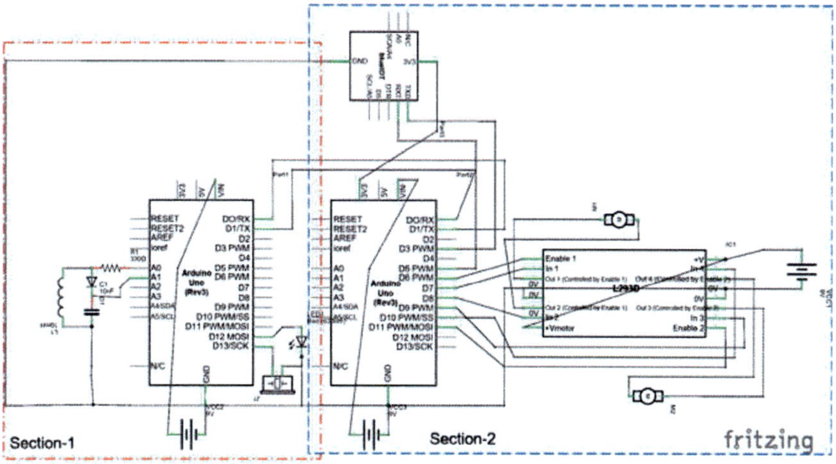

(a)

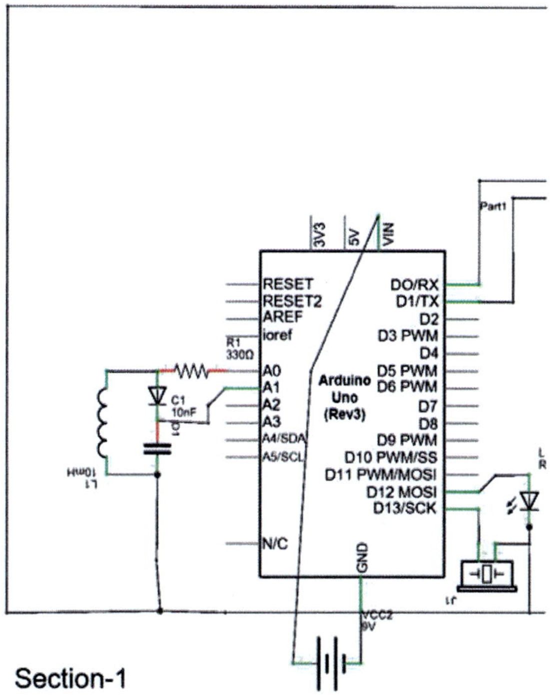

(b)

Fig. 2 **a** The whole circuit diagram of the proposed metal detecting RD, **b** magnified view of the circuit diagram of section-1, and **c** magnified view of the circuit diagram of section-2

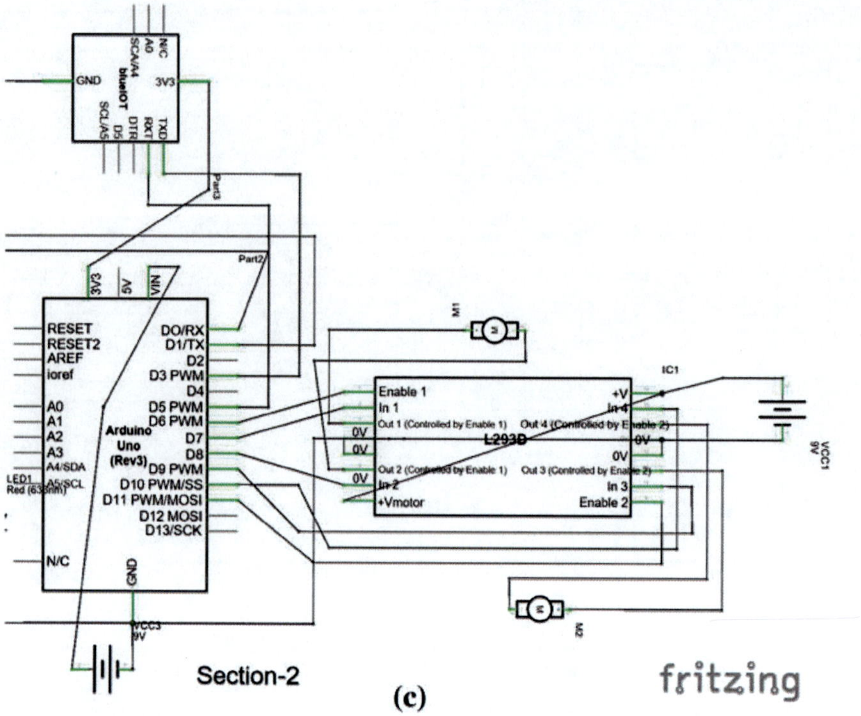

Fig. 2 (continued)

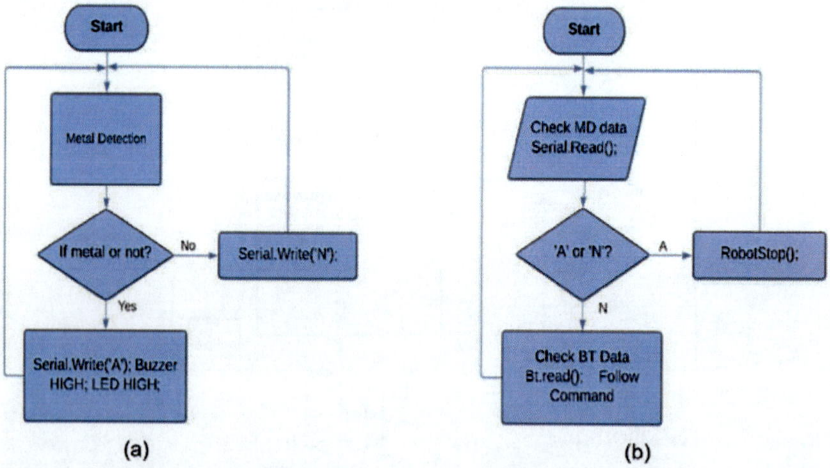

Fig. 3 Flowchart for **a** metal detecting Arduino-1 and **b** robot control Arduino-2

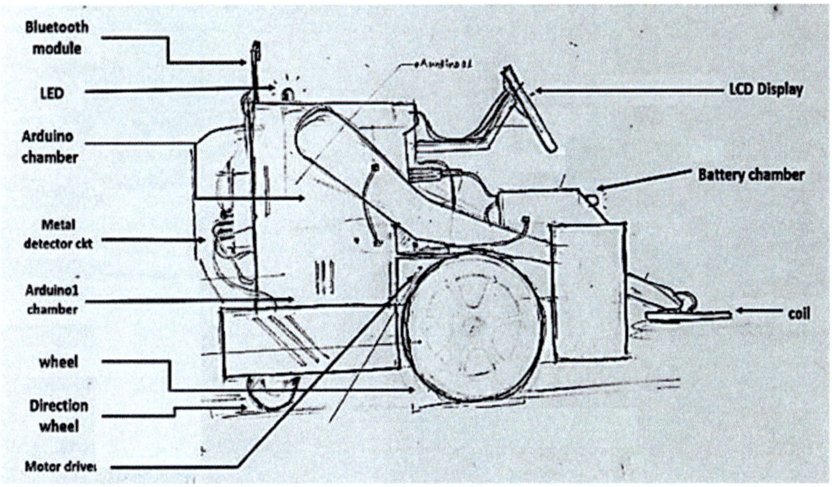

Fig. 4 Body structure of the proposed metal detecting RD

3.2 Constructional Design of the RD Body

We have tried best to keep the RD as small and lightweight as possible. Figure 4 shows the body structure of the proposed robot. The search coil is attached with a fixed hand, DC wheels are used in front side, the Arduinos, motor driver all are managed in chambers. There are chambers for battery metal detection circuits and cables also.

4 Results and Discussions

The RD is working satisfactorily. It can detect metals in its track. When the RD finds any metal, it locks movement on that place, starts playing buzzer, and turns ON the LED light as shown in Fig. 5.

We tested the performance of the RD for the operation of metal detection, Bluetooth communication, and movement control of the car. Table 1 shows the performance of the developed metal detector. It is observed that the metal detector showed a satisfactory performance. It can detect tiny object as coin. But it could not detect very thin things like pin or tiny metal piece less than of 0.5 cm. This is an agreement with the principles of electromagnetism. The field of the search coil is more concentrated at its center and decreases with the distance from it. It is found that the effect of metallic objects increases as the size of this object increases, also the effect of the metallic objects increases as the distance between it and the center of the search coil decreases [15, 16].

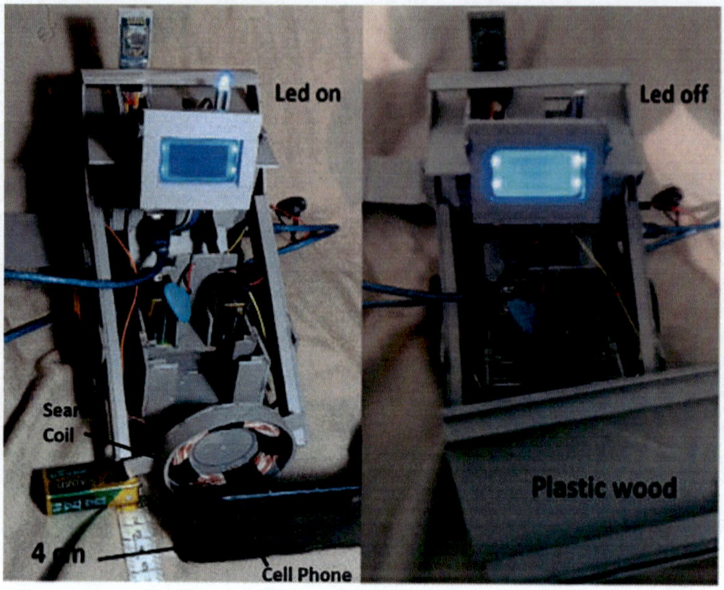

Fig. 5 Metal detection test and detection range measurement

Table 1 Metal detector performance

Sample	Size	Metallic or not?	Detected by 5 cm coil?	Detected by 10 cm coil?
Cell phone	3-in. medium	Yes	Yes	Yes
Scissor	Medium	Yes	Yes	Yes
Scale plastic	Any size	No	No	No
Scale steel	Medium	Yes	Yes	Yes
Circuits	Medium	Yes	Yes	Yes
Glass	Any size	No	No	No
Blade	Small	Yes	Yes	No
Coin	Small	Yes	Yes	No

The performance of Bluetooth communication from Android phone to RD was satisfactory. It was used a baud rate of 115,200, i.e., transmission rate was 115,200 bits/s. As a character requires 96 μs to transmit and it can send only one character for a command (F to forward, B to backward, L to left, R to right, and S to stop), so it requires a delay period of micro-seconds level which is negligible. Figure 6 shows the distance of the detection for two different size detection coils of 5 and 10 cm. It is revealed that the distance of detection is increased when the sizes of object and coil are increased. The size of the search coil needs to be optimized

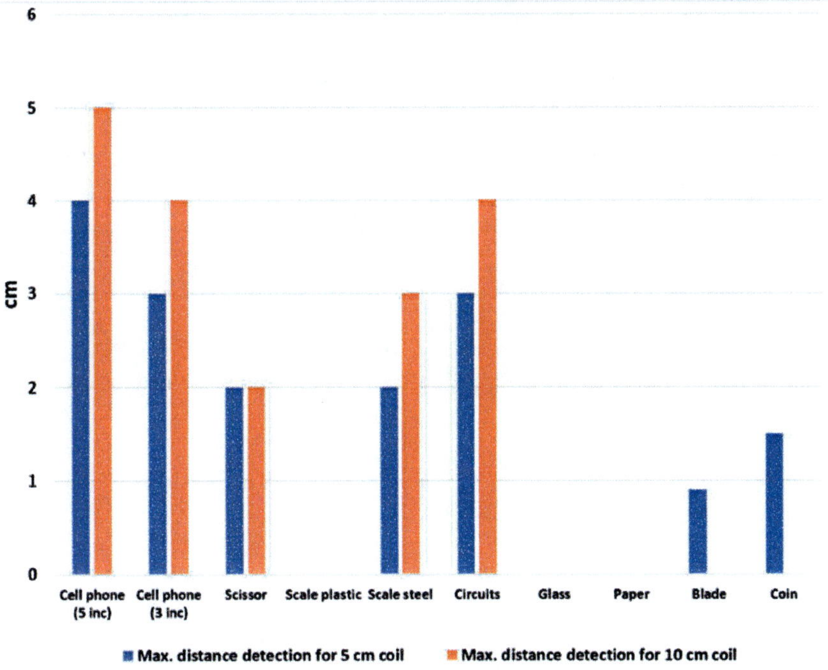

Fig. 6 Comparison of detection range of 5 and 10 cm coil

by careful consideration of design parameters of the robot system, as the large-size search coils demand more power for distance detections.

Figure 7 shows the prototype of the developed RD with metal detector system from different viewing angles and different stages. The body of the RD has been made with wooden plastic sheet. In Table 2, we represent a comparison among our proposed RD with some other related works of metal detector robots on considering the simplicity of design, performance, and cost.

Although the developed RD works perfectly, there are existence of some limitations under some application conditions. After a long time (couple of hours) running in operation, serial data communication via the Bluetooth module becomes slower. So sometimes, it may take a little time to respond. It can be programmed the RD such a way that after sending and receiving one command there is a default brake state to minimize the problem. If any inconvenience happens during Bluetooth communication like long delay or faulty command or missing any command, it would not move to any direction and lock its movement after last successful command's execution until a desired command is received. Moreover, the RD cannot detect nonmetallic explosives or chemical explosives. It cannot distinguish between several types of metals. These limitations can easily be minimized if it is designed with optimization and proper measurement and equipment. There is another limitation in this version which is marked as future scope of work, that it cannot detect obstacle or edge of any

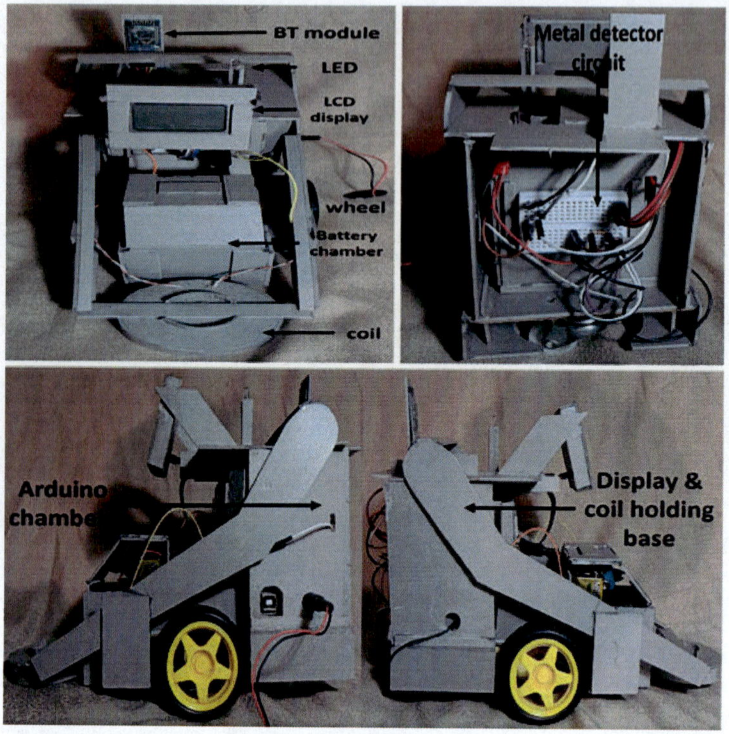

Fig. 7 Photographs of the prototype of the developed RD

Table 2 Comparison among some related works with our RD

Comparison factors	Albert et al. [5]	Suresh et al. [7]	Alauddin et al. [8]	Our proposed robotic dog (RD)
Design simplicity	Complex	Complex	Simpler	Simplest
Performance	Satisfactory	Satisfactory	Satisfactory	Satisfactory
Cost	Costly	Costly	Cheaper	Cheapest ($24)

large hole in its path. As it is a multitasking robot, there is a scope of implementation of several sensors which will add more features to it. A successful use of ultrasonic sensor or IR array sensor is enough to overcome this limitation and would add an extra feature.

5 Conclusion

The main goal of this study is to design and develop a robotic dog (RD) that acts like a sniffer barking dog, when it senses metals near its surrounding on track. The proposed RD is successfully constructed using Arduino, motor driver and Bluetooth module. The developed RD comprises a metal detector circuit that is interfaced with Arduino microcontroller-based control units that produce a buzzer sound, turns ON the LED light as an indication of danger, and stops robot movement to alert the user when a metal object is close to it. An Android application is developed to control the RD from a safe and remote distance for searching metallic objects using smartphones via Bluetooth technology. Thus, the developed robotic dog would be a potential candidate for the application field of metal searching and security.

References

1. Craig, J.J.: Introduction to Robotics: Mechanics and Control, vol. 3, pp. 48–70. Pearson/Prentice Hall, Upper Saddle River (2005)
2. Yang, G.Z., Bellingham, J., Dupont, P.E., Fischer, P., Floridi, L., Full, R., Jacobstein, N., Kumar, V., McNutt, M., Merrifield, R., Nelson, B.J.: The grand challenges of science robotics. Sci. Rob. **3**(14), eaar7650 (2018)
3. Bandyopadhyay, D., Sen, J.: Internet of things: applications and challenges in technology and standardization. Wireless Pers. Commun. **58**(1), 49–69 (2011)
4. Thierer, A.D.: The Internet of things and wearable technology: Addressing privacy and security concerns without derailing innovation. 21 Richmond J Law Technol **6** (2015)
5. Albert, F.Y.C., Mason, C.H.S., Kiing, C.K.J., Ee, K.S., Chan, K.W.: Remotely operated solar-powered mobile metal detector robot. Procedia Comput. Sci. **42**, 232–239 (2014)
6. Ambruš, D., Vasić, D., Bilas, V.: Robust estimation of metal target shape using time-domain electromagnetic induction data. IEEE Trans. Instrum. Meas. **65**(4), 795–807 (2016)
7. Suresh, K., Vidyasagar, K., Basha, A.F.: Multi directional conductive metal detection robot control. Int. J. Comput. Appl. **109**(4) (2015)
8. Alauddin, T., Islam, M.T., Zaman, H.U.: Efficient design of a metal detector equipped remote-controlled robotic vehicle. In: 2016 International Conference on Microelectronics, Computing and Communications (MicroCom), pp. 1–5. IEEE, Jan 2016
9. Boukadida, S., Gdaim, S., Mtiba, A.: Sensor fault detection and isolation based on artificial neural networks and fuzzy logic applicated on induction motor for electrical vehicle. Int. J. Power Electron. Drive Syst. (IJPEDS) **8**(2), 601–611 (2017)
10. Zin, Z.A.M., Ali, F., Ab Kadir, D.: Automatic and obstacle avoidance in metal detector robot. J. Comput. Technol. Creative Content (JTeC) **1**(1), 27–29 (2016)
11. Kasban, H., Zahran, O., Elaraby, S.M., El-Kordy, M., Abd El-Samie, F.E.: A comparative study of landmine detection techniques. Sens. Imaging Int. J. (2010)
12. Landmine Monitor 2014: Monitoring and Research Committee, ICBL-CMC Governance Board, International Campaign to Ban Landmines (2014)
13. Habib, M.K.: Humanitarian demining: reality and the challenge of technology—the state of the arts. Int. J. Adv. Rob. Syst. **4**(2), 19 (2007)
14. Larionova, S., de Almeida, A., Marques, L.: Sensor fusion for automated landmine detection with a mobile robot. In: Using Robots in Hazardous Environments: Landmine Detection, Demining and Other Applications, pp. 147–188. Woodhead Publishing (2011)

15. Hossain, M.M., Bin, W., Kong, S.H.: Large-stroke convex micromirror actuated by electromagnetic force for optical power control. Opt. Express **23**(22), 28358–28368 (2015)
16. Grant, I.S., Phillips, W.R.: Electromagnetism, 2nd edn. Wiley, Hoboken (2013)

Voice Assistant and Touch Screen Operated Intelligent Wheelchair for Physically Challenged People

Md. Shams Sayied Haque, Md. Tanvir Rahman, Risala Tasin Khan, and Mohammad Shibli Kaysar

Abstract Due to age, mishaps, or ailments, many people are suffering from temporary or permanent physical disabilities. They face a lot of difficulties in their day to day body movements. Here, Wheelchair is the mostly used assisting device for handicapped people. The initial Wheelchair (Mechanical) had some limitations as it required another person to operate. As a result, a smart Wheelchair was invented which can be managed by the patients independently. The smart Wheelchair has also provided a way to lead a more active life for impaired people. In this paper, we have proposed an Internet of Things based Intelligent Wheelchair for the physically challenged people with weak limbs and joints. The Wheelchair can be operated using the touch screen or smart voice assistants e.g. Google Assistant, Siri, Alexa, Cortana, and so on. The distinguishing feature of the wheelchair is that it can operate in both online and offline mode as well as it can send periodical and emergency notification of current and critical condition of the patients. Numerical results show the appropriate efficiency of controlling the Wheelchair using touch screen or smart voice assistants for both online and offline environments.

Keywords Smart wheelchair · IoT · Voice assistant · Sensor

Md. S. Sayied Haque (✉)
Jahangirnagar University, Savar, Dhaka, Bangladesh
e-mail: shamssayied@gmail.com
URL: https://shams.com.bd

Md. Tanvir Rahman
Mawlana Bhashani Science and Technology University, Tangail, Bangladesh

R. Tasin Khan
Jahangirnagar University, Savar, Dhaka, Bangladesh

M. Shibli Kaysar
Chittagong University of Engineering and Technology, Chittagong, Bangladesh

M. S. Kaiser et al. (eds.), *Proceedings of International Conference on Trends in Computational and Cognitive Engineering*, Advances in Intelligent Systems and Computing 1309, https://doi.org/10.1007/978-981-33-4673-4_32

1 Introduction

Internet of Things (IoT) is a network of objects with unique identifiers that can communicate in real-time with one another without any human assistance. Here, things can be gateways, variables, equipment, sensors, or any kind of device that can be connected to the Internet. IoT has become a buzzword in technology for the last couple of years and it has already spread globally [1]. Among the many application sectors of IoT and numerous more are also emerging, the healthcare is considered to be the fastest-growing sector with a very high potential rate [2]. IoT is revolutionizing the health care industry as it is providing several benefits over the traditional health care system. Now, patients can be remotely monitored, medical data and management have now become very handy, tracking of patients' records, inventory, and also early detection of any disease is now far better than before. Ramu et al. has proposed a concept where a patient from a rural area can communicate with experienced doctors from urban areas [3]. Furthermore, the authors of [4] have shown a comparative study about how the doctors can get the physiological data of a patient in real-time using wireless technologies e.g. Bluetooth, GSM, and WiFi. The wearable, commercially recognized as Fitness Band, is also another promising application of IoT in the health sector. The band comprises many important sensors such as a global positioning system sensor, heart rate sensor, blood oxygen meter sensor, temperature sensor, step counter, and a lot of other things. These features enable not only the patients to review their health but also the doctors can monitor their every patient remotely [5, 6]. Apart from telemedicine and wearables, there is a list of research associated with ingestible sensors which are an exceptional application of IoT [7–9]. Ingestible cameras, embedded into consumable tablets, are providing the doctors a way to take snaps of internal human organs and thus replacing the traditional procedures of endoscopy [10]. There is also some self-aware early warning score system for personalized healthcare. These approaches can efficiently detect any early signs of deterioration in a patient's health [11]. Physically challenged people are those who cannot do their day to day tasks by themselves. These people are highly dependent on other people or some equipment to do daily activities. In this case, where people are unable to walk due to some specific reason, the most commonly used assisting device is a wheelchair. The manual or traditional wheelchair is either operated by the user or by an external force. But this approach is not considered efficient as patients with weak limbs and bones may face difficulties. As a result, an electrical powered wheelchair is invented as an alternative to this which may resolve the issues of traditional/manual wheelchair [13].

In this paper, we have proposed an IoT based intelligent wheelchair approach for the Physically challenged people. The wheelchair can be controlled via any smart voice assistant e.g. Google Assistant, Siri, Alexa, Cortana, etc. as well as from the WiFi-powered touch screen panel. The wheelchair can continuously monitor the patient and can send regular and emergency updates.

Meanwhile, the rest of the paper is organized as follows: Sect. 2 presents an extensive review of related works, Sect. 3 focuses on the proposed model, Results are presented in Sect. 3.1. Finally, conclusions are drawn in Sect. 4.

2 Review of Related Works

Numerous research works related to a smart wheelchair can be found from a different perspective. Barriuso et al. have proposed an agent-based intelligent interface for wheelchair movement control [14]. Another intelligent wheelchair with GPS and voice recognition is proposed by Nasrin et al. so that the location of the wheelchair can be remotely monitored [15]. The above two approaches require a smartphone with an active internet connection to operate. Ghorbel et al. has also suggested a smart wheelchair with a Micro-controller and LCD for making the overall system more user friendly [16]. To have continuous monitoring of a patient, Reddy and Kumar have explained that if a wheelchair is incorporated with IoT devices along with RFID tags voice recording system and GPRS/GSM module, it is easily possible to connect the wheelchair with the smartphone of the guardian [17]. Avutu et al. has propounded an intelligent wheelchair that will be operated through voice control. The overall cost of this approach is much cheaper as it does not require any WiFi but it is only limited to small local locations [18]. Meanwhile, Raiyan et al. has also come up with another Arduino based smart wheelchair that can be operated without an internet connection. They have also added some interesting features like sending important messages to emergency contacts and auto-detecting and avoiding obstacles and conflicts using ultrasonic sensors [19]. Another obstacle avoidance wheelchair is also presented by Pu et al. which consists of RGB camera, IR camera, LIDAR, and ultrasonic sensors [20]. Masood et al. has thought about a smart wheelchair that can automatically track the sun and charge the battery by itself. The wheelchair can be controlled by both a joystick and a dependent user recognition voice system [21]. Some researchers are also thinking about controlling the wheelchair using brain waves. In that wheelchair, Electroencephalogram signals are collected from the brain and passed to Arduino which then is converted into control signals [22, 23].

3 Proposed System Model

There is a list of components that we have used to build the proposed prototype. These are Arduino, NodeMCU, Ultrasonic Sonar Sensor, Bluetooth Module, Pulse Oximeter, Motors for controlling, and Adafruit Server. To power up these devices, we have used a 9v power supply. Figure 1 shows the overall communication unit wiring. The specific modules, which are connected to Arduino, are explained as follows:

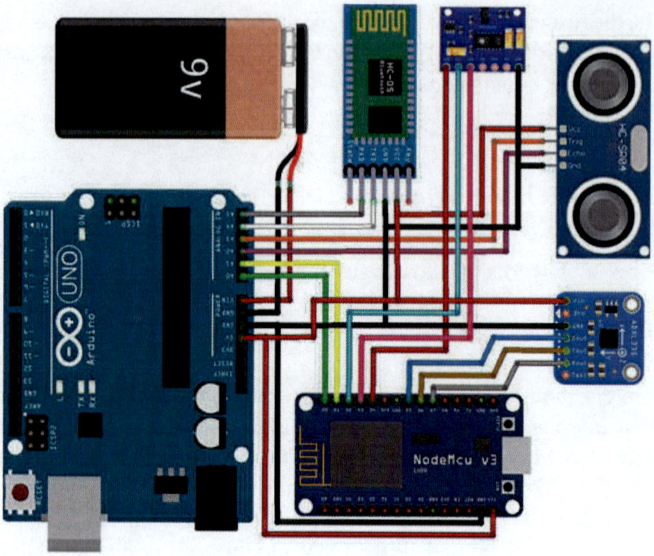

Fig. 1 Overall communication unit wiring

NodeMCU to Arduino Wiring: We have used a 5 v power supply to provide power to NodeMCU and other modules. Arduino Uno Vin supports up to 12 v so we have attached one 9 v battery to this port and Arduino +5 v connected to NodeMCU and other components. Then D0 pin (TX) of NodeMCU is connected to A0 pin (RX) of Arduino. Similarly, the D1 pin (RX) of NodeMCU is also connected to the A1 pin (TX) of Arduino. We have used this connection is to send wheelchair movement command from NodeMCU to Arduino.

Arduino Uno to Ultrasonic Sensor Wiring: The Ultrasonic sensor operates in 5 v power, so we have connected its Vin terminal to a 5 v power our Arduino port and also connected the GND into GND of Arduino. Then for the complete communication between the Arduino and the Sonar Sensor, we have connected the Trig pin and the Echo pin to the A2 and A3 pin of the Arduino respectively.

Arduino Uno to Bluetooth Module Wiring: The Bluetooth module also operates in 5 v and uses serial communication. So, as like as Arduino, the Bluetooth module is also connected as serial communication. The key and the state pin of the Bluetooth module is of no use in this proposed approach. The Bluetooth module communicates with Arduino for offline movement operation.

NodeMCU to Pulse Oximeter Wiring: The Pulse Oximeter is communicating with NodeMCU in a different serial communication way which is different than the other serial communication used in this entire circuit. This communication is called I2C communication. For the connection, the Vin and GND port of the Pulse Oximeter is connected to the 3.3 v Port and GND port of NodeMCU respectively. The SCL and

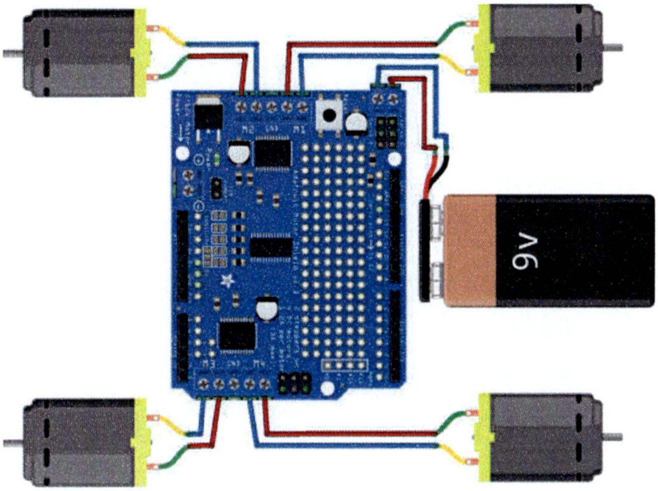

Fig. 2 Motor control unit wiring

SDA terminal of the Pulse Oximeter is also connected to the D2 and D3 terminal of NodeMCU.

Motor Control Unit Wiring: There are four motor output terminals attached to each leg of the wheelchair. These motors are used to move the wheelchair according to the command. Figure 2 shows the Motor control unit wiring.

3.1 System Workflow

Figure 3 explains how the overall system works. The step by step processes are outlined as follows:

- Step 01: The user sends a voice command
- Step 02: The command is received by a smart voice assistant
- Step 03: The voice assistant converts the command into text and compare with IFTTT server applet commands
- Step 04: If there is any match is found, the IFTTT server start running the task according to the instruction.
- Step 05: The IFTTT server then communicates with Adafruit IO
- Step 06: The Adafruit IO passes the value to NodeMCU module
- Step 07: The NodeMCU module also receives values from different sensors. Combining these data with the Adafruit IO instruction the NodeMCU sends the movement direction to Arduino.
- Step 08: Arduino then takes the decision about which motor needs to run in which direction.

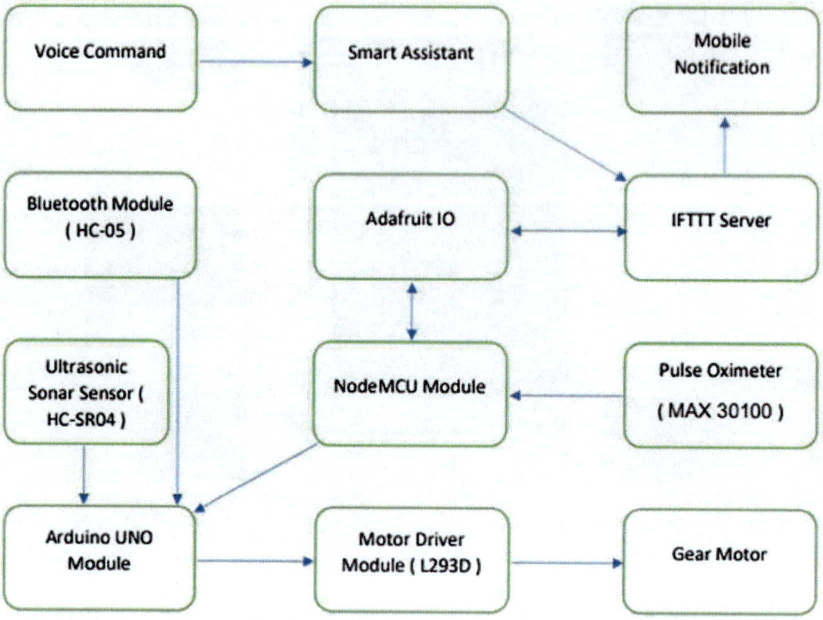

Fig. 3 Workflow of the proposed system

- Step 09: Arduino can stop running wheelchair if it detects any obstacle by the help of the Ultrasonic Sensor.

These are the predefined steps that will be happening all the time. But if the NodeMCU module receives any unstable value response from the sensors which is not present within the reference value of the sensor then it sends a warning notification to mobile by the help of the Adafruit IO and IFTTT server or does some specific tasks which are already defined for the situation. For example, if the patient's heartbeat goes out of the reference value of normal patient then it sends a warning notification to mobile. If any obstacles appear within 1 foot of the front of the wheelchair, the Arduino sends a stop command to Motor Driver Module. On the other hand, if there is no internet connection then Arduino Uno can receive wheelchair movement direction by the help of the Bluetooth module.

For the data flow and communication system, when the wheelchair starts immediately NodeMCU starts trying to communicate with the Adafruit MQTT server. If it gets failed then it tries to reconnect after 5 s. When it gets connected then it searches for data change in all server feeds link which is subscribed to the NodeMCU. If there is any change occurred then NodeMCU receives that new value and compares conditions for matching. If it gets matched then it checks data given by sensor values. After comparing all of those data it sends the final result returned from the algorithm to the Arduino module using serial communication. The latency between server and NodeMCU can be increased or decreased according to the quality of the internet connection. But communication between sensors, Bluetooth module, and Arduino module works almost instant because of serial and I2C communication. There are two sensors connected to the NodeMCU which are Pulse Oximeter and Accelerometer. One Ultrasonic Sonar sensor is attached to the Arduino as it can measure the distance. The reason for using Pulse Oximeter is to detect the patient's heart rate and oxygen level of blood. If the heart rate and oxygen level gets goes outside of the reference value then NodeMCU can send a warning notification to the family members via Adafruit and IFTTT server. We have also used Accelerometer sensors for the detection of falls or accidents of the wheelchair. We can use ECG sensors which measures heart rate and pulse quality but SpO2 can also measure the Oxygen level of blood for the patient which can give us an idea about when the patient needs oxygen support.

4 Results and Discussion

We have prepared a prototype of the intelligent wheelchair as shown in Fig. 4. The wheelchair can be controlled via WiFi and mobile data. So, firstly we have measured the latency for connecting to the Adafruit IO server while using WiFi and Mobile data considering our simulation environment. For WiFi, Fig. 5 shows that the minimum and maximum amount of latency is 275 ms and 282 ms respectively while the average latency is at 278 ms. From Fig. 6 it is seen that these values for Mobile data are always higher when compared to WiFi. For Mobile data, the minimum, maximum,

Fig. 4 Prototype of the intelligent wheelchair

Fig. 5 Ping latency to adafruit IO server using mobile wifi

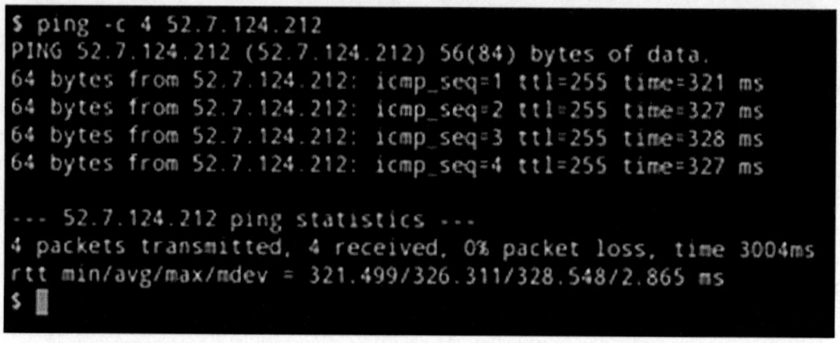

Fig. 6 Ping latency to adafruit IO server using mobile data

and average latency is 321 ms, 328 ms, and 326 ms respectively. These latency are dependent on internet connection. To control the wheelchair, the commands will be sent by the user via touch control or voice control. These two controlling methods can be initiated through the WiFi network or the internet.

Table 1 illustrates the amount of required latency for both touch control and voice control considering the connection through WiFi. While considering mobile data, the scenario is outlined in Table 2. From both of the tables, it is visible that using mobile data is requiring additional latency to communicate with the NodeMCU.

With the developed prototype wheelchair, we have also conducted some angle accuracy test. For this test, we have instructed the wheelchair to turn left or right and

Table 1 Comparison of latency required from Mobile to NodeMCU using WiFi

Trial	Touch control latency (ms)	Voice control latency (ms)
T-1	610	6230
T-2	650	5910
T-3	720	6590
T-4	630	7230

Table 2 Comparison of latency required from mobile to NodeMCU using mobile data

Trial	Touch control latency (ms)	Voice control latency (ms)
T-1	750	7910
T-2	830	7560
T-3	700	8570
T-4	770	7830

Table 3 Angle accuracy test for "Turn Left" command

Trial	Expected rotation (°)	Result of proposed wheelchair (°)	Accuracy (%)
T-1	90	81	90
T-2	90	87	96
T-3	90	76	84
T-4	90	85	94

Table 4 Angle accuracy test for "Turn Right" command

Trial	Expected rotation (°)	Result of proposed wheelchair (°)	Accuracy (%)
T-1	90	73	81
T-2	90	85	94
T-3	90	68	75
T-4	90	87	96

calculated the amount of angle the wheelchair has rotated. The overall efficiency of the wheelchair for the "Turn Left" command is explained in Table 3. Table 4 shows the efficiency of the wheelchair for "Turn Right" command. This efficiency can be increased close to 99% if we use accelerometer sensor for measuring rotation angle and stop motor rotation when the accurate angle is reached.

5 Conclusion

The wheelchair is the best assisting device for physically impaired people. If this device can be more technology-powered, the disabled person can lead a more realistic life than they live with a traditional wheelchair. Here, our proposed wheelchair can be a great solution. The proposed approach is also designed because the guardian can remotely control and monitor the person sitting onto it. The wheelchair can also be controlled without an active internet connection. The best part is anyone can connect the wheelchair with a smart assistant (Google Assistant, Siri, Alexa, Cortana, etc.).

We have used NodeMCU module which can't even run two loops or programs at the same time and it is also very slow considering the communication with the IoT server. So, in future, we can use an advanced IoT board like "Raspberry Pi" to run more than one command at a time. We can further improve the functionality of the wheelchair by integrating Google AI which may store all the movement records and can assist the voice commands later on. We may also need to change google assistant with any other voice interpreter service because it's very slow in responding. As our wheelchair is a sensitive item so it needs a very fast and accurate voice interpreter. The Stepper motor can be used to get high accuracy in the left and right movements.

References

1. Bajaj, A., Bhatnagar, M., Chauhan, A.: Recent trends in internet of medical things: a review. In: Patnaik, S., Yang, X.S., Sethi, I. (eds.) Advances in Machine Learning and Computational Intelligence. Algorithms for Intelligent Systems. Springer, Singapore (2021). https://doi.org/10.1007/978-981-15-5243-4_61
2. Doukas, C., Maglogiannis, I.: Bringing IoT and cloud computing towards pervasive healthcare. In: 2012 Sixth International Conference on Innovative Mobile and Internet Services in Ubiquitous Computing, Palermo, pp. 922–926 (2012). https://doi.org/10.1109/IMIS.2012.26.
3. Ramu, R., Kumar, A.: Real Time diagnosis of Rural Cardiac Patients Through Telemedicine: Proceedings of the International Conference on Modelling and Simulation (MS-17) (2019).. https://doi.org/10.1007/978-3-319-74808-5_44
4. Abdellatif, M., Mohamed, W.: Telemedicine: an IoT based remote healthcare system. Int. J. Online Biomed. Engi. (iJOE) **16**, 72 (2020). https://doi.org/10.3991/ijoe.v16i06.13651
5. Kassem, A., Tamazin, M., Aly, M.H.: An intelligent IoT-based wearable health monitoring system. In: Farouk, M., Hassanein, M. (eds.) Recent Advances in Engineering Mathematics and Physics. Springer, Cham.(2020). https://doi.org/10.1007/978-3-030-39847-7_29
6. Elango, K., Muniandi, K.: A Low-cost wearable remote healthcare monitoring system. In: G.R. Kanagachidambaresan (ed.) Role of edge analytics in sustainable smart city development (2020). https://doi.org/10.1002/9781119681328.ch11
7. Flore, J.: Ingestible sensors, data, and pharmaceuticals: Subjectivity in the era of digital mental health. New Media Soc. (2020). https://doi.org/10.1177/1461444820931024
8. Beardslee, L., Banis, G., Chu, S., Liu, S., Chapin, A., Stine, J., Pasricha, P., Ghodssi, R.: Ingestible sensors and sensing systems for minimally invasive diagnosis and monitoring: the next frontier in minimally invasive screening. ACS Sens. **5**(4), 891–910 (2020)
9. Kamal, S., et al.: Perceptions of people living with HIV and HIV healthcare providers on real-time measuring and monitoring of antiretroviral adherence using ingestible sensors: a qualitative study. AIDS Res. Treat. 1 (June 2020). https://doi.org/10.1155/2020/1098109
10. Hooman, H., Robertson, T.L., Moon, G.D., Au-Yeung, K.-Y., Zdeblick, M.J., Savage, G.M.: An ingestible sensor for measuring medication adherence. IEEE Trans. Biomed. Eng. **62**(1) (2014)
11. Azimi, I., Anzanpour, A., Rahmani, A.M., Liljeberg, P., Tenhunen, H.: Self-aware Early Warning Score System for IoT-Based Personalized Healthcare, in eHealth 360, pp. 49–55. Springer, Cham (2017)
12. Santos, A., Macedo, J., Costa, A., João Nicolau, M.: Internet of things and smart objects for M-health monitoring and control. Proc. Technol. **16**, 1351–1360 (2014)
13. Kedar, S., Darshitkumar, S., Anil, S., Jairaj, P., Rovina, D.: Smart wheelchair: a literature review. Int. J. Inf. Commun. Technol. (IJ-ICT) **7**(63) (2018). https://doi.org/10.11591/ijict.v7i2.pp63-66

14. Barriuso, A.L., Pérez-Marcos, J., JiménezBravo, D.M., González, G.V., De Paz, J.F.: Agent based intelligent interface for wheelchair movement control. Sensors **18**(1511), 1–31 (2018)
15. Aktar, N., Jahan, I., Lala, B.: Voice Recognition based intelligent wheelchair and GPS tracking system. In: 2019 International Conference on Electrical, Computer and Communication Engineering (ECCE), pp. 7–9 (2019)
16. Ghorbel, M., Pineau, J., Gourdeau, R., Javdani, S., Srinivasa, S.: A decision-theoretic approach for the collaborative control of a smart wheelchair. Int. J. Soc. Rob. **10**(1), 131–145 (2018)
17. Reddy, K.B.P., Kumar, K.K.: A smart wheelchair system with social media update. Ind. J. Sci. Technol. **9**(30), 1–5 (2016)
18. Avutu, S.R.: Voice control module for low cost local-map navigation based intelligent wheelchair. 2017 IEEE 7th International Advance Computing Conference, pp. 609–613 (2017)
19. Raiyan, Z., Nawaz, M.S., Adnan, A.K.M.A., Imam, M.H.: Design of an Arduino based voice-controlled automated wheelchair. In: IEEE Region 10 Humanitarian Technology Conference (R10-HTC). Dhaka, vol. 2017 (2017), pp. 267–270. https://doi.org/10.1109/R10-HTC.2017.8288954
20. Pu, J., Jiang, Y., Xie, X., Chen, X., Liu, M., Xu, S.: Low-cost sensor network for obstacle avoidance in share-controlled smart wheelchairs under daily scenarios. Microelectron. Reliab. **83**, 180–186 (2018)
21. Masood, M.R., Sufyan, A., Khalid, U., Khan, M.M., Amin, S.: Automatic sun tracking smart wheel chair with real time navigation. In: Bajwa I., Sibalija T., Jawawi D. (eds.) Intelligent Technologies and Applications. INTAP 2019. Communications in Computer and Information Science, vol. 1198. Springer, Singapore (2020). https://doi.org/10.1007/978-981-15-5232-8_5
22. Swee, S.K., You, L.Z., Kho, D.T.K.: Brainwave controlled electrical wheelchair. MATEC Web of Conferences. **54**, 03005 (2016). https://doi.org/10.1051/matecconf/20165403005
23. Satti, A.R., Coyle, D., Prasad, G.: Self-paced brain-controlled wheelchair methodology with shared and automated assistive control. In: IEEE Symposium on Computational Intelligence, Cognitive Algorithms, Mind, and Brain (CCMB). Paris, vol. 2011, pp. 1–8 (2011). https://doi.org/10.1109/CCMB.2011.5952123

Virtual Heritage of the Saith Gumbad Mosque, Bangladesh

Md. Masood Imran and Minar Masud

Abstract The notion of this paper is to introduce the Virtual Reality (VR) in archaeological research to visualise and safeguard the cultural heritage virtually. 3D visualisation of cultural heritage with scientific transparency is the ultimate motto of the London Charter and Sevilla Principle, which have been followed in this research work. Undoubtedly, 3D technology, open up an avenue to visualise, interpret and safeguard the archaeological understanding in virtual reality. In the field of archaeological research and cultural heritage studies, VR is becoming a powerful tool. Typically, archaeological understanding is the mood of prediction, which connotes the futuristic approach towards the past. They want to understand and visualise the past. In this paper, Shait Gumbad Mosque, which is the part of the World Heritage Site of Mosque Cities of Bagerhat, Bangladesh, is anticipated the VR to visualise and safeguard the Shait Gumbad Mosque by using the existing archaeological and VR technological knowledge.

Keywords Virtual reality · Virtual heritage · Shait gumbad mosque · London charter · Sevilla principle · World heritage.

1 Introduction

Worldwide armed conflicts essentialised the reasons for protecting the cultural heritage; particularly the WW II. Hague Convention (HC) of 1954 ratified to protect the cultural heritage from suffering the widespread destruction [11]. This convention highlighted that human lives and history both have lost due to the war. HC issued and ratified the first protocol to protect the cultural properties during the times of armed conflict. By this time, in 1972, UNESCO took the initiative to protect the

Md. M. Imran (✉)
Department of Archaeology, Jahangirnagar University, Dhaka, Bangladesh
e-mail: masoodimranmannu@juniv.edu

M. Masud
Institute of Information Technology, Jahangirnagar University, Dhaka, Bangladesh

M. S. Kaiser et al. (eds.), *Proceedings of International Conference on Trends in Computational and Cognitive Engineering*, Advances in Intelligent Systems and Computing 1309, https://doi.org/10.1007/978-981-33-4673-4_33

cultural heritage not only from the armed conflicts but also from the vandalism, loot, illicitly trade spoiling and started to make a World Heritage List (WHL). The second protocol has been announced after the 56 years of the first protocol of HC. It was in 1999 [5]. Due to the lack of implementation guidelines, the first protocol of HC remains neglected widely. UNESCO pointed out the problems and set a committee in 2004 and issued a guideline of implementation in 2008 [8]. It resulted to initiate a project on War Free World Heritage Listed Cites (WFWHLC) [3] funded by the European Union in 2009 [8]. These types of enterprises i.e. set-regulations, declarations, recommendations, in addition to protocols, principles, along with codes have been considered as secondary charters in many cases.

The last couple of decades, armed conflicts spread in many archaeologically sensible zones in the world. UNESCO did not initiate any protocol directly regarding computer-based visualisation of cultural heritage, which should be generated a trustworthy digital repository for conserving the human-created genius. The notion of digital heritage by UNESCO is; "digital heritage is made up of computer-based materials of enduring value that should be kept for future generations. Digital heritage emanates from different communities, industries, sectors and regions. Not all digital materials are of enduring value, but those that require active preservation approaches if continuity of digital heritage is to be maintained [4]." About protecting the archaeological heritage, there are a couple of documents from Europe, America and Asia have been played a deciding role. The computer-based visualisation of the London Charter of 2006 has been taken place to visualise and protect the cultural heritage through computer-aided simulations []. It is similar to the authentication regulated by following the Nara document of Japan [10]. Besides the London charter, there is another document developed particularly define the international borderline of Virtual-Archaeology, is acknowledged as the Seville Principles. It also maintained the usual nomenclature of ICOMOS like the London Charter. This document has been prepared as a set of principle by giving importance to the set of recommendations, which were seeking to advance the applicability of the London Charter in the study of archaeological heritage.

In this paper, the virtual reality of Shait Gumbad Mosque has been developed for visualisation and protection of listed world heritage by following the London Charter and Sevilla Principles [1]. However, both charter and principles have not been ratified by the ICOMOS general assembly yet. In 2017 UNESCO opened a poll where 77% of respondents out of 33 countries responded [8]. In fact, as a visionary document, which is the ahead of it's time to use innovative technological pieces of knowledge in the field of cultural heritage. The London charter principles aim to improve the intensities of scientific clearness present in 3D visualisations of cultural heritage. Primarily the same aim belongs to this paper. Because, the present revolution of information technology, particularly the heyday of 3D technology is a specialised way to visualise the archaeological events in virtual reality. And on the other hand, the visualization of virtual models in numerous application areas allows professionals to both study and intermingle with their study objects, while moreover authorizes the usual public to study these virtual environs, thus endorsing their social, cultural and scientific contribution. Normally archaeological understanding is a futuristic approach towards the past. It means a reverse journey. They want to understand the past. Usually, archaeological understanding is express as the model of prediction that indicates the relative potential of encountering an archaeological site. Primarily, predictive location models were attempted to locate and pattern of distribution of archaeological records [6].

Most archaeological sites encompass different structures that are brutally injured or destroyed, which does not permit to grasp them in its original excellence. The crazy visualization of their virtual models, incorporated with the authentic or remain structures, composes an essential added value acknowledging the general public to have an enhanced perception and understanding of the archaeological site.

In this paper, Shait Gumbad Mosque is inscribed the world heritage list of UNESCO as the group of buildings of Bagerhat, Bangladesh, is anticipated the VR by using the existing archaeological and technological knowledge. As new technology in archaeological heritage research of Bangladesh, VR opens a new avenue to visualise and safeguard the past. This paper shows the pathway to share the knowledge with peers as a novel manifesto for the cultural heritage related research and this work might be considered as introductory research in the evolving field of digital cultural heritage protection and restorations.

The paper of VR of Shati Gumbad Mosque has been structured in the following ways: Section two and three have been formed as a conceptual format to understand the architectural importance of Shait Gumbad Mosque and conceptualised the VR in the field of cultural heritage studies. Section 4 and 4.1 are the main part of the process of development of VR by using the application software and illustrate the final output of the VR of Shait Gumbad Mosque. The final segment is the concluding part of this paper.

2 About Shait Gumbad Mosque

The location of Shait Gumbad Mosque near to the district town of Bagerhat in Bangladesh. It is the largest mosque of Bagerhat, and this town is the famous to worldwide as Mosque city. It is a listed world heritage site by UNESCO and Shait Gumbad Mosque became identical for Mosque City of Bagerhat [6]. Architectural uniqueness made this mosque special. It has been known as Khan Jahan style, because of this city established by the warrior-saint Ulugh Khan-i-Jahan in 1459 [7]. Stylistically Shait Gumbad Mosque resembles with the sultanate style architecture. Even though, the architectural differences authenticated its excellence as the flavour of the traditional orthodox mosque plan to list it as World Heritage of UNESCO [7].

A colossal multi-domed roof characterised a mythical perspective of this mosque. Popularly people believed that because of the number of the dome, where sixty synonymous in Bangla language is Shait, therefore, they called it as Shait Gumbad means sixty-domed mosque. Interestingly, here are eighty-one domes in total. The entire roof has shaped by seventy-seven dome, where seven domes are shaped as chau-chala Bangla hut type in the middle row of domes (Fig. 1a, b). Because of that

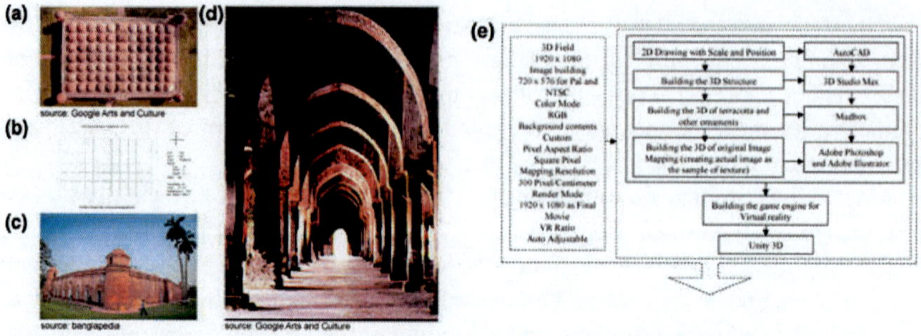

Fig. 1 **a** Birds eye view **b** Ground plan **c** Brick built massive structure **d** An aisle **e** Model

some historians speculated, it might be Sat means seven Gumbad, which colloquially pronounced as Shait Gumbad. Rest of the four domes, which comparably smaller size is on the four-corner tower. In the mosque architecture of Shait Gumbad, there are sixty pillars used to contain the seventy-seven domes. These sixty pillars have been generated another speculation that the colloquial pronunciation transcends the bangla word Khumbaz means pillar in Gumbad. Therefore, Shait Khumbaz fluidified through water-mouth as Shait Gumbad Mosque.

Fig 1b shows the ground plan, which is describing the main gateway (7.92 m × 2.44 m) in the eastern enclosure by following the cardinal central archways and another one was in the north enclosure but these days it has been disappeared. It 48.77 m long in the north to south direction and west to east length is 32.92m. It is an entirely brick-built massive structure in the Sultanate period of Bangla (Fig. 1c). There are four corner towers, which roofs are projected in the level of the main roof with small-sized domes.

Already mentioned that domes have been roofed on the sixty stone pillars, which have been distributed in the north to south as seven aisles (Fig. 1d). These have been generated eleven bays from east to west. In the central position by following the middle aisle in the west enclosure contained the central mihrab. Along with this, there are a total of ten semi-circular mihrabs have been installed. But central mihrab is big size than the others. The brick walls are again unusually thick and externally are slightly tapering. In the western wall, there is arched entrance by following the design of bays and pillar partitions. The western facade has one strange little entrance, facilitating simple entree for the prayer chief. There are two corners out of the four, are roundabout and tapering brick turrets, two of which on the eastern side contain circular staircases. From here, the faithful can be called to their prayer. The forest of stone pillars together with the eleven unusually ornated stone and terracotta mihrabs, makes the interior very special. Although this building has had a chequered history—there are reports in 1871 that the mosque was covered in jungle and in 1903 it has been considered that the mosque could only be preserved as a partial ruin—it survives in reasonable condition as a representative of an important period of architecture.

Shait Gumbad Mosque has been decorated by terracotta and brick. Stone carving exemplified in low relief. Still, some of the ornamentations have been survived in arches of doorways, mihrabs, intersecting arches of domes, the chau-chala vault's interior, in the cornices of the entire structure and in the corner towers from disappeared by ravages of time.

There are three subtly protruded horizontal bands just above the archway, the bottom one portrays a variety of descending flowers, the middle one displays lozenges alternating with small rosettes and the upper one is filled with a sequence of four-petalled flowers. There are two slightly submerged narrow panels in between those bands. The upper panel is adorned with loop-shaped scrolls illustrating lotus flowers. A series of arched tri-lobbed niches is portrayed by the remaining panel and are again enriched with designs such as palm trees, interlocking squares in the middle with small rosettes, and flowers with leafy plants. The whole composition is topped by a magnificent triangular pediment, whose embellishment has now totally vanished. More or less similar terracotta patterns have been used to design the remaining arches at the door. The spandrels and upper parts of the doorway arches were painted inside mainly with terracotta, whose motifs differ from entrance to entrance.

The qibla wall is internally embellished with ten arched-mihrabs with ornamental engravings. Like its adjoining counterparts and those of other Khan Jahan era houses, the central mihrab is constructed entirely of Gray sandstone and its ornamentation is in shallow and low relief Muslim-style carving. Many of the architectural motifs have vanished but much remains in a rotting condition. The large multi-cusped arched mihrab's arises from two faceted decorative pilasters in stone. The spandrels of this arch are also portrayed with a tree motif, which is further defined by branches with leaves and flowers growing out from the vases. A parallel band lined with a series of lozenges alternating with rosettes runs directly above the pinnacle of the arch. The semicircular mihrab niche is divided internally into two halves by a slightly elevated band carved with alternating rosettes with a frieze of lozenges. By a slightly raised band, the semicircular qibla niche is internally split into two halves and the band is carved with rosettes, alternating with a frieze of lozenges. The top portion of the niche is decorated with rows of rosettes, net patterns and lotus petals, which take the shape of a half-dome. Two horizontal rows of rectangular panels curve the lower half of the niche, which is semi-circular in form, nine in each row. Each of these panels is divided by a thin band from the other, reveals a cusped arch depicting a rosette in the middle and a tree motif interlacing identical small rosettes in the spandrels. A single sculpted chain hangs down from the centre of the splitting band of the niche ending with an oblong pendant. This pendant is illustrated with a cusped arch with spandrels, a rosette in the middle. The mihrab is enclosed inside a large rectangular boundary, now in decaying state, packed with varieties of designs. A pair of boldly sticking out mouldings carved with friezes of lotus petals, lozenges, and rosettes top this rectangular border. In the centre of this whole composition, an engraved square black stone panel crown this. There is a wide tiered rosette in the centre of this panel ringed by seven smaller ones.

The remaining nine mihrabs are made entirely of bricks which display cusping in their faces. While much of their ornamentation has vanished, it still survives enough to demonstrate that these were originally exquisitely decorated, not unlike the main mihrab, but with terracotta. The motifs and patterns used are predominantly the same but vary in their arrangement from mihrab to mihrab. With decorative cusped niches, twelve in each wall, the north and south walls are labelled inward. Each of these niches is surmounted with a few mouldings. While these mouldings display rosettes alternating with diaper motifs, the space between them is ornamented in terracotta with floral scrolls.

Above the central nave, internally decorated seven chau-chala vaults are formed by the intersection of rafters and horizontally drawn thin brick-bands, which represents the exact imitate of the bamboo frame of Bengal's chau-chala stroke cottage. The frame is further characterized by terracotta lotus flowers, each positioned at the meeting place of the rafters and horizontal brick-bands. The brick-setting of horizontal rows are arranged in a cornerwise and edgewise manner. And this setting serves the function of arches to support the dome above, which is distinctive and gives a high relief appearance of a delicate carved design. Throughout the Sultanate period, and even during the Mughal era many monuments were built in this technique and continued widely for not only constructional but also for decorative purposes.

3 The Conceptualisation of Virtual Heritage

As a term, "Virtual heritage" has been coined from Reilly's [9] works of "virtual archaeology", which contained the understanding of virtual reality in archaeology or heritage. Categorically, it can be told that virtual reality as a tool can disseminate cultural heritage. It facilitates a sensational journey towards the past where never have been experienced before. But the journey should be operated carefully, because, the first phases are going through of the new tools, therefore it should be used cautiously. It means the drawing should be done by following the basic guidelines. It would be exceptionally simple to become overly enthusiastic by its present intrigue and attempt to utilize a computer-generated genuineness to make an encounter that may be well drawn closer with a more conventional medium, or it may just replicate this real-world confinement of museum facilities, yet now in a simulated environment.

On the contrary, it enables us to take a tour, which is impossible to visit physically in the real world. Virtual reality can be contemplating the sunken excavation of the ruins of a dipped Roman Ship. Virtual reality is open up the option, where not to move physically or without permission and a visitor could be touring inside the pyramid. Visitor can easily fly over the Shait Gumbad Mosque or Paharpur Mahavihara. These entire journeys could be possible by moving the camera and light. It could be allowing the visitor to move beyond the time and space, where s/he belongs to. Any period, even the meeting with Dinosaurs is possible to recreate. Virtual reality can be possible to make a tour in any possible imaginative journey towards future-past.

However, let us pause for an instant. What is the purpose of such journeys in the context of a museum, an historic or even an archaeological site? What can be achieved? By using the VR technology in archaeology, two different things that could be done. Rebuilding the past experiences which are prohibited to access these days, especially safety or conservation measures, i.e. pyramids, cave paintings, ruins of Somapura Mahavihara of Paharpur.

The other methodology is to incorporate these encounters into the museology of heritage structures, museums and archaeological sites in a way that authorises us to give background and passionate profundity to spaces and assortments of objects.

and environment experienced actively. There is no scope to be a passive entity, which has been happened in basic cinematographic tools by framing and montage. But, in this new realm, where the relationship between the work and the spectators are a different one. There is no scope to watch the content passably, which is happening in front of them. Users are participating actively, which are taking place surrounding him/her.

During the heyday of the current information technology revolution, three-dimensional technology is undoubtedly a special way or initiative to present and explain the archaeological position in virtual reality. And the on other hand the picturing of virtual models in several application areas permits experts to both study and interact with their study objects, while additionally allows the mass people to discover these virtual environments, thus endorsing their cultural, social and scientific contribution.

Here we develop a virtual reality to visualise and conserve the Shait Gumbad Mosque. It will help out the visitor not to do a physical journey, they will be able to feel the reality of the World Heritage Site of Bangladesh. It also preserves the original detail feature of Shait Gumbad Mosque virtually.

The primary procedure has been done by collecting the field level data. Majorly the measurement and detail photographs of the studied structure. It means that the two steps methodological journey have been conducted. Here the second step was to rebuild the structure virtually with computer-aided software (e.g. AutoCAD, 3D Studio Max, Mudbox, Unity3D, Adobe Photoshop and Illustrator) (Fig. 1e).

Since a design is very important before creating any installation. Therefore, AutoCAD software, the well-known 2D architectural design tools have been used to scale the outline of Shait Gumbad Mosque (Fig. 2a). By helping of this 2D drawing software, the work of mapping, making proportional structures and making accurate measurements have been regulated by referencing of the architectural historical characteristics. Here, the AutoCAD tool serves as a pre-assisted tool in creating the structural design of the studied structure.

After preparing the initial 2D structures, the data have been exported in Autodesk 3Ds Max. It helps out respectively formulated the 3D graphics and 3D animations, developed and produced by Autodesk Media and Entertainment. For generating the 3D graphics, 3D modelling, and animations, 3D Max has been used widely as application software (Fig. 2b). Final outputs have been built by rendering tools of this software. Thus, this tool has been used for moving the objects virtually of the studied architecture. Virtually generate the background reality of Shait Gumbad Mosque (Fig. 2c), 3D Max has been used precisely means node to node.

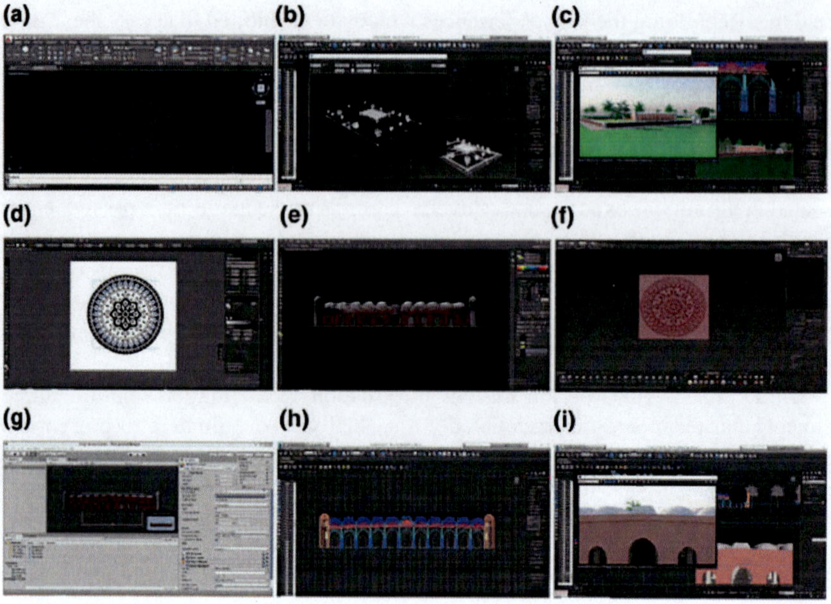

Fig. 2 **a** 2D drawing of ground plan in AutoCAD. **b** 3D wire drawing **c** Virtual background Preparation. **d** Cartographic illustration of terracotta plaque. **e** Negotiating the drawing pitfalls. **f** Mudbox preparation for 3D sculpturing the cultural objects aesthetically. **g** Preparation by unity 3D software. **h** Mesh model. **i** Final output after injecting mudbox and mesh features

4 Technological Journey to Build the Virtual Heritage of Shait Gumbad Mosque

A good map expresses a good story. Knowing what to include on a map and how to demonstrate that information is n vital skill. The map needs to be clear, accurate, oriented to the appropriate audience and the media that will be used for its presentation. Regarding this, graphics programs have been used very often in the journey to reconstruct the Shait Gumbad Mosque virtually, such as Adobe Illustrator and Photoshop, to put the final touches on objects and structures (Fig. 1g). The ending result is the product of cartography combined with some years of experience as an archaeological illustration. The created map aims to indicate the locations of the cities or ruins with minimal background disruption.

Using photographs of what can be seen today, therefore, to build pictures of what it might have looked like before. The Photoshop learning curve is difficult enough, but it might be easier if some of the pitfalls involved with the heritage sites of Shait Gumbad Mosque has been pointed out (Fig. 1h).

Mudbox has been used to generate CG sculpting programme. It also helps out in production-level texture painting program, especially the detail of brick and terracotta works. Here Fig. 9 prepared by using Adobe Illustrator for 3D sculpturing of terracotta plaque, which is showing in the figure of 11. To the professional 3D modeller, Mudbox is one of the tools to use for recreating the terracotta, brickworks and arches in the studied structure of this paper. This digital sculpting programme has been used to generate a 3D artist's workflow in the virtual heritage project of this paper, specifically working with more organic objects. It is very similar to the progression of real-life sculpting with clay, subdividing created models (essentially adding extra polygons) which permits us to add details that are far more problematic to achieve in a traditional polygonal 3D workflow. Mudbox have been used to visualize the aesthetic aspects of different constructions of the studied structure while envisaging the mosque structure. Here Mudbox has aided to make the beautiful earthenware boards, virtually created by presenting the burnt clay boards or other materials.

The fifth phase of building the VR of Shait Gumbad Mosque was utilizing Unity 3D, which has been known as the definition and modelling of such techniques in the Virtual model have been conceivable possible to the nearby joint effort of specialists in PC illustrations, art history, and ancient textiles. Furthermore, the 3D virtual demonstrating of artefacts is a common practice in cultural heritage; for instance, it has been used at the studied archaeological site, to rebuild the non-existing elements; to reconstruct the current shape of artefacts, including embellishments, brickworks, terracotta by the means of augmented reality technology.

When showing 3D renovations to the viewer, it is common to use either static images or at most models that revolve along an axis or point. However, it is possible to generate more immersive and stimulating models simply by using a navigation system that is more accessible and instinctive. In this regard, Unity3D offers the possibility of using the first-person viewpoint to navigate the site which requires little effort and no coding. It can even be used to discover a site in VR. Here we need to export the file as .obj or .fbx, which is the most suitable of texture transfer to 3D Studio Max.

At the point when we make this task in Unity3D, the format has been ensured as 3D (Fig. 2h). After that, the model files, which has been confirmed as .obj or .mtl format, has been dragged into the Unity asset panel. By following this process, some problems have arisen with the textures of the model. In this case, we reported it and copied the files directly into the project folder again. Here we importantly introduced the model file from assets into the hierarchy.

For adding components, it allows to "bump" into the model and we necessarily add some components (Fig. 2h) to the model itself. This helps out to generate a mesh of every singular object. Regarding this, we use a Mesh Collider and a Rigid body. One thing has been confirmed that the mesh collider should not be convex and untick the Use Gravity toggle in the rigid body. As a final point, the toggle was maintained as Kinematic toggle. By following this process, the meshed objects have been appeared to interact with the user. It has been generated a worth leaving the gravity on and creating it not kinematic so that the objects got liberated to move everywhere. Lights and camera works are essential to generate the environment of VR. Three-dimensional software is generous regarding solving this issue. In fact, the camera is not able to see exactly what human eyes can see with white eyes. For example, the vision of human eyes is straight and visualise as far as it can go. Along with the looking straight ahead, the vision's perspective on the right and left. The camera cannot do that. That's why virtually visualising Shait Gumbad Mosque, 3D software has been chosen because of shadowless Omni light projection and not sunlight, but this technology cannot penetrate opaque objects (Fig. 2i). In the case of cameras, all the cameras of 3D software have been used in the following Fig. 15. But these also used to do some work by increasing the fire. As an example, if the filter has been used, it takes a lot of time to match it with the light. A 3D camera, therefore, has been used, which matched automatically. Understanding the light and camera angle of the 3D software precisely is essential to bring out the real work, which has been done in the following works (Fig. 3).

5 Final Outcome of Virtual Heritage of Shait Gumbad Mosque

The above-mentioned procedure finalises the detail 3D modelling of studied Shait Gumbad Mosque. The outputs have been found in the following ways, which are described in Fig. 3. Transforming these 3D figures in virtual mood as an outcome for having the experiences beyond time and spaces, which needed some devices to enter the virtual environment, e.g. Cardboard, Mobile VR headsets, High-end VR headsets.

Cardboard glasses is a simplistic technology, which is a combination of two lenses with any mid-range smartphone can be used to experiencing the Virtual Heritage of Shati Gumbad Mosque (Fig. 19). It provides limited quality but is accessible to almost everyone, who want to join the virtual tour of Shait Gumbad Mosque. Mobile VR headsets is another VR gear of Google Daydream. A high-end mobile phone and sensors have been combined with plastic glasses, which provides a high-quality VR experience to users. A remote controller has been incorporated with this device, which allows smooth performativity action within the digital world. Last but not the least is High-end VR headsets is the HTC Vive, Oculus Rift, and Sony Play Station VR, which should be linked to a powerful computer, or in Sony's case via a PS4. It provides an engaging total experience in the virtual world. Users can, through their activities, modifications and interactivity with this world gratitude to two regulators that make it conceivable to accomplish any activities that are conceivable in reality, just as some that would be inconceivable, which previously built in the virtual heritage of Shait Gumbad Mosque.

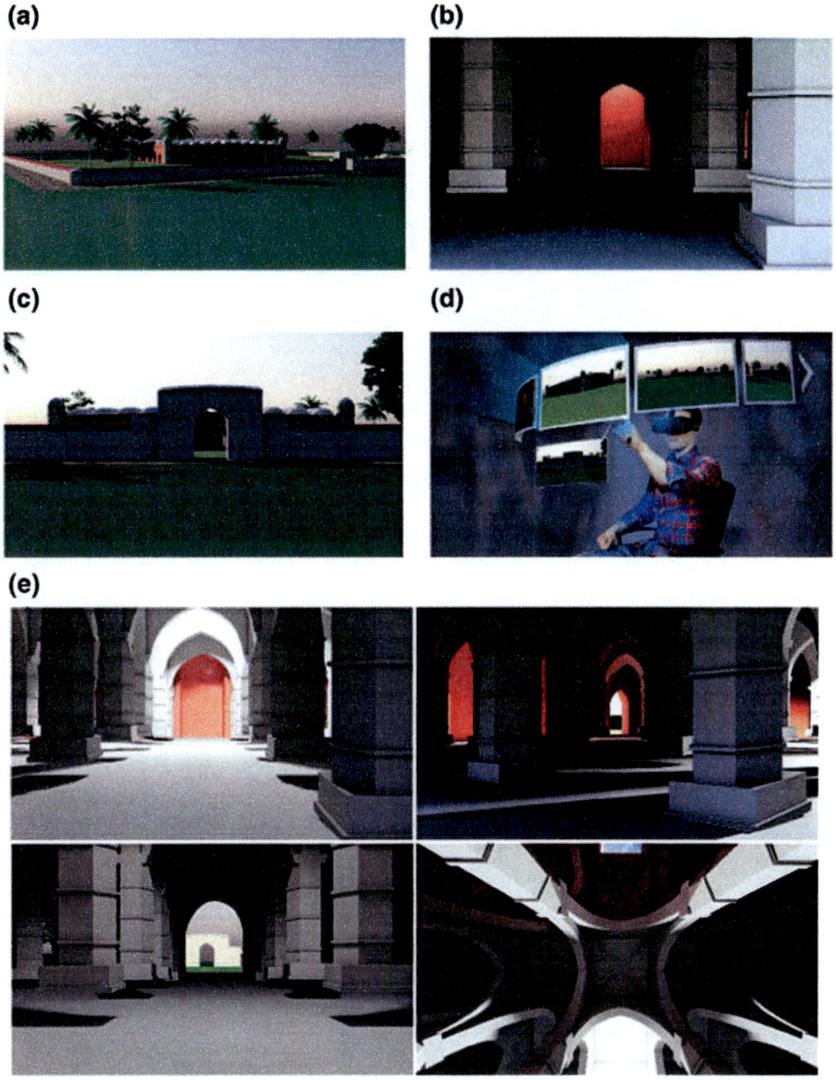

Fig. 3 Snapshot of virtual. **a** Heritage, **b** Interior, **c** Facede, **d** Illustration of shait gumbad mosque. **e** Light and camera preparation to visualise the interior of shait gumbad mosque virtually

6 Conclusion

Virtual heritage is the procedural journey beyond time and space, which not only visualise the historical structures but also preserve digitally. By following the London Charter and Sevilla Principles, the virtual reality of Shait Gumbad Mosque has been built, which could be considered as a conservation technique of the history of architecture in Bangladesh digitally. However, the geometric features are not easy enough to move besides engineering the smooth moving gematric objects. In respect of 2D, the repossession of semantic content in 3D has many canonical representational problems. Particularly, problematise the inequivalent representation, i.e. volume, surface structural representations expressed differently with the naked eye version [2]. Characterise the historical material as a subject to the sampling of built heritage is another major problem, because of the non-destructive restoration process. When this research went through the mechanism of explorative methods of social science where sketches have been incorporated the various concepts and phenomenon by following the knowledge-negotiation mechanism. Another thing is, there are so many architectural and ornamental features have been lost or partially survived. Because of all the features, a historical architectural journey has been needed. Therefore, according to the Sevilla principal, this research work has been done by the active participation of an archaeologist (first author) and professional VR expert (2nd author). The notion of achieving this is to generate some tools and methods for testing and analysing the theories and hypotheses for historical settings [12] through the use of 3D modelling tools and Virtual Reality (VR) engines in the field of the cultural heritage research in Bangladesh.

References

1. Bendicho, V.M.L.M.: International guidelines for virtual archaeology: the seville principles. In: Good Practice in Archaeological Diagnostics. Springer, pp. 269–283 (2013)
2. Chu, W.: Wir sind keine Deutschen nur dem volke nach. Multiethnic pasts and ethnic germans in the german criminal police in lodz during the second world war. ZfGen Zeitschrift für Genozidforschung **16**(1), 35–56 (2018)
3. Cimino, C.: Protection of cultural heritage in urban areas during peace and conflict times from threats to risk preparedness as a shared responsibility. In: XXIV International CIPA Symposium, Strasbourg. pp. 195–199 (2013)
4. Di Giovine, M.A.: The heritage-scape: UNESCO, world heritage, and tourism. Lexington Books (2008)
5. Henckaerts, J.M.: New rules for the protection of cultural property in armed conflict. Int'l Rev. Red Cross **81**, 593 (1999)
6. Imran, M.: Genealogy of archaeological surveying in bangladesh: Delineating peter haggett's systematic ground survey with the modification by gis in khalifatabad, bagerhat, Pratnatattva. J. Depart. Archaeol. Dhaka: Jahangirnagar University **19**, 75–87 (2013)
7. Imran, M.: Beyond Anecdotalism: questing about Ulugh khan Jahan, founder of world heritage of Shait Gumbad mosque, Bangladesh from the dominating historical and archaeological account. In: ICON: Journal of Archaeology and Culture, Wakankar Rock Art and Heritage Welfare Society, Bhopal (2016)

8. Menghi, R., Maino, G., Panebarco, M.: Virtual reality models for the preservation of the Unesco historical and artistical heritage. In: International Conference on Image Analysis and Processing. Springer, pp. 475–485 (2011)
9. Reilly, P.: Towards a virtual archaeology. In: Computer Applications in Archaeology. British Archaeological Reports, Oxford, pp. 133–139 (1990)
10. Stovel, H.: Origins and influence of the Nara document on authenticity. APT bullet. **39**(2/3), 9–17 (2008)
11. UNESCO: 1954 Hague Convention | United Nations Educational, Scientific and Cultural Organization (2017). https://rb.gy/z9ojpc. Access date: 14 Sep 2020
12. Wang, C., Miller, D., Brown, I., Jiang, Y., Castellazzi, M.: Visualisation techniques to support public interpretation of future climate change and land-use choices: a case study from ne scotland. Int. J. digit. Earth **9**(6), 586–605 (2016)

A Healthcare System for In-Home ICU During COVID-19 Pandemic

Zannatun Naiem Riya and Tanzim Tamanna Shitu

Abstract The number of reported Covid-19 cases are increasing so rapidly that even developed countries are being unable to afford to get a bed in hospitals for every Covid-19 positive patient. Shortage in number of doctors is also a problem. This is the reason why physicians encourage patients to stay at home and take care of themselves. Our intention is to formulate a system for patients with home ICU. This application will help to archive every patient's health response and alert people nearby. A threshold of each body parameter has been set so that an alarming system can call for help immediately. For constant checking of every patient's health condition for making further decision is fulfilled by developing a real time database system that helps to store and share data in remote areas. A data management system allows the clinician to access important information through one platform for avoiding crucial details. A password system makes the stored data secured for not to compromise with patients' privacy.

Keywords Patient monitoring system · Real time data · Home ICU · XAMPP · Tera term.

1 Introduction

Various disorders are related to change in various physiological measures such as [1] blood oxygen level, body temperature, respiratory rate, blood pressure, heart rate etc. In case of covid-19 patients, factors such as respiratory rate, blood oxygen level and heart rate are so important. When a person gets affected by corona virus, they can have slight to serious fluctuation in these parameters. Common or early symptoms of Covid19 affection are fever, cough, tiredness [2]. Sometimes affected people

Z. Naiem Riya (✉) · T. Tamanna Shitu
Khulna University of Engineering and Technology, Khulna, Bangladesh
e-mail: riyanaiem45@gmail.com

T. Tamanna Shitu
e-mail: shitutanzim0711@gmail.com

M. S. Kaiser et al. (eds.), *Proceedings of International Conference on Trends in Computational and Cognitive Engineering*, Advances in Intelligent Systems and Computing 1309, https://doi.org/10.1007/978-981-33-4673-4_34

with these symptoms, take this situation lightly and later suffer badly. Also, people with chronic diseases like heart diseases, diabetes, cancer are at high risk if they get affected. So during Covid-19 when social distancing is so essential, people can continuously examine their health by themselves using modern technology. Emergency patients get admitted into ICU as they need a 24/7 assertion and observation of these physiological condition at every instant. Financially unstable patients find it very expensive to be admitted into an ICU for a long time period. Doctors often can not routinely track a patient if he's out of station. The idea of home ICU is not new but in a situation of a pandemic like Covid-19, it has become a must. Vastly increased Covid-19 patient number, shortage of doctors and health staff, shortage of hospitals contribute to the concept of home ICU. An idea of home ICU can help to reduce a patients' health risk as it alarms against a patients' condition which is prone to be dangerous before being too late. During Covid19 phase, with the aid of technology, doctors can give treatment more patients than they do during usual circumstances. Physicians in the Intensive Care Unit (ICU) deal with seriously ill patients and have to take quick assessments at the stage of care. The ICU world involves an increasing array of observation and life-supporting systems that gather data but do not make it accessible automatically. In addition in [3], many patients in remote communities seek care that would not otherwise have been accessed without longer journeys or resolving other obstacles to transportation. Telemedicine facilities range from basic health care to highly advanced care provided in established accredited medical centers. Statistics shows in [4], year after year approximately 8 million people worldwide die for limited access to health treatment. It reported about 21% of males and 30% of females below 75 are resistant to basic healthcare (the comparable study is 56%). The Covid-19 fatalities rate is noted to decrease with expanded medical staff and nursing homes, based on results in [5]. It observed 0.23, 0.16 and 0.21 as a measure of its nonlinear scale involving COVID-19 mortality rate and preventative health campaign. So, to avoid the potentially ongoing COVID-19 epidemic, it may be important to upgrade health services and to implement certain safety measures within a short period of time. Satisfactory treatment is mostly out of possession whenever this comes to handling non-communicable conditions that grow relatively steadily such as diabetes, cardiac disease, cancer etc. Thus, continuous health monitoring takes the most importance so that minor changes can be detected too. Sensor based measurement in health monitoring is way more effective and popular in in-home medical check-up now-a-days as in [6]. Using the Internet of Things (IoT) and also its e-Health implementations within healthcare system of telemedicine tends to a fast and efficient transmission of data among healthcare providers and patients, making healthcare cost-effective and enhance the efficiency of care of patients in [7]. Health information technology is at a state where it is efficient in streamlining information and configured in such scenarios to be effective enough to allow doctors increase the speed and quality of service. These services are not just to enable doctors to make sure that they will be recompensed in a timely and adequate manner but also help them in evaluating patient outcomes and detecting inadequacies that may impact the security of their patient. The database management system integrates a different types of data and guarantees insights into different network applications, while trying to maintain real-time network data analysis.

Arduino IDE is good for data displaying and software such as Tera Term can turn the Arduino serial monitor data into real time data. For database creation and manipulation, XAMPP is a very easy to handle as one can create charts, import files, export them there. In case of using such database management system in medical sectors, the privacy needs to be strong enough because if any changes happen in any physiological data by mistake or intentionally, it would be life threatening. The privacy is strong sharing stuffs here in XAMPP software as password system makes it protective. Also there is scope to create modules for each patients so there is no chance of exchanging the patients' record. So, this proposed system covers the maximum technological application in medical sector in emergency and usual situations.

2 Related Work

This research examines recent advancements in mobile healthcare services and tracking in both contact and contactless techniques. Detecting physiological signals of patients with DS18B20 and MAX30102 sensors and providing cloud data via WI-FI, leading to local hospital servers and allowing physicians to collect data as require in [8]. A remotely controlled PMS (Patient Management System) is proposed in [9]. For this paper, this wireless patient screening device monitors breathing, body temperature and blood oxygen percentage. This paper addresses the development and implementation of a relatively cheap process that enables GSM technology to track a patients' medical condition. ZigBee was used as a remote device to transmit signals to PC via RS-232 serial port adapter as part of an order. The building model describes in [10] embedded web services operated by Arduino Uno with multiple sensors. Furthermore the Rich Internet Framework (RIA) Service Oriented Architecture (SOA) was used in this method. This scheme was introduced particularly to the four different sensors required for monitoring the patient, along with blood oxygen (SPO2), electrocardiogram (ECG), body temperature and pulse rate. Furthermore, system outputs show gathering real-time data, reducing manual data processing, allowing the tracking of huge patient numbers, boost high-quality care. A hybrid OCC/BLE platform is investigated in [11] to enable successful, distant, and real-time transmitting of an electrocardiogram (ECG) signal to a display. It proposes a patch circuit which integrates an LED array and a BLE transmitter module. The patch gathers the ECG data to reduce the power usage as per the patients' health condition. Then an algorithm for network selection is built for a new request for network access created within the patch system. Finally, the Fuzzy logic is used to find a particular camera for receiving data and proposes a handoff process to ensure efficient spectrum allocation given the versatility of the patient. Simulations are eventually performed to show the performance and feasibility of the measurement process. A revolutionary patient surveillance system in [12] incorporates an algorithm and wearable tools to track vital physiological information of the patient. It also includes a GPS to enable patient and medical care giver to observe each other. Their implementation has been verified, and the findings show that their proposed system can transmit high-precision critical physiological signals and also patients' movement information to a distant healthcare giver. The machine produces warning alerts for patients, also help patients to find the nearest healthcare givers. A Plethysmography system is developed in [13] which includes continuous transmitting and receiving of information about heart rate. Data is sent to an IoT gateway, Raspberry Pi. For sending information to the cloud, it has used a protocol called MQTT for updating user with his health condition with Email, SMS and also some real time plots. If required, a nearby hospital address is also mentioned. A Tele-Health application for elderly patients is described in [14]. Some physiological sensors are used to sense patients' health condition continuously and send it to required volunteer via GSM. Doctors can view and interpret information on their smartphone via developed web server where all requires patients' health information is imported.

3 Data Acquisition Scheme

Physiological information of patients is recorded with these sensors:

- **Pulse sensor**: This sensor basically monitors the difference in blood volume. It contains a light source and LED sensor. The sensor evaluates the reflected light amount when blood pumps underneath the skin past the light source. Normally the reading is collected through fingertip of the patient.
- **LM35 sensor**: To quantify an organism's infrared energy, the sensing element consists of numerous thermocouples. The outcome of this sensor is an analog signal. To produce a temperature measurement in Celsius, the given voltage outcome can be construed then.
- **SpO2 sensor**: With MAX30100, specifically blood oxygen measuring sensor, the oxygen saturation value is extracted that includes two LEDs (red and infrared). To be estimated by a photodiode, the beam travels through participant's finger. Hemoglobin with oxygen or without oxygen absorbs the LED lights at different wavelengths. MAX30100 then estimates blood oxygen level according to that particular degree of absorption.
- **Respiratory rate sensor**: The threshold of every other breath is acquired by the circulatory sensor. To predict the duration between two limits and determine the breaths per minute, a simulator is used here.

In this paper, a health care system for ICU patients is designed where patient monitoring, alarming scope, database-based data storing and medicine chart is included if it is important to observe how a patient is recovering by given treatment or medicine. e-Health applications need the data to be available on the application servers. In XAMPP, under the Apache server monitored data is imported on MySQL database so that doctors can use those data for further analytical treatment. For offline data storage, a software Tera Term is used which interface data between the Arduino

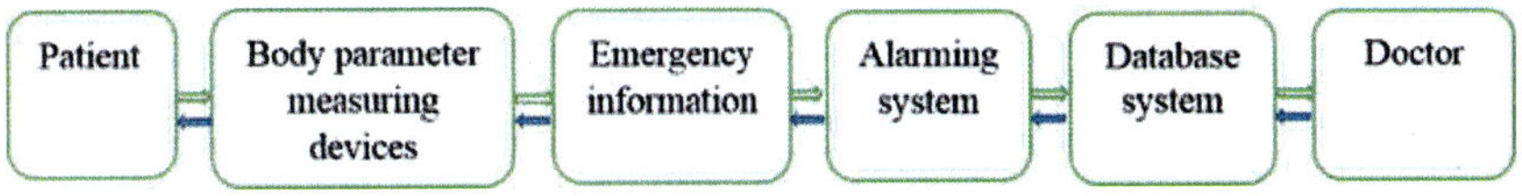

Fig. 1 Data acquisition scheme

Serial monitor and an Excel datasheet. Using the extension .csv the file is stored as a CSV file. A threshold of BPM, body temperature, blood oxygen level and respiratory rate is set. Whenever the monitoring parameter is below or above average range, a buzzer is activated which needs analysis of a patients' condition. For this, the saved CSV file is imported in MySQL database and the doctor can export the data from anywhere. The data is protected as only the desired doctor will get access by putting password. What is more important Tera Term's 'timestamp' helps the data stored in real-time so that doctors can have every patients' full record. Doctors can also send feedback and patient can export it easily. Figure 1 is an abstraction of the proposed data scheme.

4 System Architecture

The data was collected which includes heart rate, body temperature respiratory rate and blood oxygen level from around 500 patients of 2 different hospitals and checked out the alarming system of which approximately 89% cases, the system worked successfully. Figure 2 describes our proposed system architecture.

- The measured parameters are BPM, respiratory rate, blood oxygen level and body temperature. These parameters are sensed by using a pulse sensor, respiratory rate sensor, SpO2 sensor, and the body temperature sensor respectively. Using the Arduino Uno Serial monitor the measured details are collected.
- Then the measured data is converted to the real-time data via Arduino-Tera Term interfacing where the output shows the patients' condition even more specifically. The interface of this software is done by opening the Tera Term and selecting the port option as per the Arduino port.
- After the setup option is sorted, the output from the Arduino serial monitor is transferred to Tera Term where the data of each second are shown which is very useful in data recording as doctors can observe the patients' condition and come to a decision whether the patients' condition is improving or getting even more complex specifically. The data is then saved as a CSV file using an Excel datasheet.
- If the doctor is out of station, then the data is to send other places so that the doctor can give treatment via telemedicine. For this, XAMPP is used for data importation and exportation.

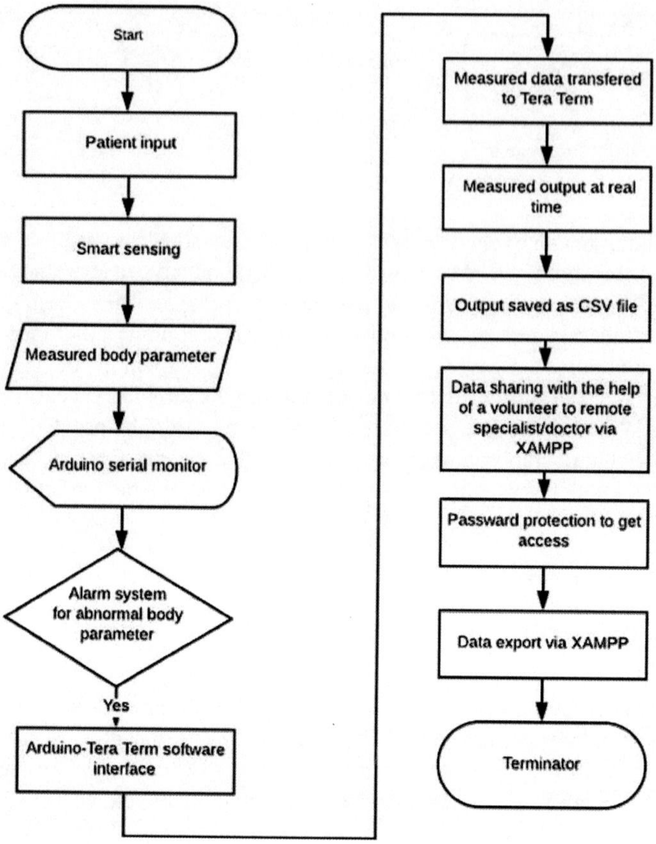

Fig. 2 System architecture

- On XAMPP, MySQL uses tables to store the information. Tables can be added to MySQL using the phpMyAdmin user interface. Here whenever a patient is released, the unnecessary data can be erased too.
- When data is imported, the measured data is observed by those who have the access to export it only. The query takes a fraction of a second to update.
- By clicking new, another patient's data can be stored. Or several charts can be created such as measured data, given medicines etc.
- For each patient, a new module is created.
- On the home page, a list of available doctors and a section of notification with important informations are to be shown like Table 1. Medical doctors in service list: 1. Dr. A to 10. Dr. X.

Table 1 Service schedule for doctor and medical staff

Time	Available medical staff in service
12.01 a.m. to 5.00 a.m.	doctor: 1, 6 and an extra intern (in case of heavy patient load)
5.01 a.m. to 10.00 a.m.	doctor: 2, 7 and an extra intern (in case of heavy patient load)
10.01 a.m. to 3.00 pm	doctor: 3, 8 and an extra intern (in case of heavy patient load)
3.01 pm to 8.00 pm	doctor: 4, 9 and an extra intern (in case of heavy patient load)
8.01 pm to 12.00 a.m.	doctor: 5, 10 and an extra intern (in case of heavy patient load)

5 Experimental Result

Normal range of body parameters for adults:

- Heart rate: 60–100 beat per minute
- SpO2 (blood oxygen saturation levels): in between 94 and 100%
- Respiratory rate: 12–20 breaths per minute
- Body temperature: 36–37 °C

Medical condition of around 500 patients have been recorded and a demo record of this work including a patients condition is categorized in 3 sections.

Category 01: Alarming system inactivates as body parameters remain within normal range. Two demo records for category 1 are shown in Table 2.

Category 02: Alarming system activates as body parameters below/above normal range. Two demo records for category 2 are shown in Table 3. Abnormal records are in bold and italic mode in table.

Category 03: Alarming system activates although body parameters remain within normal range. Two demo records for category 3 are shown in Table 4.

These categories show a demo of my work of alarming system with body parameters action. So for this an alarming buzzer is set which gives alarm whenever the measured data goes out of range. Category 01 and 02 increase the accuracy of my work. In case of category 02 there is unwanted fluctuation of parameters, the alarming system activates and the volunteer is alert to send a CSV file which was converted in an Excel datasheet to the doctor to seek for medical help via XAMPP software with

Table 2 Do not require emergency call for medical help

Time	Heart rate (beat per minute)	Respiratory rate (breathe per minute)	SpO2 (blood oxygen saturation levels) in percentage	Body temperature (degree celsius)
[year-month-date hour-min-sec]	79	16	95	36
[year-month-date hour-min-sec]	81	15	97	37

Table 3 Requires emergency call for medical help

Time	Heart rate (beat per minute)	Respiratory rate (breathe per minute)	SpO2 (blood oxygen saturation levels) in percentage	Body temperature (degree celsius)
[year-month-date hour-min-sec]	***55***	14	***87***	36
[year-month-date hour-min-sec]	80	***25***	93	***33***

Table 4 Misleading alert system

Time	Heart rate (beat per minute)	Respiratory rate (breathe per minute)	SpO2 (blood oxygen saturation levels) in percentage	Body temperature (degree celsius)
[year-month-date hour-min-sec]	78	17	94	37
[year-month-date hour-min-sec]	79	15	96	37

Apache web server and MySQL database server where the desired data is present so that doctor can detect the severity of data fluctuation and treat patient accordingly. But in category 03 alarm activates although the parameters were normal. Cable breakage, damaged cable insulation or broken connectors can lead to this malfunction.

6 Conclusion

By using such sites, doctors can treat patient of a great number in a short time in an analytic way. As PMS requires continuous monitoring and recording, so there is no chance of missing any significant information about patient condition by the doctors. Also, people from area where medical facilities are not available, do not need to come to a long path to seek for doctors help and wait for hours in front of doctor's chamber. On the home page, availability of doctors are mentioned so patients can easily choose a suitable time and physician for their treatment. What is more important as via telemedicine doctors do not have to come in direct contact with patients, for this health risk or specifically COVID19 spreading risk reduces in both doctors and patient. However there was error in minority cases in the alarming system. This can be minimized by proper system maintenance and diligent processing of data at a great pace.

References

1. Malasinghe, L.P., Ramzan, N., Dahal, K.: Remote patient monitoring: a comprehensive study. J. Ambient Intell. Human Comput. **10**, 57–76 (2019)
2. WHO emergencies diseases novel-coronavirus-2019, Q&A section, common symptoms of COVID-19
3. Sebastian, S., Jacob, N.R., Manmadhan, Y., Anand, V.R., Jayashree, M.J.: Remote patient monitoring system. Int. J. Distrib. Parall. Syst. (IJDPS) **03**(5) (2012)
4. Kruk, M., Gage, A., Joseph, N., Danaei, G., Garcia-Saiso, S., Salomon, J.: Mortality due to low-quality health systems in the universal health coverage era: a systematic analysis of amenable deaths in 137 countries. The Lancet. **392**, (2018). https://doi.org/10.1016/S0140-6736(18)31668-4
5. Abdulla, F., Nain, Z., Karimuzzaman, M., Moyazzem Hossain, M., Adhikari, U.K., Rahman, A.: Effect of preventive actions and health care factors in controlling the outbreaks of COVID-19 pandemic (2020)
6. Jagadish, D., Priya, N., Suganya, R.: Remote continuous health monitoring system for patients. Data Sci. Analyt. Appl. 129–138 (2018)
7. Saminathan, S., Geetha, K.: A survey on health care monitoring system using IOT. Int. J. Pure Appl. Mathe. **117**(17), 249–254 (2017)
8. Siam, A., Abouelazm, A., El-Bahnasawy, N., El Banby, G, Abd El-Samie, F.: Smart health monitoring system based on IoT and cloud computing (2019)
9. Chanakya Mothukuri, K.H., Prathap Kumar, M.: Patient monitoring system. Int. J. Sci. Res. (IJSR), **02**(02) (2013). India Online ISSN: 2319-7064
10. Hameed, R.T., Mohamad, O.A., Hamid, O.T., Tapus, N.: Patient monitoring system based on e-health sensors and web services. In: 8th International Conference on Electronics, Computers and Artificial Intelligence (ECAI) (2016)
11. Hasan, M.K., Shahjalal, M., Chowdhury, M.Z., Jang, Y.M.: Real-time healthcare data transmission for remote patient monitoring in patch-based hybrid OCCBLE networks. Sensors **19**, 1208 (2019)
12. Abou-ElNour, A., Tarique, M., Foyez, D., Hassan, L., Rashed, A.: Location aware health monitoring system for emergency cases. Int. J. Scient. Technol. Res. **8**(12) (2019)
13. Bansal, P., Malik, M., Kundu, R.: Smart heart rate monitoring system. In: 2018 IEEMA Engineer Infinite Conference (eTechNxT) (2018)
14. Surekha, Y., Akhil, N., Rajesh: Patient monitoring system using IoT. Int. J. Innovative Res. Adv. Eng. (IJIRAE) **V**, 176–182

Career Prediction with Analysis of Influential Factors Using Data Mining in the Context of Bangladesh

Al Amin Biswas, Anup Majumder, Md. Jueal Mia, Rabeya Basri, and Md. Sabab Zulfiker

Abstract Nowadays, the job market has become more competitive due to the advancement of higher educations. In Bangladesh, an individual selects three types of careers namely Government Job, Private Job, and Business after the completion of their graduation. Selecting a specific career is affected by many factors. These factors can be personal as well as academic. In this research, we have done career predictions and analyze several factors behind this. To accomplish the work, we have analyzed several academic and personal factors. Here, we have applied several data mining techniques for experimentation and several performance evaluation metrics to evaluate our work. Finally, we found that Part classifier outperforms the other techniques.

Keywords Prediction · Data mining · Influential factors · Career · Bangladesh

---

A. A. Biswas (✉) · Md. Jueal Mia · Md. Sabab Zulfiker
Department of CSE, Daffodil International University, Dhaka, Bangladesh
e-mail: alaminbiswas.cse@gmail.com

Md. Jueal Mia
e-mail: mjueal02@gmail.com

Md. Sabab Zulfiker
e-mail: sabab.rumc@gmail.com

A. Majumder
Department of CSE, Jahangirnagar University, Dhaka, Bangladesh
e-mail: anupmajumder@juniv.edu

R. Basri
Department of CSE, Bangladesh University of Business and Technology, Dhaka, Bangladesh
e-mail: rabeyabasri3@gmail.com

M. S. Kaiser et al. (eds.), *Proceedings of International Conference on Trends in Computational and Cognitive Engineering*, Advances in Intelligent Systems and Computing 1309, https://doi.org/10.1007/978-981-33-4673-4_35

1 Introduction

A successful and secure future largely depends on the selection of an appropriate career choice. This selection process can be influenced by some personal factors along with the individual's academic background, such as inherent talents, technical skills, academic results, research interests, and sometimes mostly on the family background as well as on the family income. Selecting an effective career path sometimes becomes difficult when an individual's interest conflicts with real circumstances in case of qualification, experience over time, economic stability, family pressure, ongoing career trends, etc. In Bangladesh, due to several factors, most of the graduate students compete for government jobs, as they consider government jobs as the most viable and secure career option through their self-observation or other family member's experiences. But in reality, everyone cannot get the opportunity to go for a government job. On the other hand, the majority among the rest, who generally belong to well-off families, tend to become self-employed by starting their businesses, whereas, youths with strong technical skills, but less institutional grades, join private jobs at several international companies in the state or abroad for better earning. Thus, depicting an accurate prediction of an individual's career is indispensable with prior knowledge about the influential factors, which have a great impact. Moreover, some individuals are unable to find out their actual interests and capabilities. In that case, career prediction will act as a guide to make an appropriate career choice. Here, we have emphasized the analysis of some empirical factors of an individual's personal and academic data to perform the career prediction. We have experimented with six popular data mining classifiers including Logistic, Multilayer Perceptron (MLP), Kstar, Part, J48, and Random Forest on our survey dataset. To determine the best classifier in the working context, we have calculated several performance evaluation metrics and a result comparison is performed here. From the analysis of the obtained result, it is confirmed that Part classifier achieves the best result in terms of metrics.

This paper is sorted as follows: Sect. 2 presents an extensive survey of related works to identify previous research gaps. Section 3 represents the research methodology including a brief description of the dataset, implementation procedure, and classifiers. Section 4 discusses the experimental result and other findings. Finally, Sect. 5 provides a conclusion.

2 Related Work

The aspects of this kind of problem are notably complex and vary in time. In this portion, we have gone through several articles to find the research gap of existing work.

Yeasin et al. [1] proposed a way to predict students' careers based on the data mining technique. This work has been done only for CS graduates. They have applied several classifiers and the accuracy varied based on several classifiers. In this work, only 506 data records have been used but no training and testing dataset has been mentioned separately. Gorad et al. [2] developed a web-based application that helps students to choose a course for their career. In this work, the dataset is not mentioned here and the work is only a developing system that suggests a student's career counseling. Rangnekar et al. [3] has drawn data from social media like Facebook to predict the student's career using data mining and linear classification technique to develop an intuitive career system. The dataset consisted of 480 entries. Assuming several parameters in this work give the accuracies of 77.41, 75.4, and 60.09%. García-Peñalvo et al. [4] built a predictive model to analyze a student's career after completing their graduation by using a machine learning-based technique. In this work, they have done a case study involving 3000 students' information from the context of Spanish Observatory for Employability and Employment. No accuracy is mentioned here to develop the model. Al-Saiyd and Al-Takrouri [5] proposed a model that supervises students about choosing IT-based works and how it is changed based on their skills over time. Only 50 student's data have been taken to implement the task using backpropagation neural networks (BPNN). Though the models are organized in a composition of 35 input factors parameters for testing, and 35 input factors for training but no accuracy is mentioned here to develop the model. Campagni et al. [6] investigated the careers of university graduate students using data mining models. A transactional database including student's information has been used and later 141 data records have been analyzed. In this work, they wanted to show different types of student behaviors and computed many types of distance between the career of any student and the ideal one. Panda et al. [7] proposed a predictive model using different data mining classifiers in the context of an Indian education system. They have proposed the approach of Random Forest to predict the career decision for the students. In this work, they have also reviewed a comparative study of the classifiers namely Iterative Dichotomiser 3 (ID3), K-Means, Naïve Bayes, and Random Forest. Roy et al. [8] introduced a career recommender system for CS graduates. Their work has been done with a huge number of records. Support Vector Machine (SVM), XG Boost, and Decision Tree is used here and SVM outperforms the other. Nie et al. [9] proposed a data-driven framework to forecast the students' career choices upon graduation based on their behavior. 4246 students' data from the same grade is used to conduct the evaluation. In this work, they have compared the result with four classification algorithms like SVM, Random Forest classifier, Logistic Regression, and Decision Tree. Among them, the Random Forest classifier showed the best result but they have not mentioned the accuracy of this task. For future career planning, Lou et al. [10] et al. applied a machine learning approach using hand rules and clustering algorithms. The data has drawn from social media like LinkedIn is about 67,000 profiles. It is mentioned that the model's logical accuracy quickly falls off when the expected length of the path like career opportunities increases. Zulfiker et al. [11] applied seven different classifiers to predict the final grade of students into four categories. Then the output of the base classifiers has been combined using the weighted voting approach to achieve better results and it outperforms the base classifier by achieving an accuracy of 81.73%. Mia et al.

[12] applied several machine learning techniques to predict the students' registration status. They have applied seven well-known classifiers. To accomplish their work, they have collected more than a thousand students' data (five attributes). They have calculated six performance evaluation metrics to evaluate their work. Finally, they found SVM outperforms others by achieving 85.76% accuracies. Biswas et al. [13] applied seven classifiers namely Multilayer Perceptron, Naïve Bayes, Logistic, Random Forest, Locally Weighted Learning, Random Tree, and Part. They have calculated the seven metrics to evaluate their work and measure the performance of the classifier. Three classifiers namely Naïve Bayes, Part, and Locally Weighted Learning perform better with 86.36% accuracy. Then from the performance evaluation metrics, it is observed that the Locally Weighted Learning classifier performed best here. In [14], to select the right career based on the individual's interest, a recommendation system is presented. A data mining approach [15] is applied to predict employability in Morocco.

3 Methodology

This section is parted into Data Description and Analysis, Classifier Description, and Implementation Procedure. This section basically presents the way we have accomplished this work. A detailed description of all the subsections is presented below.

3.1 Data Description and Analysis

Dataset used in this research is collected by a survey. The survey mainly consists of 13 questions. Among these 13 questions, both personal and academic factors are included. To do labeling all of the questions, 13 variables are used here. In the dataset, the number of independent variables is twelve, and the number of dependent variables is one. All the variables with their possible values are tabulated in Table 1. A total of 494 individual records are used here to accomplish this work. A total of 80% of data is used for the training of the classifier and 20% is employed for the testing purpose.

The correlation matrix of our working dataset is presented in Fig. 1. The linear relationship between the variables of our acquired data is represented by this correlation matrix. From Fig. 1, we have tried to show the relation between twelve independent variables (GT, FAP, MOP, FPEGJ, FPEPJ, FPEB, UT, SSCGPA, HSCGPA, GCGPA, GJRS, and MIF) with one dependent variable (CPS). From the resulting correlation matrix, we found that UT (University Type) has a strongly positive correlation with CPS (Current Profession Status) and MIF (Monthly Income of Family) has a weakly negative correlation with CPS (Current Profession Status). So, academic and personal factors are the influential parameters in this regard.

Table 1 Attribute with their possible values

Variable	Full-form	Variable type	Possible value
GT	Gender type	Independent	Female (0), Male (1)
FAP	Father's profession	Independent	Business (0), Private job (1), Government job (2), Other (3)
MOP	Mother's profession	Independent	Business (0), Private job (1), Government job (2), Housewife and Other (3)
FPEGJ	Family person engagement in government job	Independent	No (0), Yes (1)
FPEPJ	Family person engagement in private job	Independent	No (0), Yes (1)
FPEB	Family person engagement in business	Independent	No (0), Yes (1)
UT	University type	Independent	National University (0), Private University (1), Public University (2)
SSCGPA	Grade point average in secondary school certificate	Independent	Below 3.00 (0), 3.00 to 3.99 (1), 4.00 to 5.00 (2)
HSCGPA	Grade point average in higher secondary certificate	Independent	Below 3.00 (0), 3.00 to 3.99 (1), 4.00 to 5.00 (2)
GCGPA	Cumulative grade point average in graduation	Independent	Below 3.00 (0), 3.00 to 3.49 (1), 3.50 to 4.00 (2)
GJRS	Government job related to study	Independent	No (0), Yes (1)
MIF	Monthly income of family	Independent	Below or Equal 50 k (0), 51 k to 99 k (1), Equal or Above 100 k (2)
CPS	Current profession status	Dependent	Business (0), Private job (1), Government job (2)

3.2 *Classifier Description*

Logistic Regression [16] is a popular supervised learning technique that is used to predict the categorical dependent variable using a set of independent variables. This is a powerful and significant algorithm as it can give probabilities and classify new data by using both continuous and discrete data. Logistic (Sigmoid) function is a mathematical function that is used to map the values (predicted) to the probabilities. It can map any of the real value into another within a range of 0–1. The main assumption of this algorithm is that the output variable must be categorical in nature and the input variable should not have multi-collinearity. Logistic Regression can be classified into three types namely Binomial, Multinomial, and Ordinal. In binomial, there can be only two possible outcomes of the output variables such as 0 or 1. In multinomial, there can be three or more possible unordered outcome of the output variable such as "Private", "Public", or "Business". In ordinal, there can be three or

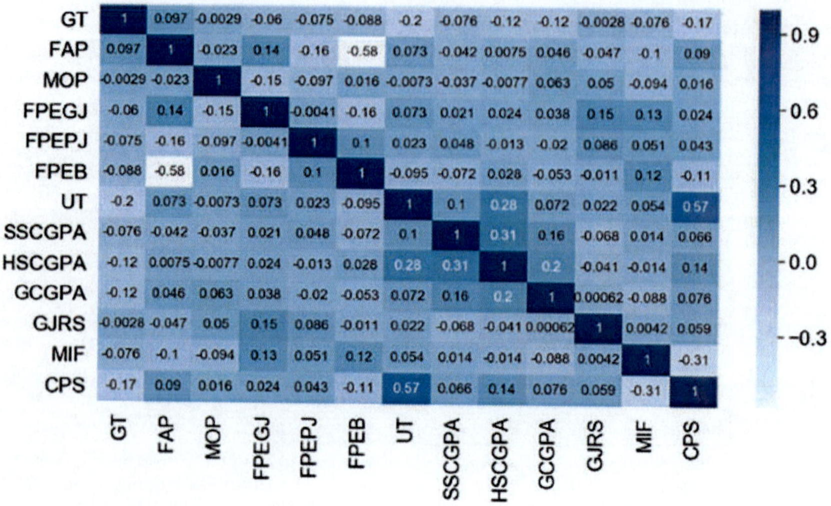

Fig. 1 Correlation matrix

more possible ordered outcome of the output variable such as “Small”, “Medium”, or “Large”.

A Multilayer Perceptron (MLP) [17] consists of an input, an output, and some number of hidden layers (one or more). A single layer perceptron can learn only linear functions whereas a multi-layer perceptron can also learn nonlinear functions. The learning process of MLP is called the Backpropagation algorithm. The input layer receives the signal and the output layer predicts a decision based on the input. The hidden layers are used as a computational engine to approximate continuous functions. In MLPs, the previous layer’s output is served as the input to the next layer. MLPs [18] are feedforward networks consist of a forward pass, where signal flows from input to output layer via hidden layers to the output layer, and a backward pass, where backpropagation is used to minimize the error by adjusting the model parameters (weights and biases) using stochastic gradient descent optimization.

KStar, known as a lazy learner is an instance-based classifier that is different from other instance-based learners as it uses an entropy-based distance function. K^* is similar to the K-Nearest-Neighbor classifier. This classifier simply stores the training data and then performs little preprocessing; finally waits for the test tuple and when it sees a test tuple, it starts classifying the tuple based on its similarity of the stored training tuples. This type of classifier does less work while training data is inputted and performs more work when the classification of a test tuple is given [19–21].

Part [21] is a rule-based classifier which obtains rules from partial decision trees. It creates the tree by using C4.5’s heuristics with the same user-defined parameters as J4.8. Because of this, J4.8 and the part classifier can give the same result for a particular dataset.

J48 is a statistical classifier that implements the C4.5 algorithm in WEKA (a widely used data mining tool) using JAVA. It is an extended version of ID3 (Iterative Dichotomiser 3), which is used to build decision trees for classification. J48 is an open-source implementation of C4.5, containing some spare features that are useful for handling pruning of decision trees, rules derivation, missing values, continuous ranges of attributes, etc. WEKA provides several options linked to the pruning of trees that is used as a precise tool to fix overfitting. This algorithm generates decision trees until it obtains optimum accuracy with flexibility [22].

Random Forest classifier, a supervised learning technique builds forests by bagging ensembles of random trees [21] and used for both regressions as well as classification. It is widely used in classification problems. It creates decision trees on data samples and then performs prediction by each of them. Then finally chooses the best solution using the voting approach. Random Forest performs better than a single decision tree as it is an ensemble method. Because of averaging the result, reduces the overfitting problem. The following steps depict the working principle of it [23].

a. Randomly selects the samples from a given dataset.
b. Construct a decision tree for every sample and perform the prediction for each of the decision trees.
c. Voting will be performed for every predicted result.
d. Select the majority voted class prediction result as the final prediction result.

3.3 Implementation Procedure

The aims of this work are to perform the career prediction and to analyze the influential factors behind choosing a particular career by an individual. Many significant parameters mostly are academic skills as well as personal factors are considered here to ensure an effective prediction. For implementing this work, we have performed several steps that are presented in Fig. 2.

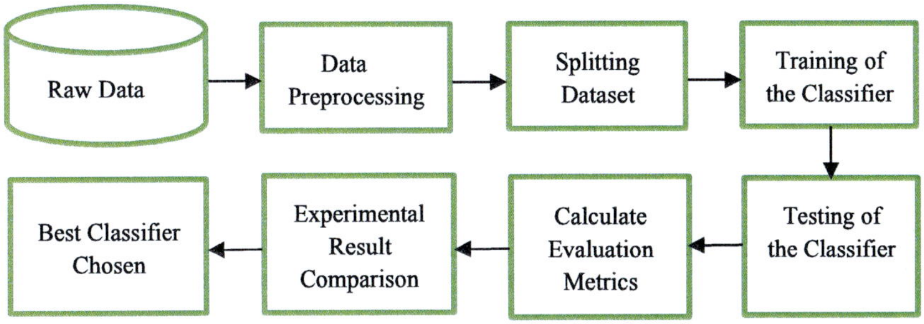

Fig. 2 Working procedure of career prediction using data mining technique in the Bangladesh context

First of all, we have created a questionnaire form which contains 13 questions including some of the academic as well as some of the personal questions. Then we have collected data from many random people through this questionnaire. Then we performed some preprocessing techniques to feed this data into the classifier. A single variable is used to label a single question. So, a total of 13 variables are used to label all the questions. Then, to label the answer to the particular question, the numbers 0–3 are used here. In our dataset, for example, the variable/attribute named "University Type (GT)" has three possible outcomes such as National University (0), Private University (1), and Public University (2). After preprocessing, our prepared data is partitioned into the training and testing set. Here, 80% of the data from the total data set has been used for training purposes. The rest 20% of the whole data set has been used for testing purposes. This splitting is done randomly. After that, the classifiers have been trained with the training data. After training the classifiers, we have used test data to predict the current professional status. Here, we have calculated some of the metrics for the performance evaluation. Using these metrics, we have determined the best classifier to predict in this context. Based on the confusion matrix generated by the classifier, several performance metrics have been calculated in percentage by using Eqs. (1–7).

$$\text{Accuracy} = \frac{\text{TP} + \text{TN}}{\text{TP} + \text{FN} + \text{FP} + \text{TN}} \times 100\% \,. \tag{1}$$

$$\text{Sensitivity or Recall or True Positive Rate(TPR)} = \frac{\text{TP}}{\text{TP} + \text{FN}} \times 100\% \,. \tag{2}$$

$$\text{Specificity or True Negative Rate(TNR)} = \frac{\text{TN}}{\text{FP} + \text{TN}} \times 100\% \,. \tag{3}$$

$$\text{False Positive Rate(FPR)} = \frac{\text{FP}}{\text{FP} + \text{TN}} \times 100\% \,. \tag{4}$$

$$\text{False Negative Rate(FNR)} = \frac{\text{FN}}{\text{FN} + \text{TP}} \times 100\% \,. \tag{5}$$

$$\text{Precision} = \frac{\text{TP}}{\text{TP} + \text{FP}} \times 100\% \,. \tag{6}$$

$$\text{F1 Score} = 2 \times \frac{\text{Precision} \times \text{Recall}}{\text{Precision} + \text{Recall}} \times 100\% \,. \tag{7}$$

4 Result and Discussion

Since this work is mainly a multiclass problem, the classifier produced a 3 * 3 confusion matrix. The generated matrix for each of the classifiers is presented in Table 2.

To evaluate this work, Accuracy, TPR, TNR, FPR, FNR, Precision, and F1 Score from the above confusion matrix is computed. The result of several performance evaluation metrics is presented in Table 3.

From Table 3, we found that the Part classifier outperforms the other five classifiers from the overall analysis of the result. The accuracy of the Part classifier for Govt. Job, Private Job, and Business class are 92.93, 87.88, and 94.95% which is the highest among all the classifiers. The F1 Score of the Part classifier for Govt. Job, Private Job, and Business class are 86.79, 90.48, and 73.69% which is also the highest among all the classifiers. And also, other results of Table 3 support the Part classifier.

Table 2 Confusion matrix for applied classifier

Logistic regression		Predicted			Multilayer perceptron		Predicted		
		Govt. job	Private job	Business			Govt. job	Private job	Business
Actual	Govt. job	23	1	1	Actual	Govt. job	20	2	3
	Private job	5	52	6		Private job	5	55	3
	Business	0	4	7		Business	0	4	7
Kstar		Predicted			Part		Predicted		
		Govt. job	Private job	Business			Govt. job	Private job	Business
Actual	Govt. job	19	5	1	Actual	Govt. job	23	2	0
	Private job	4	58	1		Private job	5	57	1
	Business	0	8	3		Business	0	4	7
J48		Predicted			Random Forest		Predicted		
		Govt. job	Private job	Business			Govt. job	Private job	Business
Actual	Govt. job	23	2	0	Actual	Govt. job	22	1	2
	Private job	5	56	2		Private job	6	54	3
	Business	0	4	7		Business	0	5	6

Table 3 Comparison of six classifier's performance

Classifier	Class name	Accuracy (%)	TPR (%)	TNR (%)	FPR (%)	FNR (%)	Precision (%)	F1 Score (%)
Logistic	Govt. job	92.93	92	93.24	6.76	8	82.14	86.79
	Private job	83.84	82.54	86.11	13.89	17.46	91.23	86.67
	Business	88.89	63.64	92.05	7.95	36.36	50	56
MLP	Govt. job	89.9	80	93.24	6.76	20	80	80
	Private job	85.86	87.3	83.33	16.67	12.7	90.16	88.71
	Business	89.9	63.64	93.18	6.82	36.36	53.85	58.34
Kstar	Govt. job	89.9	76	94.59	5.41	24	82.61	79.17
	Private job	81.82	92.06	63.89	36.11	7.94	81.69	86.57
	Business	89.9	27.27	97.73	2.27	72.73	60	37.5
Part	Govt. job	92.93	92	93.24	6.76	8	82.14	86.79
	Private job	87.88	90.48	83.33	16.67	9.52	90.48	90.48
	Business	94.95	63.64	98.86	1.14	36.36	87.5	73.69
J48	Govt. job	92.93	92	93.24	6.76	8	82.14	86.79
	Private job	86.87	88.89	83.33	16.67	11.11	90.32	89.6
	Business	93.94	63.64	97.73	2.27	36.36	77.78	70
Random forest	Govt. job	90.91	88	91.89	8.11	12	78.57	83.02
	Private job	84.85	85.71	83.33	16.67	14.29	90	87.8
	Business	89.9	54.55	94.32	5.68	45.45	54.55	54.55

5 Conclusion and Future Work

This work is mainly done to perform the prediction of the individual's career and to recognize the influential factors to choose the particular type of career by the individual. Several data mining techniques are applied here to do this. To accomplish this work, a total of 80 and 20% of data is used to train and test the classier, respectively. To evaluate the working classifier, we have measured several performance evaluation metrics. Among all the data mining techniques, we found that part classifier outperforms others. We also found that both academic factors and personal factors have a significant impact on career prediction as well as a career selection. In the future, we will work with more datasets with more attributes and apply more data mining techniques.

References

1. Arafath, M.Y., Saifuzzaman, M., Ahmed, S., Hossain, S. A.: Predicting career using data mining. In: International Conference on Computing, Power and Communication Technologies (GUCON), pp. 889–894. IEEE, (2018)

2. Gorad, N., Zalte, I., Nandi, A., Nayak, D.: Career counseling using data mining. Int. J. Eng. Sci. Comput. (IJESC) **7**(4), 10271–10274 (2017)
3. Rangnekar, R.H., Suratwala, K.P., Krishna, S., Dhage, S.: Career prediction model using data mining and linear classification. In: 4th International Conference on Computing Communication Control and Automation (ICCUBEA), pp. 1–6. Pune, India (2018)
4. García-Peñalvo, F.J., Cruz-Benito, J., Martín-González, M., Vázquez-Ingelmo, A., Sánchez-Prieto, J.C., Therón, R.: Proposing a machine learning approach to analyze and predict employment and its Factors. Int. J. Interact. Multim. Artif. Intell. (IJIMAI) **5**(2), 39–45 (2018)
5. Al-Saiyd, N.A., Al-Takrouri, A.S.: Prediction of IT jobs using neural network technique. Ubiquitous Comput. Commun. J. **9**(10), 1500–1507 (2015)
6. Campagni, R., Merlini, D., Sprugnoli, R., Verri, M.C.: Data mining models for student careers. Expert Syst. Appl. **42**(13), 5508–5521 (2015)
7. Panda, S., Pattanaik, P.A., Swarnkar, T.: A higher education predictive model using data mining techniques (2017)
8. Roy, K.S., Roopkanth, K., Teja, V.U., Bhavana, V., Priyanka, J.: Student career prediction using advanced machine learning techniques. Int. J. Eng. Technol. (IJET) **7**(2.20), 26–29 (2018)
9. Nie, M., Yang, L., Sun, J., Su, H., Xia, H., Lian, D., Yan, K.: Advanced forecasting of career choices for college students based on campus big data. Front. Comput. Sci. **12**(3), 494–503 (2018)
10. Lou, Y., Ren, R., Zhao, Y.: A Machine Learning Approach for Future Career Planning. Standford University (2010)
11. Zulfiker, M.S., Kabir, N., Biswas, A.A., Chakraborty, P., Rahman, M.M.: Predicting student's performance of the Private Universities of Bangladesh using machine learning approaches. Int. J. Adv. Comput. Sci. Appl. (IJACSA) **11**(3), 672–679 (2020)
12. Mia, M.J., Biswas, A.A., Sattar, A., Habib, M.T.: Registration status prediction of students using machine learning in the context of Private University of Bangladesh. Int. J. Innov. Technol. Explor. Eng. (IJITEE) **9**(1), 2594–2600 (2019)
13. Biswas, A.A., Majumder, A., Mia, M.J., Nowrin, I., Ritu, N.A.: Predicting the enrollment and dropout of students in the post-graduation degree using machine learning classifier. Int. J. Innov. Technol. Explor. Eng. (IJITEE) **8**(11), 3083–3088 (2019)
14. Dusane, P.D., Bhosale, N.V., Avhad, V.A., Naikwade, P.K.: Recommendation system for career path using data mining approaches. Int. J. Sci. Res. Eng. Trends **6**(2), (2020)
15. Mohamed S., Abdellah E. (2020) Data mining approach for employability prediction in Morocco. In: Bhateja, V., Satapathy, S., Satori, H. (eds.) Embedded Systems and Artificial Intelligence. Advances in Intelligent Systems and Computing, vol. 1076. Springer, Singapore. https://doi.org/10.1007/978-981-15-0947-6_69
16. Logistic Regression in Machine Learning: https://www.javatpoint.com/logistic-regression-in-machine-learning, last accessed 2020/08/25
17. A Quick Introduction to Neural Networks: https://ujjwalkarn.me/2016/08/09/quick-intro-neural-networks. Last accessed 2020/08/22
18. Multilayer Perceptron: https://medium.com/@jorgesleonel/multilayer-perceptron-6c5db6a8dfa3. Last accessed 2020/08/25
19. Class KStar: https://weka.sourceforge.io/doc.stable-3-8/weka/classifiers/lazy/KStar.html. Last accessed 2020/08/20
20. Cleary, J.G., Trigg, L.E.: K^*: an Instance-based learner using an entropic distance measure. In: 12th International Conference on Machine Learning, pp. 108–114 (1995)
21. Witten I.H., Frank, E.: Data Mining: Practical Machine Learning Tools and Techniques, 2nd edn. Morgan Kaufmann (2005)
22. Kaur, G., Chhabra, A.: Improved J48 classification algorithm for the prediction of diabetes. Int. J. Comput. Appl. (IJCA) **98**(22), 13–17 (2014)
23. Classification Algorithms—Random Forest: https://www.tutorialspoint.com/machine_learning_with_python/machine_learning_with_python_classification_algorithms_random_forest.htm. Last accessed 2020/08/20

Network and Security

Secured Smart Healthcare System: Blockchain and Bayesian Inference Based Approach

Fahiba Farhin, M. Shamim Kaiser, and Mufti Mahmud

Abstract The Internet of Healthcare Things (IoHT) is a rising smart ubiquitous structure that interconnects shrewd gadgets, devices, stakeholders (e.g., specialists, patients, caregivers, researcher, and so on.), and infrastructure utilizing smart sensors. Over the past decade, exposure to modern devices and data sensing and analysis strategies has allowed numerous analysts to develop and deliver personalized IoHT services. This brought about an impressive number of examination results tending to the applications, difficulties, and likely arrangements focusing on secured communication inside the IoHT system. Despite sizable efforts devoted to it, secured service provisioning is a first-rate mission yet. This research aims at providing a multilayer security framework for heterogeneous IoHT systems using bayesian inference based trust management with the digitally signed blockchain network that includes individual node and data protection. The simulation results reveal the feasibility and effectiveness of the proposed dynamic approach regarding malicious node detection.

Keywords Internet of Healthcare Things (IoHT) · Data analytics · Bayesian inference · Blockchain · Trust management · 5G network.

1 Introduction

The Internet of Things or usually alluded to as IoT incorporates things (for example, objects, gadgets, foundation, and so forth) associated with embedded devices via smart sensors. These inserted gadgets associate those shrewd sensors to the web.

F. Farhin (✉) · M. S. Kaiser
Institute of Information Technology, Jahangirnagar University, Dhaka, Bangladesh
e-mail: farhinfahiba@gmail.com

M. S. Kaiser
e-mail: mskaiser@juniv.edu

M. Mahmud
Department of Computer Science, Nottingham Trent University, Nottingham, UK
e-mail: mufti.mahmud@ntu.ac.uk

M. S. Kaiser et al. (eds.), *Proceedings of International Conference on Trends in Computational and Cognitive Engineering*, Advances in Intelligent Systems and Computing 1309, https://doi.org/10.1007/978-981-33-4673-4_36

In this way, IoT is considered as an incorporated piece of things to come in correspondence of the internet. In recent years, IoT has drawn considerable attention to the research communication and it provided solutions for smart cities, agriculture, industry, office, automobile, retails, security services, and healthcare.

Given that IoT related examination has sharply grown over the most recent years in various spaces, numerous interdisciplinary explorations are regulated inside the healthcare setting. The Internet of Healthcare Things (IoHT) integrates healthcare devices (e.g., imaging devices, physiological sensing devices, diagnostic devices, patient tracking, real-time patient monitoring at home/hospital, etc.), infrastructure (e.g., ambulance, wheelchair, bed, pharmacy, billing, etc.) and services (e.g., elderly care, fitness, remote monitoring, and personalised treatment delivery, etc.) enabled by the IoT [5]. The points of IoHT are diminishing healthcare cost and device downtime, upgrading the client experience, and improving personal satisfaction by enhancing the utilization of healthcare assets [6].

The IoT market in healthcare is predicted to grow USD 188$ billion by 2024, at a CAGR of 27.6% during the forecast period [5]. All the massive players during this industry try to require advantage of this growth and improving their solution or investing to adopt such technologies. Since the Internet of Things (IoT) implementation is rapidly growing, thus the IoT nodes are prone to various cyber threats. The attacker initially exploits weak points in the IoT system which is the primary step towards breaching a critical device. Securing user and data protection and privacy is one of the main problems facing the IoHT sector now.

Among numerous emerging technologies, blockchain is one of the most promising to meet recent security requirements [3]. New technology and future network generation require blockchain for 5G service deployments. From a technological viewpoint, blockchain is a distributed ledger system that was first used to support Bitcoin's public digital ledger for economic transactions [18]. Nowadays, it has been increasingly used to register, authenticate and verify properties and transactions, manage interactions, record data and manage multi-party recognition in a trustworthy, transparent and protected manner [23].

Despite the extensive effort exerted into the research on privacy and security for the IoHT frameworks, it's still in its infancy. To facilitate the mixing of security and privacy, this work proposes a two-layer security mechanism where trust management is considered for node trust and integration of blockchain for data security. It discusses an IoHT framework consists of five layers and provides a comprehensive review on privacy and security issues in those layers. The work anticipates layer-wise attacks and accordingly discusses countermeasures and eventually, main challenges and solutions are outlined at the top.

The rest of the paper describes related works in this field in Sect. 2 and proposes a security mechanism comprises IoHT framework, possible threats and solution module in 3. Simulation result of node behaviour is presented in 4. At the end, Sect. 5 is conclusion.

2 Related Work

In the past sixteen years of IoT development, many researchers concentrated on the IoT security have been published, however, the IoT model heterogeneity raised the concern of the security issue [10]. Joel et al. studied presented a study of IoHT's framework and supporting technologies to highlight the cutting-edge e-health fields examining patterns and challenges [20, 21]. Mahmoud et al. proposed the convergence of cloud computing and IoT, contributing to the Cloud of Things (CoT) in the sense of intelligent health systems that involve CoT principles, architectures, frameworks and e-health applications [14]. In hospitals with medical equipment, many patient details including main symptoms can be treated with IoHT to properly diagnose and treat patients [2].

There are numerous challenges and risks in the various context of the IoHT environment. This was explored in several research projects focusing on data preprocessing and management, interoperability, legislation, interaction between human devices and network applications [9]. Privacy and security, as anticipated at any time, anywhere and any means [5] are two key challenges in IoHT system. The survey of e-health safety and privacy by Hathaliya et al. is exhaustive [10], while Kang et al. study the latest IoHT security framework protocols in the mobile networks [12] in a systematic way. Different health data protection mechanisms have been proposed, including encryption-based, hybrid encoding for the protection of medical images in electronic health records (eHR) [8], and machine learning-based safety frameworks for the identification of malignancy in a smart health care system [19].

Trust management with different theories and techniques has been analysed in past years to protect wireless sensor nodes and networks from malicious activity [4]. Shabut *et al.* introduced energy and social aware multi-dimensional trust management model to obtained an improved quality of service (QoS) in the context of mobile ad-hoc networks [22]. Mahmoud et al. introduced an ANFIS based trust management system [1]. Authors have calculated data as well as behavioural trust values connected IoT and relay nodes in a cloud-IoT framework [16]. Later on, different machine learning-based approaches have been introduced for trust and reputation calculation. SVM based algorithm is offered to classify the extracted trust features calculated numerically and combine them to produce a final trust value to be used for decision making in [11]. Lahbib et al. executed a blockchain-based trust engineering to gather trust evidence, to characterize a trust score for every device and to safely store and offer them with different devices inside the system by implanting them into blockchain transaction [13].

3 Proposed Framework

This research offers a unique security model to ensure the end to end confidentiality and safety of the healthcare network from different types of attacks and threats. This section illustrates every step of the proposed system model.

Figure 1 shows that this study comes up with an idea to integrate a Bayesian inference based trust management incorporated in the fog controller for node security with the blockchain for data security. The proposed model can be further discussed in three modules: firstly, IoHT module covering five layers framework; secondly, attacks and vulnerabilities in concerning layers and finally, distributed security module considering nuts and bolts of proposed safety measure.

3.1 Healthcare Module

IoHT consists of a large number of connected sensors and IoT devices which are generally with limited processing power and generate plenty of data that need to be processed differently. Centralized cloud platform connected with IoT provides support in this case. However, cloud technology is most secured so the processing takes time which results in latency and power consumption issues. To solve this fact, the fog layer is introduced which adds smart IoT-Fog gateways which provide reliability and better performance at the same time. Yet, latency and QoS issues cannot be solved for delay-sensitive transmission as healthcare deals with various type of loss-sensitive and real-time data.

To solve this issue, IoHT framework needs to handle the data processing in 'on-demand' basis in different layers which introduced the mist layer in IoHT architecture. The addition of mist layer in framework adds rule-based data processing which reduces data volume transmission that will ensure better performance, low power

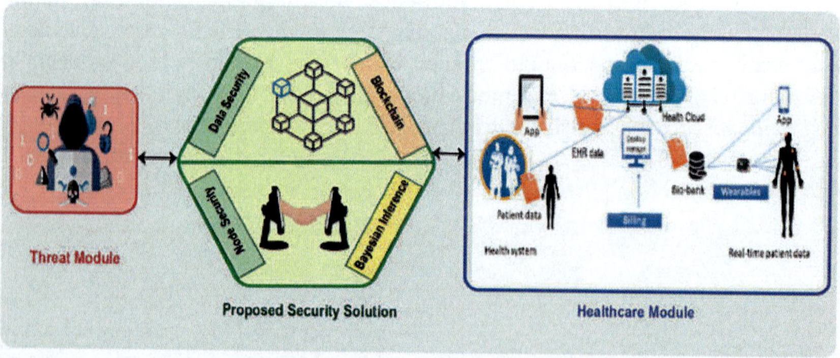

Fig. 1 Proposed system architecture comprising three module: threat, healthcare and security solution.

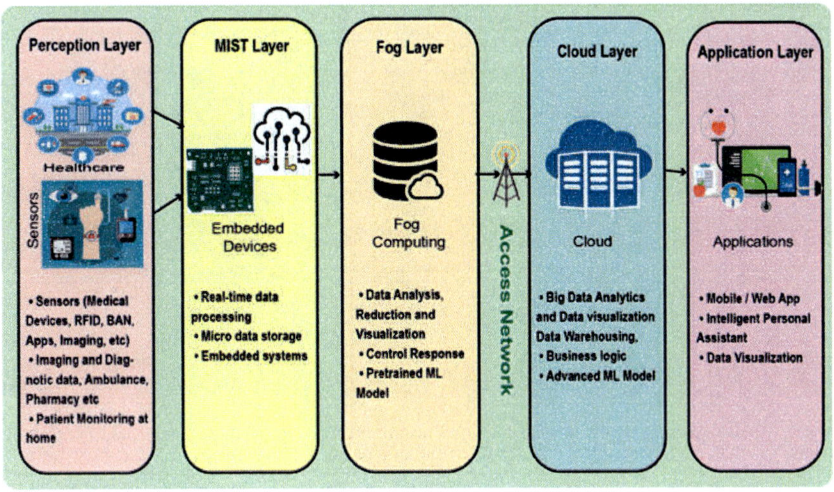

Fig. 2 Proposed five layered IoHT framework.

consumption and latency. So an IoHT framework containing five layers, namely, Perception, Mist, Fog, Cloud, and Application has been proposed (see Fig. 2) which is capable of sensing real-time health data and forward it through rule-based upper layer thus results in better care for patient [5].

3.2 *Threat Module*

Generic IoHT frameworks like the one discussed above face various challenges including security and privacy threats at devices and infrastructure levels, and lack of standardization in healthcare big data. The layer-wise relevant threats and attacks of IoHT framework are investigated and listed in Fig. 3.

At the perception layer, vulnerabilities come through the physical object as it connects people, medical devices and hospital infrastructure and collects healthcare data [5]. Mist layer work as a pathway between physical device and fog layer which contains microdata storage and embedded systems. Therefore, malicious activities by hackers can affect communication devices and/or network. Fog layer abnormalities or anomaly affects through underlined access network or intruder that results in mentioned attacks. Cloud layer is less vulnerable than the previous ones as healthcare data are stored into the cloud after being processed in the fog layers. Artificially intelligent, ML and reasoning-based algorithms are employed in the cloud for data analysis. So the attacks in the cloud layer are related to SQL injection, password stealing, cross-site scripting and similar ones. The application layer is more susceptible to attacks due to shared communication protocols and services along with various interface methods between the end-users, stakeholders and communications network [15].

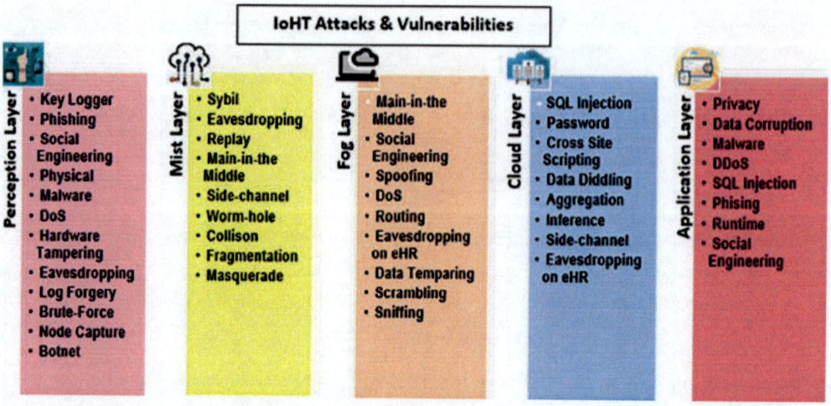

Fig. 3 Taxonomy of possible attacks and vulnerabilities on different layers of the IoHT framework.

3.3 *Security Module*

Traditional security framework generally bothers about the data security which can be efficiently mitigated by integrating blockchain in the system [7]. Since vulnerabilities or attacks may occur in physical level as smart healthcare is nothing but a hub of connected devices and sensors. So, to prevent infrastructure level problem like node capturing, eavesdropping etc., the trustworthiness of the connected node is the most considerable thing [22]. In these circumstances, a fusion of automated trust calculation to ensure node security and encryption of data in blockchain for data security can ensure the overall compactness of the system from vulnerabilities.

3.3.1 Fog Controller

The proposed system considers an IoHT environment where different stakeholders are using various sensors and medical devices which generate plenty of data. These data can be considered as packets which must go through the underlying layers to get stored in eHR. As the mentioned IoT devices are accessed through the internet, so data is passed through the Mist gateway. According to current technology, the mist layer knows about the attached devices in the physical layer. The system is considered to be tightly coupled with blockchain. Hence, the communication request of every device passing through mist gateway is considered as a request for the transaction to enter into a block.

Since the presented method transmits data packet flow to mist from IoT or mobile device, the gateway switch checks its destination address for that specific packet. When forwarding to the upper layer controller as a Tx_1 transaction, the switch's private key digitally signatures it. Figure shows the Tx_1 format Fig. 4. The controller's

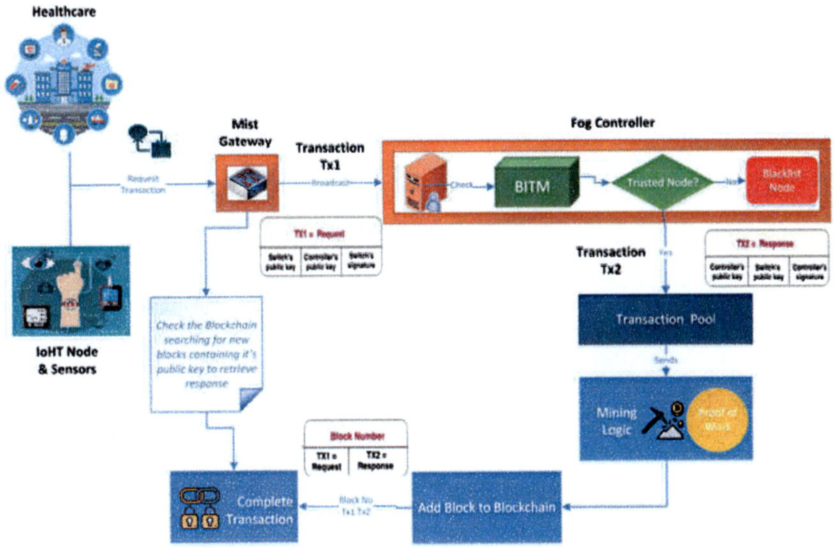

Fig. 4 Work flow of security framework that embraces machine learning based trust management in SDN controller and blockchain

public key, which controlled the switch, will also be sent to accelerate decision-making. Upon receiving Tx_1 transaction request, the controller will send it to the BITM module to verify the node's trustworthiness. Any connected device or sensor will be considered as a node as it generates or sends data. If the node is determined to be a trustworthy one, the controller will evaluate and process the Tx_1 transaction and create a new Tx_2 transaction to update the switch status. The second transaction (Tx_2) contains public keys of the controller and switch, controller's digital signature, and action answer.

3.3.2 BITM

BITM stands for Bayesian Inference based Trust Management. Different ML-based trust evaluation has been discovered in recent years [11]. A Bayesian inference based trust approach is proposed here to detect malicious node in an IoHT environment by evaluating node behaviour. Bayesian inference works on Bayes' rule to adjust the probability of an event [17]. Computing trustworthiness of connected nodes and finding the malicious one using a user-defined rule is the main goal of trust management. In the proposed system, all data packets served from a device are considered independent and so their probability of being benign or malicious can be said $\frac{1}{2}$. For instance, if N packets are considered from a device and k of them are safe, then the probability can be computed using Binomial distribution:

$$P(n(N) = k|p) = \binom{N}{k} p^k (1-p)^{N-k} \tag{1}$$

where $n(N) = k$, which shows n successes out of N Bernoulli trials, and both n and p has the same probability for the true trial.

Bayesian inference evaluates the trustworthiness on behaviour profiling of that device or node. For a given period t, a device's trust value is (t_{value}^{device}) and packet trust is (t_{value}^{packet}) which can be computed as below:

$$t_{value}^{device} = \lambda \times \frac{k_t^{data} + 1}{N_t^{data} + 2}; \tag{2}$$

$$t_{value}^{packet} = \lambda \times \frac{k_t^{packet} + 1}{N_t^{packet} + 2} \tag{3}$$

Here,N_t^{data} denotes the total number of packets where k_t^{data} are considered as benign at a certain t times and λ is the forgetting factor.

The following equation can be written for trust evaluation:

$$t_{value}^{total} = X_1 \times t_{value}^{packet} + X_2 \times t_{value}^{device} \tag{4}$$

In above Equation, X_1 and X_2 are weight values and $X_1 + X_2 = 1$. Therefore, a device will be safe and forward data packet if the trust value increase or equal to a threshold of T_{thres} which belongs to [0, 1]. To check the trustworthiness of the considered node, it has to check t_{value}^{total} periodically.

- If $t_{value}^{total} \geq T_{thres}$, then the device will be safe to transmit data packet.
- If $t_{value}^{total} < T_{thres}$, then the device will be blacklisted.

This mechanism can examine the trustworthiness of a device based totally on its long-time behaviour and offer extra flexibility for administrators to govern and control the network community in the fog controller.

3.3.3 Blockchain Environment

According to the design, the mining module is placed after the fog controller. As appeared in Fig. 4, partaking controllers will place their transactions in the transaction-pool, at that point other non-taking controllers that were not associated with the previously mentioned stage will go about as miners to gather the transactions from controllers to compare and validate them by affirming Proof-of-Work (PoW). Upon validation and achieving consensus, it will be considered as a block needed to be attached to the chain with the block number consisting of Tx_1 and Tx_2. After getting connected to the chain, the mist switch will be informed that a block is trying to retrieve the data using its public key.

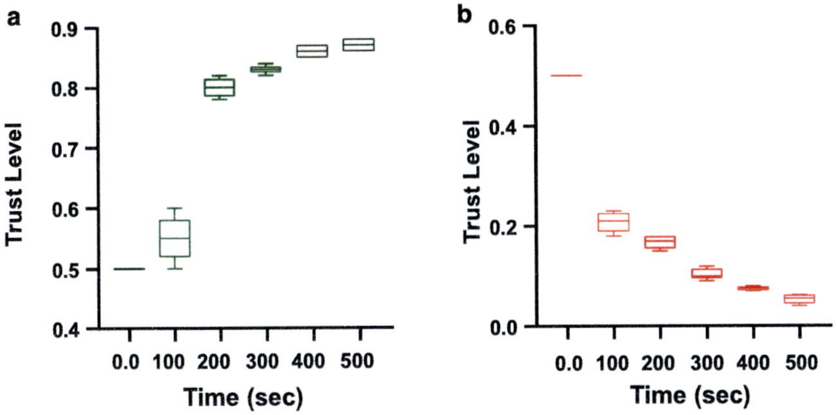

Fig. 5 Effect of node behaviour on the trust level **a** Trustworthy nodes **b** Malicious nodes

4 Simulation and Evaluation

The proposed Bayesian inference based node trust model has been simulated in NS2. This simulator considers a healthcare network consisting of various nodes that are randomly located in the simulated area. The simulation time is considered to be 500 secs. All newly-added nodes were assumed to be trustworthy with the trust value $\frac{1}{2}$. The threshold trust value is $T_{threshold} = 0.8$.

In this simulation, the trust value of every node is considered for some time. Receiving malicious packets from a node will decrease the trust value of that particular node whereas trust value will increase for sending benign packets for that period. Figure 5 shows the two scenarios of node behaviour. Initially, both incident has a trust value of 0.5. Then, in Fig. 5a, the trust level is increasing from 0.8 to above with time for behaving as a reliable one. On the contrary, in Fig. 5b, the trust level is decreasing from 0.2 to below for being a malicious one.

The data packet from nodes having a trust level more than or equal to 0.8, will be considered for sending to the transaction pool to achieve PoW and added to the new block.

5 Conclusion

Healthcare is experiencing a rapid transition from conventional to a global patient-centred approach. Advances in technologies have been boosted up this rapid transformation of healthcare which also introduces various challenges in IoHT network. This paper offers a heterogenous five-layered IoHT framework where possible vulnerabilities are also discussed. To ensure overall security, a two-layer security framework is proposed here that comprises of blockchain and bayesian inference for data and node security respectively. Healthcare data, generated from various sensors and devices, must go through the underlying layer for being stored or processed. So considering the physical device's trust have been done by behavioural profiling and then the data packet is being stored in blocks if that node is trustworthy. Future work would include the simulation of the current approach and investigate how to make it more dimensional and speedy in an even larger environment.

Acknowledgements This work is funded by the ICT Division, Government of the People's Republic of Bangladesh.

References

1. Afsana, F., Jahan, N., Sunny, F.A., Kaiser, M.S., Mamun, S.A.: Trust and energy aware cluster modeling and spectrum handoff for cognitive radio ad-hoc network. In: ICEEICT. pp. 1–6 (2015)
2. da Costa et al.: Ioht: Toward intelligent vital signs monitoring in hospital wards. Artif. Intell. Med. **89**, 61–69 (2018)
3. Arifeen, M.M., Al Mamun, A., Kaiser, M.S., Mahmud, M.: Blockchain-Enable Contact Tracing for Preserving User Privacy During Covid-19 Outbreak (2020)
4. Arifeen, M.M., et al.: Anfis based trust management model to enhance location privacy in underwater wireless sensor networks. In: ECCE. pp. 1–6. IEEE (2019)
5. Asif-Ur-Rahman, et al.: Toward a heterogeneous mist, fog, and cloud-based framework for the internet of healthcare things. IEEE Internet Things J. **6**(3), 4049–4062 (2018)
6. Biswas, S., et al.: Cloud based healthcare application architecture and electronic medical record mining: an integrated approach to improve healthcare system. In: Proc. ICCIT, pp. 286–291. IEEE (2014)
7. Chakraborty, S., Aich, S., Kim, H.C.: A secure healthcare system design framework using blockchain technology. In: 2019 ICACT, pp. 260–264. IEEE (2019)
8. Elhoseny, M., et al.: Hybrid optimization with cryptography encryption for medical image security in internet of things. Neural Comput. Appl., 1–15 (2018)
9. Farahani, B., et al.: Towards fog-driven IoT eHealth: promises and challenges of IoT in medicine and healthcare. Future Gener. Comput. Syst. **78**, 659–676 (2018)
10. Hathaliya, J.J., Tanwar, S.: An exhaustive survey on security and privacy issues in healthcare 4.0. Comput. Commun. (2020)
11. Jayasinghe, U., Lee, G.M., Um, T.W., Shi, Q.: Machine learning based trust computational model for IoT services. IEEE Trans. Sustain. Comput. **4**(1), 39–52 (2018)
12. Kang, J.J.: Systematic analysis of security implementation for internet of health things in mobile health networks. In: Data Science in Cybersecurity and Cyberthreat Intelligence, pp. 87–113. Springer, Heidelberg (2020)
13. Lahbib, A., et al.: Blockchain based trust management mechanism for IoT. In: 2019 IEEE Wirel. Commun. Netw. Conf. (WCNC), pp. 1–8 (2019)
14. Mahmoud, M.M.E., et al.: Enabling technologies on cloud of things for smart healthcare. IEEE Access **6**, 31950–31967 (2018)
15. Mahmud, M., Kaiser, M.S., Hussain, A.: Deep Learning in Mining Biological Data. arXiv:2003.00108 [cs, q-bio, stat] pp. 1–36 (Feb 2020). http://arxiv.org/abs/2003.00108, arXiv: 2003.00108
16. Mahmud, M., et al.: A brain-inspired trust management model to assure security in a cloud based IoT framework for neuroscience applications. Cogn. Comput. **10**(5), 864–873 (2018)

17. Meng, W., et al.: Towards Bayesian-based trust management for insider attacks in healthcare software-defined networks. IEEE Trans. Netw. Serv. Manag. **15**(2), 761–773 (2018)
18. Nakamoto, S., Bitcoin, A.: A peer-to-peer electronic cash system. Bitcoin (2008). https://bitcoin.org/bitcoin.pdf
19. Newaz, A.I., Sikder, A.K., Rahman, M.A., Uluagac, A.S.: HealthGuard: a machine learning-based security framework for smart healthcare systems. In: 2019 SNAMS, pp. 389–396 (2019)
20. Paul, M.C., et al.: Low cost and portable patient monitoring system for e-health services in bangladesh. In: Proc. ICCCI, pp. 1–4. IEEE (2016)
21. Rodrigues, J.J.P.C., et al.: Enabling technologies for the internet of health things. IEEE Access **6**, 13129–13141 (2018)
22. Shabut, A.M., Kaiser, M.S., Dahal, K.P., Chen, W.: A multidimensional trust evaluation model for manets. J. Netw. Comput. Appl. **123**, 32–41 (2018)
23. Zheng, Z., et al.: An overview of blockchain technology: architecture, consensus, and future trends. In: 2017 IEEE BigData Congress, pp. 557–564. IEEE (2017)

A Blockchain-Based Scheme for Sybil Attack Detection in Underwater Wireless Sensor Networks

Md. Murshedul Arifeen, Abdullah Al Mamun, Tanvir Ahmed, M. Shamim Kaiser, and Mufti Mahmud

Abstract Underwater wireless sensor network (UWSN) has become popular because of its diverse applications and massive improvements in sensing technologies in recent years. However, this sensor network is vulnerable to various types of cyber-attacks due to its inherent characteristics. Among many cyber-attacks, the Sybil attack is one of the fatal attacks and damages the network severely. In this work, we have proposed a blockchain-based Sybil attack detection scheme in UWSN. We have also integrated one of our previous trust model with the blockchain-based method to make it resilient against the attacks detection. We have conducted an experimented in Flask and discussed the implementation with code details.

Keywords Wireless sensor netwrok · UWSN · Blockchain · Sybil attack · Cluster · Postman · Flask

---

Md. M. Arifeen (✉)
Department of ICT, Bangladesh University of Professionals, Dhaka, Bangladesh
e-mail: murshedularifeendipto@gmail.com

A. Al Mamun · M. S. Kaiser
Institute of Information Technology, Jahangirnagar University, Dhaka, Bangladesh
e-mail: abdullah.iiuceee@gmail.com

M. S. Kaiser
e-mail: mskaiser@juniv.edu

T. Ahmed
Department of Mechanical and Production Engineering, Islamic University of Technology, Dhaka, Bangladesh
e-mail: tanvirahmedog@gmail.com

M. Mahmud
Department of Computing & Technology, Nottingham Trent University, Clifton Lane, UK
e-mail: muftimahmud@gmail.com

M. S. Kaiser et al. (eds.), *Proceedings of International Conference on Trends in Computational and Cognitive Engineering*, Advances in Intelligent Systems and Computing 1309, https://doi.org/10.1007/978-981-33-4673-4_37

1 Introduction

Underwater wireless sensor network (UWSN) is a network of hundreds or even thousands of tiny sensor nodes deployed in the underwater environment for various applications, for instance, oil pipeline monitoring, aquatic life monitoring, scientific investigation, underwater resource exploration, military surveillance, etc. [4, 5]. A sensor node senses its surrounding area and collects useful data depending on the application. The sensor nodes which create this large network, covering a huge sea bed, suffer from several resource constraints, such as it is limited to energy, memory, processing power, and communication range. These limitations create challenges from different aspects, especially implementing a strong security algorithm for protecting the network. This network is vulnerable to various types of cyber-attacks [19], and these attacks lead to failure of the network operation. Moreover, human intervention is not possible for monitoring the network daily due to the harsh and random nature of the underwater environment. Also, the deployment area is open, dynamic, and unattended in nature, which makes it difficult to resist the network from potential attackers physically. All these shortcomings make it easier for the attacker to intervene in the network and launch various cyber-attacks.

One of the fatal attacks in WSN is the Sybil attack, which involves attacking by replicating multiple legitimate identities. Also, it affects the network severely and imposes several threats, for instance, distributed storage, multipath routing, data aggregation, voting, fair resource allocation, and misbehavior detection [1, 18]. Therefore, the detection and prevention of the Sybil attack are one of the major challenges in UWSN. Recent studies on detection and prevention of Sybil attack are only suitable for WSN, and these methods cannot be used in terms of UWSN, since the UWSN is different in characteristics and behavior as opposed to WSN. Furthermore, the Sybil attack launches through compromising a legitimate sensor node or deploying a malicious node. The traditional cryptography and key management schemes cannot overcome the compromisation issue. In this regard, trust management can work effectively by evaluating the real-time behavior of the sensor nodes. It can eliminate internal attacks by distinguishing malicious or compromised sensor nodes from legitimate sensor nodes [15]. Storing the trust-related information along with other essential data and sharing them among the sensor nodes is another critical issue. Previous studies show that the authors introduced recommendation or indirect trust management, but it fails when the malicious nodes recommend false information [21]. Therefore, besides securing the network, it also requires to ensure confidentiality, integrity, and trustworthiness of the shared information and data. This can be achieved by introducing blockchain in UWSN.

Moving to blockchain, it is getting popular in the field of WSN and IoT for enhancing security due to its decentralized, fault-tolerant, and secure sharing of data among the distributed network nodes [2, 6, 10]. Blockchain utilizes a Hash function for connecting the blocks of the blockchain and ensures immutability. Along with these features, blockchain also offers availability due to its distributed nature and also, confidentially, integrity, and authentication. Nowadays, various applications use blockchain for many decentralized environment, such as IoT device communications, healthcare, supply chain, intrusion detection, and so on [12, 14, 24].

Therefore in this work, we are integrating blockchain with a trust management model proposed in [3] for detecting Sybil nodes in UWSN. The trust value will be evaluated by the Hidden Markov Model, which is a machine learning approach, meaning that the sensor nodes will learn through HMM about each others' behavior and store this information in the blocks of the blockchain [9, 14]. Any other sensor node can check the blockchain and can know about a sensor node's trust value without re-computing it. Also, according to the Sybil attack, any node which will attempt to access the network by coping legitimate id will get detected with the help of blockchain.

In Sect. 2, we have discussed the related works. In Sect. 3, we have explained the attack model, the trust model, and the blockchain-based Sybil attack detection model. In Sect. 4, we depicted the implementation of the proposed blockchain-based Sybil attack detection model, and finally, Sect. 5 concludes this paper.

2 Literature Review

We have reviewed the literature on Sybil attack countermeasures along with blockchain-based trust models. A message authentication and passing algorithm has been proposed in [8] for checking the trustworthiness and Sybil nodes in the network. The authors assert that if a node does not have authorization in the network, then it cannot join the network and a message authentication scheme serves this purpose. The proposed scheme is very effective in the case of Sybil attack detection. A robust and lightweight scheme based on RSSI has been proposed in [7] to detect Sybil node in WSN. The authors assert that RSSI has some demerits like it is unreliable and time-varying, but they showed that using multiple receivers and ratios of RSSI can overcome this problem. The results show a few false positives and 100% completeness.

A new detection mechanism called cooperative RSS-based Sybil detection (CRSD) has been proposed [13] for Sybil attack detection in a static wireless sensor network. It uses received signal strength (RSS) to detect the Sybil node's position. The authors have taken multiple nodes to compute RSS value in time interval T and group IDs with similar RSS to detect Sybil node. The method protects the throughput of the network. In [23], the authors have proposed a trust-based scheme to detect Sybil node in the WSN. The method first evaluates the trust value of sensor nodes and provides a certificate. Later, it measures distance and uses RSSI value to identify Sybil nodes. Blockchain-based trust model has been proposed to detect malicious nodes in wireless sensor networks, in [20, 22]. Simulation results show that the proposed methods can effectively detect malicious nodes in wireless sensor networks. The evaluation process includes processing delay, forwarding rate, and response time. The authorized nodes include this evaluation information to the blocks of the blockchain along with nodes location, ID, and state (active or inactive) for the detection of Sybil node.

To ensure secure localization of unknown sensor nodes in WSN, the authors proposed a novel trust model based on blockchain in [11]. The model includes behavior-based trust, data-based trust, and feedback-based trust. The results show that it outperforms the existing models in terms of false-positive rates, false-negative rates, and average energy consumption. The authors in [16] proposed a trust management model based on blockchain's immutable property called proof of authority (PoA). According to the model, the sink node adds a new node to the network through the PoA mechanism. The sink node has its own database to keep the transaction record. The BS issues smart contracts, issuance of activity, and process sensed data. The BS monitors the network. It removes any dead node or suspicious node. A robust, transparent, and energy-efficient distributed authentication scheme based on a blockchain has been proposed for the authentication of IoT devices in [25]. BATM [17] ensures a novel way to achieve confidentiality, authentication, and Web of trust.

From the above studies, it is clear that the recently proposed methods for mitigating Sybil attack are not suitable for UWSN, and blockchain can be used effectively to ensure the immutability, privacy, and security of the network.

3 System Model

3.1 Attack Model

The network layer works based on routing protocols. It assigns host and destination IP address to the packet header. The network layer chooses the proper and shortest routing path based on various route selection protocols or algorithms from all the available routing paths. The main operation of this layer is routing. Besides, it also performs local addressing. Most of the attacks in wireless sensor networks are involved with this layer, such as wormhole, black hole, sinkhole, Sybil, Dos attacks. But the most severe attack is the Sybil attack. Sybil attack refers to a malicious node that has several legitimate IDs. The malicious node sends packets or can attack the node of a WSN through multiple IDs. Sybil attack breaches the integrity of data, security, and resource utilization. In Fig. 1, a malicious node has cloned multiple legitimate IDs and attacks various nodes.

3.2 Trust Model

For evaluating the trust value of a sensor node, we will utilize the trust model proposed in our other work [3], where the Hidden Markov Model (HMM) is used to predict a sensor nodes trustworthiness. Since the underwater channel is dynamic, the HMM can address the dynamic nature of the sensor network. Packet loss and error are considered as observation state, and trust or malicious is considered as hidden states.

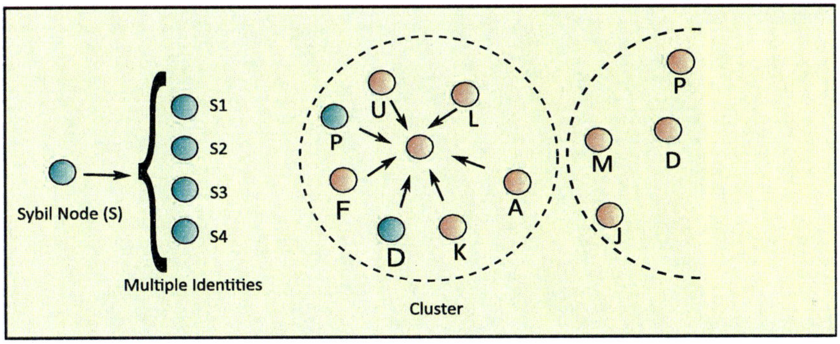

Fig. 1 A Sybil node shows multiple identities such as $S1$, $S2$, $S3$, $S4$. In the cluster, a Sybil node tries to access the network through copying legitimate id's p and D

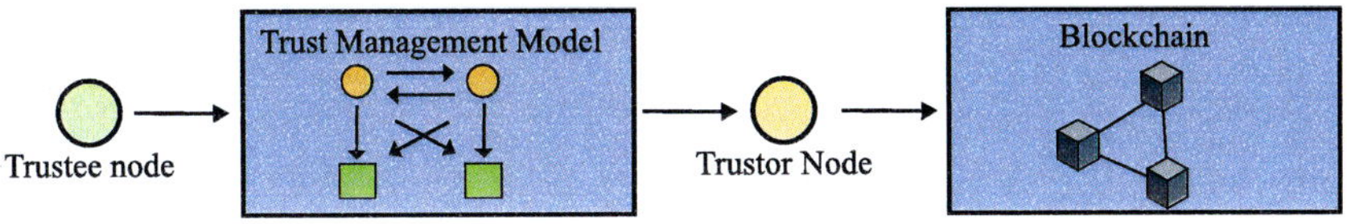

Fig. 2 Trustee nodes trust value will be evaluated through the HMM-based trust management model by the trustor node and the trust information will be stored in the blockchain for future reference

Based on the observation state, the Hidden Markov Model can successfully measure the trust and distrust values of sensor nodes. Figure 2 shows that the trustor nodes evaluate the trust value of the trustee node through the HMM-based trust model.

3.3 Blockchain-Based Sybil Attack Detection

To get access and join in the network, each sensor node will request to the CH showing its unique ID. After getting access, the sensor nodes will update its existence in the cluster after a fixed time interval periodically, since due to the dynamic nature of the environment, the sensor nodes change its position or even may out of the range. The CH will evaluate the trust values of each sensor node of its cluster based on the trust management model. If the evaluated node satisfies the trust requirements, then the CH will generate a transaction for the sensor node and add this transaction to the blockchain. Each transaction contains the status of the nodes (for instance, its ID, location) and the corresponding trust value. If the sensor node changes its location and relocates to a new cluster, then it needs to send a request message for joining the network again. The corresponding CH will check the blockchain about the previous information of the sensor node. If there is a transaction about this new node showing satisfactory trust value, then the node will get access without recomputing the trust value again. Each CH will also include the timestamp of the particular period along

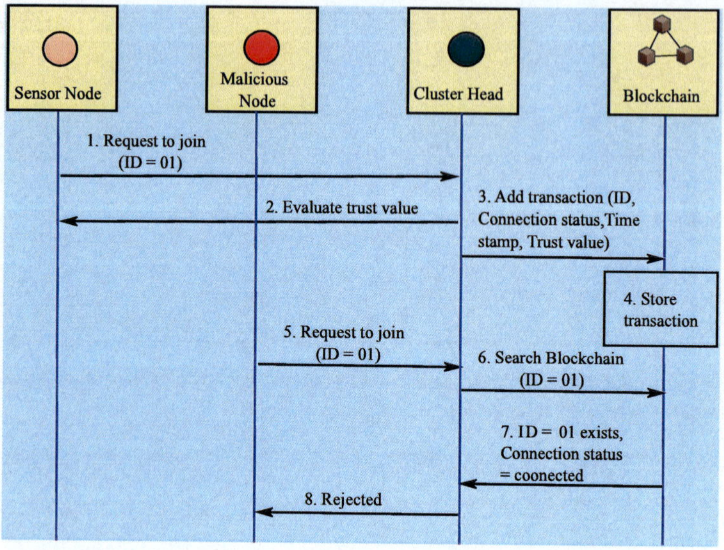

Fig. 3 A scenario which shows a sensor node requests to join in the network and the cluster head approves the joining request through the proposed scheme. Later a malicious node copies ID from the sensor node and tries to join but the cluster head rejects since from the blockchain it is evident that the requested ID is already connected in the blockchain

with each sensor's connection status that is connected or disconnected for detecting any Sybil nodes. When a malicious node copies a legitimate ID and tries to access through the copied ID, the CH will search for the ID in the blockchain, and if the ID is in the blockchain, but then with a different timestamp and the connection, status is connected and then the requesting sensor nodes is a Sybil node, that is trying to access the network. Otherwise, the node is a legitimate one.

In the flowchart shown in Fig. 3, in step 1 a, sensor node with ID = 01 requests to the CH to join in the network. The CH first evaluates the trust value through the trust management model in step 2. After measuring the trust value, the CH generates a transaction and adds this transaction to the blockchain, as shown in step 3 and step 4, respectively. The transaction contains the sensor node of ID = 01's, connection status that is connected, its trust value, and timestamp. In step 5, a malicious node copies the ID = 01 and tries to join the network by requesting to the CH. Upon getting a request, the CH searches the blockchain for the corresponding ID in step 6. However, the blockchain shows that the requested ID 01 is already connected to the network. Therefore, the CH finally rejects the new requesting sensor node with ID 01.

4 Implementation

4.1 Experimental Setup

In the proposed system, we experimented with the Windows10 operating system (version1909) built with 6 GB RAM and 1 TB HDD. We used Python (version3.7.4) as a high-level programming language and coded on Spyder IDE. Besides, we also used Flask (version0.12.2) as a Web app framework for our work. We deployed the experiment over Postman (version7.29.1) API testing.

4.1.1 Connection and Testing of Nodes

We set up the blockchain network with four individual nodes. Usually, Flask runs on $http://127.0.0.1:5000/$. However, to make the system decentralized, we made four different nodes by just adding four different ports at the end of the localhost address of Flask, which are as follows:

- $http://127.0.0.1:5007/$
- $http://127.0.0.1:5008/$
- $http://127.0.0.1:5009/$
- $http://127.0.0.1:5010/$

For the convenience of our targeted environment, we considered each node as an independent sensor network node. After compiling the codes at each node, we first checked the Genesis Block at each node, using the function *get_chain* Fig. 4a, which is a **GET** method request. *get_chain* function basically returns the updated blockchain at each node. After that, using the *connect_node* function, which is a **POST** method request, we connected all nodes successfully, where we provided the node information as input in JSON format. Now our system is ready to add transaction (nodes information) to the blockchain.

4.1.2 Adding Node's Information

We collected the trust value for each node from the aforementioned trust management model and the connection status of the node (either connected or disconnected) from the routing table. We then provided the node information into the blockchain using the *AddNewNode* Fig. 4b. The node info fed into the system using JSON formatted input. There are three mandatory information regarding a node needed to be provided at the time of AddNewNode request. The mandatory information is *node_id*, *trust_value*, and *status*. After providing this necessary info, a **POST** request was generated. The system then checks the *node_id* and the *status*. If the system found that the *node_id* and the *status* are similar, then it assumed the node is a Sybil node and hence denies the node to be added to the blockchain Fig. 4c. However, if the system found the

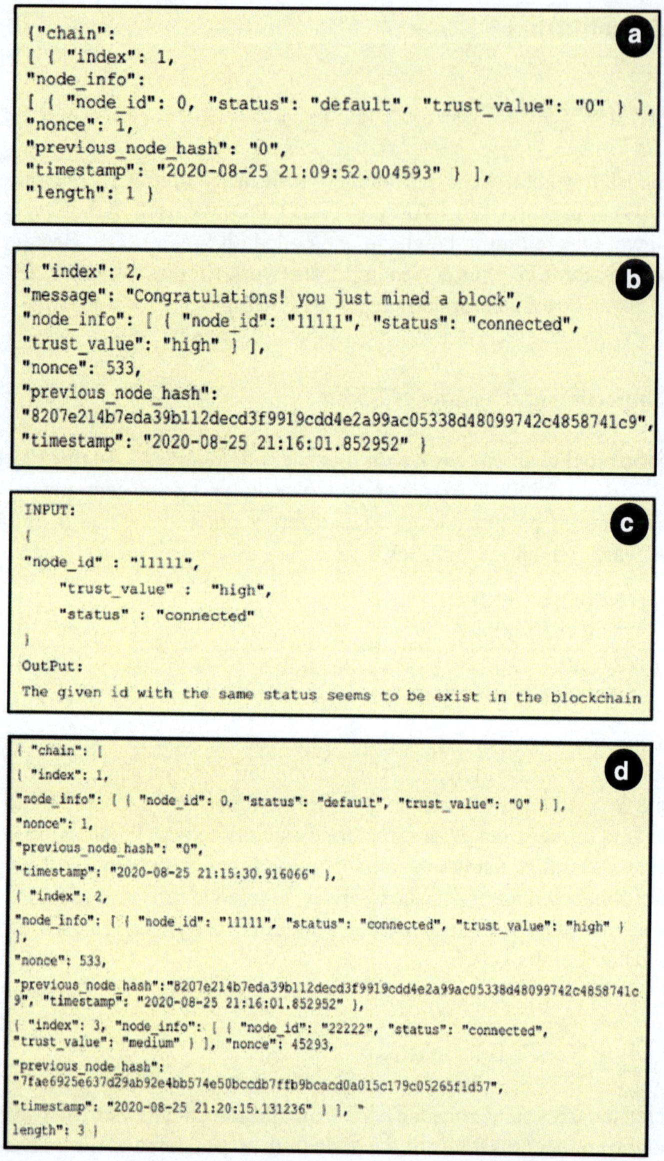

Fig. 4 Step-by-step implementation of the blockchain showing the detection of Sybil attack

node_id similarity, but the node *status* is different, then the system adds the node, as some nodes have the possibility to be disconnected at some point in time due to adverse underwater environment.

4.1.3 Updating the Blockchain

After a definite time interval, the block is updated with all new nodes information. The update of the blockchain occurs with the help of *replace_chain* function Fig. 4d. This function is a **GET** request method, which combines the updated information from each node and makes the blockchain identical at each node. Therefore, whenever a new node tries to communicate with an existing node, the node checks the information about the new node in the blockchain and can easily find out if the node is malicious.

5 Conclusion

Sybil attack is disastrous and seriously damages the network. This paper analyzes the consequences of the Sybil attack and solutions offered by several researchers. It then proposes a security framework against the Sybil attack for blockchain-focused underwater wireless sensor networks. We merged one of our previous model of trust with the blockchain, which helps with recognizing the trustworthy and malicious nodes, for creating a firewall against the Sybil attack. The research confirms that the proposed blockchain trust model will easily identify the malicious node. This work could be expanded in the future by integrating the machining learning-based trust model with blockchain to detect multichannel-based Sybil attacks.

References

1. Ahmed, M.R., Aseeri, M., Kaiser, M.S., Zenia, N.Z., Chowdhury, Z.I.: A novel algorithm for malicious attack detection in UWSN. In: 2015 International Conference on Electrical Engineering and Information Communication Technology (ICEEICT), pp. 1–6. IEEE (2015)
2. Arifeen, M.M., Al Mamun, A., Kaiser, M.S., Mahmud, M.: Blockchain-Enable Contact Tracing for Preserving User Privacy During Covid-19 Outbreak. Prrpints (2020)
3. Arifeen, M.M., Bhakta, D., Remu, S.R.H., Islam, M.M., Mahmud, M., Kaiser, M.: Hidden markov model based trust management model for underwater wireless sensor networks. In: Proceedings of the International Conference on Computing Advancements, pp. 1–5 (2020)
4. Arifeen, M.M., Rahman, M.M., Taher, K.A., Islam, M.M., Kaiser, M.S., et al.: Anfis based trust management model to enhance location privacy in underwater wireless sensor networks. In: 2019 International Conference on Electrical, Computer and Communication Engineering (ECCE), pp. 1–6. IEEE (2019)
5. Arifeen, M.M., Tasnim, Z., Mahmud, M., Kaiser, M.S.: Performance analysis of mc-cdma based underwater wireless sensor network. In: 2018 International Conference on Advances in Computing, Communications and Informatics (ICACCI), pp. 1919–1924. IEEE (2018)

6. Asif-Ur-Rahman, M., et al.: Toward a heterogeneous mist, fog, and cloud-based framework for the internet of healthcare things. IEEE Internet Things J. **6**(3), 4049–4062 (2018)
7. Demirbas, M., Song, Y.: An rssi-based scheme for sybil attack detection in wireless sensor networks. In: Proceedings of the 2006 International Symposium on on World of Wireless, Mobile and Multimedia Networks, pp. 564–570. IEEE Computer Society (2006)
8. Dhamodharan, U.S.R.K., Vayanaperumal, R.: Detecting and preventing Sybil attacks in wireless sensor networks using message authentication and passing method. Sci. World J. (2015)
9. Kaiser, M.S., et al.: Advances in crowd analysis for urban applications through urban event detection. IEEE Trans. Intell. Transp. Syst. **19**(10), 3092–3112 (2018)
10. Khan, M.I., Faisal, F., Azam, S., Karim, A., Shanmugam, B., De Boer, F.: Using blockchain technology for file synchronization. In: IOP Conference Series: Materials Science and Engineering, vol. 561, p. 012117. IOP Publishing (2019)
11. Kim, T.H., Goyat, R., Rai, M.K., Kumar, G., Buchanan, W.J., Saha, R., Thomas, R.: A novel trust evaluation process for secure localization using a decentralized blockchain in wireless sensor networks. IEEE Access **7**, 184133–184144 (2019)
12. Liang, C., Shanmugam, B., Azam, S., Karim, A., Islam, A., Zamani, M., Kavianpour, S., Idris, N.B.: Intrusion detection system for the internet of things based on blockchain and multi-agent systems. Electronics **9**(7), 1120 (2020)
13. Lv, S., Wang, X., Zhao, X., Zhou, X.: Detecting the Sybil attack cooperatively in wireless sensor networks. In: International Conference on Computational Intelligence and Security, 2008. CIS'08, Vol. 1, pp. 442–446. IEEE (2008)
14. Mahmud, M., Kaiser, M.S., Hussain, A.: Deep Learning in Mining Biological Data. arXiv (2020)
15. Mahmud, M., Kaiser, M.S., Rahman, M.M., Rahman, M.A., Shabut, A., Al-Mamun, S., Hussain, A.: A brain-inspired trust management model to assure security in a cloud based IoT framework for neuroscience applications. Cogn. Comput. **10**(5), 864–873 (2018)
16. Mateen, A., Tanveer, J., Khan, N.A., Rehman, M., Javaid, N., et al.: One step forward: towards a blockchain based trust model for wsns. In: International Conference on P2P, Parallel, Grid, Cloud and Internet Computing, pp. 57–69. Springer, Heidelberg (2019)
17. Moinet, A., Darties, B., Baril, J.L.: Blockchain Based Trust & Authentication for Decentralized Sensor Networks. arXiv preprint arXiv:1706.01730 (2017)
18. Pooja, M., Singh, D.Y.: Security issues and sybil attack in wireless sensor networks. Int. J. P2P Netw. Trends Technol. **3**(1), 7–13 (2013)
19. Rahman, A., Roy, S., Kaiser, M.S., Islam, M.S.: A lightweight multi-tier s-mqtt framework to secure communication between low-end IoT nodes. In: 2018 5th International Conference on Networking, Systems and Security (NSysS), pp. 1–6. IEEE (2018)
20. Shabut, A.M., Dahal, K., Kaiser, M.S., Hossain, M.: Malicious insider threats in tactical manet: The performance analysis of dsr routing protocol. In: 2017 IEEE Region 10 Humanitarian Technology Conference (R10-HTC), pp. 187–192. IEEE (2017)
21. Shabut, A.M., Kaiser, M.S., Dahal, K.P., Chen, W.: A multidimensional trust evaluation model for manets. J. Netw. Comput. Appl. **123**, 32–41 (2018)
22. She, W., Liu, Q., Tian, Z., Chen, J.S., Wang, B., Liu, W.: Blockchain trust model for malicious node detection in wireless sensor networks. IEEE Access **7**, 38947–38956 (2019)
23. Sujatha, V., Anita, E.M.: An efficient trust based method for Sybil node detection in mobile wireless sensor network. In: AIP Conference Proceedings, Vol. 2016, p. 020138. AIP Publishing (2018)
24. Vokerla, R.R., Shanmugam, B., Azam, S., Karim, A., De Boer, F., Jonkman, M., Faisal, F.: An overview of blockchain applications and attacks. In: 2019 International Conference on Vision Towards Emerging Trends in Communication and Networking (ViTECoN), pp. 1–6. IEEE (2019)
25. Yazdinejad, A., Parizi, R.M., Srivastava, G., Dehghantanha, A., Choo, K.K.R.: Energy efficient decentralized authentication in internet of underwater things using blockchain. In: 2019 IEEE Globecom Workshops (GC Wkshps), pp. 1–6. IEEE (2019)

An Efficient Bengali Text Steganography Method Using Bengali Letters and Whitespace Characters

Md. Shazzad-Ur-Rahman, Amit Singha, Nahid Ibne Akhtar, Md. Fahim Ashhab, and K. M. Akkas Ali

Abstract The present era is the era of information and communication. Everyday, billions of information pass from one digital device to another. Among this information and data, there are many sensitive and confidential data like banking credentials, national and international secrets, defense strategy, etc. These data must not go public and security must be ensured. It has become complicated to provide protection to these data. One solution to keep data privacy is steganography. It is a way of secure communication where a secret message is hidden within a cover object. In this paper, we propose a unique text steganography method using Bengali Unicode characters and white-space characters. In our method, we have encrypted our secret message within an ordinary plain text. It is impossible to differentiate between the normal plain text and stego-text in bear eyes. If anybody suspects the stego-text, he/she must have to break another secret pattern to decrypt the hidden message. We evaluate the efficiency of our proposed work in terms of security and embedding capacity. The comparison results suggest that our new scheme offers higher embedding capacity and more security than the existing methods.

Keywords Text steganography · Natural form · Composite form · Whitespace

M. Shazzad-Ur-Rahman: These authors have contributed equally.

Md. Shazzad-Ur-Rahman (✉) · A. Singha · N. Ibne Akhtar · Md. Fahim Ashhab
Bangladesh University of Professionals, Mirpur Cantonment, Dhaka, Bangladesh
e-mail: shazzad.bupit@gmail.com

A. Singha
e-mail: amitsinhga1996@gmail.com

N. Ibne Akhtar
e-mail: nahid.ibne.akhtar@gmail.com

Md. Fahim Ashhab
e-mail: ashhabfahim@gmail.com

K. M. A. Ali
Institute of Information Technology, Jahangirnagar University, Savar, Dhaka 1342, Bangladesh
e-mail: akkas@juniv.edu

M. S. Kaiser et al. (eds.), *Proceedings of International Conference on Trends in Computational and Cognitive Engineering*, Advances in Intelligent Systems and Computing 1309, https://doi.org/10.1007/978-981-33-4673-4_38

1 Introduction

Today's world relies on technology. We are trading a large amount of information, for example, web perusing, online exchange, mobile-banking, banking credentials, etc. through the web. In some cases, this information may be confidential, such as banking credentials, secret missions code, national secret, and personal appointments. So, these types of data must need to be secured and protected during transmission.

Steganography is a process of hiding information within a cover media so that none can realize the existence of secret data [1]. It is a process of concealing a secret message into a cover media, which can be a text, image, audio, or video file. It is a Greek term which comes from two Greek words 'Steganos' and 'Graptos' which means covered writing or secret writing [2]. It is often confused with the cryptography as both of the processes are used for hiding data but, there are some differences between them [3]. When the secret message is embedded into the cover media, the resultant output is called stego-object. When the receiver receives the stego-object then he/she can extract the hidden message through the stego-key.

The aim of steganography is hiding the data into the cover-object so that any attacker cannot even notice the existence of secret information hidden in the cover media. In a text steganography approach, a text file is used as a cover media to hide the secret data. Text steganography is very critical to implement because it has less redundant data than image, audio, and video [4]. A successful steganographic approach has four significant features which are: Invisibility, Embedding capacity, Robustness and, Security [5]. If any steganographic process can maintain all these features, that would be an ideal process.

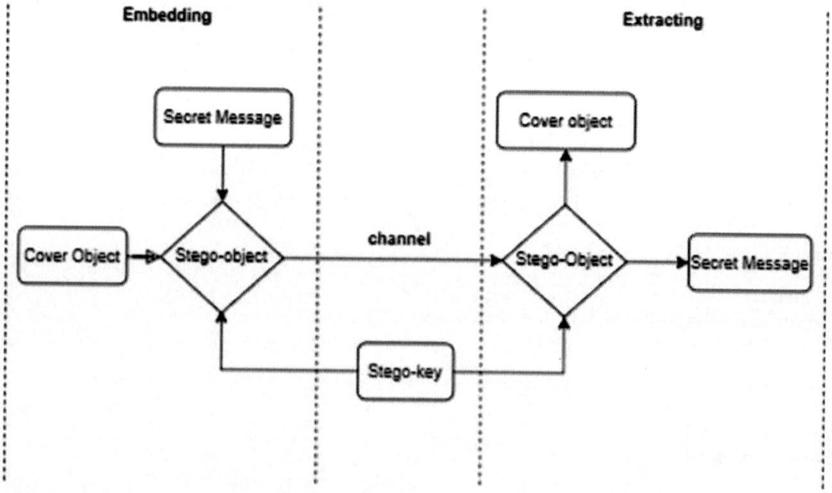

Fig. 1 Flow chart of the steganography

In our proposed scheme, we use Bengali text as the cover medium. The Bengali language is ranked as the seventh most spoken language in the world. Bengali is native language of Bangladesh and some states of India. Around 26 million people in the world speak in the Bengali language [6]. In the chart of Unicode, the location of the Bengali letter is 0980-09ff. People are also using the Bengali language in social media like Facebook, Messenger, WhatsApp, etc. So, steganography in the Bengali language will make a significant contribution to the security of information.

1.1 The Duality Form of Bengali Letter

The Bengali alphabet has some unique properties which are different from other language's alphabet. There are some diacritics (shorthand form of vowel) in Bengali letter used beside a consonant.

In contrast, some diacritics have two parts; one part is placed before the consonant, and another is after the consonant. For example, "◌ো', '◌ৌ' ' have two part. One part is placed before the consonant and another is after the consonant. These two diacritics can be written as their actual Unicode which is U+09cb for ◌ো and U+09cc for ◌ৌ. But these diacritics can be written in another way shown in Table 1.

On other hand, some consonant of the Bengali alphabet also has two parts; one part is the main body and another part is a symbol which is called 'Nukta'. 'Nukta' is basically a dot placed under the main body consonant. For example, 'র','ড়','ঢ়','য়' these are the consonants having 'Nukta'. As a character, each of them has a unique location in the Unicode chart. The main body of these characters has the same shape as another Bengali character. These characters having 'Nukta' can also be written in two ways; shown in Table 1.

Above, we discuss about the some Bengali letters, which can be written in two ways using different Unicode. This process is demonstrated at Table. 1.

2 Related Work

This section describes some important and similar methods of text steganography to the proposed scheme along with their advantages and disadvantages. Text steganography schemes can be broadly divided into three categories: format-based, NLP (Natural Language Processing), and linguistic method. In a format-based method to hide data there needs to change the formatting of the cover text. It is a very popular technique for hiding data.

There are some special Bengali letters which can be written by two different Unicode forms. These two forms can be defined as natural form and composite form, which is shown in Table 1 on page 6. The very first duality form-based Bengali text steganography method was proposed by Md. Khairullah [3]. Here, a new method of hiding bits was proposed by taking the advantage of the duality form of some special Bengali letters.

There are some Unicode space characters that look like the normal space. The very first whitespace character-based steganographic tool was proposed by Kwan which is known as SNOW. In this process, secret bit stream is embedded by using extra space character in the end of line spacing. L.Y. Por, T. F. Ang proposed a inter-word and inter-paragraph spacing technique for hiding secret message using whitespace characters [7]. This process is also known as WhiteSteg. Por [8] proposed a text-based data hiding method using Unicode space character which is known as UniSpaCh. In this approach, they encode two binary bits with combination of two whitespace character. This is one of the earliest works of text steganography using Unicode space character. Kumar and Chand [9] discusses an effective text steganography approach by using whitespace character which is very much similar to UniSpaCh [8]. Prem Singh and Raja Chaudhury proposed a null space-based text steganography method [10]. But this technique takes more space to encode a few bits.

The multilingual approach is a very a common term in text steganography. Some letters of the English language have an appearance in with another language with different Unicode. Even, some Bengali letters another language like Assam, Devanagari letters. Abdul Monem [11] proposed multilingual advantage-based text steganography process.

A color-based text steganography technique was proposed in MS Word document by Khairullah [12]. In this method, by setting foreground colors for invisible space characters or the carriage return for hiding secret bits. Samraa A. Al-Asadi and Wesam Bhaya proposed a novel steganographic method where a data hiding method was proposed by changing the font color and font type in the Excel sheet cell [1]. Wesam Bhaya proposed a novel method for hiding data by changing the font of the cover text in a Microsoft Word document [13]. Though the embedding capacity of this method is very high, it lacks the imperceptibility quality. Aruna Malik and Geeta Sikka discussed an efficient text steganographic method based on LZW compression and color coding [14].

Safia Al-Nofaie and Adnan Gutub proposed a steganographic scheme using Arabic text utilizing pseudo-spaces. Here, they proposed two technique for hiding secret message in Arabic text which are Kashida-PS and PS-betWords [15]. Vidhya P.M and Vargeshe Paul proposed an innovative text steganographic scheme using Malayalam text Unicode and alphabet indexing [16]. Malayalam is an Indian local language.

Baharudin Osman and Azman Yasin proposed an alphabet-based technique using Genetic algorithm [17].

Zhongliang and Shuyu proposed an automatically generated text steganographic model based on Markov model and Huffman coding [18].

Fig. 2 Dataset of Bengali Unicode method

আমার সোনার বাংলা, আমি তোমায় ভালোবাসি।
চিরদিন তোমার আকাশ, তোমার বাতাস, আমার প্রাণে বাজায় বাঁশি।
ও মা, ফাগুনে তোর আমের বনে ঘ্রাণে পাগল করে,
মরি হায়, হায় রে—
ও মা অঘ্রানে তোর ভরা ক্ষেতে আমি কি দেখেছি মধুর হাসি ।
কি শোভা, কী ছায়া গো, কি স্নেহ, কী মায়া গো—
কী আঁচল বিছায়েছ বটের মূলে, নদীর কূলে কূলে।
মা, তোর মুখের বাণী আমার কানে লাগে সুধার মতো,
মরি হায়, হায় রে—
মা তোর বদনখানি মলিন হলে, ও আমি নয়ন জলে ভাসি ।

3 Drawback of Previous Method

At first, Md. Khairullah proposed the duality form-based text steganography using Bengali Unicode [3]. Different methods of text steganography using whitespace character was proposed in [7–10]. In this section, we discussed about some drawbacks of previous method.

3.1 *Md. Khairullah's "Steganography in Bengali Unicode Text" [3]:*

This section discusses about two major drawbacks of this steganographic scheme.

Less Security: In this dataset, shown in Fig. 2 a secret message which is '123ab' is hidden into the cover object where the character highlighting with yellow color is written in natural form and the character highlighting with blue color is written in composite form.

If any hacker understands the duality form of the Bengali special letter, he/she can extract the secret message by only two attempts.

1st attempt: In the first attempt, we consider the special character having natural form as '1' and the special character having composite form as '0'.

So, in the first attempt, the acquired bit stream is: 11001110 11001101 11001100 10011110 10011101 If we convert the binary bit stream into their ASCII-8-character form then the probable secret message will be: ╬═╠Pts¥.

2nd attempt: In second attempt, we consider the special character having natural form as '0' and the special character having composite form as '1'.

So, in the second attempt, the acquired bit stream is: 00110001 00110010 00110011 01100001 01100010.

If we convert the binary bit stream into their ASCII-8-character form then the probable secret message will be: 123ab.

So, we can observe that the secret message can be extracted by doing only two attempts which proves that the security of this scheme is very low.

Less Embedding Capacity: This method has a very low rate of embedding capacity as it can embed only 1 bit in each embeddable location where UniSpach [8], Efficient Steganography [9] can embed 2 and 4 bits in every embeddable location respectively.

4 Proposed Method

In this paper, we propose an efficient text steganographic scheme. In our proposed method, we have selected some special Bengali letters having two forms, which are Natural form and Composite form, to embed hidden bits shown in Table 1. These letters look the same in both forms. In our proposed method, we have used five kinds of Unicode space characters shown in Table 2. Here, every space seems alike to naked eyes and can only be differentiated if there has been a Unicode analysis of the text document. But identifying the Unicode space characters will not be enough for decoding the secret message. After finding one of these five Unicode space characters there has to be a special Bengali character in the white-space character's immediately previous word.

Moreover, there have been some fake placement of these Unicode space characters, and there will not be any special characters in the previous word of some white-space characters intentionally to make the system more secure. Tables 1 and 2

Table 1 Bengali Unicode characters having two forms

Natural form		Composite form	
Char.	Unicode	Char.	Unicode
ো (ex. মো)	09cb	ো (ex. মো)	09c7 + 09be
ৌ (ex. মৌ)	09cc	ৌ (ex. মৌ)	09c7 + 09d7
র	09b0	র	09ac + 09bc
ড়	09dc	ড়	09a1 + 09bc
ঢ়	09dd	ঢ়	09a2 + 09bc
য়	09df	য়	09af + 09bc
গু	0997 + 09c1	গু	0997 + 200c + 09c1
শু	09b6 + 09c1	শু	09b6 + 200c + 09c1
হু	09b9 + 09c1	হু	09b9 + 200c + 09c1
হৃ	09b9 + 09c3	হৃ	09b9 + 200c + 09c3

Table 2 Used space and their corresponding hidden bits

Used space	Unicode	Decimal value	Hidden bits	
			Natural form	Composite form
Three-Per-Em space	U+2004	8196	000	100
Four-Per-Em space	U+2005	8201	001	101
Punctuation space	U+2008	8200	010	110
Medium mathematical space	U+205f	8287	011	111
Narrow no-break space	U+202f	8239	0	1

illustrates the special Bengali letters having the duality form and special spaces used in our method respectively.

4.1 Embedding Algorithm and Extracting Algorithm

We develop an efficient process for the embedding procedure, ensuring the bit-streams accuracy while encoding the secret message. Our extracting process is so secure that, it is almost impossible to extract the hidden message by any unauthorized accessed middle-man.

Algorithm 1 Embedding Algorithm

Input: Cover Text Document and Secret Message.
Output: Stego-object embedded with Secter Message.
Step 1: spaces = Dictionary("All possible 3 bits combinations with 0 and 1": "Spaces")
Step 2: PlainText = read(PlainText File)
Step 3: SpecialCharacters = List(Natural and composite characters)
Step 4: Patterns = Binaryform of Hidden Message
if words in PlainText **then**
  for Words in PlainText **do**
    for Character in Word **do**
      if Character in SpecialCharacters and Patterns[index: index+3] in composite form **then**
        Step 5:$Corpus = word(with\ replaced\ composite\ character) + spaces[index : index + 3]$ Character in SpecialCharacters and Patterns[index: index+3] in natural form
        Step 6:$Corpus = word + spaces[index : index + 3]$
      else
        Step 7::$Corpus = word + Regularspace$
      end if
    end for
    if patterns is empty **then**
      Step 8:: *Generate Output*
    end if
  end for
end if

Algorithm 2 Extracting Algorithm

```
Input: Stego document with the Secret Message
Output: The Secret Message in text and bits form.
Step 1: List 1 = Declaring Natural form of characters.
Step 2: List 2 = Declaring composite form of characters.
for character in Stego document do
    if special space found then
        for i = Index of(Special Space -1) to Previous Special Space do
            if Special character found then
                Step 3: Result.Append(Corresponding bits of the special character)
            else
                Step 4: Continue
            end if
            Step 5: Repeat until any special space remain.
        end for
        Step 6: Convert bits into Text
        Step 7: Print the Secret Message
    end if
end for
```

Algorithm 1 and algorithm 2 represent the secret message embedding and extracting technique of our proposed method respectively.

5 Experimental Results and Comparisons

This section analyzes the experimental result and compares it with the other method discussed in the third section. The implementation of proposed the method is made in python programming language (version 3.7) and Visual Studio Code 2020, which provides a full-featured development environment for implementing the proposal. The encryption and decryption method of our method is shown in Fig. 3

Fig. 3 Embedding and extracting process of our proposed scheme

a

করোনাভাইরাসের সংক্রমণ গোটা বিশ্বে ছড়িয়ে পড়েছে। অনেকেই মনে
করছেন, বিশ্বায়নের কারণেই এ সংক্রমণ বাড়ছে। আর এ থেকে মুক্তি
পাওয়ার একমাত্র উপায় হলো বিশ্ব থেকে বিচ্ছিন্ন থাকা। এজন্য এরই মধ্যে
অনেক দেশ ভ্রমণে এমনকি ব্যবসার ক্ষেত্রেও নিষেধাজ্ঞা জারি করেছে।
বিশ্লেষকরা বলছেন, স্বল্প-মেয়াদি সঙ্গরোধ (কোয়ারেন্টাইন) মহামারী রোধ
করতে পারে, কিন্তু এমন রোগের বিরুদ্ধে কার্যকর প্রতিরোধ না গড়ে
দীর্ঘমেয়াদী আইসোলেশনে গেলে অর্থনৈতিক পতন ঘটবে।

b

করোনাভাইরাসের সংক্রমণ গোটা বিশ্বে ছড়িয়ে পড়েছে। অনেকেই মনে
করছেন, বিশ্বায়নের কারণেই এ সংক্রমণ বাড়ছে। আর এ থেকে মুক্তি
পাওয়ার একমাত্র উপায় হলো বিশ্ব থেকে বিচ্ছিন্ন থাকা। এজন্য এরই মধ্যে
অনেক দেশ ভ্রমণে এমনকি ব্যবসার ক্ষেত্রেও নিষেধাজ্ঞা
জারি করেছে। বিশ্লেষকরা বলছেন, স্বল্প-মেয়াদি
সঙ্গরোধ (কোয়ারেন্টাইন) মহামারী রোধ করতে পারে, কিন্তু এমন
রোগের বিরুদ্ধে কারাকর পুরতিরোধ না গড়ে দীর্ঘমেয়াদী আইসোলেশনে গেলে
অর্থনৈতিক পতন ঘটবে।

Fig. 4 **a** Cover document. **b** Stego document

5.1 *Evaluation Criteria Analysis*

In this section, we evaluate the experimental results based on two major parameters which are imperceptibility and embedding capacity.

Imperceptibility The trace of a hidden message in the cover media must be invisible so that none can suspect the existence of secret message. Figure 4a, b represents the cover document and stego-document respectively.

The experimental results implies that reader cannot detect any existence of the hidden message and also cannot find any difference between the cover document and stego-document which ensures the invisibility of our proposed scheme.

Embedding Capacity and Hiding Efficiency Comparison: Our proposed method has a lucrative amount of embedding capacity as we can embed 3 bits in every embeddable location which is three times more than the previously discussed method in second section [3]. Figure 5a illustrates the embedding capacity and Fig. 5b illustrates the hiding efficiency comparison. We compare our proposed method with the existing methods, An Efficient Text Steganography [9] and Steganography in Bengali Unicode [3].

6 Discussion

Our proposed method has a lucrative amount of embedding capacity. According to the Bengali linguist Munier Chowdhury and other research works on Bengali language, the character 'র' has an appearance around 7%, 'য়' has around 1.9%, 'ো' has around 0.7%, and "ড়' has around 0.25% in a Bengali text document. This means the selected special characters for the proposed method are nearly 10% of the total characters in Bengali text which is a great advantage for our proposed scheme.

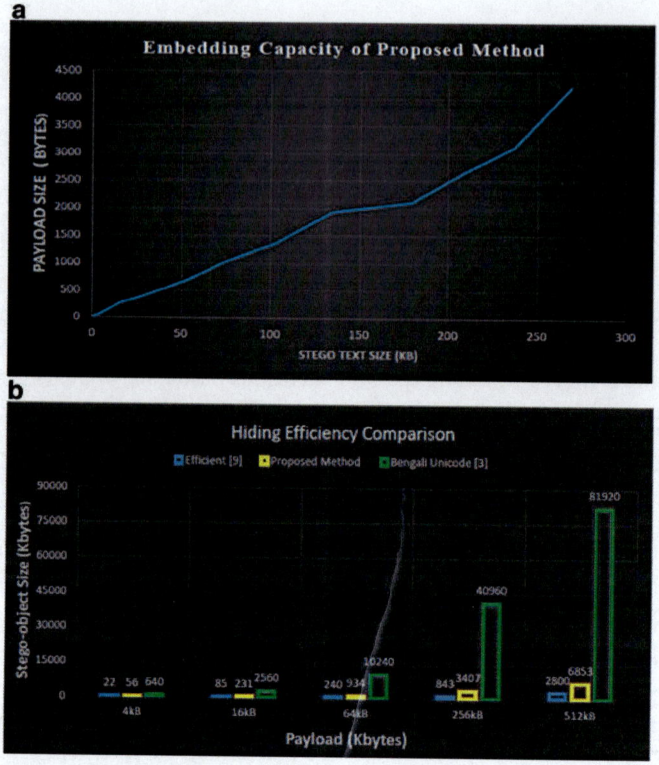

Fig. 5 **a** Embedding capacity of the proposed method. **b** Hiding efficiency comparison

7 Conclusion and Future Work

In the paper, we have proposed an efficient text steganography scheme based on Bengali Unicode characters and whitespace character for hiding information. The combination of Bengali Unicode characters and whitespace characters produce a secured encoding system which keeps the secret message protected against any kind of attack.

An error detection and correction method in the decryptor end can make the system more secure. In the future, we can use the hash function in our proposed method for mapping data of arbitrary size to bo binary bit string of fixed length.

References

1. Al-Asadi, S.A., Bhaya, W.: Text steganography in excel documents using color and type of fonts, p. 7
2. Por, L.Y., Delina, B: Information hiding: a new approach in text steganography, p. 8 (2008)
3. Ratul, A.S.: Steganography in Bengali Unicode. Text **27**(1), 11 (2017)
4. Majumder, A., Changder, S.: A novel approach for text steganography: generating text summary using reflection symmetry. Proced. Technol. **10**, 112–120 (2013)
5. Taleby Ahvanooey, M., Li, Q., Hou, J., Dana Mazraeh, H., Zhang, J.: AITSteg: an innovative text steganography technique for hidden transmission of text message via social media. IEEE Access **6**, 65981–65995 (2018)
6. What are the top 200 most spoken languages? October 2018
7. Por, L.Y., Ang, T.F., Delina, B.: WhiteSteg: a new scheme in information hiding using text steganography **7**(6), 12 (2008)
8. Yee Por, L., Wong, K., Chee, K.O.: UniSpaCh: a text-based data hiding method using Unicode space characters. J. Syst. Soft. **85**(5), 1075–1082 (2012)
9. Kumar, R., Chand, S., Singh, S.: An efficient text steganography scheme using Unicode Space Characters. Int. J. Forens. Comput. Sci. **10**(1), 8–14 (2015)
10. Singh, P.: A novel approach of text steganography based on null spaces. IOSR J. Comput. Eng. **3**(4), 11–17 (2012)
11. Rahma, A.M.S., Bhaya, W.S., Al, D.A.: Text steganography based on Unicode of characters in Multilingual **3**(4), 14 (2013)
12. Khairullah, M.: A novel text steganography system using font color of the invisible characters in microsoft word documents. In: 2009 Second International Conference on Computer and Electrical Engineering, pp. 482–484. IEEE, Dubai, UAE (2009)
13. Bhaya, W.: Text steganography based on font type in MS-word documents. J. Comput. Sci. **9**(7), 898–904 (2013)
14. Malik, A., Sikka, G., Verma, H.K.: A high capacity text steganography scheme based on LZW compression and color coding. Eng. Sci. Technol. Int. J. **20**(1), 72–79 (2017)
15. Al-Nofaie, S., Gutub, A., Al-Ghamdi, M.: Enhancing Arabic text steganography for personal usage utilizing pseudo-spaces. J. King Saud Univer. Comput. Inf. Sci. pp. S1319157819304719 (2019)
16. Vidhya, P.M., Paul, V.: A method for text steganography using Malayalam text. Proced. Comput. Sci. **46**, 524–531 (2015)
17. Osman, B., Yasin, A., Omar, M.N.: An analysis of alphabet-based techniques in text steganography **8**(10), 7
18. Zhongliang, Y., Shuyu, J., Yongfeng, H., Yujin, Z., Hui, L.: Automatically generate steganographic text based on Markov model and Huffman coding, p. 10

Enhancing Flexible Unequal Error Control Method to Improve Soft Error Tolerance

Md. Atik Shahariar, Mirazul Islam, Muhammad Sheikh Sadi, and Soumen Ghosh

Abstract Because of soft error, the reliability of modern computing systems is facing many problems. As the area of application increases, the demand for detecting and correcting soft errors increases proportionally. Several error-correcting methods have been developed to solve this problem. However, high redundancy, low correction ability, large circuit complexity, space and time overhead of most of the existing methodologies are creating the demand for enhancement in the area of soft error tolerant researches. To improve the soft error tolerability in computer systems, we have enhanced the performance of Flexible Unequal Error Control (FUEC) method in this paper. Our method has increased the correction ability for six more erroneous bits than FUEC. Moreover, the proposed method has low bit-overhead and high correction coverage compared to the previous methods.

Keywords Soft error tolerance · Reliability · Flexible unequal error control · Parity check matrix · Generator matrix

1 Introduction

With the advancement of technology, computer systems and circuits are becoming more complex [1]. The size of the computer chip is decreasing and so the circuits are becoming much denser. Although more transistors are present on a single chip,

Md. Atik Shahariar (✉) · M. Islam · M. S. Sadi · S. Ghosh
Khulna University of Engineering & Technology, Khulna 9203, Bangladesh
e-mail: shahariar1607065@stud.kuet.ac.bd

M. Islam
e-mail: islam1607062@stud.kuet.ac.bd

M. S. Sadi
e-mail: sadi@cse.kuet.ac.bd

S. Ghosh
e-mail: soumeng007@gmail.com

M. S. Kaiser et al. (eds.), *Proceedings of International Conference on Trends in Computational and Cognitive Engineering*, Advances in Intelligent Systems and Computing 1309, https://doi.org/10.1007/978-981-33-4673-4_39

there is a decrease in the supply of power and noise margins. As a result, electronic systems are becoming more vulnerable to radiation-induced errors [2], [3]. Memory devices can easily become corrupted due to this radiation induction [4]. This can lead to corruption of data or malfunction in the computer circuits [5]. Such errors can be more catastrophic in critical environments such as space stations, space shuttle, power plants, etc. [6], [7]. Nowadays, computer chips are being used in medical science to monitor patients in intensive care units, where even a little change of data can be very harmful to the patient.

As soft error poses a high threat to the system, necessary steps should be taken to make the system error resistant or less vulnerable to soft error. There are some error protection techniques based on hardware but they suffer from high area, time, and power overheads [8]. Some prior works known as Error Correction Codes (ECCs) [9] have been done for detecting and correcting soft errors. Examples of ECCs are Bose-Chaudhuri-Hocquenghem (BCH) [5], Golay codes [10], SEC–Double-adjacent Error Detection (SEC–DED) [11], Matrix Codes [12], Hamming code [13], Column–Line–Code (CLC) [14] etc. Later on, Saiz-Adalid et. al introduced FUEC [15] which shows error tolerance with lower redundant bits. However, low correction coverage, large circuit complexity, space and time overhead of most of the existing methodologies are creating the demand for enhancement in the area of soft error tolerant researches.

In this paper, we have enhanced FUEC to increase its adjacent error detection and correction capability. Our proposed method can detect and correct up to 10-bit adjacent errors. Furthermore, our method has low information overhead with a higher correction ability. It has 62.5% correction coverage and 93.75% bit overhead.

This paper is organized into the following sections. Section 2 discusses related work in the field of error correction and detection. In Sect. 3, we have discussed our methodology in brief. In Sect. 4, we have evaluated the performance of our proposed methodology. Section 5 concludes the paper.

2 Related Work

As we have mentioned before, there have been some works in the past in this field of error detection and correction. Some of these approaches are described shortly as follows.

Hamming code [13] is easy to be developed for any data length. And with parity check matrix, it can be used to detect and correct any bit error. The parity check matrix that is used with the hamming code has a unique property, Any two columns in the matrix must be pairwise linearly independent. But the bit overhead of the parity check matrix is too high.

An extended version of the hamming code is CLC [14]. CLC can be used for correcting multiple errors in memory. It can detect and correct up to a 2-bit error. It uses 24 check bits for a 16-bit data word. The bit overhead of CLC is 150% which is its biggest disadvantage.

In Matrix code [12] data of n bits is stored in a matrix form, of length l and width w, where $n = l * w$, that means if the data word is 16 bits than it is stored in 4 * 4 matrix form. Here for each row check bits are added horizontally and for each column parity bits are added vertically. It can correct a single-bit error and detect 2-bit burst error and has relatively lower redundancy than CLC.

Later SEC–DED [11] error correction code was introduced, which requires 5 check bits for a 16-bit data word. Thus, it has a low bit overhead but it can only detect and correct single-bit or two adjacent bits error and because of its low correction coverage, it has very low efficiency. After that, another technique FUEC–double adjacent error correction (DAEC) [15], FUEC–triple adjacent error correction (TAEC) [15], FUEC–quadruple adjacent error correction (QUAEC) [15] was invented which can detect and correct from any single bit error to 4-bits adjacent error. FUEC method has a very low bit overhead compared to the previously discussed methods but still, it has a low correction coverage.

3 The Proposed Approaches

We have designed several new matrices to improve the error detection and correction rate of ECCs. With the help of these matrices, our method can detect and correct up to 10 adjacent erroneous bits. We have proposed FUEC–Pente adjacent error correction (PAEC) code as shown in Fig. 1a to correct up to five-bit adjacent errors.

Using this method, we can correct from a single bit error to maximum 5-bit adjacent errors and this method is also capable of detecting up to 6-bit adjacent errors. In FUEC-PAEC, it needs a total of 10 check bits for a 16-bit data word. From Fig. 1a we can observe that the left side of the H matrix represents the check bits, which is an identity matrix and the right side of the matrix represents the data bits. Every two rows of the H matrix must be unique to each other. When the H matrix is generated, we calculate the check bits. To calculate the check bits we have performed X-OR operations. Here, C represents the check bits, A represents the actual data bits and E represents the erroneous data bits. Then the formula to calculate check bit is given below:

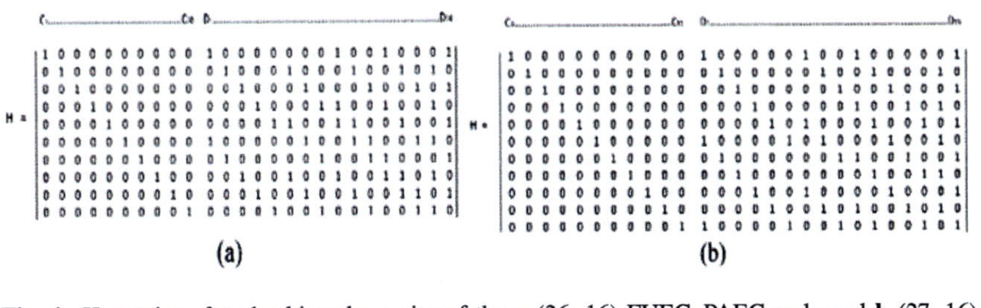

Fig. 1 H matrices for checking the parity of the **a** (26, 16) FUEC–PAEC code and **b** (27, 16) FUEC–HXAEC code

$$C[0] = A[0] \oplus A[8] \oplus A[11] \oplus A[15]$$

$$C[1] = A[1] \oplus A[5] \oplus A[9] \oplus A[12] \oplus A[14]$$

$$C[2] = A[2] \oplus A[6] \oplus A[10] \oplus A[13] \oplus A[15]$$

$$C[3] = A[3] \oplus A[7] \oplus A[8] \oplus A[11] \oplus A[14]$$

$$C[4] = A[4] \oplus A[5] \oplus A[8] \oplus A[9] \oplus A[12] \oplus A[15]$$

$$C[5] = A[0] \oplus A[6] \oplus A[9] \oplus A[10] \oplus A[13] \oplus A[14]$$

$$C[6] = A[1] \oplus A[7] \oplus A[10] \oplus A[11] \oplus A[15]$$

$$C[7] = A[2] \oplus A[5] \oplus A[8] \oplus A[11] \oplus A[12] \oplus A[14]$$

$$C[8] = A[3] \oplus A[6] \oplus A[9] \oplus A[12] \oplus A[13] \oplus A[15]$$

$$C[9] = A[4] \oplus A[7] \oplus A[10] \oplus A[13] \oplus A[14]$$

Now, we can calculate the syndrome bits by performing the X-OR operations among actual data check bits and erroneous data check bits. The formula to calculate syndrome bit is given below.

$$S[0] = C[0] \oplus E[0] \oplus E[8] \oplus E[11] \oplus E[15]$$

$$S[1] = C[1] \oplus E[1] \oplus E[5] \oplus E[9] \oplus E[12] \oplus E[14]$$

$$S[2] = C[2] \oplus E[2] \oplus E[6] \oplus E[10] \oplus E[13] \oplus E[15]$$

$$S[3] = C[3] \oplus E[3] \oplus E[7] \oplus E[8] \oplus E[11] \oplus E[14]$$

$$S[4] = C[4] \oplus E[4] \oplus E[5] \oplus E[8] \oplus E[9] \oplus E[12] \oplus E[15]$$

$$S[5] = C[5] \oplus E[0] \oplus E[6] \oplus E[9] \oplus E[10] \oplus E[13] \oplus E[14]$$

$$S[6] = C[6] \oplus E[1] \oplus E[7] \oplus E[10] \oplus E[11] \oplus E[15]$$

$$S[7] = C[7] \oplus E[2] \oplus E[5] \oplus E[8] \oplus E[11] \oplus E[12] \oplus E[14]$$

$$S[8] = C[8] \oplus E[3] \oplus E[6] \oplus E[9] \oplus E[12] \oplus E[13] \oplus E[15]$$

$$S[9] = C[9] \oplus E[4] \oplus E[7] \oplus E[10] \oplus E[13] \oplus E[14]$$

After calculating the syndrome, we need to find the positions of the erroneous bits. For doing that, we need to take five adjacent values from the right side of the *H* matrix, starting from D_1 (D_i represents the data bit part of the *H* matrix) to D_5 position and then do X-OR operation among them. The X-OR value is then compared with the 1st syndrome bit. Then we need to take the next syndrome bit and compare with the X-OR of D_1 to D_5 positional values from the 2nd-row of *H* matrix. This process keeps running until it reaches the last bit of the syndrome. When this comparison is over, we check the number of times a syndrome bit matches with the *H* matrix X-OR value.

If this number is equal to the length of syndrome array then we have 5-bit adjacent errors in the first 5 bits of the received dataword. After detecting the position of the erroneous bits, just by flipping the data bits at those positions, we can get the corrected dataword. If the number of similarities is not the same with all syndrome bits then we have to start the same process again. But at this instance, the X-OR of the *H* matrix is performed among the next 5 adjacent bits, starting from D_2 to D_6. When all the data bits (D_1 to D_{16} of each row) of *H* matrix are checked in this way and still the number of similarities is not same with syndrome bits then we can conclude that no 5-bit adjacent error exists in the 16-bit dataword.

The second approach of our proposed method is FUEC–Exi adjacent error correction (HXAEC) code which can correct from a single bit to 6-bit adjacent errors. This approach can also detect up to 7-bit adjacent errors. In FUEC-HXAEC, it needs a total 11 check bits for a 16-bit dataword. The *H* matrix for FUEC-HXAEC is shown in Fig. 1b.

In our third approach as shown in Fig. 2a, the FUEC–Epta adjacent error correction (HEPAEC) code can correct from a single bit to 7-bit adjacent errors. This method

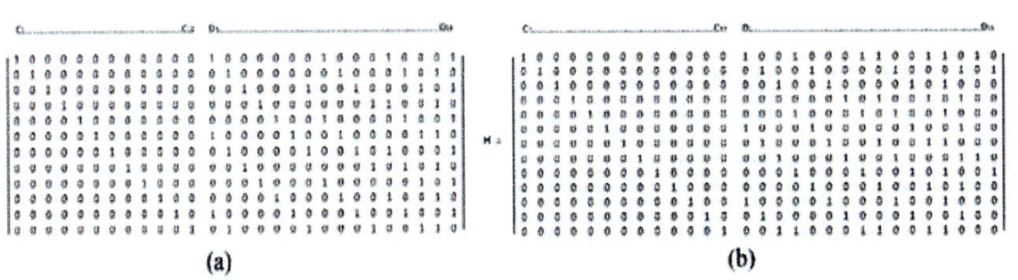

Fig. 2 *H* matrices for checking the parity of **a** (28, 16) FUEC–HEPAEC code and the **b** (29, 16) FUEC–OCAEC code

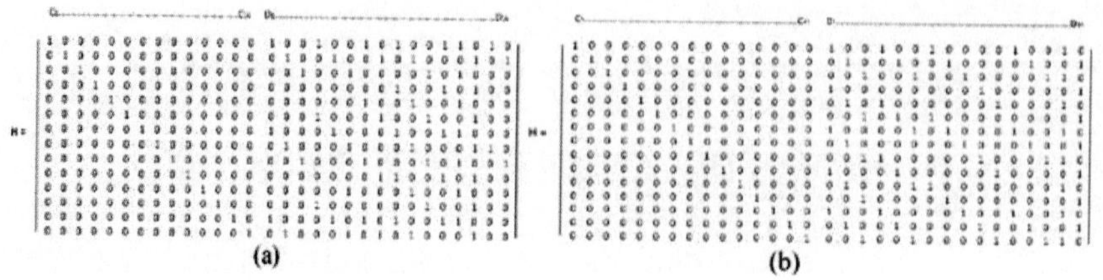

Fig. 3 *H* matrices for checking the parity of the **a** (30, 16) FUEC–ENAEC code and **b** (31, 16) FUEC–DKAEC code

can also detect up to 8-bit adjacent errors. In FUEC-HEPAEC, it needs a total of 12 check bits for a 16-bit dataword.

Our fourth approach FUEC–Octo adjacent error correction (OCAEC) code can correct from a single bit to 8-bit adjacent errors and detect a maximum of 9-bit adjacent errors. In FUEC-OCAEC, it needs a total 13 check bits for a 16-bit dataword. The *H* matrix for FUEC-OCAEC is shown in Fig. 2b.

In our fifth approach of the proposed method FUEC–Ennea adjacent error correction (ENAEC) code we can correct from a single bit to 9-bit adjacent errors and detect a maximum of 10-bit adjacent errors. In FUEC-ENAEC, it needs a total of 14 check bits for a 16-bit dataword. The *H* matrix for FUEC-ENAEC is shown in Fig. 3a. The sixth approach of our proposed method FUEC–Deka adjacent error correction (DKAEC) code can detect and correct a maximum of 10-bit adjacent errors. In FUEC-DKAEC, it needs a total of 15 check bits for a16 bit dataword. The *H* matrix for FUEC-DKAEC is shown in Fig. 3b.

3.1 *Fault Injection*

To verify our work we have considered all possible combinations of faults for a 16-bit data word and this is 2^{16} (=65,536) in total. And then we have injected all possible combinations of adjacent faults to evaluate the performance of the proposed method. This combination decreases with the increasing number of adjacent bits. For example, for 5-bit adjacent fault injection, there is $(2^{16}) * 12$ combinations and for 6-bit adjacent error, it reduces to $(2^{16}) * 11$ combinations. For 8-bit adjacent error, it becomes $(2^{16}) * 9$ combinations. In that way, we have injected all the possible combinations of the adjacent faults to evaluate the performance of the proposed method.

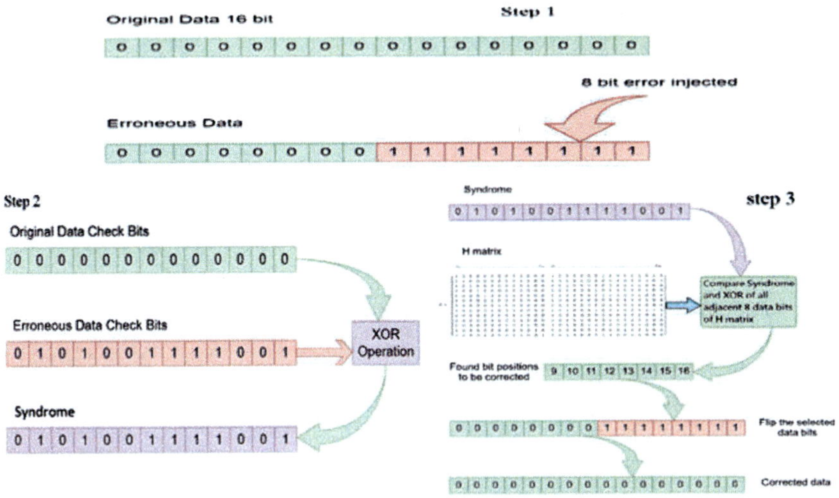

Fig. 4 Error detection and correction process for 8-bit adjacent error

3.2 A Simplified Example

The process of correcting errors using the proposed method is shown as follows. Let us consider that, 8-bit adjacent faults are injected from bit position 9 to 16 of a 16-bit dataword. Then we have calculated the check bits for original data and erroneous data using the same process as shown in FUEC–PAEC. Then we do X-OR operations between the original check bits and erroneous check bits to find the syndrome bits. Then we do X-OR operations on the data bits of the *H* matrix using only 8 adjacent data bits at a time from row 1 and then compare it with the syndrome bits. After each comparison, syndrome index increases, and the next row from the *H* matrix gets selected. By doing this process similar to FUEC-PAEC, we have found the positions of the eight adjacent erroneous data bit. After that, we need to flip the bits of the obtained positions to get the corrected 16-bit data word.

Figure 4 shows the flow diagram that illustrates the proposed approaches (since all approaches follow the same algorithmic procedure here). This flow diagram uses an example of 8-bit adjacent errors in 16-bit dataword.

4 Experimental Analysis

4.1 Experimental Setup

To evaluate the performance of our proposed method, we used a system with the following configurations.

- Intel® Core™ i5-7200U CPU @ 2.50 GHz 2.71 GHz
- Operating system Windows 10–64 bit
- RAM 8 GB
- Language C++.

4.2 Experimental Results

We have evaluated our proposed methods by running some tests on them. Firstly we have checked the bit overhead of the six approaches of the proposed method from FUEC-PAEC to FUEC-DKAEC. When we send a 16-bit data, we need to send the check bits with it as well. These extra bits are needed to verify if there is an error in the received data. And a lower number of check bits is always better as the check bits consume some extra amount of space. By using (1), we can calculate the bit overhead.

$$\text{Bit overhead} = \frac{\text{No.of check bit}}{\text{No.of data bit}} * 100\% \tag{1}$$

If the bit overhead is low then it means that the system requires less memory for the check bits. On the other hand, if the bit overhead is high, then it requires a higher amount of memory for the check bits. So it is important to minimize the bit overhead to give more memory space for data bits. Figure 5 shows the comparison between our method and existing methods with respect to bit-overhead.

From Fig. 5, it can be seen that matrix code and CLC have the highest bit overhead 100% and 150%, respectively although they have the lowest error detection and correction capability. Next, FUEC-QUAEC method has 56.25% bit overhead. After that from FUEC-PAEC to FUEC-DKAEC, the bit overhead rises with the increase of error detection and correction capability. The approaches in our proposed method FUEC-PAEC, FUEC–HXAEC, FUEC–HEPAEC, FUEC-OCAEC, FUEC-ENAEC, FUEC-DKAEC have bit overheads of 62.5%, 68.75%, 75%, 81.75%, 87.5%, and

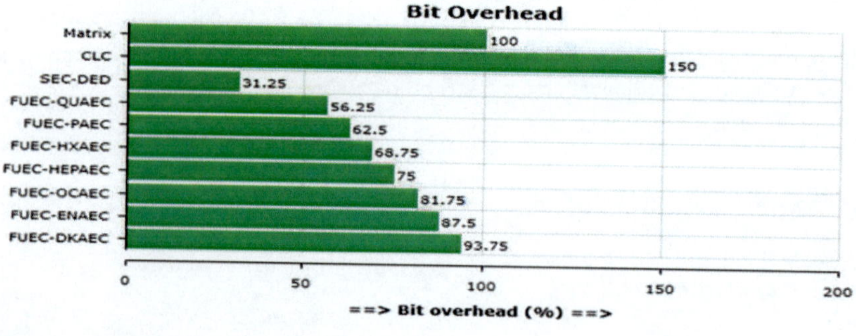

Fig. 5 Bit overhead of various methods

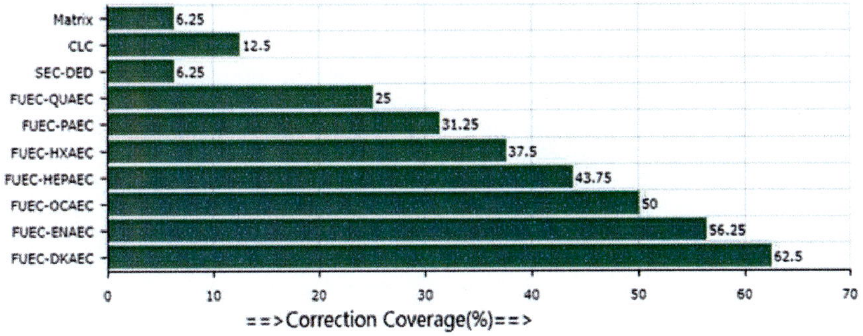

Fig. 6 Comparison of correction coverage between various methods

93.75% ,respectively, which is still much better than existing methods considering their bit overhead and error detecting and correcting capability.

Next, we have checked the correction coverage of the six approaches of the proposed method from FUEC-PAEC to FUEC-DKAEC and then compared the results with the existing methods. Now from Fig. 6, we can see that FUEC methods had a maximum of 25 % correction coverage and we have successfully increased the correction coverage in our proposed method. FUEC-PAEC, FUEC-HXAEC, FUEC-HEPAEC, FUEC-OCAEC, FUEC-ENAEC have 31.25%, 37.5%, 43.75%, 50%, 56.25% correction coverages, respectively. And our final method FUEC-DKAEC has 62.5% correction coverage which is the best result we have found yet.

The time complexity shows how fast a method can detect or correct errors. It measures the time requirement of a method. All our approaches have O(n) time complexity. Where n represents the maximum number of error correction capability. For example, FUEC-PAEC has $O(n = 5)$ time complexity as it can correct a maximum of 5-bit adjacent errors. Similarly, for FUEC-HXAEC, FUEC-HEPAEC, FUEC-OCAEC, FUEC-ENAEC, and FUEC-DKAEC the time complexities are $O(n = 6)$, $O(n = 7)$, $O(n = 8)$, $O(n = 9)$, $O(n = 10)$, respectively.

To calculate the execution time for FUEC-PAEC, we have generated a total of 786432 combinations of adjacent error inputs and then injected it into our test code. Similarly for FUEC-HXAEC, FUEC-HEPAEC, FUEC-OCAEC, FUEC-ENAEC, FUEC-DKAEC the combinations of adjacent erroneous inputs are 720896, 655360, 589824, 524288, 458752, respectively.

The time required to execute the approaches of our proposed method is shown in Fig. 7. Here we can see that for FUEC-PAEC, it needs 4008 seconds for checking total 786432 combinations which is equivalent to 5.09 ms to process each combination. Similarly for FUEC-HXAEC, FUEC-HEPAEC, FUEC-OCAEC, FUEC-ENAEC, FUEC-DKAEC the time requirement is 4.9 ms, 5 ms, 5.1 ms, 5 ms, and 5.2 ms, respectively. From this experiment, we can see that FUEC-HXAEC is a bit faster than the other methods.

The experimental results are summarized and shown in Table 1, where we can see that the number of check bits, bit overhead, correction coverage, and time complexity

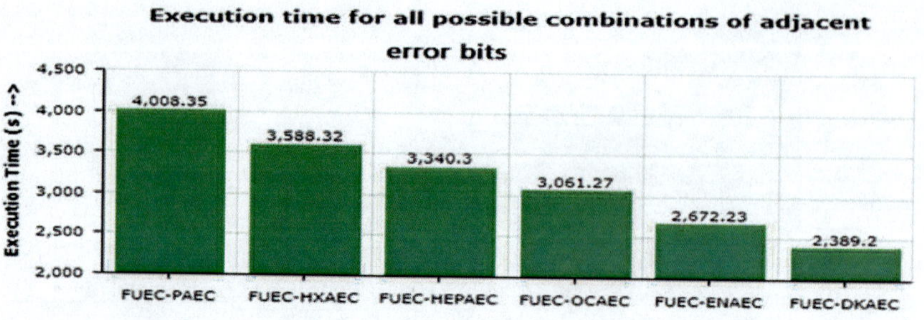

Fig. 7 Comparison of execution time requirement among different possible combinations

Table 1 Summarization of experimental results

Schemes	Required number of Check bits for a 16-bit dataword	Bit overhead (%)	Correction coverage (%)	Execution time per 16-bit dataword (ms)	Time complexity O(n=Maximum number of error correction capability)	Maximum adjacent error correction capability
FUEC-PAEC	10	62.5	31.25	5.09	$O(n = 5)$	5-bit
FUEC-HXAEC	11	68.75	37.5	4.9	$O(n = 6)$	6-bit
FUEC-HEPAEC	12	75	43.75	5	$O(n = 7)$	7-bit
FUEC-OCAEC	13	81.75	50	5.1	$O(n = 8)$	8-bit
FUEC-ENAEC	14	87.5	56.25	5	$O(n = 9)$	9-bit
FUEC-DKAEC	15	93.75	62.5	5.2	$O(n = 10)$	10-bit

increases with the increase of error correction capability but execution time remains almost the same.

5 Conclusions

In this paper, we have enhanced FUEC to increase the adjacent error detection and correction capability up to 10 bits. Though the correction coverage of our method is 62.5%, and it can detect or correct adjacent errors only, the relative correction coverage concerning the information overhead of the proposed method is much better than existing methods. In future, there are open scopes to enrich this proposed method by working with non-adjacent errors. Besides this, there is another open scope for the enhancement of this paper which is to extend the size of a data word.

References

1. Siddiqui, M., Ruchi, S., Van Le, L., Yoo, T., Chang, I., Tae-Hyoung Kim, T.: SRAM radiation hardening through self-refresh operation and error correction. IEEE Transa. Device Mater. Reliab. **20**(2), 468–474 (2020)
2. Rossi, D., Timoncini, N., Spica, M., Metra, C.: Error correcting code analysis for cache memory high reliability and performance. Des., Autom. Test Europe **2011**, 1–6 (2011)
3. Baumann, R.: Radiation-induced soft errors in advanced semiconductor technologies. IEEE Trans. Device Mater. Reliab. **5**(3), 305–316 (2005)
4. Shahriyar, S., Golap, M., Sadi. M.: An efficient error correction approach by using successive parity generation. In: 4th International Conference on Electrical Engineering and Information & Communication Technology (iCEEiCT) 2018, pp. 302–307 (2018).
5. Liu, S., Reviriego, P., Xiao, L.: Evaluating direct compare for double error correction codes. IEEE Trans. Device Mater. Reliab. **17**(4), 802–804 (2017)
6. Sadi, M., Khan, M., Uddin, M., Jürjens. J.: An efficient approach towards mitigating soft errors risks. Signal Image Process.:An Int. J. (SIPIJ) **2**(3) (2011).
7. Ibraheem, N.: Error-detecting and error-correcting using hamming and cyclic codes. IEEE Trans. Inf. Theory **51**(9), 3347–3353 (2005)
8. Rahman, M., Sadi, M., Ahammed, S.: Soft error tolerance using horizontal-vertical double-bit diagonal parity method. In: International Conference on Electrical Engineering and Information Communication Technology (ICEEICT) 2015, 1–6 (2015)
9. Koren, I., Krishna, C.: Fault-Tolerant Systems. 1st edn. Denise Penrose, 500 Sansome Street Suite 400 San Francisco CA 94111 (2007).
10. Reviriego, P., Liu, S., Xiao, L., Maestro, J.: An efficient single and double-adjacent error correcting parallel decoder for the (24,12) extended Golay code. IEEE Trans. Very Large Scale Integration (VLSI) Syst. **24**(4), 1603–1606 (2016).
11. Neale, A., Sachdev, M.: A new sec-Ded error correction code subclass for adjacent Mcu Tolerance in embedded memory. IEEE Trans. Device Mater. Reliab. **13**(3), 223–230 (2013)
12. Argyrides, C., Pradhan, D., Kocak, T.: Matrix codes for reliable and cost efficient memory chips. IEEE Trans. Very Large Scale Integr. (VLSI) Syst. **19**(3), 420–428 (2011).
13. Kumar, U., Umashankar, B.: Improved Hamming Code For Error Detection And Cor-rection. In: 2nd International Symposium On Wireless Pervasive Computing (2007)
14. Castro, H., Silveira, J., Coelho, A., Silva, F., Magalhães, P., Lima, O.: A correction code for multiple cells upsets in memory devices for space applications. In: 14th IEEE International New Circuits and Systems Conference (NEWCAS) 2016, pp. 1–4. (2016).
15. Gracia-Morán, J., Saiz-Adalid, L., Gil-Tomás, D., Gil-Vicente, P.: Improving Error Correction Codes for Multiple-Cell Upsets in Space Applications. IEEE Trans. Very Large Scale Integr. (VLSI) Syst. **26**(10), 2132–2142 (2018).

A Combined Framework of InterPlanetary File System and Blockchain to Securely Manage Electronic Medical Records

Abdullah Al Mamun, Md. Umor Faruk Jahangir, Sami Azam, M. Shamim Kaiser, and Asif Karim

Abstract Blockchain has become a popular research area since its introduction, as the benefit has been seen in various industries. It could greatly benefit the healthcare sector as it offers anonymity, immutability, and decentralization of data. Electronic Medical Record (EMR) systems face crucial problems regarding data security, accessibility, and management. A great deal of security threats relating to patient privacy involves unauthorized access to medical records, misuse of patient's disease reports, and so on. To address these issues, we have proposed a blockchain combined with the InterPlanetary File System solution framework for EMR in the healthcare industry. The aim is to implement the blockchain for EMR and provide access rules for various users of it. The proposed framework, while protecting patient privacy, allows convenient access by approved authorities such as healthcare providers to medical data.

Keywords Blockchain · IPFS · EMR · Data Security · SHA256 · Python · Postman · Flask

---

A. Al Mamun (✉) · M. S. Kaiser
Institute of Information Technology, Jahangirnagar University, Dhaka, Bangladesh
e-mail: abdullah.iiuceee@gmail.com

M. S. Kaiser
e-mail: mskaiser@juniv.edu

Md. U. Faruk Jahangir
Department of Computer Science & Engineering, Shahjalal University of Science and Technology, Sylhet, Bangladesh
e-mail: jahangircs.sust@gmail.com

S. Azam · A. Karim
College of Engineering, IT and Environment, Charles Darwin University, Casuarina, Australia
e-mail: sami.azam@cdu.edu.au

A. Karim
e-mail: asif.karim@cdu.edu.au

M. S. Kaiser et al. (eds.), *Proceedings of International Conference on Trends in Computational and Cognitive Engineering*, Advances in Intelligent Systems and Computing 1309, https://doi.org/10.1007/978-981-33-4673-4_40

1 Introduction

Technology has a significant role to play in addressing diverse issues in our everyday lives. The necessity of technological development in the medical sector is the demand of time now. The healthcare sectors produce a high volume of medical data every day that need to be stored, disseminated, and accessed daily. For example, medical data is created when a patient goes through some medical tests, such as MRI, X-Ray, and USG, or when a doctor prescribes the patient. However, healthcare data are breached every day in different parts of the world. For instance, the U.S. Department of Health and Human Services (HHS) stated in one of their reports in 2018 that 6.1 million people are victims of 229 healthcare data breaches, and the report was published in the civil rights of breach portal. In 2014, Statistica mentioned about 783 data breaches, with around 85.61 million records were compromised, which was approximately 500% since 2005 [20, 34].

Healthcare industries have management complexity and diversity of providers. Point to be noted that the medical record of a patient is stored in a hospital in electronic form, and the patient doesn't get accessed to it when taking treatment from another hospital [21, 24, 29]. People are dying due to the lack of proper infrastructure, human resources, funding, and data preservation policies [28]. Moreover, Electronic Medical Records (EMR) are important and extremely confidential private information for healthcare treatment like HIV and cancer. These data need to be shared among various peers, including healthcare providers, insurance companies, researchers, passport officials, law enforcement agencies, and relatives. Maintaining security and stacking up previous diagnosis reports of these valuable data is crucial [4, 8].

For consultation and treatment, a patient might visit multiple medical institutions or move from one hospital to another. As a result, a customer does not get all his medical data accumulated when needed at various locations. These medical records are generally too sensitive to fall into the wrong hand. Usually, these data are stored in various organizations' private databases, so these data remained centralized. Many researchers around the globe proposed several frameworks to address these issues. However, the recent solutions lack in reducing the security threats in healthcare data management [22, 30]. Hence, we proposed a framework based on the blockchain and IPFS network to tackle the problem.

The remainder of the paper is worded as follows : Section 2 outlined the relevant work on health data management and limitations. Section 3 addressed the background technology used in our proposed system. Section 4 outlined and applied the device model. Section 5 ultimately ended the paper and discussed the future continuation of the analysis.

2 Related Works

There are potential applications based on blockchain technology for healthcare data managing proposed by many researchers. Yue et al. claim to be the first to incorporate blockchain into healthcare system architecture [35]. They proposed the architecture of a gateway application for healthcare data for simple and safe monitoring and sharing of medical data between various entities. However, their proposed system has not been tested nor implemented. They only showed some possibilities of sharing data for study purposes without any security or privacy measures. Jenkins et al. suggested to use blockchain combining with supervised machine learning for medical large data analysis with some functional markers, which involves bio-metric and biomedical data [19].

MedRec [5] demonstrated how to apply decentralization principles in an EMR framework to large-scale data management. The authors attempted to illustrate an approach to accessing medical information via a detailed usage log. However, the prototype was not commercially applied. The scheme was based on the permissionless public blockchain. Dubovitskaya et al. explored possible applications of blockchain technologies in different healthcare environments [13]. The authors attempted to clarify the theoretical blockchain infrastructure to better handle healthcare records. They couldn't demonstrate any realistic application of any suggested framework. Omar et al. suggested a blockchain-based encryption network protecting patient records. The data is encrypted and processed in a federation blockchain, and to access the data [1], a data user must receive a data owner decryption key. A prototype of Hyperledger Medical data exchange between patient-doctor and pharmacist was seen in [32]. Using the AES256 encryption method, a patient encrypts the medical record, exchanging the keys with qualified users (doctors, pharmacists). However, as stated earlier, the proposed scheme was a project and had no functional implementation.

Shahnaz et al. suggested an ethereum-based blockchain algorithm to better handle EMR [31]. They wrote in Solidity the access management rules for healthcare records, measured costs, and evaluated performance for a theoretical case scenario. The authors have addressed EVM's average execution time, latency, and contract passage. They couldn't demonstrate any realistic case scenario or execution, though. The problems of treating patient medical information using blockchain are addressed of [28]. ABE and searchable ciphertext encryption technology are critical data protection, and fine-grained access control concerns technologies. A decentralized data management solution can solve the single failure point of traditional cloud storage systems. Combining Etherum and IPFS technologies may be a likely solution to storage problems [33]. People now switch to blockchain-based cloud data storage, as traditional cloud computing lacks encryption and fine-grained access control [14].

Recently, Kumar et al. suggested distributed off-chain storage of healthcare patient diagnosis reports in conjunction with IPFS. Authors suggested an idea to store medical records in the cloud with medical workers, and after some mining process, the block containing medical records would be created [23]. They have not considered, though, that anyone with IPFS hash still has access to medical records. Besides, they demonstrated no real-time implementation of the proposed method [25].

3 Background Technologies

3.1 IPFS

IPFS offers a distributed peer-to-peer storage structure to store and access encrypted massive volumes of EMR when required. When deleting redundant files from version-control history, IPFS stores files with their content-addressed hash [6] from a distributed hash table. The data contained in IPFS can only be modified by the hash value. Instead of keeping complete blockchain medical data, the IPFS uploads the hash value of the data to be transmitted. This would reduce mining latency and expense [2, 10, 12]. This stable IPFS property makes it ideal for storing vital and confidential clinical data. The cryptographic hash produced could be stored to minimize exhaustive blockchain computational operations. The IPFS protocol operates as: a unique content address is used to store and retrieve data on IPFS; The IPFS network do not allow the presence of duplicate files; and content exploration occurs using the DHTs [7].

IPFS is special in saving or accessing files over the traditional network. Although standard web uses location-based data processing addressing, IPFS uses content-based addressing. The key issue with conventional file storage or recovery is that it becomes unavailable to users if a file is deleted from the server location. However, IPFS overcome this problem by using the content address of data for accessing it. When a user saves a file in the IPFS network, IPFS splits files into several blobs, and an empty blob connects all [15] spluttered blocks (Fig. 1). Another great feature of the IPFS network is that it supports Git's document version control framework. A consumer can accurately monitor data changes over time.

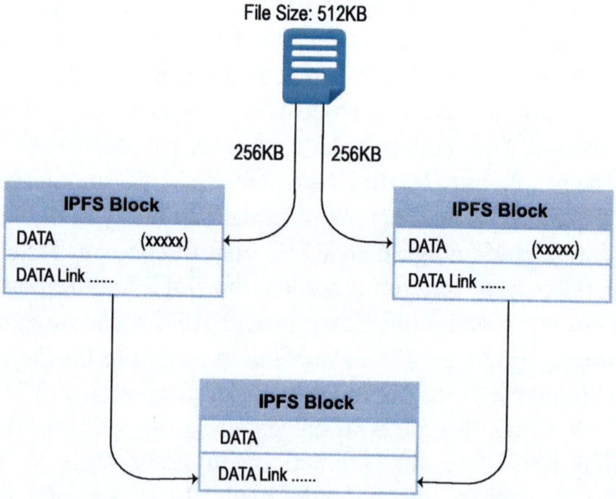

Fig. 1 Data storing process of IPFS

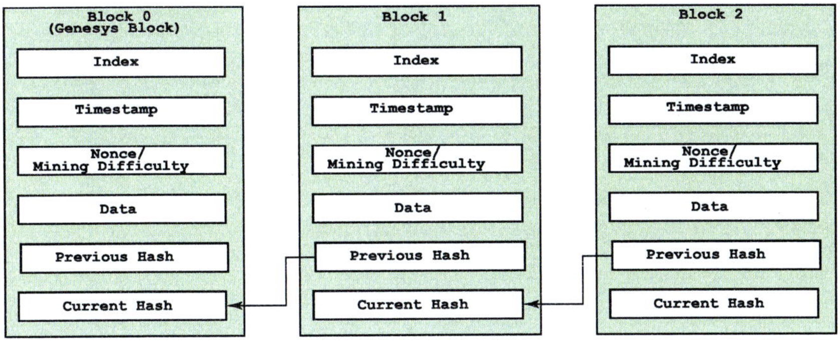

Fig. 2 Architecture of basic blockchain operation

3.2 The Blockchain

An individual (or group of people) known as Satoshi Nakamoto conceptualized the first blockchain in a digital currency form i.e. bitcoin in 2008. The primary aim was to solve the digital currency-related double-spending problem, but soon the technology began to use in several other [26] applications.

It uses a decentralized approach to disseminate and share information/data. The blockchain contains hyped, secured and peer-to-peer exchange packets. A blockchain has many advantages, including confidentiality, transparency, and data integrity, without third-party interference. These advantages make maintaining a patient's medical records equal, and technical advancement in the healthcare industry has prioritized patient health data protection. Several researchers find a viable solution [9, 17, 18] to use blockchain technologies in healthcare. Blockchain's simple architecture showed in Fig. 2 Several attributes exist, including block index, timestamp, nonce/mining complexity, details, current hash, previous hash, and more, in the blockchain.

4 System Model and Implementation

4.1 Proposed Workflow

The proposed workflow for EMR preservation and distribution has shown in Fig. 3. At first, a patient performs some medical diagnoses or tests at a healthcare center (Healthcare A) (1). Then, he receives the medical record or records from a healthcare authority (2). In step 3, the author encrypts the EMR using AES-256 algorithm [11, 16]. After that, the patient receives the IPFS hash by uploading the EMR to the IPFS network using our proposed system (4). He then uploads the hash of EMR with name in the blockchain network (5). The blockchain stores all the medical records

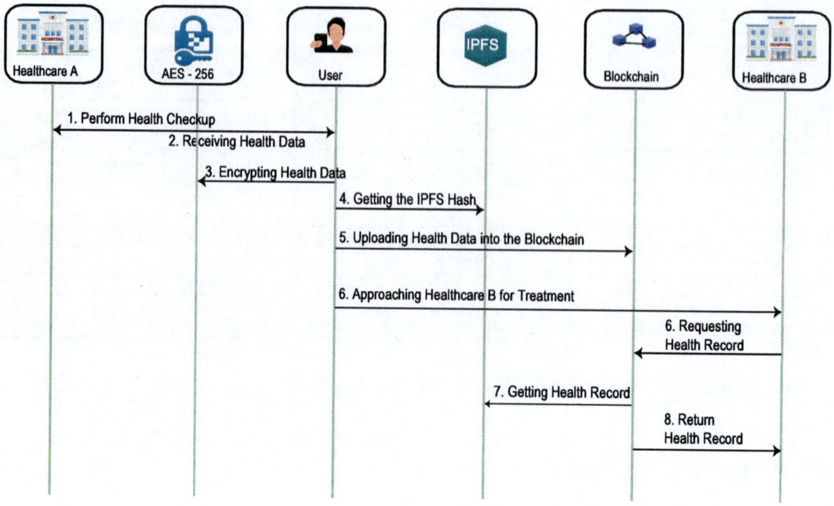

Fig. 3 Proposed workflow for secure storage and distribution of of EMR on blockchain

in different blocks. In the near future, when the patient needs to get treatment or consultation based on his recent symptoms and previous medical history, he can easily share the EMR from the blockchain with other healthcare authorities like pharmacies, doctors, or immigration agencies. At this stage, the authority will request for EMR on the blockchain using the user id and name of certificate (6). The blockchain network then retrieves the specific EMR from the IPFS network using the unique hash key for that records (7). Finally, the legitimate authority, for instance, healthcare B, will get a copy of EMR by decrypt it with the help of patient and take the necessary decision based on it (8).

4.2 *Implementation*

Experimental Setup: We used the Windows 10 operating system (version 1909) with 6 GB of RAM and 1 TB of HDD in our proposed model. We used Python (version 3.7.4) as a high-level programming language and coded it on Spyder IDE. We also used Flask (version 0.12.2) as the web application framework for our work. We deployed codes in the Postman API framework (version 7.29.1).

Connecting Peers: We had our work checked in two separate nodes Fig. 4a. Flask usually runs on http:/127.0.0.1:5000/. However, we used two separate Flask ports for our decentralization tests which are:

- http:/127.0.0.1:5002/
- http:/127.0.0.1:5003/

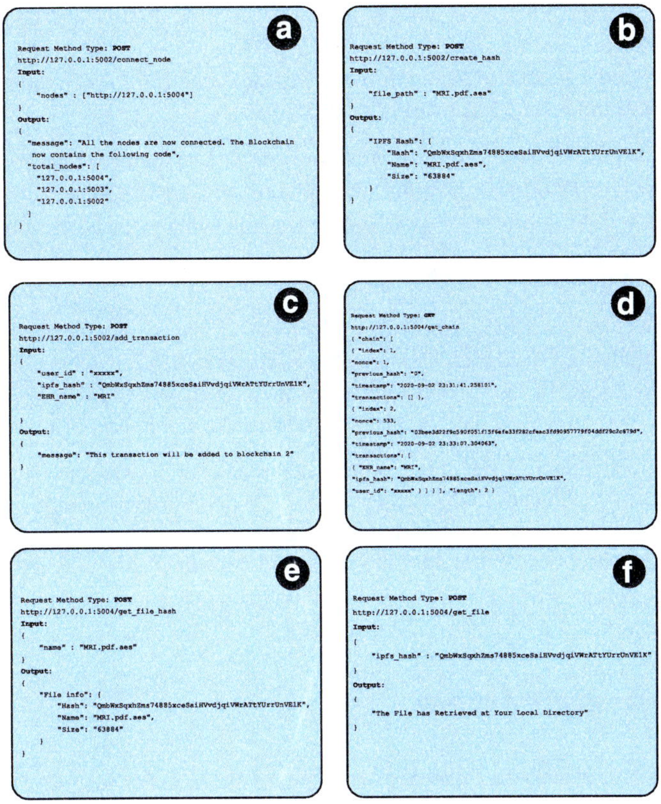

Fig. 4 Necessary coding blocks for secure EMR preservation and sharing over the blockchain network

IPFS, by default, uses the local port http:/127.0.0.1:5001/, hence we kept port 5001 for the IPFS network. We considered port 5002 for the patient and port 5003 for the healthcare authority. However, we can add as many ports for individual identity, either a patient or a legitimate user of the EMR. The function used here is *connect_node* with **POST** method request.

Getting Hash from IPFS: After receiving the EMR from healthcare A, the patient first encrypt the it and then uploads in on the IPFS network to receive a SHA256 hash key from it Fig. 4b. At this stage, the function called *create_hash* has been used, and the request was **POST** method request. The input for the function was provided in JSON format, so the user only needs to provide the directory of the EMR. After running the function, the patient will get the unique hash key with other information like file name and file size in bytes.

Adding the EMR on the Blockchain: During the uploading stage, the patient needs to provide the file name of the EMR so that using that name he can query and retrieve the EMR from the blockchain when necessary Fig. 4c. In addition to EMR name, the patient also provides his name and the unique hash key received from IPFS. Here we have uploaded a sample MRI report received from healthcare A. The function used for the operation is *add_hash* with a *POST* request.

Getting the EMR Lists: Lists of all previous EMR can be found by using the function *get_chain* Fig. 4d with a **GET** request. By initiating the function the patient can easily get the lists of all previous EMR stored on the blockchain. We have shown the list with Genesys block and an MRI test report EMR. At the end of the EMR history, the length of blocks of the blockchain has been shown.

Getting the Specific EMR from Lists: When the patient needs to retrieve a specific EMR from the lists of EMRs, he just needs to run the **POST** request with function *get_file_hash* Fig. 4e. User needs to provide the name of the EMR in JSON format. The function then returns the file info by querying the whole lists of blocks.

Getting the Targeted EMR: Finally, to get the required EMR, user need to run the *get_file* function which a *POST* request Fig. 4f. When the user provides the unique IPFS hash key for the specific EMR, the function gets the file from the IPFS network and saved it into the local directory of the user. Finally, the user can share his private key to decrypt the document to the authorized personnel.

4.3 Performance Analysis

We have analyzed the performance of the proposed model in this section. The average latency of the system, during the execution of different functions, has shown in Fig. 5. The latency for the *mine_block* and *get_file* are higher in comparison with other functions.

Security Analysis: In the proposed system, if a user wish to access the medical reports of a patient, he need to get the private key from the owner of the data, as the data is encrypted using AES-256 algorithm. Besides, the full control of the data in under

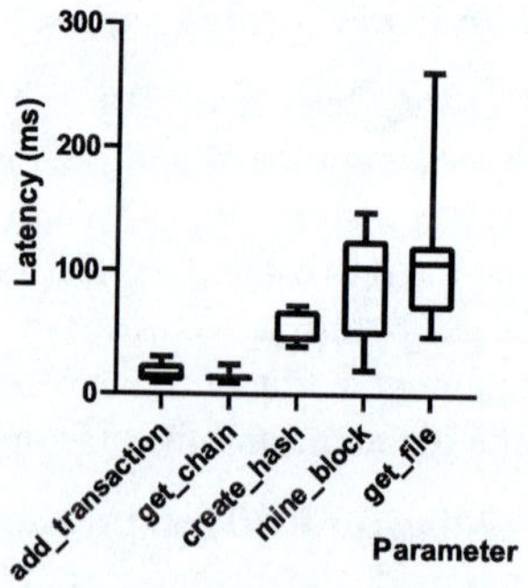

Fig. 5 Average latency of the proposed framework considering a MRI report of 62.3 KB

the control of the user, as the private blockchain was used in the proposed system. This was a unique achievement in comparison to related works mentioned in Sect. 2 [3, 27].

Scalability: The scalability in a simple term defined as the ability of the system to perform better irrespective of system data size increases or decreases. The scalability level of the proposed system is pretty good as the functions used here are lightweight and very low in size.

Integrity: The integrity of a system means the trustworthiness of the data store or transmitted in the system is unchangeable and reliable. Our system is highly reliable for patients as well for the stakeholder. The system ensures that data are immutable, using the decentralized and temper-proof nature of blockchain.

5 Conclusion and Future Work

In this article, we addressed how blockchain and IPFS technologies can be helpful to the healthcare industry, and how these can support handling of EMR. Despite advancements in the healthcare sector and technical advancement in EMR systems, they still faced some problems that this latest technology, i.e., blockchain, can tackle. Our suggested system incorporates the secure preservation of patient data with remote access guidelines for such data. It offers such a structure to promote users' usage and comprehension. The proposed system also concentrated on reducing document sizes for upload over the blockchain network. Moreover, unique IPFS hash and patient power over EMR guarantee data immutability.

We intend to introduce the platform for patients in certain local hospitals and online API. We would also attempt to simplify reimbursement for diagnostic testing and storage of EMRs. We should also define more user-based access guidelines and such strategies to follow healthcare legal expectations and values. We would also like to introduce a suitable healthcare data encryption system before sharing over a distributed network.

References

1. Al Omar, A., Rahman, M.S., Basu, A., Kiyomoto, S.: Medibchain: A blockchain based privacy preserving platform for healthcare data. In: Proceedings of SpaCCS, pp. 534–543. Springer (2017)
2. Arifeen, M.M., Al Mamun, A., Kaiser, M.S., Mahmud, M.: Blockchain-enable contact tracing for preserving user privacy during Covid-19 outbreak. Prrpints (2020)
3. Arifeen, M.M., et al.: Hidden Markov model based trust management model for underwater wireless sensor networks. In: Proceedings of ICCA, pp. 1–5 (2020)
4. Asif-Ur-Rahman, M., et al.: Toward a heterogeneous mist, fog, and cloud-based framework for the internet of healthcare things. IEEE Internet Things J. **6**(3), 4049–4062 (2018)

5. Azaria, A., Ekblaw, A., Vieira, T., Lippman, A.: Medrec: using blockchain for medical data access and permission management. In: OBD, pp. 25–30. IEEE (2016)
6. Benet, J.: Ipfs-content addressed, versioned, p2p file system. arXiv preprint arXiv:1407.3561 (2014)
7. Benet, J.: What is IPFS? (2020). https://docs.ipfs.io/concepts/what-is-ipfs/#decentralization. Last accessed 11 Aug. 2020
8. Biswas, S., Akhter, T., Kaiser, M., Mamun, S., et al.: Cloud based healthcare application architecture and electronic medical record mining: An integrated approach to improve healthcare system. In: 2014 17th International Conference on Computer and Information Technology (ICCIT), pp. 286–291. IEEE (2014)
9. Boonstra, A., Versluis, A., Vos, J.F.: Implementing electronic health records in hospitals: a systematic literature review. BMC Health Serv. Res. **14**(1), 370 (2014)
10. Chen, Y., Li, H., Li, K., Zhang, J.: An improved p2p file system scheme based on IPFS and blockchain. In: Big Data, pp. 2652–2657. IEEE (2017)
11. Daemen, J., Rijmen, V.: The Design of Rijndael: AES—The Advanced Encryption Standard. Springer (2002)
12. Dey, T., Jaiswal, S., Sunderkrishnan, S., Katre, N.: Healthsense: A medical use case of internet of things and blockchain. In: Proceedings of ICISS, pp. 486–491. IEEE (2017)
13. Dubovitskaya, A., Xu, Z., Ryu, S., Schumacher, M., Wang, F.: Secure and trustable electronic medical records sharing using blockchain. In: AMIA Annual Symposium Proceedings. vol. 2017, p. 650. American Medical Informatics Association (2017)
14. Esposito, C., De Santis, A., Tortora, G., Chang, H., Choo, K.K.R.: Blockchain: a panacea for healthcare cloud-based data security and privacy? IEEE Cloud Comput. **5**(1), 31–37 (2018)
15. Fazil, U.: A distributed file store (2020). https://medium.com/block360-labs/ipfs-a-distributed-file-store-533cda4c6047. Last accessed 22 Aug. 2020
16. Fazil, U.: Understanding AES 256 encryption (2020). https://www.solarwindsmsp.com/blog/aes-256-encryption-algorithm. Last accessed 02 Oct. 2020
17. Gordon, W.J., Catalini, C.: Blockchain technology for healthcare: facilitating the transition to patient-driven interoperability. Comput. Struct. Biotechnol. J. **16**, 224–230 (2018)
18. Gunter, T.D., Terry, N.P.: The emergence of national electronic health record architectures in the united states and Australia: models, costs, and questions. J. Med. Internet Res. **7**(1), e3 (2005)
19. Jenkins, J., et al.: Bio-mining for biomarkers with a multi-resolution block chain. In: Independent Component Analyses, Compressive Sampling, Large Data Analyses (LDA), Neural Networks, Biosystems, and Nanoengineering XIII. vol. 9496, p. 94960N. ISOP (2015)
20. Juliana, D.G.: The history of data breaches (2020). https://digitalguardian.com/blog/history-data-breaches. Last Accessed 01 Sept. 2020
21. Kaplan, R.S., Porter, M.E.: How to solve the cost crisis in health care. Harv. Bus. Rev. **89**(9), 46–52 (2011)
22. Khan, M.I., Faisal, F., Azam, S., Karim, A., Shanmugam, B., De Boer, F.: Using blockchain technology for file synchronization. In: IOP Conference Series: Materials Science and Engineering, vol. 561, p. 012117. IOP Publishing (2019)
23. Kumar, R., Marchang, N., Tripathi, R.: Distributed off-chain storage of patient diagnostic reports in healthcare system using IPFS and blockchain. In: COMSNETS, pp. 1–5. IEEE (2020)
24. Mah'd Alloubani, A., Almatari, M., Almukhtar, M.M.: Effects of leadership styles on quality of services in healthcare. Eur. Sci. J. **10**(18) (2014)
25. Mahmud, M., Kaiser, M.S., Hussain, A.: Deep learning in mining biological data. arXiv (2020)
26. Nakamoto, S.: Bitcoin: A peer-to-peer electronic cash system. Technical report, Manubot (2019)
27. Rahman, A., Roy, S., Kaiser, M.S., Islam, M.S.: A lightweight multi-tier s-MQTT framework to secure communication between low-end IOT nodes. In: Proc. NSysS, pp. 1–6. IEEE (2018)
28. Rghioui, A.: Managing patient medical record using blockchain in developing countries: challenges and security issues. In: Morgeo, pp. 1–6. IEEE (2020)

29. Sears, K., Stockley, D.: Influencing the quality, risk and safety movement in healthcare: In: Conversation with International Leaders. CRC Press (2017)
30. Semantha, F.H., Azam, S., Yeo, K.C., Shanmugam, B.: A systematic literature review on privacy by design in the healthcare sector. Electronics **9**(3), 452 (2020)
31. Shahnaz, A., Qamar, U., Khalid, A.: Using blockchain for electronic health records. IEEE Access **7**, 147782–147795 (2019)
32. Tian, H., He, J., Ding, Y.: Medical data management on blockchain with privacy. J. Med. Syst. **43**(2), 26 (2019)
33. Wang, S., Zhang, Y., Zhang, Y.: A blockchain-based framework for data sharing with fine-grained access control in decentralized storage systems. IEEE Access **6**, 38437–38450 (2018)
34. Weisbaum, H.: The total cost of a data breach-including lost business-keeps growing (2020). https://rb.gy/r1njhr. Last accessed 01 Sept. 2020
35. Yue, X., Wang, H., Jin, D., Li, M., Jiang, W.: Healthcare data gateways: found healthcare intelligence on blockchain with novel privacy risk control. J. Med. Syst. **40**(10), 218 (2016)

Evaluating Energy Efficiency and Performance of Social-Based Routing Protocols in Delay-Tolerant Networks

Md. Khalid Mahbub Khan, Muhammad Sajjadur Rahim, and Abu Zafor Md. Touhidul Islam

Abstract Energy consumption is a significant issue for data forwarding in Delay-Tolerant Network (DTN), where nodes are energy-constrained as they are driven by batteries. During their limited lifetime, these nodes have to spend a huge amount of energy since they perform forwarding, receiving, and temporarily storing messages, searching for the neighboring nodes for message forwarding, etc. The remaining energy of a node is expected to be higher as it increases the possibility of successful message delivery. Hence, awareness of energy expenditure is necessary for the routing strategies in DTN. This issue instigated us to evaluate the impact of node remaining energy on social-based routing protocols for the DTN scenario. Apart from the conventional DTN forwarding techniques, social-based forwarding protocols have involved an emerging concept that considers nodes' social relations and behaviors to make routing decisions for efficient message transmission. In this paper, we have analyzed the energy efficiency of several social-based routing approaches, such as Social-aware Content-based Opportunistic Routing Protocol (SCORP), dLife, dLifeComm, and Bubble Rap with their corresponding performance evaluations regarding the performance metrics: average remaining energy, delivery ratio, average delay, and transmission cost. Opportunistic Network Environment (ONE) simulator is used as the simulation tool, where all the simulations are performed by (i) varying the number of nodes in each group with a fixed message Time-To-Live (TTL), and (ii) changing the message TTL, while the number of nodes per group is kept constant. From the outcome of simulations, it is observed that SCORP has the highest energy efficiency and shows the best performances for average delay and transmission cost.

M. K. M. Khan · A. Z. M. T. Islam
Department of Electrical and Electronic Engineering, University of Rajshahi, Rajshahi 6205, Bangladesh
e-mail: khalidmahbub.khan@yahoo.com

A. Z. M. T. Islam
e-mail: touhid.eee@ru.ac.bd

M. S. Rahim (✉)
Department of Information and Communication Engineering, University of Rajshahi, Rajshahi 6205, Bangladesh
e-mail: sajid_ice@ru.ac.bd

M. S. Kaiser et al. (eds.), *Proceedings of International Conference on Trends in Computational and Cognitive Engineering*, Advances in Intelligent Systems and Computing 1309, https://doi.org/10.1007/978-981-33-4673-4_41

Keywords Delay-tolerant network · Social-based routing protocol · Energy efficiency · Performance evaluation · Opportunistic network environment (ONE) simulator

1 Introduction and Related Works

In wireless ad-hoc networks, the communicating nodes utilize the direct delivery link of their neighbors to route data. But in many cases, appropriate nodes cannot be addressed due to nodes' mobility, unpredictable network architecture, larger delays, etc. These networking scenarios are modeled as Delay-Tolerant Networks (DTNs), where message propagation needs to wait until a data forwarding link is available for delivering it to the next node. However, source node accumulates the message for some duration of time, carries it till the connectivity is available with the next node. This paradigm of data forwarding is known as "Store-Carry-Forward" approach. DTN can be considered as unique communication architecture as compared to other conventional TCP/IP-based networking schemes, where a dedicated communication link between source and destination is mandatory. Moreover, frequent link disruptions and longer delays are identified in DTN environment. So, "Store-Carry-Forward" approach becomes a good solution for forwarding data in DTNs [1–3]. In many disastrous circumstances, i.e., after the tsunami, earthquake, etc. the existing communication network can be broken down or completely damaged. During such situation, mobile nodes can be used to set up a DTN to forward messages. Furthermore, people living in distant and isolated areas may not find good Internet connectivity. In such case, battery-driven mobile devices of people may be utilized to establish DTN for forwarding messages successfully since they want their interest-wise data and information, such as news, music, sports, weather forecast, etc. [4].

Various routing strategies are proposed to address the forwarding problems in DTNs. Replication-based forwarding techniques are the first generation of DTN routing protocols. Uncontrolled and controlled replications of messages are the key features of such protocols. Epidemic [5], Spray and Wait [6], Spray and Focus [7], etc. are included in this class. Then, probabilistic concept is adopted in message forwarding, where node with higher probability to encounter with destined node is prioritized to forward the message by source node. Probabilistic Routing Protocol using History of Encounters and Transitivity (PRoPHET) [8] is one of the instances of such type. Eventually, next-generation forwarding strategies consider some key contexts to take routing decisions. These contexts include packet scheduling and resource allocation problems of the network. This generation of protocols includes MaxProp [9], Resource Allocation Protocol for Intentional DTN (RAPID) [10], etc. In addition, several social-based forwarding approaches are developed in this class. These routing protocols consider social interactions, relations of nodes, etc. in message forwarding to improve the performance of network. SCORP [11], dLife

[12], dLifeComm [12], and Bubble Rap [13] are included in this genre of protocols [14].

In many scenarios where nodes are distributed within numerous small-scale areas and the network topology is constantly changing, conventional DTN routing techniques do not perform well. Social-based protocols are suitable in such scenarios. These protocols exploit several social metrics like community, centrality, etc. Moreover, nodes' social characteristics, behaviors, and the reflections of their social relationships with other nodes are considered to determine the next relay node. Nonetheless, energy consumption of nodes during routing message is an important issue for such protocols in DTN environment. Notable amount of nodes' energy is spent for repeatedly transmitting and receiving message, scanning for the next relay, etc. in their limited lifetime as these nodes are powered by limited-energy batteries. As a consequence, forwarding techniques should utilize energy in the most efficient way. Less energy consumption of forwarding technique is desired for better networking performance and extending network lifetime [15, 16].

Several research works have been done regarding energy efficiency of forwarding techniques in DTNs. In [17], investigation of the energy efficiency is performed among five DTN forwarding techniques by creating simulation environment on Tirana city map. In addition, the respective performance analysis of those forwarding techniques is conducted on that scenario based on multiple performance criteria like delivery ratio, hop count, overhead ratio, etc. Also, in [18], evaluation of energy efficiency as well as performance study of six DTN forwarding techniques is reported. However, social-aware forwarding approaches are not considered in these research works. In [19], an assessment of several social-aware forwarding approaches (i.e., SCORP and dLife) is performed against the traditional DTN forwarding strategies considering metrics like delivery ratio, average latency, transmission cost, etc. But evaluation of energy efficiency of those protocols is not performed. Again, in [20], the authors include only a single social-based forwarding approach, i.e., dLife in their performance study with several existing DTN forwarding approaches. But other social-based techniques remain unexplored in terms of energy efficiency analysis. In works [21–23], the researchers have proposed various social-based forwarding approaches considering different circumstances and demonstrated the performance of their proposed protocols with respect to well-known DTN forwarding protocols. But energy efficiency investigation of social-based routing protocols, such as Bubble Rap, dLifeComm, etc. is not included in those studies. In this paper, we have considered four established social-based forwarding approaches: SCORP, dLife, dLifeComm, and Bubble Rap in DTN scenario. We intend to investigate the average remaining node energy in the mentioned social-based protocols and to evaluate their performance in terms of three metrics: delivery ratio, average delay, and transmission cost. This study is conducted by performing simulations using Opportunistic Network Environment (ONE) simulator. The structure of the remaining contents of this paper is as follows: Sect. 2 accommodates a review of the social-based protocols considered in this study. Important simulation settings and protocol requirements are explained in Sect. 3. Later, a comprehensive discussion on simulation outcomes is provided in Sect. 4. Finally, Sect. 5 gives a conclusion with motivations for some future perspectives.

2 Review of Social-Based Routing Protocols Under Investigation

In this section, a brief clarification about the basic forwarding strategies of examined social-based routing techniques is addressed. We have considered four well-known social-based routing protocols: SCORP, dLife, dLifeComm, and Bubble Rap in our study.

Social-aware Content-based Opportunistic Routing Protocol (SCORP) is a social-based forwarding approach, where the interest of message content is the key concern issue. Furthermore, as the other social-based forwarding techniques, the social interactions and behaviors of nodes within a community structure formed by the nodes are also considered in this approach. The information that is affiliated with message content includes the kind of content, interested hosts, etc. To determine the next relay node for message routing in this technique, the interests of message content get priority along with the social proximity among nodes. In this routing technique, the owner node containing message forwards it toward that particular encountered node which has the same interests about the content of message owned by the source. Otherwise, source node only forwards the message toward that node which has a higher degree of relationship with source [19, 24].

dLife is another social-aware approach which is generally based on the users' every day dynamic activity. A weighted social graph needs to be designed in order to model such dynamic activity of users. In this strategy, source routes a message copy to that neighboring node which has a strong social relation with the destination node. This relation must be higher in order as compared to that between source and that neighboring node. If the relation of the neighboring node with the destined node is unknown, then the source node's consideration of the importance of node to forward the message gets priority. Source forwards a message copy toward that node first which has the highest importance [4, 19, 23].

dLifeComm is known as the community version of dLife forwarding technique. The basic routing mechanism adopted by this technique is quite similar to that of dLife. In dLifeComm, forwarding decision is taken depending on that social strength inside a community which is evaluated from the source and encountered node toward the destined node. Moreover, the importance of node is also considered to forward the message outside the community. Here, community is the main issue instead of centrality, and these communities are detected by exploiting K-Clique algorithm in this strategy [12].

Bubble Rap protocol involves both centrality and community metrics in message forwarding decisions. A community may be formed by a number of nodes. Inside a community, there are some nodes which have more popularity than other nodes for social interactions. So, it can be considered that they have higher centrality. A source node may select a distinct node as the next relay which has higher centrality to forward the message toward the desired node. But if this protocol is implemented in a community that is spawned by the nodes having lower centrality, the message delivery ratio is expected to be low. Moreover, Bubble Rap protocol is not suited for extensive DTN architecture since the knowledge related to the entire topology of the network is required for nodes to compute the centrality. So, centrality computation may not be feasible for such large scattered networks [4, 13].

3 Simulation Scenario and Parameter Setup

Opportunistic Network Environment (ONE) is a Java-based simulator [25] that is utilized to create the proper simulation scenario for this research. Here, simulation of experimented social-based protocols is performed on an energy module that is configured in the simulator. Moreover, ONE has several attractive features, such as visualization of node movement, supporting new maps and mobility models, etc., which make it a perfect tool to simulate the social-based protocols under investigation. This research work intends to analyze the energy efficiency of simulated social-based forwarding techniques with consideration of their performance analysis. The simulation duration is kept at 6 h, and the entire area of simulation is considered as 10 km $\times$ 10 km on the map of Helsinki city. Three groups of nodes characterizing pedestrians, trams, and cars are used in the simulation. First, the number of nodes per group is varied from 30 to 80 by retaining message TTL at 300 min. Then the value of number of nodes per group is fixed at 30, while changing TTL from 120 to 300 min for further simulation. Table 1 represents some important parameters that are essential to perform simulations, whereas Table 2 lists the parameters that are essential to understand the energy settings of nodes.

4 Evaluation of Simulation Outcomes

After performing simulations of experimented social-based routing techniques in the ONE simulator, the outcomes are evaluated on the basis of the following performance metrics, such as Average Remaining Energy, Delivery Ratio, Average Delay, and Transmission Cost. A discussion related to such evaluations as well as the significance of the mentioned criteria is provided in this section.

Table 1 Necessary settings for simulation arrangements

Parameter	Value
Simulation duration	6 h
Number of nodes per group	30, 40, 50, 60,70, 80
Interface type	Bluetooth
Transmission rate	250 kbps
Transmission range	100 m
Routing protocols	SCORP, dLife, dLifeComm, Bubble Rap
Buffer capacity (MB)	2
Message TTL (min.)	120, 150, 180, 210, 240, 270, 300
Message size	500 KB–1 MB
Mobility model	Shortest path map-based movement
Simulation area size	10 km × 10 km

Table 2 Parameters pertaining to energy of nodes

Parameter	Value
Energy capacity of battery	4800 J
Scan energy	0.92 mW/s
Transmit energy	0.08 mW/s
Receive energy	0.08 mW/s
Energy recharge interval	28,000 s

4.1 Average Remaining Energy

The average amount of energy that is sustained by each node after the simulation is accomplished is known as average remaining energy, and the value of it is supposed to be high as much as possible for every node. A node is considered as dead if the energy level of it reaches zero. Figure 1a, b exhibit the variation of average remaining energy with the growth of number of nodes per group and TTL, respectively. From both the figures, it is observed that the average energy of each node is decreasing in both cases with increasing node density and message TTL. Increase in number of nodes per group results in more message transmissions, scanning, receiving, etc. through the network. So, energy consumption of each node is increased. Moreover, the lifetime of message is also extended if TTL rises and message remains active for an extra duration of time. As a result, additional energy will be consumed in each node. It is clarified in both Fig. 1a, b that SCORP exhibits the highest average remaining energy of nodes. Since in this protocol forwarding decision is taken based on the interest of message content among nodes and the relationship with source, the probability of message delivery is higher due to less amount of message wastage.

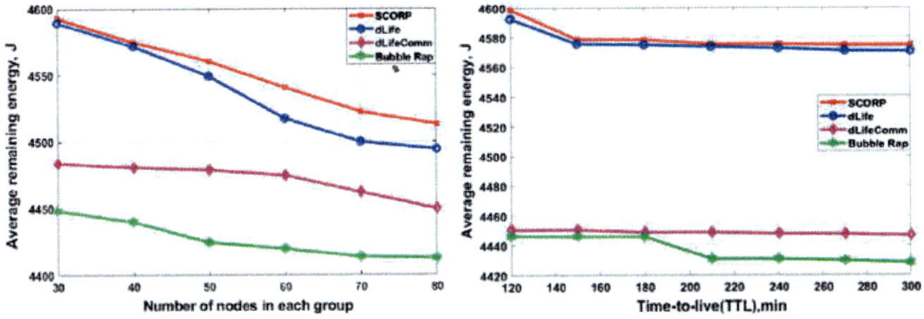

(a) (b)

Fig. 1 Average remaining energy with **a** variation of node density in each group, and **b** varying TTL

Consequently, comparatively less energy is consumed by nodes in SCORP while node density and message TTL both are increased. So, SCORP demonstrates the highest energy efficiency among the studied protocols. On the other hand, Bubble Rap protocol exerts inferior energy efficiency for both cases of varying number of nodes per group and message TTL. When both the number of nodes and TTL of a network is increased for Bubble Rap protocol, nodes need to invest more energy to calculate higher centrality nodes in order to relay messages that have longer lifetime. For this reason, the average remaining energy of node is the least for Bubble Rap protocol. However, dLife and dLifeComm protocols have higher average remaining energy than Bubble Rap but have lower average remaining energy than SCORP.

4.2 Delivery Ratio

This parameter can be determined as the ratio between the amount of messages successfully delivered to the destination and the total number of messages that are created by source. This performance metric reviews the probability of successful message delivery. Higher delivery ratio of a node means it has a higher success rate in routing messages from source to destination. From Fig. 2a, b, it is observed that dLife has the highest success rate for message delivery with increasing node density and TTL. In this case, the next relay is selected based on the relationship with destined node or in accordance with the node importance. So, the ratio of message delivery is higher than other protocols. On the other hand, Bubble Rap exhibits the lowest delivery ratio in both cases as compared to other simulated protocols due to its message forwarding mechanism. Furthermore, SCORP and dLifeComm are the medium performer for both cases in terms of delivery ratio.

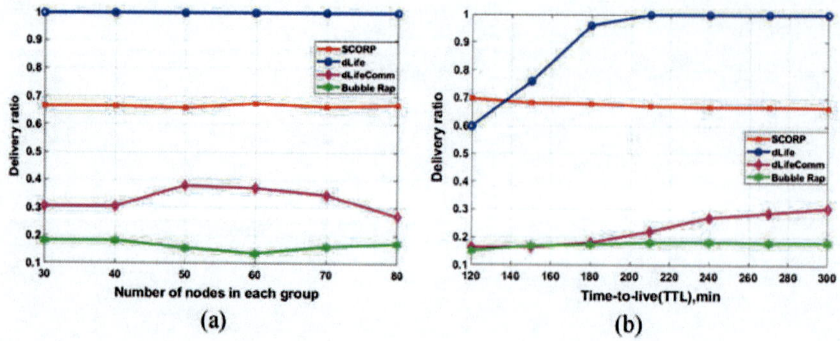

Fig. 2 Delivery ratio with **a** variation of node density in each group, and **b** varying TTL

4.3 *Average Delay*

Average delay indicates the average intermediary time that is required between the creation of a message in source and its successful reception in the destination. The value of this metric is expected to be low for effective message routing. In Fig. 3a, b, average delay is plotted with varying number of nodes in each group and TTL, respectively. From both figures, it is obtained that SCORP shows the best performance for average delay as it experiences minimum delay time for successful message delivery. In SCORP, interested node or the node which has a strong relation with source will get the message first. Thus, less time is required for perfect delivery of a single message. On the other hand, the values of average delay for dLifeComm are exceeding more than both dLife and Bubble Rap with node density and TTL. So, dLifeComm demonstrates the lowest performance than other examined protocols.

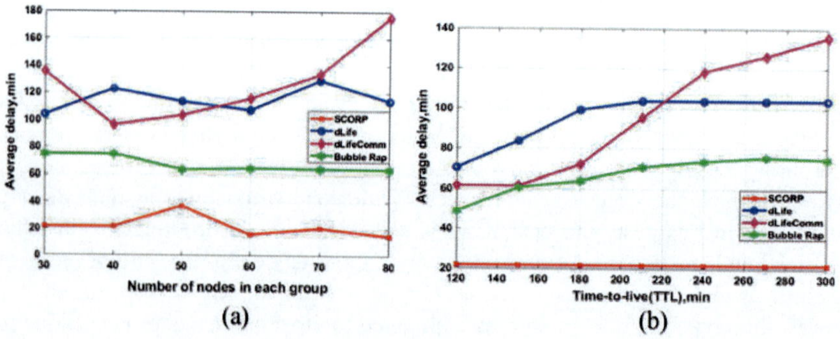

Fig. 3 Average delay with **a** variation of node density in each group, and **b** varying TTL

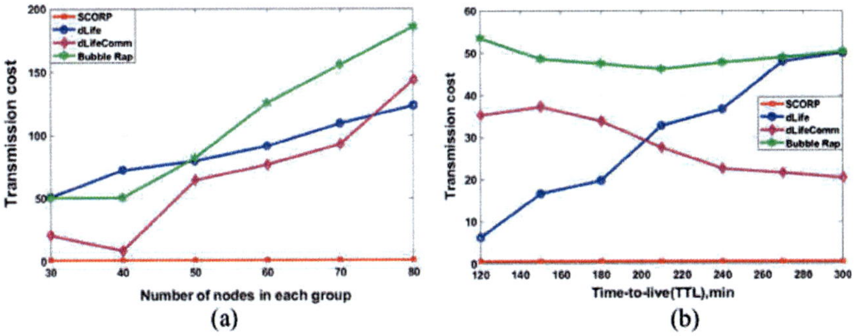

Fig. 4 Transmission cost with **a** variation of node density in each group, and **b** varying TTL

4.4 Transmission Cost

Transmission cost measures the quantity of wasted packets for a distinct packet delivery. It can be defined as the average number of packets that are forwarded for the successful end-to-end transmission of a packet. So, the transmission cost simply reflects the effectiveness of a network and it is desired to be low in value. Figure 4a, b represent the impact of transmission cost on both varying number of nodes per group and TTL, respectively. From both figures, this comes to the realization that SCORP has the minimum cost of transmission with both increasing the number of nodes within a group and TTL, whereas Bubble Rap experiences the maximum cost in those cases. In SCORP, suitable node gets the message easily that has the interest for the message as well as having strong relation with source. This is the main reason for the reduction of unnecessary message transmission for SCORP. Moreover, Bubble Rap forwarding approach becomes ineffective when both node density and message lifetime are extended. In addition, dLife and dLifeComm exhibit lesser performance than SCORP and better performance than Bubble Rap.

5 Conclusions

In this work, we have simulated four social-based routing techniques: SCORP, dLife, dLifeComm, and Bubble Rap in order to investigate their energy efficiency by considering the impact of node density and TTL. In addition, we have evaluated the performance of the mentioned routing protocols on the basis of another three metrics, such as delivery ratio, average delay, and transmission cost. After analyzing all simulation results, we finally conclude that SCORP has the maximum energy efficiency, whereas Bubble Rap is observed to be the least energy-efficient among the investigated social-based routing techniques for both cases of varying node density and message TTL. In addition, SCORP has the best performance in terms of average delay and transmission cost in those both cases. Apart from this, dLife is the best performer for delivery ratio with the variation in the number of nodes in each group and TTL, whereas Bubble Rap experiences the worst performance with respect to delivery ratio and transmission cost. In our future research, we intend to compare the observations of social-based protocols gained from this work with other DTN forwarding protocols on same simulation environment and parameter setting that would help us to contribute a better understanding of energy efficiency of diversified categories of routing protocols in DTN scenario.

Acknowledgements This work has been supported by National Science and Technology (NST) M.Phil. fellowship, granted by Ministry of Science and Technology, Government of the People's Republic of Bangladesh.

References

1. Borrego, C., Borrell, J., Robles, S.: Efficient broadcast in opportunistic networks using optimal stopping theory. Ad Hoc Netw. **88**, 5–17 (2019)
2. Wu, J., Guo, Y., Zhou, H., Shen, L., Liu, L.: Vehicular delay tolerant network routing algorithm based on bayesian network. IEEE Access **8**, 18727–18740 (2020)
3. Juyal, V., Saggar, R., Pandey, N.: On exploiting dynamic trusted routing scheme in delay tolerant networks. Wirel. Pers. Commun. **112**, 1705–1718 (2020)
4. Haq, A., Faheem, Y.: A peer-to-peer communication based content distribution protocol for incentive-aware delay tolerant networks. Wirel. Netw. **26**(1), 583–601 (2020)
5. Vahdat, A., Becker, D.: Epidemic Routing for Partially-Connected Ad Hoc Networks. Department of Computer Science, Duke University, Tech. Rep. CS-2000-06 (2000)
6. Spyropoulos, T., Psounis, K., Raghavendra, C.S.: Spray and wait: an efficient routing scheme for intermittently connected mobile networks. In: WDTN, Philadelphia, PA, pp. 252–259. ACM (2005)
7. Spyropoulos, T., Psounis, K., Raghavendra, C.S.: Spray and focus: efficient mobility-assisted routing for heterogeneous and correlated mobility. In: Fifth Annual IEEE International Conference on Pervasive Computing and Communications Workshops, White Plains, NY, pp. 79–85 (2007)
8. Lindgren, A., Doria, A., Schelén, O.: Probabilistic routing in intermittently connected networks. ACM Mobile Comput. Commun. Rev. **7**(3), 19–20 (2003)
9. Burgess, J., Gallagher, B., Jensen, D., Levine, B.N.: MaxProp: routing for vehicle-based disruption-tolerant networks. In: IEEE INFOCOM, Barcelona, Spain, pp. 1–11 (2006)
10. Balasubramanian, A., Levine, B.N., Venkataramani, A.: DTN routing as a resource allocation problem. In: SIGCOMM, Kyoto, Japan, pp. 373–384. ACM (2007)
11. Moreira, W., Mendes, P., Sargento, S.: Social-aware opportunistic routing protocol based on user's interactions and interests. In: Sherif, M., Mellouk, A., Bellavista, P. (eds.) Ad Hoc Networks. ADHOCNETS 2013. Lecture Notes of the Institute for Computer Sciences, Social Informatics and Telecommunication Engineering, vol. 129, Springer, Cham (2014)
12. Moreira, W., Mendes, P., Sargento, S.: Opportunistic routing based on daily routines. In: IEEE International Symposium on a World of Wireless, Mobile and Multimedia Networks, San Francisco, USA, pp. 1–6 (2012)
13. Hui, P., Crowcroft, J., Yoneki, E.: Bubble rap: social-based forwarding in delay-tolerant networks. IEEE Trans. Mob. Comput. **10**(11), 1576–1589 (2011)
14. Abdellaoui Alaoui, E.A., Zekkori, H., Agoujil, S.: Hybrid delay tolerant network routing protocol for heterogeneous networks. J. Netw. Comput. Appl. **148**, 102456 (2019)

15. Qirtas, M.M., Faheem, Y., Rehmani, M.H.: A cooperative mobile throwbox-based routing protocol for social-aware delay tolerant networks. Wirel. Netw. **26**, 3997–4009 (2020)
16. Zhang, X., Huang, P., Guo, L., Fang, Y.: Social-aware energy-efficient data offloading with strong stability. IEEE/ACM Trans. Netw. **27**(4), 1515–1528 (2019)
17. Spaho, E.: Energy consumption analysis of different routing protocols in a delay tolerant network. J. Ambient Intell. Humaniz. Comput. **11**, 3833–3839 (2019)
18. Khan, M.K.M., Roy, S.C., Rahim, M.S., Islam, A.Z.M.T.: On the energy efficiency and performance of delay-tolerant routing protocols. In: Bhuiyan, T., Ali, M. (eds.) Cyber Security and Computer Science. ICONCS 2020. Lecture Notes of the Institute for Computer Sciences, Social Informatics and Telecommunication Engineering, vol. 325, Springer, Cham (2020)
19. Khan, M.K.M., Rahim, M.S.: Performance analysis of social-aware routing protocols in delay tolerant networks. In: 2018 International Conference on Computer, Communication, Chemical, Material and Electronic Engineering, pp. 1–4. IEEE, Bangladesh (2018)
20. Loudari, S.E., Benamar, N.: Effects of selfishness on the energy consumption in opportunistic networks: A performance assessment. In: 2019 International Conference on Wireless Technologies, Embedded and Intelligent Systems, pp. 1–7. IEEE, Morocco (2019)
21. Kaviani, M., Kusy, B., Jurdak, R., Bergmann, N., Liu, V.: Energy-aware forwarding strategies for delay tolerant network routing protocols. J. Sens. Actuator Netw. **5**(4), 18 (2016)
22. Chilipirea, C., Petre, A.C., Dobre, C.: Energy-aware social-based routing in opportunistic networks. In: 2013 27th International Conference on Advanced Information Networking and Applications Workshops, pp. 791–796. IEEE, Spain (2013)
23. Jedari, B., Liu, L., Qiu, T., Rahim, A., Xia, F.: A game-theoretic incentive scheme for social-aware routing in selfish mobile social networks. Futur. Gener. Comput. Syst. **70**, 178–190 (2017)
24. Kuppusamy, V., Thanthrige, U.M., Udugama, A., Förster, A.: Evaluating forwarding protocols in opportunistic networks: trends, advances, challenges and best practices. Futur. Internet **11**(5), 13 (2019)
25. Opportunistic Network Environment (ONE) simulator project page. https://www.netlab.tkk.fi/tutkimus/dtn/theone. Accessed: 31 Aug 2020, 3:55:37 AM, Bangladesh Standard Time

Design a U-slot Microstrip Patch Antenna at 37 GHz mm Wave for 5G Cellular Communication

S. M. Shamim, Umme Salma Dina, Nahid Arafin, Mst. Sumia Sultana, Khadeeja Islam Borna, and Md. Ibrahim Abdullah

Abstract A U-slotted square microstrip path antenna at operating frequency 37 GHz has been proposed for millimeter wave (mmWave) fifth-generation mobile communication. The proposed antenna has a compact size of 4.35×4.35 mm^2 and printed on a FR-4 substrate material with the thickness of 0.95 mm, dielectric constant of 4.3 and loss tangent of 0.0009. A U-slotted radiating shape over the square patch has been used in the proposed antenna in order to enhance the gain and bandwidth. The result of the proposed antenna shows, return loss of −36.146 dB, gain of 5.737 dB and bandwidth of 6.17 GHz. The output of this paper also shows that proposed antenna has a good performance in terms of radiation pattern, radiation efficiency and total efficiency for the fifth-generation cellular communication at 37 GHz resonant frequency.

Keywords Microstrip path antenna · 5G cellular communication · Radiation pattern and efficiency · mmWave communication

S. M. Shamim (✉)
Department of Information and Communication Technology, Mawlana Bhashani Science and Technology University, Tangail, Dhaka, Bangladesh
e-mail: ictshamim@yahoo.com

U. S. Dina · N. Arafin · Mst. S. Sultana · K. I. Borna · Md. I. Abdullah
Department of Information and Communication Engineering, Bangladesh Army University of Engineering and Technology, Natore, Dhaka, Bangladesh
e-mail: dinahossain195@gmail.com

N. Arafin
e-mail: nahidarafin6@gmail.com

Mst. S. Sultana
e-mail: sumaiashimu1919@gmail.com

K. I. Borna
e-mail: khadeejaislamborna@gmail.com

Md. I. Abdullah
e-mail: ibrahim.risat96@gmail.com

M. S. Kaiser et al. (eds.), *Proceedings of International Conference on Trends in Computational and Cognitive Engineering*, Advances in Intelligent Systems and Computing 1309, https://doi.org/10.1007/978-981-33-4673-4_42

1 Introduction

Millimeter wave (mmWave) wireless communications are emerging as a key technology in fifth-generation (5G) cellular wireless systems due to higher potentiality to achieve their massive throughputs required by future networks [1–4]. 5G cellular systems tend to operate on the mmWave frequency band where frequency relies between 30 and 100 GHz or centimeter wave frequency band (3–30 GHz). Cellular wireless communications have dramatically contributing to the economic and social developments of both developing and developed countries in the last few decades [5]. The demanding of higher data rates to communicate and number of user are increasing day by day and wireless communication infrastructure has to develop rapidly to meet the demand. Wireless-enabled devices and data consumed rate are growing exponentially every year. 4G network technologies spectral efficiency not enough to solve the upcoming higher data rate demands. Researchers are looking for higher frequency and bandwidths as a potential solution. The emerging 5G technology has high frequency band to support multi Gbps data rates [6]. After the World Radio Communication Conference (WRC) 2015 [7], International Telecommunication Unit (ITU) released list of mmWave frequency band from 24 to 86 GHz for maintaining global standardization of mmWave frequency. In 2016, Federal Communication Commission (FCC) announced three licensed mmWave frequency band for 5G cellular communication 27 GHz (27.5–28.35 GHz), 37 GHz (37–38.6 GHz) and 39 GHz (38.6–40 GHz) [8]. Afterward different mmWave frequency band has been proposed for fifth-generation cellular communication 15, 28, 37, 60, 64, 71 and 73 GHz [9–12].

An efficient antenna must be design to incorporate with the 5G technology. Different researchers design different type's antennas to support fifth-generation cellular communication. Low profile, high gain, lightweight and simple structure antennas are the key requirement of 5G wireless technology. This is to take care of reliability, mobility and high efficiency. Microstrip patch antenna meets these entire requirements which may suite to planar as well as non-planar surfaces and their design considerations are simple. Microstrip patch antenna is extensively used due to their low cost, light weight, small size, large gain and bandwidth [13]. Circularly polarized microstrip patch antenna is using more in many wireless applications compare to linearly polarized microstrip patch antenna. It allows signal reception irrespective of the orientation of the receive antenna with respect to the transmit antenna [14]. A novel millimeter wave microstrip patch antenna is designed and analysis for 5G communication in [15]. Their proposed antenna has been described in terms of return loss, gain and impedance bandwidth. In [16], E-shaped microstrip patch antenna has been presented for fifth-generation (5G) wireless technology. Here, authors used teaching–learning-optimization (TLBO) algorithm for designing the E-shaped microstrip patch antenna. A compact substrate integrated waveguide slot-based microstrip patch antenna has been reported at 37 GHz resonant frequency for soil moisture measurement in [17]. By using high frequency structure simulator (HFSS), a low loss and high gain millimeter wave U-slotted microstrip patch antenna at 60 GHz resonant frequency has been proposed in [18]. An ultra-wideband microstrip patch antenna at 28 GHz center frequency using proximity coupling feeding technique has reported in [19] for 5G millimeter wave applications. Different researchers reported number of antenna for fifth-generation wireless network. Their reported antenna either in large size or lower gain or lower impedance bandwidth. Further studies are required to design high performance antenna to incorporate the 5G wireless communication. This paper presents a high performance U-slotted microstrip patch antenna at 37 GHz resonant frequency using EM simulation software CST microwave studio for fifth-generation wireless communications.

The rest of the papers are described as follows. The proposed antenna configurations are presented on Sect. 2. Numerical analysis described in Sect. 3 with proper equation. Section 4 descries the result and description with proper explanation. Finally, Sect. 5 concludes the paper.

2 Proposed Antenna Configuration

Proposed microstrip path antenna has been designed for resonating at 37 GHz which is applicable in 5G cellular communication as shown in Fig. 1. Copper material has used for radiating patch and ground plane due to good conducting from others. Proposed antenna built on a FR-4 substrate with height 0.95 mm, dielectric constant 4.3 and loss tangent 0.0009. Due to minimal dispersion and losses, FR-4 has selected and it also suitable for high frequency broadband application. Microstrip patch has a dimension of 4.35 mm × 4.35 mm. For ground and substrate, 8.7 mm × 8.7 mm dimension has used in simulation. A U-slot cut on the patch is enhancement of bandwidth. The microstrip feed line has length and width of 0.87 mm and 2.18 mm,

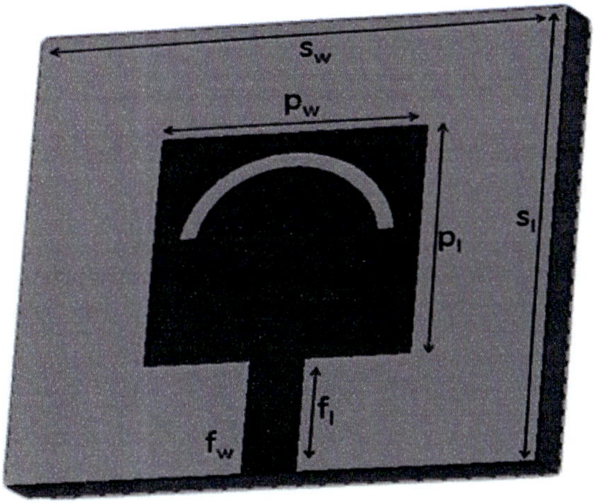

Fig. 1 Schematic structure of proposed antenna

Table 1 Proposed antenna schematic geometrical parameters

Parameter name	Parameter	Dimension (mm)
Ground length	L_g	8.7
Ground width	W_g	8.7
Patch length	L_p	4.35
Patch width	W_p	4.35
Feed length	L_f	2.18
Feed width	W_f	0.87
U-slot radius	U	0.3
U-slot outer side length	U_o	5.66
U-slot inner side length	U_i	4.71

respectively. All the geometric information regarding proposed antenna has been reported in Table 1.

3 Numerical Analysis

Microstrip patch antenna is one of the most commonly used antennas in number of wireless communication because of their low cost and easy fabrication. Different dimensions of patch antenna are calculated using the following equation. The width of microstrip antenna is calculated by Eq. (1) [20].

$$W = \frac{c}{2f_0\sqrt{((\varepsilon_r + 1)/2)}} \tag{1}$$

where w is the width of microstrip antenna, c velocity of light, f_0 resonating frequency and ϵ_r is the relative permittivity of the dielectric substrate. The length (L) of patch antenna is calculated from effective length L_{eff} subtract length extension ΔL which is presented on by Eq. (2).

$$L = L_{\text{eff}} - 2\Delta L \tag{2}$$

The effective length of patch (L_{eff}) and length extension of patch (ΔL) are calculated from Eqs. (3) to (4).

$$L_{\text{eff}} = \frac{c}{2f_0\sqrt{\varepsilon_{\text{eff}}}} \tag{3}$$

$$\Delta L = 0.421h\frac{(\varepsilon_{\text{reff}} + 0.3)\left(\frac{w}{h} + 0.264\right)}{(\varepsilon_{\text{reff}} - 0.258)\left(\frac{w}{h} + 0.8\right)} \tag{4}$$

where h thickness and $\varepsilon_{\text{reff}}$ is the effective dielectric constant for the microstrip patch antenna on dielectric substrate which is calculated by Eq. 5.

$$\varepsilon_{\text{reff}} = \frac{\varepsilon_r + 1}{2} + \frac{\varepsilon_r - 1}{2}\left[1 + 12\frac{h}{w}\right]^{-1/2} \tag{5}$$

where $\varepsilon_{\text{reff}}$ is effective relative permittivity, ε_r is the relative permittivity, h is the height of the substrate and w is the width of the patch.

4 Result and Discussion

Proposed antenna has been carried on computer simulation technology (CST) which is a microwave commercial software programs. Designed antenna operates at 37 GHz which is applicable for 5G cellular technology. The result of proposed antenna has return loss as shown in Fig. 2 for the U-slot radius $S1 = 0.3$. The upper bound and lower bound length of U-slot are 5.66 mm and 4.71 mm, respectively.

From Fig. 2, antenna has its −10 dB bandwidth microwave frequency and covers the mmWave frequency at 37 GHz. Due to U-slotted over the microstrip patch, higher bandwidth 6.3 GHz has been achieved. The return loss of proposed antenna has shown −36.146 dB. The radius between the upper bound and lower bound of U-slot has effect antenna in terms of return loss and bandwidth. The effect of U-slot radius has been analyzed which is shown in Fig. 3. Different radiuses $S1 = 0.3$, $S2 = 0.2$ and $S3 = 0.1$ used and their corresponding effect is shown. Among three

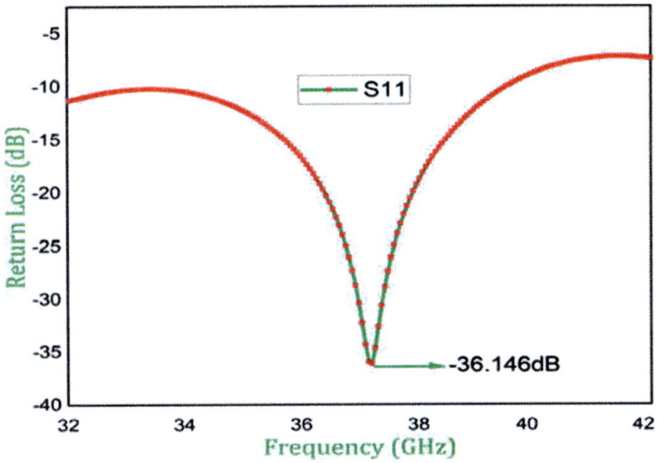

Fig. 2 Return loss versus frequency of plot of the proposed antenna

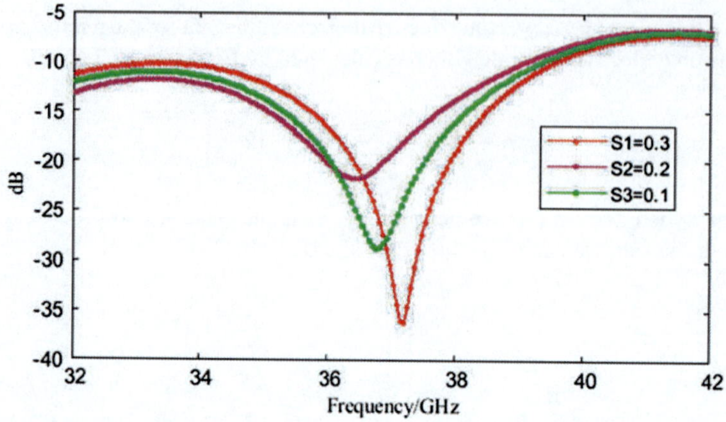

Fig. 3 Effect of U-slotted radius of proposed in terms of reflection coefficient and bandwidth

radiuses, the optimal result has achieved for the radiuses $s1 = 0.3$. For both $s2 = 0.2$ and $s3 = 0.1$ radiuses, the reflection coefficient/ return loss has below the -10 dB.

Gain versus frequency plot has been shown in Fig. 4 for three different radiuses $S1 = 0.3$, $S2 = 0.2$ and $S3 = 0.1$ in U-slot over the microstrip patch. Among the three different radiuses, highest gain 5.737 dB has achieved for $S3 = 0.3$ at operating frequency 37 GHz.

The radiation and total efficiency has been calculated after the simulation. In Fig. 5, total efficiency and radiated efficiency graph shown in terms of efficiency versus frequency. At 37 GHz, total and radiated efficiency is 59.78 and 58.528%.

The Fairfield directivity of proposed antenna at 37 GHz frequency presented in Fig. 6.

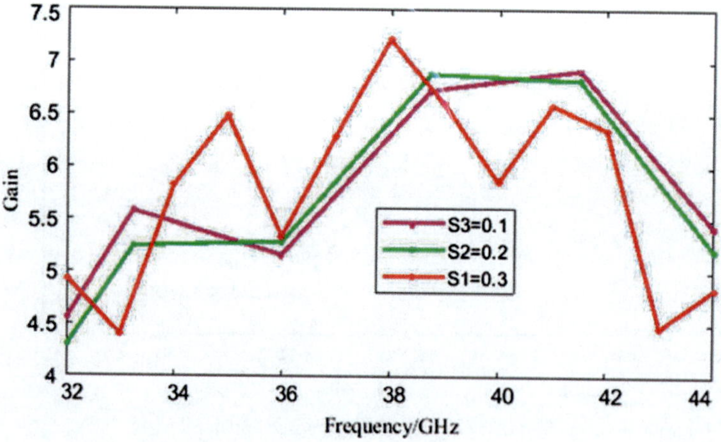

Fig. 4 Gain versus frequency graph of proposed antenna

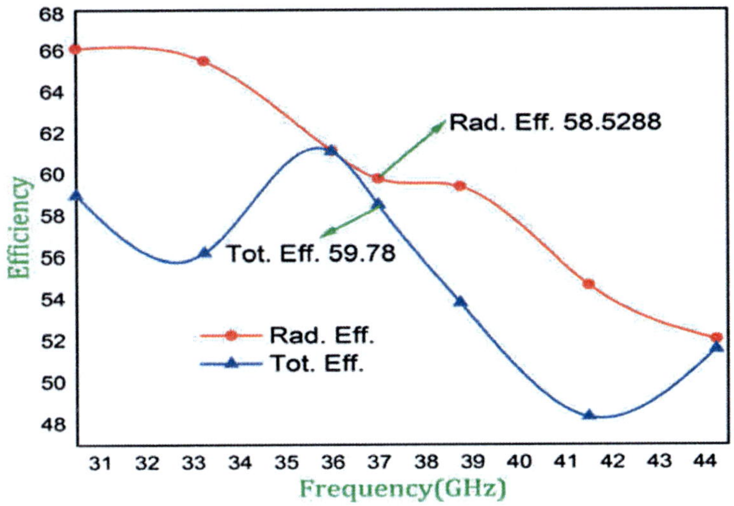

Fig. 5 Efficiency versus frequency graph of the proposed antenna

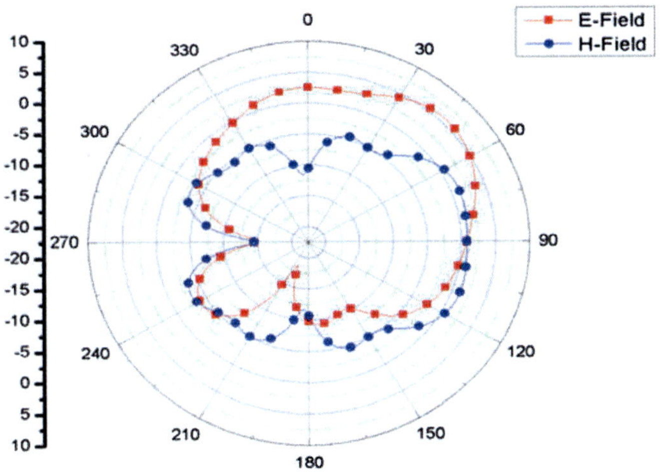

Fig. 6 E-plane and H-plane of proposed antenna

The surface current distribution of proposed antenna is presented in Fig. 7 at operating frequency 37 GHz. Larger amount of current flows below and above the U-slot in the microstrip patch, whereas almost uniform current distributed in the other areas of in this result.

At operating frequency 37 GHz, 3D radiation pattern of proposed antenna presented in Fig. 8. From the figure, the antenna directivity is 7.972 dB, and radiation efficiency and total efficiency are −2.235 dB and −2.326 dB, respectively.

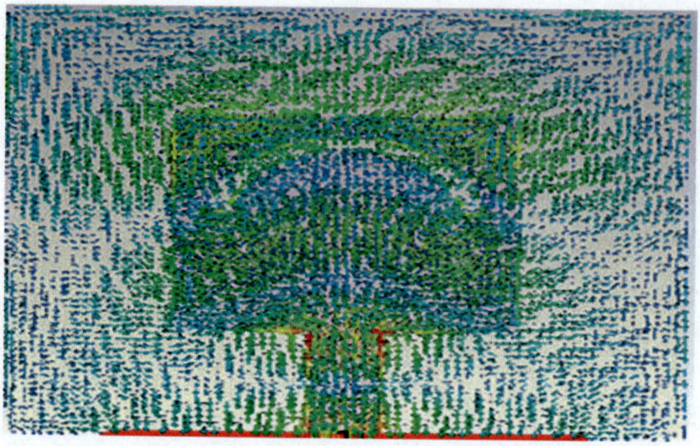

Fig. 7 Simulated current distribution of proposed antenna

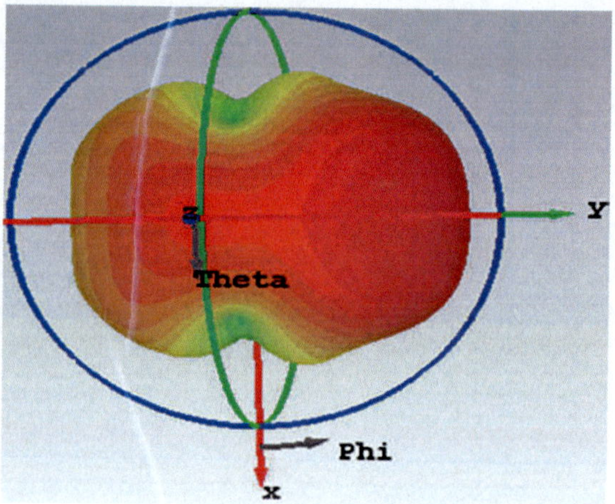

Fig. 8 3D radiation pattern of proposed antenna at operating frequency 37 GHz

Comparative analysis of proposed antenna with the previous worked has been shown in Table 2 in terms of size, resonating frequency, return loss, gain and band-

Table 2 Comparison of proposed antenna with existing design

References	Size (mm^2)	Frequency (GHz)	Return loss (dB)	Gain (dBi)	BW (GHz)
[15]	7.2 × 5	37/54	−25.8/−27.8	5.5/6	5.5/8.67
[16]	4.96 × 6.86	25/37	−25.73/−25.77	6.71/1.72	2/6
[17]	9.8 × 7	37	−36	6.7	1.4
Proposed work	4.35 × 4.35	37	−36.146	5.737	6.17

width. On comparison, we can clearly observe that the performance of the proposed antenna is better compare to reported ones.

5 Conclusion

A square microstrip patch antenna with one U-slot shape at operating frequency 37 GHz for mmWave cellular communication has been reported. Furthermore, proposed antenna has gain and bandwidths are 5.737 dB and 6.17 GHz, respectively. This designed antenna has low cost, small size and simple structure. The proposed antenna provides high efficiency, acceptable gain, higher bandwidth, good impedance matching at the desired resonate frequency 37 GHz. Proposed antenna is compact in size and it can be incorporated in many 5G communication devices, sensing, security, etc. Our future works will include different types of slot over the radiating patch to improve the antenna performances at 37 GHz resonant frequency.

References

1. Rappaport, T.S., Sun, S., Mayzus, R., Zhao, H., Azar, Y., Wang, K., et al.: Millimeter wave mobile communications for 5G cellular: it will work! IEEE Access **1**, 335–349 (2013)
2. Rangan, S., Rappaport, T.S., Erkip, E.: Millimeter Wave Cellular Wireless Networks: Potentials and Challenges (2014). arXiv preprint arXiv:1401.2560
3. Boccardi, F., Heath, R.W., Lozano, A., Marzetta, T.L., Popovski, P.: Five disruptive technology directions for 5G. IEEE Commun. Mag. **52**(2), 74–80 (2014)
4. Rappaport, T.S., Heath, R.W. Jr., Daniels, R.C., Murdock, J.N.: Millimeter Wave Wireless Communications. Pearson Education (2014)
5. Mezzavilla, M., Zhang, M., Polese, M., Ford, R., Dutta, S., Rangan, S., Zorzi, M.: End-to-end simulation of 5G mmWave networks. IEEE Commun. Surv. Tutor. **20**(3), 2237–2263 (2018)
6. Mamta, K., Singh, R.K., Sinha, N.K., Keshri, R.K.: Design and development of microstrip patch antenna for millimeter-wave application. In: Proceeding of the Second International Conference on Microelectronics, Computing & Communication Systems (MCCS 2017), pp. 79–85. Springer, Singapore (2019)
7. Resolution Com6/20, Provisional Final Acts WRC-15. WRC-15 (pp. 424–426). Geneva: ITU. https://www.itu.int/dms_pub/itu-r/opb/act/R-ACT-WRC.11-2015-PDF-E.pdf
8. Peng, M., Zhao, A.: High performance 5G millimeter-wave antenna array for 37–40 GHz mobile application. In: 2018 International Workshop on Antenna Technology (iWAT), pp. 1–4. IEEE (2018)
9. Gapeyenko, M., Samuylov, A., Gerasimenko, M., Moltchanov, D., Singh, S., Akdeniz, M.R., et al.: On the temporal effects of mobile blockers in urban millimeter-wave cellular scenarios. IEEE Trans. Veh. Technol. **66**(11), 10124–10138 (2017)
10. Naderpour, R., Vehmas, J., Nguyen, S., Järveläinen, J., Haneda, K.: Spatio-temporal channel sounding in a street canyon at 15, 28 and 60 GHz. In: 2016 IEEE 27th Annual International Symposium on Personal, Indoor, and Mobile Radio Communications (PIMRC), pp. 1–6. IEEE (2016)
11. Karttunen, A., Molisch, A.F., Hur, S., Park, J., Zhang, C.J.: Spatially consistent street-by-street path loss model for 28-GHz channels in micro cell urban environments. IEEE Trans. Wirel. Commun. **16**(11), 7538–7550 (2017)

12. Shamim, S.M., Hossain, M.S., Ta-seen, G.M.K., Miah, M.B.A., Uddin, M.S.: Performance analysis of omni-directional and directional power delay profile for millimeter wave 5G cellular networks in LOS environment. In: 2018 International Conference on Advancement in Electrical and Electronic Engineering (ICAEEE), pp. 1–4. IEEE (2018)
13. Shamim, S.M., Uddin, M.S., Hasan, M.R., Samad, M.: Design and implementation of miniaturized wideband microstrip patch antenna for high-speed terahertz applications. . Comput. Electron. 1–7 (2020)
14. Chatterjee, J., Mohan, A., Dixit, V.: Broadband circularly polarized H-shaped patch antenna using reactive impedance surface. IEEE Antennas Wirel. Propag. Lett. **17**(4), 625–628 (2018)
15. Lodro, Z., Shah, N., Mahar, E., Tirmizi, S.B., Lodro, M.: mmWave novel multiband microstrip patch antenna design for 5G communication. In: 2019 2nd International Conference on Computing, Mathematics and Engineering Technologies (iCoMET), pp. 1–4. IEEE (2019)
16. Goudos, S.K., Tsiflikiotis, A., Babas, D., Siakavara, K., Kalialakis, C., Karagiannidis, G.K.: Evolutionary design of a dual band E-shaped patch antenna for 5G mobile communications. In: 2017 6th International Conference on Modern Circuits and Systems Technologies (MOCAST), pp. 1–4. IEEE (2017)
17. Shehab, S.H., Hassan, M., Karmakar, N.: SIW slot antenna at Ka-band for soil moisture radiometer system. In: 2017 Eleventh International Conference on Sensing Technology (ICST), pp. 1–4. IEEE (2017)
18. Kumar, G., Kumar, C.: Design and analysis of u-slot mircostrip patch antenna for mobile communication at 60 ghz. Acta Technica Corviniensis-Bull. Eng. **13**(1), 35–38 (2020)
19. Sandi, E., Rusmono, A.D., Vinda, K.: Ultra-wideband microstrip array antenna for 5G millimeter-wave applications. J. Commun. **15**(2), 198–204 (2020)
20. Balanis, C.A.: Antenna Theory: Analysis and Design. Wiley (2016)

iHOREApp: A Mobile App for Hybrid Renewable Energy Model using Particle Swarm Optimization

Abrar Fahim Alam, S. M. Musfiqul Islam, Rahman Masuk Orpon, and M. Shamim Kaiser

Abstract As technology has been advanced recently and sustainability and environmental possessions of standard fuel utilization being a major concern, it is earning the world's greater interest to generate clean and usable energy in both qualitative and quantitative amounts. The goal of this paper is to build a framework that will direct cost-effective generation of renewable energy, moderately work on different environmental factors, access to offline and online information, and build a mobile platform that will enhance user experience. This is why a model built with Particle Swarm Optimization (PSO) has been proposed. With this approach, all the objectives of this paper are fulfilled and the main objective, which is to predict the best optimal way in the worst-case scenario of producing renewable energy from wind and solar energy, is also objective.

Keywords Photovoltaic · Unit-cost · Wind · Hybrid · Radiation · PSO

1 Introduction

Renewable energy emerges from renewable resources, such as sun, wind, etc., which are continuously being replenished. The abundance of these commodities on the earth's surface is not permanent, and temporal and spatial differences occur. Stimulated by recent technical developments and an increasing concern about the sustainability and environmental consequences of fossil fuel use, the possibility of producing clean, affordable electricity in sufficient quantities from renewable energy sources is of global importance.

Hybrid renewable energy (RE) systems provide more than two renewable energy resources, allow efficient use of their operational characteristics, and produce higher efficiencies than could be achieved from a single source. Many sizing and optimization tools are available for RE projects, ranging from simple thumb laws to

A. Fahim Alam (✉) · S. M. M. Islam · R. Masuk Orpon · M. Shamim Kaiser
Institute of Information Technology, Jahangirnagar University, Savar, Dhaka 1342, Bangladesh
e-mail: fahimnirzhar@gmail.com

M. S. Kaiser et al. (eds.), *Proceedings of International Conference on Trends in Computational and Cognitive Engineering*, Advances in Intelligent Systems and Computing 1309, https://doi.org/10.1007/978-981-33-4673-4_43

sophisticated software sets used extensively by RE system designers. Many software resources exist, but the adequacy of these resources is also difficult to assess [22].

Seeking optimal sizing is the biggest difficulty in introducing a hybrid renewable system. Hybrid Optimization Model for Electric Renewable (HOMER), a software that demonstrates the optimum hybrid system design and control strategy, but with this software it is not possible to evaluate optimum sizing in a mobile app-based system to satisfy both cost efficiency and optimal measurement [12].

Since various regions have different types of resources available for renewable energy generation, it is hard to foresee which technologies are more cost-effective and it is possible to hybridize inexpensive renewable energy sources to make the whole system more capable that has not yet been adopted.

Many renewable energy studies have been completed. Many experiments are solar-based, some wind and rests are biomass-based. All these processes existed in various parts of the globe. The optimal solutions will vary for various time frames and areas. Wang et al. found that the ideal summer season solutions might not work well in winter [24]. With several researchers working on a hybrid renewable energy system, predicting the most efficient way to achieve optimum performance was hard to crack. Also, optimum Hybrid Renewable Energy Systems (HRES) sizing was the most difficult. The use of a renewable wind energy system and photovoltaic system in Macada, Red Sea, Hurgada, Egypt and their results shows that the installation of such a Wind-Solar hybrid system is cheaper than the each system [10]. Saiprasad et al mentioned the possible steeplechases that India must surpass to adopt an easy model through deliberating a more developed hybrid optimization by introducing a genetic algorithm along with a software which was produced by a Spanish professor of University of Zargoza through triple bottom line analysis [19].

Previous studies of hybrid renewable energy models show that there is no interactive app-based model that can predict the best way to produce cost-efficient renewable energy.

In this paper, we proposed a mobile app focused on PSO to determine optimal sizing of hybrid RE choices. We compared the model's performance with the other current model and examined the efficacy of the proposed mobile app.

The remainder of the article is arranged as: Sect. 2 includes literature review; Sect. 3 presents the system model; Sect. 4 demonstrates a case analysis; Sect. 5 contains findings and discussion, and eventually, Sect. 6 ends the work with future work.

2 Literature Review

Understanding the underlying hybrid renewable energy generation and improving the performance of mixed lead- Solar, Wind, Biomass, etc. has become the next challenge and trend in this sector. So, here the authors present experimental and theoretical studies of renewable energy generation. Deshmukh et al. observed that modeling of the HRES reveals its popularity when it comes to satisfying basic energy requirements. HRES is mainly known for remote area power applications and are now cost-effective where grid service expansion is costly and found that to integrate HRES into the current power supply network, a detailed study should be performed on checking viability and profitability in technological matters [3]. Hansen et al. found no models which have yet been developed to the degree that they can represent all the sectors and features needed in adequate detail and have not been extended to the global level and also models of energy systems should be further developed in tandem with integrated evaluation models for a more detailed and multidisciplinary interpretation of defossilization pathways for the benefit of all the communities and stakeholders being involved [9].

Affluent of works on renewable energy forecasting have been reported in terms of different methods based on different designed algorithms, models and softwares. So, here most of the authors present more practical and mathematical studies regarding cost efficiencies. Khare et al. addressed analyzes of pre-feasibility, optimum design, modeling, control aspects, and aspects of reliability and also further discussed the application of evolutionary strategy and game theory in the area of hybrid renewable energy but they did not develop an exact optimization technique and geographical software to determine the potential of solar radiation and velocity [14]. Shivarama et al. provided a thorough analysis of the various issues modeling, analysis, optimum size, control, and energy storage approaches for hybrid systems combined with wind, solar PV, fuel cells, etc. were discussed briefly, but the most cost-effective and optimal combination could also have been listed [21]. Masud et al. adopted the homer optimization program in this research to evaluate the efficiency of a standalone PV or wind power generation system for deployment in rural communities of Sokoto state Nigeria. But the other environmental factors were missed out in this research which is not related to that zonal forecasting [18]. Gan et al. summarized and classified the optimal sizing of hybrid renewable energy system with an storage. Based on various energy storage modes of the proposed hybrid system, the software presents various combination along with optimal hybrid system [8]. The work and the proposition of Krishan et al. has recognized that the HOMER and MATLAB are synergistic for the design of the HRES [15].

In recent trends, most of the reports, journals, and publications are being released regarding machine learning and artificial intelligence. The renewable energy generation sector is not among the exceptions. Here, most authors are trying to introduce scientific theories with practical orientations in their regarded papers. Several Bio-inspired Artificial Intelligence (AI) methods like the Genetic algorithm have been used widely used by the researchers to optimize HRES. The study of Javed et al. demonstrated that GA is more predominant than HOMER as far as monetary measuring, optimization time, and system reliability in terms of autonomous hybrid solar-wind-battery systems [11, 17]. Yang et al. has applied a genetic algorithm to examine a hybrid wind-solar system on the southeast coast of china. Though the system performs effectively. The energy resources are inadequate as they have used a statistical parameter in their model. As a result, most of the time the system performance decreases [25]. Lopez et al. has used a genetic algorithm to optimize the sizing and control of several PV-Diesel Systems [1, 4].

Along with the genetic algorithm, different types of AI algorithm has been applied by the researchers. Sharifzadeh et al. has used predictive Artificial Neural Networks (ANN), Support Vector Regression (SVR), and Gaussian Process regression method to predict the wind power, solar power, and the energy demand. All of the models did well to predict solar and energy power but only ANN had successfully predicted the energy demand [16, 20]. Gampa et al. proposed a two-stage optimization algorithm based on Fuzzy logic (multi-objective) which has optimally sized distributed generations, shunt capacitors, and electric vehicle charging stations for placing distribution systems [7]. Gala et al. has predicted solar radiation values in Spain through machine learning regression methods. Though their model performs better than the stand-alone Gradient Boosted Regression and Random Forest Regression model but it doesn't perform as good as the SVR model [6]. Amer et al. has applied PSO algorithm to find the minimum Levelized Cost of Energy. But in their study, they have examined the HRES system containing only a single wind turbine or PV module [2]. Singh et al. has used an Artificial Bee Colony algorithm and observed that a grid-connected PV-Biomass hybrid system is more reliable and cost-efficient than the stand-alone systems [23]. Kefayat et al. has examined that hybrid of ant colony optimization and artificial bee colony algorithm performs better in multiple distributed energy resources allocation optimization problems [13].

3 Proposed System Model

The energy production of a solar system from the solar radiation value of a specific area and energy production of a wind turbine from the wind speed value of that area can be calculated. Then it could be found as the most optimum combination of solar and wind by using the PSO algorithm.

3.1 PV System Model

As solar energy is regarded as the most muscular available source of energy, an enlarged defined system should be utilized. PV system is that defined system that can generate that much renewable energy which is the combined system of inverter and other electrical and mechanical hardware. Its functionality is to utilize the energy produced from the sun electrically.

$$E = A \times r \times H \times PR \tag{1}$$

where A is the total solar panel area, r is the efficiency of the solar panel, H is annual average solar radiation of that area and PR is the performance ratio. If it is considered as solar panel area, efficiency and performance ratio are constant then the energy generation is proportional to solar radiation. More energy can be generated if that area has greater solar radiation.

3.2 Wind Turbine Model

Wind power being sustainable and renewable energy, it's kinetic energy is easily converted into electricity with a wind turbine. Despite it's less impact than that of burning fossils on the environment, pollutants are not generated as electrical power is generated which is renewable. So, it is hereby said as a compact and balanced design of cost and energy output.

$$P = \frac{1}{2}\rho A v^3 \tag{2}$$

where ρ is the air density value, A is the wind turbine area and v is the wind speed of that area. If it is considered as air density and wind turbine area is constant then the generated energy will depend on wind speed. As it is known that every wind turbine has a rated wind speed for which the maximum output can be provided therefore if the rated wind speed and the maximum output is known then it can easily be found the energy generation at any wind speed is quite possible for that specific wind turbine.

3.3 Proposed System Overview

In our proposed model the most optimum combination of wind and solar resources have been calculated. For that, the location has been traced by the system as it has to calculate according to solar radiation and wind speed for that specific area. So, these data from that location will be observed.The current situation will be audited and the best combination with the PSO algorithm will be analyzed. Figure 1 shows the block diagram of our proposed model. From Eq. (1) we get the daily solar energy generated

$$E_S = A \times r \times H \times PR \tag{3}$$

From Eq. (2) we can calculate the daily wind energy generated

$$E_W = 24 \times \frac{1}{2}\rho A v^3 \tag{4}$$

Now the hybridization can be calculated as

$$E = m \times E_S + n \times E_W \tag{5}$$

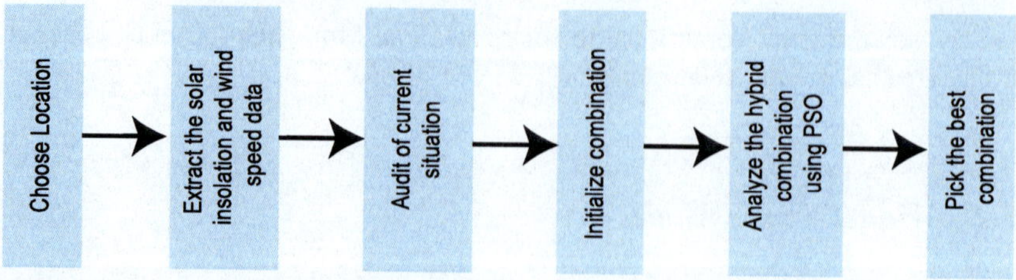

Fig. 1 The Block diagram depicts the simple illustration of our proposed hybrid optimized renewable energy model. The principle parts of our proposed model have been shown here

where m + n = 1. This hybridization will be most optimal for that values of m and n for which E is maximum at the worst-case scenario. This can be solved by using the PSO algorithm.

3.4 *PSO Based Hybrid Model*

PSO is widely used by the researchers for the optimization problem. PSO is efficient and does not have the complexity to implement like other optimization algorithms. That is why we have used PSO algorithm for our proposed hybrid optimization model. In our proposed model at first, we have calculated the monthly energy productions for each combination. Here we assigned the worst month's production as pbest. When we have found all combination's pbest value then we calculate which combination has the maximum pbest value and assigned it as gbest. For the combination, we have found the gbest has the best production in the worst-case scenario. So, the combination has been returned as the output to the user. Algorithm 1 shows the algorithm of our proposed system. Figure 2 shows the user interface of our iHOREApp.

Algorithm 1 Calculation of Most optimum combination

$P = Initialization$
for combination= 1 to 100 **do**
  for i=1 to 12 **do**
    $combination[fitness] = production(month[i])$
    if $combination[fitness] < combination[pbest]$ **then**
      $combination[pbest] = combination[fitness]$
    end if
  end for
end for
$gbest = max(combination[pbest])$
$return\ combination$

4 A Case Study

This section includes a performance comparison of the proposed iHORE app and HOMER based solution. For the study, solar radiation and wind speed of Izmir, Turkey has been considered [5]. Figure 3 shows the solar radiation and wind speed value of Izmir, Turkey.

Erken et al. has employed HOMER software and found that by mixing 55.6% solar and 44.4% wind resources the production cost will be as low as 0.064$ per KW [5]. On the other hand, for the same setting, our proposed app found the per unit energy cost will be 0.058$ per KW when the solar and wind resources are 35% solar and 65% respectively.

Figure 4 represents the energy production for each solar and wind combination. We have tried to develop a cost-effective system that will operate on multiple environmental factors. We can collect monthly solar radiation and wind speed data from an area from NASA 's website, as the app would take the time to transmit the production to the particular area and produce energy for each month.

5 Conclusion and Future Work

A hybrid framework has been intended to be built based on accumulating the selection of smarter combination for that particular area as per the solar-powered radiation and wind speed. Further, a plan has been worked out to make a cross-breed frame with sun-oriented and wind-friendly components for a good alternative and cost amicable components. A systematized solution has been outlined which will be cost-efficient, mobile-friendly, and will provide the most optimum results by the application through AI and machine learning algorithm.

The basic components like biomass, hydro, tidal etc. will further be introduced in our system which will make the hybrid system more compatible. Introducing these components will expand the opportunities of providing more optimum solutions as a suggestion in our application.

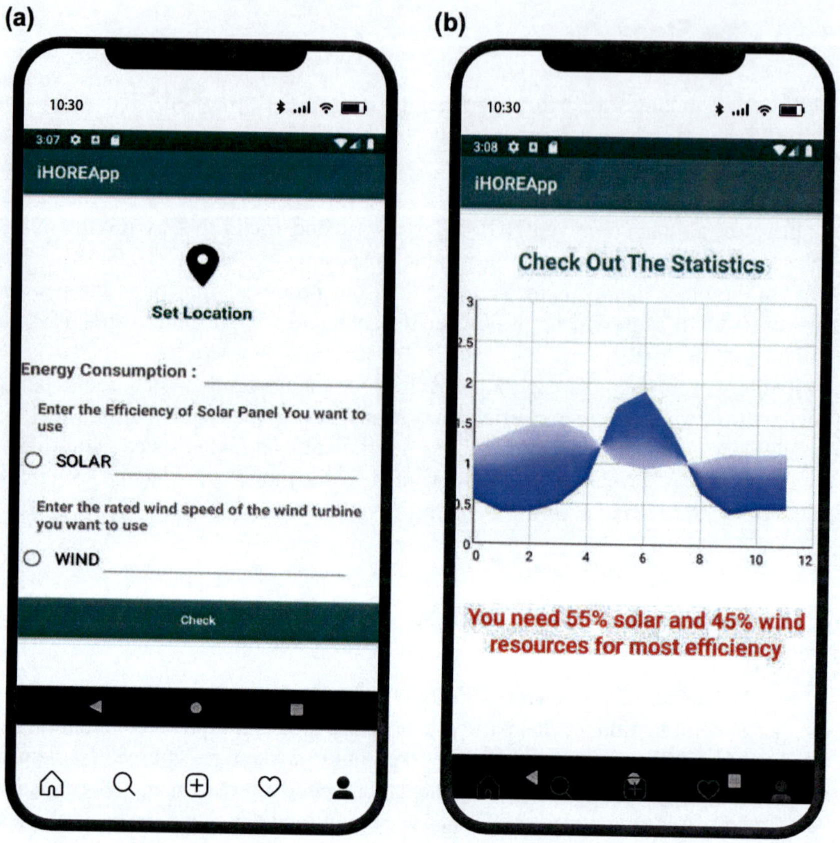

Fig. 2 Screenshot of the User interface of iHOREApp, **a** shows the input screen of our proposed application. Here the app will find the location of the user through GPS coordinates. The app will ask the user for his energy consumption. Then it will ask the efficiency of the solar panel he wants to use and the rated wind speed of the wind turbine he wants to use. **b** shows the output screen of our proposed application. Here the app will propose the optimum solution and also show the graph of monthly power generation of the hybrid system with which the user can see in which month which combination will perform better. In this graph, X-axis represents the month and Y-axis represents the energy generated by different combinations

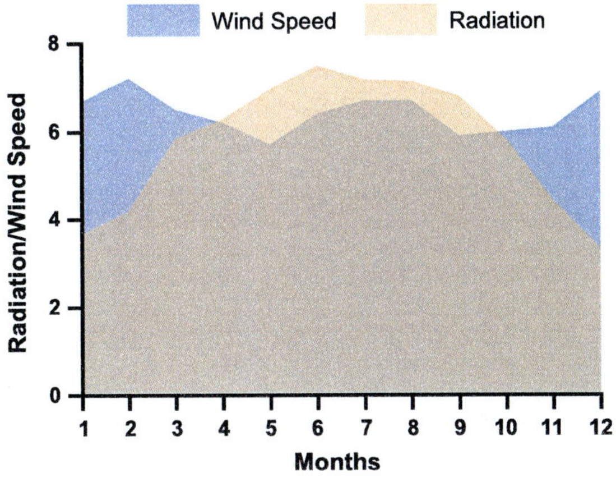

Fig. 3 Daily radiation and wind speed of Izmir, Turkey

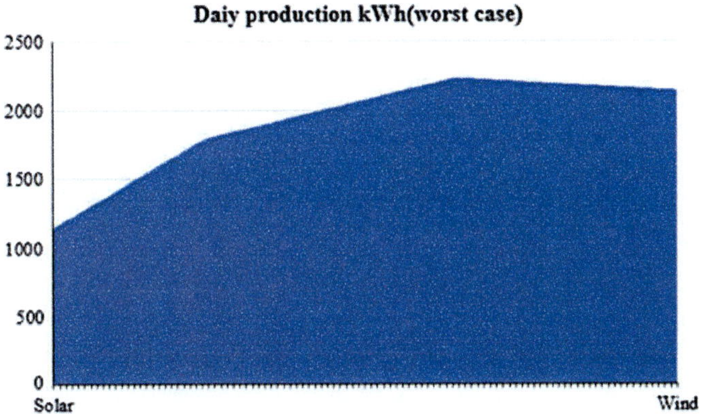

Fig. 4 Daily Energy production at worst case scenario

References

1. Al Mamun, S., Chowdhury, Z.I., Kaiser, M.S., Islam, M.S.: Techno-financial analysis and design of on-board intelligent-assisting system for a hybrid solar-deg-powered boat. Int. J. Energy Environ. Eng. **7**(4), 361–376 (2016)
2. Amer, M., Namaane, A., M'sirdi, N.: Optimization of hybrid renewable energy systems (HRES) using PSO for cost reduction. Energy Procedia **42**, 318–327 (2013)
3. Deshmukh, M., Deshmukh, S.: Modeling of hybrid renewable energy systems. Renew. Sustain. Energy Rev. **12**(1), 235–249 (2008)
4. Dufo-López, R., Bernal-Agustín, J.L.: Influence of mathematical models in design of pv-diesel systems. Energy Convers. Manage. **49**(4), 820–831 (2008)
5. Ekren, O., et al.: Sizing of a solar-wind hybrid electric vehicle charging station by using HOMER software. J. Clean. Prod. **279**, 123615 (2020)

6. Gala, Y., et al.: Hybrid machine learning forecasting of solar radiation values. Neurocomputing **176**, 48–59 (2016)
7. Gampa, S.R., Jasthi, K., Goli, P., Das, D., Bansal, R.C.: Grasshopper optimization algorithm based two stage fuzzy multiobjective approach for optimum sizing and placement of distributed generations, shunt capacitors and electric vehicle charging stations. J. Energy Storage **27**, 101117 (2020)
8. Gan, J., Li, J., Qi, W., Kurban, A., He, Y., Guo, S.: A review on capacity optimization of hybrid renewable power system with energy storage. In: E3S Web of Conferences, vol. 118, 02055. Publisher: EDP Sciences (2019)
9. Hansen, K., Breyer, C., Lund, H.: Status and perspectives on 100% renewable energy systems. Energy **175**, 471–480 (2019)
10. Hemeida, A.M., El-Ahmar, M.H., El-Sayed, A.M., Hasanien, H.M., Alkhalaf, S., Esmail, M.F.C., Senjyu, T.: Optimum design of hybrid wind/PV energy system for remote area. Ain Shams Eng. J. **11**(1), 11–23 (2020)
11. Javed, M.S., Ma, T.: Techno-economic assessment of a hybrid solar-wind-battery system with genetic algorithm. Energy Procedia **158**, 6384–6392 (2019)
12. Kaiser, M.S., Chowdhury, Z.I., Al Mamun, S., Hussain, A., Mahmud, M.: A neuro-fuzzy control system based on feature extraction of surface electromyogram signal for solar-powered wheelchair. Cognit. Comput. **8**(5), 946–954 (2016)
13. Kefayat, M., Lashkar Ara, A., Nabavi Niaki, S.A.: A hybrid of ant colony optimization and artificial bee colony algorithm for probabilistic optimal placement and sizing of distributed energy resources. Energy Convers. Manage. **92**, 149–161 (2015)
14. Khare, V., Nema, S., Baredar, P.: Solar-wind hybrid renewable energy system: a review. Renew. Sustain. Energy Rev. **58**, 23–33 (2016)
15. Krishan, O., Suhag, S.: Techno-economic analysis of a hybrid renewable energy system for an energy poor rural community. J. Energy Storage **23**, 305–319 (2019)
16. Mahmud, M., Kaiser, M.S., Hussain, A.: Deep learning in mining biological data. arXiv preprint arXiv:2003.00108 (2020)
17. Mahmud, M., et al.: Applications of deep learning and reinforcement learning to biological data. IEEE Trans. Neural Netw. Learn. Syst. **29**(6), 2063–2079 (2018)
18. Masud, A.A.: The application of homer optimization software to investigate the prospects of hybrid renewable energy system in rural communities of sokoto in Nigeria. Int. J. Electr. Comput. Eng. (**7**(2), 596 (2017)
19. Saiprasad, N., Kalam, A., Zayegh, A.: Triple bottom line analysis and optimum sizing of renewable energy using improved hybrid optimization employing the genetic algorithm: a case study from India. Energies, vol. 12(3), 349. Number: 3 Publisher: Multidisciplinary Digital Publishing Institute (2019)
20. Sharifzadeh, M., Sikinioti-Lock, A., Shah, N.: Machine-learning methods for integrated renewable power generation: A comparative study of artificial neural networks, support vector regression, and gaussian process regression. Renew. Sustain. Energy Rev. **108**, 513–538 (2019)
21. Shivarama Krishna, K., Sathish Kumar, K.: A review on hybrid renewable energy systems. Renew. Sustain. Energy Rev. **52**, 907–916 (2015)
22. Siddique, A.R.M., Kaiser, M.S.: Financial and performance analyses of microcontroller based solar-powered autorickshaw for a developing country. J. Eng. **2016**, (2016)
23. Singh, S., Kaushik, S.C.: Optimal sizing of grid integrated hybrid PV-biomass energy system using artificial bee colony algorithm. IET Renew. Power Gener. **10**(5), 642–650. Publisher: IET Digital Library (2016)
24. Wang, R., Xiong, J., He, M.f., Gao, L., Wang, L.: Multi-objective optimal design of hybrid renewable energy system under multiple scenarios. Renew. Energy **151**, 226–237 (2020)
25. Yang, H., Zhou, W., Lu, L., Fang, Z.: Optimal sizing method for stand-alone hybrid solar-wind system with LPSP technology by using genetic algorithm. Sol. Energy **82**(4), 354–367 (2008)

Cyber Threat Mitigation of Impending ADS-B Based Air Traffic Management System Using Blockchain Technology

Farah Hasin and Kazi Abu Taher

Abstract This study presents a new model to mitigate the cyber threats of the impending Automatic Dependent Surveillance-Broadcast (ADS-B) using the security features of blockchain technology. The initiatives of regional aviation authorities to upgrade the ATM system have common surveillance and data sharing equipment named ADS-B which is unencrypted. The cybersecurity threat to aviation operations has become more challenging with the up-gradation of ATM which largely depends on ADS-B. Today, the global civil aviation authorities are relying on the data-Link and IT systems for their day-to-day ATM through ADS-B and other Air Traffic Control Services (ATCS). Numerous studies and aviation enthusiasts expressed the vulnerability of the ADS-B due to unencrypted data sharing of ATCS. In this research, a blockchain framework is proposed to address the current security-related weaknesses of the ADS-B application and other data-Linked services of impending ATM. This blockchain model is designed based on the Hyperledger Fabric platform which ensures privacy, confidentiality, and origin/destination authenticity to overcome the cybersecurity threat of ADS-B and other ATCS support systems. This proposed model is significant as it leverages the synergy of multi-agent networks and approved blockchain technology like Hyperledger Fabric for a secure, stable, and unified network for collaboration. Thus, the different risks of upcoming ADS-B based ATM-like eavesdropping, message modification, injection, deletion can be combated using this proposed model.

Keywords Automatic dependent surveillance-broadcast · Hyperledger fabric · Chaincode

F. Hasin (✉) · K. A. Taher
Department of ICT, Bangladesh University of Professionals, Dhaka, Bangladesh
e-mail: farahhasin01@gmail.com

K. A. Taher
e-mail: kataher@gmail.com

M. S. Kaiser et al. (eds.), *Proceedings of International Conference on Trends in Computational and Cognitive Engineering*, Advances in Intelligent Systems and Computing 1309, https://doi.org/10.1007/978-981-33-4673-4_44

1 Introduction

In the era of globalization, the two mentionable sectors which help the world to become smaller are Information Technology (IT) and Aviation. The huge growth of both sectors brought changes in their operations with new challenges as well as opportunities. Heavy reliance on air transportation caused tremendous difficulty in Air Traffic Management (ATM) system throughout the world. Regional authority like the Federal Aviation Administration (FAA) is implementing NextGen project for safe air operations which largely depends on IT [1]. Single European Sky ATM Research (SESAR) is a collaborative project to completely overhaul European airspace and the ATM that had been in use for long [2]. Likewise; International Civil Aviation Organization (ICAO) is implementing ASIA/PACIFIC seamless ATM plan with a focus on technological and human performance upgrades [3]. ICAO expects air traffic to triple in the Asia Pacific region by 2030 [4]. Air Navigation Service Providers (ANSP) all over the world are also facing high traffic growth and growing difficulties in dealing with the rising volume of air traffic and high standards of service.

The overall easy deployment with a low budget, salient benefits to safety and capacity, harmonization with current ATM procedures of ADS-B was also discussed at the 11th ICAO Air Navigation Conference [5]. As per the Civil Aviation Authority of Bangladesh (CAAB), before the COVID-19 impact began, around 500 flights that were taking off from other nations fly over the air space of Bangladesh and the number of flights flying to and from the nation's three international airports was more than 200 [6]. In the adaptation process to digitalized ATM, ADS-B took over the place of old radar systems. Apart from ADS-B, Controller–Pilot Data Link Communications (CPDLC) is also assisting ATM in an advanced aviation environment that works mostly on an unencrypted data-Link basis. Another way of monitoring its own aircraft by the owner is Aircraft Communications Addressing and Reporting Systems (ACARS) which transmits unencrypted data to its predesignated stations. Informative services of ATM-like Flight Information Display System (FIDS) also depend on data sharing. The chances of security attacks on all these digitalized components of ATM are similar as all of them shares unencrypted data.

Cybersecurity was regarded as a high-level barrier to the execution of ICAO's Global Air Navigation Plan [7]. Historical precedent has also shown how a potential adversary can manipulate unencrypted datalinks [8]. The previous studies, presentations, and report clearly expressed about the threats, but could not provide any solution to ADS-B vulnerability directly. The number of commercial aircraft production is estimated to be twice the existing number in the 2040s, as reported by the Global Aerospace and Defence Industry [9]. In ATM, ACARS is generally prone to cybersecurity attacks. Civil Air Navigation Service Organization (CANSO) stated that nowadays it has been a major key field that demands cybersecurity concern is the introduction of increased automation and dependency on system-wide information exchange [10]. This research aims to tackle the major limitations of the ADS-B based ATM system, which is the lack of privacy and access control, by developing a scalable permissioned applying the security features of blockchain technology. To combat the different security threats of the cutting-edge technology in ATM such as eavesdropping, message injection, alteration, and masquerading. This paper presents a blockchain-based model that alleviates existing weaknesses in the ADS-B application and other ATM services that rely on data-Links. This proposed model fabricates a novel collaboration framework for distributed, multi-agent structures based on the evolving blockchain technology, especially the Hyperledger Fabric ecosystem.

The remainder of this paper is structured as follows. Section 2 discusses the related technology. Section 3 describes the new proposed model using a hyper ledger blockchain. Section 4 emphasizes detailed architecture and function. Section 5 covers the analyses of the proposed model performance, and finally Sect. 6 concludes the paper.

2 An Overview of Related Technology

2.1 ADS-B in ATM

In the impending ATM, an ADS-B enabled aircraft calculates its location using GPS or GNSS. The transponder of the aircraft then broadcasts that status at frequent time frames mentioning the identity, and salient flight conditions like altitude, speed, etc. As shown in Fig. 1 for effective aircraft tracking, field ADS-B base station continuously collects broadcasts signal and submit data to the ATC tower. Other proximate aircraft can also receive the broadcasts of the data sending aircraft. There will be very less controller's dictation to fly along a route with other multiple aircraft at proximity [11].

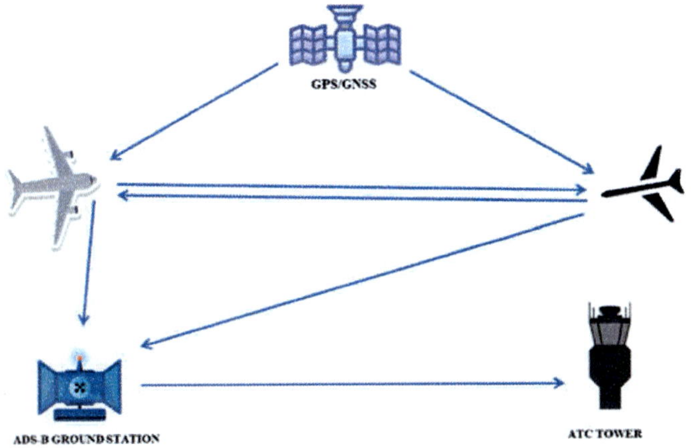

Fig. 1 Proposed ATM model which depends on ADS-B using GPS/GNSS

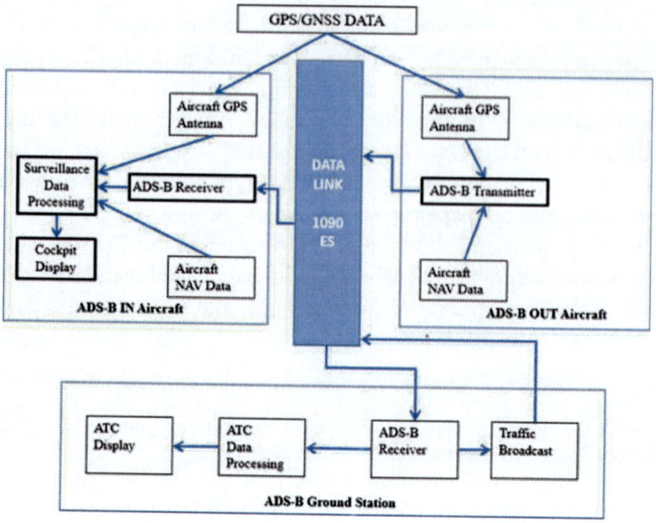

Fig. 2 ADS-B working principle with functional components and layouts

2.2 ADS-B Technology

An aircraft enabled with ADS-B and Mode S transponder, a dedicated ground station, a networked backbone between ATC Tower and ground stations are the main components of the ongoing ADS-B based ATM infrastructure. Apart from these components, ADS-B relies on GPS or GNSS. The broadcast of ADS-B along with ground stations is also received by properly equipped aircraft within range. ADS-B can perform two basic operations: ADS-B OUT and ADS-B IN optionally [12].

ADS-B OUT is a feature that permits the continuous generation of ADS-B broadcasts by aircraft and provides real-time location data to ATC. ADS-B IN helps aircraft to collect and view ADS-B OUT information from other aircraft. In addition, ADS-B IN would allow aircraft to receive ground station services (e.g., weather updates). Figure 2 describes the ADS-B system's working components and modules. Exchange of data between different components of ADS-B needs standard message formats and transmission protocols. ADS-B sends messages over 1090 MHz data-Link which is presently in use by Mode S transponder [13].

2.3 Blockchain Technology Relevant to ATM System

Hyperledger Fabric is a precise blockchain system that acts as a platform to support distributed ledger solutions organized by a specific design conveying high degrees of secrecy, versatility, and adaptability [14]. In this thesis, Hyperledger Fabric is applied because of its unique feature that is private and permissioned. In a public network, there is no privacy therefore an intruder can interpret the network system, which increases the probability of Sybil attack [15]. A permissioned blockchain network is operated by known entities as in a consortium where members or stakeholders in each business context [16]. All the members are whitelisted and obliged by strict agreement to behave properly [17]. New peers can be added by permission from the existing peers by approval or pre-defined rules or with permission from a regulator. Here a permissioned blockchain gives an approach to make sure about the connections among a gathering of elements that have a shared objective, however, which don't completely confide in one another.

3 Proposed System Model

Here in Fig. 3a Hyperledger blockchain model is proposed where nodes will be enrolled in the network through Membership Service Provider (MSP) and dedicated endorsers. It also focused on specifically about data privacy. It means some data can be exchanged and visible by all users, and other data are shielded and can only be viewed by specified entities using private channels. It has the potential to provide specific network features such as increased anonymity and security, effective communication, scalability, and a scalable framework that can be tailored for individual applications. Corporate aviation may want to keep their flight information secret with their competitive stake holders. Again, defence aviation, in general, keep their flight information hidden for security reasons.

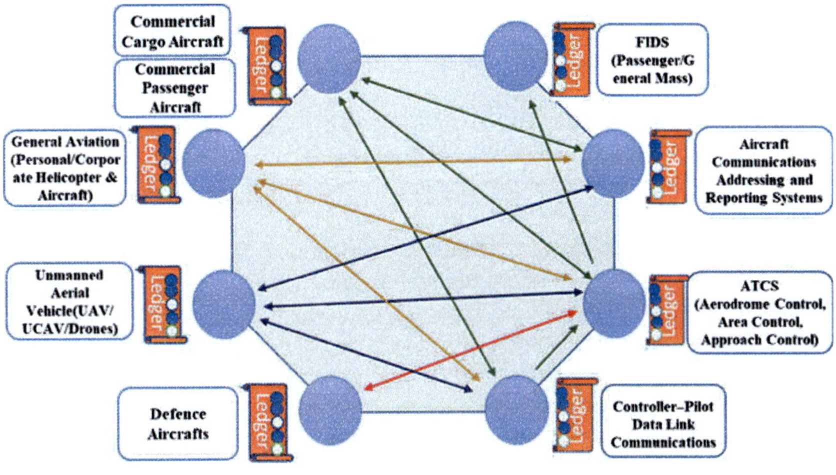

Fig. 3 Proposed blockchain framework that properly routes information between the aircraft, the ground-based ATM, and other support facilities

4 Proposed Model Architecture and Working Procedure

In this model, each type of entity—different aircraft, ATCS, FIDS, CPDLC all act as a node. Each entity enrols itself through MSP into the whole chain. Aircraft creates its own ledger which will be shared between itself and other connected peer nodes through a private channel. The ledger contains blocks of information data and each block are linked with cryptographic operations which cannot be altered. Encrypted ADS-B Out transmission data between any airplane and any approved entity of the network can be passed through private channels which is suitable to carry a private key following the particulars of the chain code. Using cyphertext that allows ADS-B consumers to preserve privacy and confidentiality while still offering a reliable and effective means of contact with designated agencies, such as ATCS. Here ATCS is the only body that knows all ADS-B information of all types of aircraft.

The ATCS creates a synopsis of all ADS-B transmissions from all aircraft and is the main approved part to get the data about every single airplane while the airplane just knows its own history. In this diagram, all components are divided into two parts. One is a ground-based operational entity which marked as zone-A and the other is different types of aircraft which is marked under zone B. Zone-A includes (i) ATCS, (ii) CPDLC, (iii) ACARS, and (iv) FIDS. On the other hand, Zone-B includes airbase components (i) Commercial and Cargo Aircraft, (ii) Personal/Corporate Helicopter & Aircraft) (iii) Unmanned Aerial Vehicle (UAV/UCAV/ Drones) and (iv) Defence Aircrafts. Here, different colored arrow indicates the relationship of consensus-service (order data exchange and provide communication channel) between different nodes of zone-A and zone-B. Respective authorized members establish one or more number of channels in which peers are separated into a subnet and private ledgers developed. Channels may be used to segment the state of the blockchain network, but there is no synchronization of agreement between channels and the overall sequence of transactions in each channel is different from the others. Such deployments where all orders find to be trustworthy can also enforce by-channel peer access control (Fig. 4).

The ledger of each channel is available only to the peer nodes of its participants. Organizations (entities) of the channel will authorize the membership of each peer at the channel. Any information that passed through that channel is routed to a specified channel and thus confidential transactions are conducted. For example, an aircraft's from zone-B physically identifiable information (height, latitude, longitude, stated airspeed, etc.) may be held safe by a private channel, whereas the common information of the aircraft (aircraft type, origin, destination, registered flight path, etc.) may be released on a channel that is available to all authorized members.

In membership management segment overall enrolment process is described. All individuals from this proposed model are required to enlist through a trusted MSP where fabric CA is an endorsement authority for this general enrolment measure. It is the answerable body for the enrolment of new peer hubs, issuance of enrolment digital certificates, and endorsement restoration or renouncement. Every CA is either a root CA or an intermediate CA running on peer hubs. MSP follows this

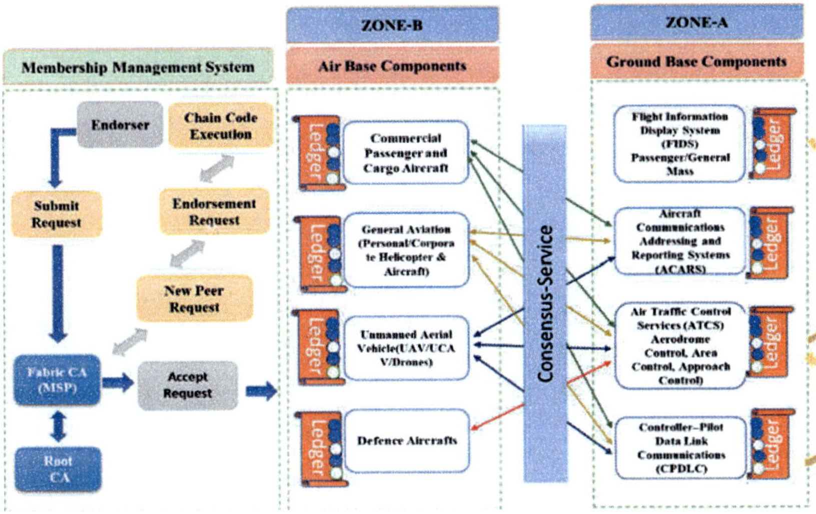

Fig. 4 Blockchain-based proposed architecture showing different components and layouts

E-Certs enrolment cycle and issues encryption and signature keys to peers. These are X.509 standard digital certificates. An aircraft which is not a part of the hyper-ledger blockchain network set a membership proposal with necessary identifiable information and parameters. Then it sends the proposal to all its endorsing peers to satisfy the endorsement policy of the given chain code. CA verifies the information transaction request based on the data and parameters received from the aircraft. Although these peer hubs are architectured to be physically long away every time it maintains its unique authenticity. The relations and various methods and techniques that peer hubs can use to get data from airplane administrators, air traffic networks, and different clients are heterogeneous and can change extraordinarily in local execution. The airplane at that point participates in the network and makes its own indistinguishable ledger. In this network, all peer hubs hold duplicates of records and chaincode, which contain an unchanging record of each exchange from each part. Here dedicated endorsers will be used for per chain code. The conventional ADS-B receivers could not detect any of these aircraft's ADS-B unless the aircraft operators decided to deliberately transmit the information in plaintext. FIDS is another entity that generates ledger by getting information from ATCS through a private channel, which in terms displays the overall departures and arrivals schedule of airline traffic and civil aircraft. This ledger is generally open to the public and therefore the data may be kept inside or outside the framework. ACARS is a special entity that creates a private channel between relevant aircraft and its intended ground stations. It is an optional service generally used by company headquarters to monitor its own aircraft. CPDLC is an additional means of service that also create a private channel between aircraft and air traffic controller via data-Link which in general passes strategic information by avoiding VHF communication to reduce congestion over the radio. Defence aircraft's operational information is highly confidential for the sake of national security. Each defence aircraft creates a ledger that is shareable only between the aircraft and ATCS. All ADS-B information is encrypted and transmitted to ATCS and ATCS is the only body that can decrypt it.

5 Performance of the Proposed Model

In Fig. 5 an aircraft A0 set a membership proposal with necessary identifiable information and parameters to the endorsing peer E0. In this case, Eo, E1, E2, and E3 will participate the endorsement process as per defined policy (chaincode) while others are not entitled to participate. Ao issues two types of certificate Tcert(N*Tcert), Ecert(1*Ecert), and E0 issue Ecert. Membership management authority accumulates all the information from A0 and endorsing peers (E0, E1, E2, E3) and hence based on the approval, it will then request to consensus service for inclusion in the ledger to achieve information read /write authority. The consensus service approves the inclusion of the transaction request into the ledger. Then A0 participates in the network and makes its own indistinguishable ledger. Here all peer hubs hold duplicates of records and chaincode. On the other hand, if subscription request is denied then the user will again attempt enrolment or discarded from the network. In this proposed

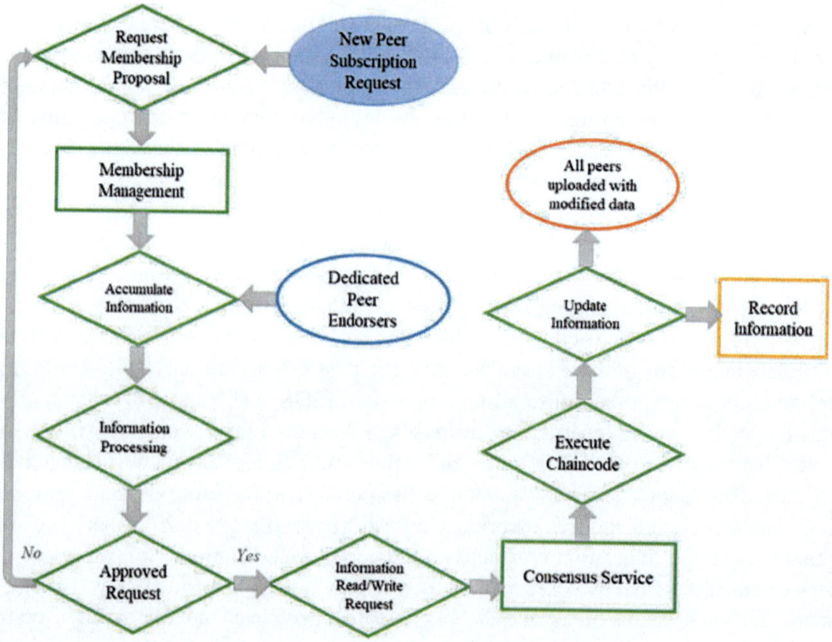

Fig. 5 Request processing flowchart of proposed model

model security services are ensured by the following four aspects: Firstly, it provides a structured PKI framework providing identification protection for aircraft operators who can obtain a synchronous cypher key supplied by PKI. The key will permit ADS-B information to be communicated to Air Traffic and close by airplane through cyphertext. The ADS-B information is confirmed by the other companion hubs who realize the key and zero-information verification separate the exchange information from on-chain records, shielding security from the hidden calculation. Secondly, the overall membership management procedure with dedicated endorsers per chain code and digital certificate management service ensures the network's legitimacy on the blockchain model. Third, the multi-channel architecture divides the details from one channel to another. Finally, data gathering on privacy also addresses the need to separate private data across different nodes within the same system. The two most distinctive methods in the above measures are the collection of data on the channel and on privacy. The channel is dedicated to protecting the privacy of the blockchain, allowing separate isolation of the data on the channel. The peer shares a record on the same channel, and the transaction peer needs to get channel acknowledgment before it can join the channel and execute with others. Different forms of security attacks are possible in ADS-B transmission vis a vis ATM and here it is illustrated how this proposed model can mitigate these vulnerabilities.

Eavesdropping: As ADS-B uses unsecured messages over an essentially transmitted channel, the eavesdrop possibility is very high. This passive attack is also called Aircraft Reconnaissance. When any signal is unencrypted, any intruder can capture the information. This is made even easier in the case of CPDLC because no current encryption is in use. For this reason, to ensure communication security and protect eavesdropping in this hyperledger fabric model, transmitted data traffic will be encrypted by creating a common symmetric key and using message authentication code. As per Fig. 4. if any external body is trying to eavesdropping in zone A or zone B deliberately his attempt will be failed as he does not have any secret key to encrypt the message and also will have no access in the entire network because of layered authentication system.

Injection: The current ADS-B system is very vulnerable for injection because of its absence of substance verification to shield against message injection from unauthorized entities. In this model, if an intruder tries to inject false messages in ATCS which belongs to zone-A, first he needs to access the network. His attempt request will be assessed through the network's validation process. His identity will be verified by the CA authority and dedicated peer endorsers and therefore his attempt will be rejected. The same scenario will be happening if any external aircraft try to enter in zone-B with inappropriate information. So, overall access control is managed through MSP and generates X.509 standard cryptographically validated digital signature which will protect any unauthorized access. Information is also controlled through dedicated Chaincode, Consensus-Service (Order management), Separate Peer to Peer Channel (Private subnet of communication), a structured PKI framework with identification check through a cipher key so it is hard to inject any new message by any intruder.

Alteration (Modification/Deletion): An alteration attack is defined when an attacker alters the original data. It shows itself as information being altered, diverted to another recipient, or deferred. The proposed model can be secured from modification, deletion, and alteration due to restriction of access in the system. Here access control through MSP, Digital signature a potential intruder will not able to access the network. Data privacy is controlled through Chaincode, Consensus-Service (Order Management), Separate Peer to Peer Channel (Private subnet of communication), a structured PKI framework with identification check through a cipher key. Likewise, general aviation information in zone-B cannot be altered because all information is kept in ledgers for individual aircraft and accessible only through a dedicated channel by defined peer nodes. A symmetric chipper key is also shared within these restricted channels.

Masquerading: Masquerading implies the interloper mimics an approved client, regardless of whether an airplane or an ATC and procures benefits that are not allowed. The masquerading attacker will therefore have the means to engage in a full conversation with his victim unless detected. Due to the strict access management process in the proposed model through MSP and Digital signature, it is difficult to impersonate someone else.

Thus, from the above discussion, it can be concluded that the proposed model ensures data privacy, confidentiality, data source/origin authenticity, and data integrity which will protect the ADS-B system from data eavesdropping, injection, alteration, and masquerading. Table 1. articulate a structural summary of the security mechanism of the proposed model against different cyber threats.

6 Conclusion

In this paper, a specific blockchain-based concept is introduced for improving the performance and safety of ATM in aviation. To address the necessities of continuous adaptation to a rapidly changing condition of ATM, the threat environment, the organizational and technological system architectures must be restructured. However, a detailed analysis of the current work is presented, particularly to ADS-B along with concepts from relevant sectors of ATM, such as CPDLC, ACARS, FIDS, and so on. In future studies, scopes are open to refining this research with a detailed view of the nodes, the data sharing, and the frameworks used to access transfer data. The ultimate goal of the paper is to introduce the recommendations and to test it for real-world information, which will be possible once seamless ATM is implemented by CAAB or ICAO in Bangladesh vis a vis Asia Pacific region.

Table 1 Security mechanism of proposed model against different attacks

Attacks	Method	Applied security service	Security mechanism of proposed model
Aircraft reconnaissance	Eavesdropping	Data privacy, confidentiality	Message encryption
Ground station target ghost injection/flooding	Message injection	Data Privacy, Confidentiality, Peer Entity Authentication, Data Origin Authentication and Data Integrity	Access control through MSP, Digital signature, Chaincode, Consensus-service, Separate peer to peer channel (private), Message encryption, Endorsement policies
Virtual aircraft hijacking	Message modification	Data Privacy, Confidentiality, Peer Entity Authentication, Data Origin Authentication and Data Integrity	Access control through MSP, Digital signature, Chaincode, Consensus-service, Separate peer to peer channel (private), Message encryption, Endorsement policies
Aircraft disappearance	Message deletion	Data Privacy, Confidentiality, Peer Entity Authentication, Data Origin Authentication and Data Integrity	Access control through MSP, Digital signature, Chaincode, Consensus-service, Separate peer to peer channel (private), Message encryption, Endorsement policies
Aircraft spoofing	Eavesdropping	Data privacy, Confidentiality	Message encryption

References

1. Martin S., Matthias S., Vincent L.: Realities and Challenges of Nextgen Air Traffic Management: The case of ADS-B, pp. 111–118, IEEE (2014). https://doi.org/10.1109/MCOM.2014.6815901
2. Chantal L.: The Single European Sky: A Window of Opportunity for EU–NATO relations, pp. 415–435, (2017). https://doi.org/10.1080/09662839.2017.1352580
3. Xiaodong, L., Kazuyuki, M., Tadashi, K., Yasuto, S.: Collaborative flight and flow Information exchange to achieve seamless air traffic management operation. IEEE (2019). https://doi.org/10.1109/ISADS45777.2019.9155791https://doi.org/10.1109/ISADS45777.2019.9155791
4. Advancing air traffic management in Asia Pacific, BRIDGING SKY, Issue 21, (2017)
5. ADS-B implementation and operations guidance document, international civil aviation organization Asia and Pacific office, 7 edn., pp. 6 (2014)
6. Mamun, S.: Bangladesh's Air Traffic System in Shambles, Dhaka Tribune (Dhaka) (2018)
7. Cyber Security For Civil Aviation, Twelfth Air Navigation Conference (2012)
8. Automatic Dependent Surveillance-Broadcast (ADS-B) Out Performance Requirements to Support Air Traffic Control (ATC) Service, Federal Aviation Administration, Final Rule, 14 CFR Part 91. Federal Reg. **75**(103) (2010)

9. Robin, L.: 2019 Global Aerospace and Defense Industry Outlook, Deloitte (2019)
10. CANSO Vision 2020. https://www.canso.org/vision-2020
11. SESAR Deployment Manager, ADS-B and other means of surveillance implementation status (2018)
12. Busyairah, A., Washington, O., Arnab, M., Wolfgang, S., Thiam, C.: ADS-B system failure modes and models. IEEE Commun. Surv. Tutor. **17**(2) (2014). https://doi.org/10.1017/S037346331400037X
13. Martin, S., Vincent, L., Ivan, M.: Security of ADS-B: state of the art and beyond. IEEE Commun. Surv. Tutor. **17**(2) (2013). https://doi.org/10.1109/COMST.2014.2365951
14. Christian, C.: Architecture of the Hyperledger Blockchain Fabric (2016)
15. Douceur J.R.: The Sybil attack. In: Revised Papers from the First International Workshop on Peer-to-Peer Systems, ser. IPTPS'01, pp. 251–260. Springer, Berlin (2002)
16. Cachin, C., Vukoli´c, M.: Blockchain Consensus protocols in the wild. In: Richa, A.W. (ed.) International Symposium on Distributed Computing (DISC), pp. 1:1–1:16 (2017)
17. Tim, S.: Consensus-as-a-Service: A Brief Report on the Emergence of Permissioned, Distributed Ledger Systems (2015)

Building Machine Learning Based Firewall on Spanning Tree Protocol over Software Defined Networking

Nazrul Islam, S. M. Shamim, Md. Fazla Rabbi, Md. Saikat Islam Khan, and Mohammad Abu Yousuf

Abstract Software-Defined Networking (SDN) is the most expected inventions for the network administrators, cloud service providers and businessmen. For its global acceptance, has to undergo security issues and get cover using a proper solution. This study implements the firewall on the Spanning Tree Protocol (STP) over SDN using Mininet emulator. The broadcast streams are controlled by the Firewall on STP based SDN without dedicated hardware support. The result focuses the Firewall functionalities over STP based SDN as well as frees the available bandwidth for traffic. Additionally, Machine Learning (ML) techniques have been applied to build up the infrastructure to get the best outcome from a model with top-notch performance. Among different machine learning models using the same NSL-KDD dataset, Random Forest Classifier algorithm stands at the top with highest accuracy in differentiating between regular communication and intrusion on the network.

Keywords Software defined networking (SDN) · Spanning tree protocol · Firewall · OpenFlow · Machine learning

---

N. Islam (✉) · S. M. Shamim · Md. Fazla Rabbi
Department of Information and Communication Technology, Mawlana Bhashani Science and Technology University, Santosh,Tangail 1902, Bangladesh
e-mail: nazrul.islam@ieee.org

S. M. Shamim
e-mail: smsict09@gmail.com

Md. Fazla Rabbi
e-mail: fazlarabbi366iou@gmail.com

Md. Saikat Islam Khan
Department of Computer Science and Engineering, Mawlana Bhashani Science and Technology University, Santosh, Tangail 1902, Bangladesh
e-mail: bappy.10.cse.mbstu@gmail.com

M. Abu Yousuf
Institute of Information Technology, Jahangirnagar University, Savar, Dhaka 1342, Bangladesh
e-mail: yousuf@juniv.edu

M. S. Kaiser et al. (eds.), *Proceedings of International Conference on Trends in Computational and Cognitive Engineering*, Advances in Intelligent Systems and Computing 1309, https://doi.org/10.1007/978-981-33-4673-4_45

1 Introduction

Software Defined Networking (SDN) replaces current network paradigm through the abstraction of central programming interface. SDN is flexible enough to let the network engineers cope up with the changing requirements. Sometimes, for adding new hardware or new program administrator requires numerous individual device configurations and even for adding QoS or security policies. As the switches and routers are from multiple vendors, each vendor has different Operating System (OS) with different configuration methods for their own switches and routers. Despite their comprehensive acceptance, network administrator faces difficulty controlling the traditional IP networks [1]. Use of machine learning in firewall helps to analyze the traffic flow, possible attacks and possibly helps to improve the QoS. Different approaches of Machine learning points to how effective and accurate, these can be when filtering the traffic of the network [2].

Moreover, Internet security and network security are on the rise as quick globalization taking place in present years. Malicious programs like worms, viruses and other digital pests from the Internet may annihilate valued information. Consequently, perfect procedure are required to keep "good" bits in and "bad" bits out. IPSec may be one of the methods to protect data among secure sites. Notwithstanding, IPSec isn't coming up with anything useful to get rid of these digital pests and intruders those get on the organization LAN. Although there's no single product or service in the market that will guarantee complete security, but a firewall may be up to stand as it's the piling of any network security strategy.

The Spanning-Tree Protocol (STP) ensures layer-2 loops do not exist when switches or bridges are connected with each other via different routes in the network. The STP does the work of exchanging Bridge Protocol Data Unit (BPDU) messages among switches and ceasing bridge interfaces prevent loops and ensures a single active path between two devices [3]. The Firewall is a security system which controls traffic flow based on a list of principles. The STP is set to block Firewall not for traffic taking in or out on the network. OpenFlow is an enabler of SDN, on the other hand remotely manages switches of various vendors so a well-organized platform between switches and SDN. OpenFlow protocol is used to communicate between controller and network devices. It is an open interface for regulating the routers, access points and forwarding tables in network switch remotely. By using this low-level primal, researchers can fabricate networks having new high-level specialty [4].

This paper proposed the SDN design that includes Spanning Tree Protocol (STP) without Firewall configuration and Spanning Tree Protocol (STP) with Firewall configuration using Mininet emulator and POX controller. The STP protocol without Firewall allows flooding and broadcast streams on the network. Besides the STP with Firewall eschews flooding and impede broadcast streams on the network and thus the available bandwidth are being free from traffic loads.

The structure of the paper follows as: Sect. 2, literature review. In Sect. 3, proposed system model. Section 4, describes the implementation of the model and Sect. 5, evaluates the results after the simulation. Section 6, conclude the paper with future outlook.

2 Literature Review

A literature review has been performed that the Firewall rules have been implemented on STP based networking topology over the Software-Defined Network (SDN). The Mininet [5] is an open-source network emulation tool which runs a number of virtual hosts, controllers, switches, and links. The minuet is developed to assist research and education in the sector of SDN. Some most popular examples of an SDN Controller on the market and research are OpenDayLight [6], Floodlight [7], RYU [8]. RYU controller can be used in the testbed wireless SDN topology for mobility solution [9, 10]. Studies show that the Python based SDN controller is POX [11]. The POX controller works on the control layer of SDN. The OpenFlow is the chosen API for POX controller to take the control over the network.

Firewall policies have composed a set of defined rules and are strictly maintained to perform a specific action on a packet. Every rule has a structure (condition, action) meaning that, if a certain condition is met, then either 'allow' or 'deny' action is taken on the data packet. In firewall policy, sometimes several rules conflict with each other, and because they may yield different decisions. For solving these issues firewall follows a mechanism naming first match mechanism. This mechanism helps to avoid the collision of rules by matching only the first rule in the order they were arranged and make a decision according to that [12].

Firewall application has implemented with a bunch of MAC address pairs. In OpenFlow table Firewall installs static flow entries to unable all the transmissions between each MAC pair at the time of establishing connection between the controller and the switch. Firewall rules have been added to certain source and destination MAC addresses. These sources and destination MAC address pairs are provided in the form of a CSV file which will be read by POX controller. The controller drop offending packets with the given MAC address pair. The simulation has been set up in a Core-i3 processor based laptop.

3 Proposed System Model

Proposed system has been implemented by using Mininet emulator and POX controller. The minuet is a network emulator that creates a network with several hosts, controllers and switches. It supports custom designed topology along with simple and comprehensible Python API for the network generation.

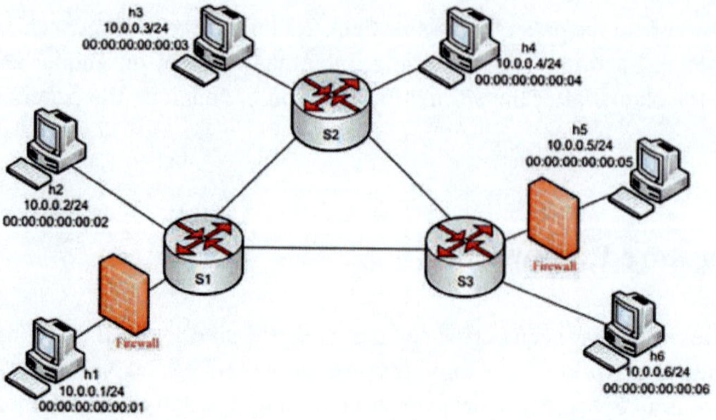

Fig. 1 Network design

The SDN controller POX is based on Python programming language. POX is bundled with Mininet emulators for SDN network. POX controller integrates OpenFlow, controllers and switches. OpenFlow which uses open protocol is the most popular in SDN. For SDN, OpenFlow counts as the first standard that executes protocols based upon specific application. OpenFlow carries the communication between open flow controller and open flow switch.

The proposed network simulation has been done in the following custom network topology in Fig. 1. The network consist of three switches, controller and six hosts. Three switches are linked in the loop and host h1 and host h2 is connected with switch s1, h3 and h4 are connected with s2, h5 and h6 are connected with switch's s3. Two firewalls have been installed in designing network where one of them is installed between host h1 and switch s1. Another firewall has been installed between switch's s3 and h5.

All of the hosts are assigned to a unique IP address and MAC address. There are six hosts. The IP addresses of the hosts start from '10.0.0.1/24' and ends at '10.0.0.1/24'. The MAC addresses are also assigned accordingly starting from '00:00:00:00:00:01' and ending at '00:00:00:00:00:06'.

Consequently, an incoming packet and outgoing packet are protected by a firewall system which comes from various sources within the SDN environment. This can also track and manage the data flow that comes from different sources within the network. The early detection of the attacks within the SDN environment required more practical approach. Machine learning approaches were widely used nowadays to identify anomalies within the SDN environment. Machine learning methods integrated with the firewall framework can be a reliable and robust model that can detect the attacks inside the SDN network early. With this motivation, machine learning methods to detect anomalies inside SDN technology is also proposed in this paper.

SDN is the key to improve network security that can inform us the risks of violating the security confidentiality of the network [13]. Using machine learning approaches in SDN can take a long stride forward. Moving to Fig. 2 shows the proposed architecture

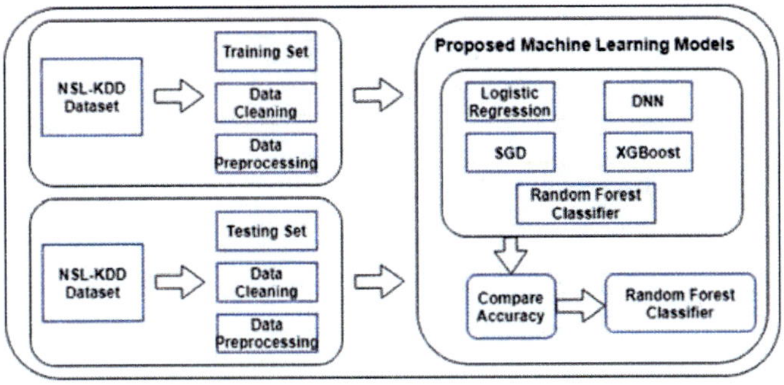

Fig. 2 Proposed machine learning based design

for detecting network intrusion within the SDN environment. The testing dataset is already broken down into a training set and test set. Because the data is not properly ordered in the dataset, the process of data cleaning was carried out. Data preprocessing techniques such as normalization and mean encoding was performed after the data were cleaned. After then, the training set was fed into the machine learning (ML) model. We apply different ML approaches such as Logistic Regression, Deep Neural Network, Stochastic Gradient Descent Classifier, XGBoost, and Random Forest Classifier on the dataset. Then we evaluate the performance of each model and compare it with each other. At the end, the best model was used to detect network intrusion in the SDN context.

4 Implementation

4.1 Spanning Tree Protocol Without Firewall Configuration

At first, Spanning Tree Protocol (STP) has been implemented without firewall configuration using Mininet. The STP protocol assures loop-free network topology for any bridged Ethernet Local Area Network (LAN). The main objective behind the STP is to impede the occurrence of the broadcast streams in a network which contains loop structure. If there is a network failure, it enables automatic path switching and ensures secure network redundancy. In spanning tree protocol each switch connects in a loop. In addition, STP generates a spanning tree inside a network which is combined with a layer-2 bridges. It assures only one enable link within more than one network nodes and drop others links that are not the portion of the spanning tree.

Moreover, STP switches have several port states blocking, forwarding, learning, listening and disabling. Blocking prevents the loop path on the network. It does not allow sending or receiving data or traffic. If some of other links fail, blocking states may go to forwarding states. In listening state switch processes Bridge Protocol

Data Units (BPDUs) and waits for probabilistic packet, which may cause back to the blocking state. In learning states, source addresses experienced from accepting frames and attach them to switching database. It does not forward frames. In forwarding states switch sends and receive data. In disable state network administrator disables a port manually and it's not strictly part of STP.

The entire connectivity of the network can be obtained by a single pingall command. The Packet Internet Grouper (PING) is used to check all possible combinations of network hosts. There are no packets will be exuded because firewall is not configured in the network. The pingall command shows that the entire hosts are completely attached and no host is unfastened from the network. Ping is used to reach one host to another host. The ping is used to check the connection from host h1 (10.0.0.1) to host h2 (10.0.0.2). Ping sends an Internet Control Message Protocol (ICMP) echo request to a destination host and waits for the destination to get an ICMP response.

The first ping always has the longest delay compared to others delay because at the beginning stream tables are vacant. Corresponding packet must be resolved by the programmable controller firstly from one host to another host before addressing command to install flow entry in order to control the packet. For every ping 5 echo request packets are sent from source host to destination host. After installing flow entry the delay is reduced after the posterior pings because switch dynamically handles the packet immediately without review the packet twice.

4.2 *Spanning Tree Protocol with Firewall Configuration*

A firewall is a network security system that manages the traffic flow of a network normally between LAN and WAN where a LAN is more secure than WAN. L2 and L3 headers (i.e. IP a MAC address) are inspected when analyzing the data packets. For filtering network traffic, the firewall does Deep Packet Inspection (DPI) for upper layer parameters like application type and services, etc. It acts as a barrier between a secure network and an unreliable network like the internet which can hardly be trusted.

A second evaluation implements layer 2 firewall on STP which runs side by side with the MAC learning module on the POX OpenFlow controller. At the time of connection establishment between controller and switch, flow rule entries are installed by the application on the OpenFlow table so that links between MAC pairs becomes inactive. The SDN controller manages and fixes how the network operates, in other words, determining the packets can be discarded or not. Firewall rules installed on host h1 (MAC = '00:00:00:00:00:01') to block packets from the other host. Firewall rules also installed on the host h5 to block packet coming from another host with the corresponding MAC address. After this the pingall command shows that the entire packets are blocked in host h2 and host h3 that is coming from host h1.

Since firewall rules have been installed on host h1 for blocking from all other hosts, the entire packet will be dropped. When host h3 ping host h1, host h3 tries to send 5 echo request packet to host h1, but all the packet is being dropped and shows a message Destination host is Unreachable. When host h4 ping host h5, host h4 sends 5 echo requests to host h5 due to firewall installation between host h5 and switch s3, the entire packet is being dropped and shows the message "Destination host is Unreachable".

Different researcher deployed firewall rules to increase security mechanism in the software defined network. To illustrate, firewall rules have been deployed in centralized controller in SDN to provide security by using Mininet emulator and POX controller [14]. In addition, a stateful firewall application has been implemented in SDN in [15] and observed that stateful firewall is more secure than the stateless. Moreover, a robust firewall, FLOWGUARD [16] has been deployed in SDN to find out firewall policy infringement while network states are refreshed automatically. This paper focuses implementing firewall functionalities on spanning tree protocol in SDN. At first spanning tree protocol has been implemented by using pox controller. After then firewall rules deployed over the spanning tree protocol in SDN.

5 Results and Analysis

Recently, every single sector of science has been using the power of machine learning to get the best out it. For a long time, machine learning is being rich with the computational resources of networking and distributed computing systems. In the future internet, SDN has become one of the new technologies which can improve our network security. Nevertheless, SDN also includes possible risks that might breach the network security confidentiality. Machine learning approaches are applied to detection of flow-based anomalies in the SDN environment.

5.1 Analyze Using Datasets

The NSL-KDD dataset has been used in this study to construct a framework that predicts whether there is an attempted attack or intrusion, or is a regular contact. The NSL-KDD data set that was obtained from KDD 99 after the cleaning operation was performed in it [17]. The dataset contains a total of 125,973 records in the training set and 22,554 records in the testing set. Total 43 attributes mentioned in the dataset among them 41 attributes are traffic data and the other two attributes are attack category and score which explained the extent of the traffic data information.

The dataset includes a total of 39 separate attacks which are further divided into four designated categories: Denial of Service (DOS), Probing Attack, Remote to Local Attack (R2L) Root Attack (R2L). The detailed categories of attacks in both the training set and the test set are out in Table 1. Although the NSL-KDD dataset is facing some problems that may not be perfect for the real networks, it can be more effective than other intrusion detection methods because of the lack of a network-based dataset. Moreover, the number of training and testing records in the NSL-KDD dataset is enormous, making the model work on a real-time network effectively.

Table 1 Category of attacks in the NSL-KDD dataset

Attack category	Training	Testing
DoS	Neptune, smurf, teardrop, back, land, pod	Neptune, pod, smurf, teardrop, back, land, mailbomb, apache2, worm, processtable, udpstorm
Probe	Portsweep, satan, ipsweep, nmap	Saint, ipsweep, satan, mscan, nmap, portsweep
R2L	Multihop, phf, spy, fpt-write, guess-passwd, warezmaste, imap, warezclient	Imap, multihop, fpt-write, guess-passwd, warezmaster, snmpguess, snmpgetattack, httptunnel, xlock, xsnoop, sendmail, phf, spy, named
U2R	Loadmodule, perl, buffer-overflow, rootkit	Buffer-overflow, sqlattack, xterm, loadmodule, perl, rootkit, ps

Table 2 Performance of the machine learning approach

Algorithm	Accuracy
Logistic regression	0.977
SGD	0.929
DNN	0.974
XGBoost	0.981
Random forest classifier	0.983

5.2 *Performance Evaluation*

The performance of the approaches to machine learning was evaluated in terms of accuracy which shows the percentage of true detection over total traffic records.

$$Accuracy = (TP + TN)/(TP + TN + FP + FN) \tag{1}$$

Where,

TP = True Positive, TN = True Negative, FP = False Positive, FN = False Negative

Table 2, note that the NSL-KDD dataset successfully employed machine learning methods. Between them, Random Forest Classifier and XGBoost methods can detect anomalies with more than 98 percent accuracy in the SDN environment. Logistic Regression, Stochastic Gradient Descent classifier and Deep Neural Network methods can also be considered as good predictive models. The network design of the work expresses significant accuracy because of the perfect data pre-processing and data cleaning operation. The comparison of the models is also demonstrated in Fig. 3.

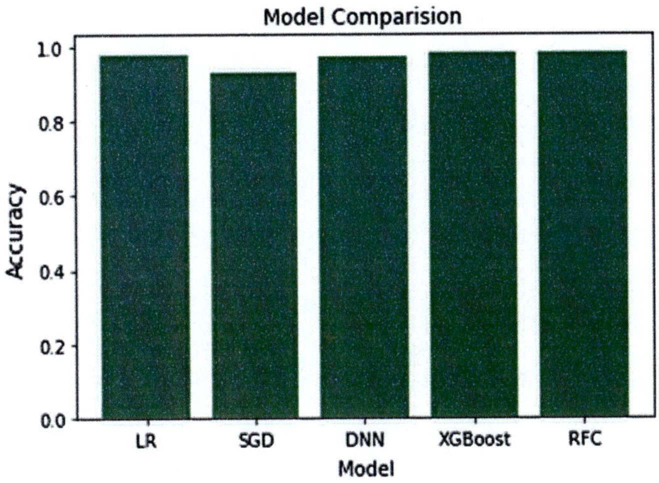

Fig. 3 Comparison among proposed models

Table 3 Comparison with other research methods

Name	Method	Accuracy
Tang et al. [18]	DNN	75.8
Abubakar et al. [19]	Pattern recognition	97.3
Meti et al. [20]	SVC	80.0
Samriddhi et al. [21]	Naïve Bayes	84.3
Proposed method	RFC	98.3

Table 2, also identify that Random Forest Classifier (RFC) displays the high accuracy. The RFC system efficiency is contrasted with the other research method found in Table 3. Abubakar et al. used the Pattern Recognition technique which shows more than 97 percent accuracy. Meti et al. also demonstrated more than 80 percent accuracy using support vector classification and Samriddhi et al. using Naïve Bayes methods. Eventually, more than 75 percent accuracy was demonstrated by Tang et al. using DNN technique. Nevertheless, our proposed RFC approach has shown more than 98 per cent accuracy in the same dataset in which they worked.

The approach also contrasts with other literature that has used the same NSL-KDD dataset with a random forest classifier. Table 4 displays that the highest accuracy is obtained by Tan et al., which is 92.3%, where Gupta and Tesfahun et al. were able to achieve almost same the efficiency for the random forest classifier. Table 4 also demonstrations that the strategies have outperformed all other literature that used the random forest as a classifier on the NSL-KDD dataset.

Table 4 Comparison with other research methods with same classifier

Name	Accuracy
Tan et al. [22]	92.3
Gupta et al. [23]	88.6
Tesfahun et al. [24]	88.5
Proposed method	98.3

6 Conclusion

This paper represents sets of observation based on the simulation and afterward data are analyzed carrying out difference Machine Learning techniques. Firstly, the STP function of the POX controller without a firewall is implemented. This shows that the network functions properly connecting all hosts without detaching a single one without any traffic loss. Secondly, the STP functions with a firewall implementation of STP based networks in SDN. The STP function eschews flooding and impede broadcast storms on the network. Traffic loss is found in applying the Firewall rules between hosts. Adopting the network parameters Firewall rules make accessible in the form of REST APIs by the POX controller.

Additionally, applying ML approaches, the structure of the network becomes safer as it shows the performance and accuracy in number. It will be interesting to evaluate and apply ML techniques in Intrusion Detection that improves the Quality of Services (QoS) and helps to figure out how the network is performing under the hood.

In future, it would be interesting to evaluate the model to a real-time SDN scenario. The efficiency of the model will be calculated using throughput and latency and cross-validation method to make the model more robust.

References

1. Xu, J., Wang, L., Xu, Z.: An enhanced saturation attack and its mitigation mechanism in software-defined networking. Comput. Netw. **169**, 107092 (2020). https://doi.org/10.1016/j.comnet.2019.107092
2. Boutaba, R., Salahuddin, M.A., Limam, N., Ayoubi, S., Shahriar, N., Estrada-Solano, F., Caicedo, O.M.: A comprehensive survey on machine learning for networking: evolution, applications and research opportunities. J. Internet Serv. Appl. **9**, 16 (2018). https://doi.org/10.1186/s13174-018-0087-2
3. Spanning Tree Protocol, http://www.cisco.com/c/en/us-/tech/lanswitching/spanning-treeprotocol/index.html. Last accessed 7 Feb. 2020
4. McKeown, N., Anderson, T., Balakrishnan, H., Parulkar, G., Peterson, L., Rexford, J., Turner, J.: OpenFlow: enabling innovation in campus networks. ACM SIGCOMM Comput. Commun. Rev. **38**(2), 69–74 (2008). https://doi.org/10.1145/1355734.1355746

5. Mininet-An Instant Virtual Network on your Laptop (or other PC)- Mininet. http://mininet.org/. Last accessed 10 Sept. 2020
6. The OpenDayLight Platforml OpenDayLight, http://www.opendaylight.org/. Last accessed 18 June 2020
7. Floodlight OpenFlow Controller-Floodlight Project, http://www.projectfloodlight.org/floodlight/. Last accessed 8 May 2020
8. Islam, M.T., Islam, N., Al Refat, M.: Node to node performance evaluation through RYU SDN controller. Wirel. Person. Commun. **1–16**, (2020). https://doi.org/10.1007/s11277-020-07060-4
9. Rahman M.H., Islam N., Swapna A.I., Habib M.A.: Analysis of Software Defined Wireless Network with IP Mobility in Multiple Controllers Domain, Cyber Security and Computer Science. (ICONCS). Lecture Notes of the Institute for Computer Sciences, Social Informatics and Telecommunications Engineering, vol 325 (2020), Springer. https://doi.org/10.1007/978-3-030-52856-0_42
10. Jany, M.H.R., Islam, N., Khondoker, R. and Habib, M.A.: December. Performance Analysis of OpenFlow based Software Defined Wired and Wireless Network. In: 20th International Conference of Computer and Information Technology (ICCIT) (pp. 1-6). IEEE, (2017). https://doi.org/10.1109/ICCITECHN.2017.8281814
11. Kaur, S., Singh, J., Ghumman, N.S.: Network programmability using POX Controller. In: International Conference on Communication, Computing and Systems (2014). https://doi.org/10.13140/RG.2.1.1950.6961
12. Hu, H., Han, W., Kyung, S., Wang, J., Ahn, G.-J., Zhao, Z., Li, H.: Towards a reliable firewall for software-defined networks. Comput. Secur. **87** (2019). https://doi.org/10.1016/j.cose.2019.101597
13. Swapna, A.I., Islam, N.: Security analysis of IEEE 802.21 standard in software defined wireless networking. In: 20th International Conference of Computer and Information Technology (ICCIT) 2017 Dec 22 (pp. 1–5). IEEE. https://doi.org/10.1109/ICCITECHN.2017.8281843
14. Mantur, B., Desai, A., & Nagegowda, K. S.: Centralized control signature-based firewall and statistical-based network intrusion detection system (NIDS) in software defined networks (SDN). In: Emerging Research in Computing, Information, Communication and Applications, pp. 497–506. Springer, (2015). https://doi.org/10.1007/978-81-322-2550-8_48
15. Kaur, K., Singh, J.: Building Stateful Firewall Over SDN. Information Systems Design and Intelligent Applications, pp. 159–168. Springer (2016). https://doi.org/10.1007/978-81-322-2752-6_15
16. Hu, H., Han, W., Ahn, G.J., Zhao, Z.: Flowguard: Building Robust Firewalls for Software-Defined Networks. Proceedings of the third workshop on Hot topics in SDN, pp. 97–102. ACM (2014). https://doi.org/10.1145/2620728.2620749
17. KDD Cup 1999, http://kdd.ics.uci.edu/databases/kddcup99/. Last accessed 04 June 2020
18. Tang, T.A., Mhamdi, L., McLernon, D., Zaidi, S.A., Ghogho, M.: Deep learning approach for network intrusion detection in SDN. In: 2016 International Conference on Wireless Networks and Mobile Communications (WINCOM), 26 Oct. 2016 (pp. 258–263). IEEE. https://doi.org/10.1109/WINCOM.2016.7777224
19. Abubakar A, Pranggono B.: Machine Learning based Intrusion Detection System for Software Defined Networks. In: 2017 seventh international conference on emerging security technologies (EST) 2017 Sep 6 (pp. 138-143). IEEE. https://doi.org/10.1109/EST.2017.8090413
20. Meti, N., Narayan, D.G., Baligar, V.P.: Detection of distributed denial of service attacks using machine learning algorithms in software defined networks. In: 2017 international Conference on Advances in Computing, Communications and Informatics (ICACCI) 13 Sep. 2017 (pp. 1366–1371). IEEE. https://doi.org/10.1109/ICACCI.2017.8126031
21. Samriddhi, V., Nithyanandam, P.: Detailed analysis of intrusion detection using machine learning algorithms. Int. J. Recent Technol. Eng. (IJRTE) **9**(1) (2020). ISSN: 2277-3878
22. Tan, X., Su, S., Huang, Z., Guo, X., Zuo, Z., Sun, X., Li, L.: Wireless sensor networks intrusion detection based on SMOTE and the random forest algorithm. Sensors **19**(1), 203 (2019). https://doi.org/10.3390/s19010203

23. Gupta GP, Kulariya M.: A Framework for Fast and Efficient Cyber Security Network Intrusion Detection using Apache Spark. Procedia Computer Science, vol.93, 2016, pp.824–31. Elsevier, (2016). https://doi.org/10.1016/j.procs.2016.07.238
24. Tesfahun, A., Bhaskari, D.L.: Intrusion detection using random forests classifier with SMOTE and feature reduction. In: 2013 International Conference on Cloud & Ubiquitous Computing & Emerging Technologies, Nov 15, 2013, (pp. 127–132). IEEE. https://doi.org/10.1109/CUBE.2013.31

A Routing Protocol for Cancer Cell Detection Using Wireless Nano-sensors Network (WNSN)

Mohammad Helal Uddin, Mohammad Nahid Hossain, and Asif Ur Rahman

Abstract Nanotechnology, being one of the buzzwords of the era, enthralled us to usher light in the fields which would have the way for implementation of this technology in medical sector, in future. The objective of this paper is to propose a routing protocol that works in a single-hop transmission process in order to establish communication between nano-devices deployed inside the human body. This paper solely focuses on the routing protocol based on a three-tier handshaking process between the nano-sensors which are dynamically movable inside body and nano-routers which are fixed at certain places in the body also. This routing protocol would operate in terahertz band so that it can take advantage of the potentials of terahertz band channel mode. The system would ensure that the data packet is received by the authorized entity through which it minimizes data loss in an efficient way where energy usage is also minimum by the system. The system uses TS-OOK communication method in order to lessen the code weight. Finally, the process ensures that the recipient receives the data from the very beginning of the process as our nano-sensors are dynamically movable inside the body.

Keywords Cancer cell detection · Nano-sensors · Routing protocol · Algorithm · Decentralized network · Terahertz

M. H. Uddin (✉) · M. N. Hossain
Department of Electronic Engineering, Kwangwoon University, Seoul, South Korea
e-mail: helaluddinsimon@gmail.com

M. N. Hossain
e-mail: nahidhossainmohammad@gmail.com

A. Ur Rahman
Department of Computer Science, American International University-Bangladesh, Dhaka, Bangladesh
e-mail: rahman.m.asifur@gmail.com

M. S. Kaiser et al. (eds.), *Proceedings of International Conference on Trends in Computational and Cognitive Engineering*, Advances in Intelligent Systems and Computing 1309, https://doi.org/10.1007/978-981-33-4673-4_46

1 Introduction

Nanotechnology expands the known fields of science to a many-fold dimension. It offers an abundance of possibilities along with a handful of challenges. One of the paradoxical challenges is its size which at the same time is a blessing. Nanotechnology encompasses the knowledge of the physics, chemistry, and biology at nanometer scale which enables us to create impact at the molecular level [1]. There are many applications of nanotechnology in different types of field. Like using nanotechnology in biomedical sector—the natural domain of the major components of living cells molecules, proteins, and also DNA is in the nanoscale. As a result, to work with this nanoscale material, a very large number of applications of nanotechnology are used in the biomedical field to satisfy its criteria [2]. Environmental sector, one of the fields of nanotechnology—in order to attract the natural predators of the insects that are attacking some of the important element of environment like trees, herbs, or bushes, or to regulate their blooming among different plantations, among others several chemical composites to the air is released [3]. To stop these insect's deployment and detect the chemical compounds that release and exchange between plants, chemical nano-sensors [4] could be used. In agriculture fields, nanotechnology already established classical sensor devices [5] using the nanoscience. The touch of the nanoscience throws a great effect on industrial field too. Flexible and stretchable electronic devices [6] to new functionalized nanomaterial for self-cleaning anti-microbial textiles can be the example of the applications of nanotechnology in the development of new industrial and consumer goods. One of the great examples of uses of nanotechnology is military and defense field. Two examples of the military applications enabled by nanotechnology are existed; one is advanced nuclear, biological, and chemical (NBC) defenses, and another one is sophisticated damage detection systems for civil structures, soldiers' armor, and military vehicles. There are too many applications using nanotechnology which existed in this science decade. Our proposal is one kind of this type of application that will be inside human body to detect the cancer cell. Here, we have proposed a model for both centralized and decentralized network which can add a new dimension for research in nanotechnology. The model will focus mainly on decentralized network with a single-hop routing framework. A customized clustering method will be used with time difference of arrival (TDOA) which is being proposed [7].

2 Related Work and Motivation

The manipulation of nano-sensors has reached the science world to the top. Several fields have included in this research. The ongoing advances and the combination of microelectro-mechanical frameworks innovation, integrated circuit advances, microprocessor equipment, and nanoinnovation, wireless communications, ad-hoc networking routing protocols, dispersed signal processing, and installed frameworks have made the idea of wireless sensor networks (WSNs) [8]. Some work has been also done for the respective field included in literature review. Centralized network for energy-aware trust-based routing scheme for WSN [9], centralized network for energy-efficient routing algorithm for WSN [10], centralized network for assigning node as a center [11] are focused on. Simple energy-aware routing protocol based on the 'backward-learning' paradigm in which simulations are carried out in ns-3 by using nano-SIM package is proposed to compare with the existing protocols [12]. An improved performance scheme of nanocommunication over terahertz bands for wireless BSNs is proposed for smart e-health applications. It contains a new energy-efficient forwarding routine or electromagnetic communication in wireless nanonetworks consisting of hybrid clusters with centralized scheduling [13, 14]. In another routing protocol, the area of candidate nodes is narrowed to control the direction of multi-hop forwarding because several nodes with low link cost have the probability to be selected as a next hop, which prolongs the lifetime of nano-sensor networks [15]. Wireless nano-sensor network (WNSNs) is proposed concerning the energy balanced clustering in the routing mechanism of WNSNs. This proposed routing framework outperforms the state-of-the-art routing mechanisms in terms of energy consumption and delay [16]. Decentralized network has been used for several research for WSN to establish better routing framework. Self-organized approach in EnvSys project [17] uses decentralized network for WSN. For efficient routing algorithm and multicast routing (geographic) decentralized preferred [18] in WSN. Working with nodes or sensors and to overcome of the challenges like optimizing the efficiency of a small group of nodes decentralized network used [19]. Extending the life time of sensor clustering is needed in WSN. To make the sensor or node into a group, several clusterings have been proposed, yet mixed nonlinear programming model clustering [20] and weight-based clustering [21] in WSN are the examples which have been done before. For body area network optimize the sensor ability and extend capability or save the energy of sensor itself in WSN hierarchical clustering come in front [22]. To communicate with sensor to another sensor by either centralized or decentralized network, routing protocol has to be chose. To apply any the algorithm on sensor and make it strong enough to work efficient, routing algorithm is needed [23]. Terahertz communication is one of the issues for communicating between nano-sensors in WSN inside the human body [24]. Highest uses of bandwidth and optimize the propagation loss terahertz band communication working [25] at a good scale. We will use statistical time division multiplexing (TDM) because in statistical TDM there will be very less chance for interference [26]. In statistical TDM, when a channel has no data to send, the channel dynamically allocates other frame for the channel who has a data to send. It will be used for bandwidth efficiency [27]. Time difference of arrival (TDOA) [28] is one of the most issues in this research. As we are using TDOA, the clock frequency should be synchronized with base station as well as cluster heads. Several works have been done in this field too. Localization in WSN nonlinear approach [29], insecurity of time of arrival [30], and so on have been done. So, efficient routing protocol [31] and clustering [32] will be the crying need to work with the nano-sensors in this sense.

3 Considered Issues in Designing Routing Protocol

Terahertz communication issues: In terahertz communication, there are some peculiarities, which are addressed in [33]. Terahertz communication allows a huge bandwidth but only in a very limited distance (around 1 m).

Sensor node deployment: The proposed model assures sensor node deployment in a self-organizing system where the sensor nodes are scattered randomly and can be stationary or moving in a limited area.

Disconnection of nano-sensors nodes: Another major challenge is sudden disconnection of the sensors due to chemical reaction or any other unexpected phenomena, which should not affect the entire system.

4 Routing Framework

4.1 System Model and Assumptions

Our model is based on a body area network where a limited area is considered to deploy nano-sensors nodes. To detect any abnormality in cells, the nano-sensors will collect data and transmit it to the nano-router. The nano-router is responsible for the transmission of the sensor data to the base station for further operations like data visualization or cancer detection. It can be a personal computer. In our case, the number of nano-router is two, which will form two clusters. The number of nano-router can be increased depending on the size of the network area. Figure 1 shows a nano-sensors network with three nano-routers.

Each nano-router has a microgateway to the base station outside the body, which can be a personal computer or wearable device. The abnormality in cell will be detected or visualized in the base station's interface. The nano-sensors can be stationary or moving. There can be high density of nano-sensors in an area. They have very limited capabilities. They are bound to do two things,

- collecting data with their sensing capabilities
- transmitting the data after a certain time schedule.

4.2 Cluster Formation

To ensure efficient communication, a clustering algorithm will be introduced where three clusters will be formed by three nano-routers. The position of the nano-routers is fixed and known to the base station. The clock frequency of them is synchronized with the nano-sensors and base station. To assign a nano-sensor in a cluster, a distance estimation will be needed. So, a multi-node time difference of arrival technique will be used which is discussed in [30]. This multi-node TDOA has been customized for

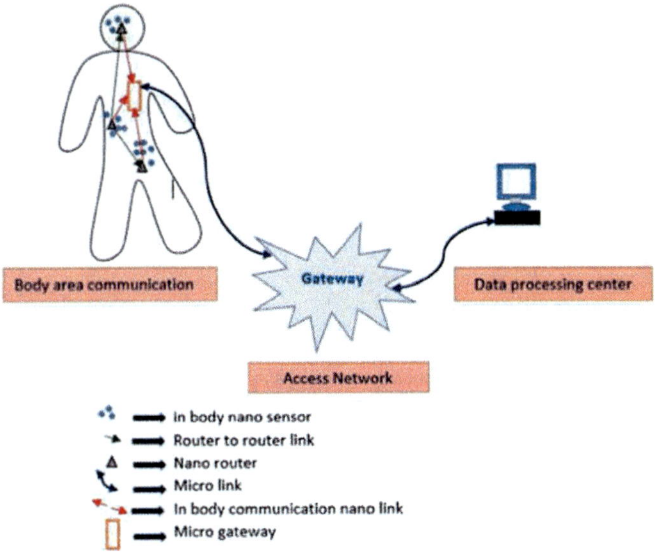

Fig. 1 Body area network using nano-sensors and nano-routers

our purpose. Let us assume three nano-routers R1, R2, and R3 which have received a cluster join request signal from a nano-sensor. The time difference of R1 and R2 is denoted by tR1-R2, which defines a hyperbola on which the nano-sensor is located. Similarly, the time difference of R2 and R3 is denoted by tR2-R3, and the time difference of R3 and R1 is denoted by tR3-R1, which defines another two hyperbolas where the nano-sensor also lies. The intersection point of these two hyperbolae is the estimated position of the sensor. Let us assume the position is Pnode = (Xnode, Ynode, Znode) and router R1′s position PR1 = (XR1,YR1,ZR1) (Fig. 2).

The nano-router with shortest Euclidean distance will accept the nano-sensor and notify the other two via a channel. The distance sometimes would not be accurate as the sensors are moving, but we only need an estimation of which nano-router is near to it.

4.3 Communication Scheme

The communication of nano-sensors and nano-routers will be single hop which will follow a dynamic time division multiple access (TDMA) which is divided into three fixed length subframes (Uplink1, Uplink2, and random link). The Uplink1 will be used by the nano-sensors to transmit data to the nano-routers. The random link will be used to receive cluster join the request by the nano-router. The Uplink 2 will be used only by the nano-routers to notify each other about the sensor connection.

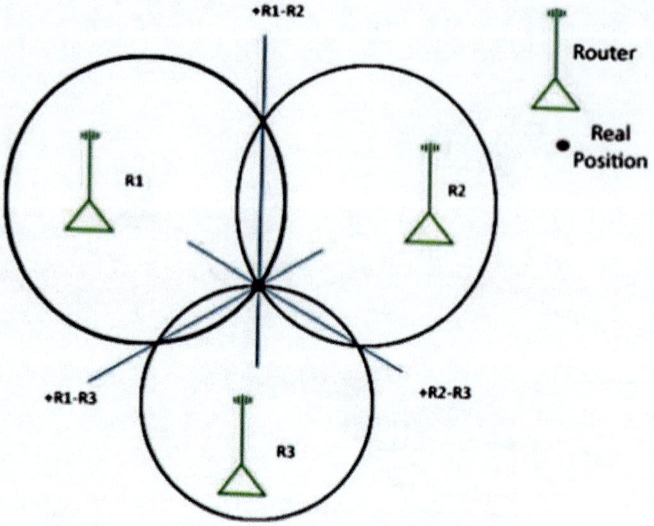

Fig. 2 Position estimation of nano-sensor using multi-node TDOA

Nano-sensor to nano-router communication: According to our proposed model, nano-sensors are deployed in the body in order to collect information from the field. These sensors can be freely movable in the body so that it increases the efficiency of the system. Nano-sensors can collect data and transmit data. This process is completed by a handshaking method with the nano-router. The nano-sensors would transmit a stream of data with its unique ID-Join Request, which would be received by the nano-routers, then nano-routers that received the data calculates the position of the nano-sensors by the method called time of arrival. And the nano-router that is close to the nano-sensors would send an acknowledgement (ACK) packet to the nano-sensors. When the nano-sensors receive the ACK packet, it sends the data packet to the nano-routers. Note to mention, the nano-sensors would only listen for the ACK packet for a certain period of time based on the location of the nano-routers. If the ACK packet is received in this time, the nano-sensor would transmit the data packet or else it would send another joint request to the nano-routers considering the previous one to be a failed attempt. The joint request would have a room of 72 bits. The main data packet of the nano-sensors would be 128 bits of data per transmission in TS-OOK communication method. According to [2], if we consider energy per pulse to transmit a data 1 pJ and target transmit distance of 10 mm, we can design an energy harvesting system that would harvest 200pJ of charge in a given time to power this process. The detailed method and calculations are stated in Sect. 5.

Nano-routers: Nano-routers are considered as the cluster head of the nano-sensors. In this process, it is responsible for creating cluster, sending ACK upon the reception of the join request by the nano-sensors, informing other routers after accepting a sensor in its cluster, receiving data packet from the nano-sensors which later would be sent them to outer body system by the nano-router. Nano-routers are set at a fixed location because of its increased size than the sensors and also for its increased energy harvesting as it would not only communicate with the nano-sensors, it would also communicate with an outer body device. That is why the nano-routers would have to be bigger in size; hence, it should be fixed in a location to avoid health hazard created by the machine. In this proposed model, nano-router works as a base station and nano-sensors are the cellular devices. Each of the nano-router works as a cluster head created by the nano-sensors s. The same energy model proposed for the nano-sensors would be used in nano-router. Energy harvesting area would be increased to meet the energy need of the device.

4.4 Cluster Formation Algorithm

In this section, the algorithm for cluster formation and the algorithm for a nano-sensors will be described.

Cluster formation algorithm: (Algorithm 1) Notation Initialization -
S_i = ID of the nano-sensors that sent a join request. i = 1,2,3.......
R_i,R_j, R_k = Router IDs
X_{Ri}, Y_{Ri}, Z_{Ri} = position of the router R_i
n = number of the total nano-sensors in cluster R_k
C (X_i, Y_i, Z_i) = position of the nano-sensors
S_t = Status of the router (online/offline)
P_{si} = Data packet of nano-sensors S_i
D_i, D_j, D_k = Nano-sensors S_i 's distance from the router R_i, R_j, R_k, respectively;
B = Outer body device
Here, $D_i = \surd((X_{Ri} - X_{Si})^2 + (Y_{Ri} - Y_{Si})^2 + (Z_{Ri} - Z_{Si})^2)$
Result: Cluster formation and data transmission to the nano-routers.

Algorithm 1 Cluster formation in router *R*

```
Select S_i after receiving join request
calculate D_i . D_j, D_k
if D_i < D_j and D_i < D_k then
  cluster R_i ←S_i
  notify R_j, R_k
else
  determine the router R with shortest distance
  if S_t of R ≠ 0 then
    break
  else
    add S_i
  end if
end if
while n > 0 do
  for each S_i do
    Ps_i ←set priority
  end for
  for each S_i do
    R ←receive Ps_i
    B ←transmit Ps_i
    S_i ←send ACK after successful transmission
  end for
  n = n - 1
end while
```

Algorithm 2 Cluster join & data transmission of S_i

```
start
t_i = 0
W = 0
while t_i < T do
  S_i ←collect and store data
end while
task
if Ps_i > 0 and W = 0 then
  send join request
  W = 1
end if
if ACK ≠ 0 then
  start transmission in assigned priority
else
  W = 0
  go to task
end if
end task
end start
while true do
  if S_i has charge then
    go to start
  else
    charge
  end if
end while
```

Cluster join & data transmission of Si (Algorithm 2) Notation Initialization - S_i = ID of the nano-sensors that sent a join request. i = 1,2,3.......
t_i = time taken for data collection of abnormality in cells
T = fixed time period for collection of data
P_{Si} = Data of nano-sensors S_i
Result: Joining cluster and data transmission by nano-sensor.

5 Conclusion and Future Work

In this paper, a routing protocol with a decentralized network inside human body for detecting cancer cell has been presented. Our challenges were to focus mainly on decentralized network and considering terahertz communication issues, energy harvesting, and disconnection of nano-sensor nodes in case of routing protocol. The most important issue is time difference of arrival (TDOA) method which has been applied for clustering nano-sensors differently. A dynamic time division multiple access (TDMA) has been followed to communicate with the nano-router from nano-sensors. In this proposal, it has been assumed that there is no chemical reaction or collision inside the human body with nano-sensors or routers.

Nowadays, image processing seems to be a very popular method for cancer cell detection [34–36]. For these kinds of methods, cancer cell-related images are mandatory as a dataset. We have presented the routing protocol for nano-sensors in this paper. Using this proposed method, nano-sensors will be able to transmit cancer cell related data to the outer data hub (such as smart watch). This data could be image and other numerical data. The data type will be determined by what kind of nano-sensors we are using. Our proposed method is able to support image processing algorithm by providing image dataset. Also, using this routing protocol, nano-sensors will be able to send different kinds of numerical data related to cancer cell which will be very helpful for traditional cancer cell analysis.

The most well-known imaging strategy used to recognize cancer and screen its spread is computed tomography (CT) [37], which gives cross-sectional imaging by computer. CT filters are made utilizing X-beams. Attractive resonance imaging (MRI) [38] is a strategy that utilizes incredibly magnetic fields. This traditional procedure produced images of suspicious cancer cell area and then analyzed it for further decision. This CT and MRI are the heavy clinical devices, and patients must need to take hospital/clinic for the procedure. Also, its time consuming. In our proposal, we have proposed a routing protocol for nano-sensors which will be able to deploy inside human body where nano-sensors will monitor the cancer cell area. It is more convenient and efficient for patients because they do not need to run hospital every time. Doctors will be able to monitor patient's cancer cell data remotely.

The proposal also has some limitations which will give the opportunity for study and research in future. As we are using TDOA, the clock frequency should be synchronized with base station as well as cluster heads. This proposal proposed that the nano-routers should be in a fixed place in the hosting body so that nano-sensors have to find those routers; it could cost some energy of nano-sensors. As the energy is a key factor for nano-devices, it could be studied in future for more improvement. As all the nano-sensors will form the cluster in the hosting body, it is really important to ensure that a moving nano-sensors must find a cluster to connect otherwise there could be some data loss. This paper proposed a well-structured routing algorithm about nano-sensors communication with base station via nano-routers. For this algorithm, nano-routers calculate some parameters and conditions which could be time consuming and also energy consuming for nano-routers.

References

1. " What is Nanotechnology?", CrNano.org, [Online], https://www.crNano.org/whatis.htm. Accessed 30 Dec. 2017
2. Nelson, D.L., Cox, M.M.: Lehninger Principles of Biochemistry. W. H. Freeman Company, Journal (2008)
3. Heil, M., Silva Bueno, J.C.: Within-plant signaling by volatiles leads to induction and priming of an indirect plant defense in nature. In: Proceedings of the National Academy of Sciences, vol. 104, no. 13, pp. 5467–5472 (2007)
4. Yonzon, C.R., Stuart, D.A., Zhang, X., McFarland, A.D., Haynes, C.L., Duyne, R.P.V.: Towards advanced chemical and biological nano-sensors s-an overview. TTalanta **67**(3), 438–448 (2005)
5. Akyildiz, I.F., Su, W., Sankarasubramaniam, Y., Cayirci, E.: Wireless sensor networks: a survey. Comput. Netw. (Elsevier) J. **38**(4), 393–422 (2002)
6. Rogers, J.: Materials and mechanics for stretchable electronics-from electronic eye cameras to conformal brain monitors. In: International Solid-State Sensors, Actuators and Microsystems Conference, pp. 1602–1603 (2009)
7. Gholami, M.R.: Positioning Algorithms for Wireless Sensor Networks (2017)
8. Biradar, R.V., Patil, V.C., Sawant, S.R., Mudholkar, R.R.: Classification and Comparison of Routing Protocols in Wireless Sensor Networks
9. Tajeddine, A., Kayssi, A., Chehab, A: A Centralized Energy- Aware Trust-Based Routing Scheme for Wireless Sensor Networks
10. Zanjireh, M.M., Larijani, H., Chehab, a.: A Survey on Centralised and Distributed Clustering Routing Algorithms for WSNs
11. Cahit, I., Cahit, R.: Structural Reliability of Centralized Communication Networks
12. Al-Turjman, F., Ihsan Kilic, K.: LaGOON: A Simple Energy-Aware Routing Protocol for Wireless Nano-Sensor Networks
13. Afsana, F., Asif-Ur-Rahman, M., Ahmed, M.R., Mahmud, M., Kaiser, M.S.: An energy conserving routing scheme for wireless body sensor nanonetwork communication. IEEE Access **6**, 9186–9200 (2018). https://doi.org/10.1109/AC-CESS.2018.2789437https://doi.org/10.1109/AC-CESS.2018.2789437
14. Yu, H., Ng, B., Winston, Seah, K.G.: Forwarding Schemes for EM-Based Wireless Nanosensor Networks in the Terahertz Band
15. Xu, J., Zhang, R., Wang, Z.: FAn Energy Efficient Multi-hop Routing Protocol for Terahertz Wireless Nanosensor Networks
16. Verma, S., Kaur, S., Dhiman, G., Kaur, A.: Design of a Novel Energy Efficient Routing Framework for Wireless Nanosensor Networks
17. Jamont, J.-P., Occello, M., Lagrèze, A.: A Decentralized Self- Organized Approach for Wireless Sensor Networks
18. Banerjee, P., Mitra, M., Barbhuiya, F.A., Chakraborty, S., Nandi, S.: An Efficient Decentralized Rekeying Scheme to Secure Hierarchical Geographic Multicast Routing in Wireless Sensor Networks

19. Mihaylov, M., Tuyls, K., Nowe, A.: Decentralized Learning in Wireless Sensor Networks
20. Zhang, Z., Wang, H., Wang, C., Fang, H.: Energy Efficient Clustering Algorithm in Wireless Sensor Networks by Cluster-based Epidemic Control Through Smartphone-based Body Area Networks
21. Tandon, R., Dey, B., Nandi, S.: Weight Based Clustering in Wireless Sensor Networks
22. Mamalis, B., Gavalas, D., Konstantopoulos, C., Pantziou, G.: Clustering in Wireless Sensor Networks
23. Chowdhury, Uddin Ahmed, T.: Energy Efficient Hierarchical Routing Protocol for Wireless Sensor Network
24. Yang, K.: Characterisation of the In-vivo Terahertz Communication Channel within the Human Body Tissues for Future Nano-Communication Networks
25. Jornet, J.: Fundamentals of electromagnetic nanonetworks in the terahertz band. Found. Trends R Network. **7**(2–3), 77–233 (2012)
26. Zakharenkov, U.A., Maurer, G.S., Leight, J.E.: Optical Time Division Multiplexing System
27. Ceylan, O., Sumbullu, D., Bulent Yagci, H.: Localization at Wireless Sensor Networks
28. Ghelichi, A., Yelamarthi, K., Abdelgawad, A.: Target Localization in Wireless Sensor Network based on Time Difference of Arrival
29. Ravindra, S., Jagadeesha, S.N.: Time of Arrival Based Localization in Wireless Sensor Networks: A Non-linear Approach
30. Glass, S., Muthukkumarasamy, V., Portmann, M.: The Insecurity of Time-of-Arrival Distance-Ranging in IEEE 802.11 Wireless Networks
31. Er. Asif Munir: Multipath Routing Protocol for Wireless Sensor Networks
32. Rashed, M.G., Hasnat Kabir, M., Sajjadur Rahim, M., Enayet Ullah, Sk.: A Cluster Based Routing Protocol for Wireless Sensor Network
33. Miquel Jornet, J., Pierobon, M., Akyildiz, I.A., Akkari, N., Almasri, S.: A routing framework for energy harvesting Wireless Nano Sensor Networks in the Terahertz Band
34. Davuluri, S., Rathna Kishore, D.: Cancer Clumps Detection using Image Processing Based on Cell Counting and Artificial Neural Network Techniques, vol. 9(2). ISSN: 2249-8958 (2019)
35. Danaee, P., Ghaeini, R., Hendrix, D.A.: A deep learning approach for cancer detection and relevant gene identification. In: Pacific Symposium on Biocomputing (2017)
36. Kadampur, M.A., Al Riyaee, S.: Skin cancer detection, applying a deep learning-based model driven architecture in the cloud for classifying dermal cell images. https://doi.org/https://doi.org/10.1016/j.imu.2019.100282
37. Griffeth, Landis, K.: Use of PET/CT scanning in cancer patients: technical and practical considerations. In: Proceedings (Baylor University. Medical Center) vol. 18(4) (2005): 321–30. https://doi.org/10.1080/08998280.2005.11928089
38. Penet, M.F., Mikhaylova, M., Li, C., et al.: Applications of molecular MRI and optical imaging in cancer. Future Med Chem. **2**(6), 975–988 (2010). https://doi.org/10.4155/fmc.10.25https://doi.org/10.4155/fmc.10.25

Signal Processing, Computer Vision and Rhythm Engineering

Cascade Classification of Face Liveliness Detection Using Heart Beat Measurement

Md. Mahfujur Rahman, Shamim Al Mamun, M. Shamim Kaiser, Md. Shahidul Islam, and Md. Arifur Rahman

Abstract Face detection and recognition is a prevalent concept in security and access control area which is commonly used in surveillance cameras at public places, attendance etc. But often this type of system can be circumvented by holding a photo or running a video of authorized person to the camera. Therefore, liveliness concept comes up with a solution to detect the person is real or spoofed. In this paper, We proposed a cascade classifier based model for detecting liveliness using deep-learning and Heart-beat measurement. Moreover, we have evaluated our model accuracy with our own dataset of real and fake videos and photos. By using our proposed model of face liveliness detection model, FPR and FNR have declined 16% and 5.22% respectively. In addition, we have also compared proposed system with other state-of-art methods. And here proposed study has achieved an accuracy of 99.46%.

Keywords Features · Face Detection · Face Liveliness · Heart Beat · PCA · FaceNet · CNN · Deep Learning

1 Introduction

Face liveliness means if the face in front of the camera belongs to the live person that helps to prevent face spoofing attacks and ensuring safe bio-metric authentication and verification. Face liveliness detection is a real-time detection using a camera and analyzes certain face characteristics to detect liveliness. Due to the rapid digitalization, nowadays, the bio-metric authentication technique is used for identity management [1–4]. Face recognition is one of the popular bio-metric technique for securing the access control of web and mobile-related software applications. Face recognition is more convenient to install than other bio-metric techniques. However,

Md. M. Rahman (✉)
Daffodil International University, Dhaka 1207, Bangladesh
e-mail: mrrajuiit@gmail.com

S. A. Mamun · M. S. Kaiser · Md. S. Islam · Md. A. Rahman
Jahangirngar University, Savar, Dhaka 1342, Bangladesh

M. S. Kaiser et al. (eds.), *Proceedings of International Conference on Trends in Computational and Cognitive Engineering*, Advances in Intelligent Systems and Computing 1309, https://doi.org/10.1007/978-981-33-4673-4_47

despite its advantage as a non-intrusive form of access, the security system might not be able to distinguish between a real person and a person's photograph. A trickster easily can get entry to the system by presenting a copy of the authenticate image to the camera. Therefore, early face recognition authentication, face liveliness detection is important to detect whether the captured face is an alive or fake captured image.

At present, many organizations or institutes deal with personal attendance using bio-metric devices [5–7]. Traditional face authentication algorithms are not able to differentiate which ones live and which are not. It is a major security issue of the Face Recognition System. Portrait photographs can be used as spoofed images while facial recognition is used for authentication process. In order to enrich security system for face authentication, liveliness detection can be robust solution for any authentication system by facial recognition. In recent years, liveliness detection has been a very active research topic in the face recognition system.

CNN improves liveliness detection to an extent which is a promising approach and offers improved user experience. Though the CNN based FaceNet model has gained much attention for face recognition, several drawbacks associated with this model has made the liveliness prone to spoofing attack. We have improved CNN based FaceNet model with the implementation of heartbeat measurement in liveliness detection. In our proposed model we used a deep learning algorithm which is MTCNN to detect Face in attendance security issues with Face. The samples input into the convolutional neural network (CNN) based FaceNet model to extract the feature and used SVM to train the data to classify the input image. The classified image can be Real or Fake. We combined CNN based FaceNet with Heartbeat Measurement Model for performing liveliness detection. The proposed model recognizes a person accurately without spoofing to provide the corresponding Name or Unknown output for accurate identity authentication. The proposed model retains all the benefits of FaceNet and also solve the shortcoming of CNN based model [8].

The remainder of this paper is organized as follows. In Sect. 2, we discuss Face Liveliness Detection Review. Our Proposed Methodologies and Model Architecture are stated in Sect. 3. In Sect. 4, Results and Discussion are shown and the details are discussed. Finally, concluding remarks are provided in Sect. 5.

2 Face Liveliness Detection Review

In this section, some of the most interesting, traditional liveliness detection techniques and deep learning on liveliness detection are presented. There are many approaches that have been proposed to detect Face Liveliness for identification systems. Gahyun Kim et al. and Jukka et al. proposed a model based on Frequency and Texture based analysis for detecting liveliness [9, 10]. This model can detect live and fake faces frequency and texture analyses. The author proposed the Local Binary Pattern (LBP) applied on the given facial images to analyze the textures. In their proposed model there are some limitations. For detecting liveliness, this research didn't deal with the attacks using portrait photographs, video and fake faces, masks or 3D models of the face. Sooyeon et al. [11] proposed the effect of defocus for face liveliness detection. The author used a feature-level fusion approach to detect spoofing images. This paper didn't consider more robust countermeasures and can not detect spoofing attacks against 3D attacks and videos by analyzing textural and temporal characteristics. Author Rabello dos et al. [12] presents two methods which are Client-definite and Client-independent methods for liveliness detection. For anti-spoofing, Client-definite methods use a Probabilistic Graphical Model framework. Some papers proposed methods based on heartbeat [13, 14], blood flow , blinking and involuntary micro-motion of facial muscles [15], to classify live and spoofed faces. Using life features, methods have a high detection accuracy but this method requires face video as the input and needs a large amount of computation. Recently, Deep learning architectures are used to detect face liveliness and proofing as an art result over the different approaches due to self-learned, robustness features, and the great power of abstraction from the training data [16, 17]. Yaman et al. proposed a deep-learning-based spoof face detection. In this model, LRF-ELM and CNN [18, 19] are used to detect liveliness. In this paper, the author just comparatively analyzed between them. Y.A.U Rehman et al. proposed a convolutional-features fusion layer which one is fully adaptive that balances the fusion for real-world face images to find out convolutional features by deep CNN during training [20]. Ranjana et al. developed deep architectures that combine CNN and texture analysis to detect the captured image as real or fake [21]. Their development greatly improved different texture in the real image. Gustavo et al. proposed a model that trained in two different parts for a task based on a novel CNN architecture and each part of the neural network analyze facial areas to learn features [22]. As a result pre-training step allows CNN to learn different local spoofing areas, the convergence speed of the final model, and improving the performance, and also performance is high to detect liveliness.

3 Model Architecture

The system provides the security of attendance application by authenticating the user with face attribute along with aliveness detection and recognition using MTCNN, CNN based face liveliness detection with Heartbeat Measurement. In illustration, Figure 1 indicate the the block model of liveliness detection model comprised with Face Detection Model and validate the result with Heart Beat Measurement for getting liveliness of the detected person to prevent spoofing in the security systems.

3.1 Face Detection Model

In face recognition model, there are three basic ways to recognize a face such as - Face Detection, Feature Extraction and Face Classification. In our proposed model for detecting a face, MTCNN is used. Moreover, facenet model is used to extract

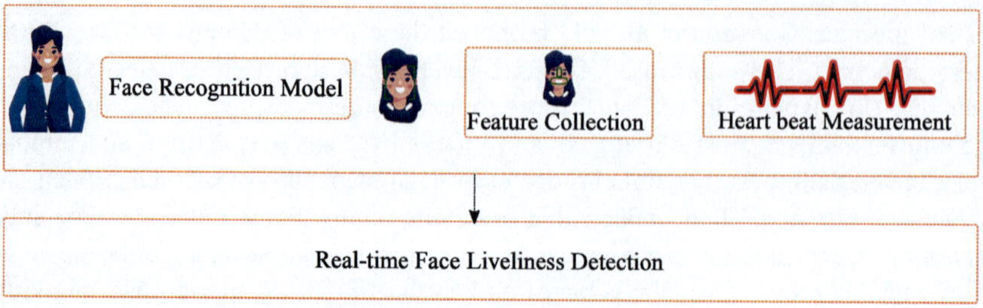

Fig. 1 Face liveliness detection block model

the feature and finally, support vector machine (SVM) is being used to classify person. MTCNN is a neural network which has three convolutional networks: P-Net, R-Net, and O-Net to perform face detection in real time environment. These convolutional networks gives a bounding box of a face. This bounding box is used as input to the FaceNet algorithm. P-Net is a fully connected network which known as Proposal Network. P-Net provides candidate frames which contains face and creates multiple scaled copies of images. Bounding box coordinates for these candidate frames. R-Net is similar to P-Net. P-Net is further optimized using R-Net. R-Net removes a large number of non-face or false frames or false candidates windows. P-Net provides candidate frames which are fed to the R-Net. R-Net can remove a huge number of false candidates, calibrates bounding boxes and performs Non-Maximum suppression(NMS) again. Finally, O-Net further reduces those bounding boxes that have low confidence scores, finds five facial landmarks, calibrates bounding boxes and landmarks further, and then does NMS. The learning objective of MTCNN is a multi-task loss and This losses are weighted and summed up in a cumulative multi-task formula. The output of MTCNN is fed to FaceNet for face recognition in the bounding box.

3.2 *CNN Model*

In this research, we use transfer learning to train our network with FaceNet to recognise a face [23]. FaceNet is a deep neural network which is used for extracting high-quality features from an image of a person's face. FaceNet is used to embeds the input image and covert into an 128 dimensional vector encoding which represent the most important features of a face. This vector encoding is called embedding. FaceNet embedding are useful for face verification, face recognition or classification and clustering. Loss function is a another important aspect of FaceNet. FaceNet uses triplet loss function. we need 3 images which can be namely anchor, positive and negative to calculate triplet loss function. The triplet loss compare face embedding for these three images. One possible way of recognising a person on an unseen image would be to calculate its embedding, calculate euclidean distances among anchor image, positive images and negative images. If euclidean distance of embedding anchor and positive image is lesser than distance of embedding anchor and negative image, then similar face will be recognized. Finally, an SVM classifier is used to identify the face in the last stage and also classify the image as fake and real.

3.3 Heart Beat Measurement

After detecting the face and recognition, we applied our heart beat measurement (HBM) to verify the face liveliness. Human heartbeat is measured in two ways: contact base and non-contact base. Contact base physically contact with human body and non-contact base associated with human faces using computer vision technology. Generally, ECG or PPG is used to measure the heart rate. But Non-contact based approached is popular due to mitigate constraint of contact-based techniques when clinical section is considered. The computer vision systems lifted the computer aided medical assistance into a new level - now we can monitor the human pulse rate without contact equipment that be very useful for diagnosing liveliness of the face. In Fig. 2, shows the major blood circulating arteries to the face. Where, our main focus the area of interest onto supratrochlear vain to detect blood circulation pulse in image green channel. Region of interest (ROI) is collected from detected face from CNN model, we covert the ROI into green channel image and calculate the intensity of the pixel of the ROI. This intensity marked as our features to measure HB. Moreover, Cardiovascular pulse signal from web cam using heartpy tools [24] which is separated from noise motion using principal component analysis method. After doing PCA method, we have extracted three to five independent components. For each component, we apply Peak Detection (PD) technique to indicate the positions of peaks shows in Fig. 3. After that best component is selected to measure the heart rate incorporate with the PD results. In cardiovascular pulse, proof frequencies is 1–2 Hz which corresponding with 40 to 120 beats per minute. After getting pulse rate, it will be checked by normal bpm. To calculate the pulse rate, we used, $heartrate = P * \frac{60}{q_2 - q_1}$. Here, P is the total number of peaks and q1 & q2 are the time variables of first and last peak points.

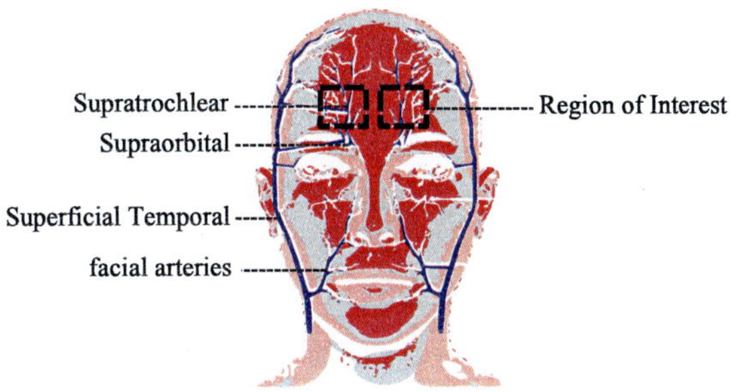

Fig. 2 Major blood supply arteries to the face and our Region of Interest (ROI)

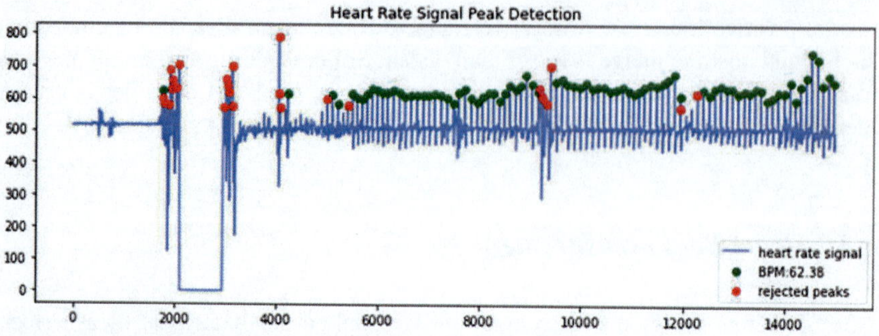

Fig. 3 Peak Detection in Heart Beat Measurement

3.4 Liveliness Detection

Proposed method uses live video or test image as an input and returns the pulse rate and identification as an output to detect a live face. Firstly, we detect a person's face using MTCNN. Then we extract the facial feature using FaceNet based CNN and recognize a face using SVM. After recognizing a face with indicating real or fake/spoof, two facial region is carried out as a key point to measure pulse rate: forehead and nose-mouth. From forehead, we can measure the pulse/heart rate using proposed heartbeat measurement. After getting pulse rate, it will be checked by normal bpm. If it is in range, then live face will be detected. A input fake image can be recognized as a real image using FaceNet Model. But Heartbeat Measurement Model will recognize as a fake image which is originally fake image. We combine the two output using sigmoid function and finally will get the desire result (See Fig. 4).

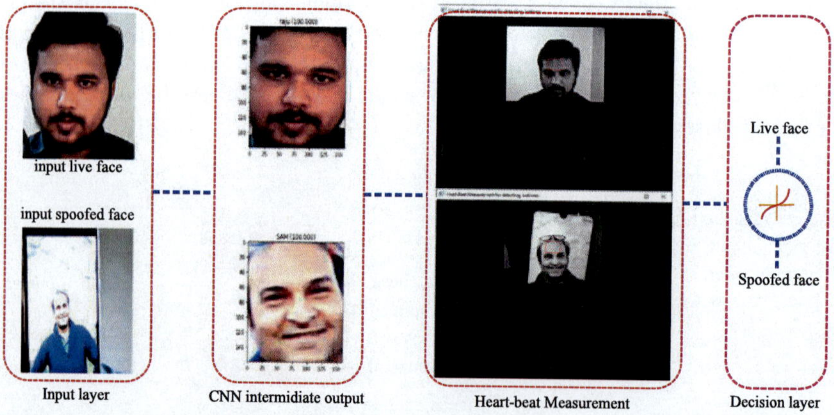

Fig. 4 Face liveliness detection output. The system has taken input as live/spoofed and detected intermediate output from facenet and finally measured heart-beat to enhance the accuracy of detecting face whether it is really a live face or not

4 Result and Discussion

4.1 Dataset Description

Dataset for liveliness detection needs diverse set of real and fake images of a person. We collected data from students of IIT, JU and students of CSE, DIU of Bangladesh and also some faculty members of those universities. In the purpose of attendance security system, we choose these universities students for better understanding. In our collected data set we have mainly two classes of images: Fake images and Real images with individual person identification. Data set are collected in two ways: captured images and video file. Fake images of a person captured from a camera while playing a video or setting up a spoofed image at screen in-front of their real face. Real images of a person captured from a selfie video/images with phone or camera and also collected heart rate of a person using garmin 235 J fitness watch which is considered as ground truth value. We collected two types of videos: Fake videos and Real Videos. one video for "real" faces and another for "fake/spoofed" faces. Finally applied face detection to both videos to extract individual ROI(Region of Interest) face for Real or Fake classes. Image dataset is available in GitHub link(https://github.com/rajuiit/livelinessdata).

4.2 Performance Analysis

Working with liveliness detection of person for maintaining security concern is a comparatively new concept because of accurate and specific results. Deep learning combined with heartbeat measurement is a new way to detect face liveliness precisely and work with this attendance or related application field. We worked with huge amount of face images data from our created dataset and worked with them with our proposed model precisely.

We represents a confusion matrices of the liveliness detection to measure the performance of our proposed model. Proposed model with the help of sigmoid function will generate the output with label "Live Face" and "Spoofed Face". Figure 5 represents the confusion matrices of CNN model. Here we took 135 live faces and 49 spoofed faces for testing purposes. First we applied this dataset in our FaceNet based CNN model. The corrected Live instances was 127 which was True Positive and incorrect instances was 8. On the other hand, The correctly identified spoofed faces 42 and incorrectly classified was 7. Our main objective was to improvement the FaceNet based CNN model. So output result of Facenet based CNN model is taken as input in Heartbeat Measurement Model. Figure 5 represents the confusion matrices of Heartbeat Measurement model. Heartbeat measurement model correctly identified all 135 live faces which was actually 135 and correctly identified 48 spoofed faces which was actually 49. We used False Positive Rate(FPR) and False Negative Rate(FNR) to evaluate the proposed model in this study which was calculated from

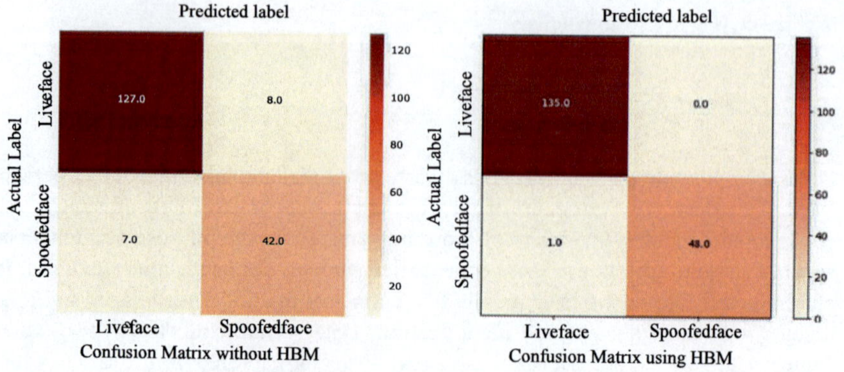

Fig. 5 Confusion Matrix without heart-beat measurement and using heart-beat measurement

confusion matrices. False Positive Rate as a performance metric is particularly useful when the goal is to reduce false positive value in the confusion matrix. In this study, false positive rate of FaceNet based CNN model is 16% which is reduced to 0% in Heartbeat Measurement to detect liveliness. On the other hand, a high false-negative rate suggests that there is an possibility that spoofed faces consider as live faces which is very risky. False negative rate of FaceNet based CNN model is 5.22% where Heartbeat measurement model reduced false-negative rate of CNN based FaceNet Model which is 0.74%. Table 1 shows details of result analysis of our dataset that has been trained with proposed model and also result has compared over an art methods which will help to verify our finding result. Proposed model seems to outperform all of existing model in Table 1.

Table 1 Performance comparison of our liveliness detection method with some state-of-arts

Refs.	Method Name	Accuracy
Gahyun Kim et al. [11]	One-class classifiers	85%
Kollreider et al. [15]	Boosting Algorithm	94%
koshy et al. [21]	Deep CNN	95.99%
Dos Anjos et al. [22]	lsCNN	95.64%
Akbulut et al. [18]	(LRF)-ELM and CNN	88.75%, 82.03%
Zukfiqar et al. [1]	Pre-trained CNN	98.76%
Sun et al. [25]	Blinking-based Conditional Random Fields	98.3%
Proposed Method	Improved CNN-Heartbeat Measurement	99.46%

5 Conclusion

This paper proposed an identity authentication for attendance system combining a CNN based FaceNet model and Heart Beat Measurement to improve face liveliness detection against spoofing attack in authentication system. Moreover, this work have achieved 99.46% of accuracy for detecting a spoofed face. In future, this work will extend to performance evaluation with standard data sets and will add some features such as background identification, context analysis for improving heart-beat calculation. Finally, we will use Bayesian Convolutional Neural Network (BayesCNN) approach to examine liveliness.

References

1. Zulfiqar, M., Syed, F., Khan, M.J., Khurshid, K.: Deep face recognition for biometric authentication. In: 2019 International Conference on Electrical, Communication, and Computer Engineering (ICECCE), pp. 1–6, July 2019
2. Mamun, S.A., Shamim Kaiser, M., Ahmed, M.R., Islam, Md. S., Islam, Md. I.: Performance analysis of optical wireless communication system employing neuro-fuzzy based spot-diffusing techniques. Commun. Netw, **5**(3), 260–265, 2013-09-27 (Number: 3 Publisher: Scientific Research Publishing)
3. Biswas, S., Kaiser, M.S., Mamun, S.A.: Applying ant colony optimization in software testing to generate prioritized optimal path and test data. In: 2015 International Conference on Electrical Engineering and Information Communication Technology (ICEEICT), pp. 1–6 (2015)
4. Shamim Kaiser, M., Chowdhury, Z.I., Al Mamun, S., Hussain, A., Mahmud, M: A neuro-fuzzy control system based on feature extraction of surface electromyogram signal for solar-powered wheelchair. Cogn. Comput. **8**(5), 946–954 (2016)
5. Fatemifar, S., Arashloo, S.R., Awais, M., Kittler, J.: Spoofing attack detection by anomaly detection. In: ICASSP 2019—2019 IEEE International Conference on Acoustics, Speech and Signal Processing (ICASSP), pp. 8464–8468, May 2019. ISSN 2379-190X
6. Al Mamun, S., Suzuki, R., Lam, A., Kobayashi, Y., Kuno. Y.: Terrain recognition for smart wheelchair. In: Huang, D., Han, K., Hussain, A. (eds.) Intelligent Computing Methodologies. Lecture Notes in Computer Science, pp. 461–470. Springer (2016)
7. Al Mamun, S., Fukuda, H., Lam, A., Kobayashi, Y., Kuno, Y.: Autonomous bus boarding robotic wheelchair using bidirectional sensing systems. In: Advances in Visual Computing. Lecture Notes in Computer Science, pp. 737–747. Springer (2018)
8. Kaiser, M.S., et al.: Advances in crowd analysis for urban applications through urban event detection. IEEE Trans. Intell. Transport. Syst. **19**(10), 3092–3112 (2017)
9. Kim, G., Eum, S., Suhr, J.K., Kim, D.I., Park, K.R., Kim, J.: Face liveness detection based on texture and frequency analyses. In: 2012 5th IAPR International Conference on Biometrics (ICB), pp. 67–72, March 2012. ISSN 2376-4201
10. Komulainen, J., Hadid, A., Pietikäinen, M.: Context based face anti-spoofing. In: 2013 IEEE Sixth International Conference on Biometrics: Theory, Applications and Systems (BTAS), pp. 1–8, Sept 2013
11. Kim, S., Ban, Y., Lee, S.: Face liveness detection using defocus. Sensors **15**(1), 1537–1563, Jan 2015 (Number: 1 Publisher: Multidisciplinary Digital Publishing Institute)
12. Chingovska, I., dos Anjos, A.R.: On the use of client identity information for face antispoofing. IEEE Trans. Inf. Forensics Secur. **10**(4), 787–796, April 2015 (Conference Name: IEEE Transactions on Information Forensics and Security)

13. Li, X., Komulainen, J., Zhao, G., Yuen, P.-C., Pietikäinen, M.: Generalized face anti-spoofing by detecting pulse from face videos. In: 2016 23rd International Conference on Pattern Recognition (ICPR), pp. 4244–4249, Dec 2016
14. Ostankovich, V., Prathap, G., Afanasyev, I.: Towards human pulse rate estimation from face video: automatic component selection and comparison of blind source separation methods. In 2018 International Conference on Intelligent Systems (IS), pp. 183–189 (2018)
15. Kollreider, K., Fronthaler, H., Faraj, M.I., Bigun, J.: Real-time face detection and motion analysis with application in "Liveness" assessment. IEEE Trans. Inf. Forensics Secur. **2**(3), 548–558, Sept 2007. (Conference Name: IEEE Transactions on Information Forensics and Security)
16. Canziani, A., Paszke, A., Culurciello, E.: An analysis of deep neural network models for practical applications. arXiv:1605.07678 [cs], Apr 2017.
17. Zulfiker, Md. S., Kabir, N., Biswas, A.A., Chakraborty, P., Rahman, Md. M.: Predicting students' performance of the private universities of Bangladesh using machine learning approaches. Int. J. Adv. Comput. Sci. Appl. **11**(3) (2020)
18. Akbulut, Y., Sengur, A., Budak, Ü., Ekici, S.: Deep learning based face liveness detection in videos. In: IDAP, pp. 1–4, Sept 2017
19. Paul, M.C., Sarkar, S., Rahman, M.M., Reza, S.M., Kaiser, M.S.: Low cost and portable patient monitoring system for e-health services in Bangladesh. In: 2016 International Conference on Computer Communication and Informatics (ICCCI), pp. 1–4 (2016)
20. Rehman, Y.A.U., Po, L.-M., Liu, M., Zou, Z., Weifeng, O., Zhao, Y.: Face liveness detection using convolutional-features fusion of real and deep network generated face images. J. Vis. Commun. Image Represent. **59**, 574–582 (2019). February
21. Koshy, R., Mahmood, A.: Optimizing deep CNN architectures for face liveness detection. Entropy **21**(4), 423, Apr 2019 (Number: 4 Publisher: Multidisciplinary Digital Publishing Institute)
22. de Souza, G.B., Papa, J.P., Marana, A.N.: On the learning of deep local features for robust face spoofing detection. arXiv:1806.07492 [cs, stat], Oct 2018.
23. Noor, M.B.T., Zenia, N.Z., Kaiser, M.S., Mamun, S.A., Mahmud, M.: Application of deep learning in detecting neurological disorders from magnetic resonance images: a survey on the detection of Alzheimer's disease, Parkinson's disease and schizophrenia. Brain Inf. **7**(1), 1–21 (2020)
24. van Gent, P., Farah, H., Nes, N., Arem, B.: Heart rate analysis for human factors: development and validation of an open source toolkit for noisy naturalistic heart rate data. In: 6th Humanist Conference, 06 2018
25. Sun, L., Pan, G., Wu, Z., Lao, S.: Blinking-based live face detection using conditional random fields. In: Lee, S.-W., Li, S.Z. (eds.) Advances in Biometrics, Lecture Notes in Computer Science, pp. 252–260. Springer, , Berlin, Heidelberg (2007)

Two-Stage Facial Mask Detection Model for Indoor Environments

Aniqua Nusrat Zereen, Sonia Corraya, Matthew N. Dailey, and Mongkol Ekpanyapong

Abstract The COVID-19 pandemic has led to many organizations around the world enforcing face mask rules for personal protection. Manual checking whether individuals entering an organization's premises are wearing masks is cumbersome and possibly confrontational. There has been relatively little work on automatic face mask rule violations thus far. We propose a system for automatic monitoring for face mask rule violations for enterprises. Our method is an efficient two-stage facial mask detection model. The first stage is based on facial landmark extraction and clustering, and the second stage analyzes the clustered nose region. A thorough accuracy evaluation on five types of sample face images (no mask, beard and mustache, single-color mask, multi-color mask, and skin-color mask) finds that the overall accuracy of the two-stage model is an excellent 97.13%, outperforming simpler single-stage models.

Keywords COVID-19 · Facial mask detection · k-means · Face landmark

1 Introduction

Globally, at the time of writing, over 36 million people are infected and over 1,056,186 deaths have been caused by the COVID-19 pandemic [1]. The World Health Organization has recommended that mask-wearing can reduce the chances of being infected or spreading COVID-19 by respiratory droplets, which constitute the main vector

A. N. Zereen (✉) · M. N. Dailey · M. Ekpanyapong
AIT Artificial Intelligence Center, Asian Institute of Technology, Pathum Thani, Thailand
e-mail: st120069@ait.ac.th

M. N. Dailey
e-mail: mdailey@ait.ac.th

M. Ekpanyapong
e-mail: mongkol@ait.ac.th

S. Corraya
Department of Computer Science and Engineering, Jagannath University, Dhaka, Bangladesh
e-mail: sonia@cse.jnu.ac.bd

M. S. Kaiser et al. (eds.), *Proceedings of International Conference on Trends in Computational and Cognitive Engineering*, Advances in Intelligent Systems and Computing 1309, https://doi.org/10.1007/978-981-33-4673-4_48

for COVID-19 virus transmission [2, 3]. For health and safety, enterprises around the world are enforcing face mask rules. Manually checking whether each individual entering an organization's premises is wearing a face mask or not is cumbersome and possibly confrontational. In this paper, we propose and evaluate a reliable and efficient model for detecting individuals wearing masks and not wearing masks in indoor environments. Up till now, as the COVID-19 pandemic situation is completely new, there has been limited research in this field. However, there have been a few studies on face-mask detection and masked face recognition using the visible area of the face, for example, iris matching and partial face matching [4]. There are several existing deep learning approaches, but they require time, effort, computational resources, and storage for model training and runtime requirements [5–7].

We propose a more lightweight approach to detecting faces and classifying them as wearing a mask or not. Our system first finds faces. S3FD [8] achieves excellent performance by applying SSD and multiple strategies to handle small faces. Dlib [9] is a toolkit for building real-world machine learning and data analysis applications in C ++. The standard Dlib detector uses histograms of oriented gradients (HOG) features with linear SVMs. Some studies have found that HOG + SVM performs better than Haar cascades or Linear Binary Patterns (LBPs) for face detection in video sequences [10, 11]. The multi-task cascaded convolutional neural network (MTCNN) model for face detection [12] consists of three convolutional networks (R-Net, P-Net, and O-Net) and is able to outperform other methods on face detection benchmarks while retaining real-time performance. In this paper, we report on the performance of Dlib and MTCNN for the face detection task with the following five types of face images: (1) no mask, (2) no mask with beard or mustache, (3) single-color mask, (4) multi-color mask, and (5) skin-color mask. After detection, we must categorize the face as to whether the individual is wearing a mask or not. We use a novel two-stage model. The first stage extracts facial landmarks then applies k-means clustering [13] to the RGB pixels in the face's nose region to get a total of four clusters, and then it analyzes the color clusters to determine whether the clustered colors fall into the skin color range or not. We use the skin color classification method proposed by Kovac et al. [14]. The RGB color vectors representing each cluster mean are converted into the YCbCr color space. Based on the cluster centers' membership in the skin region, the system can correctly distinguish faces in groups 1 + 2 + 5 above from those in groups 3 + 4, i.e. people with no mask or a skin-colored mask can be distinguished from those wearing a colorful mask. The second stage must then differentiate group 1 + 2 from group 5, i.e., distinguish no-mask individuals from those with skin-colored masks. The performance of the proposed model is assessed with a total of 5504 images. We combine existing public face mask datasets with new Web-scraped and synthetic datasets. The combined dataset includes real masked and unmasked faces as well as synthesized masked face images. The number of images in each category is presented in Table 1. Experimental results indicate that the average accuracy after stage-1 is only 83.84% across all five types of images, but with stage 1 and 2 together, the overall accuracy increases to 97.13%.

Table 1 Distribution of samples over five categories

Dataset	No mask		Mask		
	General	Beard and mustache	Single-color	Multi-color	Skin-color
Public dataset 1 [16]	592	193	996	-	392
Public dataset 2 (our Web scraping)	700	213	920	-	45
Public dataset 3 [17]	55	545	-	-	-
Our dataset	219	-	219	852	210
Total images	2517		2987		

The remainder of the paper is organized as follows: Sect. 2 describes the overall system and the two stages of the proposed model separately, with both implementation and execution flows. Section 3 describes our experimental results in detail. Finally, Sect. 4 concludes the paper with a discussion and description of planned future work.

2 Proposed Model

We first provide a high level overview of the system, then we describe the evaluation dataset, algorithm, and system runtime architecture.

2.1 *System Overview*

The system executes the following steps.

(1) Capture frames.
(2) Detect and track humans.
(3) Detect faces.
(4) Detect facial landmarks.
(5) Cluster nose region pixels.
(6) Use cluster centroids and a skin color model for stage-1 classification.
(7) Separately analyze cluster distribution over the upper and lower parts of the nose region for stage-2 classification.

2.2 *Evaluation Dataset*

We synthesize four datasets, two of which are publicly available [16, 17], one of which was constructed automatically by Web scraping, and one new dataset consisting of

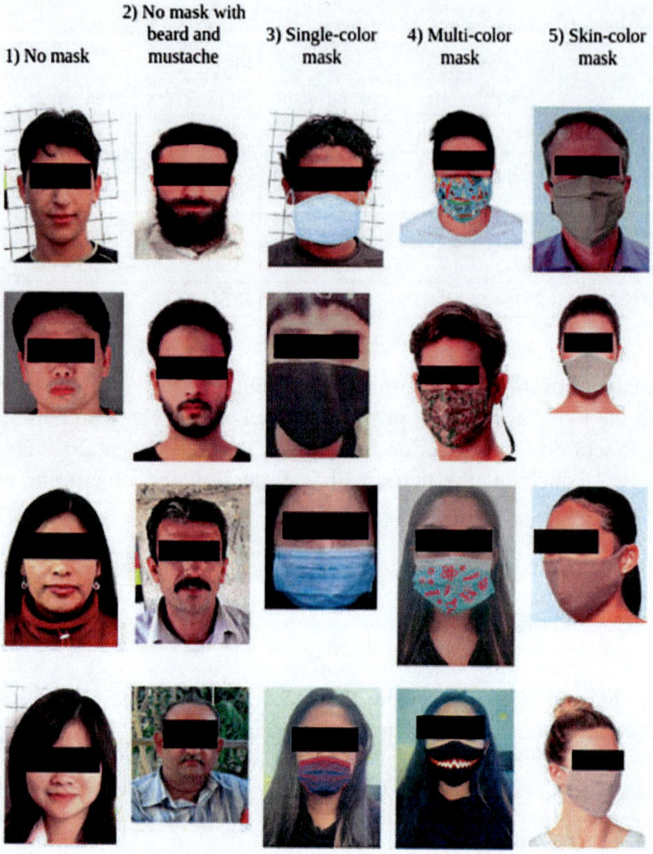

Fig. 1 Sample no-mask and masked face images from five categories

single color, multiple color, and skin-colored face mask images. To balance the data, we augmented the fourth dataset with random flips, rotations, shear, and cropping. The data comprise two categories: faces wearing masks and those wearing no mask. Examples are shown in Fig. 1. Faces without a mask include those in categories 1 and 2. Faces with masks include those in categories 3, 4, and 5. Table 1 shows the distribution of samples over the five categories.

3 Method

This section provides details of each of the blocks outlined in Sect. 2.1.

Capture Frames. For training and validation, we constructed the datasets as described previously and fed these images to the system. At runtime, the camera stream from any indoor IP or Web camera may be used. Every frame is resized to 640 × 480 prior to processing.

Detect and Track Humans. We use DeepSORT [18] with YOLO v3 [19] to track the people in the scene in real-time. DeepSORT uses the techniques of tracking-by-detection, linking YOLO detections of people using Kalman filters, deep feature extraction, a trained image patch similarity metric based on those features, and the Hungarian optimal association algorithm.

Detect Faces. We use the Dlib library or MTCNN to find faces within each detected human's bounding box. Dlib uses HOG features and a linear SVM to detect faces then performs an affine warping to align the eyes and bottom lip to the same location in each patch. MTCNN uses a cascade of three deep convolutional neural networks. First, candidate windows are produced through a fast proposal network. After that, the proposals are refined. Finally, to predict face and landmark locations, a third network generates a final bounding box and facial landmark location.

Detect Facial Landmarks. After the face has been detected, we further localize facial landmarks. We tried multiple facial landmark detection models [20, 21] and found that Dlib's 68-landmark model had sufficiently accurate performance. The essential information for mask detection is the lower part of the face region covering the lips, nose, and jawline. Dlib outperforms MTCNN, which has many false positive detections of faces on our datasets. We, therefore, run the Dlib detection method on each track, retaining the 8th, 16th, and 24th face patches and landmark sets, discarding the rest of the track [15, 22].

Cluster Nose Region Pixels. We find that the nose region of the face is most discriminative for mask / no mask classification. We use k-means clustering [13] over the nose area region of interest with k = 4.

Stage-1 Classification. In stage-1, we determine how many of the four clusters from the previous step are likely "skin" clusters. A skin cluster is a cluster in which the centroid falls within a skin region of the RGB space. If the three largest clusters are all skin clusters, we conclude that the face is either in the no-mask condition or a face with a skin-colored mask. Restricting attention to the top three clusters allows correct classification of no-mask faces with facial hair, which, if present, nearly always falls into the smallest of the four clusters.

Stage-2 Classification. In this stage, images containing three or more skin clusters are classified as to whether they contain skin-colored masks or unmasked faces. The top two of the three largest clusters are analyzed separately. We split the nose region of interest (ROI) vertically into two equal parts. We then calculate a range of expected spread of pixels in the largest two clusters over the upper and lower half of the nose ROI. First, we analyze the largest cluster. If the distribution of this cluster's pixels over the two subregions of the ROI is within range, we output "no mask." Otherwise, we consider the second cluster; if it also satisfies the condition, we output "no mask;" otherwise, we output "mask."

Figure 2 shows the overall view of the proposed system.

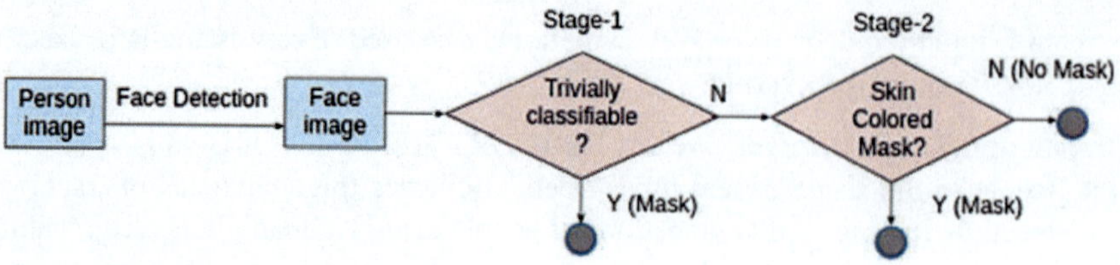

Fig. 2 Overview of the proposed system

3.1 Execution

Here we describe the workflow of the two classifier stages in more detail. The first stage determines whether the target is wearing a uniquely colored mask or required more careful consideration. The second stage considers the negatives from the first phase and determines whether the person is wearing a skin-colored mask or no mask. The final classification is based on the result of chaining the two simpler classifiers into a decision tree.

The first stage simply determines whether the top three clusters' centroids are in the skin range. We use the skin color range proposed by Kovac et al. [14] in YCbCr color space with some modification. The skin range is $77 \leq \text{Cb} \leq 127$ AND $133 \leq \text{Cr} \leq 173$. The thresholds are determined empirically. If any of the top three cluster centroids is out of range, the system outputs the "mask" class. If all three clusters do fall into the skin range, we proceed to the stage-2 classifier.

The stage-2 classifier is thus given images in which the top three clusters all fall into the skin range, and it further decides whether the person is wearing a skin-colored mask or no mask. We first split the nose region of interest vertically into two equal parts. We calculate the distribution of pixels in the top two clusters over the two subregions. We exploit the principle that real skin will have a more similar distribution over the two subregions than a skin-colored mask in the lower subregion and real skin in the upper subregion. Let α, $\beta and \gamma$ be the proportion of ROI pixels in cluster 1, cluster 2, and cluster 3, respectively. We let $L = \frac{\beta}{2}$ and let $H = \frac{2\alpha - \beta}{2}$. If the proportions α_U and α_L of ROI pixels in cluster 1 and the upper/lower regions of the ROI satisfy both $L \leq \alpha_{\text{U}} \leq H$ and $L \leq \alpha_{\text{L}} \leq H$, then we classify the image as "no mask." If cluster 1 does not satisfy the criteria, we use $L = \frac{\gamma}{2}$ and $H = \frac{2\beta - \gamma}{2}$ then test whether $L \leq \beta_U \leq H$ and $L \leq \beta_L \leq H$. If they do, the image is classified as "no mask."

Finally, if neither the first nor second-largest cluster satisfy the condition, we output "mask." This is based on the idea that if a person is not wearing a mask, the two top clusters' pixels will be relatively evenly spread over the upper and lower subregions of the nose ROI. Figures 3 and 4 show the workflow of our model.

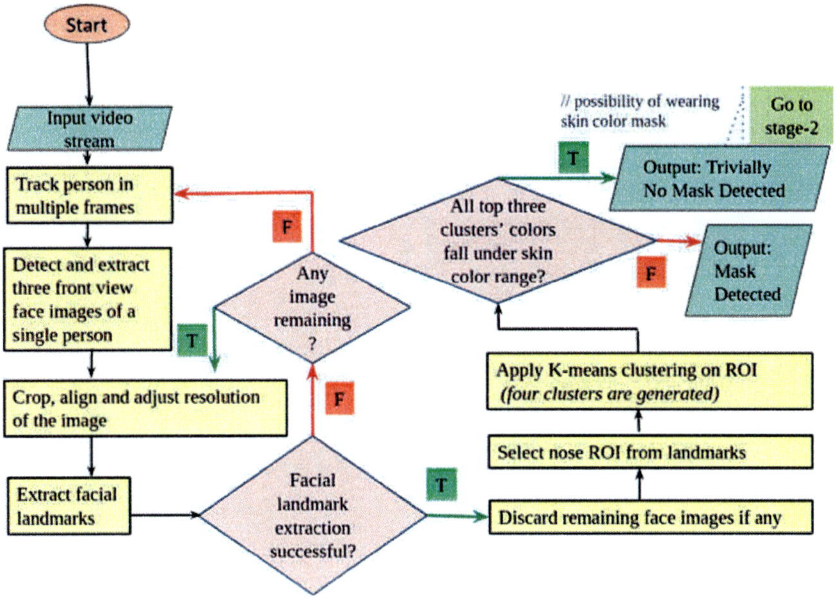

Fig. 3 Stage-1 classifier, which detects single-color and multi-color mask

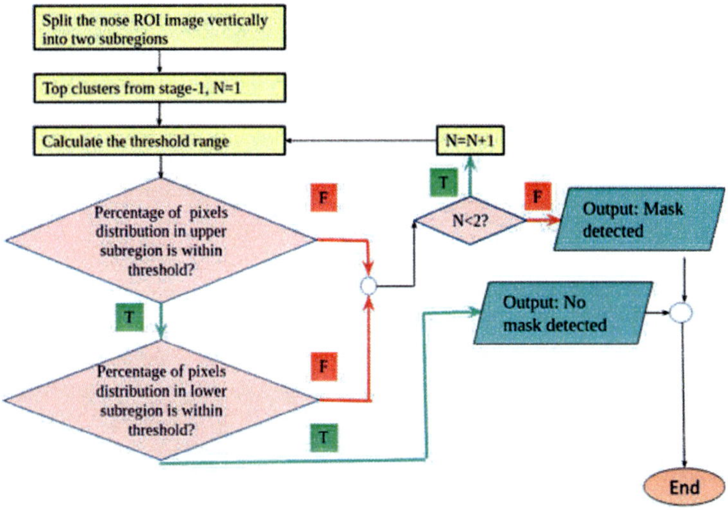

Fig. 4 Stage-2 classifier, which can distinguish skin-color masks from the no mask condition

Table 2 Face detection accuracy (%)

	No mask		Mask			Average
Category	General	Beard and mustache	Single-color	Multi-color	Skin-color	
Dlib	93	93	87	86	92	**90.2**
MTCNN	90	89	64	53	54	70

4 Experimental Results and Discussion

For face detection, we compare the accuracy of Dlib and MTCNN. Table 2 shows that Dlib performs better than MTCNN with 90.2% accuracy. MTCNN outputs many false positives, oftentimes on peoples' masks. We have also tried S3FD, which has good accuracy at multiple scales. However, in the controlled indoor environments we are targeting, we do not need multi-scale detection, so we chose Dlib as the face detector.

4.1 Stage-1 Results

Figure 5 shows the stage-1 results, where five different types of images are selected. Stage-1 alone gives an accuracy of 83.84%, shown in Table 3. Table 3 shows clearly that Stage-1 has only 32% accuracy in classifying masks that are skin-colored. Therefore, stage-2 is required to differentiate skin from skin-colored masks.

4.2 Stage-2 Results

Figure 6 shows result of Stage-2 processing of images that have all top three clusters in the skin color range (passed on from stage-1). These include faces with no mask, faces with no mask with a mustache or beard, and with skin-colored masks. The first row in the figure, consider the first image. The largest cluster contains 41.40% of pixels in the ROI. Of these pixels, the upper region has 23.13%, and the lower region has 18.27%. This satisfies the range criteria for nomask (14.71–26.69%). So we output "no mask." Similarly, the second image is also classified correctly. The third row considers an image with a skin-colored mask. The image has 55.31% of ROI pixels in the top cluster, among which the upper region has 6.055% and lower region has 49.255%, which does not satisfy the threshold range (12.32–42.98%). We then consider the second top cluster (24.65%), among which the upper region has 24.65% and lower region has 0%. This again does not satisfy the threshold range (5.03–19.62%), so the system outputs "mask." From these examples, we can conclude that

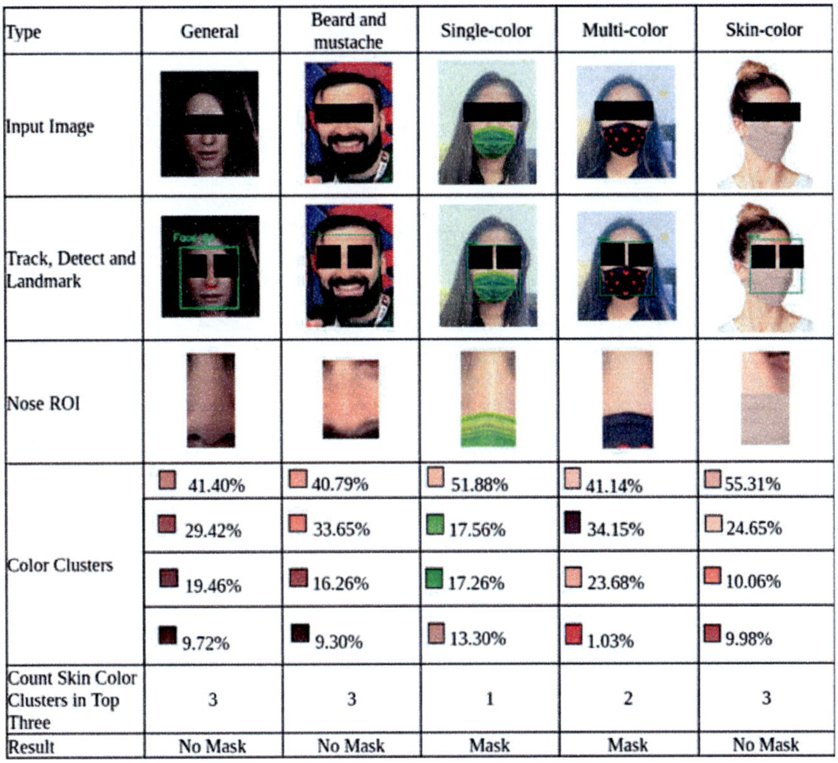

Fig. 5 Stage-1 results

Table 3 Mask and no mask detection accuracy (%)

	No mask		Mask			Average
Category	General	Beard and mustache	Single-color	Multi-color	Skin-color	
Stage-1	98.6	94.5	97.8	96.3	**32**	83.84
Stage-1 + 2	98.6	94.5	97.8	96.3	98.47	**97.13**

our system can detect skin-colored masks very efficiently. After the stage-2 process, the overall system has a high accuracy of 97.13%, as shown in Table 3.

5 Conclusion

There has been relatively little work in the area of detecting faces with and without masks. Our method efficiently classifies faces wearing more than one type of mask.

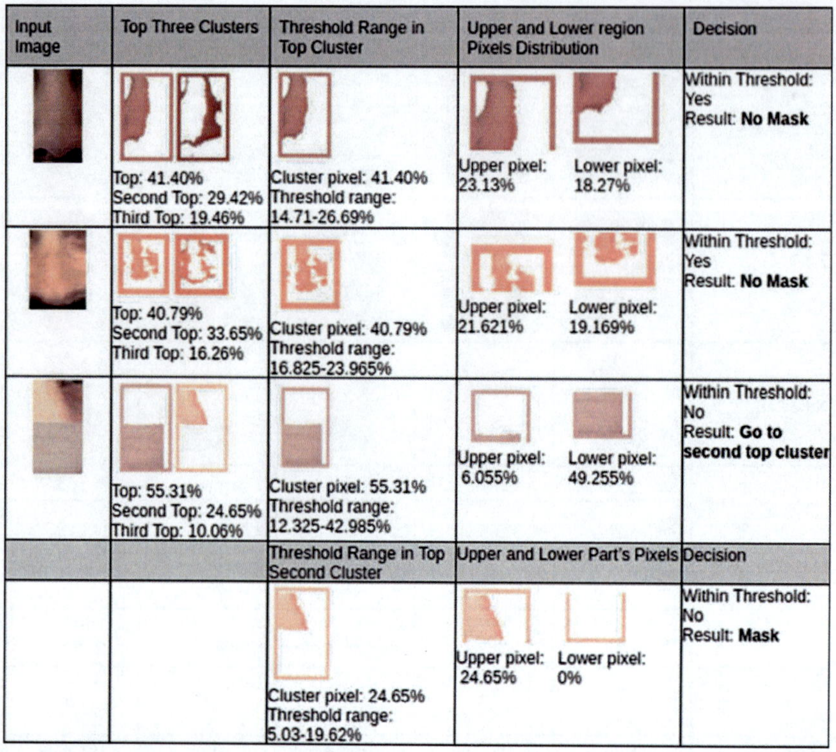

Fig. 6 Stage-2 results

Except for person detection and tracking, our system does not require expensive (in terms of runtime resources and training data) deep learning models. The system can in principle detect masks regardless of pattern or color. In particular, it overcomes the problem of differentiating people wearing skin-colored masks from those without masks. The overall performance of the model is 97.13% on a challenging dataset synthesized from multiple sources. The two-stage model outperforms a simpler single-stage model. In future work, we will consider different data distributions including faces wearing glasses and skin-colored masks covering the entire nose region of interest, which can be misclassified as no-mask faces by the current system.

References

1. Coronavirus disease 2019 (COVID-19): situation report-209. https://www.who.int/emergencies/diseases/novel-coronavirus-2019/situation-reports/. Last accessed 09 Jan 2020
2. Coronavirus disease (COVID-19) advice for the public, https://www.who.int/emergencies/diseases/novel-coronavirus-2019/advice-for-public. Last accessed 09 Jan 2020

3. Fong, S. J., Dey, N., & Chaki, J.: An introduction to COVID-19. In Artificial Intelligence for Coronavirus Outbreak (pp. 1–22). Springer, Singapore (2020)
4. Wang, Z., Wang, G., Huang, B., Xiong, Z., Hong, Q., Wu, H., ... & Chen, H.: Masked face recognition dataset and application. arXiv preprint arXiv:2003.09093 (2020)
5. Meenpal, T., Balakrishnan, A., & Verma, A.: Facial mask detection using semantic segmentation. In 2019 4th International Conference on Computing, Communications and Security (ICCCS) (pp. 1–5). IEEE (2019)
6. Jiang, M., & Fan, X.: RetinaMask: a face mask detector. arXiv preprint arXiv:2005.03950 (2020)
7. Ge, S., Li, J., Ye, Q., & Luo, Z.: Detecting masked faces in the wild with LLE-CNNS. In Proceedings of the IEEE conference on computer vision and pattern recognition (pp. 2682–2690) (2017)
8. Zhang, S., Zhu, X., Lei, Z., Shi, H., Wang, X., & Li, S. Z.: S3FD: Single shot scale-invariant face detector. In Proceedings of the IEEE international conference on computer vision (pp. 192–201) (2017)
9. King, D.E.: Dlib-ML: A Machine Learning Toolkit. the Journal of Machine Learning Research **10**, 1755–1758 (2009)
10. Adouani, A., Henia, W. M. B., & Lachiri, Z.: Comparison of Haar-like, HOG and LBP Approaches for Face Detection in Video Sequences. In 2019 16th International Multi-Conference on Systems, Signals & Devices (SSD) (pp. 266–271). IEEE (2019)
11. Adouani, A., Henia, W. M. B., & Lachiri, Z.: Comparative study of face detection methods in spontaneous videos. In 2019 International Conference on Control, Automation and Diagnosis (ICCAD) (pp. 1–6). IEEE (2019)
12. Zhang, K., Zhang, Z., Li, Z., Qiao, Y.: Joint Face Detection and Alignment using Multitask Cascaded Convolutional Networks. IEEE Signal Process. Lett. **23**(10), 1499–1503 (2016)
13. Jain, A. K., & Dubes, R. C.: Algorithms for clustering data. Prentice-Hall, Inc. (1988)
14. Kovac, J., Peer, P., & Solina, F.: 2D versus 3D Colour Space Face Detection. In Proceedings EC-VIP-MC 2003. 4th EURASIP Conference focused on Video/Image Processing and Multimedia Communications (IEEE Cat. No. 03EX667) (Vol. 2, pp. 449–454). IEEE (2003)
15. Zereen, A. N., Corraya, S.: Detecting real time object along with the moving direction for visually impaired people. In 2016 2nd International Conference on Electrical, Computer & Telecommunication Engineering (ICECTE) (pp. 1–4). IEEE (2016)
16. Face-Mask-Detection.: https://github.com/chandrikadeb7/Face-Mask-Detection. Last accessed 22 July 2020
17. Beard-Face-Dataset.: https://github.com/Keshav-Aggarwal/Beard-Face-Dataset. Last accessed 22 July 2020
18. Wojke, N., Bewley, A., & Paulus, D.: Simple online and realtime tracking with a deep association metric. In 2017 IEEE international conference on image processing (ICIP) (pp. 3645–3649). IEEE (2017)
19. Redmon, J., & Farhadi, A: YOLOV3: An Incremental Improvement. arXiv preprint arXiv:1804.02767.
20. Shape predictor.: https://github.com/codeniko/shape_predictor_81_face_landmarks. Last accessed 05 August 2020
21. 194 face landmarks, https://theonly1.tistory.com/1163,last accessed 2020/08/05.
22. Corraya, S., Uddin, J.: An efficient method for detecting electrical spark and fire flame from real time video. In International Symposium on Signal Processing and Intelligent Recognition Systems (pp. 359–368). Springer (2017)

MobileNet Mask: A Multi-phase Face Mask Detection Model to Prevent Person-To-Person Transmission of SARS-CoV-2

Samrat Kumar Dey, Arpita Howlader, and Chandrika Deb

Abstract Medical researchers around the globe provide evidence that COVID-19 pandemic diseases transmitted through droplets and respirators of respiratory aerosols and wearing a face mask is an efficient infection control recommendation process. In addition, many public and private service providers demand that consumers use the service only if they wear masks properly. However, a few research studies have been found on face mask detection based on the technology of Artificial Intelligence (AI) and Image Processing. In this article, we propose, MobileNet Mask, which is a deep learning-based multi-phase face mask detection model for preventing human transmission of SARS-CoV-2. Two different face mask datasets along with more than 5,200 images have been utilized to train and test the model for detecting with and without a face mask from the images and video stream. Experiment results show that with 770 validation samples MobileNet Mask achieves an accuracy of ~ 93% whereas with 276 validation samples it attains an accuracy of nearly ~ 100%. Lastly, we also discuss the possibility of implementing our proposed MobileNet Mask model on light-weighted computing devices such as mobile or embedded devices. Besides, this proposed model also introduces frontier technologies to support the efforts of government and public health guidelines with anticipation of implementing mandatory face mask regulations all over the world.

Keywords SARS-CoV-2 · Face mask · COVID-19 · MobileNetV2 · Deep learning

---

S. K. Dey (✉)
Dhaka International University, 66 Green Road, Dhaka 1205, Bangladesh
e-mail: sopnil.samrat@gmail.com

A. Howlader
Patuakhali Science and Technology University, Dumki 8602, Bangladesh
e-mail: arpita.cse@pstu.ac.bd

C. Deb
Cyber City Tower 2, Pune, Maharashtra, India
e-mail: chandrikadeb7@gmail.com

M. S. Kaiser et al. (eds.), *Proceedings of International Conference on Trends in Computational and Cognitive Engineering*, Advances in Intelligent Systems and Computing 1309, https://doi.org/10.1007/978-981-33-4673-4_49

1 Introduction

A few people used to wear face masks in order to protect them from air pollution. During COVID-19 epidemic, almost every healthcare professional and scientist recommended wearing face masks in public places. COVID-19 is arising issue from respiratory virus infections caused that emerging by Severe Acute Respiratory Syndrome coronavirus 2 (SARS-CoV-2). SARS-CoV-2 transmission caused through droplets that ejected from person's speaking, coughing, or sneezing with common droplet size 5–10 μm but aerosol emission and droplets emission increased when human speech and shout loudly [1]. Many laboratories setup research evidence that surgical masks were found to filter better than cloth mask filtered because large droplets evaporate and turn into minor aerosol particles such as 3 to fivefold smaller. Infected aerosol particles produced by patients coughing where this airborne virus aerosol atoms can be reduced (95% confidence interval [CI], 81–96%; $P = 0.03$) by wearing a surgical mask compared with no mask [2]. In many countries, some communities leisurely adopt any incipient convivial norm, in that case mandatory appropriate regulation of face mask implementation in public is big challenge. Mask detection on face refers to discover face position and whether face is covering with mask or not. Associating the problem narrowly to common object detection and face detection is to identify particular class of objects, i.e., face [3]. Artificial Intelligence (AI) technology support advanced video and image-based detection methods which can accurately detect objects and classify human in the camera's field of view and determine whether human face is wearing mask or not. Deep learning method with classical machine learning algorithm such as Decision Trees (DT), Support Vector Machine (SVM), and ensemble algorithm used for face mask detection using different datasets [4]. However, in recent time, some of the proposed AI technology-based methods and software tools are integrated with surveillance cameras to impede the increase of COVID-19 by detecting humans without face masks as in France (Paris Metro system). In this article, we introduce a multiple-phase face mask detector model that uses deep learning framework (MobileNetV2) to train a classifier. Two different datasets of face masks have been used to train and test the model. Apart from that, a pre-trained model of ResNet-10 architecture-based Single-Shot-Multibox Detector (SSD) has also been utilized in this exploration to detect faces from video stream. Several packages of Machine/Deep learning approaches and image processing techniques have also been followed along with the framework of OpenCV. Moreover, various steps were also employed including data augmentation, loading the classifier (MobileNetV2), building the fully-connected (FC) layer, pre-processing and loading the image data, applying classifier, training phase, validation, and testing phase. In order to provide effective monitoring and enable proactive, MobileNetV2 is an extremely effective framework for embedded systems which can also be used. The rest of this paper is organized as follows. In Sect. 2, we review relevant works based on Face Mask detection model and datasets. Description of different datasets is presented in Sect. 3. The proposed methodology is presented in Sect. 4. Section 5 describes experimental design, evaluation metrics, results, and discussion. Finally, Sect. 6 concludes the paper and discusses future work.

2 Relevant Work

Preceding many studies established the benefits of wearing a mask on face to prevent the spread of respiratory viruses of COVID-19 forasmuch large droplets dehydrate to minor aerosol particles that can hover for longer in air. Till now, there are many face recognition techniques designed based on deep/machine learning and only a few research studies have been found based on face mask detection using Artificial Intelligence (AI) technology. The author of this [5] article used YOLOv3 with the support of DarkNet-53 as backbone only for face detection. The accuracy obtained from the experiment was 93.9% with training on Large-scale CelebFaces Attributes (CelebA) and WIDER face dataset and testing on Face Detection Data Set and Benchmark Home (FDDB). In other research [6], the author combined LogitBoost (fits an additive logistic regression model) system and Gentle AdaBoost to detect face and mask, respectively, to identify whether medical staff are wearing surgical masks in the operation room or not. The system accomplished 95% accuracy with the objective of diminishing the false positive face detections. However, in this work [7] authors used Principal Component Analysis (PCA) for facial feature extraction and Viola-Jones for face detection. Accuracy of face resonation using the PCA is extremely affected by wearing masks which cause accuracy drops to less than 70%. In a very recent work [8] the authors build three face masked datasets including MFDD, RMFRD and SMFRD but the masked face images are not real rather the masked images are generated artificially. The developed datasets are widely used by the industry and academics in different study but the accuracy of face recognition is quite low and accuracy reduced further when wearing a mask initially 50%. In other work, authors have proposed a RetinaFaceMask [9] detection architecture that consist of ResNet and MobileNet as the backbone, FPN as the neck, and context attention modules as the heads initialized by kaiming's method that focuses on face and masks features.

3 Datasets Characteristics

The experimental result of this research was conducted on two different fundamental datasets. The first image dataset (IDS_1) is consists of three thousand eight hundred thirty five (3835) images belonging to two different classes including 1916 images of with_mask and 1919 images of without_mask. The sample images of IDS_1 were gathered from numerous sources including the Kaggle datasets, Bing search API, and the Real-World Masked Face Dataset (RMFD) [8].

Figure 1 illustrates the examples of faces with and without the masks. The second image dataset (IDS_2) is consists of one thousand three hundred seventy six (1376)

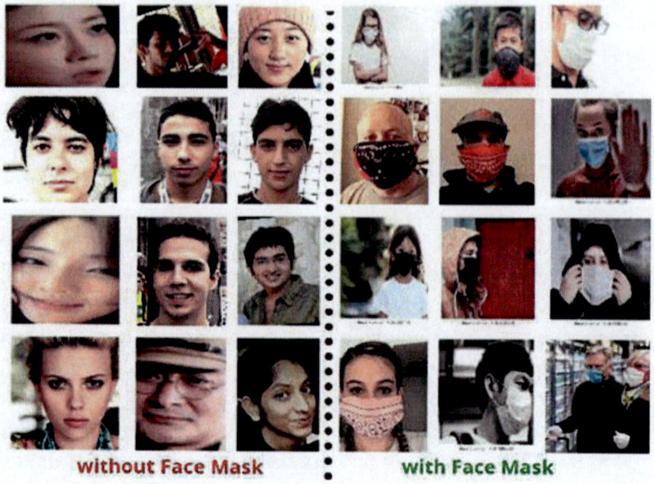

Fig. 1 Sample images of first face mask dataset (IDS_1)

images belonging to two different classes including 690 images of with_mask and 686 images of without_mask. IDS_2 is based on simulated or artificially created Masked Face Dataset (SMFD) [10]. Samples for images of the SMFD dataset are presented in Fig. 2. The first dataset (Image Dataset 1) and the SMFD (Image Dataset 2) were applied for general face recognition as well as face mask detection in training, validation, and testing phases. In terms of real-time responses, face mask detector may impede the increase of COVID-19.

Fig. 2 Sample images of face mask second dataset (IDS_2)

4 Methodology

A two-phase face mask detection model has been proposed where different datasets were reviewed to train our custom face mask detector. Finally, by using the support of the library of Keras and TensorFlow we will highlight how we have built a face mask detector model (github.com/chandrikadeb7/Face-Mask-Detection). Moreover, a two-fold approach is also implemented for detecting face masks in images as well as from real-time video streams. Following Fig. 3 highlights our proposed methodology of two phase based COVID-19 Face Mask Detector Model.

4.1 Face Mask Detector Training Phase

The methodology we proposed for the training of a custom face mask detector is split into two separate stages. One is training phase and the other is the deployment phase. In the training phase, model has been trained based on the available dataset. After finishing the training phase of face mask detector, we performed the face

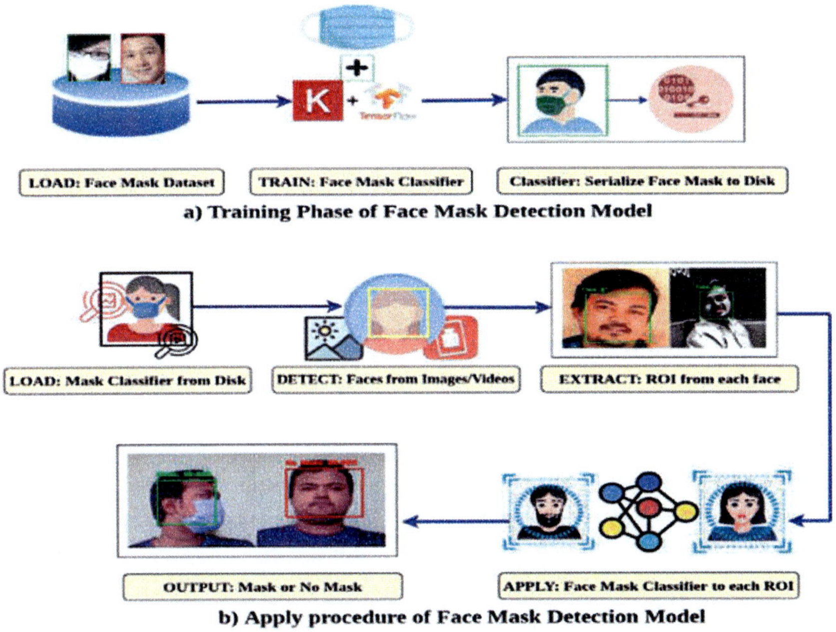

Fig. 3 Multiple stages and detailed phases for the development of a face mask detector model (images by the authors) **a** represent the training phase of face detection model whereas **b** shows how the apply procedure works in face detection model

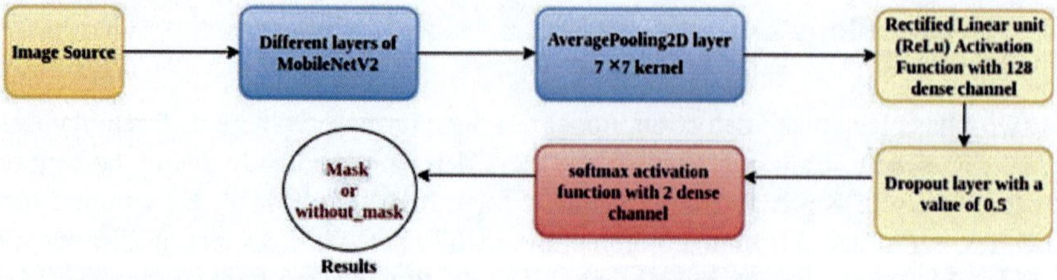

Fig. 4 Working procedure of customized fully connected layer of MobileNetV2 architecture

detection and then classified based on with_mask and without_mask. This exploration has used a deep learning system involving TensorFlow and Keras to automatically classify face mask wearing situations. To achieve this goal, a fine tuned mechanism on MobileNetV2 [11] architecture has been conducted. Several packages of machine/deep learning approaches and image processing libraries have been used in this research including OpenCV, scikit-learn, matplotlib, numpy, and many more to train the mask detector. To establish a baseline model that saves a considerable amount of time, a three-step process of fine-tuning has also been conducted to prepare the MobilenetV2 architecture. As our model is for binary classification (mask and without_mask) problem, therefore we have used binary cross-entropy, decay schedule of a learning rate, and Adam optimizer to compile our model. After the completion of the training phase, we evaluated the resulting model on the test set and generated a classification report for inspection. Finally, we serialized the face mask classification model to the disk.

4.2 MobileNetV2

The MobileNetV2 architecture is considered highly powerful and can be extended to embedded systems with minimal processing capability. Convolutional Neural Network (CNN) has a range of variants of pre-trained and well-architecture networks, such as AlexNet, ResNet, Inception Net, LeNet, MobileNet, etc. In this research, we have chosen a lightweight and very powerful mobile-oriented model which was MobileNetV2. Following Fig. 4 highlights the working procedure of MobileNetV2. Based on MobileNetV2 architecture we developed a customized fully connected layer which contains the four sequential layers on top of the architecture.

4.3 Face Mask Detector for Images

A three steps process of face mask detection from static images is employed after the training phase of the model.

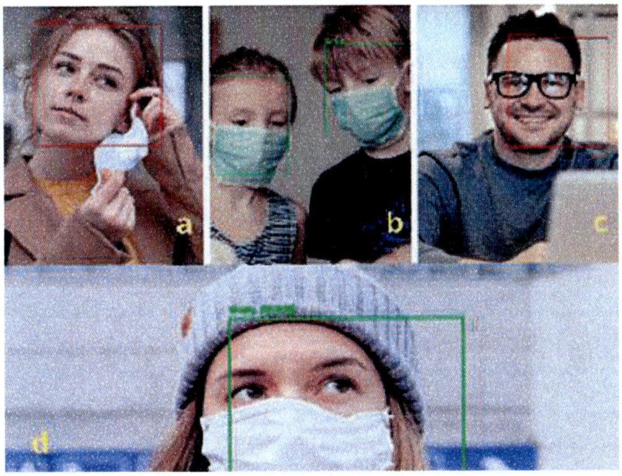

Fig. 5 Face Mask Detection from multiple static images (images free to use)

The first phase is to load an input image from the disk and second is to detect faces from the images. Figure 5 illustrates the face mask detection approach from the real images. Finally, a face mask detector was applied to identify the faces as with_mask or without_mask. Besides, the architecture of the proposed MobileNet Mask model is also capable of detecting face mask from the images of a group of people along with real-time video frame. In order to categorize whether or not a person is wearing a mask, however, face detection must be carried out first-if a face is not detected, the mask detector will not be applied.

4.4 Face Mask Detector for Webcam/Video Stream

A two-fold process of detecting masks from a webcam or video stream is also applied. In order to identify faces in the webcam a pre-trained model provided by the OpenCV framework was employed. The pre-trained model is based on Single-Shot-Multibox Detector (SSD) and utilizes the backbone of a ResNet-10 Architecture. Following Fig. 6 shows the step by step working procedure of face mask detection approach from video data.

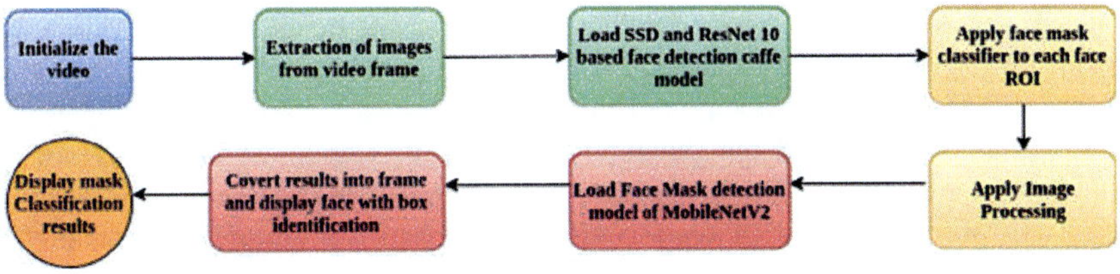

Fig. 6 Face mask detection workflow for webcam or video stream

5 Experimental Results

This section will highlight our experimental design procedure with available results. All the experimental trials have been conducted on Google Colab [12] which runs entirely in the cloud (Python 3 Google compute engine backend) and is equipped with more than 12 GB of RAM.

5.1 Evaluation Metrics

Different classifiers and performance metrics are needed to evaluate the performance to investigate the proposed model. The common performance measurement parameters are accuracy, precision, F1 score, and recall and they are accessible from the equations given below:

$$\text{Precision} = \frac{\text{TP}}{(\text{TP} + \text{FP})} \tag{1}$$

$$\text{Recall} = \frac{\text{TP}}{(\text{TP} + \text{FN})} \tag{2}$$

$$\text{Accuracy} = \frac{(\text{TP} + \text{TN})}{((\text{TP} + \text{FP}) + (\text{TN} + \text{FN}))} \tag{3}$$

$$\text{F1 score} = 2 * \frac{\text{Precision} * \text{Recall}}{(\text{Precision} + \text{Recall})} \tag{4}$$

5.2 Results Analysis

The experimental trials include two different datasets of sample images and MobileNetV2 architecture is used to conduct the experiment. For the experiment, all the datasets were split up to 80% for training and 20% for testing phase.

For the training phase, we have initialized our learning rate as (LR = 1e-4), number of training epochs were (EPOCHES = 20) and batch size was (BS = 32). In order to load and preprocess the training data, we resized our image data to 224 × 224 pixels by converting to array format and scaling the pixel intensities to the range of [−1, 1]. After that, we have encoded (using one-hot encoder method) the labels by partitioning our dataset and prepared it for data augmentation. Preprocessing of an input image is handled by the *blobFromImage* function in this experiment and resized the pixel values to 300 × 300 for performing the mean subtraction. To evaluate the performance of the classifier we have calculated the most appropriate performance measures. Figure 7 illustrates the different scenarios of detection approach of face masks or without face mask in real-time from a video stream. Also, Table 1 contains the value of performance metrics for understanding how proposed models behave during the testing phase. From the result analysis, it is evident that our proposed method of face mask detection based on different datasets performs quite significant despite having limited training data.

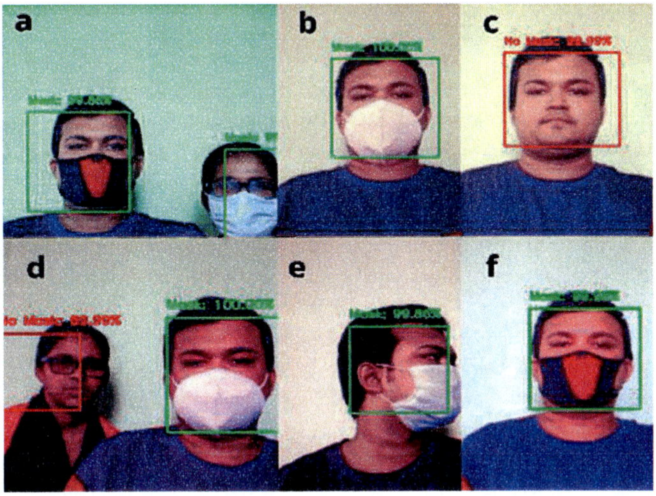

Fig. 7 Detecting face masks or without face mask in real-time from a video stream. These images contain three different categories of masks including surgical mask, KN95 mask, and homemade mask with clothes representing the different case scenario of face mask detection from a real-time video stream based on our proposed model. **a** shows the successful multiple face mask detection with more than 99% of accuracy. **b** detect the single face mask with an accuracy of 100% whereas **c** shows the successful detection of no mask in the video stream. **d** confirms the alteration in detecting mask or without mask from the video frame whereas **e** confirms that our model is capable of detecting faces from the sidewise also. However, **f** represents that our proposed model is so well trained that it can also recognize different patterns of mask successfully from a video stream (images by the authors)

Table 1 Results of different performance metrics based on the prediction on different testing dataset

Dataset category	Number of sample images	Status of images	Precision (%)	Recall (%)	F1 score (%)	Accuracy (%)	Validation sample
IDS_1	3835	with_mask	99	86	92	93	770
		without_mask	88	99	93		
IDS_2	1376	with_mask	100	99	100	100	276
		without_mask	99	100	99		

However, in IDS_1 where the data are generated from the real images of human faces, we achieved an accuracy of 93%, and in IDS_2 where the data are generated using artificial face masks produces an accuracy of almost 100% in the test set. Our model has also shown success with these findings in detecting face masks from the images outside of our training and test range.

6 Conclusion

The COVID-19 pandemic caused serious health crises as well as great loss and governments from all countries around the globe are struggling to fight against it. By means of strong encouragement of the government, the use of face masks in public with appropriate regulation can slow down the spread of the COVID-19 outbreak and diminish the hazard of the upcoming wave of the pandemic. This article proposed MobileNet Mask model that uses deep learning-based framework to train a classifier focused on the detection of face masks on two different face mask datasets. It is evident from the results; our proposed face mask detection model achieves higher accuracy than other state-of-the-art results available in the literature. In the later phase of this research, our aim is to integrate the proposed model with security camera to analyze video and images in order to provide effective monitoring and enable proactive people who are not wearing masks. Also, a real-time monitoring based mobile application can be developed through MobileNetV2 architecture.

References

1. Asadi, S., Wexler, A.S., Cappa, C.D., Barreda, S., Bouvier, N.M., Ristenpart, W.D.: Aerosol emission and superemission during human speech increase with voice loudness. Scientific Reports. **9**, 2348 (2019). https://doi.org/10.1038/s41598-019-38808-zhttps://doi.org/10.1038/s41598-019-38808-z
2. Driessche, K.V., Hens, N., Tilley, P., Quon, B.S., Chilvers, M.A., de Groot, R., Cotton, M.F., Marais, B.J., Speert, D.P., Zlosnik, J.E.A.: Surgical Masks Reduce Airborne Spread of Pseudomonas aeruginosa in Colonized Patients with Cystic Fibrosis. Am. J. Respir. Crit. Care Med. **192**, 897–899 (2015). https://doi.org/10.1164/rccm.201503-0481LEhttps://doi.org/10.1164/rccm.201503-0481LE
3. Kumar, A., Kaur, A., Kumar, M.: Face detection techniques: a review. Artif. Intell. Rev. **52**, 927–948 (2019). https://doi.org/10.1007/s10462-018-9650-2https://doi.org/10.1007/s10462-018-9650-2
4. Loey, M., Manogaran, G., Taha, M.H.N., Khalifa, N.E.M.: A hybrid deep transfer learning model with machine learning methods for face mask detection in the era of the COVID-19 pandemic. Measurement **167**, 108288 (2021). https://doi.org/10.1016/j.measurement.2020.108288https://doi.org/10.1016/j.measurement.2020.108288
5. Li, C., Wang, R., Li, J., Fei, L.: Face Detection Based on YOLOv3. In: Jain, V., Patnaik, S., Popențiu Vlădicescu, F., and Sethi, I.K. (eds.) Rec. Trends Intell. Comput. Commun. Devices. pp. 277–284. Springer, Singapore (2020). https://doi.org/10.1007/978-981-13-9406-5_34.

6. Nieto-Rodríguez, A., Mucientes, M., Brea, V.M.: System for medical mask detection in the operating room through facial attributes. In: Paredes, R., Cardoso, J.S., and Pardo, X.M. (eds.) Pattern recognition and image analysis. pp. 138–145. Springer International Publishing, Cham (2015). https://doi.org/10.1007/978-3-319-19390-8_16.
7. Ejaz, Md.S., Islam, Md.R., Sifatullah, M., Sarker, A.: Implementation of principal component analysis on masked and non-masked face recognition. In: 2019 1st international conference on advances in science, engineering and robotics technology (ICASERT). pp. 1–5 (2019). https://doi.org/10.1109/ICASERT.2019.8934543.
8. Wang, Z., Wang, G., Huang, B., Xiong, Z., Hong, Q., Wu, H., Yi, P., Jiang, K., Wang, N., Pei, Y., Chen, H., Miao, Y., Huang, Z., Liang, J.: Masked face recognition dataset and application. arXiv:2003.09093 [cs]. (2020)
9. Jiang, M., Fan, X., Yan, H.: RetinaMask: a face mask detector. arXiv:2005.03950 [cs]. (2020)
10. prajnasb: prajnasb/observations. (2020). https://github.com/prajnasb/observations. Last accessed 22 July 2020
11. Sandler, M., Howard, A., Zhu, M., Zhmoginov, A., Chen, L.-C.: MobileNetV2: inverted residuals and linear bottlenecks. In: 2018 IEEE/CVF conference on computer vision and pattern recognition. pp. 4510–4520 (2018). https://doi.org/10.1109/CVPR.2018.00474.
12. Carneiro, T., Medeiros Da NóBrega, R.V., Nepomuceno, T., Bian, G.-B., De Albuquerque, V.H.C., Filho, P.P.R.: Performance analysis of Google colaboratory as a tool for accelerating Deep Learning Applications. IEEE Access. 6, 61677–61685 (2018). https://doi.org/10.1109/ACCESS.2018.2874767

Facial Spoof Detection Using Support Vector Machine

Tandra Rani Das, Sharad Hasan, S. M. Sarwar, Jugal Krishna Das, and Muhammad Arifur Rahman

Abstract Face-based biometric authentication systems are being widely used nowadays and this use is on the rise. Because of its conveniences over the traditional access schemes, it is prevailing authentication systems. But this system has the drawback of being vulnerable to spoofing attacks. Attackers can bypass this system by presenting a printed photo or a photo in an electronic display in front of the system. In this work, we have tried to detect facial spoof. We have designed our model based on the fact that the recapturing of an image degrades the image quality. To assess the quality, we have extracted features from the images and created a feature vector. Both full reference and no reference methods have been used as an evaluator of image quality. A support vector machine (SVM) has been used for classifying the images using the feature vectors.

Keywords Spoof detection · Facial spoof · Image quality · SVM · Image classification · Feature vector · Feature extraction · Face recognition

---

T. R. Das · S. Hasan (✉) · S. M. Sarwar · J. K. Das
Department of CSE, Jahangirnagar University, Savar, Dhaka, Bangladesh
e-mail: hasansharad@gmail.com

T. R. Das
e-mail: tanudast95@gmail.com

S. M. Sarwar
e-mail: smsarwar96@gmail.com

J. K. Das
e-mail: cedas@juniv.edu

M. A. Rahman
Department of Physics, Jahangirnagar University, Savar, Dhaka, Bangladesh
e-mail: arif@juniv.edu

M. S. Kaiser et al. (eds.), *Proceedings of International Conference on Trends in Computational and Cognitive Engineering*, Advances in Intelligent Systems and Computing 1309, https://doi.org/10.1007/978-981-33-4673-4_50

1 Introduction

Biometric user authentication systems can be considered as a strong competitor of traditional systems that use username and password for authentication. Face recognition based schemes are getting much popularity, and the use of this technology has become easier because of in-built cameras in many present devices [13]. This scheme is being widely used in unlocking mobile phones, access control to secure facilities, automatic attendance, ATM identity validation, forensic investigation and so on. However, facial recognition systems are vulnerable to security attacks like facial spoofing [11]. In a spoofing attack, an imposter attempts to enter any face-based authentication system by presenting a fake face in front of the system's camera. Imposter can use a printed image of the real user's face (print attack) or a display device with a real user's image on it (replay attack) [4]. Because of the prevalent use of social media, our images are easily available on the Internet which increases the risk of facial spoof attack making it one of the strong methods of identity theft [18]. So, an efficient and secure method is necessary for facial spoof detection. In this paper, we have proposed a sophisticated,efficient and easily applicable method to detect facial spoof based on image quality analysis. That is, our model is able to classify whether an image is an original image or a spoofed image (maybe printed or on a digital screen). Our study is based on the fact that, when an image is recaptured, there occurs obvious degradation in the image quality. We have first extracted some quality measuring features from the images. After extracting the features, we have formed a feature vector. The feature vectors are fed into an SVM for classification. The speciality of our method is incorporating both full-reference and no-reference image quality based features while creating feature vector.

2 Literature Review

Unnikrishnan and Eshack [16] proposed a hybrid system for facial spoof detection that analyses image quality degradation and distortion present in spoofed images. The authors have extracted six features for distortion analysis and four features for quality assessment to create feature vectors which is then classified using neural network classifier. Their method provided better performance than the existing techniques and overcomes the problems of the ensemble classifier.

Diviya and Mishra [3] used Local Ternary Pattern (LTP) to extract textural features from facial images in their method. The extracted features are converted into upper pattern and local patterns to generate histograms. Histograms are then classified using SVM. Their LTP based approach overcomes the drawbacks of Local Binary Pattern (LBP) such as sensitivity to image noise and illumination change.

Garcia and Queiroz [4] searched for Moiré-Pattern caused by the overlapping of the digital grid of the authentication system's camera and grids present in digital media to detect 2D face spoofing. They have addressed that most of the energy of original images are present in their low frequencies but spoofed images have uncommon peaks present at higher frequencies. Authors have tried to find those unusual peaks. Their approach shows better results than the methods based on the empirical classification.

Boulkenafet et al. [18] argued that color texture can be an effective means of spoof detection because the color gamut of the various display medium (e.g. photographs, monitors) used for spoofing is different from the genuine face. Their proposed system extracts holistic texture descriptions from RGB, HSV, and CMYK color spaces and concatenates into enriched feature vectors which are then classified using SVM. Authors have claimed their approach to be more stable and promising in unknown situations compared to the existing methods.

Ruschel et al. [13] argued about the reliability of full-frame images than only face images since image environment contains rich information. To support their theory, they have compared these two approaches using a Convolutional Neural Network (CNN) model facilitated by GoogleNet architecture. The full-frame approach gained 2% higher accuracy than the only face approach, which the authors claim to be good enough considering the already high classification rate.

Tirunagari et al. [15] used Dynamic Mode Decomposition (DMD) for the first time in facial spoof detection. They have proposed a classification pipeline consisting of DMD, LBP, and SVM. Their approach captures visual dynamic cues like eye blinking, lips moving and facial dynamics using DMD. Then these are combined with LBP and classified using SVM.

Nguyen et al. [11] researched about the behavior of the distribution of noise's local variance (NLV) and found that this NLV distribution is different for fake and genuine faces. They have approximated this distribution by a Gamma distribution Γ (α, β) and exploited the parameters α and β as features in their proposed method. SVM has been used as the classifier here. The proposed method shows satisfactory result but is sensitive to the image acquisition system.

Chen et al. [2] have described that detailed facial textures are present in RGB space but it suffers from illumination sensitivity. On the other hand, facial information present in Multi-Scale Retinex (MSR) is not that detailed but it has no problem of illumination sensitivity and it captures high-frequency information effectively. In the proposed method, they have fed images from these two spaces to a two stream CNN to learn the discriminating features for spoof detection. They have used an attention-based fusion method to combine the features.

Killioğlu et al. [5] tried to detect spoofed faces with liveliness detection by developing a pupil direction observing system using basic hardware. Their system takes several snapshots from a live video. They have used the Haar-Cascade classifier to detect eye region. Kanade-Lucas-Tomasi (KLT) algorithm has been used to find feature points in the eye area and track the direction of eye area.

3 Methodology

In this study, we have proposed an effective and efficient method to distinguish real face images and spoofed face images based on the quality of the image. Our work can be divided into two parts. In the first part, we have extracted features associated with the image quality from the images and created a feature vector for each image. In the second part, we have fed the feature vectors into a support vector machine and classified the images as real or spoofed. The extracted features are peak signal to noise ratio (PSNR), mean square error (MSE), structural similarity (SSIM), blurriness, sharpness, wavelet coefficients, blind/reference less image spatial quality evaluator (BRISQUE), natural image quality evaluator (NIQE) and perception-based image quality evaluator (PIQE).

3.1 Feature Extraction

In total, we have extracted 9 features from the images associated with image quality. Both full reference and no-reference methods have been used to have the advantage of feature diversity and robust performance.

(i) Peak signal to noise ratio (PSNR): PSNR is the ratio of the maximum possible intensity of an image and the power of noise that corrupts an image and affects quality. PSNR has been calculated by comparing the image with a ground truth image.
(ii) Mean square error (MSE): The MSE of an image is calculated by taking the ground truth image as the ideal image and computing the cumulative squared error between our reference image and the ideal image. The formula is:

$$MSE = \frac{\sum_{M,N}[I_1(m,n) - I_2(m,n)]^2}{M * N} \tag{1}$$

Here, M and N represents the number of rows and columns of our subject image and the ground truth image.

(iii) Sharpness: Sharpness of the image is derived by calculating the probability of blurriness in the image edges and accumulating it for all the edges [10]. Based on this blurriness probability, the sharpness of the image is calculated.
(iv) Blurriness: To calculate the blurriness of the image we have blurred the image using Gaussian blur with sigma 4.2. Then the difference between the original image and the blurred image is calculated and the mean is taken. The formula is:

$$Blurriness = \frac{\sum_{M,N}[I(m,n) - B(m,n)]}{M * N} \tag{2}$$

Here, I is our subject image and B is the blurred version of the image. M and N represents the number of rows and columns of the images.

(v) Structural similarity (SSIM): Structural similarity between two images is the perceptual difference between the images. We compared both the real and fake images with our ground truth image.
(vi) Wavelet coefficients: We have taken the discrete wavelet transform of the image. Four wavelet coefficients are extracted. These coefficients are arrays of numbers. We have taken the average of the coefficients and stored them for training our model.
(vii) Blind/reference less image spatial quality evaluator (BRISQUE): It is calculated by statistical properties of normalized luminance coefficients and measuring the loss in natural properties of the image [9].
(viii) Natural image quality evaluator (NIQE): To calculate this index, the difference of the image from the natural statistical properties of an image is calculated [8].
(ix) Perception-based image quality evaluator (PIQE): This method is applied by extracting the local features for quality evaluation and mimicking the human perception of image quality [17].

3.2 Classification Using SVM

After extracting the features, we have concatenated those features and formed a feature vector for each of the images in our dataset. Then the feature vectors are fed into a support vector machine which is a popular supervised machine learning algorithm. We have used it for classifying real and spoofed images based on the extracted features. The classification problem we are dealing with is linearly separable, for which SVM is suitable. Besides, feature engineering is the cornerstone of SVM and we are working with extracted features from the images instead of directly using the images. That's why SVM is chosen for this classification problem. The support vector machine tries to find an optimal hyperplane among the classes and distinguish them. The closest points to the hyperplane are called the support vectors. The distances among the support vectors and the hyperplane are calculated. This distance is the margin for our classification. SVM tries to find a optimal hyperplane for which the margin is maximum.

A generalized block diagram for our model is in Fig. 1.

4 Data Set

We have used NUAA Photograph Imposter Dataset [14] for our study. It contains images of 15 people. All images are taken using a webcam having a resolution of 640 X 480 pixels. To formulate the spoofing attack, high-quality picture of the subject is

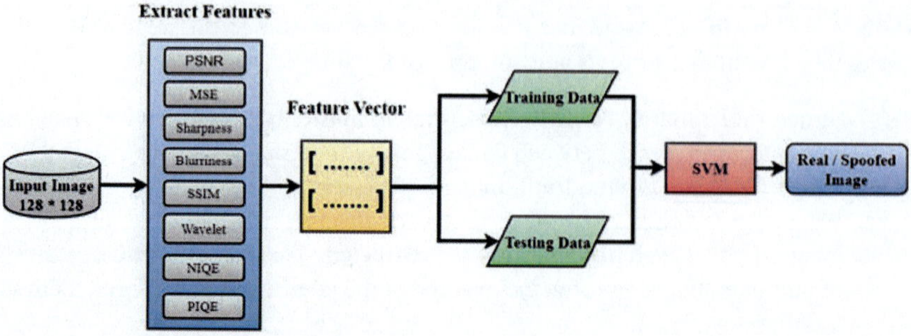

Fig. 1 Generalized block diagram of our model

taken with a Canon camera and it is printed on both traditional photography paper and normal A4 papers. The sample of each object is provided in separate directories. In total the dataset contains 5105 images of genuine face and 7504 images of fake faces.

5 Results and Analysis

We have applied our model on the NUAA Photograph Imposter Database. We have preprocessed the images and rescaled those into 128x128 pixels. 80% of the samples have been used to train our model and the rest 20% have been used for testing purpose. Some example images from the dataset we used are shown in Fig.2.

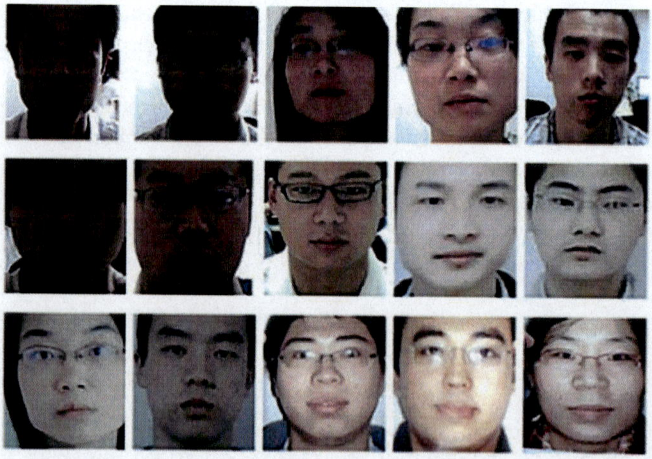

Fig. 2 Sample images from the dataset. The dataset has 5105 real images and 7509 spoofed images

Our proposed model has been able to efficiently distinguish fake and real faces. For NUAA Photograph Imposter Database our model has got 91% accuracy. It separates the fake and real faces successfully because we have used both full reference and no-reference methods for assessing the image quality. When an image is recaptured and printed in paper or displayed in the monitor of an electrical device, the quality of the recaptured image is degraded than the original ones. Using both full reference and no-reference image quality measurement techniques help in a better assessment of quality. As ground truth image for full reference methods, we have used one of the client images from the database. The following formula has been used to calculate the accuracy of our model:

$$Accuracy = \frac{tp + tn}{tp + tn + fp + fn} \tag{3}$$

where

tp = true positive, tn = true negative, fp = false positive, fn = false negative.

Figure 3 shows us the confusion matrix for our model. Here, class '1' is for real face and class '0' is for the fake face. The confusion matrix shows us that, among 1004 real faces of the test set, our model has classified 893 real faces correctly (true positive) and in the case of 111 faces the classification was wrong (false negative). Among 1519 fake faces, our model was able to identify 1395 fake faces correctly (true negative) and the other 124 was incorrect classification (false positive). The probability of the model to classify real and fake faces accurately can be understood from the receiver operating characteristics (ROC) curve given in Fig. 4. ROC curve is a metric for evaluating the classification performance. The area under the curve (AUC) specifies the ability of the model to correctly classify the instances. Higher the AUC, the better the model in distinguishing between fake and real faces.

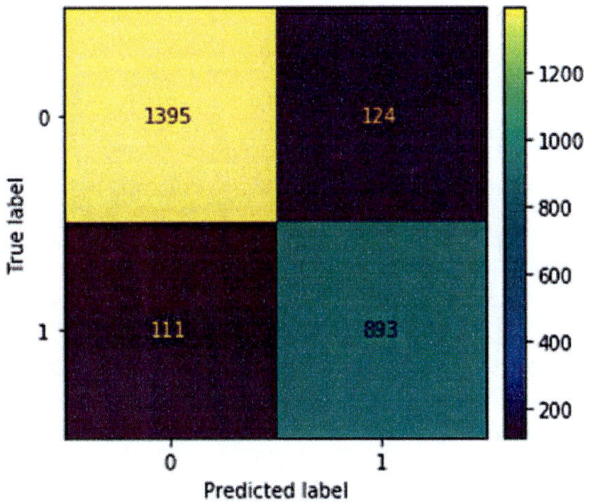

Fig. 3 Confusion matrix of our model. It is a summary of the prediction made by the model

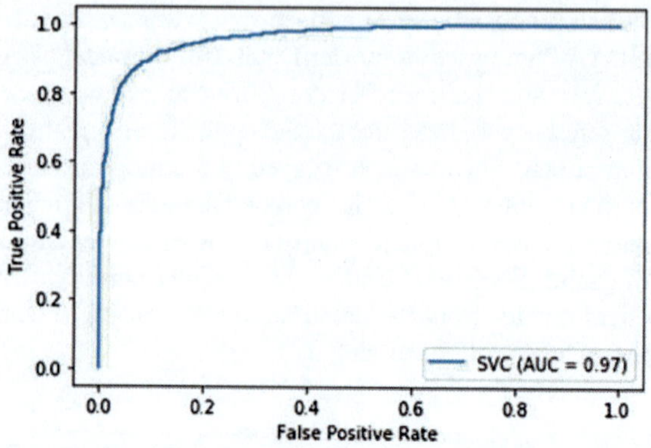

Fig. 4 Receiver operating characteristics (ROC) curve. As much the curve is close to the top left corner, it is more accurate in classifying samples. Area Under the Curve (AUC) shows how much the model is capable of distinguishing classes

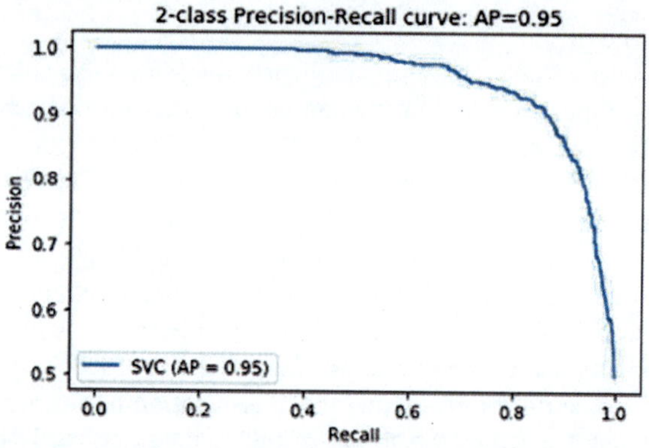

Fig. 5 Precision-Recall curve of our model. It plots the precision and recall for the different threshold values. The value of average precision (AP) summarizes the plot as a single value. For a no-skill classifier, AP is almost 0 and for a skilled classifier, it closes to 1

We can see, the ROC curve is near to the top left corner which indicates the performance of our model is better. We can see the AUC is 0.97. It is the probability that our model will highly rank a random positive sample than a random negative sample. In this case, it is pretty much high.

There are some other metrics like precision and recall to evaluate the classification performance of our model. Precision is the ratio of correct positive predictions to the total number of positive predictions. Recall is the proportion of actual positive instance that was identified correctly by the model. The precision-recall curve is presented in Fig. 5.

To determine the usability of our model, we have compared it with state of the art method. For the basis of comparison, we have taken the method proposed by Akbulut et al. [1]. Deep learning-based models [7] have been used here. Two deep models namely, CNN and Local Respective Fields – Extreme Learning Machine (LRF-ELM) are applied on the NUAA photograph imposter database. The comparison result is presented in Fig. 6.

Here, our SVM based model has outperformed CNN and LRF-ELM for this dataset mainly because of our selection of specific image quality based feature instead

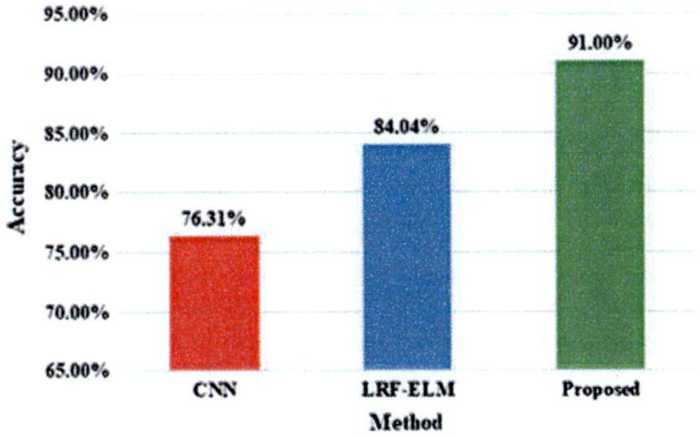

Fig. 6 This figure shows the comparative result of the spoof detection performance of our proposed image quality based method and the deep learning-based method [15]

Table 1 Summary of the detection performance of our model using different metrics. F1 score is the weighted average of the precision and recall which tries to seek the balance between these two. HTER is the average of false negative rate and false positive rate

Class	Precision	Recall	F1	HTER (%)	Accuracy
0 (Fake)	0.91	0.90	0.90	9.5	0.91
1 (Real)	0.86	0.86	0.86		

of automatic feature selection in deep learning based methods. The size of NUAA photograph imposter dataset is another reason. Deep learning based models require a huge amount of data. But SVM works well when dataset size is moderate. The dataset used in this study is not that big. So our SVM based method has outperformed CNN and LRF-ELM. Using both full-reference and no-reference method is another reason.

The precision, recall, f1 score, half the error rate (HTER) and the accuracy of the model is summarized in Table 1.

6 Discussion

As a summary, we can say, our model is good enough in distinguishing between real and fake faces. In 91% cases, it provides an accurate result. But there is still some challenges and scope of improvement. The devices for recapturing images are getting improved day by day and having the ability to reproduce high quality images. When the image is recaptured in a controlled and professional environment, the quality of original and recaptured images is almost the same and in that case, it becomes hard for the model to classify accurately. There is another type of spoofing technique called 3D mask attack which has not been covered in this study. We plan to incorporate this in future improvement. We have another plan to think this problem from non parametric modelling's i. e. Gaussian process [6, 12] perspective and evaluate the outcome.

7 Conclusion

When an image is recaptured, at least a minimum distortion occurs in the image, and the image quality is degraded. We have tried to implement this quality-based approach to distinguish between real and spoofed images. For quality measurement, 9 types of quality associated features are extracted. Both full reference and no-reference methods have been used to incorporate feature diversity and it has given a robust performance. For classification purpose, support vector machine has been used which finds the optimal hyperplane among the classes and classifies the images as real or spoofed. Our model has been 91% accurate in detecting spoofed faces for the NUAA photograph imposter dataset. According to the ROC curve, our model has 97% probability to accurately distinguish between real and fake images. Our model has outperformed some of the present states of the art methods.

References

1. Akbulut, Y.: Deep learning based face liveness detection in videos. In: 2017 International Artificial Intelligence and Data Processing Symposium (IDAP), vol. **2017**, pp. 1–4. IEEE (2017)
2. Chen, H., et al.: Attention-based two-stream convolutional networks for face spoofing detection. IEEE Trans. Inf. Forensics Secur. **15**, 578–593 (2019)
3. Diviya, M., Mishra, S.: A novel approach for detecting facial image spoofing using local ternary pattern. In: 2016 Second International Conference on Science Technology Engineering and Management (ICON-STEM), pp. 61–66. IEEE (2016)
4. Garcia, D.C., de Queiroz, R.L.: Face-spoofing 2D-detection based on moiré-pattern analysis. In: IEEE Trans. inf. Forensics Secur. **10**(4), 778–786 (2015)
5. Killioğlu, M., Taşkiran, M., Kahraman, N.: Anti-spoofing in face recognition with liveness detection using pupil tracking. In: IEEE 15th International Symposium on Applied Machine Intelligence and Informatics (SAMI), vol. 2017, pp. 000087–000092. IEEE (2017)
6. Lawrence, N.: Probabilistic non-linear principal component analysis with Gaussian process Latent variable models. J. Mach. Learn. Res. **6**, 1783–1816 (2005). ISSN 1532-4435. http://dl.acm.org/citation.cfm?id=1046920.1194904
7. Mahmud, M., Shamim Kaiser, M., Hussain, A.: Deep learning in mining biological data. In: arXiv (2020). eprint 2003. 00108 (q-bio.QM) (2020)
8. Mittal, A., Soundararajan, R., Bovik. A.C.: Making a completely blind image quality analyzer. IEEE Signal Process. Lett. **20**(3), 209–212 (2012)
9. Mittal, A.: Moorthy A. Bovik. In: No-Reference Image Quality Assessment in the Spatial Domain, pp. 4695–4708
10. Narvekar, N.D., Karam, L.J.: A no-reference image blur metric based on the cumulative probability of blur detection (CPBD). IEEE Trans. Image Process. **20**(9), 2678–2683 (2011)
11. Nguyen, H.P., et al.: Face spoofing attack detection based on the behavior of noises. In: 2016 IEEE Global Conference on Signal and Information Processing (GlobalSIP), pp. 119–123. IEEE (2016)
12. Rahman, M.A.: Gaussian process in computational biology: covariance functions for transcriptomics. Ph.D. thesis. The University of Sheffield (2018)
13. Ruschel, R., Schardosim, L.R., Scharcanski, J.: On the data conditioning for facial spoofing attacks detection using deep learning. In: 2019 IEEE International Instrumentation and Measurement Technology Conference (I2MTC), vol. **2019**, pp. 1–6. IEEE (2019)

14. Tan, X.: et al.: Face liveness detection from a single image with sparse low rank bilinear discriminative model. In: European Conference on Computer Vision, pp. 504–517. Springer (2010)
15. Tirunagari, S., et al.: Detection of face spoofing using visual dynamics. IEEE Trans. inf. Forensics Secur. **10**(4), 762–777 (2015)
16. Unnikrishnan, S., Eshack, A.: Face spoof detection using image distortion analysis and image quality assessment. In: 2016 International Conference on Emerging Technological Trends (ICETT), pp. 1–5. IEEE (2016)
17. Venkatanath, N., et al.: Blind image quality evaluation using perception based features. In: 2015 Twenty First National Conference on Communications (NCC), pp. 1–6. IEEE (2015)
18. Zinelabidine, B., Jukka, K., Abdenour, H.: Face spoofing detection using color texture analysis. In: IEEE Trans. Inf. Forensics Secur. **11**(8), 1818–1830 (2016)

Machine Learning Approach Towards Satellite Image Classification

Humayra Ferdous, Tasnim Siraj, Shifat Jahan Setu, Md. Musfique Anwar, and Muhammad Arifur Rahman

Abstract Classification of the image is an integral part of the digital picture and plays a very significant role in the development of remote sensing technologies. Thus the need to find sophisticated algorithms and methods have shown great interest over the years in solving classification issues. Remote sensing was implemented globally for the local usage of sophisticated satellite networks and sensors. However, the requirement to provide data and decision-taking was a major obstacle. It would be vitally necessary to implement machine learning methods for classification purposes to support the Graphics Processing Unit (GPU) systems to work faster. This paper proposes the supervised technique of machine learning systems, such as K-nearest neighbor (KNN), Artificial Neural Network (ANN), and Decision Tree, in this regard. KNN algorithm is based on matching and averaging non-local neighborhoods. Neural nets are inspired by the learning process that takes place inside human brains while instances are categorized in the decision tree by sorting them down the tree from the root to some leaf nodes. Preliminary findings by performing extensive experiments on satellite image dataset suggest that the proposed classification system may be a competitive alternative in terms of classification and accuracy over current feature-based extraction schemes.

Keywords Image Classification · Machine learning · Supervised Technique · KNN · Decision tree · Neural Network · Support Vector Machine

H. Ferdous (✉) · T. Siraj · S. J. Setu · Md. M. Anwar
Department of CSE, Jahangirnagar University, Dhaka, Bangladesh
e-mail: humayratasnu@gmail.com

T. Siraj
e-mail: tasnimsiraj07@gmail.com

S. J. Setu
e-mail: sjahans.cseju@gmail.com

Md. M. Anwar
e-mail: manwar@juniv.edu

M. A. Rahman
Department of Physics, Jahangirnagar University, Dhaka, Bangladesh
e-mail: arif@juniv.edu

M. S. Kaiser et al. (eds.), *Proceedings of International Conference on Trends in Computational and Cognitive Engineering*, Advances in Intelligent Systems and Computing 1309, https://doi.org/10.1007/978-981-33-4673-4_51

1 Introduction

Over the last decades, improvements in digital technology have made a great impact on storage devices as significant amounts of data can be stored. This huge amount of data includes not only texts but also images and videos. If multiple images have to be managed, as is the case with recordings of files or satellite images, it is supreme to set up data processing systems to help the consumer to locate relevant content. Classification and analysis of satellites have multi-dimensional uses in many specific fields including protection, geography, forestry, tracking of natural disaster damages, etc. Classifying of the images also plays a salient role in mapping pixels, objects, and other features found in satellite images. Over the years, remote sensing has allowed academics, society, policy, and other impacted citizens to benefit from its use in different fields of research. Over the years, remote sensing has gained much attention due to the development of image processing of satellite imagery. The human eye alone cannot cover large areas for analysis. In the event where the strategic development is required and if the information derived from the image is mistakenly labeled such as water instead of soil or vegetation, then the information would provide incorrect feedback for decision making. This type of problem shows that highly required techniques are capable of classifying images accurately. Therefore, it is very important to apply efficient techniques to classify images correctly. Nevertheless, machine learning can generate approaches that can collocate images. In this paper, we present a framework based on machine learning techniques that can automatically classify satellite images. Another notable contribution in our paper will be the performance evolution of different machine learning techniques using different datasets.

2 Literature Review

In the past, numerous researchers have established several classifiers. Tao et al. [11] proposed an unsupervised multi-layer feature learning-based solution that mines the framework details automatically from basic to complex, which fits the human visual cortex's hierarchical perceptual process. The modular design of this multi-layer feature extraction architecture is another significant property that can be easily applied to further layers. A Convolutional Neural Network (CNN) architecture-based classifier was proposed by Thao Nguyen et al. [16] that deals with the classification of satellite images which can be defined as the representation of the earth part. Dhingra et al. [8] have used the future lies dataset of satellite images to achieve better accuracy. They also focused to minimize classification error rate by applying different machine learning-based classifiers which are done on the basis of several types of bands that exist in the satellite images. A deep learning based approach Temporal Convolutional Neural Networks (TempCNNs) was presented by Charlotte et al. [19] which was used to learn temporal features automatically. The visual quality of the land cover maps provided by TempCNNs was also assessed by their suggested model. TempCNNs immersed themselves in their research as a very strong learner. The downside of their classification, however, was the incidence of salt and pepper noise and as a result, the spatial dimension needs to be taken into account to resolve the issue. A comparison between various classifications was performed in [24]. The authors randomly divided their training sample data into 14 separate datasets for evaluating the impact of training sample sizes and the performance of classification methods on the accuracy of classification. However, different machine learning classifiers perform equally in different methods.

Liu Qun, et al. considered the CNN architecture method with handcrafted features [13]. This architecture is used to solve a variety of problems in image classification. Using this framework, they have tried to fused spatial information obtained from handcrafted features as well as CNN feature maps where the accuracies were 99.90% and 99.84%, respectively. A technique to accurately select the appropriate scale for feeding the convolutionary neural network was developed by Noureldin et al [17]. Initially, the models with chosen scales for a dummy time were used in training CNNs to detect the best output. The next step is to perform a full training phase for the CNN model after choosing the best scale with the highest precision from the previous step. Again, embedding an adaptive scale mechanism in this step will result in more precision. Using deep satellite image analysis architectures, an unsupervised ordered clustering algorithm was suggested by Block et al. [5]. In their method, they promised to analyze images that have weak socio-economic and demographic ground, but the main downside was the requirement of an immense amount of computing power and storage. Tuba et al. has focused a recent bare bone fireworks algorithm based on k-means optimization which is used for image classification [25]. They represented that the proposed approach is better for image classification compared to the original k-means algorithm.

Authors in [21] observed that classifiers like Naive Bayes and Multilayer Perceptron performed well for classifying atmospheric images using satellite imagery corrected by the Atmospheric & Topographic Correction (ATCOR) method. They mainly analyzed coffee berry, which is located in the southern region of Brazil. They have performed 3 tests of the algorithm and the global accuracy, user accuracy, producer accuracy have been calculated. Le Hoang Thai et al. [23] applied Artificial Neural Network (ANN) and Support Vector Machine (SVM) techniques for the Roman numerals recognition application. Here, the preprocessing square image was used as input image, and 10-dimensional vector was considered. The ten elements and ten Roman numerals are corresponding to the output vector. Soo et al.[18] proposed the classification of images based on a neural network where the classification of features was form-based texture features that were derived from wavelet-transformed images. The neural network classifier was built using the backpropagation algorithm and other tasks such as data preprocessing, classification and feature extraction were employed to gain a classification rate of 81.7%. Biplab Banerjee et al [3] proposed a cluster-ensemble-based method using an unsupervised iterative expectation-maximization algorithm following a statistical model. They achieved better performance producing reduced overlapping among different clusters. This technique gave a result similar to a supervised technique, but it received less data collection costs. A review study [12] was conducted on a learning algorithm for the decision tree and other common algorithms. The research concentrated on evaluating features in the training process and extracting rules for evaluating features in the validation phase. Sunitha Abburu et al. [2] classified the images into three categories based on image pixel values. The results found by these categories were mainly dependent on datasets, machine learning algorithms as well as meaningful grouping of pixel values. The authors in [7] focused mainly on SVM, KNN and Naïve Bayes techniques on binary classification tasks. The results showed the comparable performance of the classifiers on most of the problems. They presented the SVM based linear kernel and found a large error e. They also showed that every result was highly dependent on the considered methodology.

3 Methodology

Our proposed method first performs the preprocessing of the images [1]. Next, feature extraction is performed for dimension reduction purposes using high priority features. We have adopted a supervised learning algorithm for training and testing instances. Classification algorithms based on the supervised technique are revised and applied as well. For comparison, we have employed some of the performance measurement techniques that can differentiate from other existing techniques. The techniques are listed below:

- Image analysis-converting images to digit values.
- Classification Techniques- KNN, ANN and Decision Tree.
- Performance Measurement- Accuracy Testing, Elapsed Time calculation.
- Finally, testing the algorithms on different datasets to see the consistency of performance.

3.1 Classification Methods

The problem of object classification is specified as an obstacle to classify the category or class to which the new observations belong to based on a training dataset comprising of observations whose category or class is identified.

3.1.1 K-Nearest Neighbor

KNN is an object-based classifying tool. This tool is used to train the closest instances of the domain. KNN considers storing only feature vectors and the training image labels. The unlabeled query is authorized to the label of it's K nearest neighbors in the classification process. [9].

When K = 1, the object is simply considered as the nearest object class. K must be an odd integer as there are only two classes. Although K is an odd integer, conducting multi-classification may still exist. After converting each image to a fixed-length vector with real numbers, we have applied Euclidean distance which is the most common distance function used in KNN. Given two points such as $a = (a_1, a_2, ..., a_n)$ and $b = (b_1, b_2, ..., b_n)$, the Euclidean distance d from either side can be represented as,

$$d(a, b) = \sqrt{\sum_{i=1}^{n} (b_i - a_i)^2} \quad (1)$$

3.1.2 Artificial Neural Network

ANN is an important tool in machine learning classification for the development of image classification. It is also used in other research areas such as in speech recognition, computer vision, Data Science applications like trust management model [14] etc. ANN is a graph interconnecting with different elements known as the nodes of the graph and these nodes have different numerical values known as weights. ANN graph has three main layers such as an input layer, a hidden layer and an output layer. The middle layer lied between the input and output layer known as the hidden layer. The hidden layer is also known as an individual's feature detector, recognizer and complex pattern solver [22]. The Back-propagation and K-means algorithm are two algorithms used in ANN. In this work, we have applied different number rounds to show better results of image classification.

3.1.3 Decision Tree

Decision Tree is generated using an algorithmic method that determines ways to isolate a dataset based on multiple circumstances. The purpose is to create a model that predicts the value of a target variable by learning simple rules of decisions inferred from the data characteristics. The target variable in tree models can take a distinct set of values which are called classification trees. In such tree structures, the leaves denote the class labels and branches depict conjunctions of attributes that correspond to individual class label. Each tree leaf is labeled with a class or a distribution of probability over the classes. The formula for finding the Gini impurity score of a particular level is (Fig. 1).

$$Gini = 1 - \sum_{p=1}^{q} (w_p)^2 \quad (2)$$

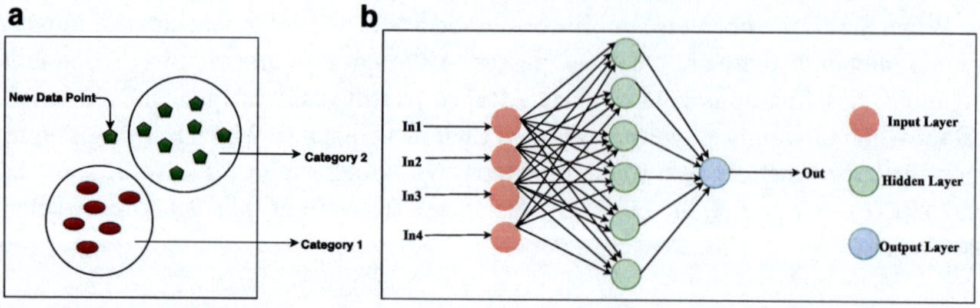

Fig. 1 (a) KNN classification with different values of K (b) Classification using Artificial Neural Network (ANN) which is represented with Input Layer, Hidden Layer, and Output Layer

Here, w_p is the ratio of class p in the node whose *Gini* index is being calculated. There is a total of q classes.

4 Data Set

We have used the satellite dataset [1] named as Statlog (Landsat Satellite) Data Collection. It includes data for pixel detection in satellite images. 4435 objects have been used for training and 2000 objects have been deployed for testing purposes. There are 36 dimensions for each object introducing 6 class labels. The last column is used to store the labels of the classes. The pendigits dataset [1] with 7494 training objects and 3498 test objects which covered 10 classes. The yeast dataset [1] has 484 test objects and 1000 training objects (Fig. 2).

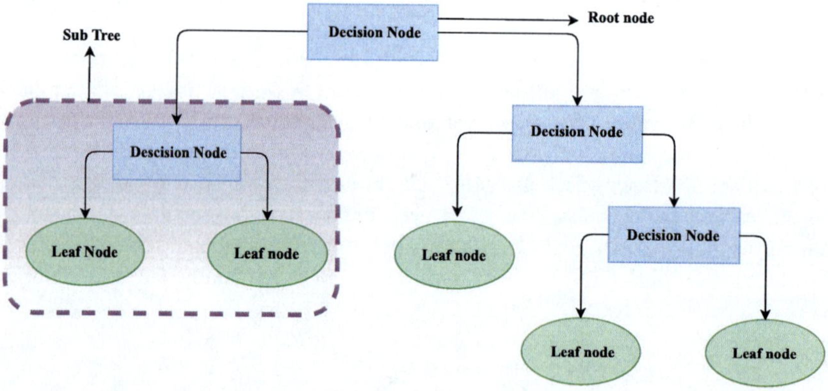

Fig. 2 Decision Tree for classification [6]

5 Results and Analysis

5.1 Result Using KNN

For the KNN classification, Euclidean distance has been employed. At first, the *mean* has been calculated for 36 dimensions and later, the *standard deviation* has been calculated. Since this algorithm relies on distance for classification, normalizing the training data instances improves its accuracy dramatically. We choose the values of K ranging from 1 to 10. The best classification accuracy of 90.50% is achieved at K = 7. Choosing the optimal K-value is salient as the large K-value overfits the model, whereas a smaller value of K underfits. Generally, for most datasets, the optimal value of K lies between 2–10. It produces better results than taking a single value of K, which is 1NN. Figure 3(a) summarizes the accuracy of KNN for different values of K.

5.2 Result Using ANN

For ANN, we can see that it performs better by using 2 layers, where the number of rounds is 120. We have also taken a different number of rounds and summarized the result in Fig. 3(b).

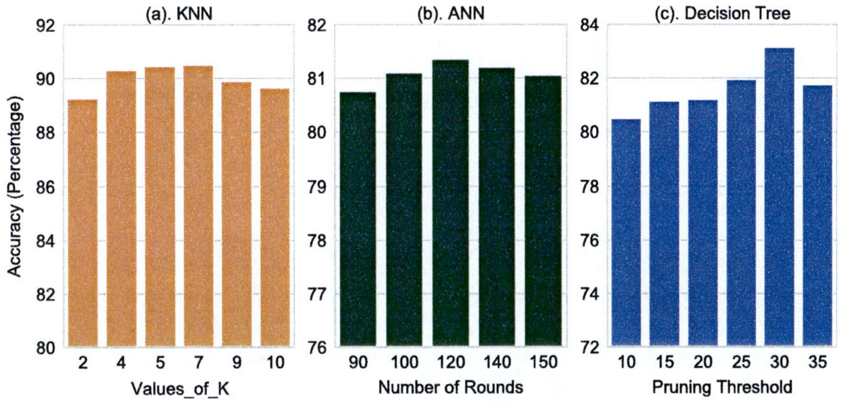

Fig. 3 (a) Accuracy of KNN for different values of K ranging from 2 to 10, plotted in the x-axis and accuracy in y-axis ranging from 80% to 92%. (b) Accuracy of ANN at the different number of rounds ranging from 90 to 150 plotted in the x-axis and accuracy ranging from 76% to 82% placed in the y-axis. (c) Accuracy of Decision Tree at different pruning thresholds ranging from 10 to 50 placed in the x-axis and accuracy ranging from 72% to 84% in the y-axis

5.3 Result Using Decision Tree

For the Decision tree, we have employed 'Information Gain' – a metric that is applied to quantify how good a split is and worked to train the Decision Tree. So far, by computing the information gain for each attribute, the best feature is selected as root. We used 'unique' thresholds that produce different splits. So thresholds have been used to prune trees. In machine learning, pruning is familiar to a search algorithm and a problem-solving optimizer. It is applied to reduce the final classifier complexity. We have adopted a pruning threshold ranging from 10 to 50. From Fig. 3(c), we can see a snapshot of the accuracy of the Decision tree. Here, the best accuracy is obtained by using pruning threshold = 30, giving an accuracy of 83.15%.

5.4 Comparison Among KNN, ANN and Decision Tree with Respect to Accuracy and Elapsed Time

After taking different values of parameters in KNN, ANN and Decision Tree, we evaluated five different values of K for KNN, a different number of rounds for ANN and five different values of thresholds for the Decision Tree. We measured the average accuracy for each classification method and the result is summarized in Fig. 4. According to Fig. 4(a), we can see that KNN has a maximum accuracy of 90.50% while the ANN has the least accuracy because KNN does not use assumptions about data.

From Fig. 4(b), KNN makes predictions just-in-time by calculating the similarity between an input sample and each training instance. The computation time in the Decision tree is faster compared to KNN. But the decision tree is prone to outliers and loses valuable information while handling continuous variables. ANN lags in

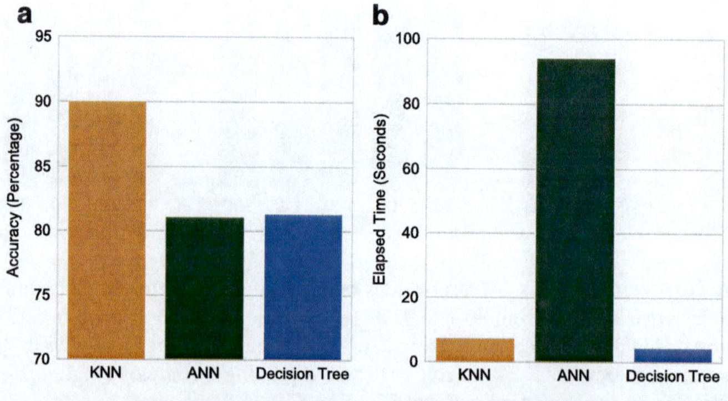

Fig. 4 (a) Accuracy of KNN, ANN and Decision Tree (b) Elapsed time of KNN, ANN and Decision Tree

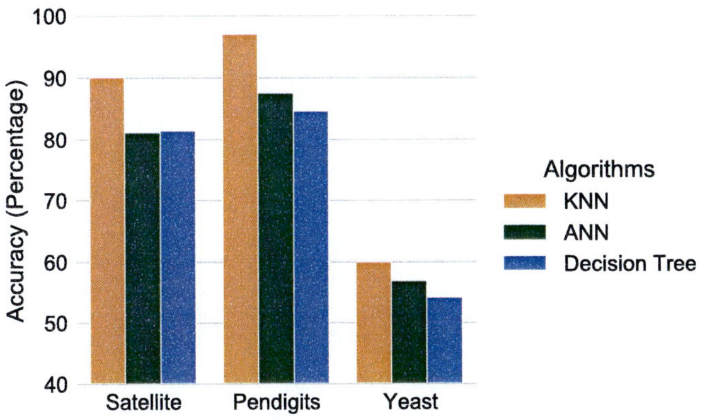

Fig. 5 Accuracy of KNN, ANN and Decision Tree on different datasets

terms of accuracy as well as elapsed time. ANN needs larger training data compared to KNN for achieving sufficient accuracy. Decision Trees are better than ANN, but ANN outperforms decision trees when there is adequate training data. So overall we can come to the conclusion that for satellite image classification, KNN works better in comparison to the other two algorithms.

5.5 Performance of Different Algorithms on Different Datasets

Here we can see that for the satellite image dataset, KNN worked better than the other two methods. We performed the same operation for the other two datasets. For pendigits and yeast datasets, the result is summarized in Fig. 5. Here we can also see that KNN outperforms the other two classification algorithms. So KNN works as a better classifier.

6 Conclusion

In this paper, we remarked that classification using KNN is giving much better results than classification using Decision Tree and ANN. Previously in the Kernel-based learning approach for satellite image classification, the accuracy of classification was around 90-92% [15] as their training data was not large enough. Our study shows a result of around 91% with a much bigger dataset. Therefore, KNN clearly outperforms the other two techniques. In future work, we would like to focus on developing better features that classify images more effectively and enhance our system by building an improved SVM model using more huge datasets or use a non-parametric framework like the Gaussian process [10, 20] to acquire higher classification results.

References

1. Index of / athitsos/courses/cse6363_spring2017/assignments/uci_datasets. http://vlm1.uta.edu/~athitsos/courses/cse6363_spring2017/assignments/uci_datasets/
2. Abburu, S., Golla, S.B.: Satellite image classification methods and techniques: a review. Int. J. Comput. Appl. **119**(8) (2015)
3. Banerjee, B., Bovolo, F., Bhattacharya, A., Bruzzone, L., Chaudhuri, S., Mohan, B.K.: A new self-training-based unsupervised satellite image classification technique using cluster ensemble strategy. IEEE Geosci. Remote Sens. Lett. **12**(4), 741–745 (2014)
4. Bishop, C.M.: Pattern Recognition and Machine Learning (Information Science and Statistics). Springer (2006)
5. Block, J., Yazdani, M., Nguyen, M., Crawl, D., Jankowska, M., Graham, J., DeFanti, T., Altintas, I.: An unsupervised deep learning approach for satellite image analysis with applications in demographic analysis. In: 2017 IEEE 13th International Conference on e-Science (e-Science). pp. 9–18. IEEE (2017)
6. Cavaioni, M.: Machine learning: decision tree classifier—By Michele Cavaioni—Machine learning bites—Medium. https://medium.com/machine-learning-bites/machine-learning-decision-tree-classifier-9eb67cad263e
7. Colas, F., Brazdil, P.: Comparison of SVM and some older classification algorithms in text classification tasks. In: IFIP International Conference on Artificial Intelligence in Theory and Practice, pp. 169–178. Springer (2006)
8. Dhingra, S., Kumar, D.: A review of remotely sensed satellite image classification. Int. J. Electr. Comput. Eng. **9**(3), 2088-8708 (2019)
9. Kim1, J., Kim, B., Savarese, S.: Comparing image classification methods: K-nearest-neighbor and support-vector-machines. In: Proceedings of the 6th WSEAS international conference on Computer Engineering and Applications, and Proceedings of the 2012 American conference on Applied Mathematics. vol. 1001, pp. 48109–2122 (2012)
10. Lawrence, N.: Probabilistic non-linear principal component analysis with Gaussian process latent variable models. J. Mach. Lear. Res. **6**, 1783–1816 (2005). http://dl.acm.org/citation.cfm?id=1046920.1194904
11. Li, Y., Tao, C., Tan, Y., Shang, K., Tian, J.: Unsupervised multilayer feature learning for satellite image scene classification. IEEE Geosci. Remote Sens. Lett. **13**(2), 157–161 (2016)
12. Liu, H., Cocea, M., Ding, W.: Decision tree learning based feature evaluation and selection for image classification. In: 2017 International Conference on Machine Learning and Cybernetics (ICMLC). vol. 2, pp. 569–574. IEEE (2017)
13. Liu, Q., Basu, S., Ganguly, S., Mukhopadhyay, S., DiBiano, R., Karki, M., Nemani, R.: Deepsat v2: feature augmented convolutional neural nets for satellite image classification. Remote Sens. Lett. **11**(2), 156–165 (2020)
14. Mahmud, M., Kaiser, M.S., Rahman, M.M., Rahman, M.A., Shabut, A., Al-Mamun, S., Hussain, A.: A brain-inspired trust management model to assure security in a cloud based iot framework for neuroscience applications. Cogn. Comput. **10**, 864–873 (2018)
15. Moorthi, S.M., Misra, I., Kaur, R., Darji, N.P., Ramakrishnan, R.: Kernel based learning approach for satellite image classification using support vector machine. In: 2011 IEEE Recent Advances in Intelligent Computational Systems. pp. 107–110. IEEE (2011)
16. Nguyen, T., Han, J., Park, D.C.: Satellite image classification using convolutional learning. In: AIP Conference Proceedings. vol. 1558, pp. 2237–2240. American Institute of Physics (2013)

17. Noureldin Laban, B.A., Ebied, H.M., Shedeed, H.A., Tolba, M.F.: Multiscale satellite image classification using deep learning approach. Mach. Learn. Data Mining Aerosp. Technol. **836**, 165 (2019)
18. Park, S.B., Lee, J.W., Kim, S.K.: Content-based image classification using a neural network. Pattern Recogn. Lett. **25**(3), 287–300 (2004)
19. Pelletier, C., Webb, G.I., Petitjean, F.: Temporal convolutional neural network for the classification of satellite image time series. Remote Sens. **11**(5), 523 (2019)
20. Rahman, M.A.: Gaussian process in computational biology: covariance functions for transcriptomics. Ph.D. thesis, University of Sheffield (2018)
21. da Rocha Miranda, J., de Carvalho Alves, M., Pozza, E.A., Neto, H.S.: Detection of coffee berry necrosis by digital image processing of Landsat 8 Oli satellite imagery. Int. J. Appl. Earth Observ. Geoinf. **85**, 101983 (2020)
22. Sootla, S., Matiisen, T.: Artificial Neural Network for Image Classification. Univ, Tartu (2015)
23. Thai, L.H., Hai, T.S., Thuy, N.T.: Image classification using support vector machine and artificial neural network. Int. J. Inf. Technol. Comput. Sci. **4**(5), 32–38 (2012)
24. Thanh Noi, P., Kappas, M.: Comparison of random forest, k-nearest neighbor, and support vector machine classifiers for land cover classification using sentinel-2 imagery. Sensors **18**(1), 18 (2018)
25. Tuba, E., Jovanovic, R., Tuba, M.: Multispectral satellite image classification based on bare bone fireworks algorithm. In: Information and Communication Technology for Sustainable Development, pp. 305–313. Springer (2020)

Anonymous Person Tracking Across Multiple Camera Using Color Histogram and Body Pose Estimation

Tasnuva Tabassum, Nusrat Tasnim, Nusaiba Nizam, and Shamim Al Mamun

Abstract Tracking anonymous persons is a necessary domain of the computer vision sector for security purposes. Nowadays, most of the important areas are covered with video surveillance cameras to tackle unwanted occurrences. Previous learning-based studies overlooked the checking of similarities of features of a person among multiple cameras for person detection and tracking. Only color feature-based studies may fail to recognize the same dressed different persons. In this paper, we are proposing a method for tracking and re-identification of a person in multiple cameras using color-based features with posture estimation in real-time scenarios. For this, we have used YOLOv3 for person tracking and OpenCV and OpenPose libraries for feature collection [16]. Model accuracy is tested on experiment video of single person multi-camera environmental scenario. In our proposal, we have taken HSV values of the most dominant color (DMC) for a targeted person on a camera as well as the fast Fourier transformed spectrum magnitude (FSM) as our feature vector to compare with another subject on different cameras. Our experiment shows that the re-identification of a person in different camera locations can be done successfully using cosine similarity and one ID is to be assigned to the same person and different IDs to different persons among the cameras to make the security issue more reliable, accurate, and faster.

Keywords FSM · FFT · Tracking · Color Histogram · Re-identification · Pose Estimation

1 Introduction

Person tracking and re-identification [19, 22, 24], a striking scope in computer vision, is an approach to test the positions to monitor a state's residents or transitory occupants, which depends on exceptional identifiers that are transiently or for all time

T. Tabassum · N. Tasnim · N. Nizam (✉) · S. Al Mamun
Institute of Information Technology, Jahangirngar University, Savar, Dhaka 1342, Bangladesh
e-mail: saranusaibanizam@gmail.com

M. S. Kaiser et al. (eds.), *Proceedings of International Conference on Trends in Computational and Cognitive Engineering*, Advances in Intelligent Systems and Computing 1309, https://doi.org/10.1007/978-981-33-4673-4_52

allocated to people like individual identifiers. The reasons for doing so are for government assistance and open security to mass observation. The absolute most normal inspiration for individual tracking items incorporates checking kids' whereabouts, babies in medical clinic nurseries, older family members, and crime control issues. Even though in the following reason there are examines where only color [2] or only posture [29] is utilized independently, a mixture of these two procedures is not frequently found. So, consolidating both color and posture for tracking persons in multiple cameras is the novelty of this study. This study aims at working on the problem of detection and tracking a person from multiple camera videos automatically by analyzing the features of the person.We have used YOLOv3 [19] algorithm and OpenCV to detect person, combining color histogram for extracting color features, OpenPose [10] for gesture pose estimation to track the person and cosine similarity for re-identification of a person.

A literature study on related work using single camera, multiple cameras, and re-identification studies are presented in section two, section three contains our proposed system model, feature extraction method, the method of pose estimation, the result of our proposed model are discussed in section four and conclusion is presented in section five.

2 Person Re-identification Review

2.1 *Single Camera Orientation*

In this section, we summarize a few works on single-camera based person tracking. These works take previously detected objects [19, 21, 24–28] as inputs using some object detection algorithm like YOLOv3, faster R-CNN [20], etc. [25] proposes a novel graph-based a formulation for linking and clustering person hypotheses over time by solving an instance of a minimum cost lifted multi-cut problem generalizing the mathematical model of [24] to express the fact that similar looking people are considered as the same person only if they are connected by at least one feasible track. High-performance detection and deep learning-based appearance feature lead to significantly better MOT results in an online and offline setting taking both detections and tracklets as input and calculating the affinity between tracklets and detections [7, 15, 17, 18, 28]. A person tracking algorithm fusing shadow signals in the RGB image with skeleton data, both of which are captured solely by a single RGB-D camera is proposed in [12]. Arbués-Sangüesa et al. [9] presents a novel multi-scale detection method to extract geometric and content features, resulting in a person tracking system for single-camera systems and low-powered embedded devices. Pedestrians are detected using the YOLOv3 algorithm and the detected target is tracked and predicted using the Kalman filter algorithm and the Hungarian algorithm in [27]. In single-camera tracking, correlations are poor when there is a significant pose change, significant occlusion, or abrupt motion, resulting in low identity recall [21], which can be solved by using Multi-Camera tracking.

2.2 Multi-camera Orientation

In this section, we summarize a few works on multiple-camera based person tracking. Color histogram depicts the picture from an alternate point of view by tallying comparative pixels and storing it and gives a distinctive outcome for the recovery task when the picture's direction and position or scales are changed in [2, 14]. An objective is set to foresee 1 second before the motion incorporating basic movements such as hand signal and walking movement utilizing OpenPose from OpenCV and YOLOv3 to remove features of a human body pose and to snip the main features in the bounding boxes with the movement of the person is anticipated with 98% of precision in [29]. Convolution neural networks (CNNs) is utilized to evaluate human posture in recordings using time-related data from numerous edges, prompting better execution, indicating that pre-sectioning the frontal area of the video improves execution and showing that even without frontal view divisions, the system figures out how to digest away from the background and can approximate the posture even within the sight of a composite, shifting circumstances in [3]. A suspicious tracking mechanism is designed across multiple cameras based on correlation filters and implemented using YOLO object detection framework, a camera correlation model, and a two-stage gait recognition strategy to deal with problems of re-identification across multiple cameras in [5, 6]. An intelligence video surveillance system is proposed that would use multiple cameras to track a person's movement and analysis that information to determine if their movement is deemed threatening or not detecting the abnormal behavior of that person by using three architectures for object detection, template matching, and decision making. Motion detection and recognition, trajectory analysis, pattern matching, and understanding the actions and interactions of the observed objects based on the information acquired by sensors are proposed to be used in [8].

A feature extraction network named NCA-Net developed by combining deep feature learning and metric learning to acquire a high-level representation of pedestrian is designed to improve the performance of multiple objects tracking across multiple cameras in [23]. In [13], a deep Multi-Shot neural model is presented for measuring the Degree of Appearance Similarity (MS-DoAS) between person observations. For dealing with complex person identification to obtain robust tracking in variate real-world scenarios, the MS-DoAS model has been trained. This model relies on the temporal consistency of the agent's appearance by analyzing the visual features measured in previous frames and also shows versatility and robustness while tested with multiple complex scenarios. A weighted two-stream preparing target function is proposed in [11] which consolidates the Siamese expense of the spatial and temporal streams to foresee an individual's personality by removing the features for each picture in the grouping and calculating the average which is utilized by the metric learning techniques.

Table 1 Object Detection and Tracking Methods

Ref.	Methods used for detection and tracking
[28]	MOT result, Tracklets, Affinity calculation
[9]	Geometric andcontent features, Single-camera systems
[27]	YOLOv3 algorithm, Kalman filter algorithm, Hungarian algorithm
[5]	YOLO algorithm, Correlation filter, Two stage gait analysis
[8]	Motion detection and recognition, Trajectory analysis, Pattern matching
[23]	NCA-Net, Deep feature learning, Metric learning

2.3 Person Re-identification

A re-identification approach is developed by using the information contained in automatically detected attributes in [22]. Based on Convolutional Neural Network (CNN), an attribute classifier on separate data is trained for this purpose which includes its responses into the training process of this person re-id model. This approach achieved state-of-the-art results on three public datasets for re-identifying persons. Tracklets are introduced to their system for reducing the state space but mistakes introduced by the tracklet generation are likely to be propagated to the final solution in [26, 28]. A tracking method is proposed utilizing tracklets called TrackletNet Tracker that combines temporal and appearance information as a unified framework in [26]. Both tracklets and detection are used as input and new tracklets and detection is produced in [28]. Here, in Table 1 some methods of object detection and tracking are listed.

3 Proposed Model Architecture

The proposed system for tracking anonymous persons across multiple cameras can largely be divided into three subsections of person detection, feature extraction and tracking the detected person.

Figure 1 represents the proposed system from which it's seen that video sequences from two of the cameras are passed through pre-trained YOLOv3 architecture through which persons are detected with bounding boxes. For collecting features from these detected persons, color extraction and pose estimation are used in the next stage. Finally, in the classification section cosine similarity of the extracted features is calculated. A higher value of cosine similarity indicates that the persons of different cameras are the same where a lower value indicates the opposite. Based on this decision, the same IDs are assigned to the same person and different IDs to different persons.

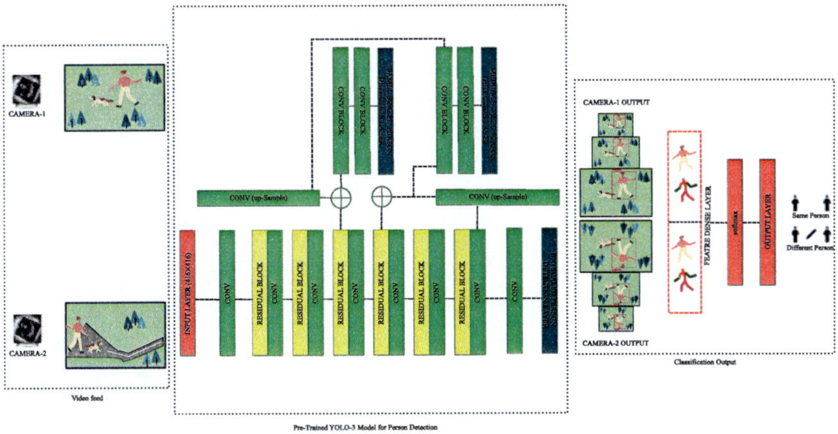

Fig. 1 System Model for Person tracking with re-identification in multiple cameras

3.1 Color Features

The color histogram represents the color distribution in a digital color image which can be visualized as a graph or plot. Figure 2 represents the dominant color bar for an image on the left side.

The dominant color bar of Fig. 2 shows the three most dominant colors for the persons' image. To get this dominant color bar, the k-means clustering algorithm is applied to cluster the pixels in an image where the same kind of pixels is stored within a cluster. HSV (Hue, Saturation, Value) values from every cluster are used in this proposed model for further matching purposes. In our study, we have used a dominant color bar to extract color features from the detected person which can give a clue for tracking the same person across multiple cameras.

Fig. 2 Most dominant color bar (right) for the image (left)

Fig. 3 Joint points used for pose estimation

3.2 *Pose Estimation Features*

Human pose estimation refers to the process of predicting the positions of a person's joints i.e. elbow or knee joints in an image or video. Figure 3 shows joints that are normally used for pose estimation. 2D pose estimation simply estimates the x and y coordinates for the location of key points in 2D space relative to an image or video frame. The pose can be estimated both in the top-down and bottom-up approaches. The top-down approach will be applied in this study starting with identifying and localizing a person instances using object detection first and then estimate the pose of a person. Angles of both ankles and elbows are calculated from the corresponding coordinates of the joint points. After plotting the angle values in a graph, it is converted into its spectrum magnitude by the Fast Fourier Transform algorithm. Estimated pose using OpenPose will be another feature set for tracking the same person across multiple cameras in this study.

4 Result and Discussion

4.1 *Data Collection and Prepossessing*

For evaluating our proposed model, we captured some videos from a busy road. Videos were captured using two different cameras and from two opposite sides of a road for getting both the front and backside view of the same pedestrians. The reason behind using multiple cameras for capturing videos from two different sides of the road was for evaluating if our proposed model can track anonymous persons from different angles. Before using the videos in the evaluation we extracted the unnecessary portions of the video to reduce the duration of a video. For evaluating purposes, we used Python IDE, Python programming language, and its libraries.

4.2 *Extracting Persons Using YOLOv3*

To extract a person from the captured video, the YOLOv3 algorithm was applied by which persons were detected with a rectangular bounding box with random IDs. After detecting persons, every bounding box was cropped which is shown in Fig. 4 to extract color features from it. The cropping area was determined by using the upper-left coordinate, width, and height of the rectangle. For cropping the bounding boxes and plotting the cropped area 'OpenCV' and 'matplotlib' library were used.

4.3 *Matching Color Feature with Pose Feature*

To preserve color features from the cropped bounding boxes of persons, dominant color bars were used. We needed OpenCV, Python, and k-means clustering algorithm for getting this dominant color bars. K-means clustering algorithm takes the pixel values of the image as data points and clusters them in three sections where the number of the section was specified by us. Pixels clustered in one region are more similar in color than pixels in other regions. For showing the clusters in a color bar,

Fig. 4 Person detection with unique ID using YOLOv3 and bounding box cropping

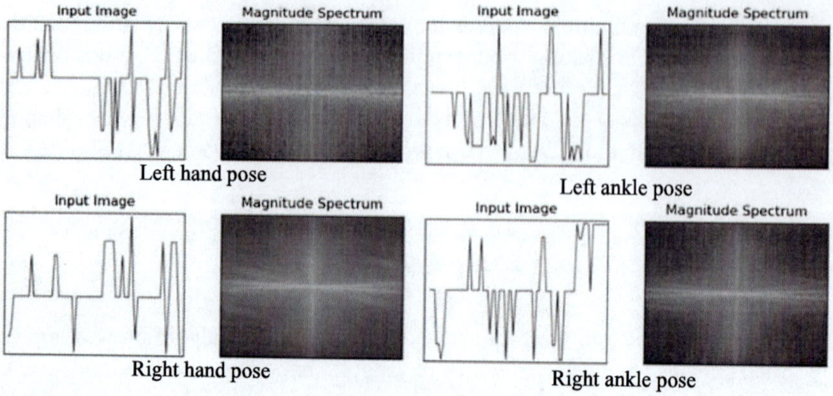

Fig. 5 Pose detection of a subject and it's spectrum magnitude

a centroid histogram was created from the number of pixels assigned to each cluster. Finally, the dominant color bar was shown using the 'matpoltlib' library. Figure 2 shows dominant color bar for a persons. HSV(Hue, Saturation, Value) values of this dominant color bar are preserved in an array for calculating cosine similarity.

We used OpenPose for pose feature estimation of pedestrians. A deep learning model pre-trained Dataset(MPII) [4] and COCO dataset [1] was used for this purpose. Figure 5 shows the graph of detected pose for both hands and ankles with the spectrum magnitude of a person. The spectrum magnitude for the graphs was calculated using the Fast Fourier Transform algorithm. Spectrum magnitude returns an array of values which was also preserved for cosine similarity.

Finding the cosine similarity of these preserved arrays of dominant color and spectrum magnitude for different persons decides if two persons are the same or different who were captured from multiple cameras. A similar ID will be provided for the same person that appeared in multiple cameras decided by the cosine similarity. The time needed in the detection stage was related to the number of persons in the frame. The total processing time can be varied in case of working with a good processor and also with the length of the captured video.

4.4 Experiment

We have taken five videos with two cameras. The color and pose features were from one camera and matched the features in another one. We have tested our system with a single person re-identification with color-based histogram and pose spectrum magnitude features. Figure 6 shows the confusion matrix of our result, where we got overall accuracy is 81.4%. for single person re-identification.

Fig. 6 Confusion Matrix

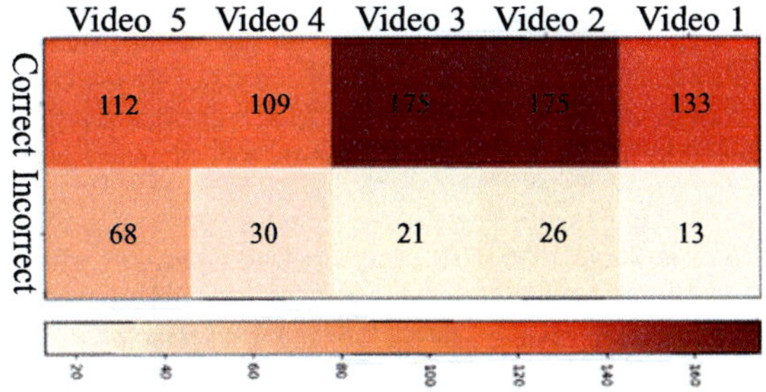

5 Conclusion

Our proposed model can track and re-identify people in a multi-camera setting using the color histogram and OpenPose. We have used both color features and the skeleton feature to increase the accuracy of our model. We used multiple cameras to detect and track people from different angles in a wide area. In this paper, we are proposing a model of person re-identification and tracking throughout multi-camera to evaluate the accuracy level by using cosine similarity matching. Though our proposed system is for a simple single person based scenario, in the future we want to do it for more complex scenes with a comparison with other existing systems. We also wish to apply low-level face recognition with this model as an enhancement to make the system more robust and reliable.

References

1. COCO—Common Objects in Context. https://cocodataset.org/#keypoints-2018
2. Color feature with spatial information extraction methods for CBIR: a review
3. Deep Convolutional Neural Networks for Efficient Pose Estimation in Gesture Videos. https://link.springer.com/chapter/10.1007/978-3-319-16865-4_35
4. MPII Human Pose Database. http://human-pose.mpi-inf.mpg.de/
5. STAM-CCF: Suspicious tracking across multiple camera based on correlation filters. https://www.ncbi.nlm.nih.gov/pmc/articles/PMC6651151/
6. Al Banna, M.H., et al.: Camera model identification using deep cnn and transfer learning approach. In: Proceedings of ICREST, pp. 626–630. IEEE (2019)
7. Al Mamun, S., Fukuda, H., Lam, A., Kobayashi, Y., Kuno, Y.: Autonomous bus boarding robotic wheelchair using bidirectional sensing systems. In: Bebis, G., Boyle, R., Parvin, B., Koracin, D., Turek, M., Ramalingam, S., Xu, K., Lin, S., Alsallakh, B., Yang, J., Cuervo, E., Ventura, J. (eds.) Advances in Visual Computing. Lecture Notes in Computer Science, pp. 737–747. Springer, Cham (2018)
8. Alshammari, A., Rawat, D.B.: Intelligent multi-camera video surveillance system for smart city applications. In: 2019 IEEE 9th Annual Computing and Communication Workshop and Conference (CCWC), pp. 0317–0323 (2019)
9. Arbués-Sangüesa, A., Haro, G., Ballester, C.: Multi-person tracking by multi-scale detection in Basketball scenarios. arXiv:1907.04637 [cs] (2019)

10. Cao, Z., Hidalgo, G., Simon, T., Wei, S.E., Sheikh, Y.: OpenPose: realtime multi-person 2D pose estimation using part affinity fields. arXiv:1812.08008 [cs] (2019)
11. Chung, D., Tahboub, K., Delp, E.J.: A Two stream Siamese convolutional neural network for person re-identification. In: 2017 IEEE International Conference on Computer Vision (ICCV). pp. 1992–2000. IEEE, Venice http://ieeexplore.ieee.org/document/8237480/ (2017)
12. Gai, W., Qi, M., Ma, M., Wang, L., Yang, C., Liu, J., Bian, Y., de Melo, G., Liu, S., Meng, X.: Employing shadows for multi-person tracking based on a single RGB-D camera. Sensors **20**(4), 1056 (2020). https://www.mdpi.com/1424-8220/20/4/1056 (number: 4 Publisher: Multidisciplinary Digital Publishing Institute)
13. Gómez-Silva, M.J.: Deep multi-shot network for modelling appearance similarity in multi-person tracking applications. arXiv:2004.03531 [cs] (2020).
14. Kaiser, M.S., Chowdhury, Z.I., Mamun, S.A., Hussain, A., Mahmud, M.: A neuro-fuzzy control system based on feature extraction of surface electromyogram signal for solar-powered wheelchair. Cognit. Comput. **8**(5), 946–954 (2016). https://doi.org/10.1007/s12559-016-9398-4
15. Kaiser, M.S., et al.: Advances in crowd analysis for urban applications through urban event detection. IEEE Trans. Intell. Syst. **19**(10), 3092–3112 (2017)
16. Mahmud, M., Kaiser, M.S., Hussain, A.: Deep learning in mining biological data. arXiv:2003.00108 (2020)
17. Mahmud, M., Kaiser, M.S., Hussain, A., Vassanelli, S.: Applications of deep learning and reinforcement learning to biological data. IEEE Trans. Neural Netw. Learn. Syst. **29**(6), 2063–2079 (2018)
18. Mamun, S.A., Lam, A., Kobayashi, Y., Kuno, Y.: Single laser bidirectional sensing for robotic wheelchair step detection and measurement. In: Huang, D.S., Hussain, A., Han, K., Gromiha, M.M. (eds.) Intelligent Computing Methodologies. Lecture Notes in Computer Science, pp. 37–47. Springer International Publishing, Cham (2017)
19. Redmon, J., Farhadi, A.: YOLOv3: An Incremental Improvement, p. 6
20. Ren, S., He, K., Girshick, R., Sun, J.: Faster R-CNN: towards real-time object detection with region proposal networks. arXiv:1506.01497 [cs] (2016)
21. Ristani, E., Tomasi, C.: Features for multi-target multi-camera tracking and re-identification. In: 2018 IEEE/CVF Conference on Computer Vision and Pattern Recognition, pp. 6036–6046 (2018). ISSN 2575-7075
22. Schumann, A., Stiefelhagen, R.: Person re-identification by deep learning attribute-complementary information. In: 2017 IEEE Conference on Computer Vision and Pattern Recognition Workshops (CVPRW), pp. 1435–1443. IEEE, Honolulu, HI, USA (2017). http://ieeexplore.ieee.org/document/8014920/
23. Tan, Y., Tai, Y., Xiong, S.: NCA-net for tracking multiple objects across multiple cameras. Sensors **18**, 3400 (2018)
24. Tang, S., Andres, B., Andriluka, M., Schiele, B.: Multi-person tracking by multicut and deep matching. arXiv:1608.05404 [cs] (2016)
25. Tang, S., Andriluka, M., Andres, B., Schiele, B.: Multiple people tracking by lifted multicut and person re-identification. In: 2017 IEEE Conference on Computer Vision and Pattern Recognition (CVPR). pp. 3701–3710. IEEE, Honolulu, HI (2017). http://ieeexplore.ieee.org/document/8099877/
26. Wang, G., Wang, Y., Zhang, H., Gu, R., Hwang, J.N.: Exploit the connectivity: multi-Object tracking with TrackletNet. arXiv:1811.07258 [cs] (2018)
27. Yi, Z., Shen, Y., Zhao, Q.: Multi-person tracking algorithm based on data association. Optik **194**, 163124 (2019). http://www.sciencedirect.com/science/article/pii/S0030402619310150
28. Yu, F., Li, W., Li, Q., Liu, Y., Shi, X., Yan, J.: POI: multiple object tracking with high performance detection and appearance feature. arXiv:1610.06136 [cs] (2016)
29. Yunus, A.P., Shirai, N.C., Morita, K., Wakabayashi, T.: Human motion prediction by 2D human pose estimation using OpenPose (2020). https://easychair.org/publications/preprint/8P4x

Quantification of Groundnut Leaf Defects Using Image Processing Algorithms

Check for updates

Ashraf Mahmud, Balasubramaniam Esakki, and Sankarasrinivasan Seshathiri

Abstract Identification, classification, and quantification of crop defects are of paramount interest to the farmers for preventive measures and decrease the yield loss through necessary remedial actions. Due to the vast agricultural field, manual inspection of crops is tedious and time-consuming. UAV based data collection, observation, identification, and quantification of defected leaves area are considered to be an effective solution. The present work attempts to estimate the percentage of affected groundnut leaves area across five regions of Andhra Pradesh using image processing techniques. The proposed method involves color space transformation combined with a thresholding technique to perform the segmentation. The calibration measures are performed during acquisition with respect to UAV capturing distance, angle, and other relevant camera parameters. Finally, our method can estimate the consolidated leaves and defected area. The image analysis results across these five regions reveal that around 14–28% of leaves area is affected across the groundnut field, and thereby yield will be diminished correspondingly. Hence, it is recommended to spray the pesticides on the areas affected alone across the field to improve the plant growth and thereby yield will be increased.

Keywords UAV · Image processing · Groundnut crop · Thresholding · HSV

A. Mahmud (✉)
Department of Mechanical Engineering, Chittagong University of Engineering and Technology, Raozan, Chittagong, Bangladesh
e-mail: rayed.ugresearch@gmail.com

B. Esakki
Department of Mechanical Engineering, Vel Tech Rangarajan Dr. Sagunthala R and D Institute of Science and Technology, 400 Feet Outer Ring Road, Avadi, Chennai, Tamil Nadu 600 062, India
e-mail: balasubramaniane@veltech.edu.in

S. Seshathiri
Department of Electrical Engineering, National Taiwan University of Science and Technology, Taipei, Taiwan
e-mail: sankarsrin@gmail.com

M. S. Kaiser et al. (eds.), *Proceedings of International Conference on Trends in Computational and Cognitive Engineering*, Advances in Intelligent Systems and Computing 1309, https://doi.org/10.1007/978-981-33-4673-4_53

1 Introduction

Image processing is considered to be a proficient and versatile tool to analyze the microscale parts of an image in many industrial and agricultural sectors [1]. The detection of plant diseases across a large scale farming area in an early stage is a viable solution to predict the distress of the plants. This helps to forecast crop production and initiate necessary remedial actions such as pesticide spraying, soil testing, seasonal crop rotation, and maintaining proper water drainage. To detect the plant diseases and classify them into different stages, agriculturalists inspect them manually and make subjective decisions. However, this kind of monitoring is time-consuming, laborious, and often inaccessible to adequately cover large agricultural fields [2]. With advancements in digital cameras and computer vision techniques, the prediction of plant disease becomes more comfortable and efficient to achieve precision agriculture.

The agricultural monitoring systems are mainly based on geospatial technology, automated systems, and other sustainable approaches. The optimal deployment methods are often chosen based on infrastructure and resource availability such as electricity, networking, accessibility, safety measurements, and so on. As most of the agricultural fields in developing countries lack most of these facilities, more sophisticated solutions are of immediate requisite. The recent advances in aerial vehicles paved the way for finding novel solutions towards agricultural monitoring, especially in rural and developing areas. Specifically, the unmanned aerial vehicles (UAV) can offer a perfect solution towards agricultural control for its easy transportation, independent operation and monitoring, and superior manoeuvring capacity. The paper deals with the deployment of UAVs to monitor such remote agricultural fields and assessment schemes to obtain the maximum yield.

2 Literature Review

The recent works on image-based analysis on agricultural crops and irrigation planning to maximize the yield are discussed. Rastogi et al. utilized K-MEANS-based segmentation, feature extraction algorithms to classify the leaf damage and develop a neural network model to determine the defected leaf area [3]. Vijay et al. have classified plant leaf into different species. The classification is done with the image segmentation technique to detect the diseases automatically [4]. Jayalakshami et al. developed an image-based diagnostic method to determine the deficiency symptoms, and it is useful to the farmers for taking necessary actions earlier for increasing the yield [5]. Shanwen et al. segmented the defected leaf image with hybrid clustering. The leaf pictures are divided by superpixels clustering using the Expectation Maximization algorithm [6]. Rupali Mahajan et al. developed a genetic algorithm to classify different types of leaf disease. Speed and accuracy of detection are improved and the unhealthy leaf can be separated easily using this technique [7].

Pejman et al. introduced a scatter transform algorithm for the extraction of weed. The texture classification is performed with multiple scattering [8]. Hanuman et al. demonstrated to detect the diseases at microscale level and classify them into various stages of diseases at an early stage to help the farmers. The segmentation is carried out using the HIS algorithm to classify the diseases [9]. Mehmet et al. utilized the UAV to detect the defective tomatoes. The algorithm provided 90% accurate results, which helps the farmers to maximize the yield [10]. Yash et al. developed an image processing algorithm to utilize the collected RGB and thermal images for calculating the temperature of the upper part of the leaves [11]. Aditya Khamparia et al. used a hybrid convolutional neural network and auto encoder for the classification of diseases. They have used a huge data set of images of a leaf, detect the crop diseases by cropping them with auto encoder [12]. Poornima et al. [13] utilized edge and color-based approaches to segment the disease potion of the leaves and classified the diseases using a support vector machine algorithm. Dat et al. [14] performed segmentation, labeling, size filtering to determine the boundaries of banana leaves and extracted the color characteristics to identify the defects. Arivazhagan et al. [2] used a transformation approach to convert RGB into grayscale and then applied a threshold to segment the leaf defects. Singh and Misra [4] used an image algorithm to segment the leaf images for automatic detection and classification of leaf diseases. Dhingra et al. [15] discussed the various detection and classification image algorithms to identify and classify leaf defects. Wang et al. [16] used a deep learning approach to differentiate various leaf features and defects with the incorporation of VGG16, VGG19, Inception-v3, and ResNet50 models. Camargo et al. [17] developed an image algorithm to convert RGB images into H, I3a, and I3b color transformations to identify and classify plant diseases.

In contrast to the existing approaches, the proposed method aims to perform leaf disease detection and quantification using a simple image processing method. The main advantage is that the proposed detection method can be easily realizable in hardware and can also be deployed in futuristic UAV agricultural real-time systems. In this work, groundnut crop images are collected using UAV equipped with a high definition camera across five regions of Andhra Pradesh, India. The various defects of the leaves are identified, and image analysis is performed to estimate the area of the defected leaf. Further, for each region group of images are selected based on their high resolution, and the total leaves area is calculated. The developed image algorithm is utilized to quantify the total defected area of leaves, and thus, the percentage of defected leaves area is estimated.

3 Collection of Groundnut Field Data Using UAV

A multi-rotor UAV with a payload of 1 kg and an endurance of 30 min is utilized to collect the images using a high definition camera. The UAV is flown (Fig. 1) at an altitude of 5 m with a speed of 1 m/sec in a stable manner to collect the groundnut field data. Groundnut crop field data is collected at various regions of Kadappa district

Fig. 1 Collection of groundnut field data using UAV

Fig. 2 Samples of collected images of groundnut leaves

such as Vedururu, Vallur, Gondhipalle, and Ananthapur district across Rugubanpalli and Vitlampalli of Andhra Pradesh (Fig. 2).

4 Proposed Method and Analysis of Groundnut Leaf

The image processing and analysis are divided into three phases, as shown in Fig. 3.

- Collection of images, selection of good quality of images, and training.
  - A UAV mountable GoPro HERO6 camera with 4 K video and 12MP sensor resolution was used to collect the groundnut image data. The images were

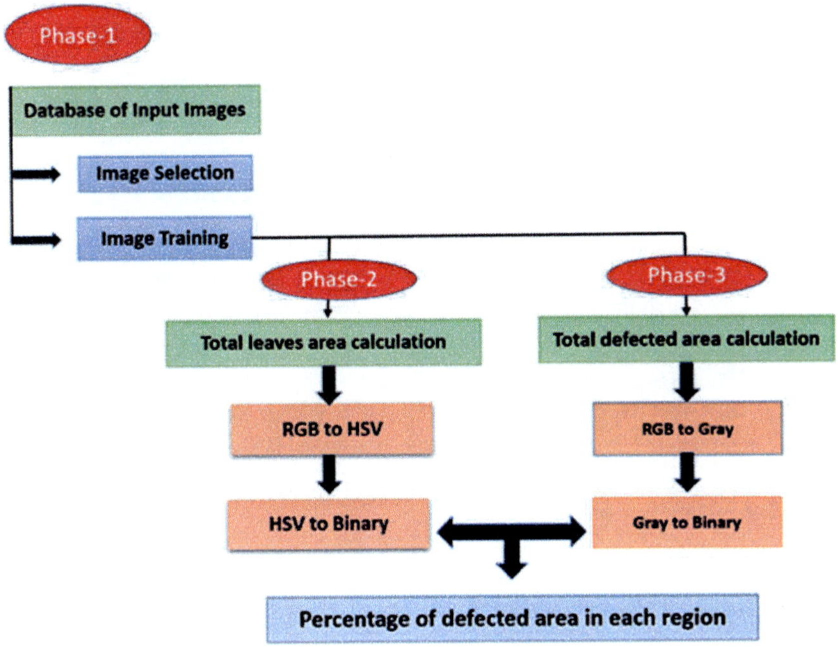

Fig. 3 Proposed approach using image processing

gathered during the day time of 11.00 h (IST) in a bright sunlight condition having an illumination of approximately 10,000 lx.

- The UAV was flown at a speed of 1 m/sec at an altitude of 5 m and the camera was positioned perpendicular to the ground surface. UAV with a battery capacity of 5000mAh was utilized and it was flown for 20 min duration for each flight. A 32 GB micro SD card was used to store the collected data.

- Leaf area calculation, which includes soil removing, conversion of RGB to HSV, convert grayscale into binary images.
- Quantification of defected leaf areas through pixel calculations.

In order to train the defected leaves, the leaf samples with defects are considered for the analysis (Fig. 4).

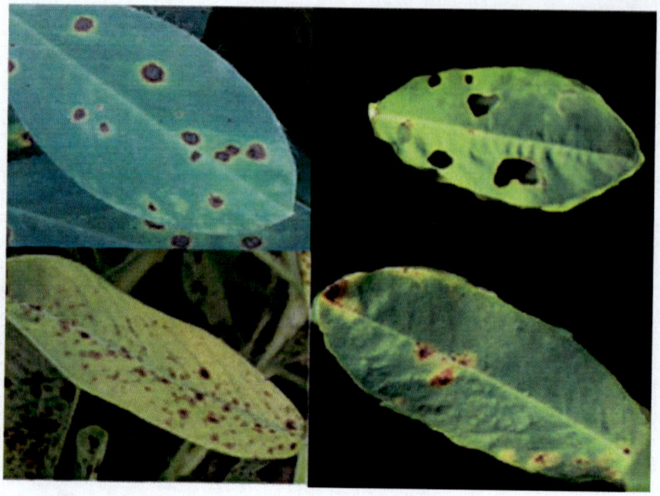

Fig. 4 Leaf samples with defects

4.1 Color Space Transformation

A. **Conversion of RGB to HSV**
The collected images in the RGB scale are converted into HSV color space to remove the soil part and obtain the green part alone. The sample input and output are shown in Table 1.

B. **Conversion of HSV to Binary**
Further, the conversion from HSV to binary is performed to detect the green part of the leaves with a threshold value, and the sample image is converted into binary as given in Table 2. The threshold value is obtained by manually inspecting the

Table 1 Conversion of RGB to HSV

Input	Output

Table 2 Conversion from HSV to Binary

Input	Output
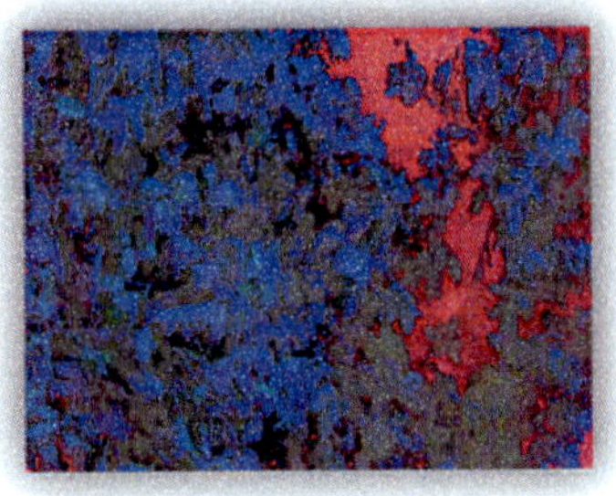	

result and it varies with respect to crop type and field. As HSV thresholding is performed it is illumination independent and for these leaf samples, the hue values are set in the range of 0.35 to 0.45.

C. **Conversion of RGB to Grayscale**
Furthermore, the transformation of images from RGB to Grayscale is per-formed to detect the defected part of the leaves with a threshold value, as shown in Table 3. Here, for detection of the defected part using the HSV scale is not possible. No unique threshold value is found to detect the defected part along with the green part.

D. **Calculation of Leaf Area**

The leaf area is calculated by converting the HSV images into binary format. The total number of pixels available with respect to the green segment alone will be counted, and it will be converted as a leaf area using the pixel to area conversion.

Based on the conversions from RGB to binary, the defected area of the leaf is calculated. Few samples of the defected leaf are given as an input, and they are

Table 3 Conversion from RGB to Grayscale

Input	Output
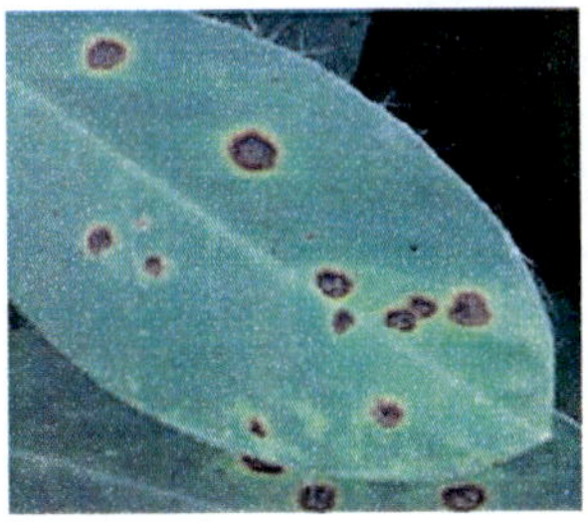	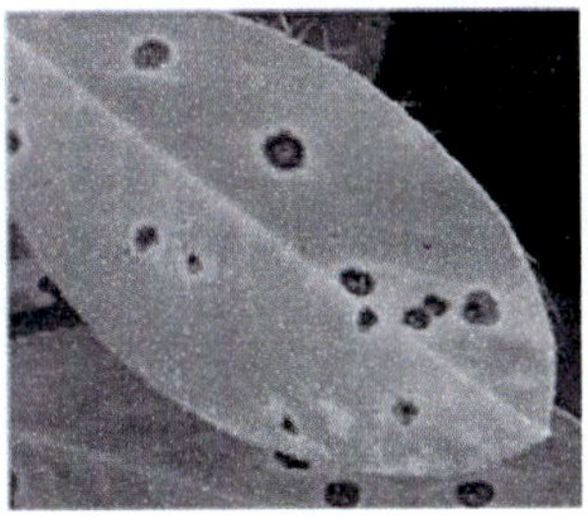

converted into the binary percentage of leaf defected area is calculated, and they are given in Table 4.

The developed image processing algorithms are applied to the set of collected images of groundnut field data across five regions of Andhra Pradesh to quantify the leaves' defected area. In this work, high-quality images alone are considered for each region, and it is varied for each region. The total leaves area in which defected leaves area is calculated, and then the percentage of defect is identified. It gives an

Table 4 Quantification of leaf defected area

Input	Output	Percentage of the defected area
	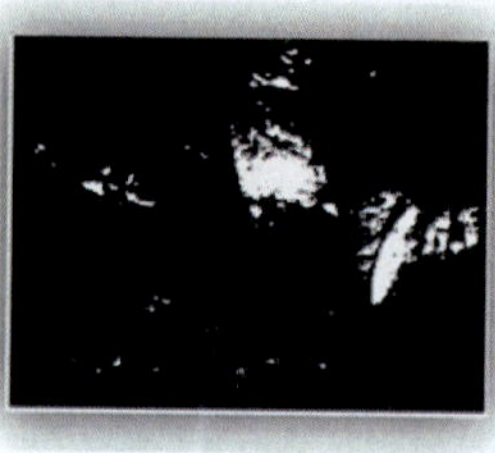	9.5%
	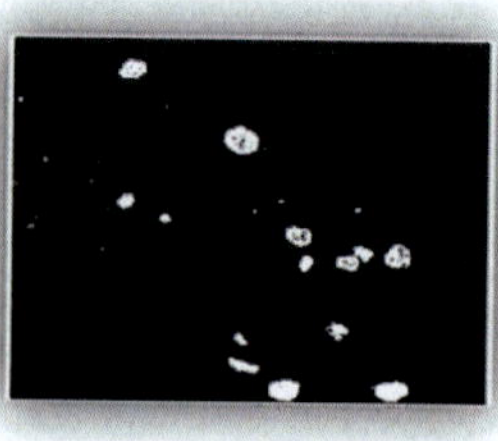	3.3%
	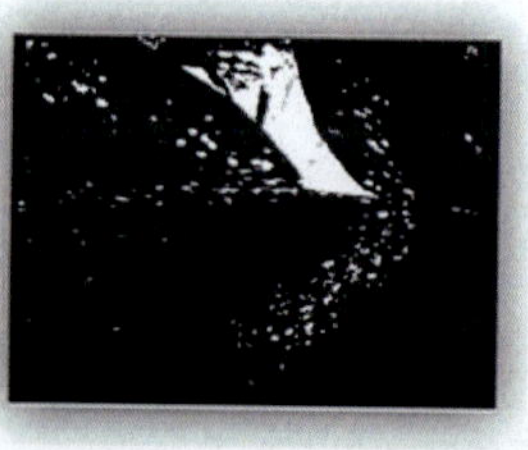	14.3%
	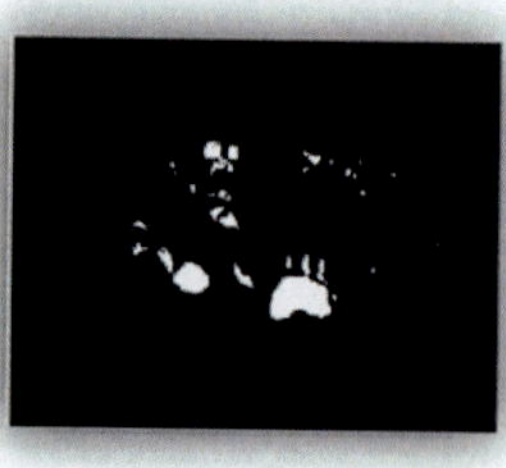	9.3%

Table 5 Quantification of the defected area of leaves in various regions

Sl. no	Region name	Total number of images	Total leaf area (mm^2)	Total defected area (mm^2)	Percentage of defect (%)
1	Vedururu	33	1587.76	275.77	17.37
2	Vallur	32	1661.37	459.68	27.67
3	Gondhipalle	33	1740.76	258.45	14.85
4	Vitlampalli	32	1711.21	389.94	22.79
5	Rughubanpalli	33	1842.39	307.45	19.93

idea to the farmers that approximately 14–28% of regions may get affected due to pest attacks and plant distress, and correspondingly, the yield will be deduced. In addition, it is observed from Table 5 that, region Vallur has predominantly affected in comparison to other regions. The experimentation is performed using Intel(R) Core(TM) i5-8300H CPU @ 2.30 GHz 2.30 GHz with 8 GB RAM, equipped with graphics card NVIDIA GeForce GTX 1050 Ti. The execution time for each image sample is around 1.5 s. In the future development, the algorithm is directly embedded into the UAV provided with image processing modules.

To validate the algorithm, few leaf samples are manually inspected and their defected area is measured using scales. The algorithm is tested with respect to manually obtained defected area and found to have obtained consistent performance.

5 Conclusion

UAV based collection of groundnut field data and image processing analysis is performed to quantify the defected area of groundnut leaves across five regions of Andhra Pradesh. The conversion of RGB image to HSV, grayscale, and then into binary format yielded initial segregation of removing the soil, identification of defected portions of leaves, and further calculation of total pixels into leaf area. Image analysis considering the high-quality images alone gathered using UAV among the five regions resulted in quantifying the percentage of defected leaves area, and thereby, tentative estimation of yield loss is determined. It is observed that around $14 - 28\%$ of leaves area are affected across these regions, and Vallur is affected to the maximum amongst all the regions considered for the present study. In the future, machine learning-based algorithms will be considered to estimate the defected leaves area, and precise locations across the groundnut field will be identified using UAV based geo-referencing.

Acknowledgements We would like to thank the Department of Bio-Technology, Govt. of India for providing the funding support under the project Expansion of Activities of Biotech- KISAN Hub in Three Aspirational Districts (Kadapa, Vizaingaram, and Visakhapatnam) of Andhra Pradesh.

References

1. Al-Hiary, H., Bani-Ahmad, S., Reyalat, M., Braik, M., Alrahamneh, Z.: Fast and accurate detection and classification of plant diseases. Int. J. Comput. Appl. **17**(1), 31–38 (2011)
2. Arivazhagan, S., Shebiah, R.N., Ananthi, S., Varthini, S.V.: Detection of unhealthy region of plant leaves and classification of plant leaf diseases using texture features. Agricultural Engineering International: CIGR J. **15**(1), 211–217 (2013)
3. Rastogi, A., Arora, R., & Sharma, S.: Leaf disease detection and grading using computer vision technology & fuzzy logic. In 2015 2nd international conference on signal processing and integrated networks (SPIN) (pp. 500–505). IEEE (2015)
4. Singh, V., Misra, A.K.: Detection of plant leaf diseases using image segmentation and soft computing techniques. Inf. Process. Agric. **4**(1), 41–49 (2017)
5. Jeyalakshmi, S., & Radha, R.: A review on diagnosis of nutrient deficiency symptoms in plant leaf image using digital image processing. ICTACT J. Image Video Process. 7(4) (2017)
6. Zhang, S., You, Z., Wu, X.: Plant disease leaf image segmentation based on superpixel clustering and EM algorithm. Neural Comput. Appl. **31**(2), 1225–1232 (2019)
7. Sahithya, V., Saivihari, B., Vamsi, V. K., Reddy, P. S., & Balamurugan, K.: GUI based detection of unhealthy Lleaves using image processing techniques. In 2019 International Conference on Communication and Signal Processing (ICCSP) (pp. 0818–0822). IEEE (2019)
8. Rasti, P., Ahmad, A., Samiei, S., Belin, E., Rousseau, D.: Supervised image classification by scattering transform with application to weed detection in culture crops of high density. Remote Sensing **11**(3), 249 (2019)
9. Reddy, N. H., Kumar, E. R., Reddy, M. V., Reddy, K. R., & Valli, G. S.: Bioinformatics and image processing—detection of plant diseases. In First International Conference on Artificial Intelligence and Cognitive Computing (pp. 149–154). Springer, Singapore (2019)
10. Altınbaş, M. D., & Serif, T.: Detecting Defected Crops: Precision Agriculture Using Haar Classifiers and UAV. In International Conference on Mobile Web and Intelligent Information Systems (pp. 27–40). Springer, Cham (2019)
11. Osroosh, Y., Khot, L.R., Peters, R.T.: Economical thermal-RGB imaging system for monitoring agricultural crops. Comput. Electron. Agric. **147**, 34–43 (2018)
12. Khamparia, A., Saini, G., Gupta, D., Khanna, A., Tiwari, S., de Albuquerque, V.H.C.: Seasonal crops disease prediction and classification using deep convolutional encoder network. Circuits Syst. Signal Process. **39**(2), 818–836 (2020)
13. Poornima, S., Kavitha, S., Mohanavalli, S., & Sripriya, N.: Detection and classification of diseases in plants using image processing and machine learning techniques. In AIP Conference Proceedings (Vol. 2095, No. 1, p. 030018). AIP Publishing LLC (2019)
14. Dat, D. T., Hai, N. D. X., & Thinh, N. T.: Detection and classification defects on exported banana leaves by computer vision. In 2019 International Conference on System Science and Engineering (ICSSE) (pp. 609–613). IEEE (2019)
15. Dhingra, G., Kumar, V., Joshi, H.D.: Study of digital image processing techniques for leaf disease detection and classification. Multimedia Tools Appl. **77**(15), 19951–20000 (2018)
16. Wang, G., Sun, Y., Wang, J.: Automatic image-based plant disease severity estimation using deep learning, Computational intelligence and neuroscience (2017)
17. Camargo, A., Smith, J.S.: An image-processing based algorithm to automatically identify plant disease visual symptoms. Biosys. Eng. **102**(1), 9–21 (2009)

Performance Analysis of Different Loss Function in Face Detection Architectures

Rezowan Hossain Ferdous, Md. Murshedul Arifeen, Tipu Sultan Eiko, and Shamim Al Mamun

Abstract Masked face detection is a challenging task due to the occlusions created by the masks. Recent studies show that deep learning models can achieve effective performance for not only occluded faces but also for unconstrained environments, illuminations or various poses. In this study, we have addressed the problem of occlusion due to wearing masks in masked face detection technique in deep transfer learning method. We have also reviewed the recent deep learning models for face detection and considered VGG16, VGG19, MobileNet and DenseNet as our underlying masked face detection models. Moreover, we have prepared a dataset containing masked face and without mask from 120 individuals and enhanced the dataset using augmentation. After training the deep learning models with our own dataset, we have analysed the performance of the deep learning models for several types of loss functions. From the experiment, it is clear that all the deep learning models perform well in terms of classification losses like categorical cross entropy loss and KL divergence loss.

Keywords Face detection · VGG face · Deep learning

R. H. Ferdous (✉) · Md. M. Arifeen · T. S. Eiko
Department of ICT, Bangladesh University of Professionals, Dhaka, Bangladesh
e-mail: rezosam0@gmail.com

Md. M. Arifeen
e-mail: murshedularifeendipto@gmail.com

T. S. Eiko
e-mail: tipusultan50@gmail.com

S. A. Mamun
Institute of Information Technology, Jahangirnagar University, Dhaka, Bangladesh
e-mail: shamim@juniv.edu

M. S. Kaiser et al. (eds.), *Proceedings of International Conference on Trends in Computational and Cognitive Engineering*, Advances in Intelligent Systems and Computing 1309, https://doi.org/10.1007/978-981-33-4673-4_54

1 Introduction

Face detection can be considered as a computer vision and pattern recognition task where various face detection algorithms try to identify and determine faces including their positions, sizes and poses from the input image. Face detection involves three principle steps [24] selecting a probable area on the input image which is also known as observation window. Then from that window several features are extracted and finally based on these features it is determined by a face detection algorithm that whether there is a face or not. Face detection is the underlying mechanism of various computer vision problems such as—face recognition, verification, tracking, behavior analysis, attribute recognition, emotion detection, gender classification, synthesis and sentiment analysis [2, 3, 9, 11, 19]. Face detection is the very first step towards various face based technologies like mentioned above. Moreover, face detection itself is essential for various purposes like photo shooting [1, 10, 12]. Taking photo using camera utilizes face detection technology to move the lens in the right position. Another useful uses of face detection is head pose estimation which is significant in drivers drowsiness detection, behaviour analysis, gesture detection and gaze estimation. Also, face detection is used in mobile devices for unlocking the device or for security purpose. Bio metric security is another research field where face detection is used enormously.

Face detection and recognition is a challenging task for several years but now a days researchers have come up with various ideas and utilization of deep learning in face detection problems make this field one step ahead. But still there are lots of issues regarding face detection which needs to be considered. In case of bio metric security applications, occluded face or partially occluded face creates difficulty for the automated systems to detect and recognize the intended person effectively. Furthermore in case of pandemic situation like COVID-19 when everybody wears masks for their health safety, it is necessary to adapt the existing face detection algorithms to identify the persons with masks without any difficulty. The performance of the existing algorithms decrease sharply when the input images contain masked faces. The main problem is the occlusion which is generated from partially covering the face with mask. Other challenging tasks like illumination, pose variation problems are addressed in various works but masked face detection is not considered well enough for various applications like video surveillance system. Therefore, the performance of the algorithms decline drastically due to masked faces which creates occlusion. Furthermore, there are lack of proper dataset containing masked faces with augmentation. If we want to train a deep learning model for classifying masked faces or face without mask, it is required to construct a proper dataset with a reasonable amount of images containing both masked faces and faces with no mask. The existing dataset is imbalanced and do not provide sufficient amount of images to train a deep learning model for the purpose of classification.

In this work, we have performed a comparative study on the performance of various deep learning model and find out that VGG can work well in case of masked face classification. We have constructed a dataset balancing masked face and face with no mask of 200 people and augmented the images into various types. Then, we performed an experiment on VGG for various loss functions and determined which loss function performs better in case of masked face classification task (Table 1).

2 Deep Learning in Face Detection

A multitask convolutional neural network is proposed in [24] for face detection and facial landmark localization at the same time. Integrating these two tasks not only improve detection accuracy but also reduces the running time. The face detection method proposed in [2] is based on mobile net. The architecture is lightweight and applicable for mobile applications. Since mobile net uses depth wise convolution so it is more effective in terms of computational costs rather than general convolution operation. For detecting and aligning face in unconstrained environment, the authors in [23] proposed a deep learning based multitask cascaded joint face detection method known as MTCNN which shows effective results in terms of illuminations or occlusions. This network involves proposal network, refinement network and output network for the generation of the outputs. To increase the efficiency and more effective results than AlexNet, the authors in [18] proposed a deep CNN based method known as VGG for object detection. The convolutional layers of VGG network consists of 3×3 filter and also a 1×1 filter working as linear transformation with 1 pixel convolutional stride followed by ReLU function. A fast face detection method based on DCF and deep CNN has been proposed in [8]. The proposed method consists of 2 key components. The first one is the nonlinear mapping function which itself includes convolution layer and max pooling layer. The second component is the sparse discriminative features which enhances linear separability of the features on the output feature maps of the nonlinear mapping function. The recently proposed face detection methods use single stage network which lacks the generalization ability as well as require fixed size images. In [15] the authors proposed a cascaded deep convolutional network to overcome these challenges. The proposed method performs two tasks in detection stage-face classification and bounding box regression. The accuracy of the model is enhanced through cascading convolutional neural network. The work in [6] focuses on improving the accuracy of face detection using YOLO method. The proposed network includes seven convolution layer and a max pooling layer. The authors in [21] focused on facial attributes and proved that facial attributes based supervision can effectively increase the occlusions handling capability of face detection network. They have proposed a deep convolutional neural network based face detection method by including facial attributes based supervision. Similarly, various methods and classification tasks are proposed in various literature's for face detection.

Table 1 Comparisons of different deep learning models

Methods	Filter size	Loss function	Data set	Input image size	Performance
MTCNN [23]	3 × 3	Cross-entropy loss (classification) Euclidian loss (bounding box regression)	Wider face, CelebA	n/a	16 fps on CPU 99 fps on GPU
VGG [18]	3 × 3, 1 × 1	Euclidian Loss	ILSVRC-2012	224 × 224 RGB	n/a
DCF [8]	5 × 5	n/a	CAS-PEAL-R1,VOC2012	32 × 32	n/a
Cascade [15]	1 × 1	Cross-entropy loss (classification), Euclidian loss (bounding box regression)	Wider face (training + test), FDDB (test)	n/a	28 fps
Mobile [2]	1 × 1		FDDB	224 × 224 RGB	n/a
DAFE-FD [19]	3 × 3	Multitask loss	WIDER, FDDB, Pascal faces	400 × 800, 1200 × 1600	Computational time is 50 ms/image and 190 ms/image
Yolo based [6]			FDDB	448 × 448	10–25 fps,26–40
Faceness net [21]	n/a	Cross entropy loss (classification)	CelebA	n/a	14 fps 100 fps GPU
Hyperface [16]	n/a	Euclidian loss (landmark localization)	AFLW	n/a	n/a
Masked face [4]	n/a	n/a	WIDER FACE, MASKED FACE	12 × 12, 24 × 24	n/a
Multiscale feature [5]	n/a	Cross-entropy loss	FDDB, AFLW	24 × 24	n/a
Tiny face [22]	n/a	Regression loss	FDDB, wider face	200 × 200	45 fps
Densenet [13]	1 × 1	(Focal loss function), Progressive Anchor loss	WIDER FACE	640 × 640	n/a
Deep cascade CNN [14]	1 × 1	Cross entropy loss and L1 loss	FDDB, WIDER face	24 × 24	Validation accuracy is above 94%

3 Loss Functions

3.1 *Cross Entropy Loss*

Cross entropy is a type of measurement used in information theory based on entropy theory and computes the differences between two probability distribution. This loss is also known as log loss. Cross entropy loss is used to measure the performance of a machine learning classification model which gives output as a probability. This loss increases with the deviation of the predicted probability value from the actual label value. Cross entropy loss is also known as softmax loss and effectively used in face detection or face recognition task. It can be defined as [20]

$$\mathrm{CE} = -\frac{1}{N}\sum_{i=1}^{N}\log\frac{\mathrm{e}^{W_{y_i}^T x_i + b_{y_i}}}{\sum_{j=1}^{n}\mathrm{e}^{W_{y_j}^T x_i + b_{y_j}}}$$

where, the symbol W represents the weight matrix, b is the bias term, x_i is the ith training sample, y_i denotes class label for ith training sample, N defines the total number of samples. Finally, W_j and W_{yi} are jth and y_ith column of W.

3.2 *KL Divergence Loss*

This loss function is used to measure the difference between two probability distributions and shows how one probability distribution differs from another one. Lets say we have a probability distribution $P(X)$ and we want to replace this distribution with a known probability distribution say $Q(X)$. In this scenario, the KL divergence loss measure the difference of the approximate distribution $Q(X)$ from the true distribution $P(X)$. Mathematically, it can be defined as [17],

$$KL(P||Q) = E_{x\sim P}\left[\log\frac{P(X)}{Q(X)}\right]$$

Intuitively, this loss defines that how much information is lost or gained when we approximate a probability distribution with another probability distribution.

3.3 *Mean Squared Error Loss*

This loss in the average of the squared error that is used as the machine learning loss function for least squares regression. This loss can be defined mathematically as [7]

$$MAE = \frac{\sum_{i=1}^{n}|y_i - y_i^p|}{n}$$

This function performs sum over all the data points, of the square of the difference between the predicted value and target values and the whole is divided by the total number of data points. This loss function is one of the most simplest loss function used in Machine Learning.

3.4 Huber Loss

Huber loss function includes mean squared error loss and mean absolute error loss function and provides the best results by balancing those two loss functions. It can be defined as [7]

$$L(y, y') = \begin{cases} \frac{1}{2}(y - y')^2 & \text{if } \|y - y'\| \leq \delta \\ \delta\|y - y'\| - \frac{1}{2}\delta^2 & \text{otherwise} \end{cases}$$

From this equation, we can understand that if the loss vales is less than δ, mean squared error loss will be used and mean squared absolute error will be used otherwise.

4 Experimental Setup

The experiment has been conducted on Google Colab. We have used Tesnsorflow and Keras libraries for our implementation. We have also used the MTCNN library to extract faces from the input images. Firstly, we have loaded the models from Keras without the fully connected layers, application model is initialized with imagenet weights and proper input shape was given as parameters. Then we build our custom fully connected layers with Dense and Flatten function was made, 10% Dropout is used here to prevent overfitting and relu is used as activation function. The input layer of keras model api is used as the input parameter of our final model and our fully connected layer is set as output parameter. The VGG face weight is loaded for tuning the model. After that we have fitted our dataset for further training. Lastly we have compiled the new model with different loss functions and optimizers (sgd, adam).

4.1 Datasets

Since there are no balanced masked face datasets available in terms of skin tone, aging factor, orientation and gradient, therefore we have prepared our own dataset containing masked and non-masked faces. The dataset contains 720 face images belonging to 120 individuals. To obtain all the features of a face, we have taken front

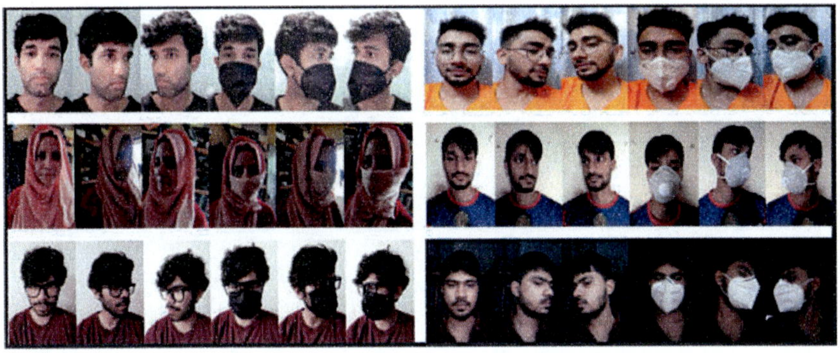

Fig. 1 Sample of our dataset which shows each individuals six different input images including masked faces and without masked faces. Among the dataset of each individual, one is front view and the other two is side views

view and both side views of each individual wearing having mask and without masks. Moreover, there are 6 images for each person which was used image data generator from Keras library to augment on those images. Furthermore, 6 images of each person is augmented to 24 images through changing various factors for instance—brightness, rotation, shifting, flipping, sharing and zooming. We have separated 18 images for training purpose, 4 images for validation purpose and 2 images for testing purpose from the augmented 24 images of each individual. We divided our dataset 80:20 ration that is the total number of training images is 1724, and for validation and testing rest of the 432 images are being used for getting performance accuracy. Figure 1 shows the sample of our prepared dataset.

5 Result and Discussion

Machine learning or deep learning models learn from the given datasets using the loss functions. Through loss function, we can understand that how a deep learning algorithms models the given data. If the predicted output of a given deep learning model deviates too much from the actual data or output, loss function will show high error value. In this study, through our experiment we have showed the best loss function which can be chosen for masked face detection tasks. We have used classification losses such as Categorical cross entropy loss, and KL divergence loss, regression losses such as Mean squared error loss and Huber loss for our experiment. Figure 2 shows the experimental results. Classification losses are probabilistic loss and they predict the output based on the highest probability it gets from the last layer of the deep learning models. On the contrary, regression losses predict continuous values. From our experimental results, we can see that, all the models perform better in case of classification loss that is for Categorical cross entropy (CCE) loss and

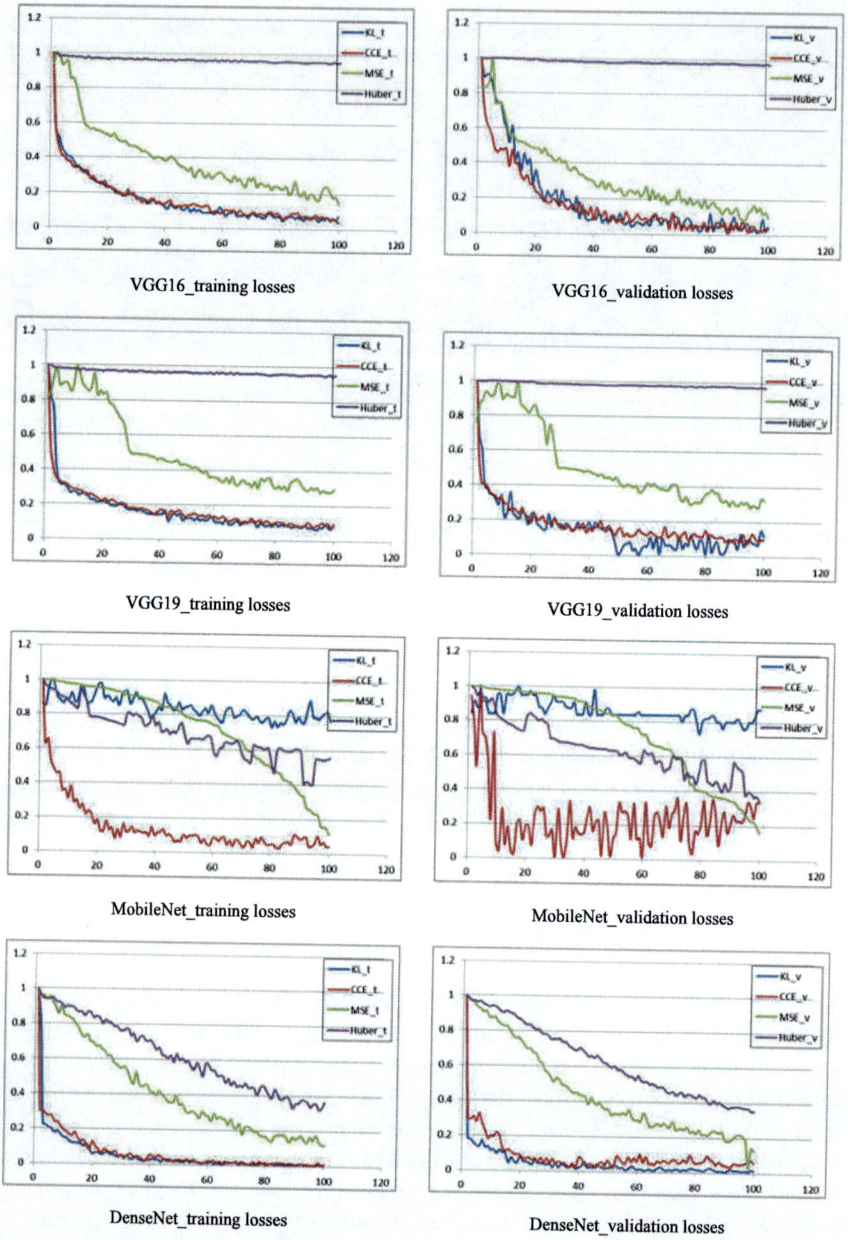

Fig. 2 Four different loss functions KL divergence loss (KL), categorical cross entropy loss (CCE), mean squared error loss (MSE) and huber loss (Huber) against epochs for four different deep learning models (VGG16, VGG19, MobileNet and DenseNet)

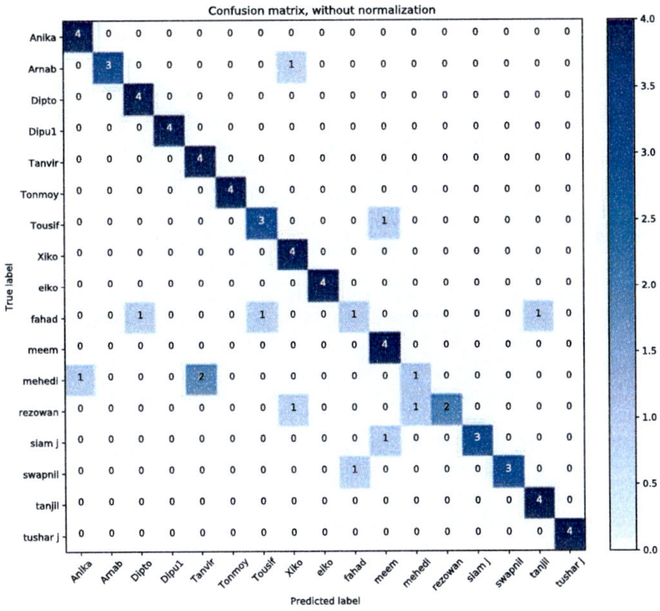

Fig. 3 Confusion matrix for test data

Kl divergence (Kld)loss. From the figures it is evident that for all the deep learning models that is for VGG16, VGG19, MobileNet and DenseNet, the loss function curves tend towards zero for only for CCE and Kld losses. Other losses do not converge towards zero. Hence, we can say that the CCE and Kld loss performs well in case of deep learning based face detection models. To visualize the performance, we have rebuilt our model with 17 classes and plot the confusion matrix shown in Fig. 3. We have tested the model with 1 frontal and side view images for both masked and non-masked in total 4 images of an individual. VGG16 with CCE loss is used for this Confusion matrix shown in Fig. 3.

6 Conclusion

In this work, we have addressed the problem of occlusion due to wearing masks in case of masked face detection and analysed the performance of various types loss functions for different deep learning models. We have prepared our own dataset for the experimental purpose which is enhanced through augmentation technique. For the experiment, we have considered four different types of deep learning models for instance VGG16, VGG19, MobileNet and DenseNet. Experimental results show that the deep learning models performs better for classification losses such as—Categorical cross entropy loss and KL divergence loss.

References

1. Al Mamun, S., Fukuda, H., Lam, A., Kobayashi, Y., Kuno, Y.: Autonomous bus boarding robotic wheelchair using bidirectional sensing systems. In: Bebis, G., Boyle, R., Parvin, B., Koracin, D., Turek, M., Ramalingam, S., Xu, K., Lin, S., Alsallakh, B., Yang, J., Cuervo, E., Ventura, J. (eds.) Advances in Visual Computing, Lecture Notes in Computer Science, pp. 737–747. Springer International Publishing (2018). https://doi.org/10.1007/978-3-030-03801-4_64
2. Almadhor, A.: Deep learning based face detection algorithm for mobile applications. In: TENCON 2018–2018 IEEE Region 10 Conference, pp. 1158–1162. IEEE (2018)
3. Biswas, S., Kaiser, M.S., Mamun, S.A.: Applying ant colony optimization in software testing to generate prioritized optimal path and test data. In: 2015 International Conference on Electrical Engineering and Information Communication Technology (ICEEICT), pp. 1–6 (2015). https://doi.org/10.1109/ICEEICT.2015.7307500
4. Bu, W., Xiao, J., Zhou, C., Yang, M., Peng, C.: A cascade framework for masked face detection. In: 2017 IEEE International Conference on Cybernetics and Intelligent Systems (CIS) and IEEE Conference on Robotics, Automation and Mechatronics (RAM), pp. 458–462. IEEE (2017)
5. Chen, Q., Meng, X., Li, W., Fu, X., Deng, X., Wang, J.: A multi-scale fusion convolutional neural network for face detection. In: 2017 IEEE International Conference on Systems, Man, and Cybernetics (SMC), pp. 1013–1018. IEEE (2017)
6. Garg, D., Goel, P., Pandya, S., Ganatra, A., Kotecha, K.: A deep learning approach for face detection using yolo. In: 2018 IEEE Punecon, pp. 1–4. IEEE (2002)
7. Grover, P.: 5 regression loss functions all machine learners should know (2018), https://heartbeat.fritz.ai/5-regression-loss-functions-all-machine-learners-should-know-4fb140e9d4b0. Last accessed 11 Oct 2020
8. Guo, G., Wang, H., Yan, Y., Zheng, J., Li, B.: A fast face detection method via convolutional neural network. Neurocomputing (2019)
9. Kaiser, M.S., Chowdhury, Z.I., Mamun, S.A., Hussain, A., Mahmud, M.: A neuro-fuzzy control system based on feature extraction of surface electromyogram signal for solar-powered wheelchair. Cogn. Comput. **8**(5), 946–954 (2016). https://doi.org/10.1007/s12559-016-9398-4
10. Liu, H.: Face technologies on mobile devices. In: Facial Detection and Recognition on Mobile Devices, pp. 11–38. Elsevier. https://doi.org/10.1016/b978-0-12-417045-2.00002-1
11. Mamun, S.A., Kaiser, M.S., Ahmed, M.R., Islam, M.S., Islam, M.I.: Performance analysis of optical wireless communication system employing neuro-fuzzy based spot-diffusing techniques. Commun. Netw. **5**(3), 260–265 (2013). https://doi.org/10.4236/cn.2013.53B2048, http://www.scirp.org/Journal/Paperabs.aspx?paperid=39354 (Scientific Research Publishing)
12. Mamun, S.A., Suzuki, R., Lam, A., Kobayashi, Y., Kuno, Y.: Terrain recognition for smart wheelchair. In: Huang, D.S., Han, K., Hussain, A. (eds.) Intelligent Computing Methodologies, pp. 461–470. Lecture Notes in Computer Science. Springer International Publishing (2016). https://doi.org/10.1007/978-3-319-42297-8_43
13. Nandy, A.: A densenet based robust face detection framework. In: Proceedings of the IEEE International Conference on Computer Vision Workshops, p. 0 (2019)
14. Peng, B., Gopalakrishnan, A.K.: A face detection framework based on deep cascaded full convolutional neural networks. In: 2019 IEEE 4th International Conference on Computer and Communication Systems (ICCCS), pp. 47–51. IEEE (2019)
15. Qi, R., Jia, R.S., Mao, Q.C., Sun, H.M., Zuo, L.Q.: Face detection method based on cascaded convolutional networks. IEEE Access **7**, 110740–110748 (2019)
16. Ranjan, R., Patel, V.M., Chellappa, R.: Hyperface: a deep multi-task learning framework for face detection, landmark localization, pose estimation, and gender recognition. IEEE Trans. Pattern Anal. Mach. Intell. **41**(1), 121–135 (2017)
17. Seo, G., Yoo, J., Choi, J., Kwak, N.: Kl-divergence-based region proposal network for object detection. arXiv preprint arXiv:2005.11220 (2020)
18. Simonyan, K., Zisserman, A.: Very deep convolutional networks for large-scale image recognition. arXiv preprint arXiv:1409.1556 (2014)

19. Sindagi, V., Patel, V.: Dafe-fd: density aware feature enrichment for face detection. In: 2019 IEEE Winter Conference on Applications of Computer Vision (WACV), pp. 2185–2195. IEEE (2019)
20. Srivastava, Y., Murali, V., Dubey, S.R.: A performance comparison of loss functions for deep face recognition. arXiv preprint arXiv:1901.05903 (2019)
21. Yang, S., Luo, P., Loy, C.C., Tang, X.: Faceness-net: face detection through deep facial part responses. IEEE Trans. Pattern Anal. Mach. Intell. **40**(8), 1845–1859 (2017)
22. Ye, F., Ding, M., Gong, E., Zhao, X., Hang, L.: Tiny face detection based on deep learning. In: 2019 14th IEEE Conference on Industrial Electronics and Applications (ICIEA), pp. 407–412. IEEE (2019)
23. Zhang, K., Zhang, Z., Li, Z., Qiao, Y.: Joint face detection and alignment using multitask cascaded convolutional networks. IEEE Signal Process. Lett. **23**(10), 1499–1503 (2016)
24. Zhang, Y., Wang, H., Xu, F.: Face detection based on multi task learning and multi layer feature fusion. In: 2017 6th International Conference on Computer Science and Network Technology (ICCSNT), pp. 289–293. IEEE (2017)

Convid-Net: An Enhanced Convolutional Neural Network Framework for COVID-19 Detection from X-Ray Images

Sabbir Ahmed, Md. Farhad Hossain, and Manan Binth Taj Noor

Abstract This article aims to demonstrate a deep convolutional neural network (CNN) framework namely $Convid\text{-}Net$ based on a combination of residual network and parallel convolution ($CONV$) to detect COVID-19 from chest X-ray images. The proposed architecture can choose optimum features from different parallel $CONV$ and residual connection increasing overall accuracy with less computational expenses. A custom dataset has been created for this work which consists of total 1440 images of COVID-19, 2470 normal images and 2407 chest X-ray images of viral and bacterial pneumonia; collected from different publicly available sources. Augmentation and preprocessing have been applied as well to increase the number of data for better training purposes. $Convid\text{-}Net$ has been trained and tested on a prepared augmented dataset which achieved accuracy of 97.99%. The promising result of the proposed system shows that it converges to an overall higher accuracy and can be a very useful method for physicians and radiologists to assist them in rapid detection and diagnosis of COVID-19 from radiography images. These results also indicate that $Convid\text{-}Net$ architecture can further be used in other image based classification tasks.

Keywords Convid-net · Coronavirus · Covid-19 · Deep learning · X-ray images · CNN · LSTM · Residual network

---

S. Ahmed (✉) · Md. F. Hossain · M. B. T. Noor
Institute of Information Technology, Jahangirnagar University, Dhaka, Bangladesh
e-mail: sabbir.iit.ju@gmail.com

Md. F. Hossain
e-mail: farhadiitju@gmail.com

M. B. T. Noor
e-mail: manan.noor@juniv.edu

M. S. Kaiser et al. (eds.), *Proceedings of International Conference on Trends in Computational and Cognitive Engineering*, Advances in Intelligent Systems and Computing 1309, https://doi.org/10.1007/978-981-33-4673-4_55

1 Introduction

Coronavirus disease 2019 (COVID-19) is characterized as a disease caused by a novel coronavirus now known as Severe Acute Respiratory Syndrome Coronavirus 2 (SARS-CoV-2; formerly known as 2019-nCoV), first identified as a result of an outbreak of respiratory disease in Wuhan City, Hubei, China. Globally, as of 26 August 2020, there have been 23,752,965 confirmed cases of COVID-19, including 815,038 deaths, reported to the World Health Organization (WHO) [1]. Currently, Reverse Transcription Polymerase Chain Reaction (RT-PCR) and antibody testing are prominent methods of testing people for COVID-19 across global healthcare systems. In medical science, different test-based techniques are used for diagnosis, which require expert opinion. From that perspective, exploring data strategies for medical image is a task in terms of obtaining insight, then assessing and diagnosing a particular disease. Following that, in order to track the diagnosis of SARS-CoV-2 infections, there are some medical imaging methods utilizing radiological images; such as chest X-rays or Computed Tomography (CT). Former experiments [2] indicate that COVID-19 in the context of ground-glass opacities causes abnormalities that are noticeable in the chest X-rays and CT images. Classification of medical images is a tedious task for handcrafted features, thus recent trends are focused on deep learning approaches for higher recognition rate. A survey summarizes over 300 contributions in deep learning for medical image classification, object detection, segmentation, registration, and other tasks discussed in [3]. Deep learning-based image classification can also be conducted on X-ray and CT images to detect COVID-19 patterns. However, datasets of X-ray images are more available than CT images and less time-consuming in terms of processing the image [4].

In this article, multiple parallel $CONV$ with different $CONV$ filters and residual layer-based architecture has been proposed to classify X-ray images into COVID-19, Pneumonia and Normal classes. Dataset has been split into a testing set and training set before augmentation for correctly measuring classifier performance. The initial dataset is created by combining several other publicly available datasets which contain different number of COVID-19, normal and pneumonia chest X-ray images. Since, the initial dataset generated for this work is imbalanced, the same number of data have been generated for each class with augmentation for this work. Parallel $CONV$ layer has been used for feature extraction and data complexity reduction. Residual layer with varying parallel $CONV$ has been applied to deal with vanishing gradient problems. Finally the trained networks later can be used for COVID-19 detection from X-ray images in general, thus larger dataset can be collected by crowdsourcing. Flow diagram of the proposed approach is shown in Fig. 1.

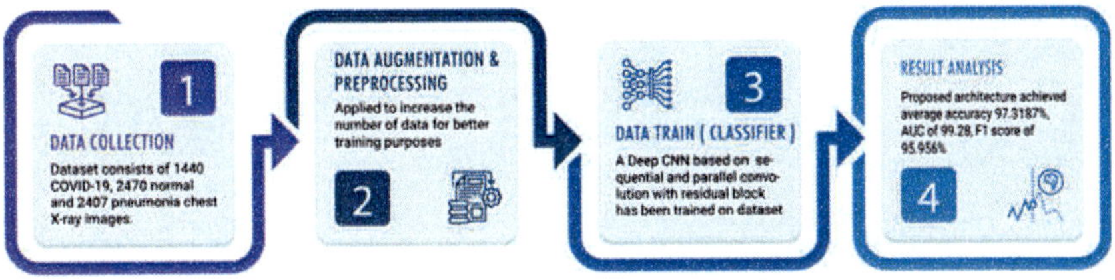

Fig. 1 Flow diagram of the proposed system

2 Related Studies

This section discusses the newly developed COVID-19 detection systems based on deep learning techniques. Islam et al. proposed a CNN for deep feature extraction with Long Short-Term Memory ($LSTM$) for detection model [5]. A patch-based CNN approach has been developed for COVID-19 diagnosis with clinically interpretable saliency maps with limited data by [6]. However, their datasets only contain 180 chest X-ray (CXR) of COVID-19 patients. In another work Wang et al. introduced COVID-Net to identify distinctive anomalies in CXR images [7]. Overall this method has been hampered by training on imbalance data, lack of rigorous preprocessing of data, and inadequate simulation of decisions. Transfer learning-based CNN has also been implied in [8] for the diagnosis of infected, coronavirus and pneumonia patients using chest X-ray radiographs with fivefold cross validation. Then, Resnet-18 with Decompose, Transfer and Compose has been developed in [9] for detection of COVID-19 from chest X-ray images. But their validation accuracy severely exceeded train accuracy which is a result of overfitting. Bayesian Deep Learning classifier was trained on COVID-19 X-ray images utilizing transfer learning method for estimating model uncertainty by Ghoshal et al. [10]. A Generative Adversarial Networks for a pneumonia chest X-ray dataset creation and AlexNet, GoogLeNet, Squeeznet, ResNet-18 for detection is presented by Khalifa et al. [11]. Ozturk et al. on the other hand used handcrafted feature extraction, SMOTE algorithm-based data processing and stacked auto-encoder for classifying imbalance COVID-19 chest X-ray [12]. Four X-ray based detection methods [13], 48 Diagnostic imaging [2], 69 Models [14] based on X-ray images and CT have been studied from the preprocessing perspective, extraction of the characteristics and classification. In contrast to the studies above, the proposed methodology enforces a customized larger dataset incorporating a newer and effective version of CNN; $Convid\text{-}Net$ which tends to probe optimum solution for covid-19 detection.

3 Methodology

3.1 Dataset

Table 1 contains the information of those publicly available datasets which have been used to create the customized dataset for the proposed architecture. Collected data set contains total 1440 chest X-ray images of COVID-19, 2470 normal chest X-ray images and 2407 chest X-ray images of pneumonia. Since this amount of data is insufficient for neural network training, augmentation has been applied in this context to generate a larger dataset. Online augmentation with neural networks takes more time than offline augmentation. If the train and test data are augmented from the same source, then this undermines the performance of neural network. For this reason, each of the three classes: COVID-19, Normal, pneumonia has been split into testing set and training set. 30% of the total data of each class is reserved for testing and 70% for training purposes. Furthermore, as the combined dataset contains images from different sources; randomly selecting 30% images can result in erroneous accuracy. Because, testing sets may contain completely different images than training images, or testing sets may be the same as training sets. To overcome this problem, images of each class have been ordered. Then every third image is selected sequentially for the test set; rest are selected for the train set. Thus both test and train set contain images from every collected dataset which in turns creates a balanced dataset.

3.2 Data Preprocessing and Augmentation

While classifying using deep learning algorithms, the mismatch ratio of dataset class distribution has an immense effect on model output after splitting data into test and train directories; thus image augmentation has been applied to create a larger dataset. In this case, the number of COVID-19 images are significantly lower than in other classes. To solve this issue, augmentation has been applied to generate

Table 1 Public datasets used to create the customized dataset for this work

Dataset	Covid-19	Normal	Pneumonia
COVID-19 X-ray [15]	94	94	0
COVID-19 patients lungs X-ray images 10,000 [16]	70	20	0
COVID-19 radiography database [17]	219	1341	1345
IEEE8023/covid-chestxray-dataset [18]	180	0	0
COVID-19 detection X-ray dataset [19]	60	880	1072
Actualmed COVID-19 chest X-ray dataset [20]	58	127	0
BIMCV covid19+ [21]	805	0	0

Table 2 Dataset details

	Non augmented			Augmented		
	Train	Test	Total	Train	Test	Total
COVID-19	1008	432	1440	7052	3022	10,074
Normal	1729	741	2470	7092	3004	10,096
Pneumonia	1685	722	2407	6925	2928	9903
Total augmented data						30,073

same number of images for each class. Again, the dimension of images varies in the collected dataset. Thus, after resizing the images at 256 $*$ 256 pixels; the following augmentation techniques have been applied:

- **Feature wise center** Normalizing input by subtracting mean value of the input distribution.
- **Rotation** Rotates image in certain direction (clockwise and counterclockwise) at most 20°.
- **Zoom** Scaling up images at most 10%.
- **Width shift** Moving image to a certain direction along the horizontal axis at most 15%.
- **Shear range** Displaces each point in the vertical direction by a sum commensurate with its distance from the picture edge, at most 10%.
- **Horizontal Flip** Making mirror of input image by flipping horizontal axis.
- **Fill mode nearest** Points outside the boundaries of the input are filled according to nearest color.

Table 2 shows the train and test data distribution of augmented and non-augmented data used for the dataset, created for this work.

3.3 *Convid-Net*

Proposed deep learning architecture $Convid\text{-}Net$ consists of the following building blocks:
$(CONV)$ layer Set of learnable filter (kernel) which calculates the output of neurons attached to local input areas, each calculating a dot product between their weights and a specific area related to them in the input volume.

Sub sampling (pooling) layer Pooling layer takes maximum (Max Pooling (MP)) or average (Average Pooling) value from each $CONV$ feature matrix input.

Batch normalization (BN) layer A process used to render neural networks quickly in a robust way by re-centering and re-scaling of the input layer.

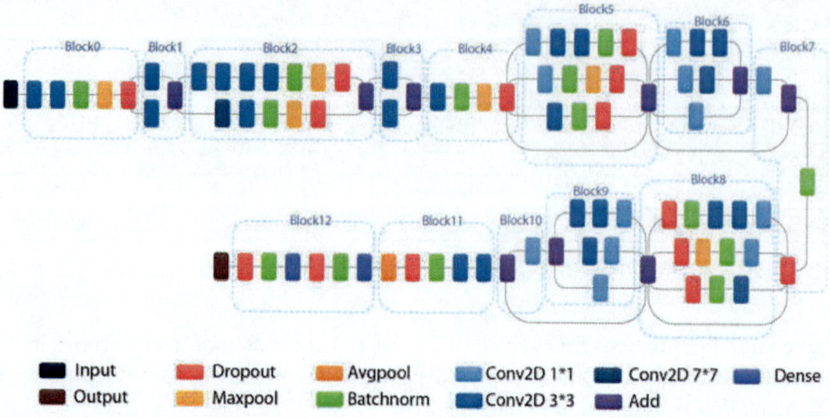

Fig. 2 *Convid-Net* architecture

Dropout (DO) **layer** A regularization method to temporarily remove units, including all its incoming and outgoing connections. This regularization prevents overfitting of the network.

Add layer Add same dimension input tensor by allowing skipping connection (residual) from previous layers and also parallel $CONV$, which in turns creates an alternative route for gradients to pass.

Dense layer Completely Connected layers in a neural network where all inputs from one layer are linked to each activation unit of the previous and following layer. Softmax is generally used as an activation function in the dense layer.

Figure 2 shows the overall *Convid-Net* architecture.

Here, deep CNN architecture has been proposed with various degrees of reduction based valid padding parallel $CONV$ and reduction less same padding parallel $CONV$ with residual block. This network subsequently expands into different $CONV$, pooling, BN layers and then fuses those layers for different feature extraction. Valid padding $CONV$ reduces image dimension (height and width) while extracting features (Block 0–4 in Fig. 2). Subsequent $CONV$ layers have been used in sub-blocks for complex feature extraction. Since, the valid padding $CONV$ layers reduce input dimension at a constant rate, it constrains the number of convolutions applied to a specific image input. The same padding $CONV$ with a residual connection from the previous layer has also been applied so that input images get fully covered by filters and output preserves the dimension of the input image (Block 5–10 in Fig. 2). Subsequent $CONV$ layer(Block 0, 2 in Fig. 2) has been used in this architecture because larger kernel size $CONV$ is computationally expensive. Instead of using subsequent smaller kernel size $CONV$ also results in the same feature extraction and output dimension with less computation. While creating the subsequent layer, learning of the previous layer may disappear. Parallel $CONV$ allows a network to choose relevant filter size $CONV$. To reduce overfitting BN and DO have been added either in each parallel $CONV$ or at the end of the concatenation of parallel layers. Parallel $CONV$ have been used with a residual block (Block 5, 7, 8, 10 of Fig. 2) to prevent vanishing gradient. Thus the network can select relevant features from different $CONV$ filters or input from the previous block. To minimize the number of trainable parameters and computational complexity; reduction based blocks have been used. Block 0–4 reduces the image dimension from $120 * 120$ to $10 * 10$ pixels. Then more complex parallel $CONV$ (block 5–10) were applied for feature extraction. To further reduce dimension, for being able to feed into dense layer, sequential $CONV$-MP blocks were placed after block 10. Following that, since flatten turns a multidimensional vector into a one dimensional vector without any calculation; thus global average pooling has been used in this case for additional special dimensions and overfitting depletion. A Summary of the proposed architecture is shown in Fig. 3. To compile the given architecture, *Adam* has been used as optimizer and categorical cross entropy as loss function with batch size 64. Total parameters calculated as 4,292,067 for the proposed architecture.

4 Experimental Analysis

The experiments are performed in anaconda environment using python language and TensorFlow 2.1 running on a machine equipped with intel Core i5 processor, 8 GB ram, 2GB Nvidia 920mx GPU. All models trained on a total 21,096 and tested on a total 8954 chest X-ray images. $120 * 120$ dimensional images have been used as input in all networks. DO of 0.25 and 0.5 have been applied in the $CONV$ and dense/$LSTM$ portion respectively. Learning rate of 0.001 has been utilized for each model. For comparison a less deep version of proposed architecture and a CNN-$LSTM$ model have also been designed and evaluated on the same dataset. These model details are following:

Model 1 Model 1 consists of 8 blocks with less $CONV$ layer than the proposed architecture. Only two parallel $CONV$ residual blocks have been used for it. Total parameters used in this model are 271,638.

Model 2 A sequential $CONV$ layer with MP, BN and DO has been utilized in model 2. The feature extraction CNN is 18 layer deep. Flatten parameters fed into 512 units of $LSTM$ layer with DO. Subsequently, similar $LSTM$ and DO layer have been applied again before the dense layer. Two hidden dense layers with 256,256 neuron units have been applied before the output layer. Total parameters used in this model are 11,140,483.

Model 3 Model 3 comprises the proposed architecture. Model 3 has been created from model 1 by optimizing parameters and layers

Figures 4 and 5 shows the graphical representation of different performance metrics; evaluated for different models of this work. In order to evaluate and quantify the efficiency of the proposed architecture *Convid-Net* on the dataset; several evaluation metrics have been utilized. Accuracy, Recall, Precision, F1-Score, Sensitivity,

	Layer(type)	Output Shape	Param	Connected to
	Image (Input Layer)	(120, 120, 3)	0	
Block_0	Conv2D_0	(118, 118, 32)	896	Image
Block_0	Conv2D_1	(116, 116, 32)	9248	Conv_0
Block_0	BatchNorm_0	(116, 116, 32)	128	Conv_1
Block_0	MaxPooling_0	(58, 58, 32)	0	BatchNorm_0
Block_0	DropOut_0	(58, 58, 32)	0	MaxPooling_0
Block_1	Conv2D_2	(56, 56, 32)	9248	DropOut_0
Block_1	Conv2D_3	(56, 56, 32)	9248	DropOut_0
Block_1	Add_0	(56, 56, 32)	0	Conv2D_2 Conv2D_3
Block_2	Conv2D_6	(54, 54, 64)	18496	Add_0
Block_2	Conv2D_7	(52, 52, 64)	36924	Conv2D_6
Block_2	Conv2D_4	(50, 50, 64)	100416	Add_0
Block_2	Conv2D_8	(50, 50, 64)	36928	Conv2D_7
Block_2	Conv2D_5	(48, 48, 64)	36928	Conv2D_4
Block_2	Conv2D_9	(48, 48, 64)	36928	Conv2D_8
Block_2	BatchNorm_1	(48, 48, 64)	256	Conv2D_5
Block_2	BatchNorm_2	(48, 48, 64)	256	Conv2D_9
Block_2	MaxPooling_1	(24, 24, 64)	0	BatchNorm_1
Block_2	MaxPooling_2	(24, 24, 64)	0	BatchNorm_2
Block_2	DropOut_1	(24, 24, 64)	0	MaxPooling_1
Block_2	DropOut_2	(24, 24, 64)	0	MaxPooling_2
Block_2	Add_1	(24, 24, 64)	0	DropOut_1 DropOut_2
Block_3	Conv2D_10	(22, 22, 64)	36928	Add_1
Block_3	Conv2D_11	(22, 22, 64)	36928	Add_1
Block_3	Add_2	(22, 22, 64)	0	Conv2D_10 Conv2D_11
Block_4	Conv2D_12	(20, 20, 128)	73856	Add_2
Block_4	BatchNorm_3	(20, 20, 128)	512	Conv2D_12
Block_4	MaxPooling_3	(10, 10, 128)	0	BatchNorm_3
Block_4	DropOut_3	(10, 10, 128)	0	MaxPooling_3
Block_5	Conv2D_15	(10, 10, 128)	16512	DropOut_3
Block_5	Conv2D_13	(10, 10, 128)	16512	DropOut_3
Block_5	Conv2D_16	(10, 10, 128)	147584	Conv2D_15
Block_5	BatchNorm_4	(10, 10, 128)	512	Conv2D_13
Block_5	Conv2D_14	(10, 10, 128)	147584	DropOut_3
Block_5	Conv2D_17	(10, 10, 128)	147584	Conv2D_16
Block_5	MaxPooling_4	(10, 10, 128)	0	BatchNorm_4
Block_5	BatchNorm_5	(10, 10, 128)	512	Conv2D_14
Block_5	BatchNorm_6	(10, 10, 128)	512	Conv2D_17
Block_5	DropOut_4	(10, 10, 128)	0	MaxPooling_4
Block_5	DropOut_5	(10, 10, 128)	0	BatchNorm_5
Block_5	DropOut_6	(10, 10, 128)	0	BatchNorm_6
Block_5	Add_3	(10, 10, 128)	0	DropOut_3 DropOut_4 DropOut_5 DropOut_6
Block_6	Conv2D_21	(10, 10, 128)	16512	Add_3
Block_6	Conv2D_19	(10, 10, 128)	16512	Add_3
Block_6	Conv2D_22	(10, 10, 128)	147584	Conv2D_21

	Layer(type)	Output Shape	Param	Connected to
Block_6	Conv2D_18	(10, 10, 128)	16512	Add_3
Block_6	Conv2D_20	(10, 10, 128)	147584	Conv2D_19
Block_6	Conv2D_23	(10, 10, 128)	147584	Conv2D_20
Block_6	Add_4	(10, 10, 128)	0	Conv2D_18 Conv2D_20 Conv2D_23
Block_7	Conv2D_24	(10, 10, 128)	16512	Add_4
Block_7	Add_5	(10, 10, 128)	0	Conv2D_24 Add_3
Block_7	BatchNorm_7	(10, 10, 128)	512	Add_5
Block_7	DropOut_7	(10, 10, 128)	0	BatchNorm_7
Block_8	Conv2D_27	(10, 10, 128)	16512	DropOut_7
Block_8	Conv2D_25	(10, 10, 128)	16512	DropOut_7
Block_8	Conv2D_28	(10, 10, 128)	147584	Conv2D_27
Block_8	BatchNorm_8	(10, 10, 128)	512	Conv2D_25
Block_8	Conv2D_26	(10, 10, 128)	147584	DropOut_7
Block_8	Conv2D_29	(10, 10, 128)	147584	Conv2D_28
Block_8	MaxPooling_5	(10, 10, 128)	0	BatchNorm_8
Block_8	BatchNorm_9	(10, 10, 128)	512	Conv2D_26
Block_8	BatchNorm_10	(10, 10, 128)	512	Conv2D_29
Block_8	DropOut_8	(10, 10, 128)	0	MaxPooling_5
Block_8	DropOut_9	(10, 10, 128)	0	BatchNorm_9
Block_8	DropOut_10	(10, 10, 128)	0	BatchNorm_10
Block_8	Add_6	(10, 10, 128)	0	DropOut_7 DropOut_8 DropOut_9 DropOut_10
Block_9	Conv2D_33	(10, 10, 128)	16512	Add_6
Block_9	Conv2D_31	(10, 10, 128)	16512	Add_6
Block_9	Conv2D_34	(10, 10, 128)	147584	Conv2D_33
Block_9	Conv2D_30	(10, 10, 128)	16512	Add_6
Block_9	Conv2D_32	(10, 10, 128)	147584	Conv2D_31
Block_9	Conv2D_35	(10, 10, 128)	147584	Conv2D_34
Block_9	Add_7	(10, 10, 128)	0	Conv2D_30 Conv2D_32 Conv2D_35
Block_10	Conv2D_36	(10, 10, 128)	16512	Add_7
Block_10	Add_8	(10, 10, 128)	0	Conv2D_36 Add_6
Block_11	Conv2D_37	(8, 8, 256)	295168	Add_8
Block_11	Conv2D_38	(6, 6, 512)	1180160	Conv2D_37
Block_11	BatchNorm_11	(6, 6, 512)	2048	Conv2D_38
Block_11	DropOut_11	(6, 6, 512)	0	BatchNorm_11
Block_11	GlobalAvgPool	(512)	0	DropOut_11
Block_12	Dense_0	(512)	262656	GlobalAvgPool
Block_12	BatchNorm_12	(512)	2048	Dense_0
Block_12	DropOut_12	(512)	0	BatchNorm_12
Block_12	Dense_1	(256)	131328	DropOut_12
Block_12	BatchNorm_13	(256)	1024	Dense_1
Block_12	DropOut_13	(256)	0	BatchNorm_13
	Dense_2 (Output layer)	(3)	771	DropOut_13

Fig. 3 Summary of *Convid-Net* architecture

Specificity, Area Under Curve (AUC) are calculated for each of the models to distinguish the performance measures among the models.

Different evaluation metrics show that Model 3 performs better than Model 1 and Model 2 owing to the presence of residual connections; nonetheless the number of parameters in Model 3 are nearly one-third of Model 2. Since test accuracy is lower than train accuracy and test accuracy closely follow train accuracy it can be stated that proposed model does not suffer from any overfitting or underfitting problem; which further validates dataset creation logic. Proposed architecture achieved average

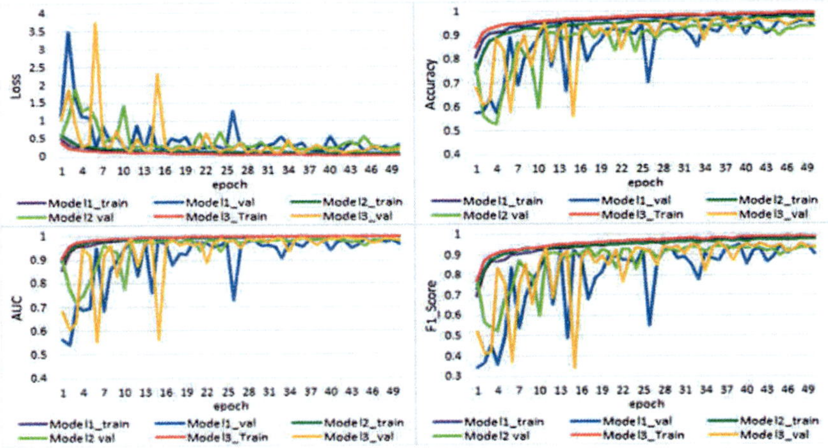

Fig. 4 AUC, loss, F1-score and accuracy measurements of Model 1, 2 and 3 of train and test dataset

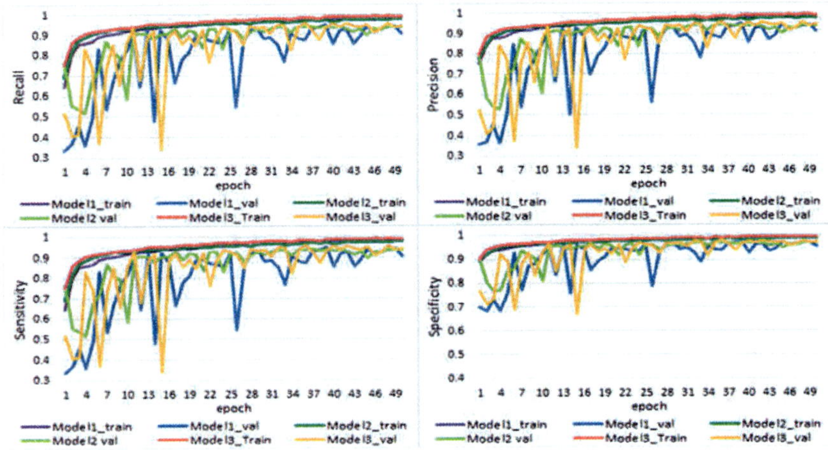

Fig. 5 Precision, recall, sensitivity and specificity measurements of Model 1, 2 and 3 of train and test dataset

training accuracy of 97.3187 and 89.37% testing accuracy with maximum 99.47% training and 97.4% testing accuracy in only 50 epochs. Average training recall of 95.78%, precision of 96.13% and AUC of 99.28, F1 score of 95.956%, specificity of 98.08%, sensitivity of 95.78% has also been achieved by the proposed architecture.

5 Conclusion

As COVID-19 emerges all over the world; more tests are needed to detect and track the spread of COVID-19. PCR test kit shortage is one of the main obstacles for rapid testing. From that perspective, a deep CNN framework ($Convid\text{-}Net$) has been proposed with parallel $CONV$ and residual connection for the detection of COVID-19 from X-ray images for this work. Proposed model achieves great performance in the given augmented dataset. Since data have been collected from various sources and augmentation has been used to generate enough data to train the networks, shortage of COVID-19 chest X-ray data is a major limitation. Though $Convid\text{-}Net$ has achieved admissible performance in classifying X-ray images, usage of the architecture in other imaging techniques such as CT scan, ultrasonic imaging is yet to be determined. In future, an openly available prediction application can be built for general use and to crowdsource COVID-19 X-ray images.

References

1. WHO Coronavirus Disease (COVID-19) Dashboard. https://covid19.who.int
2. Wynants, L, et al.: Prediction models for diagnosis and prognosis of covid-19 infection: systematic review and critical appraisal. BMJ **369** (2020)
3. Litjens, G., Kooi, T., Bejnordi, B.E., Setio, A.A.A., Ciompi, F., Ghafoorian, M., Van Der Laak, J.A., Van Ginneken, B., Sánchez, C.I.: A survey on deep learning in medical image analysis. Med. Image Anal. **42**, 60–88 (2017)
4. Khan, A.I., Shah, J.L., Bhat, M.M.: Coronet: a deep neural network for detection and diagnosis of covid-19 from chest X-ray images. In: Computer Methods and Programs in Biomedicine, p. 105581 (2020)
5. Islam, M.Z., Islam, M.M., Asraf, A.: A combined deep CNN-LSTM network for the detection of novel coronavirus (covid-19) using X-ray images. In: Informatics in Medicine Unlocked, p. 100412 (2020)
6. Oh, Y., Park, S., Ye, J.C.: Deep learning covid-19 features on CXR using limited training data sets. IEEE Trans. Med. Imag. (2020)
7. Wang, L., Wong, A.: Covid-net: a tailored deep convolutional neural network design for detection of covid-19 cases from chest X-ray images. arXiv preprint arXiv:2003.09871 (2020)
8. Narin, A., Kaya, C., Pamuk, Z.: Automatic detection of coronavirus disease (covid-19) using X-ray images and deep convolutional neural networks. arXiv preprint arXiv:2003.10849 (2020)
9. Abbas, A., Abdelsamea, M.M., Gaber, M.M.: Classification of covid-19 in chest X-ray images using detrac deep convolutional neural network. arXiv preprint arXiv:2003.13815 (2020)
10. Ghoshal, B., Tucker, A.: Estimating uncertainty and interpretability in deep learning for coronavirus (covid-19) detection. arXiv preprint arXiv:2003.10769 (2020)
11. Khalifa, N.E.M., Taha, M.H.N., Hassanien, A.E., Elghamrawy, S.: Detection of coronavirus (covid-19) associated pneumonia based on generative adversarial networks and a fine-tuned deep transfer learning model using chest X-ray dataset. arXiv preprint arXiv:2004.01184 (2020)
12. Ozturk, S., Ozkaya, U., Barstugan, M.: Classification of coronavirus images using shrunken features. medRxiv (2020)
13. Shi, F., Wang, J., Shi, J., Wu, Z., Wang, Q., Tang, Z., He, K., Shi, Y., Shen, D.: Review of artificial intelligence techniques in imaging data acquisition, segmentation and diagnosis for covid-19. IEEE reviews in biomedical engineering (2020)

14. Chen, D., Ji, S., Liu, F., Li, Z., Zhou, X.: A review of automated diagnosis of covid-19 based on scanning images (2020)
15. COVID-19 X ray dataset (Train & test sets). https://kaggle.com/khoongweihao/covid19-xray-dataset-train-test-sets
16. COVID-19 patients lungs X ray images 10000. https://kaggle.com/nabeelsajid917/covid-19-x-ray-10000-images
17. COVID-19 radiography database. https://kaggle.com/tawsifurrahman/covid19-radiography-database
18. Cohen, J.P.: IEEE8023/covid-chestxray-dataset, Sept 2020. https://github.com/ieee8023/covid-chestxray-dataset, original-date: 2020-02-14T23:22:23Z
19. COVID-19 detection X-ray dataset. https://kaggle.com/darshan1504/covid19-detection-xray-dataset
20. Agchung: Agchung/Actualmed-COVID-chestxray-dataset, Sept 2020. https://github.com/agchung/Actualmed-COVID-chestxray-dataset, original-date: 2020-04-17T15:45:49Z
21. BIMCV Medical Imaging Databank of the Valencia Region, Pertusa, A., de la Iglesia Vaya, M.: BIMCV-COVID19+ (May 2020). OSF. https://doi.org/10.17605/OSF.IO/NH7G8. https://osf.io/nh7g8/

Predicting Level of Visual Focus of Human's Attention Using Machine Learning Approaches

Partha Chakraborty, Mohammad Abu Yousuf, and Saifur Rahman

Abstract Attention is a human's concentration of awareness on a specific moment or information. The level of attention describes how deep the focus is on that specific moment or information. The level of attention generally depends on the environment and the mental condition of a human. This paper has offered various machine learning approaches for predicting the level of visual focus of human's attention. Using data collected from survey reports (environmental data) and eyeball data (eyeball movement, reading time, and head turn) while reading an article, a dataset with each participant's attention level was formed. Eight different classifiers were trained, namely: Logistic Regression, Support Vector Machine (SVM), Decision Tree, *K*-Nearest Neighbor (KNN), AdaBoost, Multilayer Perceptron (MLP), Extra Tree Classifier, and Voting Classifier for classifying the participant's attention level into three classes: High, Average, and Low. The Logistic Regression achieved the highest accuracy of 96% beating others in predicting these classes. The aggregated weighted Voting Classifier attained an accuracy of 95%.

Keywords LVFOA · Prediction · Eyeball detection · Weighted voting approach · Machine learning

P. Chakraborty (✉) · S. Rahman
Department of Computer Science and Engineering,
Comilla University, Cumilla 3506, Bangladesh
e-mail: partha.chak@cou.ac.bd

S. Rahman
e-mail: saifurrahmany43@gmail.com

M. A. Yousuf
Institute of Information Technology, Jahangirnagar University,
Savar, Dhaka 1342, Bangladesh
e-mail: yousuf@juniv.edu

M. S. Kaiser et al. (eds.), *Proceedings of International Conference on Trends in Computational and Cognitive Engineering*, Advances in Intelligent Systems and Computing 1309, https://doi.org/10.1007/978-981-33-4673-4_56

1 Introduction

'He who works, he who rises'—this proverb has been recited by our parents since childhood. The problem is all of us in one place; all kinds of thoughts come to mind when anyone sits down to read. Who knows where the attention disappears! Before we do anything, we have to prove to ourselves that we can. In Biology, it has been proven that if it is done with confidence before doing any work, then the body's cells give much more effort, the results of which are multiplied. So we have to be confident and relieve stress. That is, fear or narrowness is the main obstacle to success. If the mind is immersed in the surroundings' thoughts, our mind becomes normal to perform poorly.

The analysis shows that participants like students may be under mental pressure for a variety of reasons. Maybe it's a quiz or mid or final exam tomorrow, but they are unable to complete their studies, or a lot of reading is stuck together. The study is not going well but he is very talented. Due to poor financial condition, he has to continue his tuition near the university or in the city and support himself and his family at the same time. For these reasons, he lacks concentration in his studies. When the teachers are teaching in the classroom or when a student is reading, he is engrossed in various thoughts. So he seems to be reading as he sees at the book, but his attention is on the other side. Surveys determine his surroundings during data analysis, but the eyeball movement issue is also taken seriously during his reading.

In order to provide higher education, about 44 public universities have been established in Bangladesh, where most of the students in these universities are from middle or lower middle-class households. They have to study at the university by facing various problems around them. Despite having a lot of talent due to the pressure of their surroundings at different times, they have to face many obstacles in their studies due to which their lack of attention affects the results.

In this case, various personal features were collected through survey reports and other eyeball data collected through a small experiment using our 'PC Eyeball Tracker' software. Finally, this joint part (survey report and eyeball data) gives complete attention information. Therefore, various statistical analysis and machine learning approaches have been applied to predict the level of visual focus of human attention by rubbing the data of different factors to the total participants of a department.

The original purpose of this paper is not only to predict the level of visual focus of human's attention as low, average, and high categories using various machine learning approaches but also to compare various machine learning approaches to find out which approach performs better to give accuracy.

The entire manuscript is organized in such a way that the literature review is done in Sect. 2, the analysis of attention level detection is done in Sect. 3, the system architecture in Sect. 4, the experimental details in Sect. 5, the discussion is in Sect. 6 and the final Sect. 7 is in conclusion.

2 Literature Review

Caye et al. [1] developed various risk models: Stochastic Gradient Boosting, Random Forest, and Artificial Neural Network which predicts adult attention-deficit/hyperactivity disorder (ADHD). A human-robot interaction system was established by Chakrabarty et al. [2] to detect the visual focus of attention (VFOA) based on human attention. The system detects the person's current attention and calculates the level by detecting gaze pattern calculation. A deep neural network model was trained by Zhang et al. [3] with high-quality gaze data of people playing Atari games to predict human focus positions and visual interest with high precision. A database was collected by studying subject's head movement in panoramic video sequences in paper [4] to form a DRL-based HM prediction (DHP) approach. The behavioral patterns of students were compared in terms of their suitability for predicting the course outcomes of students participating in MOOCs [5]. Using the current state of game variables, a classifier [6] was developed and then used at runtime to predict object category and gaze during gameplay with more accurately than previous approaches. In [7], a query-guided spatial attention model was developed that searches for regions related to either individual words or the whole query, repeating the procedure over several recurrent steps. Han et al. [8] showed a comparative study of spatial and temporal visual attention models for videos, sequentially by collecting human gaze [9] data.

A comparison study of various visual attention modeling was presented in paper [10]. A taxonomy of almost 65 models was presented in the paper, which offers a crucial comparison of methods, their strengths, and their drawbacks. Heller et al. [11] developed a highly accurate predictive algorithm in correctly predicting diagnoses based on gameplay of "Groundskeeper". A human attention prediction model [12] was developed considering the sentiment or emotion of the object. A new evaluation metric (AttI) was formed and then trained a deep neural network for human attention prediction. Zaletelj and Košir [13] suggested an innovative approach to the automated calculation of students' concentration during classroom lectures. For both facial and body movement data gathered by the Kinect One sensor, various machine learning algorithms [14] have been trained for prediction.

3 Analysis of Attention Level Detection

Before the experiment, we recorded videos of participants while studying very intensely, just to analyze how long a participant's attention has been on them and how it is done without letting them know. Each video is carefully watched from the 1st second to the end, where the average video length was three to five minutes long, and the attention fluctuates in different ways over time.time. Table 1. represent the activities of a participant to calculate the attention level based on Fig. 1 that represent the analysis of various level of attention of a participant over times. For the convenience

Table 1 Activities of a participant to calculate attention level

Time (min)	Activities	Figures	Time (min)	Activities	Figures
0–0.22	1. Has studied continuously with full attention 2. The eyeball moves from left to right 3. Has studied by uttering words	1a	1.11–1.19	1. Has studied continuously with full attention again 2. The eyeball moves from left to right	1e
0.23–0.26	1. Suddenly she stopped studying by uttering words 2. The direction of the eyeball has moved away from the book 3. Scratching her cheek	1b	1.20–1.42	1. Suddenly the sound of reading faded 2. Folded her fingers and put them in her mouth and repeatedly brushing his teeth with her fingers 3. Looking at the same place at a glance, scratching her leg and touching her hair	1f
0.27–1.02	1. Has studied continuously with full attention again 2. The eyeball moves from left to right 3. Studied by uttering words	1c	1.43–1.59	1. Deep focus on reading again 2. Eyeball moves from left to right 3. Study is going on by marking with a pen	1g
1.02–1.10	1. Suddenly the sound of reading faded 2. She put her finger in her mouth 3. She turned the pages of the book while reading 4. She shook hands and opened it again	1d	2.00–2.10	1. Started scratching her head with her hands 2. Eyes moved from the book to the fingers	1h
2.11–3.00	1. The eyeball has moved toward he book again 2. The eyeball moves from lef to right 3. Studied by uttering words				1i

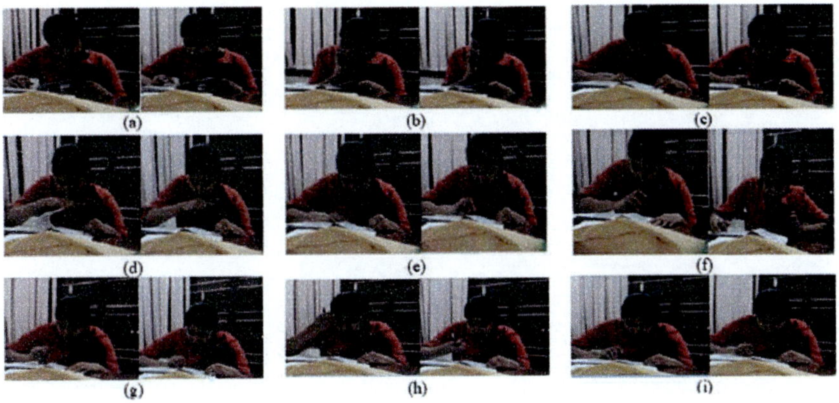

Fig. 1 Analysis of attention level calculation

of calculation, information is given in one video file through which it can be seen that the level of attention fluctuates in different ways over time. After the video was recorded, they were asked what it means to increase or decrease concentration. They talk about various family problems that are faced with their life. After observing all the videos in this way, it can be concluded that the level of attention in reading may be low or high or average, which is the same as shown in the above participant's case. Every participant is taken in both the survey and the eyeball data calculation method.

4 System Architecture

After creating the dataset based on analysis part, it was spited into train and test parts. A total number of eight classifiers were used in the study. These classifiers were trained with the training dataset. Finally, each classifier was evaluated with the test dataset by forming some evaluation matrix. The full proposed system architecture is divided into two parts named data collection and classification model which is shown in Fig. 2.

5 Experimental Details

5.1 Data Collection

The survey part has been obtained by studying a total of 400 participants from a department of a reputated public university. This dataset contains a total of 23 param-

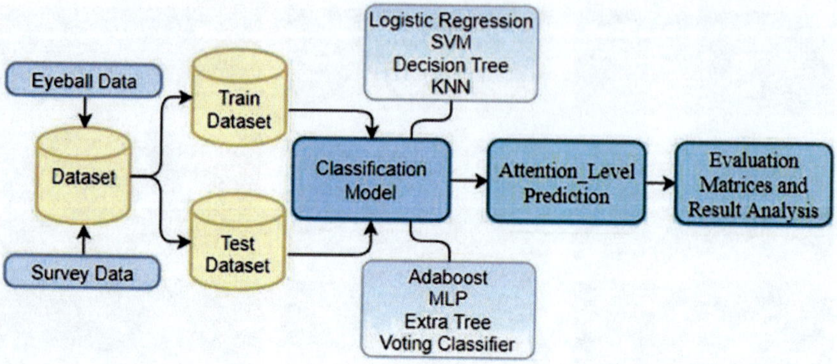

Fig. 2 Proposed system architecture

eters. These are Participant_ID, Name, Age, Gender, Mental Pressure, Mark_MP as M_p, Study Pressure, Mark_SP as S_p, Exam Pressure, Mark_EP as E_p, Tuition Profession, Mark_TP as T_p, Economic Status, Mark_ES as E_s, Residence Status, Mark_RS as R_s, Room Environment, Mark_RE as R_e, Time_taken_to_read_the_article(minute) as T_t, Eyeball_movement as E_m, Number_of_head_turn as H_t, Total_Marks as T_m, and Attention_Level as A_l. The whole process of creating a dataset has been divided into two parts:

I. Value obtained from survey report: All the columns in the survey part were not used as features in our study; the main seven featured are used to calculate the total survey data, SD_t:

$$SD_t = M_p + S_p + E_p + T_p + E_s + R_s + R_e \tag{1}$$

The attributes described were just not labeled as 'Yes' or 'No'. Each feature of the survey is marked based on a level of 0–10 where 0 is the lowest mark and 10 being the highest mark. It was done to get a clear picture between a particular attribute and attention for a particular student. All features are also described as the same. A complete survey collected these fields and data.

II. Value obtained from the eyeball movement:
The last three features are used to calculate the total eyeball data, ED_t:

$$ED_t = T_t + E_m + H_t \tag{2}$$

These features were collected through a small experiment using our 'PC Eyeball Tracker' software. Participants were instructed to read a short article (analysis show that it takes roughly one minute to read the selected article) by sitting in front of a computer. Only one standard laptop and webcam is enough to use the software. The software is programmed with Python, where various packages of OpenCV are used. The process of PC eyeball tracker is described below:

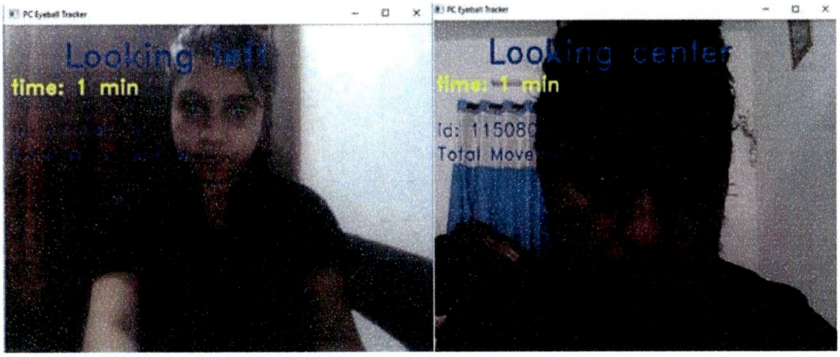

Fig. 3 Collect eyeball data from PC eyeball tracker soft

1. Reading the face from webcam
2. Detect face using dlib frontal face detector method
3. Binarizing, isolating eyes and detecting the left and right eyes and the blinking
4. Detect the iris of the eyes and estimating the position of the iris by calculating the centroid
5. If pupils are located, then the methods return the co ordinates (x, y) of each eye, the left or right position of each eye
6. Taking input of every Participant_ID
7. Counting the movement of the eyeball data from left-to-right or right-to-left in one minute in Fig. 3
8. Record the data in a csv file with Participant_ID, Eyeball_Movement, Number_of_head_turn, time(min).

The last column, 'Attention_Level' from the dataset is our target class. 'Attention_Level' for a participant was calculated by adding total of 10 features mark where the total marks as T_m were calculated in 100.

$$T_m = SD_t + ED_t \tag{3}$$

Table 2 shows how 'Attention_Level as A_t' of a participant was calculated in dataset. Participant's dataset has been created by sampling a total of 400 data where 118 participants were found highly attentive, 130 participants average and the rest 152 were found low attentive.

Table 2 Example of attention_level calculation for two participant

Participant_ID	SD_t	ED_t			T_m	A_l	Remarks
		T_t	E_m	H_t			
11508035	53	1	4	2	60	Average	Low: 1–39
11508022	62	1	8	5	76	High	High: 70–100 Average: 40–69

5.2 *Classification Model*

Seven base classifiers were used in the study for prediction purposes. Logistic Regression, Support Vector Machine, *K*-Nearest Neighbours, Adaboost Classifier [15], Multi-layer Perception Classifier, and Extra Tree Classifier [16]. A weighted voting classifiers was also used to aggregate the results of these base classifiers. The dataset was split into two parts, training and testing before fitting into the models. The training dataset was 75, and 25% were used for testing. All the classifiers were trained separately, and their accuracy were also measured separately. Then a weighted voting classifier was used for classification. Weights were added to the classifiers based on the performance of that classifier. If a particular classifier achieved better accuracy than others, then a higher weight was given to that classifier.

5.3 *Result Analysis*

Five evaluation matrix was used to measure the performance of each classifier. These are precision, recall, f1-score, accuracy, and Area Under the Curve (AUC). 25% of data, which is 100 samples in number, was used for testing purposes. In the test dataset, there were 45 students with low attention, 29 with average, and 26 with a high attention level. A confusion matrix was made for each classifier from the predicted data and actual data shown in Table 3.

Low attention class was predicted almost rightly for all the classifiers. But other classes were hard to catch for most of the classifiers. From the confusion matrix, all the evaluation matrices mentioned earlier were calculated as shown in the classification report Table 4. It reveals that Adaboost Classifier has the lowest 84% accuracy, and Logistic Regression achieved the maximum accuracy of 96%. The second highest accuracy of 95% is attained by two classifiers, the KNN classifier, and Weighted Voting Classifier. AUC was also used for classifier evaluation as it informs how much a model can differentiate between classes. The higher the AUC (maximum value of 1) of a model, the better the model is on class prediction. From the eight classifiers, the maximum AUC 0.966 value was achieved by Logistic Regression. The

Table 3 Confusion matrix

Classifiername	Actual →	High	Average	Low
Logistic	Predicted ↓			
regression	High	25	1	0
	Average	2	26	1
	Low	0	0	45
SVM	High	21	5	0
	Average	0	28	1
	Low	0	0	45
Decision tree	High	23	3	0
	Average	6	22	1
	Low	0	2	43
KNN	High	23	3	0
	Average	1	27	1
	Low	0	0	45
AdaBoost	High	18	8	0
	Average	7	21	1
	Low	0	0	45
MLP	High	23	3	0
	Average	9	19	1
	Low	0	0	45
Extra tree	High	23	3	0
	Average	7	21	1
	Low	0	1	44
Voting classifier	High	22	4	0
	Average	0	28	1
	Low	0	0	45

second highest AUC 0.956 was also attained by two classifiers as Accuracy, KNN, and Voting Classifier, followed by SVM in the third position with an AUC value of 0.947. AdaBoost Classifier achieved the lowest AUC value. Table 4 and earlier analysis has shown that Logistic Regression achieved the highest accuracy of 96% among the eight classifiers. It also achieved the highest AUC means the probability of distinguishing classes is higher than others in Logistic Regression. Its weighted average of f1-score was 0.96.

6 Discussion

After collecting data from participants, the proposal system has performed much better, especially for those who are under a lot of stress, whether personal or family

Table 4 Result analysis

Classifier name	Class label	Precision	Recall	F1-score	Accuracy (%)	AUC
Logistic regression	High	0.93	0.96	0.94	96	0.966
	Average	0.96	0.90	0.93		
	Low	0.98	1.00	0.99		
SVM	High	1.00	0.81	0.89	94	0.947
	Average	0.85	0.97	0.90		
	Low	0.98	1.00	0.99		
Decision tree	High	0.79	0.88	0.84	88	0.904
	Average	0.81	0.76	0.79		
	Low	0.98	0.96	0.97		
KNN	High	0.96	0.88	0.92	95	0.956
	Average	0.90	0.93	0.92		
	Low	0.98	1.00	0.99		
AdaBoost	High	0.72	0.69	0.71	84	0.865
	Average	0.72	0.72	0.72		
	Low	0.98	1.00	0.99		
MLP	High	0.72	0.88	0.79	87	0.893
	Average	0.86	0.66	0.75		
	Low	0.98	1.00	0.99		
Extra tree	High	0.77	0.88	0.82	88	0.903
	Average	0.84	0.72	0.78		
	Low	0.98	0.98	0.98		
Voting classifier	High	1.00	0.85	0.92	95	0.956
	Average	0.88	0.97	0.92		
	Low	0.98	1.00	0.99		

or in some way, have a big impact on attention in reading. But, working with data from all the university students is very challenging if we build a dataset and learn it by machine learning approaches where most of the students of public universities are not available online. The possibilities of getting personal data collection of the participants was very difficult for us.

7 Conclusion

It can be said that the proposed study calculated not only to predict the level of visual focus of human's attention as low, average, and high categories but also showed a clear comparison between various machine learning models with the dataset. Although the logistic regression classifier topped the others with 96% accuracy, the voting classifier also performed better than the rest of the others.

The limitation of this system is that there are only three levels of attention class, and the dataset is created using a total of 400 participants in a department of a public university, and no comparison is made with the work of any researchers.

In the future, we must not only be limited to one department of the public university but also work with all the participants mainly students of all the departments of public and private universities to create datasets. Apart from this, increasing the class of attention and train that data with neural network model could be a challenging task.

References

1. Caye, A., Agnew-Blais, J., Arseneault, L., Gonçalves, H., Kieling, C., Langley, K., Menezes, A.M.B., Moffitt, T.E., Passos, I.C., Rocha, T.B., Sibley, M.H., Swanson, J.M., Thapar, A., Wehrmeister, F., Rohde, L.A.: A risk calculator to predict adult attention-deficit/hyperactivity disorder: generation and external validation in three birth cohorts and one clinical sample. In: Epidemiology and Psychiatric Sciences (2020)
2. Chakraborty, P., Yousuf, M.A., Zahidur Rahman, M., Faruqui, N.: How can a robot calculate the level of visual focus of human's attention. In: Algorithms for Intelligent Systems. Springer, Singapore (2020)
3. Zhang, R., Liu, Z., Zhang, L., Whritner, J., Muller, K., Hayhoe, M., Ballard, D.: AGIL: learning attention from human for visuomotor tasks. In: 15th European Conference. Munich, Germany, 8–14 Sept 2018
4. Xu, M., Song, Y., Wang, J., Qiao, M., Huo, L., Wang, Z.: Predicting head movement in panoramic video: a deep reinforcement learning approach. IEEE Trans. Pattern Anal. Mach. Intell. 1 (2018)
5. AL-Shabandar, R., Hussain, A., Laws, A., Keight, R., Lunn, J., Radi, N.: Machine learning approaches to predict learning outcomes in Massive open online courses. In: International Joint Conference on Neural Networks (IJCNN), pp. 713-720 (2017)
6. Koulieris, G., Drettakis, G., Cunningham, D., Mania, K.: Gaze prediction using machine learning for dynamic stereo manipulation in games. In: IEEE Virtual Reality (VR), pp. 113–120 (2016)
7. Xu, H., Saenko, K.: Ask, Attend and answer: exploring question-guided spatial attention for visual question answering. In: Information Security and Privacy, pp. 451–466 (2016)
8. Han, J., Sun, L., Hu, X., Han, J., Shao, L.: Spatial and temporal visual attention prediction in videos using eye movement data. Neurocomputing **145**, 140–153 (2014)
9. Chakraborty, P., Roy, D., Zahid, Z.R., Rahman, S.: Eye gaze controlled virtual keyboard. Int. J. Rec. Technol. Eng. (IJRTE) **8**(4), 3264–3269 (2019)
10. Borji, A., Itti, L.: State-of-the-art in visual attention modeling. IEEE Trans. Pattern Anal. Mach. Intell. **35**, 185–207 (2013)
11. Heller, M.D., Roots, K., Srivastava, S., Schumann, J., Srivastava, J., Hale, T.: A machine learning-based analysis of game data for attention deficit hyperactivity disorder assessment. Games Health J. **2**(5), 291–8 (2013)

12. Cordel, M.O., Fan, S., Shen, Z., Kankanhalli, M.: Emotion-aware human attention prediction. In: 2019 IEEE/CVF Conference on Computer Vision and Pattern Recognition (CVPR), pp. 4021–4030 (2019)
13. Zaletelj, J., Košir, A.: Predicting students' attention in the classroom from Kinect facial and body features. J. Image Video Proc. 80 (2017)
14. Zulfiker, M.S., Kabir, N., Biswas, A.A., Chakraborty, P., Rahman, M.M.: Predicting students' performance of the private universities of Bangladesh using machine learning approaches. Int. J. Adv. Comput. Sci. Appl. **11**(3), 672–679 (2020)
15. Freund, Y., Schapire, R.: A short introduction to boosting. J. Jpn. Soc. Artif. Intell. **14**(5), 771–780 (1999)
16. Geurts, P., Ernst, D., Wehenkel, L.: Extremely randomized trees. Mach. Learn. **63**, 3–42 (2006)

An Integrated CNN-LSTM Model for Bangla Lexical Sign Language Recognition

Nanziba Basnin, Lutfun Nahar, and Mohammad Shahadat Hossain

Abstract The development of Human-Computer Interaction (HCI) will not only empower the disabled with access to technology but also help to build tools for assisting the disable and their caregivers. Some disabled people are deaf, they use signs to communicate. Signs are emblematic forms which convey a specific meaning. These emblems construct the sign language, which is a non-verbal language used by the deaf community. This research attempts to build a Bangla Sign language recognition system that can recognize signs of both hands. Hands vary in shapes and sizes, as well as signs vary in orientations and motions. Accurate feature extraction is necessary for such systems. For such purposes, deep learning approaches can prove to be effective for the classification and feature engineering of images. In the beginning, an integrated CNN-LSTM model is proposed for building a Sign language Recognition System, a Bangla Sign Language (BSL) dataset consisting of Bangla lexical signs is considered. This dataset consists of 13,400 images comprising thirty-six classes of Bangla lexical signs. The model is trained using a CNN-LSTM model. This model produces a training accuracy of 90% and a testing accuracy of 88.5%. The proposed model is compared to CNN and other CNN state of the art models, namely VGG16, VGG9 and MobileNet. The CNN model and other CNN state undergo the problem of overfitting as their training accuracy is greater than the testing accuracy.

Keywords CNN-lSTM model · BSL dataset · Lexical signs

1 Introduction

A language is a structured tool for communication which can be verbal and non-verbal. Non-verbal language is known as Sign Language. From 1960, Sign language was recognized as a non-verbal language used by the deaf community. In general,

N. Basnin · L. Nahar
International Islamic University Chittagong, Chittagong, Bangladesh

M. S. Hossain (✉)
University of Chittagong, Chittagong, Bangladesh
e-mail: hossain_ms@cu.ac.bd

M. S. Kaiser et al. (eds.), *Proceedings of International Conference on Trends in Computational and Cognitive Engineering*, Advances in Intelligent Systems and Computing 1309, https://doi.org/10.1007/978-981-33-4673-4_57

each spoken language has a Sign Language. Recent studies by the World Health Organization (WHO) show that 5% of the world population suffer from impaired speech and hearing disabilities [8]. In Bangladesh, 13 million people have hearing and speech disabilities [25].

Bangla Sign language (BSL) gained recognition in 2000 by the Center for disability in development (CDD) as a language for the Bangladeshi deaf community. This Bangla Sign Language (BSL) has complex meaningful semantics formed from the combination of emblematic elements such as hand forms, orientation, location, movement, and non-manual markers. Most of the lexical signs in Bangla are configured from the arrangement of two hands. As a result, it is difficult for the deaf to communicate or teach their language to hearing people. Moreover, Sign Language is perceived as sophisticated and complex because of its unique phonology, morphology, grammar and lexical level which is different from that of spoken language. Therefore, to help the deaf community interact with society, it is important to develop an interpreter that translates sign language into text or speech [14, 20, 28]. To build such an interpreter, it is required to develop a method for classification of Bangla Sign Language. Classification can be carried out in two ways. A sensor-based approach or a computer vision-based approach can be implied to automatically recognize signs in sign language.

This research focuses on building an effective tool for the recognition of sign language, by introducing an integrated CNN-LSTM (Convolution Neural Network—Long Short Term Memory) model. The CNN model is considered because it can handle large amounts of raw data with comparatively less pre-processing effort, while LSTM can optimise the backpropagation [4]. Therefore, the integration of CNN with LSTM would play an important role to achieve better model accuracy. To perform this a local dataset with augmentation is used to train the CNN-LSTM model. This local dataset consists of thirty-six classes of Bangla Sign Language Dataset made up of lexical signs. The proposed model is then compared to CNN and other CNN state of the art models.

The rest of the paper is ordered as follows. Section 2 explores the existing work related to Sign Language recognition. Section 3 describes the methodology used to build the proposed model. Section 4 confers the results. While Sect. 5 draws a conclusion as well as outlines the future work.

2 Related Work

There exist few research works, when it comes to Bangla Sign Language Recognition using Bangla Sign alphabets. In [26] the recognition system developed for Bangla Sign Language (BSL) of alphabets use SVM to classify the classes in the dataset. The static dataset of 2400 images is pre-processed using HSV colour spacing. The model produced an accuracy of 97.7%. However, BSL alphabet target classes generally overlap with each other, this makes SVM unsuitable for recognizing each class accurately.

In [27] the dataset a YCBr Algorithm is used for the detection of skin. The total size of Bangla Sign Language for lexical signs consists of only 330 images. The segmented Principal Component Analysis (PCA) dataset were both trained using Artificial Neural Network. The training accuracy was 36.54%, the validation accuracy was 73.69% while the testing accuracy was 73.57%. From here we can observe that there was overfitting in the model.

In [22] the dataset was collected in real-time, by using Haar-like feature-based cascade classifiers to detect the region of hand in each frame. The dataset consisted of 3600 images. This dataset was trained using K-nearest neighbours (KNN) method. The Bangla vowels resulted in an accuracy of 98.2% whereas the Bangla consonants resulted in an accuracy of 94.75%. The model could not recognize a few alphabets of the Bangla Sign Language (BSL), because of overlapping target classes.

In [16], 47 signs of single-handed signs are incorporated into the Bangladeshi Sign Language dataset (BDSL), where each class contains 5 samples of images. So that the total number of the dataset contains 235 images. The model was trained using a different number of neural networks in NCL. Combining feature extraction with NCL achieved far better accuracy of 93%. Although the resulting accuracy is great, choosing the correct topology of NN in NCL to fit any other dataset can be tricky. As the model of NCL is sensitive to learning perimeters, initial weights, number of training samples and training methods.

In [13] uses a Dataset of Bangla sign alphabets composed of thirty-seven classes of Bengali letters, where each class consists of thirty-one images. The total dataset constitutes 1147 images. To train this dataset, VGG16 was used. VGG 16 is a pre-trained CNN model. The resulting training accuracy was 96.33% while the testing accuracy is 84.68%. From the above result, it can be deduced that the model suffered from the problem of overfitting.

The above research works [22, 26, 27] shows that Machine Learning is a good choice when it comes to working with small datasets. However, if Machine Learning is used to train large datasets it becomes computationally expensive because of feature extraction. Although in [13] VGG 16 is computationally cheaper the model experiences the problem of overfitting. As increasing the number of CNN layers decreases accuracy, because of saturated neurons and the problem of vanishing gradient. This can be overcome by integrating CNN with an LSTM layer in order to optimize backpropagation. Thus a CNN-LSTM integrated model is proposed in this research which will be described in the following section.

3 Methodology

The methodology implemented for this research is demonstrated in terms of its components. Each component illustrates how the research was carried out to identify the classes of signs (Fig. 1).

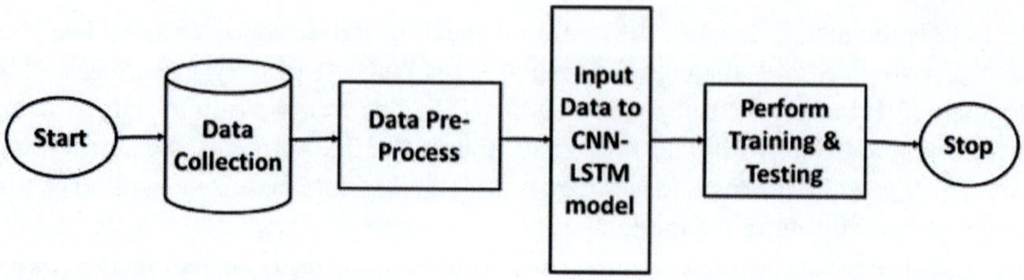

Fig. 1 Methodology of system

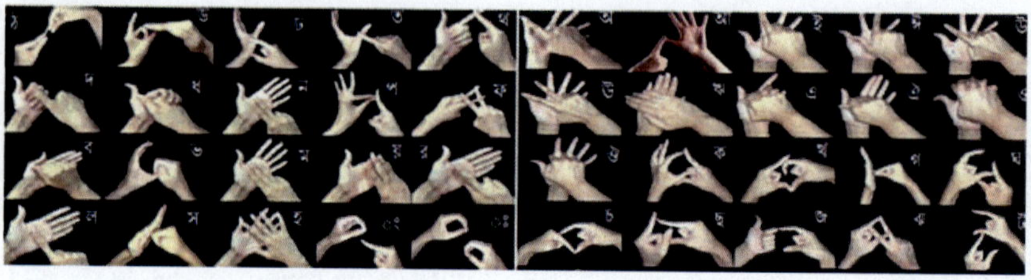

Fig. 2 Sample of local dataset [26]

3.1 Data Collection

The images in the dataset were collected using a webcam. With ethical approval, ten subjects were used to build the dataset. The dataset contains thirty-six classes of lexical signs formed by using two hands as shown in Fig. 2. Lexical signs contain Bangla consonants and vowels which represent the Bangla alphabets. For training and test, the dataset is split in a ratio of 8:2. As a result, the training set includes 10,800 images so that each class is made up of 300 images. In contrast, the testing set includes 2160 images so that each class consists of 60 images.

3.2 Data Pre-processing

The local dataset for Bangla Sign Language is pre-processed before passing it to the CNN-LSTM model. The data pre-processing goes through five different stages. These stages are namely, background subtraction, grayscale conversion, morphological erosion, median filtering and resize, as shown in Fig. 3. Initially, background subtraction is applied to each image. This method is used to generate the foreground mask. Grayscale conversion gives the image a single channel property. Next, morphological erosion is carried out which is a set of non-linear operations that take place for the shape of the image [10]. To reduce the resultant noise in images, median filtering is employed [7]. Afterwards, the image is resized into a pixel size of 60×60.

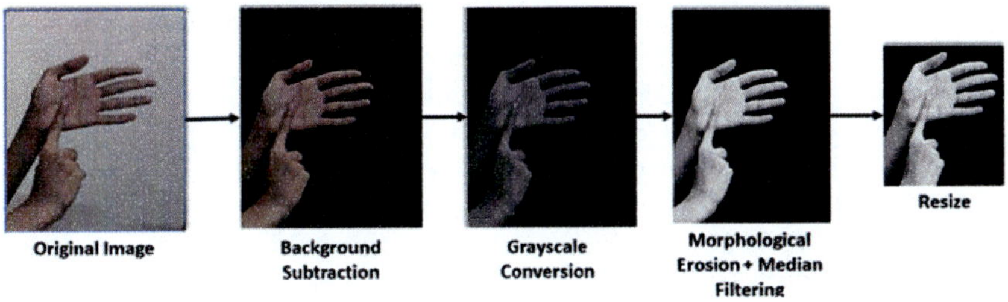

Fig. 3 Steps of data pre-processing

3.3 CNN-LSTM Integrated Model

The CNN-LSTM model shown in Table 1 consists of two convolution layers, followed by two max-pooling layers and two drop-out layers. The CNN model is separated from the LSTM layer using a Time Distributed Flattening layer. The LSTM layer is accompanied by a dropout layer. The output layer is followed by parameters such as optimizer functions, learning rates and matrices for loss as well as accuracy.

Convolution layers are layers used to generate feature maps, which are then passed on to the next layers [21]. Here, the first convolution layer, namely the input layer, consists of 64 filters which have a kernel size of three by three. The input layer is also responsible to receive images of a pixel size of sixty by sixty. Besides, the convolution layers include an activation function called Rectified Linear Unit (ReLU). ReLU

Table 1 CNN-LSTM Integrated configuration

Model content	Details
1st convolution layer 2D	Input size 60 × 60, 64 filters of kernel size 3 × 3, ReLU
1st max pooling layer	Pool size 2 × 2
Dropout layer	Randomly deactivates 25% neurons
2nd convolution layer 2D	64 filters of kernel 3 × 3, ReLU
2nd max pooling layer	Pool size 2 × 2
Dropout layer	Randomly deactivates 25% neurons
Dropout layer	Randomly deactivates 25% neurons
Flattening layer	Time Distributed
LSTM layer	200 nodes
Dropout layer	Randomly deactivates 25% neurons
Output layer	36 nodes, SoftMax
Optimization function	Stochastic Gradient Descent (SGD)
Learning rate, momentum, Nestorov	0.001, 0.9, True
Matrix	Loss, accuracy

benefits the model by reducing the problem of vanishing gradients. After each convolution layer, a max-pooling and dropout layers are added. The maximum features of the feature map are extracted using a max-pooling layer. The dropout layer excludes 25% of the nodes to prevent the model from overfitting [3, 17, 24]. This flattening layer flattens the two-dimensional feature maps into one-dimensional linear arrays. The LSTM layers include 200 nodes. It is used to interpret the features which were extracted by the CNN layer, resulting in a well defined perceptual representation of images in the convolution. Lastly, the model is embedded in an output layer. The output layers contain thirty-six nodes. A softmax activation function is also included in the output layer.

Optimizers named as Stochastic Gradient Descent (SGD) [18, 19, 23] is used as an optimizer, including perimeters such as learning rate, momentum and Nesterov. The learning rate allows the model to converge faster, providing better testing accuracy. Momentum encourages SGD to head towards a direction of reduced loss. Nestorove solves the oscillatory problem around the minima when momentum is unable to slow down, corresponding to a faster rate of model convergence.

3.4 System Implementation

The collection of images which constitutes the local dataset was carried out in real-time using the Jupyter IDE. The trained model is implemented in Google Colab. Google Colab provides access to free Graphics Processing Unit (GPU) which makes the system computationally faster. As a result libraries such as Tensor Flow, Keras, OpenCV, Sklearn, Matplotlib, Tkinter, PIL, Pandas and Numpy can be supported. Keras libraries are used to develop the CNN-LSTM model. On the other hand, TensorFlow supports the backend of the system. OpenCV is responsible for the processing of images in the dataset [1]. The graphical representations for model loss and model accuracy are generated using Matplotlib. For a real-time collection of data Tkinter is used. Numpy supports the operations performed on arrays.

For the purpose of regularization, callbacks are included while training the CNN-LSTM model. Callbacks help the model avoid the problem of overfitting and underfitting [6]. Early stopping, checkpoints and reducing learning rate on the plateau are provided as perimeters to the callback function. The loss of the model is monitored and the best model is saved by using checkpoints. Whenever the loss in validation converges, reducing learning rate on the plateau is invoked.

Before the dataset is passed onto the model, data augmentation [2] is performed on the images. The built-in data augmentation API in Keras allows augmentation of data. The perimeters applied to augment the images are shifting, rotation, normalization and zoom.

4 Result and Discussion

Sections 4.1 and 4.2 evaluates the performance of the deep learning models. The performance of each model is graphically illustrated in terms of its model accuracy and model loss. Section 4.3 displays the real-time validation carried out on the proposed CNN-LSTM model.

4.1 CNN-LSTM Model on Bangla Sign Language (BSL) Dataset

The proposed CNN-LSTM model depicted in Sect. 3.3, is trained using the BSL dataset with data augmentation and perimeters to optimize model performance. For regularization, of the model callbacks are implemented as mentioned in Sect. 3.4.

Figures 4 and 5 illustrates the accuracy and loss of the model. In Fig. 4 it is observed that at 25 epochs, there is a sudden significant increase in both testing and training accuracy. The accuracy model converges after 80 epochs before experiencing an early stopping at 99 epochs. The model reaches a training accuracy of 90.0% and a testing accuracy of 88.5%. Thus, it can be deduced that the model is well fit. In Fig. 5 a training and testing loss descends steeply till 80 epochs until it reaches a constant value after 85 epochs. This constant value for training is 0.5 while for testing it is 1.2. Thus between testing and training, there is a 0.7 loss.

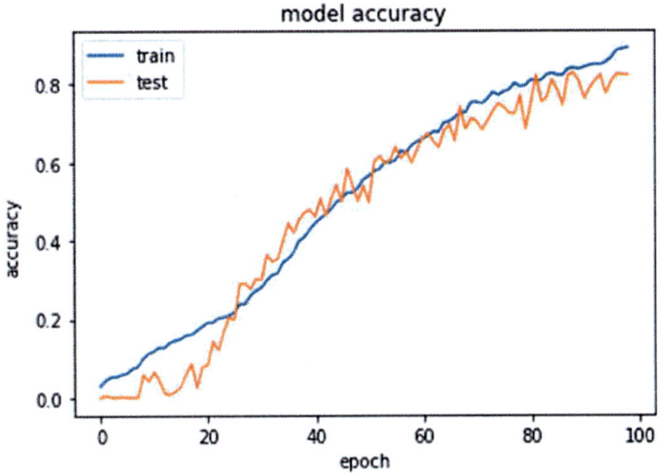

Fig. 4 Accuracy of model

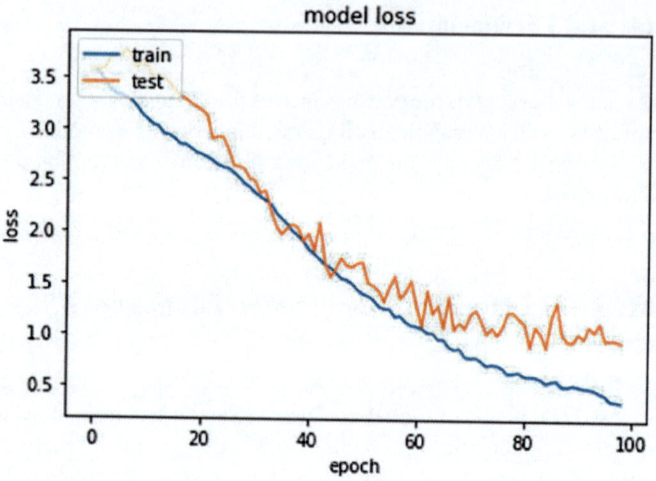

Fig. 5 Loss of model

4.2 *CNN and Other CNN State of the Art Models on BSL Dataset*

The BSL dataset with data augmentation is applied to other CNN excluding LSTM and other CNN state of the art models, namely VGG16, VGG9 and MobileNet.

Figure 6a, b represents the accuracy and loss model of CNN. In Fig. 6a. The training accuracy is 85% and the testing accuracy is 78% at 60 epochs when early stopping is invoked. This means the training accuracy is 7% larger than the testing accuracy. Thus the CNN model experiences the problem of overfitting. In Fig. 6b, the loss of the model gradually reaches a constant value after 30 epochs. The training loss becomes 0.3 while the testing loss is 1.7.

Figure 6c, d represents the model accuracy and model loss of the VGG 16 model. The model experiences an early stopping at 60 epochs. From the accuracy of the model in Fig. 6c it is observed that the training accuracy is greater than the testing accuracy. After 30 epoch, the training accuracy converges to an 85% whereas the testing accuracy fluctuates between 65 and 70%. Thus, it can be deduced that the model is overfitting. In contrast, when it comes to the model loss of VGG16 as illustrated in Fig. 6d the curve for testing increases the size of fluctuation after 50 epochs and the training loss resorts to a constant value of 1 after 30 epochs.

Figure 6e, f shows the model accuracy and loss of VGG9 model. In Fig. 6e, the training accuracy of the model is 85%. The testing accuracy fluctuates is between 70 and 72%. It can be inferred that the training accuracy is greater than the testing accuracy, corresponding to overfitting. Besides, the training loss is less than the testing loss in Fig. 6f.

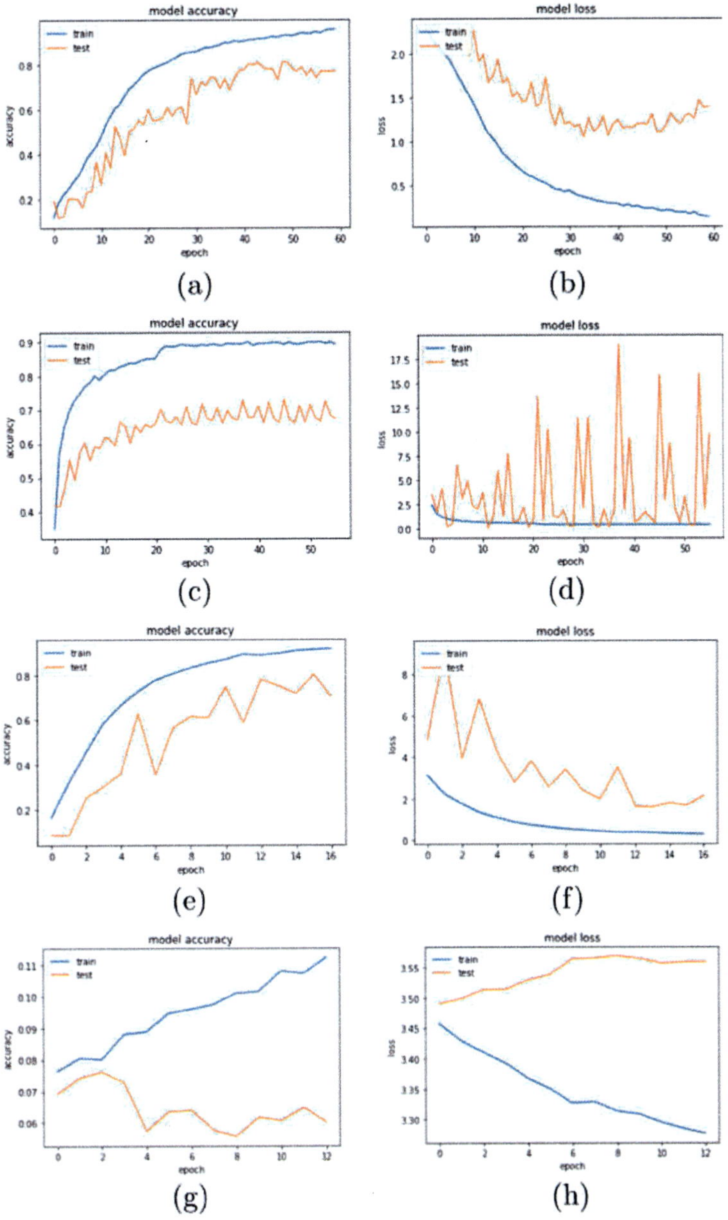

Fig. 6 **a** CNN model accuracy **b** CNN model loss **c** VGG16 model accuracy **d** VGG16 model loss **e** VGG9 model accuracy **f** VGG9 Model loss **g** MobileNet model accuracy **h** MobileNet model loss

Figure 6g, h illustrates the model accuracy and loss of the MobileNet model. It is observed in Fig. 6g, the model experienced an early stopping after 12 epochs. The training accuracy is 12% whereas the testing accuracy is 0.08%. Thus, the MobileNet model is overfitting. In Fig. 6h the curve for training loss gradually descends while the curve for model testing loss gradually ascends.

4.3 Comparison of Proposed CNN-LSTM with Other Models

According to Fig. 7, the proposed CNN-LSTM model outperforms CNN and other models with an accuracy of 88.5% as well as it does not undergo the problem of overfitting.

4.4 Real Time Validation

Real-Time Validation for Bengali Sign language is carried out using a webcam to collect real-time data as shown in Fig. 8.

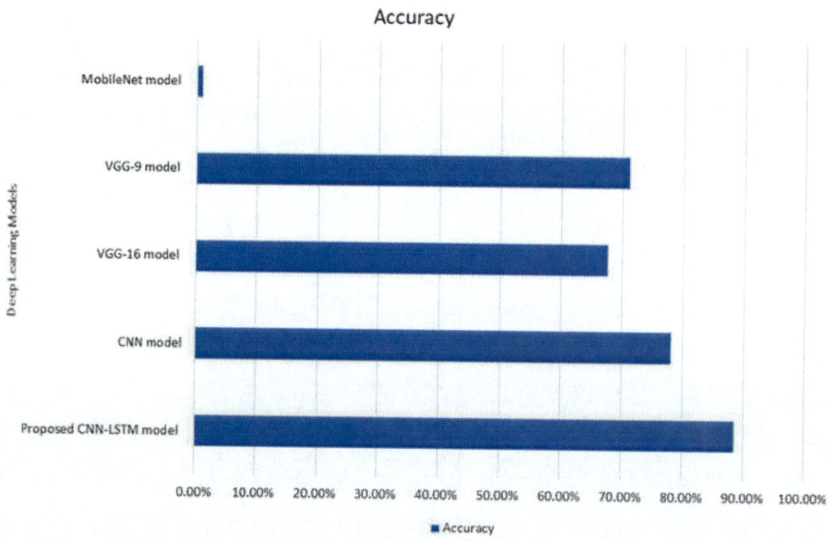

Fig. 7 Comparing the accuracy of each model

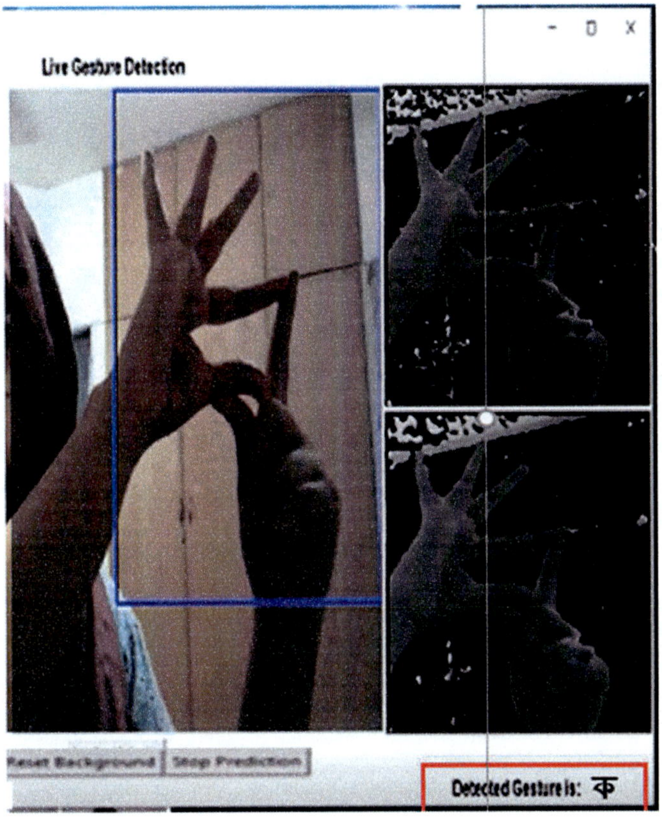

Fig. 8 BSL real time validation

5 Conclusion and Future Work

The Bengali Sign Language dataset was collected at real-time. This dataset was pre-processed using background subtraction, grey-scale conversion and morphological erosion, before being fed to the proposed CNN-LSTM model. The result of CNN-LSTM model was compared with CNN excluding LSTM and other CNN state of the art models, namely, VGG16, VGG9 and MobileNet. It was observed that the proposed CNN-LSTM model did not experience the problem of overfitting unlike the other models used in this research for comparison. Further, real-time validation is performed.

While the method provides commendable results in recognizing Bangla lexical signs, there exists no opportunity for detecting lexical signs occurring at a sequence. The recognition of sequences of lexical signs that correspond to words can enhance the research further in future. However, to achieve the recognition Bangla lexical sign sequences requires a complex methodology which will involve complex architecture.

Moreover, a dynamic dataset will need to be built comprising more number of subjects and different environmental factors. Additional classes such as numeric digits of Bangla signs can be considered in this robust dataset. Indeed the performance of such a model with a large number of classes of Bangla signs will exhibit interesting phenomenons. Besides, the complexity of the methodology can be removed if we consider a knowledge-driven approach such as Belief Rule Base (BRB). BRB's are an efficient approach when it comes to dealing with uncertain problems [5, 11, 12, 15]. Another future scope can be considered, where the output of the BSL recognition system is converted to the English Language [9].

References

1. Abedin, M.Z., Nath, A.C., Dhar, P., Deb, K., Hossain, M.S.: License plate recognition system based on contour properties and deep learning model. In: 2017 IEEE Region 10 Humanitarian Technology Conference (R10-HTC), pp. 590–593 (2017)
2. Ahmed, T.U., Hossain, S., Hossain, M.S., ul Islam, R., Andersson, K.: Facial expression recognition using convolutional neural network with data augmentation. In: 2019 Joint 8th International Conference on Informatics, Electronics Vision (ICIEV) and 2019 3rd International Conference on Imaging, Vision Pattern Recognition (icIVPR), pp. 336–341 (2019)
3. Al Banna, M.H., Ali Haider, M., Al Nahian, M.J., Islam, M.M., Taher, K.A., Kaiser, M.S.: Camera model identification using deep CNN and transfer learning approach. In: 2019 International Conference on Robotics, Electrical and Signal Processing Techniques (ICREST), pp. 626–630 (2019)
4. Al Mamun, S., Fukuda, H., Lam, A., Kobayashi, Y., Kuno, Y.: Autonomous bus boarding robotic wheelchair using bidirectional sensing systems. In: Advances in Visual Computing, pp. 737–747. LNCS, Springer International Publishing (2018)
5. Andersson, K., Hossain, M.S.: Smart risk assessment systems using belief-rule-based DSS and WSN technologies. In: 2014 4th International Conference on Wireless Communications, Vehicular Technology, Information Theory and Aerospace & Electronic Systems (VITAE), pp. 1–5. IEEE (2014)
6. Bour, A., Castillo-Olea, C., Garcia-Zapirain, B., Zahia, S.: Automatic colon polyp classification using convolutional neural network: a case study at basque country. In: 2019 IEEE International Symposium on Signal Processing and Information Technology (ISSPIT), pp. 1–5. IEEE (2019)
7. Bovik, A., Huang, T., Munson, D.: A generalization of median filtering using linear combinations of order statistics. IEEE Trans. Acoust. Speech Signal Process. **31**(6), 1342–1350 (1983)
8. Deafness: https://www.who.int/news-room/facts-in-pictures/detail/deafness
9. Gupta, D., Hossain, E., Hossain, M.S., Andersson, K., Hossain, S.: A digital personal assistant using bangla voice command recognition and face detection. In: 2019 IEEE International Conference on Robotics, Automation, Artificial-intelligence and Internet-of-Things (RAAICON), pp. 116–121 (2019)
10. Haralick, R.M., Sternberg, S.R., Zhuang, X.: Image analysis using mathematical morphology. IEEE Trans. Pattern Anal. Mach. Intell. **4**, 532–550 (1987)
11. Hossain, M.S., Khalid, M.S., Akter, S., Dey, S.: A belief rule-based expert system to diagnose influenza. In: 2014 9th International Forum on Strategic Technology (IFOST), pp. 113–116. IEEE (2014)
12. Hossain, M.S., Rahaman, S., Kor, A.L., Andersson, K., Pattinson, C.: A belief rule based expert system for datacenter PUE prediction under uncertainty. IEEE Trans. Sustain. Comput. **2**(2), 140–153 (2017)

13. Hossen, M., Govindaiah, A., Sultana, S., Bhuiyan, A.: Bengali sign language recognition using deep convolutional neural network. In: 2018 Joint 7th International Conference on Informatics, Electronics & Vision (ICIEV) and 2018 2nd International Conference on Imaging, Vision & Pattern Recognition (icIVPR), pp. 369–373. IEEE (2018)
14. Kaiser, M.S., et al.: Advances in crowd analysis for urban applications through urban event detection. IEEE Trans. Intell. Transp. Syst. **19**(10), 3092–3112 (2018)
15. Karim, R., Andersson, K., Hossain, M.S., Uddin, M.J., Meah, M.P.: A belief rule based expert system to assess clinical bronchopneumonia suspicion. In: 2016 Future Technologies Conference (FTC), pp. 655–660. IEEE (2016)
16. Karmokar, B.C., Alam, K.M.R., Siddiquee, M.K., et al.: Bangladeshi sign language recognition employing neural network ensemble. Int. J. Comput. Appl. **58**(16), 43–46 (2012)
17. Mahmud, M., Kaiser, M.S., Hussain, A.: Deep learning in mining biological data. arXiv preprint arXiv:2003.00108 (2020)
18. Mahmud, M., et al.: A brain-inspired trust management model to assure security in a cloud based IoT framework for neuroscience applications. Cogn. Comput. **10**(5), 864–873 (2018)
19. Mahmud, M., Kaiser, M.S., Hussain, A., Vassanelli, S.: Applications of deep learning and reinforcement learning to biological data. IEEE Trans. Neural Netw. Learn. Syst. **29**(6), 2063–2079 (2018)
20. Mamun, S.A., et al.: Terrain recognition for smart wheelchair. In: Intelligent Computing Methodologies, pp. 461–470. LNCS, Springer International Publishing (2016)
21. Noor, M.B.T., et al.: Application of deep learning in detecting neurological disorders from magnetic resonance images: a survey on the detection of Alzheimer's disease, Parkinson's disease and schizophrenia. Brain Inf. **7**(1), 1–21 (2020)
22. Rahaman, M.A., Jasim, M., Ali, M.H., Hasanuzzaman, M.: Real-time computer vision-based Bengali sign language recognition. In: 2014 17th International Conference on Computer and Information Technology (ICCIT), pp. 192–197. IEEE (2014)
23. Ruder, S.: An overview of gradient descent optimization algorithms. arXiv preprint arXiv:1609.04747 (2016)
24. Srivastava, N., Hinton, G., Krizhevsky, A., Sutskever, I., Salakhutdinov, R.: Dropout: a simple way to prevent neural networks from overfitting. J. Mach. Learn. Res. **15**(1), 1929–1958 (2014)
25. Sultana, Z.: Agony of persons with disability—a comparative study of Bangladesh. J. Polit. Law **3**, 212 (2010)
26. Uddin, M.A., Chowdhury, S.A.: Hand sign language recognition for Bangla alphabet using support vector machine. In: 2016 International Conference on Innovations in Science, Engineering and Technology (ICISET), pp. 1–4. IEEE (2016)
27. Yasir, R., Khan, R.A.: Two-handed hand gesture recognition for Bangla sign language using LDA and ANN. In: The 8th International Conference on Software, Knowledge, Information Management and Applications (SKIMA 2014), pp. 1–5. IEEE (2014)
28. Zisad, S.N., Hossain, M.S., Andersson, K.: Speech emotion recognition in neurological disorders using convolutional neural network. In: International Conference on Brain Informatics, pp. 287–296. Springer (2020)

Deep Learning-Based Algorithm for Skin Cancer Classification

M. Afzal Ismail, Nazia Hameed, and Jeremie Clos

Abstract Skin cancer is a disease that affects numerous people all over the world annually. It is the UK's most frequently diagnosed cancer. If not treated timely, malignant skin cancer, such as melanoma, can be extremely dangerous and even fatal as the cancer may spread to organs that are vital. Melanoma accounts for 4% of the new cancer cases diagnosed in the UK annually. This figure keeps on increasing every year. This paper proposes a classification framework using the innovations made in deep learning. This can be of great benefit to places where medical resources are minimal. For their diagnosis, it can be of great help to doctors and can give them a better chance in providing the patients with appropriate treatment. This research aims at investigating the efficiency of pre-trained convolutional neural networks and the transfer learning technique for classifying images of skin lesions. In image feature extraction, these neural networks have demonstrated to be very effective. A dataset of images, with seven different types of skin lesions and a total of 10,015 images, was separated into two classes. Skin lesions of the seven forms were classified either malignant or benign. By augmenting the number of images, the dataset was balanced. Numerous pre-trained neural networks were trained on this dataset; ResNet50, VGG16, VGG19, MobileNet, DenseNet, Inception V3. A classification framework was proposed based on their results. The framework suggested combining the three best pre-trained networks to enhance the classification accuracy. The combined model was made up of ResNet50, VGG16, and DenseNet. The addition of convolution layers to each model, that were trainable, further improved the classification accuracy. The overall accuracy achieved was 84.01% for this final proposed framework.

M. A. Ismail (✉) · N. Hameed · J. Clos
University of Nottingham, Jubilee Campus, Nottingham NG8 1BB, UK
e-mail: psxmi4@nottingham.ac.uk

N. Hameed
e-mail: Nazia.hameed@nottingham.ac.uk

J. Clos
e-mail: Jeremie.clos@nottingham.ac.uk

M. S. Kaiser et al. (eds.), *Proceedings of International Conference on Trends in Computational and Cognitive Engineering*, Advances in Intelligent Systems and Computing 1309, https://doi.org/10.1007/978-981-33-4673-4_58

Keywords Skin cancer classification · Deep learning · Pre-trained convolutional neural networks · Transfer learning · Skin lesion classification

1 Introduction

The main objective of this research work is to create a classification framework that will help to diagnose and ultimately identify skin lesions. One of the most frequently diagnosed forms of cancer annually in the UK is skin cancer. One particular malignant skin cancer, melanoma, is ranked fifth in the most frequently diagnosed type of cancer [1]. It represents 4% of the estimated number of new cases of cancer each year. 15,906 new melanoma cases were discovered in 2015, and this figure is projected to increase by 7% by 2035 [1]. It can be fatal for the patient if diagnosis or treatment of such skin cancer is delayed [2, 3]. If the cancer is diagnosed rapidly and treatment is given without delay to the patient, the chances of survival are nearly 100%. If diagnosis or treatment is delayed, however, the cancer can be fatal to the patient as it may spread to other major body parts [4]. 152,000 new cases of non-melanoma skin cancer are estimated to be diagnosed annually [5]. Over the last ten years, this number has risen by 56%. Up to 60% of the UK population will suffer from skin diseases during their lifetime [6]. This has led to an increase in the number of suicide attempts from patients suffering from such diseases [7].

It can be difficult for dermatologists to identify the skin condition from which a patient suffers [8]. This is because most of either malignant or benign lesions have similar appearances and can be distinguished only by some imperceptible characteristics. There are scoring rules that they can employ to arrive at a diagnosis, but they are not always accurate. In certain occasions, they can result in a misdiagnosis of the lesion. A biopsy can also be performed (skin tissue sample taken from patient) to assess the type of lesion. The laboratory results can take several days or weeks to be ready, however.

This paper aims to make use of developments in machine learning techniques in the extraction of features and classification of images as a helping tool to dermatologists. In underprivileged areas where resources are scarce, this can be of great benefit as this can help dermatologists in starting the appropriate treatment procedures. The main contribution of this research work is a classification framework that can distinguish between malignant and benign images of skin lesions.

Due to advancements made in the machine learning field, especially with regards to those based on image feature extraction and classification, interest in this field of research has only recently increased. In recent years, the advent of deep convolutional neural networks (CNNs) has been helpful in the development of medical image processing. Compared with other conventional machine learning approaches, CNNs have shown to have a better performance in image analysis and classification. We will be using pre-trained CNNs for this research work.

The remainder of the paper is structured according to this. A literature analysis of prior work performed in this area in Sect. 2. The methodology to train the models used is demonstrated in Sect. 3. The results are illustrated and explained in Sect. 4. The conclusion of the work carried out is given in Sect. 5.

2 Literature Review

Several methods and techniques for classifying skin lesions were proposed in the literature. Some researchers have used the traditional machine learning [8–10], while others used deep learning [11]. Some researchers also proposed classification techniques for mobile devices [12, 13].

Mahbod et al. [14] proposed the use of a combination of AlexNet, VGG16, and ResNet-18, for training and classification. They classified Nevus, seborrheic keratosis, and melanoma. The three models were combined in pairs and one where the outputs of all three were combined. The model with all three combined was the most accurate, with an accuracy of 90.69%. Al-masni et al. [15] proposed to initially pass the images into a full resolution convolutional network (FrCN) to segment the lesion from the surrounding skin. Afterwards, feature extraction and classification were done using ResNet-50. The accuracy achieved was 81.57%. The images were not balanced in the dataset. Barata and Marques [16] developed a hierarchical model. The images are segmented and initially classified as being either melanoma or not. Once this is completed, the image is further classified. The DenseNet121 architecture was used for feature extraction and classification. This approach performed better than a multi-class approach. They achieved an average recall value of 70.0%.

Yap et al. [17] proposed to use both lesion images and patients' personal data and passed into two ResNet-50 networks for feature extraction and classification. The addition of the patients' data proved to increase the accuracy of some lesions but decreased others. Bassi and Gomekar [18] trained a modified VGG16 network to classify seven different classes of skin lesions. The images in the dataset were balanced. An overall accuracy of 82.8% was obtained.

Younis et al. [20] used MobileNet CNN for classification of skin cancer. They trained their model on HAM1000 skin lesion dataset and 97% accuracy was achieved. Lau and his colleague [21] used three-layer neural network for classification of skin cancer using a traditional machine learning approach. They achieved an accuracy of 89.%. Hameed et al. [8] used traditional machine learning as well as deep learning for classification of skin lesions. They proposed a multi-class multi-level classification algorithm where skin cancer was classified at the last stage of the proposed model. They evaluated their proposed algorithm on dataset obtained from different sources and developed a desktop application using the proposed algorithm.

The above literature shows that more research work is required for the classification of skin lesions using state-of-the-art machine learning techniques.

Table 1 Number of images in each lesion class

Lesion class	Number of images
Actinic keratoses	327
Basal cell carcinoma	514
benign keratosis-like lesions	1099
Dermatofibroma	115
Melanoma	1113
Melanocytic nevi	6705
Vascular lesions	142

3 Dataset

The dataset used in this paper is the Harvard Dataverse HAM10000 dataset. It consists of seven groups of distinct skin lesions. A total of 10,015 images are on the dataset. It also has a CSV file that contains unique image IDs for each image and the class of lesions to which they belong. The images were initially split into seven different folders according to their lesion class. The seven lesion datasets are not balanced, i.e., they do not have the same number of images in each class (Table 1).

4 Methodology

4.1 Data Augmentation

Data augmentation has been used in the data set to balance the two classes. This was done on images in the malignant class. This has been done to increase its size and make it match the size of the benign class. Creating new images from the original dataset is the idea behind data augmentation. Each image in the malignant dataset undergone a series of transformations to accomplish this. This series of transformations consisted of rotating each image in the clockwise direction at an angle of 90°, 180°, and 270°. Through this rotation series, three new images were created from each original image. Furthermore, the images in the original dataset were horizontally flipped, and these were in turn rotated through the same sequence of angles previously described. From each of the original images, four new images are produced. This brings the total number of new created images to seven from each of the original images. The size of the dataset was increased by an eight factor, from 1113 to 8904 images. The size of the malignant dataset was brought up to approximately the same number as the benign dataset (8902 images). The entire dataset was divided into 0.8 for training, and 0.2 for testing.

4.2 Implementation and Training

The algorithms mentioned in this paper, including the pre-trained neural networks, were implemented and initialized using the Keras API, version 2.4.3, in Python 3.7. The training was done on a Windows 10 PC, with a core i5 2.4 GHz, 8 GB of RAM and a GTX 1050 GPU. To prepare and feed the images into the CNNs, the ImageDataGenerator class in the Keras library was used. The images were originally of dimensions 600 × 450. In order to use the images as inputs for the pre-trained models, they were resized to 224 × 224. The ImageDataGenerator will not feed the models with the original images. Each image was transformed randomly, instead. The pictures were randomly sheared through a 0.2 range, zoomed in through a 0.2 range, rotated up to 40°, and horizontally flipped at random. For each epoch, the models were trained, and these transformations were applied to the images. This ensured that during each epoch, the model saw almost new and different versions of the images, thereby reducing the odds of overfitting.

ResNet50, VGG16, VGG19, Inception V3, MobileNet, and DenseNet were the pre-trained neural networks used for this work. The networks have been initialized using the weight values from 'imagenet'. The 'imagenet' dataset is a compilation of over 14 million images on which these networks have been trained. Using these weight values, rather than random ones, has proven to improve the efficiency for extraction of features, as the weights are already good at identifying features. This method is called 'transfer learning'. The overall methodology is graphically presented in Fig. 1.

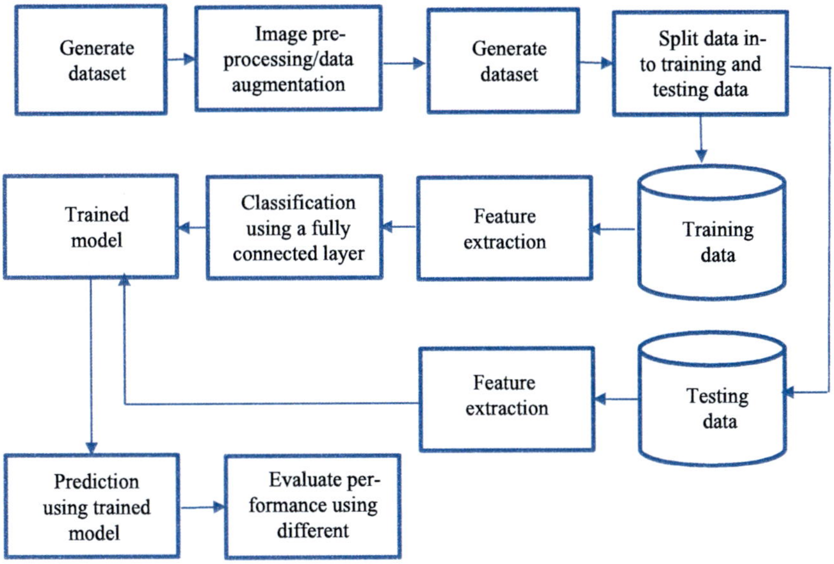

Fig. 1 Methodology diagram to train the pre-trained networks

The top layers were excluded from the pre-trained networks. They were replaced with a fully connected layer with a size of 512, with the activation function 'ReLU'. Following this was added a dropout layer with a value of 0.3. A last fully connected layer with the 'softmax' activation was added. This was to classify the images into two classes.

The networks were compiled using the stochastic gradient descent (SGD), with a learning rate of 0.01. The loss function used was 'categorical crossentropy'. The models were trained for 10 epochs.

4.3 Proposed Framework

We proposed a classification framework that combined the performance of three of the pre-trained neural networks that had the best accuracy. ResNet50, VGG16, and DenseNet were the three highest performing algorithms. To have a more efficient feature extraction from the training dataset, we proposed running and training the three networks in parallel and combining their outputs.

A combination of 2 out of the 3 pre-trained models were initially used. Both architectures have a shared input layer. The networks will be fed the same images and trained in parallel. After each network's convolution layers, one fully connected layer, with an output size of 512, was added. A concatenated layer then connects to the two fully connected layers. This layer connects the inputs and gives a single output. It will simply stack the outputs from the two fully connected layers, in order to create a single, which will have twice the width of that of the inputs. The output will be of different dimensions as the inputs, as it will not add the values.

Afterwards, the proposed approach was changed, and all three models were used. This was to assess whether the addition of one more pre-trained model, the accuracy of the feature extraction will increase.

The final architecture we proposed was based on the work described above. We proposed to improve the architecture by adding more layers for extraction of features. The three architectures ran in parallel and their layers were frozen. Their weights during training remain unchanged, as we used transfer learning. In each of the three networks, three convolution layers, each of dimensions $7 \times 7 \times 512$, were stacked after each pre-trained network. In order to prevent overfitting, a fully connected layer was added to each network, with a dropout value of 0.3. The outputs were then fed into the concatenate layer from each fully connected layer. The final architecture for the proposed model is shown in Fig. 2.

5 Results

The results for the pre-trained networks are shown in Table 2.

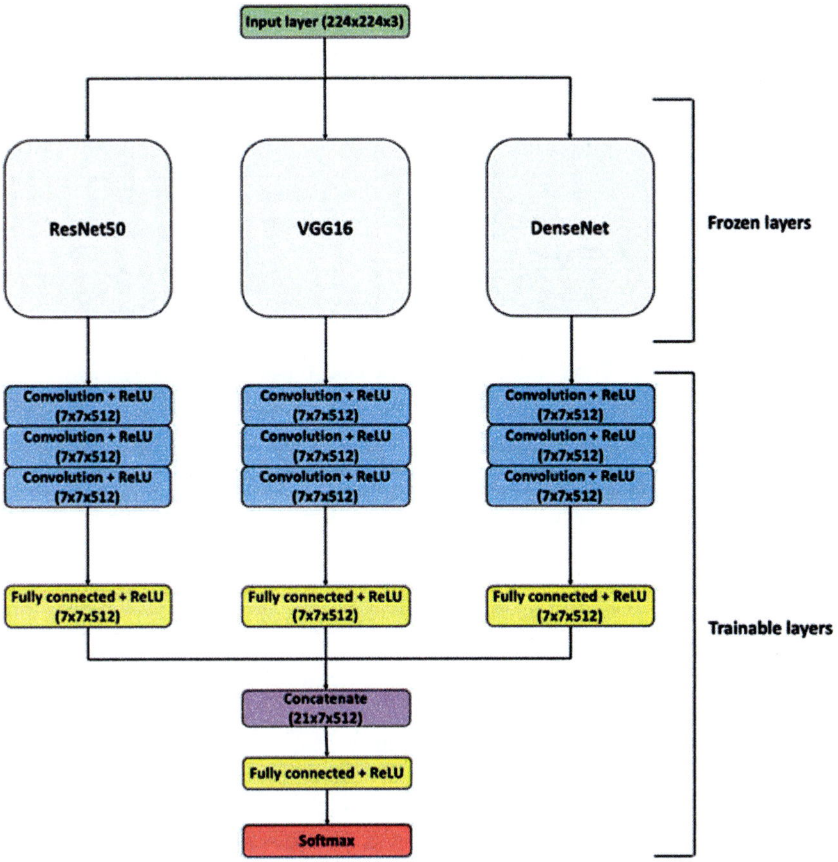

Fig. 2 Proposed architecture

Table 2 Performance of different models

Pre-trained model	Accuracy (%)
ResNet50	**81.05**
VGG16	77.51
VGG19	76.41
Inception V3	50.08
MobileNet	77.31
DenseNet	79.39

Table 3 Results of combination of pre-trained networks

Combination of pre-trained models	Accuracy (%)	Sensitivity (%)	Specificity (%)
ResNet50 + VGG16	81.67	78.52	82.5
ResNet50 + DenseNet	81.64	78.31	82.42
VGG16 + DenseNet	79.08	76.53	81.65
ResNet50 + VGG16 + DenseNet	82.74	81.34	83.32
ResNet50 + VGG16 + DenseNet + additional trainable convolution layer	**84.01**	**83.55**	**84.32**

The highest performing pre-trained model was ResNet50, with an accuracy of 81.05%. From these observations, the three best pre-trained performances were combined for the proposed architecture. They were ResNet50 (81.05%), DenseNet (79.39%), and VGG16 (77.51%).

The findings shown in Table 3 are when 2 out of the 3 models chosen were combined. It also shows the outcomes of combining the three networks and the effects of applying, to each network, the three trainable convolution layers. The findings, in Table 3, show that combining multiple pre-trained models for feature extraction increases the accuracy. We can observe that the specificity is always higher as compared to sensitivity, suggesting that all models can identify benign lesion better than malignant ones. Combining ResNet50 with any other pre-trained, either VGG16 or DenseNet improves accuracy by about 0.6% (81.67% and 81.64%, respectively) compared to ResNet50 alone (81.05%).

There is no successful effect on increasing accuracy from the combination of VGG16 and DenseNet. A positive impact is noted on the accuracy when all 3 models are combined. This value, 82.74%, is greater than 1% higher than the highest accuracy when combining two models, ResNet50 and VGG16 (accuracy 81.67%). A positive improvement in sensitivity can also be observed. This means that relative to the previous versions, the model is better at identifying malignant lesions.

Adding the trainable convolution layers into each model has significantly enhanced the accuracy. This final proposed architecture was the model with the best accuracy, sensitivity, and specificity discussed in this paper. The accuracy achieved was 84.01%. Graph showing the training and validation is graphically presented in Fig. 3.

6 Conclusion and Future Work

The main contribution of this research work is a classification framework that can distinguish between malignant and benign images of skin lesions. The proposed algorithm has performed better than some of the papers reviewed in Sect. 2. These works were proposed by Al-masni et al. [19] with an accuracy of 81.57%, Barata and Marques [16] with a recall of 70% and Bassi and Gomekar [18] with an accuracy of

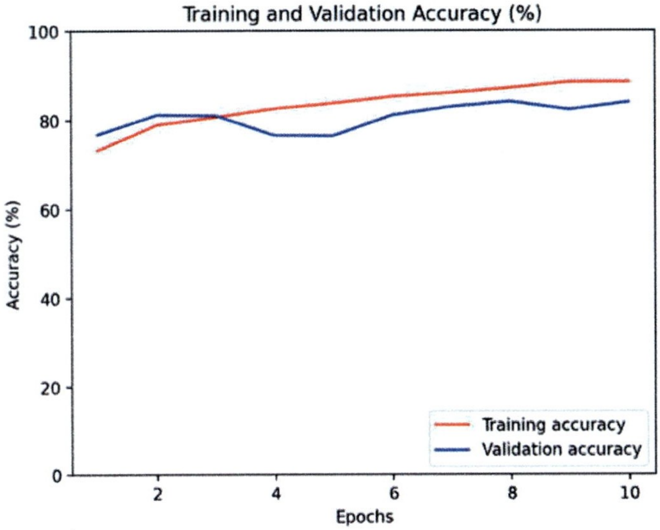

Fig. 3 Training and validation accuracy of the proposed classification framework

82.8%. In the first two works, smaller training datasets were used, with only 2000 images, compared to 17,800 for this project. This makes the proposed architecture in this paper more accurate for the prediction of new images. On the same dataset as the one used in this paper, the third work was trained on. However, they predicted more classes than we did for the proposed architecture (Table 4).

This research work shows the performance of using only deep learning to classify skin lesion images. Of all the models discussed in this project, the final proposed algorithm, with the trainable convolution layers, had the best results. For the final version of the proposed algorithm, an accuracy of 84.01% was achieved. We believe this can be of great help to doctors to diagnose malignant skin lesions, especially in regions where there is scarce access to medical services.

For future works, we believe the accuracy of the architecture by optimizing the hyperparameter values, such as the size of the kernels, size of fully connected layers, and size of inputs and outputs of the layers. This can be achieved by using machine learning techniques such as genetic algorithms.

Table 4 Comparison of the proposed framework with existing literature

References	Results
[19]	Acc: 81.57%
[16]	Re: 70%
[18]	Acc: 82.8%
Proposed algorithm	Acc: 84.01%

Furthermore, the algorithm can be expanded to classify more classes of lesions in a hierarchical manner. This means, initially classifying the image as either malignant or benign and then classifying it to its appropriate lesion class. Redesigning of the proposed framework for mobile platform is also part of our future work.

References

1. Mayers, K.: What is the scale of the skin cancer problem in the UK?J. Aesth. Nurs. (2018)
2. Taufiq, M.A., Hameed, N., Anjum A., Hameed, F.: m-Skin Doctor: a mobile enabled system for early melanoma skin cancer detection using support vector machine. In: Giokas, K., Bokor, L., Hopfgartner, F. (eds.) eHealth 360°. Lecture notes of the institute for computer sciences, social informatics and telecommunications engineering, vol. 181. Springer, Berlin (2017)
3. Skin cancer (melanoma),2020. [Online]. Available: https://www.nhs.uk/conditions/melanoma-skin-cancer/
4. Hameed, N., Hassan, K.A., Hossain, M.A.: A comprehensive survey on image-based computer-aided diagnosis systems for skin cancer. In: 2016 10th International Conference on Software, Knowledge, Information Management and Applications (SKIMA), Chengdu, China, pp 205–214 (2016)
5. Non-melanoma skin cancer statistics, [Online]. Available: https://www.cancerresearchuk.org/health-professional/cancer-statistics/statistics-by-cancer-type/non-melanoma-skin-cancer#heading-Zero
6. British Skin Foundation,2018. [Online]. Available: https://www.britishskinfoundation.org.uk/Home.aspx. Accessed 20 May 2018
7. Picardi, A.: Suicide risk in skin disorders.In: Clinics in Dermatology, vol. 31 (2013)
8. Hameed, N., Shabut, A., Gosh, M., Hossain, M.A.: Multi-class multi-level classification algorithm for skin lesions classification using machine learning techniques. Exp. Syst. Appl. 141 (2019)
9. Hameed, N., Shabut, A.M., Hossain, M.A.:A computer-aided diagnosis system for classifying prominent skin lesions using machine learning. In: 2018 10th Computer Science and Electronic Engineering (CEEC), Colchester, United Kingdom, pp 186–191 (2018)
10. Hameed, N., Hameed, F., Shabut, A., Khan, S., Cirstea, S., Hossain, A.: An intelligent computer-aided scheme for classifying multiple skin lesions. Computers **8**, 62 (2019)
11. Hameed, N., Shabut, A., Hossain, M.A.: Multi-class skin diseases classification using deep convolutional neural network and support vector machine. In: 12th International Conference on Software, Knowledge, Information Management and Applications, Phnom Penh, Cambodia (2018)
12. Hameed, N., Hameed, F., Shabut, A., Cirstea, S., Hossain, A.: An intelligent inflammatory skin lesions classification scheme for mobile devices. In: IEEE International Conference on Computing, Electronics and Communications Engineering (2019)
13. Hameed, N, Shabut, A., Hameed, F., Cirstea, S., Harriet, S., Hossain, A.: Mobile-based skin lesions classification using convolution neural network. Ann .Emerg. Technol. Comput. (AETiC) **4**(2), 12
14. Mahbod, A., Schaefer, G., Wang, C., Ecker, R., Ellinger, I.: Skin lesion classification using hybrid deep neural networks.In: International conference on acoustics, speech, and signal processing (2019)
15. Al-masni, M.A., Al-antari, M.A., Park, H.M., Park, N.H., Kim, T.: A deep learning model integrating FrCN and residual convolutional networks for skin lesion segmentation and classification.In: 2019 IEEE Eurasia Conference on Biomedical Engineering, Healthcare and Sustainability (ECBIOS), Okinawa, Japan (2019)

16. Barata, C., Marques, J.S.: Deep learning for skin cancer diagnosis with hierarchical architectures.In: 2019 IEEE 16th International Symposium on Biomedical Imaging (ISBI 2019), Venice, Italy (2019)
17. Yap, J., Yolland, W., Tschandl, P.: Multimodal skin lesion classification using deep learning.Exp. Dermatol. (2018)
18. Bassi, Gomekar: Deep learning diagnosis of pigmented skin lesions.In: 2019 10th International Conference on Computing, Communication and Networking Technologies (ICCCNT), Kanpur, India (2019)
19. Al-masni, M., Al-antari, M.A., Park, H.M., Park, N.H., Kim, T.: A deep learning model integrating FrCN and residual convolutional networks for skin lesion segmentation and classification.In: 2019 IEEE Eurasia Conference on Biomedical Engineering, Healthcare and Sustainability (ECBIOS), Okinawa, Japan (2019)
20. Younis, H., Bhatti, M.H., Azeem, M.: Classification of skin cancer dermoscopy images using transfer learning. In: 2019 15th International Conference on Emerging Technologies (ICET), Peshawar, Pakistan, pp. 1–4 (2019). https://doi.org/10.1109/ICET48972.2019.8994508
21. Lau, H.T., Al-Jumaily, A.: Automatically early detection of skin cancer: study based on neural network classification. In: 2009 International Conference of Soft Computing and Pattern Recognition, Malacca, pp. 375–380 (2009). https://doi.org/10.1109/SoCPaR.2009.80

Author Index

M. S. Kaiser et al. (eds.), *Proceedings of International Conference on Trends in Computational and Cognitive Engineering*, Advances in Intelligent Systems and Computing 1309, https://doi.org/10.1007/978-981-33-4673-4